Student Solutions Manual

Elementary Linear Algebra with Applications

Ninth Edition

Bernard Kolman

Drexel University

David R. Hill

Temple University

PEARSON

Prentice Hall

Upper Saddle River, NJ 07458

Editorial Director, Computer Science, Engineering, and Advanced Mathematics: *Marcia J. Horton*
Senior Editor: *Holly Stark*
Editorial Assistant: *Jennifer Lonschein*
Senior Managing Editor/Production Editor: *Scott Disanno*
Art Director: *Juan López*
Supplement Cover Designer: *Daniel Sandin*
Art Editor: *Thomas Benfatti*
Manufacturing Buyer: *Lisa McDowell*
Marketing Manager: *Tim Galligan*

Printed in the United States of America
10 9 8 7 6 5 4 3 2 1

ISBN 0-13-229656-X
    978-0-13-229656-4

Pearson Education, Ltd., *London*
Pearson Education Australia PTY. Limited, *Sydney*
Pearson Education Singapore, Pte., Ltd
Pearson Education North Asia Ltd, *Hong Kong*
Pearson Education Canada, Ltd., *Toronto*
Pearson Educación de Mexico, S.A. de C.V.
Pearson Education—Japan, *Tokyo*
Pearson Education Malaysia, Pte. Ltd
Pearson Education, Inc., *Upper Saddle River, New Jersey*

Contents

Preface

This manual is to accompany the Ninth Edition of Bernard Kolman and David R. Hill's *Elementary Linear Algebra with Applications*. Detailed solutions to all odd numbered exercises are included. It was prepared by Dennis Kletzing, Stetson University. It contains many of the solutions found in the Eighth Edition, as well as solutions to new exercises included in the Ninth Edition of the text.

Chapter 0

Introduction to Proofs

In many instances a first course in linear algebra is used to introduce students to active participation in abstract mathematics. That is, not only will new concepts and proofs of such concepts be given in the text, but the student is expected to create proofs for related concepts. In many cases a student is not prepared for this "jump" in mathematical sophistication. Most high school mathematics and calculus courses emphasize computation and manipulation based on the ideas introduced, rather than on building new ideas (proofs) from previous ideas. It is not that the "jump" is so hard, rather it is often the case that there has been very little foundation laid to understand the language of the statement of theorems and the nature of the techniques of proof.

Instead of blaming the system or previous instructors, what is developed here is a short summary and explanation of fundamental notions that are valuable on two fronts:

1. Reading and interpreting statements in the language of mathematics.

2. Developing proofs.

In order to use this material effectively, you must actively participate in it. One way to do this is to record your work in a notebook, both scratch work, which often contains initial ideas and attempts, and any refined versions. That way you can compare your work with others and with the solutions supplied here. Moreover, you (and your instructor) will be able to see that the strategies you have employed will vary and mature as you work throughout your linear algebra course. Remember that proofs presented in books are polished versions that have undergone numerous refinements. Such proofs are meant to be read by those who understand the language of mathematics. You should not expect your proofs to appear as slick or succinct as those in a book. With practice and experience your proofs will improve.

Throughout your previous mathematical experiences you have been introduced to some of the vocabulary of mathematics. Such terms as variable, equation, and real number should be quite familiar. In fact, there is a set of mathematical terminology that has become almost automatic due to your courses in algebra, trigonometry, and calculus. As you progress through your linear algebra course your mathematical vocabulary will expand. What is important in abstract mathematics is the way that the mathematical vocabulary is woven into the grammar used to describe concepts and make statements about concepts. We must communicate these in an unambiguous way; such precision is a crucial aspect of mathematics. The verification (proof) of a statement must be done so that there is no ambiguity present, for only then, will there be no doubt about its correctness.

In order to develop the precision of exposition required, we include a discussion of logic. Our focus will be to use the structure of logic as part of the method of mathematical proofs and their validity. We introduce a symbolic language, with symbols having precise meanings and uses. In this way known facts can be combined to prove new facts in a systematic fashion. We also avoid the imprecise overtones of everyday language that can cloud our reasoning. The use of connectives for building compound statements and the use of conditional statements is developed.

This Introduction to Proofs is meant to be for self-study. It has been the authors' experience that small groups of students working together can greatly aid one another when using this material. It is recommended that this material be studied early in conjunction with the first few sections of Chapter 1. Thus the ideas will be available as needed for proofs as you progress through Chapter 1 and in later chapters which are more theoretically oriented.

Section 0.1 Logic

Logic is an analytical theory of the art of reasoning. We shall introduce symbols which facilitate the development of rules and techniques for determining the validity of an argument. Most of what we will prove or that you are asked to prove in linear algebra appear in the form of a declarative sentence. That is, mathematical language that makes a statement. We adopt the following terminology.

Definition 1. A **statement** or **proposition** is a declarative sentence that is either true or false.

Example 1. Each of the following is a statement.

    (a)   Nero is dead.

    (b)   Every rectangle is a square.

    (c) $4 + 3 = 7$.

    (d) $x^2 \geq 0$, for every real number x.

    (e) $\sqrt{x^2} = x$, for every real number x.

Parts (a), (c) and (d) are true statements, while (b) and (e) are false. (In (e), $\sqrt{(-3)^2} = \sqrt{9} = 3 \neq -3$, so it is false.). ∎

Example 2. Each of the following is not a statement because it is not a declarative sentence or its truth value is not known.

    (a) $x^2 - 3x + 2 = 0$ (It is a declarative mathematical sentence, but its truth value depends on the value of x.)

    (b)   Can you read Russian?   (This is a question, not a statement.)

    (c)   Call me tomorrow.   (This is a command, not a statement.)   ∎

In algebra we studied equations for which our focus was to determine a solution. The mathematical formulation involved using a variable, often represented by $x, y, z, \ldots$, to represent a real number solution of the equation. In (mathematical) logic we represent statements using letters $p, q, r, \ldots$. In this case the letters $p, q, r, \ldots$, are variables representing statements. Thus we can assign

$$p: \quad \text{Six is prime.}$$
$$q: \quad \text{The earth is larger than the moon.}$$

Hence in the context of the problem where these assignments are made we can use symbols p and q rather than the explicit verbal statements.

In algebra we manipulate numbers and variables representing numbers by using the arithmetic operations $+$, $-$, $\times$, $\div$, and exponentiation. In logic we manipulate statements and variables representing statements using connective operations and conditional operations. We consider three connectives which are defined next.

Definition 2.   Given statements p and q.

(a)   The statement "p and q" is denoted $p \wedge q$ and is called the **conjunction** of p and q. Statement $p \wedge q$ is true when both p and q are true.

(b)   The statement "p or q" is denoted $p \vee q$ and is called the **disjunction** of p and q. Statement $p \vee q$ is true when either p or q or both are true.

(c)   The statement "not p" is denoted $\sim p$ and is called the *negation* of p. Statement $\sim p$ is true when p is false.

The use of connective operations builds **compound statements** from p and q, the original statements. The relationship between the statements p and q and compound statements $p \wedge q$, $p \vee q$, and $\sim p$ is conveniently displayed using a **truth table**. A truth table is a visual display of the truth value of the compound statement given the various possibilities for the truth of the original statements. Directly from Definition 2 we have the following truth tables.

p	q	$p \wedge q$
T	T	T
T	F	F
F	T	F
F	F	F

Table 1.

p	q	$p \vee q$
T	T	T
T	F	T
F	T	T
F	F	F

Table 2.

p	$\sim p$
T	F
F	T

Table 3.

The use of logic and connectives for aid in proofs requires that we recognize the verbal equivalents of the symbols $\wedge$, $\vee$, and $\sim$. We present some common verbal forms next:

$\wedge$ and, together with, combined with, all of
$\vee$ or, one of, any
$\sim$ not, opposite

In most cases you will see the concise verbal equivalents "and", "or", and "not".

Several different compound statement forms often appear. Examples 3 through 6 briefly discuss these forms using truth tables.

Example 3.   Given statements p and q determine the truth table of compound statement $\sim (p \wedge q)$. (In words, not $(p$ and $q)$.)

p	q	$p \wedge q$	$\sim (p \wedge q)$
T	T	T	F
T	F	F	T
F	T	F	T
F	F	F	T

Table 4.

We see that column 3 comes directly from Table 1 and then column 4 is obtained by negating (finding the opposite) of the truth values of column 3. As Table 4 shows $\sim (p \wedge q)$ is false only when both p and q are true.   ∎

Example 4. Given statements p and q determine the truth table of compound statement $\sim (p \vee q)$. (In words, not (p or q).)

p	q	$p \vee q$	$\sim (p \vee q)$
T	T	T	F
T	F	T	F
F	T	T	F
F	F	F	T

Table 5.

We see that column 3 comes directly from Table 2 and then column 4 is obtained by negating column 3. As Table 5 shows $\sim (p \vee q)$ is false when either p or q is true. ■

Example 5. Given statements p and q determine the truth table of compound statement $(\sim p) \vee (\sim q)$. (In words, not p or not q.)

p	q	$\sim p$	$\sim q$	$(\sim p) \vee (\sim q)$
T	T	F	F	F
T	F	F	T	T
F	T	T	F	T
F	F	T	T	T

Table 6.   ■

Example 6. Given statements p and q determine the truth table of compound statement $(\sim p) \wedge (\sim q)$. (In words, not p and not q.)

p	q	$\sim p$	$\sim q$	$(\sim p) \wedge (\sim q)$
T	T	F	F	F
T	F	F	T	F
F	T	T	F	F
F	F	T	T	T

Table 7.   ■

From the truth tables in Tables 4, 5, 6, and 7 we make the following observations.

(a) The truth table for $\sim (p \wedge q)$ is identical to the truth table for $(\sim p) \vee (\sim q)$. (We mean that the final columns in Tables 4 and 6 are the same.)

(b) The truth table for $\sim (p \vee q)$ is identical to the truth table for $(\sim p) \wedge (\sim q)$. (We mean that the final columns in Tables 5 and 7 are the same.)

Definition 3. Two statements are **equivalent** provided they have the same truth table.

Thus we can rephrase (a) and (b) above as follows.

    (a) $\sim (p \wedge q)$ is equivalent to $(\sim p) \vee (\sim q)$.

    (b) $\sim (p \vee q)$ is equivalent to $(\sim p) \wedge (\sim q)$.

As far as proofs are concerned we can use an equivalent statement to replace a given statement without changing the problem. Hence we can substitute an equivalent statement for a given statement to help simplify things or make things easier or just to change the point of view. You did this type of maneuver many times in algebra as the following illustrates:

▶ Fraction $\frac{5}{3}$ is equivalent to mixed number $1\frac{2}{3}$.

▶ Multiplying by 1 is equivalent to multiplying by $\frac{5}{5}$, or by $\dfrac{x-2}{x-2}$, $x \neq 2$.

▶ Dividing by 2 is equivalent to multiplying by $\frac{1}{2}$.

▶ Simplifying or putting in lowest terms.

We give further illustrations of the power and versatility of equivalent statements later. The use of equivalent statements is an important technique in constructing proofs.

    Another way to construct compound statements is to use conditional operations. The simplest of these operations is defined as follows.

Definition 4. Given statements p and q, the **conditional statement** or **implication** is

$$\text{if } p \text{ then } q$$

which is denoted symbolically as

$$p \Longrightarrow q$$

This is read "p implies q".

    The conditional statement is the most important kind of statement in mathematics. You have used conditional statements many times as illustrated in the following example.

Example 7.

▶ If $\underbrace{\text{two lines are parallel}}_{p}$, then $\underbrace{\text{the lines do not intersect}}_{q}$.

▶ If $\underbrace{\text{it does not rain today}}_{p}$, $\underbrace{\text{I will cut the grass}}_{q}$.
    (Sometimes, "then" does not appear explicitly.)

▶ $\underbrace{\text{Quantity } \dfrac{-b + \sqrt{b^2 - 4ac}}{2a} \text{ is a complex number}}_{q}$ if $\underbrace{b^2 - 4ac < 0}_{p}$.   ■

To appropriately use the compound statement $p \Longrightarrow q$ we must identify statements p and q correctly. The following terminology is used in this regard:

    The *if-statement* p is called the **antecedent** or **hypothesis**.

    The *then-statement* q is called the **consequent** or **conclusion**.

In this manual we use the terms *hypothesis for the if-statement* and *conclusion for the then-statement*.
    Mathematicians have agreed that the truth value of a conditional statement will be given as follows.

Definition 4. The conditional statement $p \Longrightarrow q$ is true whenever the hypothesis is false or the con-
(continued)   clusion is true.

Thus we have the following truth table.

p	q	$p \Longrightarrow q$
T	T	T
T	F	F
F	T	T
F	F	T

Table 8.

An unexpected result from Table 8 is that conditional statements may be true even when there is no connection between the hypothesis and conclusion. For example,

$$\text{Eagles can fly} \Longrightarrow 1 + 1 = 2.$$

We will have no occasion to exploit this curiosity.

A primary objective in our work will be to construct a proof of $p \Longrightarrow q$. This means we want to show

$$\boxed{\text{If } p \text{ is true, then } q \text{ is true.}}$$

In order to construct a proof, a logical argument in the language of mathematics, we must identify the hypothesis and conclusion. A conditional statement can appear disguised in a number of different ways. The following expressions are all equivalent to $p \Longrightarrow q$.

if p then q	q whenever p
p implies q	p only if q
q if p	p is sufficient for q
q provided that p	q is necessary for p
	from p it follows that q

Example 7 gives several illustrations. Exercise 6 at the end of the section provides an opportunity to identify the hypothesis and conclusion.

We digress briefly to comment on two forms equivalent to $p \Longrightarrow q$ which occur frequently in mathematics. The form

$$p \text{ is sufficient for } q$$

can be thought of in the following way. It is sufficient to show that statement p is true in order to conclude that statement q is true. We sometimes say **p is a sufficient condition for q**. The form

$$q \text{ is necessary for } p$$

can be thought of in the following way. It necessarily follows that if statement p is true then statement q is true. We sometimes say **q is a necessary consequence of p**.

In constructing proofs for conditional statements, $p \Longrightarrow q$, it is sometimes simpler to prove an equivalent statement called the **contrapositive**.

Definition 5. Given statements p and q the **contrapositive** of $p \Longrightarrow q$ is $(\sim q) \Longrightarrow (\sim p)$.

Verification that $p \Longrightarrow q$ and $(\sim q) \Longrightarrow (\sim p)$ are equivalent is done by showing that their truth tables are identical. See Exercise 7.

Example 8. Form the contrapositive of each of the following conditional statements.

(a)   Statement $p \Longrightarrow q$:   If two lines are parallel, then the lines do not intersect.

Contrapositive $(\sim q) \Longrightarrow (\sim p)$:   If two lines intersect, then the lines are not parallel.

(b)   Statement $p \Longrightarrow q$:   If a quadratic equation has two distinct real roots, then its graph crosses the x-axis.

Contrapositive $(\sim q) \Longrightarrow (\sim p)$:   If the graph of a quadratic equation fails to cross the x-axis, then the equation does not have two distinct real roots.

An effort should be made to make the contrapositive statement grammatically correct.   ■

Another type of construction that is related but not equivalent to a conditional statement is defined next.

Definition 6.   Given statements p and q, the **converse** of $p \Longrightarrow q$ is $q \Longrightarrow p$.

Note that the converse merely interchanges the roles of the hypothesis and the conclusion. To show that the converse is not equivalent to the conditional statement we construct the truth tables in Table 9. We see that the third and fourth columns are not identical hence a conditional statement and its converse are not equivalent.

p	q	$p \Longrightarrow q$	$q \Longrightarrow p$
T	T	T	T
T	F	F	T
F	T	T	F
F	F	T	T

Table 9.

Warning: In proofs we can *not* substitute the converse for the original statement.

Example 9.   Construct the converse of each of the following conditional statements.

(a)   Statement $p \Longrightarrow q$:   If numbers a and b are positive, then ab is positive.

Converse $q \Longrightarrow p$:   If ab is positive, then a and b are positive.

Note that the original statement $p \Longrightarrow q$ is true while its converse $q \Longrightarrow p$ is false because $ab > 0$ for $a = -1$, $b = -2$.

(b)   Statement $p \Longrightarrow q$:   If $n + 1$ is odd, then n is even.

Converse $q \Longrightarrow p$:   If n is even, then $n + 1$ is odd.

Note that in this case both $p \Longrightarrow q$ and its converse $q \Longrightarrow p$ are true.   ■

There is another type of conditional operation that is closely related to a conditional statement $p \Longrightarrow q$.

Definition 7.   Given statements p and q, the **biconditional statement** is

$$p \text{ if and only if } q$$

which is denoted symbolically as

$$p \Longleftrightarrow q$$

A biconditional is true when p and q have the same truth value.

It follows directly from Definition 7 that a biconditional has the following truth table.

p	q	$p \Longleftrightarrow q$
T	T	T
T	F	F
F	T	F
F	F	T

Table 10.

Because a biconditional statement $p \Longleftrightarrow q$ is true precisely when the truth values of p and q are the same, it follows from Definition 3 that a biconditional can be used to test whether p and q are equivalent. Thus from Examples 3 and 5 we have

$$\sim (p \wedge q) \Longleftrightarrow (\sim p) \vee (\sim q)$$

and from Examples 4 and 6 we have

$$\sim (p \vee q) \Longleftrightarrow (\sim p) \wedge (\sim q).$$

Example 10. Each of the following is a biconditional statement.

(a)   An algebraic equation represents a line if and only if it can be put into the form $ax + by = c$ for some constants a, b, c.

(b) $a > b$ if and only if $a - b > 0$.

(c)   A function $f(x)$ is a polynomial if and only if it can be put into the form $a_n x^n + \cdots + a_1 x + a_0$.

(d)   An integer $n > 1$ is prime if and only if its only divisors are 1 and itself.   ∎

A convenient way to think of a biconditional is

> $p \Longleftrightarrow q$ is true exactly when p and q are equivalent.

Several additional comments concerning biconditionals are useful for learning to read abstract mathematics and learning to do proofs. As you have seen in this brief discussion of logic, definitions give us a set of (elementary) facts which can be used to build more complicated statements or relationships. Any definition supplies us with an equivalence or biconditional statement. Hence definitions give us items which can be used to substitute for one another. In the text you will see many definitions which have the following form:

> **Definition:**   A matrix is called   ‘some name’   if   ‘algebraic type expression’ .

Because a definition is an equivalence (a biconditional), any time the ‘some name’ appears we can substitute the corresponding ‘algebraic type expression’ if it is more useful in a given circumstance. Thus we often can convert a verbal description to an equivalent algebraic form. This is quite valuable when justifying a particular step of a proof.

From Definition 7, $p \Longleftrightarrow q$ is read "p if and only if q". Some times "if and only if" is abbreviated "iff" and in such cases the biconditional would appear in the form p iff q. A biconditional is actually a conjunction of the two conditional statements $p \Longrightarrow q$ and $q \Longrightarrow p$. To verify this we show that the truth table for compound statement $(p \Longrightarrow q) \wedge (q \Longrightarrow p)$ is the same as Table 10. (See Exercise 13.) Thus it follows that

> To prove $p \Longleftrightarrow q$ we must show *both* $p \Longrightarrow q$ and (its converse) $q \Longrightarrow p$ are true.

Keep in mind that to prove an if and only if statement there are two things that must be proved. We will return to this in the section on proof techniques.

In the language of mathematics, a biconditional ($p \Longleftrightarrow q$) or an if and only of statement is often referred to as **a set of necessary and sufficient conditions**. This is just another way of saying statements p and q are equivalent. The search for alternative sets of necessary and sufficient conditions is very important in mathematics. With several such sets, we have flexibility in substituting equivalent expressions which in many cases permit us to change the point of view adopted in a proof. You will see that the text emphasizes this point by presenting sets of equivalent properties in summaries of important concepts.

Exercises 0.1

1. Determine which of the following are statements. For those which are statements give its truth value.

   (a) $5 - 1 = 8$

   (b) The diagonals of a square are perpendicular.

   (c) $x^2 - 1 = 0$

   (d) Are we having fun yet?

   (e) $\sqrt{x^2} = |x|$, for any real number x.

   (f) If $p(x)$ is a polynomial, then it is a function whose domain is the set of all real numbers.

2. Given statements p, q, and r. Write out each of the following compound statements.

   $$\begin{aligned} p&: \quad \text{The door is open.} \\ q&: \quad \text{The tables are set.} \\ r&: \quad \text{The food is ready.} \end{aligned}$$

   (a) $p \wedge q$ (b) $q \vee (\sim r)$ (c) $p \wedge (\sim q)$ (d) $p \wedge q \wedge r$

3. Given statements p, q, and r. Express each of the following symbolically.

   $$\begin{aligned} p&: \quad \text{The car will not start.} \\ q&: \quad \text{It is freezing outside.} \\ r&: \quad \text{I am going back to bed.} \end{aligned}$$

   (a) The car will start and it is freezing outside.

   (b) Either I am going back to bed or it is freezing outside.

   (c) I am not going back to bed and the car won't start.

   (d) I am staying up.

4. Construct a truth table for each of the following statements.

   (a) $p \wedge (\sim q)$ (b) $p \vee (q \wedge (\sim p))$ (c) $p \wedge (q \vee (\sim p))$ (d) $p \vee (\sim q)$

5. Formulate an (algebraically) equivalent statement for each of the following.

   (a) Multiply both sides of the equation by $\frac{1}{3}$.

   (b) Subtract 3 from both sides of the equation.

   (c) $5 - 3$

   (d) $\dfrac{x^2 + x - 2}{x^2 - 1}$

6. Identify the hypothesis p and conclusion q in each of the following conditional statements.

   (a) I will go to the movies if it rains.

   (b) If the diagonals of a rectangle are not perpendicular, it is not a square.

   (c) If $x > 0$ and $y > 0$, then $x + y > 0$.

   (d) $f(x)$ is a function on $[a, b]$ implies that $f(x)$ is continuous on $[a, b]$.

   (e) I will pay \$5.00 for a cup of coffee only if elephants fly.

   (f) All sides are equal provided that T is an equilateral triangle.

   (g) Parallel opposite sides is sufficient for a quadrilateral to be a parallelogram.

7. Show using truth tables that $p \implies q$ and the contrapositive $(\sim q) \implies (\sim p)$ are equivalent statements.

8. Form the contrapositive of each of the following statements.

   (a) If two lines are perpendicular, then they intersect at right angles.

   (b) Function $f(x)$ is differentiable provided that $f(x)$ is a polynomial.

9. Form the converse of each of the statements in Exercise 8.

10. Show that $\sim (p \implies q) \iff p \wedge (\sim q)$.

11. Show that $\sim (p \wedge q) \iff (p \implies (\sim q))$.

12. Complete the following biconditional statements.

   (a) Positive integer n is even $\iff$ _____.

   (b) Angles a and b are complementary if and only if _____.

   (c) $x^2 \leq 4$ iff _____.

   (d) $n + 1$ is even $\iff$ _____.

   (e) Parabola $y = ax^2 + 1$ opens upward iff _____.

13. Show that $(p \implies q) \iff ((p \implies q) \wedge (q \implies p))$.

Section 0.2   Techniques of Proof

In this section we discuss techniques for constructing proofs of conditional statements $p \implies q$ and biconditionals $p \iff q$. As you progress through your linear algebra course the material changes, hence the content of statements p and q will change. Once p and q have been determined you should identify the type of material related to p and q. In other words, the context or mathematical area that is relevant to the problem. This helps you focus on the area from which you may be able to use definitions, previous theorems, or even previous problems as justifications or reasons (replacements) for steps in a proof.

   To prove $p \implies q$ we must show that whenever p is true it follows that q is true using a logical argument in the language of mathematics. The construction of this logical argument may be quite elusive, while the logical argument itself is the thing we call the **proof**. Conceptually the proof that p implies q is a sequence of steps that logically connect p to q. Each step in the "connection" must be justified or have a reason for its validity which is usually a previous definition, a property or axiom that is known to be true, a previously proven theorem or problem, or even a previously verified step in the current proof. Thus we connect p and q by logically building blocks of known (or accepted) facts. Often it is not clear what building blocks (facts) to use, and especially how to get started on a fruitful path. In many cases *the first step of the proof is crucial*. Unfortunately we have no explicit guidelines in this area other than to recommend that you *carefully read the hypothesis p and conclusion q* in order to clearly understand them. Only in this way can you begin to seek relationships (connections) between them. Experience and practice are a must in developing proof skills.

   The construction of a proof requires us to build a step-by-step connection (a logical bridge) between p and q. If we let $b_1, b_2, \ldots, b_n$ represent logical building blocks then conceptually our proof appears as

$$p \implies b_1 \implies b_2 \implies \cdots \implies b_n \implies q,$$

where each conditional must be justified. This approach is known as a direct proof. We illustrate this in Examples 1 and 2.

Example 1. Prove: If m and n are even integers, then $m + n$ is even.

Proof: Let p: m and n are even integers
q: $m + n$ is even

Note that the mathematical area is numbers, particularly even numbers. Reasons will appear in braces $\{\cdots\}$.

To start we ask ourselves, *assuming p is true*, what facts do we know that can lead to q. Since both p and q include the use of even numbers, for building block b_1 we try for something involving even numbers. We know that an even number is a multiple of 2. Try the following.

$$p \implies \underbrace{m = 2k, n = 2j}_{b_1} \text{ for some integers } k \text{ and } j \quad \{\text{property of even numbers}\}.$$

Since q involves the sum $m + n$, we try involving the sum in b_2

$$p \implies \underbrace{m + n = 2k + 2j = 2(k + j)}_{b_2} \quad \{\text{properties of arithmetic}\}.$$

We observe that b_2 implies that the sum $m + n$ is a multiple of 2. Hence $m + n$ is even. This is just q, hence we have

$$b_2 \implies q$$

In summary $p \implies b_1 \implies b_2 \implies q$. ∎

Example 2. Let x be a real number.
Prove: $|x + 2| < 1$ implies $-3 < x < -1$.

Proof: Let p: $|x + 2| < 1$
q: $-3 < x < -1$

The mathematical area involves inequalities.

To start we ask ourselves, *assuming p is true*, what facts do we know that can lead to q. The conclusion involves an expression that has x isolated in the middle. Thus we will try an algebraic rearrangement of p.

$$p \implies \underbrace{-1 < x + 2 < 1}_{b_1} \quad \{\text{properties of absolute value and inequalities}\}$$

Next try an arithmetic operation to isolate x in the middle; subtract 2 from each piece of the string of inequalities.

$$b_1 \implies \underbrace{-3 < x < -1}_{q}$$

In summary $p \implies b_1 \implies q$. ∎

In both Examples 1 and 2 we began with

> "What fact(s) concerning hypothesis p can be used to start a bridge to q?"

We obtained a building block b_1 such that $p \implies b_1$. We proceeded by asking ourselves

> "What fact(s) concerning hypothesis b_1 can be used to continue the bridge to q?"

This scenario was continued until q was obtained. Examples 1 and 2 proceeded forward from p to q, building logical connections. We call this **forward building**.

Alternatively we could ask ourselves

> "What fact(s) must be known to conclude that q is true?"

Call this b_1. Continue backwards from b_1 by asking

> "What fact(s) must be known to conclude that b_1 is true?"

Call this b_2. Continue in this fashion until we reach p. Such a logical bridge is called **backward building**. Proofs can sometimes be built either way, but we have no way to determine in a particular instance which technique may be more fruitful. As you might expect, the two techniques can be combined. Build forward a few steps, build backward a few steps and strive to logically join the two ends. Often it is an individual's choice based on intangibles that determines the approach used. Most proofs in books appear to use forward building, but remember such proofs are the result of many refinements. Any of these constructions is called a direct proof.

Often building at both ends is used in proving trigonometric identities since substitutions that express both sides in terms of sines and cosines are easy to make. See Example 3. Another interesting combination of forward and backward building occurs in calculus limit proofs involving the definition of limits in terms of epsilon and delta. Students are often amazed at how it was known what δ (in terms of ϵ) should be chosen so that the proof works out perfectly. The answer is that what you are reading is a refined version of a proof in which the δ was not known but determined from the manipulation of the proof and then used to make a "clean" looking proof for publication. See Example 4.

Example 3. Prove the trigonometric identity $\cos^2 \theta = \dfrac{1 + \cos 2\theta}{2}$.

Proof: Let p: $\cos^2 \theta$ and q: $\dfrac{1 + \cos 2\theta}{2}$

Starting forward from p using the Pythagorean identity we have

$$p \Longrightarrow 1 - \sin^2 \theta, \quad \text{that is, } \cos^2 \theta = 1 - \sin^2 \theta$$

Starting backward from q using the identity for the cosine of a sum of two angles we have

$$q \Longrightarrow \frac{1 + \cos(\theta + \theta)}{2} \Longrightarrow \frac{1 + \cos^2 \theta - \sin^2 \theta}{2}$$

Replacing $\cos^2 \theta$ by $1 - \sin^2 \theta$ as in the forward step we have

$$\frac{1 + \cos^2 \theta - \sin^2 \theta}{2} \Longrightarrow \frac{1 + (1 - \sin^2 \theta) - \sin^2 \theta}{2} \Longrightarrow \frac{2(1 - \sin^2 \theta)}{2}$$
$$\Longrightarrow 1 - \sin^2 \theta$$

Connecting the forward and backward building we conclude that $p \Longrightarrow q$. Actually an identity is a biconditional. Since algebraic operations and substitutions are really equivalences, each of the preceding $\Longrightarrow$ can be replaced by $\Longleftrightarrow$. Thus we have simultaneously proved $q \Longrightarrow p$. Note that this is a special case and in general we can not replace $\Longrightarrow$ by $\Longleftrightarrow$. $\blacksquare$

Warning: Example 4 is more difficult than any of the preceding examples and uses theory that appears in calculus. This example can be omitted without loss of continuity in the remaining discussions in this chapter. If you choose to omit this example, then Exercise 7 should also be omitted.

Example 4. Prove $\lim\limits_{x \to 3} x^2 = 9$.

Proof: According to the definition of a limit we proceed as follows. Let $\epsilon > 0$ be given. We must find $\delta > 0$ such that whenever x is chosen such that $0 < |x - 3| < \delta$, then we have $|x^2 - 9| < \epsilon$. Let

p: $0 < |x - 3| < \delta$ for some appropriate choice of δ

q: $|x^2 - 9| < \epsilon$

In order to determine δ, which will depend on ϵ, we start backwards.

$$q : \; |x^2 - 9| < \epsilon \iff \underbrace{|x + 3|\,|x - 3| < \epsilon}_{b_1}$$

Note that the term $|x - 3|$ which is important to p has appeared in this equivalent form of q. Near $x = 3$, $|x+3|$ is near 6 and $|x-3|$ is near 0. The next step is to move backwards from b_1 so that we leave the term $|x-3|$ alone but replace the term $|x+3|$ with a constant. Eventually as $x \to 3$ we can consider x in a small interval around 3. Suppose we take $2.9 < x < 3.1$ which is equivalent to saying $|x - 3| < 0.1$ and which is equivalent to $5.9 < x + 3 < 6.1$. For $2.9 < x < 3.1$, $x+3 = |x+3|$ so $|x+3| < 6.1$. Thus we can see that b_1 is implied by b_2 where

$$b_1 \Longleftarrow \underbrace{6.1|x - 3| < \epsilon \text{ when } 0 < |x - 3| < 0.1}_{b_2}$$

We see that

$$b_2 \Longrightarrow \underbrace{|x - 3| < \frac{\epsilon}{6.1} \text{ when } 0 < |x - 3| < 0.1}_{b_3}$$

In b_3 we have two conditions that $|x - 3|$ is to satisfy in order to imply b_2 which in turn implies b_1 which in turn implies q. We combine these conditions into one as follows. Let

$$\delta = \min\{0.1, \epsilon/6.1\},$$

then

$$b_3 \Longleftarrow \underbrace{0 < |x - 3| < \delta}_{p}$$

In summary, if we choose δ as defined above (or smaller), then $p \Longrightarrow q$.   ∎

The proof of a biconditional requires that we prove both $p \Longrightarrow q$ and $q \Longrightarrow p$. If in proving $p \Longrightarrow q$ each building block used is an equivalence then we have simultaneously proved $q \Longrightarrow p$. See Example 3.

Warning: You must check to determine that each building block in the bridge from p to q is an equivalence. You can *not* assume that it is.

Another proof technique replaces the original statement $p \Longrightarrow q$ by an equivalent statement and then proves the new statement. Such a procedure is called an **indirect method of proof**. One indirect method uses the equivalence between $p \Longrightarrow q$ and its contrapositive $(\sim q) \Longrightarrow (\sim p)$. The proof of $(\sim q) \Longrightarrow (\sim p)$ is done directly. We call this **proof by contrapositive**. Unfortunately there is no specific indicator in a conditional that informs you that an indirect proof by contrapositive may be fruitful. Sometimes the appearance of the word not in the conclusion q is a *suggestion* to try this method. There are no guarantees that it will work. We illustrate the use of proof by contrapositive in Examples 5 and 6.

Example 5.   Use proof by contrapositive to show:

If $3m$ is an odd number, then m is odd.

Proof: Let p: $3m$ is odd, and q: m is odd. The contrapositive, $(\sim q) \Longrightarrow (\sim p)$ is

If $\underbrace{m \text{ is even}}_{(\sim q)}$, then $\underbrace{3m \text{ is even}}_{(\sim p)}$.

Using forward building we have

$$(\sim q) \Longrightarrow \underbrace{m = 2k, \text{ for some integer } k}_{b_1} \quad \{\text{a meaning of even}\}$$

We see that $(\sim p)$ involves $3m$, so we arrange to get $3m$ into the act.

$$b_1 \Longrightarrow \underbrace{3m = 3(2k) = 2(3k)}_{b_2} \quad \{\text{by algebra}\}$$

Thus $3m$ is expressed as 2 times another number, hence $3m$ is even. That is,

$$b_2 \Longrightarrow (\sim p)$$

In summary, $(\sim q) \Longrightarrow (\sim p)$ which is equivalent to $p \Longrightarrow q$.   ■

Example 6.   Use proof by contrapositive to show:

If a and b are positive real numbers such that $4ab$ is not equal to $(a + b)^2$ then a is not equal to b.

Proof: Let p: $4ab \neq (a + b)^2$, and q: $a \neq b$. The contrapositive $(\sim q) \Longrightarrow (\sim p)$ is

$$\text{If } a = b, \text{ then } 4ab = (a + b)^2.$$

Using backward building we have

$$(\sim p) \Longleftarrow \underbrace{4ab = (a + b)^2 = a^2 + 2ab + b^2}_{b_1} \quad \{\text{by algebra}\}$$

$$b_1 \Longleftarrow \underbrace{0 = a^2 - 2ab + b^2}_{b_2} \quad \{\text{by algebra}\}$$

$$b_2 \Longleftarrow \underbrace{0 = (a - b)^2}_{b_3} \quad \{\text{by algebra}\}$$

$$b_3 \Longleftarrow \underbrace{0 = |a - b|}_{b_4} \quad \{\text{take square root of both sides}\}$$

$$b_4 \Longleftarrow \underbrace{a = b}_{(\sim q)} \quad \{\text{by algebra}\}$$

In summary, $(\sim p) \Longleftarrow b_1 \Longleftarrow b_2 \Longleftarrow b_3 \Longleftarrow b_4 \Longleftarrow (\sim q)$ and the equivalence of conditional and contrapositive tells us $p \Longrightarrow q$. In this proof each $\Longleftarrow$ can be replaced by $\Longleftrightarrow$ because the rules of algebra employed are equivalences. Hence the original statement could have been phrased as an if and only if statement.   ■

A second indirect method of proof, called **proof by contradiction**, uses the equivalence

$$(p \Longrightarrow q) \Longleftrightarrow ((p \wedge (\sim q)) \Longrightarrow c)$$

where c is a statement that is always false. (See Exercise 10.) The motivation for this procedure can be seen by referring to Table 8 in Section 1. the only way $p \Longrightarrow q$ can be false is if hypothesis p is true and the conclusion q is false. Proof by contradiction assumes "p is true and q is false" and then attempts to build a logical bridge to a statement that is known never to be true. When this is done we say we have reached a contradiction, hence our additional hypothesis "q is false" must be incorrect. The preceding equivalence then implies that $p \Longrightarrow q$ is true. If we are unable to build our bridge to some always false statement, then this technique of proof just fails to verify that $p \implies q$ is true. We can not claim $p \implies q$ is false. Possibly we were not clever enough to build the bridge. In such cases an alternate proof technique may yield the result. As with proof by contrapositive, unfortunately, there is no specific indicator within $p \implies q$ that informs you that proof by contradiction may be fruitful. Sometimes the appearance of the word not in the conclusion q is a suggestion to try this method. Another possible indicator is if a is one of two possible alternatives like even/odd, rational/irrational, or prime/composite (having factors other than itself and 1). (One indicator of this type in linear algebra is, "a square matrix is either singular or nonsingular.")

A seemingly added complication in proof by contradiction is that we must be on the lookout for an "always false statement". The specific always false statement varies from proof to proof. Things like $0 = 1$ or a number is both rational and irrational are obviously false. In linear algebra your logic bridge starting with hypothesis $p \wedge (\sim q)$ may lead to some absurdity dealing with matrices or other linear algebra notion.

If we are using proof by contradiction, then we begin by assuming both p and $(\sim q)$ are true. This gives us an additional hypothesis to work with. We use forward building taking both p and $(\sim q)$ into account in order to use equivalences and implications to obtain simpler statements. Details are dependent upon the situation. We give two classic illustrations of proof by contradiction in Examples 7 and 8.

Example 7. Prove by contradiction:

$$\text{If } x = \sqrt{2}, \text{ then } x \text{ is irrational.}$$

Proof: Let p: $x = \sqrt{2}$, and let q: x is irrational.

Assume $p \wedge (\sim q)$ are true; that is, $x = \sqrt{2}$ and x is rational. The assumption x is rational implies that there is a fraction n/d such that

$$\sqrt{2} = \frac{n}{d}$$

We know that every fraction can be reduced to lowest terms so we can take the fraction n/d to be in lowest terms which is equivalent to saying that n and d have no common factors other than 1. Hence

$$p \wedge (\sim q) \implies \underbrace{\sqrt{2} = \frac{n}{d}}_{b_1}$$

Using algebra we have

$$b_1 \iff 2 = \frac{n^2}{d^2} \iff \underbrace{2d^2 = n^2}_{b_2}$$

Statement b_2 implies n^2 is even. The only way n^2 is even is for n to be even. So let $n = 2k$ and we have

$$b_2 \implies 2d^2 = n^2 = (2k)^2 \iff 2d^2 = 4k^2 \iff \underbrace{d^2 = 2k^2}_{b_3}$$

$$b_3 \implies d^2 \text{ is even and as before this implies that } d \text{ is even.}$$

We now have that both n and d are even which implies that they have a common factor of 2. This contradicts the fact that n/d is in lowest terms. Thus our assumption $(\sim q)$ is invalid and it follows that $p \implies q$. ∎

Example 8. Prove by contradiction:

$$\text{If } S \text{ is the set of all prime numbers, then } S \text{ has infinitely many members.}$$

Proof: Let p: S is the set of *all* prime numbers
        q: S has infinitely many members

Assume $p \wedge (\sim q)$ is true; that is, assume that S is the set of all primes and has only finitely many members. Since S has only a finite number of members, there is a largest member, call it n.

$$p \wedge (\sim q) \implies S \text{ contains all primes and there is the largest member } n \text{ of } S.$$

This suggests that we try to construct a prime bigger than n or one that is not in S. We proceed as follows. Form the number

$$k = (n \times (n-1) \times \cdots \times 2 \times 1) + 1 = n! + 1$$

Number k is not evenly divisible by any of the numbers $2, 3, 4, \ldots, n$ since k divided by any of these will leave a remainder of 1. This implies k has no prime factors which are in S since the largest prime is n. The number k is either a prime greater than n or else k is not prime in which case it is divisible by a prime greater than n. In either case we contradict that S contains all the primes. Thus our assumption $(\sim q)$ is invalid and it follows that $p \implies q$. ∎

There is an important difference between direct proofs and indirect proofs. A direct proof (in the forward direction) starts with the hypothesis p and step-by-step builds conclusion q. We call this a **constructive proof** because q is actually built (by logical arguments). However, an indirect proof, say by contradiction, provides an argument that q follows from p without actually building q. We call this an **existence proof**. Note that in Example 7 we did not explicitly show $\sqrt{2}$ had a nonterminating nonrepeating decimal expansion, which is what a construction of an irrational number should look like. Similarly in Example 8 the infinite set of primes was not constructed. That is, we did not list its elements or give a way to determine all of its elements. (At present, such a construction is not known.) To see some of the power in an indirect proof, try to prove either Example 7 or 8 directly. A number of existence proofs appear in calculus.

Beware of the way that certain conditionals are stated. Often it is not explicitly stated that some statement is true for all objects of a certain type. For instance, the conditional statement in Example 1 could have been phrased,

If m and n are even, then $m + n$ is even.

From the usual meaning of even, you are to infer that m and n are to be considered integers. In Example 2, the description "Let x be a real number." could have been omitted and you are then to infer from context (and experience) that we want real numbers not just integers. In Example 3, the result is to be valid for every angle θ. You cannot verify that the expression is true for a particular choice, say $\theta = 0$ or $\theta = \pi/2$, and then claim that it holds for all values of θ.

Most "proof situations" that you will encounter explicitly are of the form "prove $p \implies q$" or "prove $p \iff q$". The particular statement may be labeled a Theorem, Corollary, or Exercise. However, occasionally you may encounter a situation like

"Show that statement p is false."

In this case you need only find one example where statement p fails to be true. We call this determining a **counterexample** to p. There could be many counterexamples to p each as good as any other. Example 9 gives a classic use of a counterexample from calculus.

Example 9.   Show that not every continuous function is differentiable.

Discussion: A simple counterexample used in calculus is to consider $f(x) = |x|$ on $[-1, 1]$. Function f is continuous at every point of $[-1, 1]$, but fails to be differentiable at $x = 0$. See your calculus book for details. ∎

A related situation is,

Prove or disprove $p \implies q$.

A fundamental decision that you must make is, do you think the statement is true or false. If you think it is false, then you try to construct a counterexample. If you think it is true, then you try to prove it using techniques discussed in this section. While there are no explicit guidelines for making this decision there are several things to keep in mind.

▶ Think of equivalences for p.

▶ Is this the contrapositive of a previous result?

▶ Is this the converse of a previous result?

▶ Does the statement relate to any special cases discussed in the text?

▶ The appearance of words like every, all, never, and always are a signal to consider the corresponding statement very carefully.

Exercises 0.2

1. Use forward building to prove:

   If m and n are odd integers, then $m + n$ is even.

2. Use forward building to prove:

   If m, n, and k are integers and m divides n and m divides k, then m divides $n + k$.

   (Hint: x divides $y \iff y = sx$, for some integer s.)

3. Use backward building to prove:

   $$|x - 3| < 2 \quad \text{implies that} \quad 1 < x < 5.$$

4. Prove: $|a - b| \geq |a| - |b|$ for any real numbers a and b.
   (Hint: Assume that you know $|a + b| \leq |a| + |b|$ and start with $|a| = |a + (b - b)|$.)

5. Prove: $\dfrac{\sin\theta}{\sec\theta} = \dfrac{1}{\tan\theta + \cot\theta}$

6. Prove: $\dfrac{1 - \cos\theta}{\sin\theta} = \dfrac{\sin\theta}{1 + \cos\theta}$.

7. Prove: $\lim\limits_{x \to 0} \dfrac{1}{x + 1} = 1$ using a δ-ϵ approach as in Example 4.

8. Use proof by contradiction to prove:

   If m^2 is odd, then m is odd.

9. The **inverse** of $p \implies q$ is $\sim p \implies \sim q$.

   (a) Show that $p \implies q$ and its inverse are not equivalent.

   (b) State a relationship between the inverse, converse, and contrapositive of $p \implies q$.

10. Prove: $(p \implies q) \iff ((p \wedge (\sim q)) \implies c$, where c is any statement that is always false.

11. Prove by contradiction that

    $$\text{If } x > 0, \text{ then } \frac{1}{x} > 0.$$

12. Prove by contradiction that

    If x is irrational, then $3x$ is irrational.

13. We have used the fundamental properties of even and odd integers several times. Prove by contradiction that

    If 2 divides integer m, then m is even.

14. Prove or disprove: There are no consecutive integers m, n, and k so that $m^2 + n^2 = k^2$.

15. Let $\{a_n\}$ represent sequence $a_1, a_2, \ldots$. In addition suppose that a_i is positive for every i and that $\{a_n\}$ is convergent. Prove or disprove: $\lim\limits_{n \to \infty} a_n = 0$.

16. Prove or disprove: If n is an even integer, then $n^2 + n + 4$ is even.

Section 0.3 Sets

In Appendix A.1 of the text there is a brief discussion of sets. We will assume that you have read Appendix A.1 or are familiar with the concepts of **sets**, **subsets**, **equal sets**, and the **empty set**. In this manual we will use the following terminology and notation.

▶ Capital letters denote sets.

▶ Lower case letters denote elements or members of a set.

▶ The equivalent statements "x belongs to A", "x is an element of A", and "x is in A" are denoted

$$x \in A$$

▶ The equivalent statements "x does not belong to A", "x is not an element of A", and "x is not in A" are denoted

$$x \notin A$$

▶ The statement "B is a subset of A" means each element of B is in A and is denoted

$$B \subseteq A$$

▶ The statement "B is not a subset of A" means there is at least one element of B that is not in A. We denote this by

$$B \nsubseteq A$$

▶ The empty set is denoted $\varnothing$.

▶ A set A equals a set B, denoted $A = B$, if each element of A is also in B and each element of B is also in A.

In our study of linear algebra we will use ideas about subsets and equal sets. Using the language of logic from Section 2 we have

$$\boxed{B \subseteq A} \text{ is equivalent to } \boxed{\text{if } x \in B, \text{ then } x \in A}$$

Hence to show that B is a subset of A we need to prove that every element of B is in A. The building blocks for such a proof will involve the definitions for membership to sets A and B. This type of proof will be used in Chapter 2 with vector spaces and subspaces. We illustrate the concept of subsets with Examples 1–5, which should be familiar from algebra and calculus.

Example 1. Let

$$N = \text{all positive integers}$$
$$Z = \text{all integers}$$
$$Q = \text{all rational numbers}$$
$$W = \text{all irrational numbers}$$
$$R = \text{all real numbers}$$

The following set relations are well known.

(a) $N \subseteq Z$ (b) $Z \subseteq Q$ (c) $Q \subseteq R$ (d) $W \subseteq R$

(e) $W \nsubseteq Q$, see the proof of Example 7 in Section 2

(f) $N \subseteq Z \subseteq Q \subseteq R$, follows from (a), (b), and (c). ∎

Example 2. Let k be a nonnegative integer and P_k denote the set of polynomials of degree k or less.

(a) $P_1 \subseteq P_2$

    Proof: We show that

$$\text{if } \underbrace{f(x) \in P_1}_{p} \text{ then } \underbrace{f(x) \in P_2}_{q}.$$

$$p \Longleftrightarrow f = a_0 + a_1 x, \text{ for any real numbers } a_0, a_1$$
$$\Longleftrightarrow f(x) = a_0 + a_1 x + 0 x^2$$
$$\Longrightarrow f(x) \in P_2, \text{ since it is of the form } a_0 + a_1 x + a_2 x^2,$$
$$\text{where } a_0,\ a_1,\ a_2 \text{ are real numbers}$$

    Thus $p \Longrightarrow q$ or $P_1 \subseteq P_2$.

(b) $P_2 \subset P_3$ (Verify by following the pattern of proof in (a).)

(c) $P_1 \subseteq P_2 \subseteq \cdots \subseteq P_k \subseteq P_{k+1} \subseteq \cdots$ ∎

Example 3. Let

$$C(a, b) = \text{set of all continuous functions over interval } (a, b)$$
$$C^1(a, b) = \text{set of all differentiable functions over interval } (a, b)$$

Let P_k be defined as in Example 2.

(a) $P_k \subseteq C(a, b)$ for any k and any interval (a, b)
That is, every polynomial (of degree k or less) is continuous over any interval (a, b). This is shown in calculus.

(b) $P_k \subseteq C^1(a, b)$, for any k and any interval (a, b)
That is, every polynomial (of degree k or less) has a derivative at each point of (a, b). This is shown in calculus.

(c) $C^1(a, b) \subseteq C(a, b)$
That is, every function that is differentiable over (a, b) is continuous over (a, b). This is shown in calculus.

(d) $C(a, b) \nsubseteq C^1(a, b)$
That is, there exists a continuous function over (a, b) that is not differentiable at some point of (a, b). See Example 9 in Section 2 for a particular case. ∎

Example 4. Let $A = $ all lines not parallel to $x + y = 1$ and let $B = $ all lines of slope $m = 1$. Prove $B \subseteq A$.

    Proof: We show that if line $\underbrace{\ell \in B}_{p}$ then $\underbrace{\ell \in A}_{q}$.

$$p \Longrightarrow \underbrace{\text{slope of line } \ell \text{ is } 1}_{b_1} \Longrightarrow \underbrace{\ell \text{ does not have slope } -1}_{b_2}$$

    $b_2 \Longrightarrow q$ since $x + y = 1 \Longleftrightarrow y = -x + 1$ which means members of A have slope $\neq -1$.

    Thus $p \Longrightarrow q$ or $B \subseteq A$. ∎

Example 5. Let $B = $ the intercepts of $y = x^2 - 1$ and $A = $ the points on circle $x^2 + y^2 = 1$. Prove $B \subseteq A$.

Proof: We show that each element in B belongs to A. From algebra $B = \{(0, -1), (1, 0), (-1, 0)\}$. Using algebra it is easy to show that $(0, -1) \in A$, $(1, 0) \in A$, and $(-1, 0) \in A$. Thus $B \subseteq A$. ∎

We continue our illustrations of subsets using concepts that appear in Chapter 1 in the text. For each of these examples we specify the section to which it is related.

Example 6. (Section 1.1) Let A be the set of all systems of 2 equations in 2 unknowns and let B be the set of pairs of perpendicular lines. We have $B \subseteq A$ since a pair of perpendicular lines can be represented by system

$$\begin{cases} y = mx + b \\ y = (-1/m)x + c \end{cases}, \quad m \neq 0 \qquad \text{or} \qquad \begin{cases} y = b \\ x = c \end{cases}, \quad m = 0$$

Such representations are systems of 2 equations in 2 unknowns. ∎

Example 7. (Section 1.1) Let A be the set of all consistent systems of m equations in n unknowns, for any choice of positive integers m and n. Let B be the set of all pairs of distinct parallel lines. We have $B \nsubseteq A$. We see this as follows. A pair of distinct lines is represented by a system of 2 equations in 2 unknowns. Since the lines are parallel they do not intersect; hence there is no solution. ∎

Example 8. (Section 1.1) Let A be the set of all consistent systems of m equations in n unknowns, for any choice of positive integers m and n. Let B be the set of all homogeneous systems of equations regardless of size. We have $B \subseteq A$ since every homogeneous system has the trivial solution. Hence we can say *every homogeneous system is consistent*. ∎

Example 9. (Section 1.4) Let D be the set of $n \times n$ diagonal matrices and let S be the set of $n \times n$ symmetric matrices. We have $D \subseteq S$ since $D^T = D$. This result is true for each positive integer n, thus we can say *every diagonal matrix is symmetric*. ∎

Example 10. (Section 1.4) Let S be the set of $n \times n$ symmetric matrices and let N be the set of $n \times n$ nonsingular (or invertible) matrices. Is $S \subseteq N$?

Discussion: We are asked whether $S \subseteq N$ is true or false. If we suspect it is true, then we must produce a proof. However, if we suspect it is false, then we must find a counterexample. The counterexample in this case would be an $n \times n$ symmetric matrix A for which there is no matrix B (of the same size) such that $AB = I_n$. Note that O_n is symmetric, but

$$O_n \times (\text{any } n \times n \text{ matrix}) = O_n \neq I_n.$$

Thus we have a counterexample. Hence $S \nsubseteq N$. See also Example 11 in Section 1.4. ∎

Example 10 illustrates the following:

$$\boxed{B \nsubseteq A} \iff \text{there is at least one } x \in B \text{ such that } x \notin A$$

A useful set of equivalences for equality of sets follows.

$$\begin{aligned} A = B &\iff (\text{every element in } A \text{ is in } B) \wedge (\text{every element in } B \text{ is in } A) \\ &\iff (x \in A \implies x \in B) \wedge (x \in B \implies x \in A) \\ &\iff (A \subseteq B) \wedge (B \subseteq A) \end{aligned}$$

Thus we see that to prove equality of sets there are two things to be shown:

$$1. \ A \subseteq B \qquad \text{and} \qquad 2. \ B \subseteq A$$

Only if both are verified to be true is $A = B$.

The following discussion assumes that you have read Section 1.4.

A major goal in linear algebra is the study of linear systems of equations in which the coefficient matrix is nonsingular. It is important to be able to identify nonsingular matrices. What is obtained in parts of Chapters 2–6 are various alternative descriptions of the set of nonsingular matrices. The proof that we have a set of matrices equal to the set of nonsingular matrices will involve the two steps described above. The text emphasizes such "equivalent descriptions" in highlighted areas. Be on the look out for such equivalences. They should be studied carefully for they play a major role in theoretical as well as computational concepts.

Section 0.4 Mathematical Induction (Optional)

In Section 1 we introduced the ideas of elementary logic to provide a basis upon which to introduce techniques of proof. The discussion in Section 0.2 of techniques of proof can be succinctly summarized as follows:

From a stated hypothesis reason to a valid conclusion.

The term "reason" can be thought of as building a logic bridge. Our building blocks were other truths. The connection of one truth to the next formed our bridge from the hypothesis to the conclusion. This procedure is called **deductive reasoning** and is generally what is meant by a formal proof in mathematics. A new truth is deduced from other truths.

Another type of reasoning is known as **inductive reasoning**. Inductive reasoning involves collecting evidence from experiments or observations and using this information to formulate a general law or principle. Inductive reasoning attempts to go from the specific to the general, but even with large quantities of evidence the conclusion is not guaranteed. In general, mathematics rejects direct inductive reasoning. However, mathematics often uses an inductive process to formulate conjectures which are then subjected to rigorous deductive reasoning before they are accepted. Remember that a conjecture is a conclusion based on incomplete evidence—that is, a guess. Much can be gained from using experimental evidence to suggest conjectures and then applying deductive arguments to determine the truth or falsity of the conjecture. A number of important advances in science and engineering evolved in just this way.

The cycle of

> **experiment(s) — conjecture — check by deductive reasoning**

is very important in mathematics. Here we introduce one method for performing the "deductive check" on certain conjectures that can be derived from experiments that involve using the natural numbers. (The natural numbers N are the set $\{1, 2, 3, \dots\}$.) The process we describe below is called **Mathematical Induction**, but it is really a particular deductive checking procedure *not* reasoning by induction.

Experiments are often performed in mathematics to determine patterns of behavior. Patterns of behavior are often reformulated into mathematical properties and mathematical theorems. Experiments in mathematics often take the form of looking at special cases and trying to see some common pattern in order to derive a conjecture. The special cases in the patterns we explore are obtained by using the first "few" natural numbers.

Example 1.   Determine a conjecture for a formula to compute the sum of consecutive natural numbers.

Discussion: The experiments for the cases of 1, 2, 3, 4, and 5 consecutive natural numbers are shown next.

$$
\begin{aligned}
1 &= 1 \\
1 + 2 &= 3 \\
1 + 2 + 3 &= 6 \\
1 + 2 + 3 + 4 &= 10 \\
1 + 2 + 3 + 4 + 5 &= 15
\end{aligned}
$$

Looking for a pattern in the sums is not easy. Try to connect the sum to the largest integer used in the set of addends. (We do not always have to start looking for a pattern at the "beginning".) Looking at

$$1 + 2 + \boxed{3} \qquad = \boxed{6}$$
$$1 + 2 + 3 + \boxed{4} = 10$$

we see that $\boxed{3}\,\boxed{4} = 12 = 2\,\boxed{6}$. Looking at

$$1 + 2 + 3 + \boxed{4} \qquad = \boxed{10}$$
$$1 + 2 + 3 + 4 + \boxed{5} = 15$$

we see that $\boxed{4}\,\boxed{5} = 20 = 2\,\boxed{10}$. Checking the first two sums we see that the same pattern holds. For

$$\boxed{1} \qquad\qquad = \boxed{1}$$
$$1 + \boxed{2} \qquad = 3$$

$\boxed{1}\,\boxed{2} = 2 = 2\,\boxed{1}$. Checking the second pair of sums we see that the same pattern holds. (Verify.) Next we adjoin another row to the experiment table above:

$$1 + 2 + 3 + 4 + 5 + 6 = 21$$

Checking the last pair of rows:

$$\boxed{5}\,\boxed{6} = 2\,\boxed{15}$$

so the pattern holds here also. To formulate a conjecture we proceed as follows. Let the largest natural number used in a sum of consecutive natural numbers be k. Then we have

$$1 + 2 + \cdots + k - 1 + \boxed{k} \qquad\qquad = S_k$$
$$1 + 2 + \cdots + k - 1 + k + \boxed{k+1} = S_{k+1}$$

The pattern above suggests that

$$k(k+1) = 2S_k$$

or equivalently

$$S_k = \frac{k(k+1)}{2}$$

Using summation notation (see Section 1.2) we have

$$S_k = 1 + 2 + \cdots + k = \sum_{i=1}^{k} i$$

Hence our conjecture is

If we form the sum S_k of the first k consecutive natural numbers then $S_k = \displaystyle\sum_{i=1}^{k} = \frac{k(k+1)}{2}$.

We have only done the experiment and conjecture steps of our cycle. Before we can claim that the sum of the first k consecutive natural numbers is $k(k+1)/2$ we must use deductive reasoning. (See Example 3.) ∎

Example 2. Determine a conjecture for a relationship between the expressions $(1+x)^k$ and $1 + kx$ where x is such that $1 + x > 0$ and k is any natural number.

Discussion: We experiment with the first few natural numbers k and various values of x.

Case $k = 1$: x any real number so that $x > -1$
$$(1 + x)^1 = 1 + 1x$$

Case $k = 2$: $x = -0.5$ $(1 + (-0.5))^2 = 0.25$ $1 + 2(-0.5) = 0$

$x = 0$ $(1 + 0)^2 = 1$ $1 + 2(0) = 1$

$x = 1$ $(1 + 1)^2 = 4$ $1 + 2(1) = 3$

$x = 15$ $(1 + 15)^2 = 256$ $1 + 2(15) = 31$

Summary: for the few cases above, $(1 + x)^2 > 1 + 2x$

Case $k = 3$: $x = -0.75$ $(1 + (-0.75))^3 = 0.015625$ $1 + 3(-0.75) = -1.25$

$x = -0.1$ $(1 + (-0.1))^3 = 0.970299$ $1 + 3(-0.1) = 0.7$

$x = 2$ $(1 + 2)^3 = 27$ $1 + 3(2) = 7$

$x = 7$ $(1 + 7)^3 = 512$ $1 + 3(7) = 22$

$x = 20$ $(1 + 20)^3 = 9261$ $1 + 3(20) = 61$

Summary: for the few cases above, $(1 + x)^3 > 1 + 3x$

This limited evidence suggests that we form the conjecture

If k is any natural number and $x > -1$, then $(1 + x)^k \geq 1 + kx$.

Before we can claim that this relationship is true we must use deductive reasoning to supply a proof. (See Example 4.) ∎

Note that the conjecture in both examples is stated as a conditional $p \implies q$. The technique of Mathematical Induction which we define next is a method of proof for the special kind of conditional statements which appear in Examples 1 and 2. The principle of **Mathematical Induction** is stated as follows:

If S is a set of natural numbers with
    (a) $1 \in S$
and

    (b) $n \in S \implies n + 1 \in S$ for each natural number n,

then $S = N$, the set of all natural numbers.

The set S is the set of natural numbers for which the conjecture $p \implies q$ is true. The method of mathematical induction says

first: show $p \implies q$ for $n = 1$
next: assume $p \implies q$ for arbitrary natural number n and
     prove that $p \implies q$ is true for natural number $n + 1$.

If both steps are successful the principle of mathematical induction guarantees that $p \implies q$ is true for *every* natural number.

Warning: The proof for part (b) depends on the contents of the implication $p \implies q$ and many times requires ingenuity.

To illustrate the principle of mathematical induction we prove the conjectures developed in Examples 1 and 2. (Terminology: The name mathematical induction is often shortened to just **induction**.)

Example 3. Apply induction to prove the conjecture developed in Example 1. Let S be the set of all natural numbers k for which

$$\sum_{i=1}^{k} = \frac{k(k+1)}{2}$$

From the experiments in Example 1, we have 1, 2, 3, 4, 5 $\in S$. The principle of induction says assume that $n \in S$; that is,

$$\sum_{i=1}^{n} i = 1 + 2 + \cdots + n = \frac{n(n+1)}{2} \tag{1}$$

Then we must verify that $n + 1 \in S$; that is, we must show

$$\sum_{i=1}^{n+1} i = 1 + 2 + \cdots + n + n + 1 = \frac{(n+1)(n+2)}{2} \tag{2}$$

(Note that n in the right-hand side of (1) was replaced by $n + 1$ to obtain the right-hand side of (2).) In order to verify that (2) is true we use (1) which is assumed to be true. (The expression in (1) is called the **induction hypothesis** for this conjecture.) Starting with the expression

$$\sum_{i=1}^{n+1} i$$

we must show algebraically that we can produce the formula

$$\frac{(n+1)(n+2)}{2}.$$

We have

$$\sum_{i=1}^{n+1} i = \sum_{i=1}^{n} i + (n+1) \qquad \{\text{by properties of sums}\}$$

$$= \frac{n(n+1)}{2} + (n+1) \qquad \{\text{by (1)}\}$$

$$= \frac{n(n+1) + 2(n+1)}{2} \qquad \{\text{by algebra}\}$$

$$= \frac{(n+1)(n+2)}{2} \qquad \{\text{by factoring}\}$$

Hence we have shown that if $n \in S$ then $n + 1 \in S$. Thus the principle of induction implies that $S = N$; that is, the formula

$$\sum_{i=1}^{k} i = \frac{k(k+1)}{2}$$

is valid for all natural numbers.   ■

The deductive reasoning in the principle of mathematical induction is the proof *we must supply* to show part

$$\text{(b): } n \in S \implies n + 1 \in S.$$

This step is itself a conditional statement that must be proven using its hypothesis, $n \in S$, and appropriate building blocks. The inductive hypothesis, $n \in S$, does *not* mean we are assuming what we want to prove. We are not assuming what we want to prove because we must supply a proof that $n + 1 \in S$. Thus to prove (b) we must adhere to the rules about proving conditionals which are stated in Table 8 in Section 0.1. From Table 8 we see that (b) is false only if n is in S but $n + 1$ is not. Hence if we can show that whenever $n \in S$ that it must follow that $n + 1 \in S$, then the conditional (b) is always true. In summary, *assuming $n \in S$ does not assume what we must prove, namely that $n + 1 \in S$.*

Example 4. Apply induction to prove the conjecture developed in Example 2. Let S be the set of all natural numbers k such that

$$(1 + x)^k \geq 1 + kx, \quad \text{whenever } 1 + x > 0$$

From the experiments in Example 2 we have that $1 \in S$ and we suspect that 2 and 3 belong to S. It is only a suspicion since we have not verified the conjecture for all $x > -1$ when $k = 2$ or 3. We next verify the conditional in (b). Assume that $n \in S$; that is,

$$(1 + x)^n \geq 1 + nx \quad \text{for all } x > -1$$

We must prove deductively that $n + 1 \in S$: that is, prove that

$$(1 + x)^{n+1} \geq 1 + (n + 1)x \quad \text{for all } x > -1$$

We proceed as follows

$$
\begin{aligned}
(1 + x)^{n+1} &= (1 + x)^n(1 + x) & \{\text{by algebra}\} \\
&\geq (1 + nx)(1 + x) & \{\text{by the inductive hypothesis}\} \\
&= 1 + nx + x + nx^2 & \{\text{by algebra}\} \\
&= 1 + (n + 1)x + nx^2 & \{\text{by algebra}\} \\
&\geq 1 + (n + 1)x & \{\text{since } nx^2 > 0\}
\end{aligned}
$$

Hence by the principle of induction

$$(1 + x)^k \geq 1 + kx$$

for all $x > -1$ and for all natural numbers k.   ∎

Most cases in which you may need to use induction in linear algebra are already phrased as prove: $p \implies q$. The experiments and conjecture formulation stages of the process we have described have been done. You must recognize that the special structure of inductive proofs is present. That is, the natural numbers play a role in $p \implies q$. The building blocks in proving part (b) of the principle of induction will most likely be facts about matrices or other linear algebra concepts rather than ordinary algebra facts as in Examples 3 and 4. Proof by induction appears in only a few places in this manual but it is an important mathematical technique. Places where induction can be used are listed next:

> Section 1.4 Exercise 7;
> Supplementary Exercises for Chapter 1 Exercises 1(d) and 15;
> Section 4.4 Exercise 23;
> Supplementary Exercises for Chapter 5 Exercise 3.

For further reading on induction see the following sources

[1] H. Burrows, et al, Mathematical Induction, FIAM Module, COMAP, 1989.

[2] S. Lay, **Analysis, An Introduction to Proof**, Prentice Hall, 1986.

[3] L. Swanson and R. Hansen, Mathematical Induction or "What Good is All This Stuff if We Are Going to Assume It's True Anyway?", Two Year College Mathematics Journal, v.12, 1981, pp. 8–12.

[4] B. Youse, **Mathematical Induction**, Prentice Hall, 1964.

Exercises 0.4

1. Determine a conjecture for a formula to compute the sum of consecutive odd integers. Try to prove your conjecture by induction.

2. Determine a conjecture for a relationship between 2^k and $(k+1)!$ for natural numbers k. Try to prove your conjecture by induction.

3. Prove by induction that $\dfrac{d}{dx}(x^k) = kx^{k-1}$ for all natural numbers k.

4. Prove by induction that for every natural number k the expression $k^2 + k$ is divisible by 2.

5. Prove by induction that

$$\frac{1}{(1)(2)} + \frac{1}{(2)(3)} + \frac{1}{(3)(4)} + \cdots + \frac{1}{k(k+1)} = \frac{k}{k+1}$$

for all natural numbers k.

Solutions to Exercises

Section 0.1

1. (a) It is a statement which is false.

   (b) It is a true statement.

   (c) Not a statement since its truth depends on the value of x.

   (d) This is a question not a declarative sentence.

   (e) It is a true statement.

   (f) It is a true statement.

2. (a) The door is open and the tables are set.

   (b) The tables are set or the food is not ready.

   (c) The door is open and the tables are not set.

   (d) The door is open and the tables are set and the food is ready.

3. (a) $(\sim p) \wedge q$ (b) $r \vee q$ (c) $(\sim r) \wedge p$ (d) $\sim r$

4. (a)

p	q	$\sim q$	$p \wedge (\sim q)$
T	T	F	F
T	F	T	T
F	T	F	F
F	F	T	F

(b)

p	q	$\sim p$	$q \wedge (\sim p)$	$p \vee (q \wedge (\sim p))$
T	T	F	F	T
T	F	F	F	T
F	T	T	T	T
F	F	T	F	F

(c)

p	q	$\sim p$	$q \vee (\sim p)$	$p \wedge (q \vee (\sim p))$
T	T	F	T	T
T	F	F	F	F
F	T	T	T	F
F	F	T	T	F

(d)

p	q	$\sim q$	$p \vee (\sim q)$
T	T	F	T
T	F	T	T
F	T	F	F
F	F	T	T

5. (a) Divide both sides of the equation by 3.

   (b) Add -3 to both sides of the equation.

   (c) 2

   (d) $\dfrac{(x-1)(x+2)}{(x-1)(x+1)} = \dfrac{x+2}{x+1}, \; x \neq 1$

6. (a) p: it rains q: I will go to the movies

   (b) p: the diagonal of a rectangle are not perpendicular
   q: it is not a square

   (c) p: $x > 0$ and $y > 0$ q: $x + y > 0$

(d) p: $f(x)$ is a function on $[a, b]$
 q: $f(x)$ is continuous on $[a, b]$

(e) p: I will pay \$5.00 for a cup of coffee
 q: elephants fly

(f) p: T is an equilateral triangle q: all sides are equal

(g) p: opposite sides are parallel q: a quadrilateral to be a parallelogram

7.

p	q	$\sim q$	$\sim p$	$(\sim q) \Longrightarrow (\sim p)$
T	T	F	F	T
T	F	T	F	F
F	T	F	T	T
F	F	T	T	T

The last column is identical to the last column of Table 8, hence the statements are equivalent.

8. (a) hypothesis p: two lines are perpendicular
 conclusion q: intersect at right angles
 contrapositive: If two lines do not intersect at right angles, then they are not perpendicular.

(b) hypothesis p: $f(x)$ is a polynomial
 conclusion q: $f(x)$ is differentiable
 contrapositive: If $f(x)$ is not differentiable, then $f(x)$ is not a polynomial.

9. (a) If two lines intersect at right angles, then they are perpendicular.

(b) If $f(x)$ is differentiable, then $f(x)$ is a polynomial.

10. Show that the truth tables for $\sim (p \Longrightarrow q)$ and $p \wedge (\sim q)$ are identical. Using Table 8 we have

$\sim (p \Longrightarrow q)$
F
T
F
F

and

p	q	$\sim q$	$p \wedge (\sim q)$
T	T	F	F
T	F	T	T
F	T	F	F
F	F	F	F

Since the final columns are identical, the statements are equivalent.

11. Show that the truth tables for $\sim (p \wedge q)$ and $p \Longrightarrow (\sim q)$ are identical.

p	q	$p \wedge q$	$\sim (p \wedge q)$
T	T	T	F
T	F	F	T
F	T	F	T
F	F	F	T

p	q	$\sim q$	$p \Longrightarrow (\sim q)$
T	T	F	F
T	F	T	T
F	T	F	T
F	F	T	T

Since the final columns are identical, the statements are equivalent.

12. More than one equivalent statement may be possible. We present one such statement.

(a) evenly divisible by 2

(b) their sum is 90°.

(c) $|x| \leq 2$.

(d) n is odd.

(e) $a > 0$.

13. We construct the truth table for each part of the biconditional.

p	q	$p \Longleftrightarrow q$	$p \Longrightarrow q$	$q \Longrightarrow p$	$(p \Longrightarrow q) \wedge (q \Longrightarrow p)$
T	T	T	T	T	T
T	F	F	F	T	F
F	T	F	T	F	F
F	F	T	T	T	T

Since columns 3 and 6 are identical biconditional $p \Longleftrightarrow q$ is equivalent to conjunction $p \Longrightarrow q$ and $q \Longrightarrow p$.

Section 0.2

1. Prove $p \Longrightarrow q$ where p: m and n are odd, q: $m + n$ is even.

   Proof: $p \Longrightarrow \underbrace{m = 2k + 1 \; n = 2j + 1}_{b_1}$ for some integers k and j {since m and n are odd}

   $b_1 \Longrightarrow \underbrace{m + n = (2k + 1) + (2j + 1) = 2k + 2j + 2 = 2(k + j + 1)}_{b_2}$ {by algebra}

   $b_2 \Longrightarrow \underbrace{m + n \text{ is even}}_{q}$ {since it is a multiple of 2}

2. Prove $p \Longrightarrow q$ where p: $(m$ divides $n) \wedge (m$ divides $k)$, q: m divides $n + k$

   Proof: $p \Longrightarrow \underbrace{(n = ma) \wedge (k = mb)}_{b_1}$ for some integers a and b {using the hint}

   $b_1 \Longrightarrow \underbrace{n + k = ma + mb = m(a + b)}_{b_2}$ {by algebra}

   $b_2 \Longrightarrow \underbrace{m \text{ divides } n + k}_{q}$ {by the hint}

3. Prove $p \Longrightarrow q$, where p: $|x - 3| < 2$ and q: $1 < x < 5$.

   Proof: Use backward building.

   What implies $1 < x < 5$, keeping in mind we want to involve the quantity $x - 3$?

   Starting with $1 < x < 5$, to obtain an equivalent expression involving $x - 3$, subtract 3 from each piece of the inequality and simplify. Thus we have

   $$q \Longleftarrow \underbrace{1 - 3 < x - 3 < 5 - 3 \Longleftrightarrow -2 < x - 3 < 2}_{b_1}$$

   Using properties of absolute values we have

   $$b_1 \Longleftarrow \underbrace{|x - 3| < 2}_{p}$$

4. Use forward building starting with the hint:

   $$|a| = |a - 0| = |a - (b - b)|$$

   The objective is to show an inequality involving $|a - b|$, so we try to involve this quantity:

   $$|a - (b - b)| = |(a - b) + b|$$

The hint also implies that we can use $|a + b| \le |a| + |b|$. We can interpret this as saying

$$|\text{quantity } \#1 - \text{quantity } \#2| \le |\text{quantity } \#1| + |\text{quantity } \#2|$$

Do not let the symbols a and b prejudice you into thinking you can *only* use this hint with a and b. Let $a - b = \text{quantity } \#1$ and let $b = \text{quantity } \#2$. Then apply the hint. We obtain

$$|(a - b) + b| \le |a - b| + |b|$$

Putting things together we have

$$|a| \le |a - b| + |b|$$

but by algebra this is equivalent to

$$|a| - |b| \le |a - b| \iff |a - b| \ge |a| - |b|$$

This proof is more involved (tricky) than our examples or preceding exercises. Note that the first step is not obvious without the hint. Also note that we use another result, $|a + b| \le |a| + |b|$, called the **triangle inequality**. We had to interpret the meaning of it so we did not let the "generic use" of symbols a and b cloud the issue.

5. Using forward building we have

$$\frac{\sin \theta}{\sec \theta} \iff \frac{\sin \theta}{1/\cos \theta} \iff \sin \theta \cos \theta$$

Using backward building we have

$$\frac{1}{\tan \theta + \cot \theta} \iff \frac{1}{\dfrac{\sin \theta}{\cos \theta} + \dfrac{\cos \theta}{\sin \theta}} \iff \frac{1}{\dfrac{\sin^2 \theta + \cos^2 \theta}{\sin \theta \cos \theta}} \iff \sin \theta \cos \theta$$

Since we used (algebraic) equivalence at each step the biconditional is proved.

6. To start with

$$\frac{1 - \cos \theta}{\sin \theta}$$

and eventually obtain an expression with $1 + \cos \theta$ in the denominator, we must determine a way to introduce $1 + \cos \theta$ into the denominator. Here we try the equivalence of multiplying by 1 in the disguise $\dfrac{1 + \cos \theta}{1 + \cos \theta}$.

$$\frac{1 - \cos \theta}{\sin \theta} \iff \frac{(1 - \cos \theta)(1 + \cos \theta)}{\sin \theta \,(1 + \cos \theta)} \iff \frac{1 - \cos^2 \theta}{\sin \theta \,(1 + \cos \theta)}$$

$$\iff \frac{\sin^2 \theta}{\sin \theta \,(1 + \cos \theta)} \iff \frac{\sin \theta}{1 + \cos \theta}$$

Since we used (algebraic) equivalence at each step the biconditional is proved.

7. Let $\epsilon > 0$ be given. We must find $\delta > 0$ so that whenever x is chosen such that $0 < |x - 0| = |x| < \delta$,

$$\left| \frac{1}{1 + x} - 1 \right| < \epsilon.$$

Here we let

$$p: \quad 0 < |x| < \delta \text{ for some appropriate choice of } \delta$$
$$\text{(that in general depends on } \epsilon)$$

$$q: \quad \left| \frac{1}{1 + x} - 1 \right| < \epsilon$$

We start backwards to determine a δ that depends on ϵ.

$$q: \quad \left| \frac{1}{1+x} - 1 \right| < \epsilon \quad \Longleftrightarrow \quad \underbrace{\frac{|x|}{|x+1|} < \epsilon}_{b_1}$$

Note that the term $|x|$, which is important to p, has appeared in this equivalent form of q. Near $x = 0$, $|x+1|$ is near 1 and $|x|$ is near zero, of course. The next step is to move backwards from b_1 so that we leave term $|x|$ alone but replace the term $1/|x+1|$ with a constant. Eventually as $x \to 0$ we can consider x in a small interval about 0. Suppose we take $-0.01 < x < 0.01$ which is equivalent to $|x| < 0.01$. For $-0.01 < x < 0.01$, $x + 1 = |x + 1|$. Hence we have $0.99 < |x + 1| < 1.01$ which is equivalent to

$$\frac{1}{1.01} < \left| \frac{1}{x+1} \right| < \frac{1}{0.99}$$

and implies

$$\left| \frac{1}{1+x} \right| < \frac{1}{0.99} \quad \text{(for } |x| < 0.01)$$

Thus we can see that b_1 is implied by b_2 where

$$b_1 \Longleftarrow \underbrace{\frac{1}{0.99}|x| < \epsilon \quad \text{when } 0 < |x| < 0.01}_{b_2}$$

We see that

$$b_2 \Longleftrightarrow \underbrace{|x| < 0.99\epsilon \quad \text{when } 0 < |x| < 0.01}_{b_3}$$

In b_3 we have two conditions that $|x|$ is to satisfy in order to imply b_2 which in turn implies b_1 which in turn implies q. We combine the two conditions into one as follows. Let $\delta = \min\{0.01, 0.99\epsilon\}$, then

$$b_3 \Longleftarrow \underbrace{0 < |x| < \delta}_{p}$$

In summary, if we choose δ as defined above (or smaller), then $p \Longrightarrow q$.

8. Let p: m^2 is odd and q: m is odd. Prove $(\sim q) \Longrightarrow (\sim p)$.

   Proof: $(\sim q) \Longleftrightarrow m$ is even $\Longrightarrow m = 2k$, for some integer k

   $$\Longrightarrow m^2 = (2k)^2 = 4k^2 = 2(2k^2)$$

   $$\Longrightarrow m^2 \text{ is even} \Longleftrightarrow (\sim p)$$

   The reasons have been purposefully omitted. You supply reasons for each step.

9. (a)

p	q	$p \Longrightarrow q$	$(\sim p)$	$(\sim q)$	$(\sim p) \Longrightarrow (\sim q)$
T	T	T	F	F	T
T	F	F	F	T	T
F	T	T	T	F	F
F	F	T	T	T	T

   Since columns 3 and 6 are not identical, $p \Longrightarrow q$ is not equivalent to its inverse.

   (b) The converse of the inverse of $p \Longrightarrow q$ is the contrapositive of $p \Longrightarrow q$.

10.

p	q	$p \Longrightarrow q$	p	$(\sim q)$	$p \wedge (\sim q)$	c	$(p \wedge (\sim q)) \Longrightarrow c$
T	T	T	T	F	F	F	T
T	F	F	T	T	T	F	F
F	T	T	F	F	F	F	T
F	F	T	F	T	F	F	T

   Since columns 3 and 8 are identical we have $(p \Longrightarrow q) \Longleftrightarrow ((p \wedge (\sim q)) \Longrightarrow c)$.

11. Prove by contradiction that $p \implies q$ where p: $x > 0$ and q: $1/x > 0$.

   Proof: Assume $p \wedge (\sim q)$. That is, $(x > 0) \wedge (1/x \leq 0)$.

$$p \wedge (\sim q) \quad \implies \quad \text{since } x > 0 \text{ we can multiply each side of } 1/x \leq 0 \text{ by } x$$
$$\text{and preserve the inequality; } x(1/x) \leq x(0) = 0$$

$$\implies \quad 1 \leq 0, \text{ which is clearly a contradiction}$$

   It follows that assumption $(\sim q) \iff 1/x \leq 0$ is false, so $1/x > 0$. That is, $p \implies q$.

12. Prove by contradiction that $p \implies q$, where p: x is irrational and q: $3x$ is irrational.

   Proof: Assume $p \wedge (\sim q)$. That is, assume x is irrational and $3x$ is rational.

$$p \wedge (\sim q) \implies x \text{ is irrational and } 3x = m/n, \text{ for some integers } m \text{ and } n$$
$$\implies x \text{ is irrational and } x = m/3n$$
$$\implies x \text{ is irrational and rational, which is a contradiction}$$

   It follows that assumption $(\sim q) \iff 3x$ is rational is false, so $3x$ must be irrational. That is, $p \implies q$.

13. Prove by contradiction that $p \implies q$ where p: 2 divides m and q: m is even.

   Proof: Assume $p \wedge (\sim q)$. That is, assume 2 divides m and m is odd.

$$p \wedge (\sim q) \implies (m = 2k, \text{ for some } k) \wedge (m = 2j + 1, \text{ for some } j)$$
$$\implies 2k = 2j + 1 \quad \{\text{Note we cannot say even } = \text{ odd here because we are}$$
$$\text{trying to prove the characterization of even that says}$$
$$\text{it is a multiple of 2.}\}$$
$$\implies 2k - 2j = 1 \implies 2(k - j) = 1$$
$$\implies \text{the left side is negative (if } j > k) \text{ or it is } \geq 2;$$
$$\text{in any case this is a contradiction that it is equal to 1}$$

   It follows that assumption, $(\sim q) \iff m$ is odd, is false. Hence m must be even. That is $p \implies q$.

14. The expression $m^2 + n^2 = k^2$ looks like something derived from the Pythagorean theorem for right triangles. A familiar right triangle is a 3, 4, 5 triangle. We see that $3^2 + 4^2 = 5^2$. Hence we have a counterexample.

15. A decreasing positive sequence $\{a_n\}$ could get closer and closer to a number L, but stay above it. Think about an easy number other than 0. Consider $L = 1$. We want a sequence to converge to 1, but stay above it. Hence $a_n = 1 + $ a little bit. Try something easy like $a_n = 1 + 1/n$. Check things carefully and you find that $\{1 + 1/n\}$ is decreasing, has positive terms, and converges to 1. Thus we have a counterexample.

16. To show something is even we show that it is a multiple of 2. We have

$$\begin{array}{ccccc} n^2 & + & n & + & 4 \\ & & \uparrow & & \uparrow \\ & & \text{even} & & \text{even} \end{array}$$

   If n^2 is even we have sum of evens which is even because each term would have a factor of 2. Thus try to prove if n is even so is n^2. (See Exercise 8 for a similar problem.) Once we have done this, the result follows from the preceding argument.

Section 0.4

1. Experiments:

$$
\begin{aligned}
1 &= 1 \\
1 + 3 &= 4 \\
1 + 3 + 5 &= 9 \\
1 + 3 + 5 + 7 &= 16
\end{aligned}
$$

The sums can be written as 1^2, 2^2, 3^2, 4^2 respectively.

Conjecture: If S_k is the sum of the first k odd integers then

$$S_k = 1 + 3 + \cdots + (2k - 1) = k^2$$

Proof by induction: Let S be the set of all natural numbers k for which the conjecture is true. From the experiments $1 \in S$. To show (b), we assume $n \in S$; that is,

$$S_n = 1 + 3 + \cdots + 2n - 1 = n^2$$

We must show that $n + 1 \in S$. We have

$$
\begin{aligned}
S_{n+1} &= 1 + 3 + \cdots + 2n - 1 + 2(n+1) - 1 \\
&= (1 + 3 + \cdots + 2n - 1) + 2(n+1) - 1 \quad &\{\text{by algebra}\} \\
&= n^2 + 2(n+1) - 1 \quad &\{\text{by the induction hypothesis}\} \\
&= n^2 + 2n + 1 \quad &\{\text{by algebra}\} \\
&= (n+1)^2
\end{aligned}
$$

Thus by induction $S = N$.

2. Experiments:

$$
\begin{array}{llll}
n = 1 & 2^1 = 1 & (1+1)! = 2 \\
n = 2 & 2^2 = 4 & (2+1)! = 6 \\
n = 3 & 2^3 = 9 & (3+1)! = 24 \\
n = 4 & 2^4 = 16 & (4+1)! = 120
\end{array}
$$

Conjecture: If k is any natural number $2^k \le (k+1)!$
Proof by induction: Let S be the set of all natural numbers k for which the conjecture is true. From the experiments $1 \in S$. To show (b), we assume $n \in S$; that is

$$2^n \le (n+1)!$$

We must show that $n + 1 \in S$. We have

$$
\begin{aligned}
2^{n+1} &= 2^n(2) \quad &\{\text{by algebra}\} \\
&\le (n+1)! \, 2 \quad &\{\text{by the induction hypothesis}\} \\
&\le (n+1)! \, (n+2) \quad &\{\text{since } 2 \le n+1 \text{ when } n \ge 1\} \\
&= (n+2)!
\end{aligned}
$$

Thus by induction $S = N$.

3. Let S be the set of all natural numbers for which it is true that

$$\frac{d}{dx}(x^k) = kx^{k-1}$$

From calculus we have

$$\frac{d}{dx}x^1 = \frac{d}{dx}x = 1 = 1x^0$$

Hence $1 \in S$. Assume that $n \in S$; that is,

$$\frac{d}{dx}x^n = nx^{n-1}$$

We must show that $n + 1 \in S$; that is, prove

$$\frac{d}{dx}(x^{n+1}) = (n+1)x^n$$

We have

$$\frac{d}{dx}x^{n+1} = \frac{d}{dx}(x^n x) \qquad \text{\{by algebra\}}$$

$$= (x^n)\left[\frac{d}{dx}x\right] + \left[\frac{d}{dx}x^n\right]x \quad \text{\{product rule\}}$$

$$= (x^n)(1) + (nx^{n-1})(x) \qquad \text{\{since } 1 \in S \text{ and the induction hypothesis\}}$$

$$= x^n + nx^n \qquad \text{\{by algebra\}}$$

$$= (n+1)x^n$$

Thus by induction $S = N$.

4. Let S be the set of all natural numbers k for which it is true that $k^2 + k$ is divisible by 2. If $k^2 + k$ is divisible by 2 then there exists some number m such that $k^2 + k = 2(m)$.

If $k = 1$, then $1^2 + 1 = 2$, which is certainly divisible by 2. Hence $1 \in S$. Assume $n \in S$; that is

$$n^2 + n = 2(q)$$

for some number q. We must show that $n + 1 \in S$; that is, prove there exists a number r such that

$$(n+1)^2 + (n+1) = 2(r)$$

We have

$$(n+1)^2 + (n+1) = n^2 + 2n + 1 + n + 1 \qquad \text{\{by algebra\}}$$
$$= (n^2 + n) + (2n + 2) \qquad \text{\{by algebra\}}$$
$$= 2q + 2(n+1) \qquad \text{\{by the induction hypothesis and algebra\}}$$
$$= 2(q + n + 1)$$

Hence $(n+1)^2 + (n+1)$ is divisible by 2 and then by induction $S = N$.

5. Let S be the set of all natural numbers k for which it is true that

$$\frac{1}{(1)(2)} + \frac{1}{(2)(3)} + \frac{1}{(3)(4)} + \cdots + \frac{1}{k(k+1)} = \frac{k}{k+1}$$

If $k = 1$, then

$$\frac{1}{1(1+1)} = \frac{1}{2} = \frac{1}{1+1}$$

Hence $1 \in S$. Assume that $n \in S$; that is

$$\frac{1}{(1)(2)} + \frac{1}{(2)(3)} + \frac{1}{(3)(4)} + \cdots + \frac{1}{n(n+1)} = \frac{n}{n+1}$$

We must show that $n + 1 \in S$; that is, prove

$$\frac{1}{(1)(2)} + \frac{1}{(2)(3)} + \frac{1}{(3)(4)} + \cdots + \frac{1}{n(n+1)} + \frac{1}{(n+1)(n+2)} = \frac{n+1}{n+1+1}$$

We have

$$\frac{1}{(1)(2)} + \frac{1}{(2)(3)} + \frac{1}{(3)(4)} + \cdots + \frac{1}{n(n+1)} + \frac{1}{(n+1)(n+2)}$$

$$= \left[\frac{1}{(1)(2)} + \frac{1}{(2)(3)} + \frac{1}{(3)(4)} + \cdots + \frac{1}{n(n+1)}\right] + \frac{1}{(n+1)(n+2)}$$

$$= \frac{n}{n+1} + \frac{1}{(n+1)(n+2)} = \frac{n(n+2)+1}{(n+1)(n+2)}$$

$$= \frac{n^2 + 2n + 1}{(n+1)(n+2)} = \frac{(n+1)^2}{(n+1)(n+2)} = \frac{n+1}{n+2}$$

Thus by induction $S = N$.

Chapter 1

Linear Equations and Matrices

Section 1.1, p. 8

1. $x + 2y = 8$
   $3x - 4y = 4$

   Add 2 times the first equation to the second one to obtain $5x = 20$. Therefore $x = 4$. Substituting this value into the first equation, we obtain:

   $$4 + 2y = 8 \implies y = 2.$$

   The solution is $x = 4$, $y = 2$.

3. $3x + 2y + z = 2$
   $4x + 2y + 2z = 8$
   $x - y + z = 4$

   To eliminate z, add -2 times the first equation to the second one and -1 times the first equation to the third to obtain the system

   $$\begin{aligned} -2x - 2y &= 4 \\ -2x - 3y &= 2 \, . \end{aligned} \tag{1.1}$$

   To eliminate x from (1.1), add -1 times the first equation to the second to obtain

   $$-y = -2 \implies y = 2.$$

   Substitute $y = 2$ into the first equation of (1.1) to obtain $-2x - 2(2) = 4$. Therefore $x = -4$. Finally, substitute $x = -4$, $y = 2$ into the first equation of the given system to get $3(-4) + 2(2) + z = 2$. Therefore $z = 10$. The solution is $x = -4$, $y = 2$, $z = 10$.

5. $2x + 4y + 6z = -12$
   $2x - 3y - 4z = 15$
   $3x + 4y + 5z = -8$

   To simplify the system, multiply the first equation by $\frac{1}{2}$ to obtain the system

   $$\begin{aligned} x + 2y + 3z &= -6 \\ 2x - 3y - 4z &= 15 \\ 3x + 4y + 5z &= -8 \, . \end{aligned} \tag{1.2}$$

To eliminate x, add -2 times the first equation to the second one and -3 times the first equation to the third one to obtain

$$-7y - 10z = 27$$
$$-2y - 4z = 10.$$

(1.3)

Multiply the second equation by $\frac{1}{2}$:

$$-7y - 10z = 27$$
$$-y - 2z = 5.$$

(1.4)

Add -5 times the second equation in (1.4) to the first one to obtain

$$-2x = 2 \quad \Longrightarrow \quad x = -1.$$

Substitute $x = -1$ into the second equation in (1.4) to obtain

$$-y - 2(-1) = 5 \quad \Longrightarrow \quad y = -3.$$

Substituting the values $x = -1$, $y = -3$ into the second equation in (1.3), we obtain

$$-2(-1) - 4z = 10 \quad \Longrightarrow \quad z = -2.$$

The solution is $x = 2$, $y = -1$, $z = -2$.

7.　$x + 4y - z = 12$
　　$3x + 8y - 2z = 4$

To eliminate x, add -3 times the first equation to the second one to obtain

$$-4y + z = -32.$$

We now solve this equation for y in terms of z:

$$y = \frac{1}{4}z + 8.$$

The variable z can be chosen to be any real number. To find x in terms of z, substitute the expression for y into the first equation:

$$x + 4\left[\frac{1}{4}z + 8\right] - z = 12 \quad \Longrightarrow \quad x = -20.$$

The solution is $x = -20$, $y = \frac{1}{4}z + 8$, $z =$ any real number.

9.　$x + y + 3z = 12$
　　$2x + 2y + 6z = 6$

To eliminate x, add -2 times the first equation to the second one to obtain $0 = -18$. This makes no sense. Therefore the linear system has no solution. It is inconsistent.

11.　$2x + 3y = 13$
　　　$x - 2y = 3$
　　　$5x + 2y = 27$

To eliminate x, add -2 times the second equation to the first equation and -5 times the second equation to the third one to obtain

$$7y = 7$$
$$12y = 12.$$

(1.5)

Therefore $y = 1$. Substituting the value $y = 1$ into the second equation, we obtain

$$x - 2(1) = 3 \quad \Longrightarrow \quad x = 5.$$

The solution is $x = 5$, $y = 1$.

13. $x + 3y = -4$
     $2x + 5y = -8$
     $x + 3y = -5$

The first and third equations have the same left side but different right sides. Therefore the system has no solution. It is inconsistent.

15. $2x - y = 5$
     $4x - 2y = t$

(a) We first eliminate y from this system by adding -2 times the first equation to the second to obtain

$$0 = -10 + t.$$

Therefore, to be a consistent system we must have $t = 10$.

(b) The system is inconsistent if t has any value other than 10. For example, if $t = 3$, we are led to the conclusion that $0 = 10 - t = 7$, which makes no sense.

(c) There are an infinite number of choices for t that result in an inconsistent system.

17. $x + \quad 2y = 10$
     $3x + (6 + t)y = 30$

(a) Eliminate x from this system by adding (-3) times the first equation to the second to obtain $ty = 0$. Thus, for the system to have infinitely many solutions, $t = 0$. Then any real number y, and $x = 10 - 2y$, is a solution.

(b) If $t \neq 0$, then $y = 0$ so $x = 10$. For example, if $t = 1$, the unique solution is $x = 10$, $y = 0$.

(c) Infinitely many.

19. $2x + 3y - z = 0$
     $x - 4y + 5z = 0$

(a) Substituting $x_1 = 1$, $y_1 = -1$, and $z_1 = -1$ into both equations we obtain

$$2(1) + 3(-1) - (-1) = 0$$
$$(1) - 4(-1) + 5(-1) = 0.$$

Therefore $x_1 = 1$, $y_1 = -1$, $z_1 = -1$ is a solution to the system.

(b) Substituting $x_2 = -2$, $y_2 = 2$, $z_2 = 2$ into the both equations we obtain

$$2(-2) + 3(2) - (2) = 0$$
$$(-2) - 4(2) + 5(2) = 0.$$

Therefore $x_2 = -2$, $y_2 = 2$, $z_2 = 2$ is a solution to the system.

(c) Substituting $x = x_1 + x_2 = -1$, $y = y_1 + y_2 = 1$, $z = z_1 + z_2 = 1$ into the both equations we obtain

$$2(-1) + 3(1) - (1) = 0$$
$$(-1) - 4(1) + 5(1) = 0.$$

Therefore $x = x_1 + x_2 = -1$, $y = y_1 + y_2 = 1$, $z = z_1 + z_2 = 1$ is a solution to the system.

(d) Multiplying each of the values in part (c) by 3 gives $x = 3$, $y = -3$, $z = -3$. Substituting these values into both equations we obtain

$$2(3) + 3(-3) - (-3) = 0$$
$$(3) - 4(-3) + 5(-3) = 0.$$

Therefore these are also a solution to the system.

21. $$4x \qquad\qquad = \quad 8$$
$$-2x + 3y \qquad = -1$$
$$3x + 5y - 2z = \quad 11$$

Solving the first equation for x we obtain $x = 2$. Substitute $x = 2$ into the second equation to obtain

$$-2(2) + 3y = -1 \quad \Longrightarrow \quad y = 1.$$

Finally, substitute $x = 2$, $y = 1$ into the third equation to get

$$3(2) + 5(1) - 2z = 11 \quad \Longrightarrow \quad z = 0.$$

The solution is $x = 2$, $y = 1$, $z = 0$.

23. $$3x \qquad\quad - 2z = \quad 4$$
$$x - 4y + \quad z = -5$$
$$-2x + 3y + 2z = \quad 9$$

If such a value exists then $x = r$, $y = 2$, $z = 1$ must satisfy the first equation. Substituting these values into the first equation we obtain

$$3r - 2(1) = 4 \quad \Longrightarrow \quad r = 2.$$

Now substitute $x = r = 2$, $y = 2$, $z = 1$ into the second and third equations to see if it is a solution:

$$(2) - 4(2) + (1) = -5$$
$$-2(2) + 3(2) + 2(1) = \quad 4.$$

The third equation does not check. Therefore there is no value of r for which $x = r$, $y = 2$, $z = 1$ is a solution to the system.

25. If $x_1 = s_1$, $x_2 = s_2$, ..., $x_n = s_n$ is a solution to (2), then the pth and qth equations are satisfied. That is,

$$a_{p1}s_1 + \cdots + a_{pn}s_n = b_p$$
$$a_{q1}s_1 + \cdots + a_{qn}s_n = b_q.$$

Thus, for any real number r,

$$(a_{p1} + ra_{q1})s_1 + \cdots + (a_{pn} + ra_{qn})s_n = b_p + rb_q.$$

Then if the qth equation in (2) is replaced by the preceding equation, the values $x_1 = s_1$, $x_2 = s_2$, ..., $x_n = s_n$ are a solution to the new linear system since they satisfy each of the equations.

To show the converse, we must show that any solution $x_1 = s_1$, $x_2 = s_2$, ..., $x_n = s_n$ to the new linear system is also a solution to the original linear system. Since the pth equation is still in the new linear system we have

$$a_{p1}s_1 + \cdots + a_{pn}s_n = b_p.$$

Now,
$$(a_{p1} + ra_{q1})s_1 + \cdots + (a_{pn} + ra_{qn})s_n = b_p + rb_q$$
since this is the new qth equation. But by rearranging this equation we obtain
$$(a_{p1} + ra_{q1})s_1 + \cdots + (a_{pn} + ra_{qn})s_n = (a_{p1}s_1 + \cdots + a_{pn}s_n) + r(a_{q1} + \cdots + a_{qn}s_n)$$
$$= b_p + r(a_{q1} + \cdots + a_{qn}s_n).$$

Therefore $b_p + r(a_{q1} + \cdots + a_{qn}s_n) = b_p + rb_q$ which implies that $a_{q1} + \cdots + a_{qn}s_n = b_q$. That is, the qth equation of (2) is satisfied by any solution of the new linear system. It follows that the two linear systems are equivalent.

27. (a) No points simultaneously lie in all three planes.

    (b) There are infinitely many points.

    (c) No points simultaneously lie in all three planes.

29. The two spheres may have no points of intersection, one point of intersection, or infinitely many points of intersection:

 ;   ;

		Intersection is a circle	Intersection is S_1 $(=S_2)$
No points lie on both spheres.	One point lies on both spheres.	Infinitely many points lie on both spheres.	

31. Let the amount of regular plastic be denoted by x and the amount of special plastic denoted by y. Then

$$\text{plant A requirements:} \quad 2x + 2y = 8$$
$$\text{plant B requirements:} \quad 5x + 3y = 15.$$

To eliminate x we multiply the first equation by 5 and add (-2) times the second equation to it to obtain
$$4y = 10 \quad \Longrightarrow \quad y = \frac{10}{4} = 2.5.$$

To solve for x, substitute $y = 2.5$ into either of the original equations. Using the plant A requirements equation, we have
$$2x + 2(2.5) = 8 \quad \Longrightarrow \quad 2x = 3 \quad \Longrightarrow \quad x = \frac{3}{2} = 1.5.$$

Thus, we can make 1.5 tons of regular plastic and 2.5 tons of special plastic.

33. Let the amount of 2-minute developer be denoted by x, the amount of 6-minute developer by y, and the amount of 9-minute developer by z. Then:

$$\text{plant A requirements:} \quad 6x + 12y + 12z = (10)(60) = 600$$
$$\text{plant B requirements:} \quad 24x + 12y + 12z = (16)(60) = 960.$$

(Note that the left side of each of these equations expresses the total time in minutes; therefore, on the right, must convert the 10 hours and 16 hours to minutes by multiplying by 60.)

To eliminate y and z, we multiply the first equation by 1 and add (-1) times the second equation to it to obtain
$$-18x = -360 \quad \Longrightarrow \quad x = \frac{360}{18} = 20.$$

To solve for y, substitute $x = 20$ into either of the original equations. Using the plant A requirements, we obtain
$$6(20) + 12y + 12z = 600 \quad \Longrightarrow \quad 12y + 12z = 480.$$

If we substitute $x = 20$ into the plant B requirements equation, we obtain

$$24(20) + 12y + 12z = 960 \quad \Longrightarrow \quad 12y + 12z = 480,$$

which is the same equation. Thus, there is no unique solution for y and z. But

$$12y + 12z = 480 \quad \Longrightarrow \quad y + z = 40.$$

Therefore, the best we can say is that 20 tons of the 2-minute deleoper should be made, and that the toal amount of 6-minute developer and 12-minute developer should be 40 tons.

35. Let the amount of money in the first trust be x, the amount in the second trust y, and the amount in the third trust z. Then:

$$\begin{array}{lrrrr}
\text{inheritance requirements:} & x + & y + & z = & 24,000 \\
\text{second trust requirements:} & & y & = & 2x \\
\text{inheritance requirements:} & 0.09x + & 0.10y + & 0.06z = & 2210.
\end{array}$$

Substituting the second equation into the first and third equations, we have:

$$x + (2x) + z = 24,000 \quad \Longrightarrow \quad 3x + z = 24,000$$
$$0.09x + 0.10(2x) + 0.06z = 2210 \quad \Longrightarrow \quad 0.29x + 0.06z = 2210$$

To eliminate z, multiply the first equation by 0.06 and add (-1) times the second equation to it to obtain

$$-0.11x = -770 \quad \Longrightarrow \quad x = 7000.$$

Substituting $x = 7000$ into the second original equation, we obtain

$$y = 2(7000) = 14,000.$$

Finally, substitute $x = 7000$, $y = 14,000$ into the first original equation to obtain

$$7000 + 14,000 + z = 24,000 \quad \Longrightarrow \quad z = 3000.$$

Therefore, $7000 was invested in the first trust, $14,000 in the second trust, and $3000 in the third trust.

Section 1.2, p.19

1. (a) $a_{12} = -3$, $a_{22} = -5$, $a_{23} = 4$.
   (b) $b_{11} = 4$, $b_{31} = 5$.
   (c) $c_{13} = 2$, $c_{31} = 6$, $c_{33} = -1$.

3. (a) (b)

5. Two matrices are equal if corresponding entries agree. Hence,

$$\begin{bmatrix} a + 2b & 2a - b \\ 2c + d & c - 2d \end{bmatrix} = \begin{bmatrix} 4 & -2 \\ 4 & -3 \end{bmatrix}$$

implies that we have two systems of equations:

$$\begin{array}{rrr} a + 2b = & 4 \\ 2a - b = & -2 \end{array} \quad \text{and} \quad \begin{array}{rrr} 2c + d = & 4 \\ c - 2d = & -3. \end{array}$$

We solve the system on the left by multiplying the first equation by (-2) and adding it to the second equation to obtain

$$-5b = -10 \implies b = 2.$$

Substituting $b = 2$ into the first equation, we find that

$$a + 2(2) = 4 \implies a = 0.$$

Similarly, to solve the system on the right for c, we multiply the first equation by 2 and add it to the second equation to obtain

$$5c = 5 \implies c = 1.$$

Substituting $c = 1$ into the first equation, we obtain

$$2(1) + d = 4 \implies d = 2.$$

Thus, $a = 0$, $b = 2$, $c = 1$, and $d = 2$.

7. (a) $3D + 2F = 3\begin{bmatrix} 3 & -2 \\ 2 & 4 \end{bmatrix} + 2\begin{bmatrix} -4 & 5 \\ 2 & 3 \end{bmatrix} = \begin{bmatrix} 9 & -6 \\ 6 & 12 \end{bmatrix} + \begin{bmatrix} -8 & 10 \\ 4 & 6 \end{bmatrix} = \begin{bmatrix} 1 & 4 \\ 10 & 18 \end{bmatrix}.$

   (b) $3(2A) = 3\left(2\begin{bmatrix} 1 & 2 & 3 \\ 2 & 1 & 4 \end{bmatrix}\right) = 3\begin{bmatrix} 2 & 4 & 6 \\ 4 & 2 & 8 \end{bmatrix} = \begin{bmatrix} 6 & 12 & 18 \\ 12 & 6 & 24 \end{bmatrix}.$

   $6A = 6\begin{bmatrix} 1 & 2 & 3 \\ 2 & 1 & 4 \end{bmatrix} = \begin{bmatrix} 6 & 12 & 18 \\ 12 & 6 & 24 \end{bmatrix}.$

   Note: $3(2A) = 6A$.

   (c) $3A + 2A = 3\begin{bmatrix} 1 & 2 & 3 \\ 2 & 1 & 4 \end{bmatrix} + 2\begin{bmatrix} 1 & 2 & 3 \\ 2 & 1 & 4 \end{bmatrix} = \begin{bmatrix} 5 & 10 & 15 \\ 10 & 5 & 20 \end{bmatrix}.$

   $5A = 5\begin{bmatrix} 1 & 2 & 3 \\ 2 & 1 & 4 \end{bmatrix} = \begin{bmatrix} 5 & 10 & 15 \\ 10 & 5 & 20 \end{bmatrix}.$

   Note: $3A + 2A = 5A$.

   (d) $2(D + F) = 2\left(\begin{bmatrix} 3 & -2 \\ 2 & 4 \end{bmatrix} + \begin{bmatrix} -4 & 5 \\ 2 & 3 \end{bmatrix}\right) = 2\begin{bmatrix} -1 & 3 \\ 4 & 7 \end{bmatrix} = \begin{bmatrix} -2 & 6 \\ 8 & 14 \end{bmatrix}.$

   $2D + 2F = 2\begin{bmatrix} 3 & -2 \\ 2 & 4 \end{bmatrix} + 2\begin{bmatrix} -4 & 5 \\ 2 & 3 \end{bmatrix} = \begin{bmatrix} 6 & -4 \\ 4 & 8 \end{bmatrix} + \begin{bmatrix} -8 & 10 \\ 4 & 6 \end{bmatrix} = \begin{bmatrix} -2 & 6 \\ 8 & 14 \end{bmatrix}.$

   Note: $2(D + F) = 2D + 2F$.

   (e) $(2 + 3)D = 5\begin{bmatrix} 3 & -2 \\ 2 & 4 \end{bmatrix} = \begin{bmatrix} 15 & -10 \\ 10 & 20 \end{bmatrix}.$

   $2D + 3D = 2\begin{bmatrix} 3 & -2 \\ 2 & 4 \end{bmatrix} + 3\begin{bmatrix} 3 & -2 \\ 2 & 4 \end{bmatrix} = \begin{bmatrix} 6 & -4 \\ 4 & 8 \end{bmatrix} + \begin{bmatrix} 9 & -6 \\ 6 & 12 \end{bmatrix} = \begin{bmatrix} 15 & -10 \\ 10 & 20 \end{bmatrix}.$

   Note: $(2 + 3)D = 2D + 3D$.

   (f) $3(B + D)$ is undefined since B is 3×2 and D is 2×2.

9. (a) $(2A)^T = \left(2\begin{bmatrix} 1 & 2 & 3 \\ 2 & 1 & 4 \end{bmatrix}\right)^T = \begin{bmatrix} 2 & 4 & 6 \\ 4 & 2 & 8 \end{bmatrix}^T = \begin{bmatrix} 2 & 4 \\ 4 & 2 \\ 6 & 8 \end{bmatrix}.$

   Note: $(2A)^T = 2A^T$.

   (b) $(A - B)^T$ is undefined since A is 2×3 and B is 3×2.

   (c) $(3B^T - 2A)^T = \left(3\begin{bmatrix} 1 & 0 \\ 2 & 1 \\ 3 & 2 \end{bmatrix}^T - 2\begin{bmatrix} 1 & 2 & 3 \\ 2 & 1 & 4 \end{bmatrix}\right)^T$

   $= \left(3\begin{bmatrix} 1 & 2 & 3 \\ 0 & 1 & 2 \end{bmatrix} - 2\begin{bmatrix} 1 & 2 & 3 \\ 2 & 1 & 4 \end{bmatrix}\right)^T = \begin{bmatrix} 1 & 2 & 3 \\ -4 & 1 & -2 \end{bmatrix}^T = \begin{bmatrix} 1 & -4 \\ 2 & 1 \\ 3 & -2 \end{bmatrix}.$

(d) $(3A^T - 5B^T)^T$ is undefined since A^T is 3×2 and B^T is 2×3.

(e) $(-A)^T = \begin{bmatrix} -1 & -2 & -3 \\ -2 & -1 & -4 \end{bmatrix}^T = \begin{bmatrix} -1 & -2 \\ -2 & -1 \\ -3 & -4 \end{bmatrix}$; $-(A^T) = -\begin{bmatrix} 1 & 2 \\ 2 & 1 \\ 3 & 4 \end{bmatrix} = \begin{bmatrix} -1 & -2 \\ -2 & -1 \\ -3 & -4 \end{bmatrix}$.

Note: $(-A)^T = -(A^T)$.

(f) $(C + E + F^T)^T$ is undefined since $C + E$ is 3×3 and F^T is 2×2.

11. No. If there are scalars x_1 and x_2 such that

$$x_1 \begin{bmatrix} 1 & 0 \\ 0 & 1 \end{bmatrix} + x_2 \begin{bmatrix} 1 & 0 \\ 0 & 0 \end{bmatrix} = \begin{bmatrix} 4 & 1 \\ 0 & -3 \end{bmatrix},$$

then equating the (1,2) entries gives $0 = 1$, a contradiction.

13. 0. To justify this answer, let $A = \begin{bmatrix} a_{ij} \end{bmatrix}$ be an $n \times n$ matrix. Then $A^T = \begin{bmatrix} a_{ji} \end{bmatrix}$. Thus, the (i,i)th entry of $A - A^T$ is $a_{ii} - a_{ii} = 0$. Therefore, all entries on the main diagonal of $A - A^T$ are 0.

15. If $A = A^T$, then the matrix A must be symmetric about the main diagonal. To show this, let $A = \begin{bmatrix} a_{ij} \end{bmatrix}$. Then $A = \begin{bmatrix} a_{ij} \end{bmatrix} = A^T = \begin{bmatrix} a_{ji} \end{bmatrix}$, so $a_{ij} = a_{ji}$ for all i, j, $1 \le i \le n$, $1 \le j \le n$. Since a_{ji} is the entry a_{ij} reflected about the main diagonal, A is symmetric about the main diagonal.

17. (a) $\displaystyle\sum_{i=1}^{n}(r_i + s_i)a_i = (r_1 + s_1)a_1 + (r_2 + s_2)a_2 + \cdots + (r_n + s_n)a_n$

$$= r_1 a_1 + s_1 a_1 + r_2 a_2 + s_2 a_2 + \cdots + r_n a_n + s_n a_n$$

$$= (r_1 a_1 + r_2 a_2 + \cdots + r_n a_n) + (s_1 a_1 + s_2 a_2 + \cdots + s_n a_n) = \sum_{i=1}^{n} r_i a_i + \sum_{i=1}^{n} s_i a_i$$

(b) $\displaystyle\sum_{i=1}^{n} c(r_i a_i) = cr_1 a_1 + cr_2 a_2 + \cdots + cr_n a_n = c(r_1 a_1 + r_2 a_2 + \cdots + r_n a_n) = c\sum_{i=1}^{n} r_i a_i.$

19. (a) True. $\displaystyle\sum_{i=1}^{n}(a_i + 1) = \sum_{i=1}^{n} a_i + \sum_{i=1}^{n} 1 = \sum_{i=1}^{n} a_i + n.$

(b) True. $\displaystyle\sum_{i=1}^{n}\left(\sum_{j=1}^{m} 1\right) = \sum_{i=1}^{n} m = mn.$

(c) True. $\displaystyle\left[\sum_{i=1}^{n} a_i\right]\left[\sum_{j=1}^{m} b_j\right] = a_1 \sum_{j=1}^{m} b_j + a_2 \sum_{j=1}^{m} b_j + \cdots + a_n \sum_{j=1}^{m} b_j$

$$= (a_1 + a_2 + \cdots + a_n)\sum_{j=1}^{m} b_j$$

$$= \sum_{i=1}^{n} a_i \sum_{j=1}^{m} b_j = \sum_{j=1}^{m}\left(\sum_{i=1}^{n} a_i b_j\right)$$

21. The average daily price of stock is $\frac{1}{2}(\mathbf{t} + \mathbf{b})$.

Section 1.3, p. 30

1. (a) $\mathbf{a} \cdot \mathbf{b} = (1)(4) + (2)(-1) = 2.$

(b) $\mathbf{a} \cdot \mathbf{b} = (-3)(1) + (-2)(-2) = 1.$

(c) $\mathbf{a} \cdot \mathbf{b} = (4)(1) + (2)(3) + (-1)(6) = 4.$

(d) $\mathbf{a \cdot b} = (1)(1) + (1)(0) + (0)(1) = 1.$

3. We have $\mathbf{a \cdot b} = (-3)(-3) + (2)(2) + (x)(x) = 13 + x^2 = 17.$ It follows that $x^2 = 4.$ Hence $x = 2$ or $x = -2.$

5. We have

$$\mathbf{v \cdot w} = \begin{bmatrix} x \\ 1 \\ y \end{bmatrix} \cdot \begin{bmatrix} 2 \\ -2 \\ 1 \end{bmatrix} = 2x - 2 + y = 0$$

$$\mathbf{v \cdot u} = \begin{bmatrix} x \\ 1 \\ y \end{bmatrix} \cdot \begin{bmatrix} 1 \\ 8 \\ 2 \end{bmatrix} = x + 8 + 2y = 0$$

Solving these two equations, we find $x = 4,\ y = -6.$

7. $\mathbf{w \cdot w} = \sin^2 \theta + \cos^2 \theta = 1.$

9. We have

$$\mathbf{v \cdot v} = \begin{bmatrix} \frac{1}{2} \\ -\frac{1}{2} \\ x \end{bmatrix} \cdot \begin{bmatrix} \frac{1}{2} \\ -\frac{1}{2} \\ x \end{bmatrix} = \frac{1}{4} + \frac{1}{4} + x^2 = 1 \quad \Longrightarrow \quad x^2 = \frac{1}{2} \quad \Longrightarrow \quad x = \pm\sqrt{\frac{1}{2}} = \pm\frac{1}{2}\sqrt{2}.$$

11. (a) $AB = \begin{bmatrix} 1 & 2 & 3 \\ 2 & 1 & 4 \end{bmatrix} \begin{bmatrix} 1 & 0 \\ 2 & 1 \\ 3 & 2 \end{bmatrix} = \begin{bmatrix} 14 & 8 \\ 16 & 9 \end{bmatrix}$

(b) $BA = \begin{bmatrix} 1 & 0 \\ 2 & 1 \\ 3 & 2 \end{bmatrix} \begin{bmatrix} 1 & 2 & 3 \\ 2 & 1 & 4 \end{bmatrix} = \begin{bmatrix} 1 & 2 & 3 \\ 4 & 5 & 10 \\ 7 & 8 & 17 \end{bmatrix}$

(c) $F^T E = \begin{bmatrix} -1 & 0 & 3 \\ 2 & 4 & 5 \end{bmatrix} \begin{bmatrix} 2 & -4 & 5 \\ 0 & 1 & 4 \\ 3 & 2 & 1 \end{bmatrix} = \begin{bmatrix} 7 & 10 & -2 \\ 19 & 6 & 31 \end{bmatrix}.$

(d) $CB + D = \begin{bmatrix} 3 & -1 & 3 \\ 4 & 1 & 5 \\ 2 & 1 & 3 \end{bmatrix} \begin{bmatrix} 1 & 0 \\ 2 & 1 \\ 3 & 2 \end{bmatrix} + \begin{bmatrix} 3 & -2 \\ 2 & 5 \end{bmatrix} = \begin{bmatrix} 10 & 5 \\ 21 & 11 \\ 13 & 7 \end{bmatrix} + \begin{bmatrix} 3 & -2 \\ 2 & 5 \end{bmatrix};$ impossible.

(e) $AB + D^2 = \begin{bmatrix} 1 & 2 & 3 \\ 2 & 1 & 4 \end{bmatrix} \begin{bmatrix} 1 & 0 \\ 2 & 1 \\ 3 & 2 \end{bmatrix} + \begin{bmatrix} 3 & -2 \\ 2 & 5 \end{bmatrix} \begin{bmatrix} 3 & -2 \\ 2 & 5 \end{bmatrix} = \begin{bmatrix} 19 & -8 \\ 32 & 30 \end{bmatrix}$

13. (a) $FD - 3B = \begin{bmatrix} -1 & 2 \\ 0 & 4 \\ 3 & 5 \end{bmatrix} \begin{bmatrix} 3 & -2 \\ 2 & 5 \end{bmatrix} - 3 \begin{bmatrix} 1 & 0 \\ 2 & 1 \\ 3 & 2 \end{bmatrix} = \begin{bmatrix} 1 & 12 \\ 8 & 20 \\ 19 & 19 \end{bmatrix} - \begin{bmatrix} 3 & 0 \\ 6 & 3 \\ 9 & 6 \end{bmatrix} = \begin{bmatrix} -2 & 12 \\ 2 & 17 \\ 10 & 13 \end{bmatrix}$

(b) $AB - 2D = \begin{bmatrix} 1 & 2 & 3 \\ 2 & 1 & 4 \end{bmatrix} \begin{bmatrix} 1 & 0 \\ 2 & 1 \\ 3 & 2 \end{bmatrix} - 2 \begin{bmatrix} 3 & -2 \\ 2 & 5 \end{bmatrix} = \begin{bmatrix} 14 & 8 \\ 16 & 9 \end{bmatrix} - \begin{bmatrix} 6 & -4 \\ 4 & 10 \end{bmatrix} = \begin{bmatrix} 8 & 12 \\ 12 & -1 \end{bmatrix}$

(c) $F^T B + D = \begin{bmatrix} -1 & 0 & 3 \\ 2 & 4 & 5 \end{bmatrix} \begin{bmatrix} 1 & 0 \\ 2 & 1 \\ 3 & 2 \end{bmatrix} + \begin{bmatrix} 3 & -2 \\ 2 & 5 \end{bmatrix} = \begin{bmatrix} 8 & 6 \\ 25 & 14 \end{bmatrix} + \begin{bmatrix} 3 & -2 \\ 2 & 5 \end{bmatrix} = \begin{bmatrix} 11 & 4 \\ 27 & 19 \end{bmatrix}$

(d) $2F - 3(AE) = 2\begin{bmatrix} -1 & 2 \\ 0 & 4 \\ 3 & 5 \end{bmatrix} - 3\begin{bmatrix} 1 & 2 & 3 \\ 2 & 1 & 4 \end{bmatrix}\begin{bmatrix} 2 & -4 & 5 \\ 0 & 1 & 4 \\ 3 & 2 & 1 \end{bmatrix} = \begin{bmatrix} -2 & 4 \\ 0 & 8 \\ 6 & 10 \end{bmatrix} - \begin{bmatrix} 33 & 12 & 48 \\ 48 & 3 & 54 \end{bmatrix}$;

impossible.

(e) $BD + AE = \begin{bmatrix} 1 & 0 \\ 2 & 1 \\ 3 & 2 \end{bmatrix}\begin{bmatrix} 3 & -2 \\ 2 & 5 \end{bmatrix} + \begin{bmatrix} 1 & 2 & 3 \\ 2 & 1 & 4 \end{bmatrix}\begin{bmatrix} 2 & -4 & 5 \\ 0 & 1 & 4 \\ 3 & 2 & 1 \end{bmatrix} = \begin{bmatrix} 3 & -2 \\ 8 & 1 \\ 13 & 4 \end{bmatrix} + \begin{bmatrix} 11 & 4 & 16 \\ 16 & 1 & 18 \end{bmatrix}$;

impossible.

15. (a) $A^T = \begin{bmatrix} 1 & 2 & 3 \\ 2 & 1 & 4 \end{bmatrix}^T = \begin{bmatrix} 1 & 2 \\ 2 & 1 \\ 3 & 4 \end{bmatrix}$

(b) $(A^T)^T = \begin{bmatrix} 1 & 2 \\ 2 & 1 \\ 3 & 4 \end{bmatrix}^T = \begin{bmatrix} 1 & 2 & 3 \\ 2 & 1 & 4 \end{bmatrix}$

(c) $(AB)^T = \begin{bmatrix} 14 & 8 \\ 16 & 9 \end{bmatrix}^T = \begin{bmatrix} 14 & 16 \\ 8 & 9 \end{bmatrix}$

(d) $B^T A^T = \begin{bmatrix} 1 & 2 & 3 \\ 0 & 1 & 2 \end{bmatrix}\begin{bmatrix} 1 & 2 \\ 2 & 1 \\ 3 & 4 \end{bmatrix} = \begin{bmatrix} 14 & 16 \\ 8 & 9 \end{bmatrix}$

(e) $(C + E)^T = \left(\begin{bmatrix} 3 & -1 & 3 \\ 4 & 1 & 5 \\ 2 & 1 & 3 \end{bmatrix} + \begin{bmatrix} 2 & -4 & 5 \\ 0 & 1 & 4 \\ 3 & 2 & 1 \end{bmatrix}\right)^T = \begin{bmatrix} 5 & -5 & 8 \\ 4 & 2 & 9 \\ 5 & 3 & 4 \end{bmatrix}^T = \begin{bmatrix} 5 & 4 & 5 \\ -5 & 2 & 3 \\ 8 & 9 & 4 \end{bmatrix}$;

$(C + E)^T B = \begin{bmatrix} 5 & 4 & 5 \\ -5 & 2 & 3 \\ 8 & 9 & 4 \end{bmatrix}\begin{bmatrix} 1 & 0 \\ 2 & 1 \\ 3 & 2 \end{bmatrix} = \begin{bmatrix} 28 & 14 \\ 8 & 8 \\ 38 & 17 \end{bmatrix}$

$C^T B + E^T B = \begin{bmatrix} 3 & 4 & 2 \\ -1 & 1 & 1 \\ 3 & 5 & 3 \end{bmatrix}\begin{bmatrix} 1 & 0 \\ 2 & 1 \\ 3 & 2 \end{bmatrix} + \begin{bmatrix} 2 & 0 & 3 \\ -4 & 1 & 2 \\ 5 & 4 & 1 \end{bmatrix}\begin{bmatrix} 1 & 0 \\ 2 & 1 \\ 3 & 2 \end{bmatrix} = \begin{bmatrix} 28 & 14 \\ 8 & 8 \\ 38 & 17 \end{bmatrix}$

(f) $A(2B) = \begin{bmatrix} 1 & 2 & 3 \\ 2 & 1 & 4 \end{bmatrix}\left(2\begin{bmatrix} 1 & 0 \\ 2 & 1 \\ 3 & 2 \end{bmatrix}\right) = \begin{bmatrix} 1 & 2 & 3 \\ 2 & 1 & 4 \end{bmatrix}\begin{bmatrix} 2 & 0 \\ 4 & 2 \\ 6 & 4 \end{bmatrix} = \begin{bmatrix} 28 & 16 \\ 32 & 18 \end{bmatrix}$;

$2(AB) = 2\left(\begin{bmatrix} 1 & 2 & 3 \\ 2 & 1 & 4 \end{bmatrix}\begin{bmatrix} 1 & 0 \\ 2 & 1 \\ 3 & 2 \end{bmatrix}\right) = 2\begin{bmatrix} 14 & 8 \\ 16 & 9 \end{bmatrix} = \begin{bmatrix} 28 & 16 \\ 32 & 18 \end{bmatrix}$

17. (a) $(AB)_{12} = \text{row}_1(A) \cdot \text{col}_2(B) = \begin{bmatrix} 2 & 3 \end{bmatrix}\begin{bmatrix} -1 \\ 2 \end{bmatrix} = (2)(-1) + (3)(2) = 4.$

(b) $(AB)_{23} = \text{row}_2(A) \cdot \text{col}_3(B) = \begin{bmatrix} -1 & 4 \end{bmatrix}\begin{bmatrix} 3 \\ 4 \end{bmatrix} = (-1)(3) + (4)(4) = 13.$

(c) $(AB)_{31} = \text{row}_3(A) \cdot \text{col}_1(B) = \begin{bmatrix} 0 & 3 \end{bmatrix}\begin{bmatrix} 3 \\ 1 \end{bmatrix} = (0)(3) + (3)(1) = 3.$

(d) $(AB)_{33} = \text{row}_3(A) \cdot \text{col}_3(B) = \begin{bmatrix} 0 & 3 \end{bmatrix}\begin{bmatrix} 3 \\ 4 \end{bmatrix} = (0)(3) + (3)(4) = 12.$

19. We find that

$$AB = \begin{bmatrix} 1 & 2 \\ 3 & 2 \end{bmatrix} \begin{bmatrix} 2 & -1 \\ -3 & 4 \end{bmatrix} = \begin{bmatrix} -4 & 7 \\ 0 & 5 \end{bmatrix} \quad \text{and} \quad BA = \begin{bmatrix} 2 & -1 \\ -3 & 4 \end{bmatrix} \begin{bmatrix} 1 & 2 \\ 3 & 2 \end{bmatrix} = \begin{bmatrix} -1 & 2 \\ 9 & 2 \end{bmatrix}.$$

Therefore $AB \neq BA$.

21. (a) The first column of AB is

$$A \cdot \mathrm{col}_1(B) = \begin{bmatrix} 1 & -1 & 2 \\ 3 & 2 & 4 \\ 4 & -2 & 3 \\ 2 & 1 & 5 \end{bmatrix} \begin{bmatrix} 1 \\ 3 \\ 4 \end{bmatrix} = \begin{bmatrix} 6 \\ 25 \\ 10 \\ 25 \end{bmatrix}.$$

(b) The third column of AB is

$$A \cdot \mathrm{col}_3(B) = \begin{bmatrix} 1 & -1 & 2 \\ 3 & 2 & 4 \\ 4 & -2 & 3 \\ 2 & 1 & 5 \end{bmatrix} \begin{bmatrix} -1 \\ -3 \\ 5 \end{bmatrix} = \begin{bmatrix} 12 \\ 11 \\ 17 \\ 20 \end{bmatrix}.$$

23. $A\mathbf{c} = \begin{bmatrix} 2 & -3 & 4 \\ -1 & 2 & 3 \\ 5 & -1 & -2 \end{bmatrix} \begin{bmatrix} 2 \\ 1 \\ 4 \end{bmatrix} = \begin{bmatrix} (2)(2) + (-3)(1) + (4)(4) \\ (-1)(2) + (2)(1) + (3)(4) \\ (5)(2) + (-1)(1) + (-2)(4) \end{bmatrix}$

$$= \begin{bmatrix} (2)(2) \\ (-1)(2) \\ (5)(2) \end{bmatrix} + \begin{bmatrix} (-3)(1) \\ (2)(1) \\ (-1)(1) \end{bmatrix} + \begin{bmatrix} (4)(4) \\ (3)(4) \\ (-2)(4) \end{bmatrix}$$

$$= 2 \begin{bmatrix} 2 \\ -1 \\ 5 \end{bmatrix} + 1 \begin{bmatrix} -3 \\ 2 \\ -1 \end{bmatrix} + 4 \begin{bmatrix} 4 \\ 3 \\ -2 \end{bmatrix}.$$

25. (a) $AB = \begin{bmatrix} 2 & -3 & 1 \\ 1 & 2 & 4 \end{bmatrix} \begin{bmatrix} 3 \\ 5 \\ 2 \end{bmatrix} = \begin{bmatrix} -7 \\ 21 \end{bmatrix}$;

$$3\mathbf{a}_1 + 5\mathbf{a}_2 + 2\mathbf{a}_3 = 3 \begin{bmatrix} 2 \\ 1 \end{bmatrix} + 5 \begin{bmatrix} -3 \\ 2 \end{bmatrix} + 2 \begin{bmatrix} 1 \\ 4 \end{bmatrix} = \begin{bmatrix} -7 \\ 21 \end{bmatrix} = AB$$

(b) $(\mathrm{row}_1(A))B = \begin{bmatrix} 2 & -3 & 1 \end{bmatrix} \begin{bmatrix} 3 \\ 5 \\ 2 \end{bmatrix} = \begin{bmatrix} -7 \end{bmatrix}$; $(\mathrm{row}_2(A))B = \begin{bmatrix} 1 & 2 & 4 \end{bmatrix} \begin{bmatrix} 3 \\ 5 \\ 2 \end{bmatrix} = \begin{bmatrix} 21 \end{bmatrix}$;

$$\begin{bmatrix} (\mathrm{row}_1(A)) \\ (\mathrm{row}_2(A)) \end{bmatrix} = \begin{bmatrix} -7 \\ 21 \end{bmatrix} = AB$$

27. $AB^T = \begin{bmatrix} 1 & r & 1 \end{bmatrix} \begin{bmatrix} -2 \\ 2 \\ s \end{bmatrix} = \begin{bmatrix} -2 + 2r + s \end{bmatrix} = \begin{bmatrix} 0 \end{bmatrix} \implies 2r + s = 2.$ One possible solution is $r = 1,\ s = 0.$

29. $A^T A = \begin{bmatrix} -3 & 4 \\ 2 & 5 \\ 1 & 0 \end{bmatrix} \begin{bmatrix} -3 & 2 & 1 \\ 4 & 5 & 0 \end{bmatrix} = \begin{bmatrix} 25 & 14 & -3 \\ 14 & 29 & 2 \\ -3 & 2 & 1 \end{bmatrix}$. Therefore

$$\begin{bmatrix} \mathbf{a}_1^T \mathbf{a}_1 & \mathbf{a}_1^T \mathbf{a}_2 & \mathbf{a}_1^T \mathbf{a}_3 \\ \mathbf{a}_2^T \mathbf{a}_1 & \mathbf{a}_2^T \mathbf{a}_2 & \mathbf{a}_2^T \mathbf{a}_3 \\ \mathbf{a}_3^T \mathbf{a}_1 & \mathbf{a}_3^T \mathbf{a}_2 & \mathbf{a}_3^T \mathbf{a}_3 \end{bmatrix} = \begin{bmatrix} \begin{bmatrix} -3 & 4 \end{bmatrix} \begin{bmatrix} -3 \\ 4 \end{bmatrix} & \begin{bmatrix} -3 & 4 \end{bmatrix} \begin{bmatrix} 2 \\ 5 \end{bmatrix} & \begin{bmatrix} -3 & 4 \end{bmatrix} \begin{bmatrix} 1 \\ 0 \end{bmatrix} \\ \begin{bmatrix} 2 & 5 \end{bmatrix} \begin{bmatrix} -3 \\ 4 \end{bmatrix} & \begin{bmatrix} 2 & 5 \end{bmatrix} \begin{bmatrix} 2 \\ 5 \end{bmatrix} & \begin{bmatrix} 2 & 5 \end{bmatrix} \begin{bmatrix} 1 \\ 0 \end{bmatrix} \\ \begin{bmatrix} 1 & 0 \end{bmatrix} \begin{bmatrix} -3 \\ 4 \end{bmatrix} & \begin{bmatrix} 1 & 0 \end{bmatrix} \begin{bmatrix} 2 \\ 5 \end{bmatrix} & \begin{bmatrix} 1 & 0 \end{bmatrix} \begin{bmatrix} 1 \\ 0 \end{bmatrix} \end{bmatrix}$$

$$= \begin{bmatrix} 25 & 14 & -3 \\ 14 & 29 & 2 \\ -3 & 2 & 1 \end{bmatrix} = A^T A$$

31. $\begin{aligned} -2x_1 - x_2 \quad\quad + 4x_4 &= 5 \\ -3x_1 + 2x_2 + 7x_3 + 8x_4 &= 3 \\ x_1 \quad\quad\quad + 2x_4 &= 4 \\ 3x_1 \quad\quad + x_3 + 3x_4 &= 6 \end{aligned}$

33. $\begin{bmatrix} 2 & 3 & 0 \\ 0 & 3 & 1 \\ 2 & -1 & 0 \end{bmatrix} \begin{bmatrix} x_1 \\ x_2 \\ x_3 \end{bmatrix} = \begin{bmatrix} 0 \\ 0 \\ 0 \end{bmatrix}$

35. The linear systems are equivalent. That is, they have the same solutions.

37. (a) $\begin{aligned} 2x_1 + x_2 &= 4 \\ 3x_2 &= 2 \end{aligned}$

(b) $\begin{aligned} x_1 \quad\quad + 3x_3 + x_4 &= 2 \\ 2x_1 + x_2 + 4x_3 + 3x_4 &= 5 \\ -x_1 + 2x_2 + 5x_3 + 4x_4 &= 8 \end{aligned}$

39. (a) Let $A = \begin{bmatrix} 1 & 2 & 1 \\ -3 & 6 & -3 \\ 0 & 1 & -1 \end{bmatrix}$. We must determine if $\begin{bmatrix} 0 \\ 0 \\ 0 \end{bmatrix}$ is a linear combination of the columns of A:

$$c_1 \operatorname{col}_1(A) + c_2 \operatorname{col}_2(A) + c_3 \operatorname{col}_3(A) = c_1 \begin{bmatrix} 1 \\ -3 \\ 0 \end{bmatrix} + c_2 \begin{bmatrix} 2 \\ 6 \\ 1 \end{bmatrix} + c_3 \begin{bmatrix} 1 \\ -3 \\ -1 \end{bmatrix}$$

$$= \begin{bmatrix} c_1 + 2c_2 + c_3 \\ -3c_1 + 6c_2 - 3c_3 \\ c_2 - c_3 \end{bmatrix} = \begin{bmatrix} 0 \\ 0 \\ 0 \end{bmatrix}.$$

This leads to the linear system:

$$\begin{aligned} c_1 + 2c_2 + c_3 &= 0 \\ -3c_1 + 6c_2 - 3c_3 &= 0 \\ c_2 - c_3 &= 0. \end{aligned}$$

Add 3 times the first equation to the second equation:

$$12c_2 \qquad = 0$$
$$c_2 - c_3 = 0.$$

Therefore, $c_2 = 0$, $c_3 = c_2 = 0$ and, from the first equation, $c_1 = 0$. Therefore the system $A\mathbf{x} = \mathbf{b}$ has only one solution: $\mathbf{x} = \begin{bmatrix} 0 \\ 0 \\ 0 \end{bmatrix}$.

(b) We observe that the vector

$$\mathbf{x} = \begin{bmatrix} 2 \\ 2 \\ 2 \\ 2 \end{bmatrix}$$

is a sollution to the linear system because the sum of the coefficients of each equation is 10 (multiply by $\begin{bmatrix} 1 \\ 1 \\ 1 \\ 1 \end{bmatrix}$) and hence doubling these entries gives a solution. To obtain another solution, first obtain the reduced row echelon formof the augmented matrix:

$$\begin{bmatrix} 1 & 0 & 0 & -1 & -20 \\ 0 & 1 & 0 & 9 & 20 \\ 0 & 0 & 1 & -1 & 0 \end{bmatrix}.$$

From the last equation, $x_3 = x_4$. Choose the value $x_3 = x_4 = 1$. Using back substitution, we find from the second equation that $x_2 = 20 - 9(1) = 11$ and, from the first equation, that $x_1 = -20 + 11(1) = -9$. Therefore

$$\begin{bmatrix} -9 \\ 11 \\ 1 \\ 1 \end{bmatrix}$$

is another solution.

41. We have

$$\mathbf{u} \cdot \mathbf{v} = \sum_{i=1}^{n} u_i v_i = \begin{bmatrix} u_1 & u_2 & \cdots & u_n \end{bmatrix} \begin{bmatrix} v_1 \\ v_2 \\ \vdots \\ v_n \end{bmatrix} = \mathbf{u}^T \mathbf{v}.$$

43. (a) $\mathrm{Tr}(cA) = \displaystyle\sum_{i=1}^{n} c a_{ii} = c \sum_{i=1}^{n} a_{ii} = c \, \mathrm{Tr}(A).$

(b) $\mathrm{Tr}(A + B) = \displaystyle\sum_{i=1}^{n} (a_{ii} + b_{ii}) = \sum_{i=1}^{n} a_{ii} + \sum_{i=1}^{n} b_{ii} = \mathrm{Tr}(A) + \mathrm{Tr}(B).$

(c) Let $AB = C = \begin{bmatrix} c_{ij} \end{bmatrix}$. Then

$$\mathrm{Tr}(AB) = \mathrm{Tr}(C) = \sum_{i=1}^{n} c_{ii} = \sum_{i=1}^{n} \sum_{k=1}^{n} a_{ik} b_{ki} = \sum_{k=1}^{n} \sum_{i=1}^{n} b_{ki} a_{ik} = \mathrm{Tr}(BA).$$

(d) Since $a_{ii}^T = a_{ii}$, $\text{Tr}(A^T) = \sum\limits_{i=1}^{n} a_{ii}^T = \sum\limits_{i=1}^{n} a_{ii} = \text{Tr}(A)$.

(e) Let $A^T A = B = \begin{bmatrix} b_{ij} \end{bmatrix}$. Then

$$b_{ii} = \sum_{j=1}^{n} a_{ij}^T a_{ji} = \sum_{j=1}^{n} a_{ji}^2 \implies \text{Tr}(B) = \text{Tr}(A^T A) = \sum_{i=1}^{n} b_{ii} = \sum_{i=1}^{n}\sum_{j=1}^{n} a_{ij}^2 \geq 0.$$

Hence, $\text{Tr}(A^T A) \geq 0$.

45. We have $\text{Tr}(AB - BA) = \text{Tr}(AB) - \text{Tr}(BA) = 0$, while $\text{Tr}\left(\begin{bmatrix} 1 & 0 \\ 0 & 1 \end{bmatrix} \right) = 2$.

47. Let $A = \begin{bmatrix} a_{ij} \end{bmatrix}$ and $B = \begin{bmatrix} b_{ij} \end{bmatrix}$ be $m \times n$ and $n \times p$, respectively. Then the jth column of AB is

$$(AB)_j = \begin{bmatrix} a_{11}b_{1j} + \cdots + a_{1n}b_{nj} \\ \vdots \\ a_{m1}b_{1j} + \cdots + a_{mn}b_{nj} \end{bmatrix}$$

$$= b_{1j} \begin{bmatrix} a_{11} \\ \vdots \\ a_{m1} \end{bmatrix} + \cdots + b_{nj} \begin{bmatrix} a_{1n} \\ \vdots \\ a_{mn} \end{bmatrix}$$

$$= b_{1j}\text{Col}_1(A) + \cdots + b_{nj}\text{Col}_n(A).$$

Thus the jth column of AB is a linear combination of the columns of A with coefficients the entries in $\mathbf{b}_j$.

49. $AB = \begin{bmatrix} 2 & 2 \\ 3 & 4 \end{bmatrix} \begin{bmatrix} 9 & 10 \\ 10 & 12 \end{bmatrix} = \begin{bmatrix} 38 & 44 \\ 67 & 78 \end{bmatrix}$.

The entries of the matrix product AB tell the manufacturer the total cost of producing chairs and tables in each city.

$$B = \begin{pmatrix} \begin{array}{cc} \text{Salt Lake} & \\ \text{City} & \text{Chicago} \\ 38 & 44 \\ 67 & 78 \end{array} \end{pmatrix} \begin{array}{l} \text{Chair} \\ \text{Table.} \end{array}$$

51. (a) No. If $\mathbf{x} = (x_1, x_2, \ldots, x_n)$, then $\mathbf{x} \cdot \mathbf{x} = x_1^2 + x_2^2 + \cdots + x_n^2 \geq 0$.

(b) $\mathbf{x} = \mathbf{0}$.

53. The i, ith element of the matrix AA^T is

$$\sum_{k=1}^{n} a_{ik} a_{ki}^T = \sum_{k=1}^{n} a_{ik} a_{ik} = \sum_{k=1}^{n} (a_{ik})^2.$$

Thus if $AA^T = O$, then each sum of squares $\sum\limits_{k=1}^{n} (a_{ik})^2$ equals zero, which implies $a_{ik} = 0$ for each i and k. Thus $A = O$.

Section 1.4, p. 40

1. Let $A = \begin{bmatrix} a_{ij} \end{bmatrix}$, $B = \begin{bmatrix} b_{ij} \end{bmatrix}$, $C = \begin{bmatrix} c_{ij} \end{bmatrix}$. Then the (i, j) entry of $A + (B + C)$ is $a_{ij} + (b_{ij} + c_{ij})$ and that of $(A + B) + C$ is $(a_{ij} + b_{ij}) + c_{ij}$. By the associative law for addition of real numbers, these two numbers are equal.

3. $A(BC) = \begin{bmatrix} 1 & 3 \\ 2 & -1 \end{bmatrix} \left(\begin{bmatrix} -1 & 3 & 2 \\ 1 & -3 & 4 \end{bmatrix} \begin{bmatrix} 1 & 0 \\ 3 & -1 \\ 1 & 2 \end{bmatrix} \right) = \begin{bmatrix} 1 & 3 \\ 2 & -1 \end{bmatrix} \begin{bmatrix} 10 & 1 \\ -4 & 11 \end{bmatrix} = \begin{bmatrix} -2 & 34 \\ 24 & -9 \end{bmatrix};$

$(AB)C = \left(\begin{bmatrix} 1 & 3 \\ 2 & -1 \end{bmatrix} \begin{bmatrix} -1 & 3 & 2 \\ 1 & -3 & 4 \end{bmatrix} \right) \begin{bmatrix} 1 & 0 \\ 3 & -1 \\ 1 & 2 \end{bmatrix} = \begin{bmatrix} 2 & -6 & 14 \\ -3 & 9 & 0 \end{bmatrix} \begin{bmatrix} 1 & 0 \\ 3 & -1 \\ 1 & 2 \end{bmatrix} = \begin{bmatrix} -2 & 34 \\ 24 & -9 \end{bmatrix}$

5. $C(A + B) = \begin{bmatrix} 1 & -3 \\ -3 & 4 \end{bmatrix} \left(\begin{bmatrix} 2 & -3 & 2 \\ 3 & -1 & -2 \end{bmatrix} + \begin{bmatrix} 0 & 1 & 2 \\ 1 & 3 & -2 \end{bmatrix} \right)$

$= \begin{bmatrix} 1 & -3 \\ -3 & 4 \end{bmatrix} \begin{bmatrix} 2 & -2 & 4 \\ 4 & 2 & -4 \end{bmatrix} = \begin{bmatrix} -10 & -8 & 16 \\ 10 & 14 & -28 \end{bmatrix};$

$CA + CB = \begin{bmatrix} 1 & -3 \\ -3 & 4 \end{bmatrix} \begin{bmatrix} 2 & -3 & 2 \\ 3 & -1 & -2 \end{bmatrix} + \begin{bmatrix} 1 & -3 \\ -3 & 4 \end{bmatrix} \begin{bmatrix} 0 & 1 & 2 \\ 1 & 3 & -2 \end{bmatrix}$

$= \begin{bmatrix} -7 & 0 & 8 \\ 6 & 5 & -14 \end{bmatrix} + \begin{bmatrix} -3 & -8 & 8 \\ 4 & 9 & -14 \end{bmatrix} = \begin{bmatrix} -10 & -8 & 16 \\ 10 & 14 & -28 \end{bmatrix}$

7. Let $A = \begin{bmatrix} a_{ij} \end{bmatrix}$. Then

$$CA = \begin{bmatrix} c_1 & c_2 & \cdots & c_m \end{bmatrix} \begin{bmatrix} a_{11} & a_{12} & \cdots & a_{1n} \\ \vdots & \vdots & & \vdots \\ a_{m1} & a_{12} & \cdots & a_{mn} \end{bmatrix}$$

$$= \begin{bmatrix} c_1 a_{11} + \cdots + c_m a_{m1} & \cdots & c_1 a_{1n} + \cdots + c_m a_{mn} \end{bmatrix}$$

$$= \begin{bmatrix} c_1 a_{11} & \cdots & c_1 a_{1n} \end{bmatrix} + \cdots + \begin{bmatrix} c_m a_{m1} & \cdots & c_m a_{mn} \end{bmatrix}$$

$$= c_1 \begin{bmatrix} a_{11} & \cdots & a_{1n} \end{bmatrix} + \cdots + c_m \begin{bmatrix} a_{m1} & \cdots & a_{mn} \end{bmatrix}$$

$$= c_1 A_1 + \cdots + c_m A_m = \sum_{j=1}^{m} c_j A_j$$

9. One such pair is $A = \begin{bmatrix} 1 & 1 \\ 2 & 2 \end{bmatrix}$ and $B = \begin{bmatrix} 1 & 1 \\ -1 & -1 \end{bmatrix}$.

11. One such pair is $A = \begin{bmatrix} 1 & 1 \\ 0 & 1 \end{bmatrix}$ and $B = \begin{bmatrix} 1 & -1 \\ 0 & 1 \end{bmatrix}$.

13. To prove that $r(sA) = (rs)A$, we must show that corresponding entries of these two matrices are equal. Let $A = \begin{bmatrix} a_{ij} \end{bmatrix}$. Then $sA = \begin{bmatrix} sa_{ij} \end{bmatrix}$, $r(sA) = \begin{bmatrix} r(sa_{ij}) \end{bmatrix}$, and $(rs)A = \begin{bmatrix} (rs)a_{ij} \end{bmatrix}$. It follows from properties of real numbers that $r(sa_{ij}) = (rs)a_{ij}$. Thus the corresponding entries of $r(sA)$ and $(rs)A$ are equal. Therefore $r(sA) = (rs)A$.

15. $(r + s)A = (4 + (-2))A = 2A = \begin{bmatrix} 4 & -6 \\ 8 & 4 \end{bmatrix}$

$rA + sA = 4A + (-2)A = \begin{bmatrix} 8 & -12 \\ 16 & 8 \end{bmatrix} + \begin{bmatrix} -4 & 6 \\ -8 & -4 \end{bmatrix} = \begin{bmatrix} 4 & -6 \\ 8 & 4 \end{bmatrix}$

Thus $(r + s)A = rA + sA$.

17. $r(A + B) = (-3) \left(\begin{bmatrix} 4 & 2 \\ 1 & -3 \\ 3 & 2 \end{bmatrix} + \begin{bmatrix} 0 & 2 \\ 4 & 3 \\ -2 & 1 \end{bmatrix} \right) = (-3) \begin{bmatrix} 4 & 4 \\ 5 & 0 \\ 1 & 3 \end{bmatrix} = \begin{bmatrix} -12 & -12 \\ -15 & 0 \\ -3 & -9 \end{bmatrix}$

$$rA + rB = (-3)\begin{bmatrix} 4 & 2 \\ 1 & -3 \\ 3 & 2 \end{bmatrix} + (-3)\begin{bmatrix} 0 & 2 \\ 4 & 3 \\ -2 & 1 \end{bmatrix} = \begin{bmatrix} -12 & -6 \\ -3 & 9 \\ -9 & -6 \end{bmatrix} + \begin{bmatrix} 0 & -6 \\ -12 & -9 \\ 6 & -3 \end{bmatrix} = \begin{bmatrix} -12 & -12 \\ -15 & 0 \\ -3 & -9 \end{bmatrix}$$

Thus $r(A + B) = rA + rB$.

19. $A(rB) = \begin{bmatrix} 1 & 3 \\ 2 & -1 \end{bmatrix}\left((-3)\begin{bmatrix} -1 & 3 & 2 \\ 1 & -3 & 4 \end{bmatrix}\right) = \begin{bmatrix} 1 & 3 \\ 2 & -1 \end{bmatrix}\begin{bmatrix} 3 & -9 & -6 \\ -3 & 9 & -12 \end{bmatrix} = \begin{bmatrix} -6 & 18 & -42 \\ 9 & -27 & 0 \end{bmatrix}$

$r(AB) = (-3)\left(\begin{bmatrix} 1 & 3 \\ 2 & -1 \end{bmatrix}\begin{bmatrix} -1 & 3 & 2 \\ 1 & -3 & 4 \end{bmatrix}\right) = (-3)\begin{bmatrix} 2 & -6 & 14 \\ -3 & 9 & 0 \end{bmatrix} = \begin{bmatrix} -6 & 18 & -42 \\ 9 & -27 & 0 \end{bmatrix}$

$(rA)B = \left((-3)\begin{bmatrix} 1 & 3 \\ 2 & -1 \end{bmatrix}\right)\begin{bmatrix} -1 & 3 & 2 \\ 1 & -3 & 4 \end{bmatrix} = \begin{bmatrix} -3 & -9 \\ -6 & 3 \end{bmatrix}\begin{bmatrix} -1 & 3 & 2 \\ 1 & -3 & 4 \end{bmatrix} = \begin{bmatrix} -6 & 18 & -42 \\ 9 & -27 & 0 \end{bmatrix}$

It follows that $A(rB) = r(AB) = (rA)B$.

21. (a) Use $k = 3$ to get

$$3A = \begin{bmatrix} 3\cos 0° & 3\sin 0° \\ 3\cos 1° & 3\sin 1° \\ \vdots & \vdots \\ 3\cos 359° & \sin 359° \end{bmatrix}.$$

(b) The ordered pairs $(3\cos i, 3\sin i)$, where $i = 0°, 1°, \ldots, 359°$, lie on the circle $x^2 + y^2 = 9$ of radius 3 since

$$x^2 + y^2 = (3\cos i)^2 + (3\sin i)^2 = 9(\cos^2 i + \sin^2 i) = 9.$$

23. $A\mathbf{x} = \begin{bmatrix} 1 & 2 & -1 \\ 1 & 0 & 1 \\ 4 & -4 & 5 \end{bmatrix}\begin{bmatrix} -\frac{1}{2} \\ \frac{1}{4} \\ 1 \end{bmatrix} = \begin{bmatrix} -1 \\ \frac{1}{2} \\ 2 \end{bmatrix} = r\mathbf{x} = r\begin{bmatrix} -\frac{1}{2} \\ \frac{1}{4} \\ 1 \end{bmatrix} = \begin{bmatrix} -\frac{1}{2}r \\ \frac{1}{4}r \\ r \end{bmatrix}$

Equating corresponding elements gives $-1 = -\frac{1}{2}r$ and $\frac{1}{2} = \frac{1}{4}r$, which implies that $r = 2$.

25. Let $A\mathbf{x} = r\mathbf{x}$. Then to find a scalar s such that $A^2\mathbf{x} = s\mathbf{x}$ we proceed as follows:

$$A^2\mathbf{x} = A(A\mathbf{x}) = A(r\mathbf{x}) = r(A\mathbf{x}) = r(r\mathbf{x}) = r^2\mathbf{x}.$$

If $s = r^2$, then we have the desired result.

27. Theorem 1.4(b): $(A + B)T = A^T + B^T$

Proof: We show that corresponding entries of $(A + B)^T$ and $A^T + B^T$ are equal. Let $A = [\ a_{ij}\ ]$ and $B = [\ b_{ij}\ ]$. Then
$$A + B = [\ a_{ij} + b_{ij}\ ].$$

Therefore
$$(A + B)^T = [\ (a_{ij} + b_{ij})^T\ ] = [\ a_{ji} + b_{ji}\ ].$$

Now $A^T = [\ a_{ij}^T\ ] = [\ a_{ji}\ ]$ and $B^T = [\ b_{ij}^T\ ] = [\ b_{ji}\ ]$. Therefore

$$A^T + B^T = [\ a_{ji}\ ] + [\ b_{ji}\ ] = [\ a_{ji} + b_{ji}\ ].$$

It follows that $(A + B)^T = A^T + B^T$ since corresponding entries of these matrices are equal.

Theorem 1.4(d): $(rA)^T = r(A^T)$

Proof: We show that corresponding entries of $(rA)^T$ and $r(A^T)$ are equal. Let $A = [\ a_{ij}\ ]$. Then

$$rA = r[\ a_{ij}\ ] = [\ ra_{ij}\ ].$$

Therefore

$$(rA)^T = \left[\ (ra_{ij})^T\ \right] = \left[\ ra_{ji}\ \right].$$

Now

$$r(A^T) = r\left[\ a_{ij}^T\ \right] = \left[\ ra_{ji}\ \right].$$

It follows that $(rA)^T = r(A^T)$.

29. $(AB)^T = \left(\begin{bmatrix} 1 & 3 & 2 \\ 2 & 1 & -3 \end{bmatrix} \begin{bmatrix} 3 & -1 \\ 2 & 4 \\ 1 & 2 \end{bmatrix}\right)^T = \begin{bmatrix} 11 & 15 \\ 5 & -4 \end{bmatrix}^T = \begin{bmatrix} 11 & 5 \\ 15 & -4 \end{bmatrix}$

$B^T A^T = \begin{bmatrix} 3 & -1 \\ 2 & 4 \\ 1 & 2 \end{bmatrix}^T \begin{bmatrix} 1 & 3 & 2 \\ 2 & 1 & -3 \end{bmatrix}^T = \begin{bmatrix} 3 & 2 & 1 \\ -1 & 4 & 2 \end{bmatrix} \begin{bmatrix} 1 & 2 \\ 3 & 1 \\ 2 & -3 \end{bmatrix} = \begin{bmatrix} 11 & 5 \\ 15 & -4 \end{bmatrix} = (AB)^T$

31. $(kA)^T(kA) = \begin{bmatrix} -2k \\ k \\ -k \end{bmatrix}^T \begin{bmatrix} -2k \\ k \\ -k \end{bmatrix} = \begin{bmatrix} -2k & k & -k \end{bmatrix} \begin{bmatrix} -2k \\ k \\ -k \end{bmatrix} = 4k^2 + k^2 + k^2 = 6k^2 = 1$

Hence $k^2 = \frac{1}{6}$ and $k = \pm\sqrt{\frac{1}{6}}$. Thus there are two values of k such that $(kA)^T(kA) = 1$.

33. If $cA = O$, then each element of matrix cA is zero. That is, $ca_{ij} = 0$. Since c and a_{ij} are real numbers, this means that either $c = 0$ or $a_{ij} = 0$ for every i and j. It follows that either $c = 0$ or matrix $A = O$.

35. $(A - B)^T = (A + (-1)B)^T$

$\qquad = A^T + ((-1)B)^T \quad$ by Theorem 1.4(b)

$\qquad = A^T + (-1)B^T \quad$ by Theorem 1.4(d)

$\qquad = A^T - B^T$

37. Let $\mathbf{x}_1$ and $\mathbf{x}_2$ be solutions to the linear system $A\mathbf{x} = \mathbf{b}$. Then $A\mathbf{x}_1 = \mathbf{b}$ and $A\mathbf{x}_2 = \mathbf{b}$. Now, let r be any real number and let $s = 1 - r$. Let $\mathbf{x}_3 = r\mathbf{x}_1 + s\mathbf{x}_2$. Then

$$A\mathbf{x}_3 = A(r\mathbf{x}_1 + s\mathbf{x}_2) = A(r\mathbf{x}_1) + A(s\mathbf{x}_2) = r(A\mathbf{x}_1) + s(A\mathbf{x}_2) = r\mathbf{b} + s\mathbf{b} = (r + s)\mathbf{b} = \mathbf{b}.$$

Therefore $\mathbf{x}_3$ is a solution to the linear system $A\mathbf{x} = \mathbf{b}$. Thus there are infinitely many solutions to this system since there are infinitely many choices for r.

Section 1.5, p. 52

1. (a) To show that $I_m A = A$ we show that corresponding entries are equal. Let $A = \left[\ a_{ij}\ \right]$ and $I_m = \left[\ d_{ij}\ \right]$, where $d_{ij} = 0$ for $i \neq j$ and $d_{ii} = 1$. Then

    $$(i,j) \text{ entry of } I_m A = d_{i1}a_{1j} + \cdots + d_{ii}a_{ij} + \cdots + d_{im}a_{mj}.$$

    Since $d_{ik} = 0$ for $k \neq i$ and $d_{ii} = 1$, this sum is equal to $(1)(a_{ij}) = a_{ij}$. Therefore $I_m A = A$. Similarly,

    $$(i,j) \text{ entry of } AI_n = a_{i1}d_{1j} + \cdots + a_{ij}d_{jj} + \cdots + a_{in}d_{nj} = a_{ij}.$$

    Therefore $AI_n = A$.

    (b) A scalar matrix $A = \left[\ a_{ij}\ \right]$ is a diagonal matrix whose diagonal elements are equal. Hence

    $$a_{ij} = \begin{cases} r, & i = j \\ 0, & i \neq j \end{cases}$$

    for some value r. Therefore $A = rI_n$.

3. Let $A = [\ a_{ij}\ ]$ and $B = [\ b_{ij}\ ]$ be $n \times n$ diagonal matrices. Then $a_{ij} = 0$ and $b_{ij} = 0$ whenever $i \neq j$. To show that $AB = BA$ we show that corresponding entries of these matrices are equal. Now,

$$(i, j) \text{ entry of } AB = a_{i1}b_{1j} + \cdots + a_{ik}b_{kj} + \cdots + a_{in}b_{nj} = \begin{cases} a_{ii}b_{ii}, & \text{if } i = j \\ 0, & \text{if } i \neq j \end{cases}$$

and

$$(i, j) \text{ entry of } BA = b_{i1}a_{1j} + \cdots + b_{ik}a_{kj} + \cdots + b_{in}a_{nj} = \begin{cases} b_{ii}a_{ii}, & \text{if } i = j \\ 0, & \text{if } i \neq j \end{cases}$$

Since $a_{ii}b_{ii} = b_{ii}a_{ii}$ for all i, it follows that $AB = BA$.

5. A matrix A that is both upper and lower triangular has

$$a_{ij} = 0 \text{ for } i > j \quad \text{and} \quad a_{ij} = 0 \text{ for } i < j.$$

These two conditions are equivalent to $a_{ij} = 0$ for $i \neq j$, which implies that A is a diagonal matrix.

7. (a) $A^2 = \begin{bmatrix} 1 & 0 & -1 \\ 2 & 1 & 1 \\ 3 & 1 & 0 \end{bmatrix} \begin{bmatrix} 1 & 0 & -1 \\ 2 & 1 & 1 \\ 3 & 1 & 0 \end{bmatrix} = \begin{bmatrix} -2 & -1 & -1 \\ 7 & 2 & -1 \\ 5 & 1 & -2 \end{bmatrix}$

$A^3 = A^2 \cdot A = \begin{bmatrix} -2 & -1 & -1 \\ 7 & 2 & -1 \\ 5 & 1 & -2 \end{bmatrix} \begin{bmatrix} 1 & 0 & -1 \\ 2 & 1 & 1 \\ 3 & 1 & 0 \end{bmatrix} = \begin{bmatrix} -7 & -2 & 1 \\ 8 & 1 & -5 \\ 1 & -1 & -4 \end{bmatrix}$

(b) $B^2 = \begin{bmatrix} 0 & 0 & 1 \\ -1 & 1 & 1 \\ 2 & 0 & 1 \end{bmatrix} \begin{bmatrix} 0 & 0 & 1 \\ -1 & 1 & 1 \\ 2 & 0 & 1 \end{bmatrix} = \begin{bmatrix} 2 & 0 & 1 \\ 1 & 1 & 1 \\ 2 & 0 & 3 \end{bmatrix}$

(c) $AB = \begin{bmatrix} 1 & 0 & -1 \\ 2 & 1 & 1 \\ 3 & 1 & 0 \end{bmatrix} \begin{bmatrix} 0 & 0 & 1 \\ -1 & 1 & 1 \\ 2 & 0 & 1 \end{bmatrix} = \begin{bmatrix} -2 & 0 & 0 \\ 1 & 1 & 4 \\ -1 & 1 & 4 \end{bmatrix}$

$(AB)^2 = \begin{bmatrix} -2 & 0 & 0 \\ 1 & 1 & 4 \\ -1 & 1 & 4 \end{bmatrix} \begin{bmatrix} -2 & 0 & 0 \\ 1 & 1 & 4 \\ -1 & 1 & 4 \end{bmatrix} = \begin{bmatrix} 4 & 0 & 0 \\ -5 & 5 & 20 \\ -1 & 5 & 20 \end{bmatrix}$

$(AB)^3 = (AB)^2(AB) = \begin{bmatrix} 4 & 0 & 0 \\ -5 & 5 & 20 \\ -1 & 5 & 20 \end{bmatrix} \begin{bmatrix} -2 & 0 & 0 \\ 1 & 1 & 4 \\ -1 & 1 & 4 \end{bmatrix} = \begin{bmatrix} -8 & 0 & 0 \\ -5 & 25 & 100 \\ -13 & 25 & 100 \end{bmatrix}$

9. If $p = 1$ the result is true since $AB = BA$. Now assume the result is true for $p = k$. That is, assume that $(AB)^k = A^k B^k$. We now verify that the result is true for $p = k + 1$:

$$(AB)^{k+1} = (AB)^k(AB) = A^k B^k (AB) \quad \text{by our assumption}$$
$$= A^k(B^k A)B$$
$$= A^k(B^{k-1}BA)B = A^k(B^{k-1}AB)B = A^k(B^{k-2}BAB)B$$
$$= A^k(B^{k-2}ABB)B = A^k(B^{k-2}AB^2)B = \cdots = A^k(BAB^{k-1})B$$
$$= A^k(ABB^{k-1})B = A^k(AB^k)B = A^{k+1}B^{k+1}$$

Hence by the induction principle, the result is true for any positive integer p.

11. If $p = 0$ the statement is true since $(A^T)^0 = I = (A^0)^T$. Now assume the statement is true for $p = k$: $(A^T)^k = (A^k)^T$. Then for $p = k + 1$, we have:

$$(A^T)^{k+1} = (A^T)^K A^T$$
$$= (A^k)^T A^T$$
$$= (AA^k)^T \quad \text{by Theorem 1.4(c)}$$
$$= (A^{k+1})^T.$$

Hence by the induction principle, the result is true for any nonnegative integer p.

13. We have

$$(kA)\left(\frac{1}{k}A^{-1}\right) = \left(k \cdot \frac{1}{k}\right)(AA^{-1}) = 1 \cdot I = I$$

and

$$\left(\frac{1}{k}A^{-1}\right)(kA) = \left(\frac{1}{k}k\right)(A^{-1}A) = 1 \cdot I = I.$$

Therefore, $(kA)^{-1} = \frac{1}{k}A^{-1}$.

15. Let $B = \begin{bmatrix} a & b \\ c & d \end{bmatrix}$. We compute AB and BA and equate corresponding entries of these two matrices:

$$AB = \begin{bmatrix} 1 & 2 \\ 2 & 1 \end{bmatrix}\begin{bmatrix} a & b \\ c & d \end{bmatrix} = \begin{bmatrix} a+2c & b+2d \\ 2a+c & 2b+d \end{bmatrix} = BA = \begin{bmatrix} a & b \\ c & d \end{bmatrix}\begin{bmatrix} 1 & 2 \\ 2 & 1 \end{bmatrix} = \begin{bmatrix} a+2b & 2a+b \\ c+2d & 2c+d \end{bmatrix}.$$

Then

$$\begin{aligned} a+2c &= a+2b & \Longrightarrow & \quad c = b \\ b+2d &= 2a+b & \Longrightarrow & \quad d = a \\ 2a+c &= c+2d & \Longrightarrow & \quad a = d \\ 2b+d &= 2c+d & \Longrightarrow & \quad b = c. \end{aligned}$$

It follows that $B = \begin{bmatrix} a & b \\ b & a \end{bmatrix}$, where a and b are arbitrary. For example, $B = \begin{bmatrix} 1 & 3 \\ 3 & 1 \end{bmatrix}$ is a matrix such that $AB = BA$.

17. The statement is false. For example, let $A = \begin{bmatrix} 1 & 2 \\ 3 & 4 \end{bmatrix}$. Then

$$AA^T = \begin{bmatrix} 1 & 2 \\ 3 & 4 \end{bmatrix}\begin{bmatrix} 1 & 3 \\ 2 & 4 \end{bmatrix} = \begin{bmatrix} 5 & 11 \\ 11 & 25 \end{bmatrix} \quad \text{and} \quad A^T A = \begin{bmatrix} 1 & 3 \\ 2 & 4 \end{bmatrix}\begin{bmatrix} 1 & 2 \\ 3 & 4 \end{bmatrix} = \begin{bmatrix} 10 & 14 \\ 14 & 18 \end{bmatrix}.$$

19. We are given that A is symmetric; that is, $A^T = A$. We are to show that A^T is symmetric. We proceed to show that $(A^T)^T = A^T$. From Theorem 1.4(a) and the fact that $A = A^T$ we have that $(A^T)^T = A = A^T$. Hence A^T is symmetric.

21. To show that AA^T is symmetric we show that $(AA^T)^T = AA^T$. Now:

$$(AA^T)^T = (A^T)^T A^T \quad \text{by Theorem 1.4(c)}$$

$$= AA^T \qquad\quad \text{by Theorem 1.4(a)}$$

Therefore AA^T is symmetric. Similarly, for the matrix $A^T A$ we obtain

$$(A^T A)^T = A^T(A^T)^T \quad \text{by Theorem 1.4(c)}$$

$$= A^T A \qquad\quad \text{by Theorem 1.4(a)}$$

Hence $A^T A$ is symmetric.

23. If $k = 1$, the result is true since A is symmetric. Now assume the result is true for $k = n$. That is, assume that A^n is symmetric. We now verify that the result is true for $k = n+1$:

$$(A^{n+1})^T = (AA^n)^T$$

$$= (A^n)^T A^T \quad \text{Theorem 1.4(c)}$$

$$= A^n A \qquad\quad \text{since } A \text{ and } A^n \text{ are symmetric by assumption}$$

$$= A^{n+1}$$

Therefore A^{n+1} is symmetric. Hence A^k is symmetric for all positive integers k.

25. (a) Let $A = \begin{bmatrix} a_{ij} \end{bmatrix}$ be upper triangular, so that $a_{ij} = 0$ for $i > j$. Since $A^T = \begin{bmatrix} a_{ij}^T \end{bmatrix}$, where $a_{ij}^T = a_{ji}$, we have $a_{ij}^T = 0$ for $j > i$, or $a_{ij}^T = 0$ for $i < j$. Hence A^T is lower triangular.

　　(b) Proof is similar to that for (a).

27. If A is skew symmetric, $A^T = -A$. Thus $a_{ii} = -a_{ii}$, so $a_{ii} = 0$.

29. Assume there is such a decomposition $A = S + K$, where S is symmetric and K is skew symmetric. We shall determine S and K. We have

$$A^T = (S + T)^T = S^T + K^T = S - K$$

using Theorem 1.4(b) and the assumptions about S and K. Thus we have the expressions

$$A = S + K$$
$$A^T = S - K.$$

Treat these expressions as a pair of matrix equations in the unknown matrices S and K. If we add the two equations we eliminate K to get

$$A + A^T = 2S \quad \Longrightarrow \quad S = \frac{1}{2}(A + A^T).$$

Then substituting for S in the equation $A = S + K$ and solving for K we have

$$A = \frac{1}{2}(A + A^T) + K \quad \Longrightarrow \quad K = \frac{1}{2}(A - A^T).$$

Thus we have a unique solution to the pair of matrix equations. The matrix S is symmetric since

$$S^T = \left(\frac{1}{2}(A + A^T) \right)^T = \frac{1}{2}\left(A^T + (A^T)^T \right) = \frac{1}{2}(A^T + A) = S$$

and the matrix K is skew symmetric since

$$K^T = \left(\frac{1}{2}(A - A^T) \right)^T = \frac{1}{2}\left(A^T - (A^T)^T \right) = \frac{1}{2}(A^T - A) = -\frac{1}{2}(A - A^T) = -K.$$

31. To show that $A = \begin{bmatrix} 2 & 3 \\ 4 & 6 \end{bmatrix}$ is singular we follow Definition 1.10 and show that there is no matrix $B = \begin{bmatrix} a & b \\ c & d \end{bmatrix}$ such that $AB = BA = I_2$. From the requirement that

$$AB = \begin{bmatrix} 2 & 3 \\ 4 & 6 \end{bmatrix} \begin{bmatrix} a & b \\ c & d \end{bmatrix} = \begin{bmatrix} 2a + 3c & 2b + 3d \\ 4a + 6c & 4b + 6d \end{bmatrix} = \begin{bmatrix} 1 & 0 \\ 0 & 1 \end{bmatrix}$$

upon equating corresponding entries we obtain equations

$$2a + 3c = 1 \qquad 2b + 3d = 0$$
$$4a + 6c = 0 \qquad 4b + 6d = 1.$$

Attempting to solve the first pair of equations for a and c we add -2 times the first equation to the second equation to obtain $0 = -2$, which makes no sense. Hence there is no matrix B such that $AB = I_2$ and thus A is singular.

33. It is implied that the matrices are nonsingular. Thus we follow the procedure in Example 11.

(a) Let $A = \begin{bmatrix} 1 & 3 \\ 5 & 2 \end{bmatrix}$ and $A^{-1} = \begin{bmatrix} a & b \\ c & d \end{bmatrix}$. Then we must have

$$AA^{-1} = \begin{bmatrix} 1 & 3 \\ 5 & 2 \end{bmatrix} \begin{bmatrix} a & b \\ c & d \end{bmatrix} = \begin{bmatrix} a + 3c & b + 3d \\ 5a + 2c & 5b + 2d \end{bmatrix} = \begin{bmatrix} 1 & 0 \\ 0 & 1 \end{bmatrix}.$$

Equating corresponding entries we have the systems

$$\begin{aligned} a + 3c &= 1 & b + 3d &= 0 \\ 5a + 2c &= 0 & 5b + 2d &= 1 \end{aligned}$$

Solving the systems we obtain $a = -\frac{2}{13}$, $c = \frac{5}{13}$, $b = \frac{3}{13}$, and $d = -\frac{1}{13}$. Then

$$\begin{bmatrix} -\frac{2}{13} & \frac{3}{13} \\ \frac{5}{13} & -\frac{1}{13} \end{bmatrix} \begin{bmatrix} 1 & 3 \\ 5 & 2 \end{bmatrix} = I_2 = \begin{bmatrix} 1 & 3 \\ 5 & 2 \end{bmatrix} \begin{bmatrix} -\frac{2}{13} & \frac{3}{13} \\ \frac{5}{13} & -\frac{1}{13} \end{bmatrix}$$

and therefore

$$A^{-1} = \begin{bmatrix} -\frac{2}{13} & \frac{3}{13} \\ \frac{5}{13} & -\frac{1}{13} \end{bmatrix}.$$

(b) Let $A = \begin{bmatrix} 1 & 2 \\ 2 & 1 \end{bmatrix}$ and $A^{-1} = \begin{bmatrix} a & b \\ c & d \end{bmatrix}$. Then we must have

$$AA^{-1} = \begin{bmatrix} 1 & 2 \\ 2 & 1 \end{bmatrix} \begin{bmatrix} a & b \\ c & d \end{bmatrix} = \begin{bmatrix} a + 2c & b + 2d \\ 2a + c & 2b + d \end{bmatrix} = \begin{bmatrix} 1 & 0 \\ 0 & 1 \end{bmatrix}.$$

Equating corresponding entries we obtain the systems

$$\begin{aligned} a + 2c &= 1 & b + 2d &= 0 \\ 2a + c &= 0 & 2b + d &= 1 \end{aligned}$$

Solving the systems we obtain $a = -\frac{1}{3}$, $c = \frac{2}{3}$, $b = \frac{2}{3}$, and $d = -\frac{1}{3}$. Then

$$\begin{bmatrix} -\frac{1}{3} & \frac{2}{3} \\ \frac{2}{3} & -\frac{1}{3} \end{bmatrix} \begin{bmatrix} 1 & 2 \\ 2 & 1 \end{bmatrix} = I_2 = \begin{bmatrix} 1 & 2 \\ 2 & 1 \end{bmatrix} \begin{bmatrix} -\frac{1}{3} & \frac{2}{3} \\ \frac{2}{3} & -\frac{1}{3} \end{bmatrix}$$

and therefore

$$A^{-1} = \begin{bmatrix} -\frac{1}{3} & \frac{2}{3} \\ \frac{2}{3} & -\frac{1}{3} \end{bmatrix}.$$

35. By Theorem 1.6, $(AB)^{-1} = B^{-1}A^{-1} = \begin{bmatrix} 2 & 5 \\ 3 & -2 \end{bmatrix} \begin{bmatrix} 3 & 2 \\ 1 & 3 \end{bmatrix} = \begin{bmatrix} 11 & 19 \\ 7 & 0 \end{bmatrix}.$

37. $\mathbf{x} = C^{-1}A^{-1}\mathbf{b} = \begin{bmatrix} 2 & 1 \\ 1 & 2 \end{bmatrix} \begin{bmatrix} 2 & 1 \\ -1 & 1 \end{bmatrix} \begin{bmatrix} 2 \\ 3 \end{bmatrix} = \begin{bmatrix} 3 & 3 \\ 0 & 3 \end{bmatrix} \begin{bmatrix} 2 \\ 3 \end{bmatrix} = \begin{bmatrix} 15 \\ 9 \end{bmatrix}.$

39. $\mathbf{x} = (A^T)^{-1}\mathbf{b} = (A^{-1})^T\mathbf{b} = \begin{bmatrix} 4 & 1 \\ 1 & 0 \end{bmatrix} \begin{bmatrix} 1 \\ -2 \end{bmatrix} = \begin{bmatrix} 2 \\ 1 \end{bmatrix}.$

41. We use Equation (2) in the book and follow Example 13. We have from Exercise 33(a) that that

$$A^{-1} = \begin{bmatrix} -\frac{2}{13} & \frac{3}{13} \\ \frac{5}{13} & -\frac{1}{13} \end{bmatrix}.$$

(a) $\mathbf{x} = A^{-1}\mathbf{b} = \begin{bmatrix} -\frac{2}{13} & \frac{3}{13} \\ \frac{5}{13} & -\frac{1}{13} \end{bmatrix} \begin{bmatrix} 3 \\ 4 \end{bmatrix} = \begin{bmatrix} \frac{6}{13} \\ \frac{11}{13} \end{bmatrix}$

(b) $\mathbf{x} = A^{-1}\mathbf{b} = \begin{bmatrix} -\frac{2}{13} & \frac{3}{13} \\ \frac{5}{13} & -\frac{1}{13} \end{bmatrix} \begin{bmatrix} 5 \\ 6 \end{bmatrix} = \begin{bmatrix} \frac{8}{13} \\ \frac{19}{13} \end{bmatrix}$

43. One way to solve this problem is to find two matrices whose sum is O, since O is singular. Let A be any 2×2 nonsingular matrix and let $B = -A$. Then $A + B = O$.

45. To show that A^{-1} is nonsingular we find a matrix B such that $A^{-1}B = BA^{-1} = I_2$. Note that since A is nonsingular, $AA^{-1} = A^{-1}A = I_2$, so $B = A^{-1}$ satisfies the preceding relation. Since the inverse of a matrix, if it exists, is unique by Theorem 1.5, it follows that $(A^{-1})^{-1} = A$.

47. Let

$$ A = \begin{bmatrix} a_{11} & 0 & 0 & \cdots & 0 \\ 0 & a_{22} & 0 & \cdots & 0 \\ \vdots & \vdots & \vdots & & \vdots \\ 0 & 0 & 0 & \cdots & a_{nn} \end{bmatrix} \quad \text{and} \quad B = \begin{bmatrix} \dfrac{1}{a_{11}} & 0 & 0 & \cdots & 0 \\ 0 & \dfrac{1}{a_{22}} & 0 & \cdots & 0 \\ \vdots & \vdots & \vdots & & \vdots \\ 0 & 0 & 0 & \cdots & \dfrac{1}{a_{nn}} \end{bmatrix} $$

where $a_{ii} \neq 0$ for all i. Then $AB = \begin{bmatrix} c_{ij} \end{bmatrix}$ where

$$ c_{ij} = \text{row } i \text{ of } A \times \text{column } j \text{ of } B = \begin{cases} 0, & i \neq j \\ 1, & i = j \end{cases} $$

Thus $AB = I_n$ and in a similar manner we can show that $BA = I_n$. It follows that $B = A^{-1}$.

49. Let

$$ A = \begin{bmatrix} a_{11} & 0 & \cdots & 0 \\ 0 & a_{22} & \cdots & 0 \\ \vdots & \vdots & \vdots & \vdots \\ 0 & 0 & \cdots & a_{nn} \end{bmatrix}. $$

Then, for any nonnegative integer p:

$$ A^p = \begin{bmatrix} a_{11}^p & 0 & \cdots & 0 \\ 0 & a_{22}^p & \cdots & 0 \\ \vdots & \vdots & \vdots & \vdots \\ 0 & 0 & \cdots & a_{nn}^p \end{bmatrix}. $$

51. We are given that A is nonsingular and that $AB = O$. Multiply both sides of this relation by A^{-1} and simplify:
$$ A^{-1}AB = A^{-1}O \quad \Longrightarrow \quad I_n B = O \quad \Longrightarrow \quad B = O. $$

53. We are given that A is nonsingular and that $A\mathbf{x} = \mathbf{0}$. Multiply both sides of this relation by A^{-1} and simplify:
$$ A^{-1}A\mathbf{x} = A^{-1}\mathbf{0} \quad \Longrightarrow \quad I_n\mathbf{x} = \mathbf{0} \quad \Longrightarrow \quad \mathbf{x} = \mathbf{0}. $$

55. In order to add partitioned matrices, each matrix must be partitioned in the same manner. We partition matrices A and B as follows:

$$ A = \left[\begin{array}{c|cc} 1 & 3 & -1 \\ \hline 2 & 1 & 0 \\ 2 & -3 & 1 \end{array} \right] = \begin{bmatrix} A_{11} & A_{12} \\ A_{21} & A_{22} \end{bmatrix}, \quad B = \left[\begin{array}{c|cc} 3 & 2 & 1 \\ \hline -2 & 3 & 1 \\ 4 & 1 & 5 \end{array} \right] = \begin{bmatrix} B_{11} & B_{12} \\ B_{21} & B_{22} \end{bmatrix}. $$

Then

$$ A + B = \begin{bmatrix} A_{11} + B_{11} & A_{12} + B_{12} \\ A_{21} + B_{21} & A_{22} + B_{22} \end{bmatrix} = \left[\begin{array}{c|cc} 4 & 5 & 0 \\ \hline 0 & 4 & 1 \\ 6 & -2 & 6 \end{array} \right]. $$

Alternatively we could partition A and B as follows:

$$A = \begin{bmatrix} 1 & 3 & -1 \\ 2 & 1 & 0 \\ 2 & -3 & 1 \end{bmatrix} = \begin{bmatrix} A_{11} & A_{12} \\ A_{21} & A_{22} \end{bmatrix}, \qquad B = \begin{bmatrix} 3 & 2 & 1 \\ -2 & 3 & 1 \\ 4 & 1 & 5 \end{bmatrix} = \begin{bmatrix} B_{11} & B_{12} \\ B_{21} & B_{22} \end{bmatrix}.$$

Then

$$A + B = \begin{bmatrix} A_{11} + B_{11} & A_{12} + B_{12} \\ A_{21} + B_{21} & A_{22} + B_{22} \end{bmatrix} = \begin{bmatrix} 4 & 5 & 0 \\ 0 & 4 & 1 \\ 6 & -2 & 6 \end{bmatrix}.$$

57. A symmetric matrix. To show this, let $A_1, \ldots, A_n$ be symmetric matrices and let $x_1, \ldots, x_n$ be scalars. Then $A_1^T = A_1, \ldots, A_n^T = A_n$. Therefore

$$\begin{aligned} (x_1 A_1 + \cdots + x_n A_n)^T &= (x_1 A_1)^T + \cdots + (x_n A_n)^T \\ &= x_1 A_1^T + \cdots + x_n A_n^T \\ &= x_1 A_1 + \cdots + x_n A_n. \end{aligned}$$

Hence the linear combination $x_1 A_1 + \cdots + x_n A_n$ is symmetric.

59. (a) $\mathbf{w}_1 = A\mathbf{w}_0 = \begin{bmatrix} 5 & -6 \\ 1 & 0 \end{bmatrix} \begin{bmatrix} 1 \\ 0 \end{bmatrix} = \begin{bmatrix} 5 \\ 1 \end{bmatrix}$

$\mathbf{w}_2 = A\mathbf{w}_1 = \begin{bmatrix} 5 & -6 \\ 1 & 0 \end{bmatrix} \begin{bmatrix} 5 \\ 1 \end{bmatrix} = \begin{bmatrix} 19 \\ 5 \end{bmatrix}$

$\mathbf{w}_3 = A\mathbf{w}_2 = \begin{bmatrix} 5 & -6 \\ 1 & 0 \end{bmatrix} \begin{bmatrix} 19 \\ 5 \end{bmatrix} = \begin{bmatrix} 65 \\ 19 \end{bmatrix}$

$u_2 = 5u_1 - 6u_0 = 5(1) - 6(0) = 5$

$u_3 = 5u_2 - 6u_1 = 5(5) - 6(1) = 19$

$u_4 = 5u_3 - 6u_2 = 5(19) - 6(5) = 65$

(b) $\mathbf{w}_{n-1} = A\mathbf{w}_{n-2} = A(A\mathbf{w}_{n-3}) = A^2\mathbf{w}_{n-3}$. In general, $\mathbf{w}_{n-1} = A^{n-1}\mathbf{w}_0$. This is true if $n = 1$ since $\mathbf{w}_0 = A^0\mathbf{w}_0 = I\mathbf{w}_0$. Assuming that it is true for $n = k$, that is, $\mathbf{w}_{k-1} = A^{k-1}\mathbf{w}_0$, we have for $n = k + 1$:

$$\mathbf{w}_k = \mathbf{w}_{n-1} = A\mathbf{w}_{n-2} = A\mathbf{w}_{k-1} = A(A^{k-1}\mathbf{w}_0) = (AA^{k-1})\mathbf{w}_0 = A^k\mathbf{w}_0.$$

Hence $\mathbf{w}_{n-1} = A^{n-1}\mathbf{w}_0$ for every integer $n \geq 1$.

Section 1.6, p. 62

1. Since $f: R^2 \to R^2$ is defined by $f(x, y) = (x, -y)$, we have $f(2, 3) = (2, -3)$.

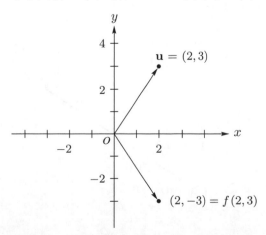

3. Since $f\colon R^2 \to R^2$ is defined by

$$f(x,y) = \begin{bmatrix} \cos\phi & -\sin\phi \\ \sin\phi & \cos\phi \end{bmatrix} \begin{bmatrix} x \\ y \end{bmatrix} \qquad \text{(see Example 8)}$$

we have

$$f(x,y) = \begin{bmatrix} \cos 30° & -\sin 30° \\ \sin 30° & \cos 30° \end{bmatrix} \begin{bmatrix} x \\ y \end{bmatrix} = \begin{bmatrix} \frac{\sqrt{3}}{2} & -\frac{1}{2} \\ \frac{1}{2} & \frac{\sqrt{3}}{2} \end{bmatrix} \begin{bmatrix} x \\ y \end{bmatrix}.$$

Thus

$$f(-1,3) = \begin{bmatrix} \frac{\sqrt{3}}{2} & -\frac{1}{2} \\ \frac{1}{2} & \frac{\sqrt{3}}{2} \end{bmatrix} \begin{bmatrix} -1 \\ 3 \end{bmatrix} = \left(\frac{-3-\sqrt{3}}{2}, \frac{-1+3\sqrt{3}}{2} \right) \approx (-2.366, 2.098).$$

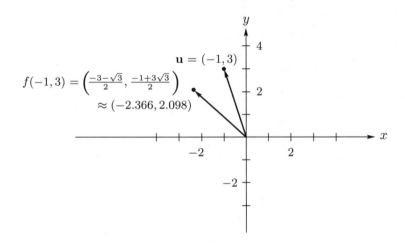

5. Since $f\colon R^2 \to R^2$ is defined by $f(\mathbf{u}) = -\mathbf{u}$, we have $f(3,2) = (-3,-2)$.

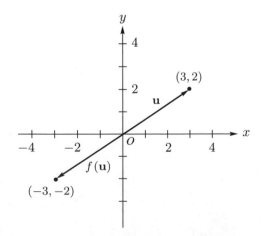

7. Since $f: R^3 \rightarrow R^2$ is defined by $f\left(\begin{bmatrix} x \\ y \\ z \end{bmatrix}\right) = \begin{bmatrix} x \\ x-y \\ 0 \end{bmatrix}$, we have $f\left(\begin{bmatrix} 2 \\ -1 \\ 3 \end{bmatrix}\right) = \begin{bmatrix} 2 \\ 3 \\ 0 \end{bmatrix}$.

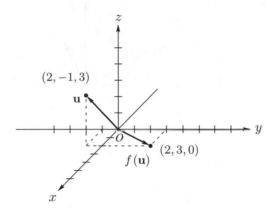

9. Yes; $\mathbf{x} = \begin{bmatrix} 3 \\ -1 \end{bmatrix}$

$$f(\mathbf{x}) = \begin{bmatrix} 1 & 2 \\ 0 & 1 \\ 1 & 1 \end{bmatrix} \begin{bmatrix} 3 \\ -1 \end{bmatrix} = \begin{bmatrix} 1 \\ -1 \\ 2 \end{bmatrix} = \mathbf{w}$$

11. Yes; $\mathbf{x} = \begin{bmatrix} 0 \\ 0 \end{bmatrix}$

$$f(\mathbf{x}) = \begin{bmatrix} 1 & 2 \\ 0 & 1 \\ 1 & 1 \end{bmatrix} \begin{bmatrix} 0 \\ 0 \end{bmatrix} = \begin{bmatrix} 0 \\ 0 \\ 0 \end{bmatrix} = \mathbf{w}$$

13. No. If $\mathbf{x} = \begin{bmatrix} x \\ y \end{bmatrix}$ is such that

$$f(\mathbf{x}) = A \begin{bmatrix} x \\ y \end{bmatrix} = \begin{bmatrix} 1 & 2 \\ 0 & 1 \\ 1 & 1 \end{bmatrix} \begin{bmatrix} x \\ y \end{bmatrix} = \begin{bmatrix} x+2y \\ y \\ x+y \end{bmatrix} = \begin{bmatrix} 1 \\ 4 \\ 2 \end{bmatrix},$$

then $y = 4$ and $x + y = 2$, so $x = -2$. But then $x + 2y = 6 \neq 1$. Thus, there is no such vector $\mathbf{w}$.

15. (a) Reflection about the y-axis.

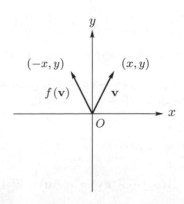

(b) Counterclockwise rotation through $\pi/2$ radians.

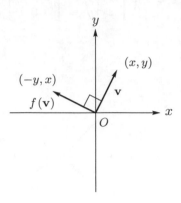

17. (a) For any vector $\mathbf{u} = \begin{bmatrix} u_1 \\ u_2 \end{bmatrix}$ in R^2,

$$f(\mathbf{u}) = A\mathbf{u} = \begin{bmatrix} 1 & 0 \\ 0 & 0 \end{bmatrix} \begin{bmatrix} u_1 \\ u_2 \end{bmatrix} = \begin{bmatrix} u_1 \\ 0 \end{bmatrix},$$

which is the projection of $\mathbf{u}$ onto the x-axis. Therefore, f is projection onto the x-axis.

(b) For any vector $\mathbf{u} = \begin{bmatrix} u_1 \\ u_2 \end{bmatrix}$ in R^2,

$$f(\mathbf{u}) = A\mathbf{u} = \begin{bmatrix} 0 & 0 \\ 0 & 1 \end{bmatrix} \begin{bmatrix} u_1 \\ u_2 \end{bmatrix} = \begin{bmatrix} 0 \\ u_2 \end{bmatrix},$$

which is the projection of $\mathbf{u}$ onto the y-axis. Therefore, f is projection onto the y-axis.

19. (a) $A^2 = \begin{bmatrix} \cos\phi & -\sin\phi \\ \sin\phi & \cos\phi \end{bmatrix} \begin{bmatrix} \cos\phi & -\sin\phi \\ \sin\phi & \cos\phi \end{bmatrix} = \begin{bmatrix} \cos^2\phi - \sin^2\phi & -2\sin\phi\cos\phi \\ 2\sin\phi\cos\phi & \cos^2\phi - \sin^2\phi \end{bmatrix}.$

$$= \begin{bmatrix} \cos 2\phi & -\sin 2\phi \\ \sin 2\phi & \cos 2\phi \end{bmatrix} = \begin{bmatrix} \cos 60° & -\sin 60° \\ \sin 60° & \cos 60° \end{bmatrix}$$

Thus, T_1 rotates $\mathbf{u}$ $60°$ counterclockwise.

(b) $A^{-1} = \begin{bmatrix} \cos\phi & \sin\phi \\ -\sin\phi & \cos\phi \end{bmatrix} = \begin{bmatrix} \cos(-\phi) & -\sin(-\phi) \\ \sin(-\phi) & \cos(-\phi) \end{bmatrix} = \begin{bmatrix} \cos(30°) & -\sin(-60°) \\ \sin(-30°) & \cos(-30°) \end{bmatrix}.$

Thus, T_2 rotates $\mathbf{u}$ $-30°$ counterclockwise or, equivalently, $30°$ clockwise.

(c) From the result in part (a), it is clear that $A^k\mathbf{u}$ rotates $\mathbf{u}$ $30k°$ counterclockwise. For this to equal $\mathbf{u}$, we must have $30k° = 360°$. Therefore, $k = 12$.

21. For any real numbers c and d, we have

$$f(c\mathbf{u} + d\mathbf{v}) = A(c\mathbf{u} + d\mathbf{v}) = A(c\mathbf{u}) + A(d\mathbf{v}) = c(A\mathbf{u}) + d(A\mathbf{v}) = cf(\mathbf{u}) + df(\mathbf{v}) = c\mathbf{0} + d\mathbf{0} = \mathbf{0} + \mathbf{0} = \mathbf{0}.$$

Section 1.7, p. 70

1. We have

$$f\left(\begin{bmatrix} 1 \\ 1 \end{bmatrix}\right) = \begin{bmatrix} -1 & 0 \\ 0 & 1 \end{bmatrix}\begin{bmatrix} 1 \\ 1 \end{bmatrix} = \begin{bmatrix} -1 \\ 1 \end{bmatrix}$$

$$f\left(\begin{bmatrix} 2 \\ 1 \end{bmatrix}\right) = \begin{bmatrix} -1 & 0 \\ 0 & 1 \end{bmatrix}\begin{bmatrix} 2 \\ 1 \end{bmatrix} = \begin{bmatrix} -2 \\ 1 \end{bmatrix}$$

$$f\left(\begin{bmatrix} 1 \\ 3 \end{bmatrix}\right) = \begin{bmatrix} -1 & 0 \\ 0 & 1 \end{bmatrix}\begin{bmatrix} 1 \\ 3 \end{bmatrix} = \begin{bmatrix} -1 \\ 3 \end{bmatrix}$$

$$f\left(\begin{bmatrix} 2 \\ 3 \end{bmatrix}\right) = \begin{bmatrix} -1 & 0 \\ 0 & 1 \end{bmatrix}\begin{bmatrix} 2 \\ 3 \end{bmatrix} = \begin{bmatrix} -2 \\ 3 \end{bmatrix}$$

The image of R is a rectangle with vertices $(-1, 1)$, $(-2, 1)$, $(-1, 3)$, and $(-2, 3)$, as shown in the following figure.

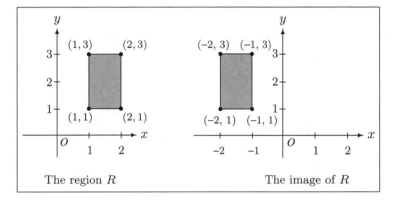

The region R The image of R

3. Here $f\left(\begin{bmatrix} a_1 \\ a_2 \end{bmatrix}\right) = \begin{bmatrix} 1 & 0 \\ -2 & 1 \end{bmatrix}\begin{bmatrix} a_1 \\ a_2 \end{bmatrix}$. The vertices of the image of R are therefore

$$f\left(\begin{bmatrix} 1 \\ 1 \end{bmatrix}\right) = \begin{bmatrix} 1 & 0 \\ -2 & 1 \end{bmatrix}\begin{bmatrix} 1 \\ 1 \end{bmatrix} = \begin{bmatrix} 1 \\ -1 \end{bmatrix} \qquad f\left(\begin{bmatrix} 1 \\ 4 \end{bmatrix}\right) = \begin{bmatrix} 1 & 0 \\ -2 & 1 \end{bmatrix}\begin{bmatrix} 1 \\ 4 \end{bmatrix} = \begin{bmatrix} 1 \\ 2 \end{bmatrix}$$

$$f\left(\begin{bmatrix} 3 \\ 1 \end{bmatrix}\right) = \begin{bmatrix} 1 & 0 \\ -2 & 1 \end{bmatrix}\begin{bmatrix} 3 \\ 1 \end{bmatrix} = \begin{bmatrix} 3 \\ -5 \end{bmatrix} \qquad f\left(\begin{bmatrix} 3 \\ 4 \end{bmatrix}\right) = \begin{bmatrix} 1 & 0 \\ -2 & 1 \end{bmatrix}\begin{bmatrix} 3 \\ 4 \end{bmatrix} = \begin{bmatrix} 3 \\ -2 \end{bmatrix}$$

5. Here $f\left(\begin{bmatrix} a_1 \\ a_2 \end{bmatrix}\right) = \begin{bmatrix} 2 & 0 \\ 0 & 1 \end{bmatrix}\begin{bmatrix} a_1 \\ a_2 \end{bmatrix}$. Therefore the vertices of the image of the unit square are

$$f\left(\begin{bmatrix} 0 \\ 0 \end{bmatrix}\right) = \begin{bmatrix} 2 & 0 \\ 0 & 1 \end{bmatrix}\begin{bmatrix} 0 \\ 0 \end{bmatrix} = \begin{bmatrix} 0 \\ 0 \end{bmatrix} \qquad f\left(\begin{bmatrix} 1 \\ 0 \end{bmatrix}\right) = \begin{bmatrix} 2 & 0 \\ 0 & 1 \end{bmatrix}\begin{bmatrix} 1 \\ 0 \end{bmatrix} = \begin{bmatrix} 2 \\ 0 \end{bmatrix}$$

$$f\left(\begin{bmatrix} 0 \\ 1 \end{bmatrix}\right) = \begin{bmatrix} 2 & 0 \\ 0 & 1 \end{bmatrix}\begin{bmatrix} 0 \\ 1 \end{bmatrix} = \begin{bmatrix} 0 \\ 1 \end{bmatrix} \qquad f\left(\begin{bmatrix} 1 \\ 1 \end{bmatrix}\right) = \begin{bmatrix} 2 & 0 \\ 0 & 1 \end{bmatrix}\begin{bmatrix} 1 \\ 1 \end{bmatrix} = \begin{bmatrix} 2 \\ 1 \end{bmatrix}$$

7. The images of the vertices are

$$f\left(\begin{bmatrix} 5 \\ 0 \end{bmatrix}\right) = \begin{bmatrix} -2 & 1 \\ 3 & 4 \end{bmatrix}\begin{bmatrix} 5 \\ 0 \end{bmatrix} = \begin{bmatrix} -10 \\ 15 \end{bmatrix}$$

$$f\left(\begin{bmatrix} 0 \\ 3 \end{bmatrix}\right) = \begin{bmatrix} -2 & 1 \\ 3 & 4 \end{bmatrix}\begin{bmatrix} 0 \\ 3 \end{bmatrix} = \begin{bmatrix} 3 \\ 12 \end{bmatrix}$$

$$f\left(\begin{bmatrix} 2 \\ -1 \end{bmatrix}\right) = \begin{bmatrix} -2 & 1 \\ 3 & 4 \end{bmatrix}\begin{bmatrix} 2 \\ -1 \end{bmatrix} = \begin{bmatrix} -5 \\ 2 \end{bmatrix}.$$

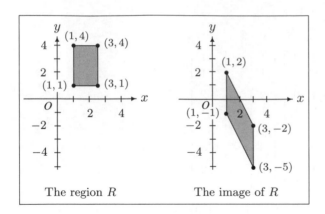

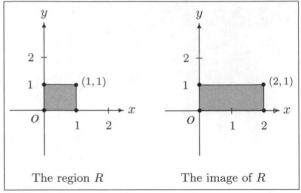

Exercise 3   Exercise 5

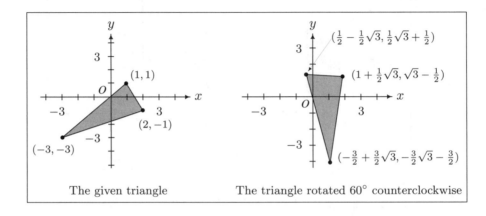

Exercise 9

The coordinates are therefore $(-10, 15)$, $(3, 12)$, and $(-5, 2)$

9. Using the rotation matrix in Example 8, with $\varphi = 60°$, we obtain

$$f\left(\left[\begin{array}{c} a_1 \\ a_2 \end{array}\right]\right) = \left[\begin{array}{cc} \cos 60° & -\sin 60° \\ \sin 60° & \cos 60° \end{array}\right] = \left[\begin{array}{cc} \frac{1}{2} & -\frac{1}{2}\sqrt{3} \\ \frac{1}{2}\sqrt{3} & \frac{1}{2} \end{array}\right].$$

The images of the vertices of the triangle are therefore

$$f\left(\left[\begin{array}{c} 1 \\ 1 \end{array}\right]\right) = \left[\begin{array}{cc} \frac{1}{2} & -\frac{1}{2}\sqrt{3} \\ \frac{1}{2}\sqrt{3} & \frac{1}{2} \end{array}\right]\left[\begin{array}{c} 1 \\ 1 \end{array}\right] = \left[\begin{array}{c} \frac{1}{2} - \frac{1}{2}\sqrt{3} \\ \frac{1}{2}\sqrt{3} + \frac{1}{2} \end{array}\right] \approx \left[\begin{array}{c} -0.366 \\ 1.366 \end{array}\right]$$

$$f\left(\left[\begin{array}{c} -3 \\ -3 \end{array}\right]\right) = \left[\begin{array}{cc} \frac{1}{2} & -\frac{1}{2}\sqrt{3} \\ \frac{1}{2}\sqrt{3} & \frac{1}{2} \end{array}\right]\left[\begin{array}{c} -3 \\ -3 \end{array}\right] = \left[\begin{array}{c} -\frac{3}{2} + \frac{3}{2}\sqrt{3} \\ -\frac{3}{2}\sqrt{3} - \frac{3}{2} \end{array}\right] \approx \left[\begin{array}{c} 1.098 \\ -4.098 \end{array}\right]$$

$$f\left(\left[\begin{array}{c} 2 \\ -1 \end{array}\right]\right) = \left[\begin{array}{cc} \frac{1}{2} & -\frac{1}{2}\sqrt{3} \\ \frac{1}{2}\sqrt{3} & \frac{1}{2} \end{array}\right]\left[\begin{array}{c} 2 \\ -1 \end{array}\right] = \left[\begin{array}{c} 1 + \frac{1}{2}\sqrt{3} \\ \sqrt{3} - \frac{1}{2} \end{array}\right] \approx \left[\begin{array}{c} 1.866 \\ 1.232 \end{array}\right]$$

11. We compute the image of the vertices under the transformation $f(\mathbf{x}) = A\mathbf{x}$.

$$f\left(\begin{bmatrix} 1 \\ 1 \end{bmatrix}\right) = \begin{bmatrix} 1 & 2 \\ 2 & 4 \end{bmatrix}\begin{bmatrix} 1 \\ 1 \end{bmatrix} = \begin{bmatrix} 3 \\ 6 \end{bmatrix}$$

$$f\left(\begin{bmatrix} -3 \\ -3 \end{bmatrix}\right) = \begin{bmatrix} 1 & 2 \\ 2 & 4 \end{bmatrix}\begin{bmatrix} -3 \\ -3 \end{bmatrix} = \begin{bmatrix} -9 \\ -18 \end{bmatrix}$$

$$f\left(\begin{bmatrix} 2 \\ -1 \end{bmatrix}\right) = \begin{bmatrix} 1 & 2 \\ 2 & 4 \end{bmatrix}\begin{bmatrix} 2 \\ -1 \end{bmatrix} = \begin{bmatrix} 0 \\ 0 \end{bmatrix}.$$

Observe that $\begin{bmatrix} 3 \\ 6 \end{bmatrix}$ and $\begin{bmatrix} -9 \\ -18 \end{bmatrix}$ are scalar multiples of each other. Thus the image of the triangle T in Exercise 8 under f is the line segment connecting the points $(3, 6)$, $(-9, -18)$, and $(0, 0)$.

13. We have

$$f\left(\begin{bmatrix} 0 \\ 0 \end{bmatrix}\right) = \begin{bmatrix} 1 & -1 \\ 2 & 3 \end{bmatrix}\begin{bmatrix} 0 \\ 0 \end{bmatrix} = \begin{bmatrix} 0 \\ 0 \end{bmatrix}$$

$$f\left(\begin{bmatrix} 1 \\ 0 \end{bmatrix}\right) = \begin{bmatrix} 1 & -1 \\ 2 & 3 \end{bmatrix}\begin{bmatrix} 1 \\ 0 \end{bmatrix} = \begin{bmatrix} 1 \\ 2 \end{bmatrix}$$

$$f\left(\begin{bmatrix} 1 \\ 1 \end{bmatrix}\right) = \begin{bmatrix} 1 & -1 \\ 2 & 3 \end{bmatrix}\begin{bmatrix} 1 \\ 1 \end{bmatrix} = \begin{bmatrix} 0 \\ 5 \end{bmatrix}$$

$$f\left(\begin{bmatrix} 0 \\ 1 \end{bmatrix}\right) = \begin{bmatrix} 1 & -1 \\ 2 & 3 \end{bmatrix}\begin{bmatrix} 0 \\ 1 \end{bmatrix} = \begin{bmatrix} -1 \\ 3 \end{bmatrix}$$

The image of the rectangle is a parallelogram with vertices $(0, 0)$, $(1, 2)$, $(0, 5)$, and $(-1, 3)$, as shown in the following figure.

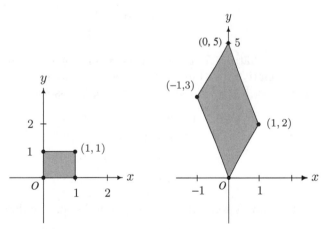

The region R The image of R

15. (a) Possible answer: Use f_1 with $\phi = 90°$, then f_3.

(b) Possible answer: Use f_1 with $\phi = -135°$.

17. Let $f\colon R^2 \to R^2$ be the matrix transformation defined by $f(\mathbf{v}) = A\mathbf{v}$, where

$$A = \begin{bmatrix} \cos\theta_1 & -\sin\theta_1 \\ \sin\theta_1 & \cos\theta_1 \end{bmatrix}$$

and let $g\colon R^2 \to R^2$ be the matrix transformation defined by $g(\mathbf{v}) = B\mathbf{v}$, where

$$B = \begin{bmatrix} \cos(-\theta_2) & -\sin(-\theta_2) \\ \sin(-\theta_2) & \cos(-\theta_2) \end{bmatrix} = \begin{bmatrix} \cos\theta_2 & \sin\theta_2 \\ -\sin\theta_2 & \cos\theta_2 \end{bmatrix}.$$

Then f represents a counterclockwise rotation through θ_1 radians and g represents a counterclockwise rotation through $-\theta_2$ radians. The composition $g \circ f \colon R^2 \to R^2$ represents a counterclockwise rotation through $\theta_1 + (-\theta_2) = \theta_1 - \theta_2$ radians. Now, $(g \circ f)(\mathbf{v}) = g(f(\mathbf{v})) = B(A\mathbf{v}) = (BA)\mathbf{v}$. The matrix for the matrix transformation $g \circ f$ is

$$\begin{bmatrix} \cos(\theta_1 - \theta_2) & -\sin(\theta_1 - \theta_2) \\ \sin(\theta_1 - \theta_2) & \cos(\theta_1 - \theta_2) \end{bmatrix}.$$

Since the matrix representing $g \circ f$ is BA, we have that

$$BA = \begin{bmatrix} \cos\theta_2 & \sin\theta_2 \\ -\sin\theta_2 & \cos\theta_2 \end{bmatrix} \begin{bmatrix} \cos\theta_1 & -\sin\theta_1 \\ \sin\theta_1 & \cos\theta_1 \end{bmatrix} = \begin{bmatrix} \cos(\theta_1 - \theta_2) & -\sin(\theta_1 - \theta_2) \\ \sin(\theta_1 - \theta_2) & \cos(\theta_1 - \theta_2) \end{bmatrix}.$$

Therefore

$$\begin{bmatrix} \cos(\theta_1 - \theta_2) & -\sin(\theta_1 - \theta_2) \\ \sin(\theta_1 - \theta_2) & \cos(\theta_1 - \theta_2) \end{bmatrix} = \begin{bmatrix} \cos\theta_2 \cos\theta_1 + \sin\theta_2 \sin\theta_1 & -\cos\theta_2 \sin\theta_1 + \sin\theta_2 \cos\theta_1 \\ -\sin\theta_2 \cos\theta_1 + \cos\theta_2 \sin\theta_1 & \sin\theta_2 \sin\theta_1 + \cos\theta_2 \cos\theta_1 \end{bmatrix}.$$

Equating the $(1, 1)$ entry and the $(2, 1)$ entry, we obtain

$$\cos(\theta_1 - \theta_2) = \cos\theta_1 \cos\theta_2 + \sin\theta_1 \sin\theta_2$$
$$\sin(\theta_1 - \theta_2) = \sin\theta_1 \cos\theta_2 - \cos\theta_1 \sin\theta_2.$$

Section 1.8, p. 79

1. We first compute the angles corresponding to each of the correlation coefficients:

$$\arccos(0.97) = 14.1°$$
$$\arccos(0.93) = 21.6°$$
$$\arccos(0.88) = 28.4°$$
$$\arccos(0.76) = 40.5°$$

From these angles we infer that the first correlation coefficient represents highly positively correlated data, the second less correlated, the third, less yet, and the fourth, the least correlated. The figures suggest that Figure C is most highly correlated, A less so, D less yet, and B the least correlated. Thus:

$$\begin{array}{ccc} \text{Figure A} & \longleftrightarrow & 0.93 \\ \text{Figure B} & \longleftrightarrow & 0.76 \\ \text{Figure C} & \longleftrightarrow & 0.97 \\ \text{Figure D} & \longleftrightarrow & 0.88 \end{array}$$

3. We must first center the data by computing the mean of distance and amplitude:

$$\text{mean of distance} = 455.56 \quad \text{mean of amplitude} = 9.71.$$

Subtracting the means from the data, we obtain the table of centered data:

Centered distance	Centered amplitude
−255.56	2.89
−255.56	10.19
−55.56	−0.41
−55.56	−0.21
44.44	−1.81
44.44	−1.91
44.44	−1.71
244.44	−3.71
244.44	−3.31

Let

$$
\mathbf{v} = \begin{bmatrix} -255.56 \\ -255.56 \\ -55.56 \\ -55.56 \\ 44.44 \\ 44.44 \\ 44.44 \\ 244.44 \\ 244.44 \end{bmatrix} \quad \text{and} \quad \mathbf{w} = \begin{bmatrix} 2.89 \\ 10.19 \\ -0.41 \\ -0.21 \\ -1.81 \\ -1.91 \\ -1.71 \\ -3.71 \\ -3.31 \end{bmatrix}.
$$

Then

$$
Cor(\mathbf{v}, \mathbf{w}) = \frac{\mathbf{v} \cdot \mathbf{w}}{\|\mathbf{v}\| \, \|\mathbf{w}\|} = -0.8482.
$$

Finally, $\arccos(-0.85) = 2.5834$ radians $= 148.02°$. Thus the distance and amplitude data are moderately negatively correlated.

Supplementary Exercises for Chapter 1, p. 80

1. (a) $2 + 1 = 3$ (b) $3 + 2 + 1 = 6$ (c) $4 + 3 + 2 + 1 = 10$

   (d) $n + (n-1) + (n-2) + \cdots + 2 + 1 = \dfrac{n(n+1)}{2}$

3. Let $A = \begin{bmatrix} a & b \\ 0 & c \end{bmatrix}$. Then

$$
A^2 = \begin{bmatrix} a & b \\ 0 & c \end{bmatrix} \begin{bmatrix} a & b \\ 0 & c \end{bmatrix} = \begin{bmatrix} a^2 & ab + bc \\ 0 & c^2 \end{bmatrix} = \begin{bmatrix} 1 & 0 \\ 0 & 1 \end{bmatrix} = I_2
$$

and equating corresponding entries gives

$$
a^2 = 1, \quad ab + bc = b(a+c) = 0, \quad c^2 = 1.
$$

To satisfy the second equality we have either $b = 0$ or $a = -c$.

Case $b = 0$: $a = \pm 1$ and $c = \pm 1$; thus we have

$$
A = \begin{bmatrix} 1 & 0 \\ 0 & 1 \end{bmatrix} \quad \text{or} \quad A = \begin{bmatrix} 1 & 0 \\ 0 & -1 \end{bmatrix} \quad \text{or} \quad A = \begin{bmatrix} -1 & 0 \\ 0 & 1 \end{bmatrix} \quad \text{or} \quad A = \begin{bmatrix} -1 & 0 \\ 0 & -1 \end{bmatrix}.
$$

Case $a = -c$: Then $b = r$, where r is any real number, and we have

$$
A = \begin{bmatrix} 1 & r \\ 0 & -1 \end{bmatrix} \quad \text{or} \quad A = \begin{bmatrix} -1 & r \\ 0 & 1 \end{bmatrix}.
$$

5. (a) $(A^T A)_{ii} = (\text{row}_i A^T) \times (\text{col}_i A) = (\text{col}_i A)^T \times (\text{col}_i A)$

   (b) From part (a)

$$
(A^T A)_{ii} = \begin{bmatrix} a_{1i} & a_{2i} & \cdots & a_{ni} \end{bmatrix} \times \begin{bmatrix} a_{1i} \\ a_{2i} \\ \vdots \\ a_{ni} \end{bmatrix} = \sum_{j=1}^{n} a_{ji}^2 \ge 0.
$$

   (c) $A^T A = O_n$ if and only if $(A^T A)_{ii} = 0$ for $i = 1, \ldots, n$. But this is possible if and only if $a_{ij} = 0$ for $i = 1, \ldots, n$ and $j = 1, \ldots, n$

7. Let A be a symmetric upper (lower) triangular matrix. Then $a_{ij} = a_{ji}$ and $a_{ij} = 0$ for $j > i$ ($j < i$). Thus, $a_{ij} = 0$ whenever $i \neq j$, so A is diagonal.

9. We are asked to prove an "if and only if" statement. Hence two things must be proved.

   (a) If A is nonsingular, then $a_{ii} \neq 0$ for $i = 1, \ldots, n$.

   Proof: If A is nonsingular then A is row equivalent to I_n. Since A is upper triangular, this can occur only if we can multiply row i by $1/a_{ii}$ for each i. Hence $a_{ii} \neq 0$ for $i = 1, \ldots, n$. (Other row operations will be needed to get I_n.)

   (b) If $a_{ii} \neq 0$ for $i = 1, \ldots, n$ then A is nonsingular.

   Proof: Just reverse the steps given above in part (a).

11. Using the definition of trace and Supplementary Exercise 5(a), we find that

$$\text{Tr}(A^T A) = \text{sum of the diagonal entries of } A^T A \qquad \text{(definition of trace)}$$

$$= \sum_{i=1}^{n} (A^T A)_{ii} = \sum_{i=1}^{n} \left[\sum_{j=1}^{n} a_{ji}^2 \right] \qquad \text{(Supplementary Exercise 5(a))}$$

$$= \text{sum of the squares of all entries of } A$$

Thus the only way $\text{Tr}(A^T A) = 0$ is if $a_{ij} = 0$ for $i = 1, \ldots, n$ and $j = 1, \ldots, n$. That is, if $A = O$.

13. Let $A = \begin{bmatrix} 1 & \frac{1}{2} \\ 0 & \frac{1}{2} \end{bmatrix}$. Then

$$A^2 = \begin{bmatrix} 1 & \frac{1}{2} + \left(\frac{1}{2}\right)^2 \\ 0 & \left(\frac{1}{2}\right)^2 \end{bmatrix} \quad \text{and} \quad A^3 = \begin{bmatrix} 1 & \frac{1}{2} + \left(\frac{1}{2}\right)^2 + \left(\frac{1}{2}\right)^3 \\ 0 & \left(\frac{1}{2}\right)^3 \end{bmatrix}.$$

Following the pattern for the elements we have

$$A^n = \begin{bmatrix} 1 & \frac{1}{2} + \left(\frac{1}{2}\right)^2 + \cdots + \left(\frac{1}{2}\right)^n \\ 0 & \left(\frac{1}{2}\right)^n \end{bmatrix}.$$

A formal proof by induction can be given.

15. Since A is skew symmetric, $A^T = -A$. Therefore,

$$A[-(A^{-1})^T] = -A(A^{-1})^T = A^T(A^{-1})^T = (A^{-1}A)^T = I^T = I$$

and similarly, $[-(A^{-1})^T]A = I$. Hence $-(A^{-1})^T = A^{-1}$, so $(A^{-1})^T = -A^{-1}$, and therefore A^{-1} is skew symmetric.

17. If $A\mathbf{x} = \mathbf{x}$ for all $n \times 1$ matrices X, then $AE_j = E_j$, where E_j is column j of I_n. Since

$$AE_j = \begin{bmatrix} a_{1j} \\ a_{2j} \\ \vdots \\ a_{nj} \end{bmatrix} = E_j$$

it follows that $a_{ij} = 1$ if $i = j$ and 0 otherwise. Hence $A = I_n$.

19. (a) $I_n^2 = I_n$ and $O^2 = O$

   (b) One such matrix is $\begin{bmatrix} 0 & 0 \\ 0 & 1 \end{bmatrix}$ and another is $\begin{bmatrix} 1 & 0 \\ 0 & 0 \end{bmatrix}$.

(c) If $A^2 = A$ and A^{-1} exists, then $A^{-1}(A^2) = A^{-1}A$ which simplifies to give $A = I_n$.

21. (a) We prove this statement using induction. The result is true for $n = 1$. Assume it is true for $n = k$ so that $A^k = A$. Then
$$A^{k+1} = AA^k = AA = A^2 = A.$$
Thus the result is true for $n = k + 1$. It follows by induction that $A^n = A$ for all integers $n \geq 1$.

(b) $(I_n - A)^2 = I_n^2 - 2A + A^2 = I_n - 2A + A = I_n - A$.

23. $\mathbf{w} = \begin{bmatrix} 1 \\ 1 \\ 1 \\ 1 \\ 1 \end{bmatrix}$. If $\mathbf{v}$ is an n-vector, $\mathbf{w}$ is the n-vector all of whose components are equal to 1.

25. (a) $\text{Mcd}(cA) = \displaystyle\sum_{i+j=n+1} (ca_{ij}) = c \sum_{i+j=n+1} a_{ij} = c\,\text{Mcd}(A)$

(b) $\text{Mcd}(A + B) = \displaystyle\sum_{i+j=n+1} (a_{ij} + b_{ij}) = \sum_{i+j=n+1} a_{ij} + \sum_{i+j=n+1} b_{ij} = \text{Mcd}(A) + \text{Mcd}(B)$

(c) $\text{Mcd}(A^T) = (A^T)_{1n} + (A^T)_{2\,n-1} + \cdots + (A^T)_{n1} = a_{n1} + a_{n-1\,2} + \cdots + a_{1n} = \text{Mcd}(A)$

(d) Let $A = \begin{bmatrix} 7 & -3 \\ 0 & 0 \end{bmatrix}$ and $B = \begin{bmatrix} 1 & 1 \\ -1 & 1 \end{bmatrix}$. Then

$$AB = \begin{bmatrix} 10 & 4 \\ 0 & 0 \end{bmatrix} \quad \text{with } \text{Mcd}(AB) = 4$$

and

$$BA = \begin{bmatrix} 7 & -3 \\ -7 & 3 \end{bmatrix} \quad \text{with } \text{Mcd}(BA) = -10.$$

27. Let
$$A = \begin{bmatrix} 0 & a \\ -a & 0 \end{bmatrix} \quad \text{and} \quad B = \begin{bmatrix} 0 & b \\ -b & 0 \end{bmatrix}.$$

Then A and B are skew symmetric and

$$AB = \begin{bmatrix} -ab & 0 \\ 0 & -ab \end{bmatrix}$$

which is diagonal. The result is not true for $n > 2$. For example, let

$$A = \begin{bmatrix} 0 & 1 & 2 \\ -1 & 0 & 3 \\ -2 & -3 & 0 \end{bmatrix}.$$

Then

$$A^2 = \begin{bmatrix} 5 & 6 & -3 \\ 6 & 10 & 2 \\ -3 & 2 & 13 \end{bmatrix}.$$

29. Let

$$A = \begin{bmatrix} A_{11} & A_{12} \\ O & A_{22} \end{bmatrix}$$

where A_{11} is $r \times r$ and A_{22} is $s \times s$. Let

$$B = \begin{bmatrix} B_{11} & B_{12} \\ B_{21} & B_{22} \end{bmatrix}$$

where B_{11} is $r \times r$ and B_{22} is $s \times s$. Then

$$AB = \begin{bmatrix} A_{11}B_{11} + A_{12}B_{21} & A_{11}B_{12} + A_{12}B_{22} \\ A_{22}B_{21} & A_{22}B_{22} \end{bmatrix} = \begin{bmatrix} I_r & O \\ O & I_s \end{bmatrix}.$$

We have $A_{22}B_{22} = I_s$, so $B_{22} = A_{22}^{-1}$. We also have $A_{22}B_{21} = O$, and multiplying both sides of this equation by A_{22}^{-1}, we find that $B_{21} = O$. Thus $A_{11}B_{11} = I_r$, so $B_{11} = A_{11}^{-1}$. Next, since

$$A_{11}B_{12} + A_{12}B_{22} = O$$

then

$$A_{11}B_{12} = -A_{12}B_{22} = -A_{12}A_{22}^{-1}$$

Hence,

$$B_{12} = -A_{11}^{-1}A_{12}A_{22}^{-1}.$$

Since we have solved for B_{11}, B_{12}, B_{21}, and B_{22}, we conclude that A is nonsingular. Moreover,

$$A^{-1} = \begin{bmatrix} A_{11}^{-1} & -A_{11}^{-1}A_{12}A_{22}^{-1} \\ O & A_{22}^{-1} \end{bmatrix}.$$

31. Let $X = \begin{bmatrix} 1 & 5 \end{bmatrix}^T$ and $Y = \begin{bmatrix} 4 & -3 \end{bmatrix}^T$. Then

$$XY^T = \begin{bmatrix} 1 \\ 5 \end{bmatrix} \begin{bmatrix} 4 & -3 \end{bmatrix} = \begin{bmatrix} 4 & -3 \\ 20 & -15 \end{bmatrix} \quad \text{and} \quad YX^T = \begin{bmatrix} 4 \\ -3 \end{bmatrix} \begin{bmatrix} 1 & 5 \end{bmatrix} = \begin{bmatrix} 4 & 20 \\ -3 & -15 \end{bmatrix}.$$

It follows that XY^T is not necessarily the same as YX^T.

33. We obtain

$$AB = \begin{bmatrix} 1 & 7 \\ 3 & 9 \\ 5 & 11 \end{bmatrix} \begin{bmatrix} 2 & 4 \\ 6 & 8 \end{bmatrix} = \begin{bmatrix} 44 & 60 \\ 60 & 84 \\ 76 & 108 \end{bmatrix}.$$

Now, the outer product of $\text{col}_1(A)$ with $\text{row}_1(B)$ is

$$\begin{bmatrix} 1 \\ 3 \\ 5 \end{bmatrix} \begin{bmatrix} 2 & 4 \end{bmatrix} = \begin{bmatrix} 2 & 4 \\ 6 & 12 \\ 10 & 20 \end{bmatrix};$$

the outer product of $\text{col}_2(A)$ with $\text{row}_2(B)$ is

$$\begin{bmatrix} 7 \\ 9 \\ 11 \end{bmatrix} \begin{bmatrix} 6 & 8 \end{bmatrix} = \begin{bmatrix} 42 & 56 \\ 54 & 72 \\ 66 & 88 \end{bmatrix}.$$

Therefore

$$\sum_{i=1}^{2} \text{outer product of } \text{col}_i(A) \text{ with } \text{row}_i(B) = \begin{bmatrix} 2 & 4 \\ 6 & 12 \\ 10 & 20 \end{bmatrix} + \begin{bmatrix} 42 & 56 \\ 54 & 72 \\ 66 & 88 \end{bmatrix} = \begin{bmatrix} 44 & 60 \\ 60 & 84 \\ 76 & 108 \end{bmatrix} = AB.$$

35. (a) $\begin{bmatrix} 1 & 2 & 3 \\ 3 & 1 & 2 \\ 2 & 3 & 1 \end{bmatrix}$ (b) $\begin{bmatrix} 1 & 2 & 5 & -1 \\ -1 & 1 & 2 & 5 \\ 5 & -1 & 1 & 2 \\ 2 & 5 & -1 & 1 \end{bmatrix}$ (c) $\begin{bmatrix} 1 & 0 & 0 & 0 & 0 \\ 0 & 1 & 0 & 0 & 0 \\ 0 & 0 & 1 & 0 & 0 \\ 0 & 0 & 0 & 1 & 0 \\ 0 & 0 & 0 & 0 & 1 \end{bmatrix} = I_5$

$$(d) \quad \begin{bmatrix} 1 & 2 & 1 & 0 & 0 \\ 0 & 1 & 2 & 1 & 0 \\ 0 & 0 & 1 & 2 & 1 \\ 1 & 0 & 0 & 1 & 2 \\ 2 & 1 & 0 & 0 & 1 \end{bmatrix}$$

37. $C\mathbf{x} = \left(\sum_{i=1}^{n} c_i \right) \mathbf{x}$

Chapter Review for Chapter 1, p. 83

True or False

1. False.   2. False.   3. True.   4. True.   5. True.

6. True.   7. True.   8. True.   9. True.   10. True.

Quiz

1. Add 3 times the second equation to the first equation to get $10x = 20$. Therefore, $x = 2$. Substitute $x = 2$ into the second equation to get $y = -4$. Thus $\mathbf{x} = \begin{bmatrix} 2 \\ -4 \end{bmatrix}$.

2. Substitute $x = 1$, $y = -1$, $z = r$ into the equations:

$$\begin{aligned} 1 + 2 + 3r &= 3 &&\implies& r &= 0 \\ 4 - 5 - r &= -1 &&\implies& r &= 0 \\ 6 - 1 + 5r &= 5 &&\implies& r &= 0 \end{aligned}$$

Thus, $r = 0$.

3. Multiplying the matrices on the left and equating entries, we obtain

$$\begin{bmatrix} -5 & b+2 \\ 3a & ab \end{bmatrix} = \begin{bmatrix} -5 & 6 \\ 12 & 16 \end{bmatrix} \implies \begin{aligned} b + 2 &= 6 \\ 3a &= 12 \\ ab &= 16 \end{aligned} \implies \begin{aligned} b &= 4 \\ a &= 4 \end{aligned}$$

Thus, $a = b = 4$.

4. (a) $LU = \begin{bmatrix} 2 & 0 & 0 \\ 1 & -2 & 0 \\ a & 1 & 3 \end{bmatrix} \begin{bmatrix} 1 & 4 & b \\ 0 & -1 & 5 \\ 0 & 0 & c \end{bmatrix} = \begin{bmatrix} 2 & 2a & 2b \\ 1 & 6 & b-10 \\ a & 4a-1 & ab+5+3c \end{bmatrix}$. Setting $4a - 1 = 7$, we find that $a = 2$.

    (b) Setting $b - 10 = 0$, we find that $b = 10$, $c =$ any real number.

5. $\mathbf{u} = \begin{bmatrix} 3 \\ r \end{bmatrix}$, where r is any real number.

Chapter 2

Solving Linear Systems

Section 2.1, p. 94

1. (a) One possible solution:

$$B = A_{-\mathbf{r}_1 \to \mathbf{r}_1} = \begin{bmatrix} 1 & -2 & 5 \\ 2 & -1 & 6 \\ 2 & -2 & 7 \end{bmatrix}$$

$$C = B_{\substack{\mathbf{r}_2 - 2\mathbf{r}_1 \to \mathbf{r}_2 \\ \mathbf{r}_3 - 2\mathbf{r}_1 \to \mathbf{r}_3}} = \begin{bmatrix} 1 & -2 & 5 \\ 0 & 3 & -4 \\ 0 & 2 & -3 \end{bmatrix}$$

$$D = C_{\frac{1}{3}\mathbf{r}_2 \to \mathbf{r}_2} = \begin{bmatrix} 1 & -2 & 5 \\ 0 & 1 & -\frac{4}{3} \\ 0 & 2 & -3 \end{bmatrix}$$

$$E = D_{\mathbf{r}_3 - 2\mathbf{r}_2 \to \mathbf{r}_3} = \begin{bmatrix} 1 & -2 & 5 \\ 0 & 1 & -\frac{4}{3} \\ 0 & 0 & -\frac{1}{3} \end{bmatrix}$$

$$F = E_{-3\mathbf{r}_3 \to \mathbf{r}_3} = \begin{bmatrix} 1 & -2 & 5 \\ 0 & 1 & -\frac{4}{3} \\ 0 & 0 & 1 \end{bmatrix}$$

Matrix F is a row echelon form of A. In summary,

$$F = A_{\substack{-\mathbf{r}_1 \to \mathbf{r}_1 \\ \mathbf{r}_2 - 2\mathbf{r}_1 \to \mathbf{r}_2 \\ \mathbf{r}_3 - 2\mathbf{r}_1 \to \mathbf{r}_3 \\ \frac{1}{3}\mathbf{r}_2 \to \mathbf{r}_2 \\ \mathbf{r}_3 - 2\mathbf{r}_2 \to \mathbf{r}_3 \\ -3\mathbf{r}_3 \to \mathbf{r}_3}} = \begin{bmatrix} 1 & -2 & 5 \\ 0 & 1 & -\frac{4}{3} \\ 0 & 0 & 1 \end{bmatrix}.$$

(b) One possible solution:

$$A_{\substack{\mathbf{r}_2 - 3\mathbf{r}_1 \to \mathbf{r}_2 \\ \mathbf{r}_3 - 5\mathbf{r}_1 \to \mathbf{r}_3 \\ \mathbf{r}_4 + 2\mathbf{r}_1 \to \mathbf{r}_4 \\ \mathbf{r}_3 - \mathbf{r}_2 \to \mathbf{r}_3}} = \begin{bmatrix} 1 & 1 & -1 \\ 0 & 1 & 2 \\ 0 & 0 & 0 \\ 0 & 0 & 0 \end{bmatrix}.$$

3. (a) $A_{\substack{\mathbf{r}_2 + 2\mathbf{r}_3 \to \mathbf{r}_2 \\ \mathbf{r}_1 - 4\mathbf{r}_3 \to \mathbf{r}_1 \\ \mathbf{r}_1 - 2\mathbf{r}_2 \to \mathbf{r}_1}} = \begin{bmatrix} 1 & 0 & 0 \\ 0 & 1 & 0 \\ 0 & 0 & 1 \end{bmatrix} = I_3.$

   (b) $A_{\substack{\mathbf{r}_2 + 4\mathbf{r}_3 \to \mathbf{r}_2 \\ \mathbf{r}_1 - 5\mathbf{r}_3 \to \mathbf{r}_1 \\ \mathbf{r}_1 - 3\mathbf{r}_2 \to \mathbf{r}_1}} = \begin{bmatrix} 1 & 4 & 0 & 0 \\ 0 & 0 & 1 & 0 \\ 0 & 0 & 0 & 1 \\ 0 & 0 & 0 & 0 \end{bmatrix}.$

5. (a) $A_{\substack{\mathbf{r}_2 + 2\mathbf{r}_1 \to \mathbf{r}_2 \\ \mathbf{r}_3 - 3\mathbf{r}_1 \to \mathbf{r}_3 \\ \mathbf{r}_3 - 2\mathbf{r}_2 \to \mathbf{r}_3}} = \begin{bmatrix} 1 & 0 & -2 \\ 0 & 1 & 5 \\ 0 & 0 & 0 \end{bmatrix}.$

   (b) $A_{\substack{\mathbf{r}_2 - 2\mathbf{r}_3 \to \mathbf{r}_2 \\ \mathbf{r}_4 - 7\mathbf{r}_3 \to \mathbf{r}_4 \\ -\mathbf{r}_2 \to \mathbf{r}_2 \\ \mathbf{r}_1 - \mathbf{r}_2 \to \mathbf{r}_1 \\ \mathbf{r}_4 + 3\mathbf{r}_2 \to \mathbf{r}_4 \\ \mathbf{r}_3 \to \mathbf{r}_2 \\ \mathbf{r}_2 \to \mathbf{r}_3}} = \begin{bmatrix} 1 & 0 & 0 \\ 0 & 1 & 0 \\ 0 & 0 & 1 \\ 0 & 0 & 0 \end{bmatrix}.$

7. (a) N

   (b) REF

   (c) RREF

9. Let A be an $n \times n$ matrix in reduced row echelon form and $A \neq I_n$. Then at least one row of A does not have a leading 1. From Definition 2.1, this row must contain all zeros.

11. (a) $A_{\substack{\mathbf{c}_2 - 2\mathbf{c}_1 \to \mathbf{c}_2 \\ \mathbf{c}_3 + 3\mathbf{c}_1 \to \mathbf{c}_3 \\ \mathbf{c}_4\mathbf{c}_1 \to \mathbf{c}_4 \\ \frac{1}{2}\mathbf{c}_2 \to \mathbf{c}_2 \\ \frac{1}{2}\mathbf{c}_3 \to \mathbf{c}_3 \\ \mathbf{c}_4 - 5\mathbf{c}_2 \to \mathbf{c}_4 \\ \frac{7}{2}\mathbf{c}_3 + \mathbf{c}_4 \to \mathbf{c}_4 \\ \frac{1}{8}\mathbf{c}_4 \to \mathbf{c}_4}} = \begin{bmatrix} 1 & 0 & 0 & 0 \\ -1 & 1 & 0 & 0 \\ 0 & \frac{1}{2} & 1 & 0 \\ 2 & -\frac{1}{2} & 3 & 1 \end{bmatrix}.$

   (b) Beginning with the echelon form in part (a), we apply the indicated column operations to obtain the reduced column echelon form:

   $$\begin{bmatrix} 1 & 0 & 0 & 0 \\ -1 & 1 & 0 & 0 \\ 0 & \frac{1}{2} & 1 & 0 \\ 2 & -\frac{1}{2} & 3 & 1 \end{bmatrix} \begin{array}{l} \mathbf{c}_1 + \mathbf{c}_2 \to \mathbf{c}_1 \\ \mathbf{c}_2 - \frac{1}{2}\mathbf{c}_3 \to \mathbf{c}_2 \\ \mathbf{c}_1 - \frac{1}{2}\mathbf{c}_4 \to \mathbf{c}_1 \\ \mathbf{c}_2 + 2\mathbf{c}_4 \to \mathbf{c}_2 \\ \mathbf{c}_3 - 3\mathbf{c}_4 \to \mathbf{c}_3 \end{array} = \begin{bmatrix} 1 & 0 & 0 & 0 \\ 0 & 1 & 0 & 0 \\ 0 & 0 & 1 & 0 \\ 0 & 0 & 0 & 1 \end{bmatrix}.$$

13. $\begin{bmatrix} \cos\theta & \sin\theta \\ -\sin\theta & \cos\theta \end{bmatrix} \mathbf{r}_2 + \frac{\sin\theta}{\cos\theta}\mathbf{r}_1 \to \mathbf{r}_2 \quad = \begin{bmatrix} 1 & 0 \\ 0 & 1 \end{bmatrix}.$
   $(\cos\theta)\mathbf{r}_2 \to \mathbf{r}_2$
   $\mathbf{r}_1 - (\sin\theta)\mathbf{r}_2 \to \mathbf{r}_1$
   $\frac{1}{\cos\theta}\mathbf{r}_1 \to \mathbf{r}_1$

Section 2.2, p. 113

1. We use back substitution to solve each system.

   (a) $x + 2y - z = 6$
   $y + z = 5$
   $z = 4$

   Substitute $z = 4$ into the second equation to get $y + 4 = 5$ and hence $y = 1$. Now substitute these values into the first equation to get $x + 2(1) - 4 = 6$ and therefore $x = 8$. The solution is $x = 8$, $y = 1$, $z = 4$.

   (b) $x - 3y + 4z + w = 0$
   $z - w = 4$
   $w = 1$

   Substitue $w = 1$ into the second equation: $z - 1 = 4$, so $z = 5$. Then, from the first equation,

   $$x - 3y + 4(5) + 1 = 0 \quad \Longrightarrow \quad x - 3y = -21.$$

   Let $y = t$ be any real number. Then $x - 3t = -21$ so $x = -21 + 3t$. The solution is $x = -21 + 3t$, $y = t$, $z = 5$, $w = 1$, where t is any real number.

3. (a) $x + y = 2$
   $ z + w = -3$

   Let $y = s$, $w = t$, where s and t are any real numbers. Then $x + s = 2 \Longrightarrow x = 2 - s$ and $z + t = -3 \Longrightarrow z = -3 - t$. The solution is $x = 2 - s$, $y = s$, $z = -3 - t$, $w = t$, where s and t are any real numbers.

   (b) The solution is $x = 3$, $y = 0$, $z = 1$.

5. (a) To use the Gaussian elimination method we form the augmented matrix and use row operations to obtain an equivalent system in which the coefficient matrix is in row echelon form. Then we apply back substitution to determine the solution(s).

   $$\begin{bmatrix} 1 & 1 & 2 & | & -1 \\ 1 & -2 & 1 & | & -5 \\ 3 & 1 & 1 & | & 3 \end{bmatrix} \to \begin{bmatrix} 1 & 1 & 2 & | & -1 \\ 0 & -3 & -1 & | & -4 \\ 0 & -2 & -5 & | & 6 \end{bmatrix} \to \begin{bmatrix} 1 & 1 & 2 & | & -1 \\ 0 & 1 & \frac{1}{3} & | & \frac{4}{3} \\ 0 & -2 & -5 & | & 6 \end{bmatrix}$$
   $$\begin{array}{l}(-1)\mathbf{r}_1 + \mathbf{r}_2 \to \mathbf{r}_2 \\ (-3)\mathbf{r}_1 + \mathbf{r}_3 \to \mathbf{r}_3\end{array} \qquad \left(-\frac{1}{3}\right)\mathbf{r}_2 \to \mathbf{r}_2 \qquad \begin{array}{l}2\mathbf{r}_2 + \mathbf{r}_3 \to \mathbf{r}_3 \\ \mathbf{r}_2 + \mathbf{r}_4 \to \mathbf{r}_4\end{array}$$

   $$\to \begin{bmatrix} 1 & 1 & 2 & | & -1 \\ 0 & 1 & \frac{1}{3} & | & \frac{4}{3} \\ 0 & 0 & -\frac{13}{3} & | & \frac{26}{3} \end{bmatrix} \to \begin{bmatrix} 1 & 1 & 2 & | & -1 \\ 0 & 1 & \frac{1}{3} & | & \frac{4}{3} \\ 0 & 0 & 1 & | & -2 \end{bmatrix}$$
   $$\left(-\frac{3}{13}\right)\mathbf{r}_3 \to \mathbf{r}_3$$

   Thus we have the equivalent linear system

   $$\begin{aligned} x + y + 2z &= -1 \\ y + \tfrac{1}{3}z &= \tfrac{4}{3} \\ z &= -2 \end{aligned}$$

Applying back substitution gives

$$z = -2$$
$$y = \tfrac{4}{3} - \tfrac{1}{3}z = \tfrac{4}{3} - \tfrac{1}{3}(-2) = 2$$
$$x = -1 - 2z - y = -1 - 2(-2) = 1$$

Thus the system has a unique solution $x = 1$, $y = 2$, $z = -2$.

(b) To use the Gauss-Jordan method we form the augmented matrix and use row operations to obtain an equivalent system in which the coefficient matrix is in reduced row echelon form. Then the solution(s) are determined from the final augmented column without the need for back substitution. We start with the equivalent system obtained in part (a) that is in row echelon form and use row operations to put zeros above leading 1's.

$$\begin{bmatrix} 1 & 1 & 2 & | & -1 \\ 0 & 1 & \frac{1}{3} & | & \frac{4}{3} \\ 0 & 0 & 1 & | & -2 \end{bmatrix} \rightarrow \begin{bmatrix} 1 & 0 & \frac{5}{3} & | & -\frac{7}{3} \\ 0 & 1 & \frac{1}{3} & | & \frac{4}{3} \\ 0 & 0 & 1 & | & -2 \end{bmatrix} \rightarrow \begin{bmatrix} 1 & 0 & 0 & | & 1 \\ 0 & 1 & 0 & | & 2 \\ 0 & 0 & 1 & | & -2 \end{bmatrix}$$

$$(-1)\mathbf{r}_2 + \mathbf{r}_1 \rightarrow \mathbf{r}_1 \qquad \begin{array}{l} \left(-\frac{1}{3}\right)\mathbf{r}_3 + \mathbf{r}_2 \rightarrow \mathbf{r}_2 \\ \left(-\frac{5}{3}\right)\mathbf{r}_3 + \mathbf{r}_1 \rightarrow \mathbf{r}_1 \end{array} \qquad \text{reduced row echelon form}$$

Hence we have the unique solution $x = 1$, $y = 2$, $z = -2$.

7. (a) We put the given augmented matrix into reduced row echelon form:

$$\begin{bmatrix} 1 & 1 & 1 & | & 0 \\ 1 & 1 & 0 & | & 3 \\ 0 & 1 & 1 & | & 1 \end{bmatrix} \rightarrow \begin{bmatrix} 1 & 1 & 1 & | & 0 \\ 0 & 0 & -1 & | & 3 \\ 0 & 1 & 1 & | & 1 \end{bmatrix} \rightarrow \begin{bmatrix} 1 & 1 & 1 & | & 0 \\ 0 & 1 & 1 & | & 1 \\ 0 & 0 & -1 & | & 3 \end{bmatrix}$$

$$(-1)\mathbf{r}_1 + \mathbf{r}_2 \rightarrow \mathbf{r}_2 \qquad\qquad \mathbf{r}_3 \rightarrow \mathbf{r}_2 \qquad\qquad (-1)\mathbf{r}_2 + \mathbf{r}_1 \rightarrow \mathbf{r}_1$$

$$\rightarrow \begin{bmatrix} 1 & 0 & 0 & | & -1 \\ 0 & 1 & 1 & | & 1 \\ 0 & 0 & -1 & | & 3 \end{bmatrix} \rightarrow \begin{bmatrix} 1 & 0 & 0 & | & -1 \\ 0 & 1 & 0 & | & 4 \\ 0 & 0 & 1 & | & -3 \end{bmatrix}$$

$$\begin{array}{l} 1\mathbf{r}_3 + \mathbf{r}_2 \rightarrow \mathbf{r}_2 \\ (-1)\mathbf{r}_3 \rightarrow \mathbf{r}_3 \end{array} \qquad\qquad \text{reduced row echelon form}$$

Thus there is a unique solution $x = -1$, $y = 4$, $z = -3$.

(b) We put the given augmented matrix into reduced row echelon form:

$$\begin{bmatrix} 1 & 2 & 3 & | & 0 \\ 1 & 1 & 1 & | & 0 \\ 1 & 1 & 2 & | & 0 \end{bmatrix} \rightarrow \begin{bmatrix} 1 & 2 & 3 & | & 0 \\ 0 & -1 & -2 & | & 0 \\ 0 & -1 & -1 & | & 0 \end{bmatrix} \rightarrow \begin{bmatrix} 1 & 2 & 3 & | & 0 \\ 0 & 1 & 2 & | & 0 \\ 0 & -1 & -1 & | & 0 \end{bmatrix}$$

$$\begin{array}{l} (-1)\mathbf{r}_1 + \mathbf{r}_2 \rightarrow \mathbf{r}_2 \\ (-1)\mathbf{r}_1 + \mathbf{r}_3 \rightarrow \mathbf{r}_3 \end{array} \qquad (-1)\mathbf{r}_2 \rightarrow \mathbf{r}_2 \qquad \begin{array}{l} 1\mathbf{r}_2 + \mathbf{r}_3 \rightarrow \mathbf{r}_3 \\ (-2)\mathbf{r}_2 + \mathbf{r}_1 \rightarrow \mathbf{r}_1 \end{array}$$

$$\rightarrow \begin{bmatrix} 1 & 0 & -1 & | & 0 \\ 0 & 1 & 2 & | & 0 \\ 0 & 0 & 1 & | & 0 \end{bmatrix} \rightarrow \begin{bmatrix} 1 & 0 & 0 & | & 0 \\ 0 & 1 & 0 & | & 0 \\ 0 & 0 & 1 & | & 0 \end{bmatrix}$$

$$\begin{array}{l} 1\mathbf{r}_3 + \mathbf{r}_1 \rightarrow \mathbf{r}_1 \\ (-2)\mathbf{r}_3 + \mathbf{r}_2 \rightarrow \mathbf{r}_2 \end{array} \qquad\qquad \text{reduced row echelon form}$$

There is a unique solution $x = 0$, $y = 0$, $z = 0$.

(c) Row reducing the augmented matrix we obtain:

$$\begin{bmatrix} 1 & 2 & 3 & | & 0 \\ 1 & 1 & 1 & | & 0 \\ 5 & 7 & 9 & | & 0 \end{bmatrix} \rightarrow \begin{bmatrix} 1 & 2 & 3 & | & 0 \\ 0 & -1 & -2 & | & 0 \\ 0 & -3 & -6 & | & 0 \end{bmatrix}$$

$$\begin{array}{l} (-1)\mathbf{r}_1 + \mathbf{r}_2 \rightarrow \mathbf{r}_2 \\ (-5)\mathbf{r}_1 + \mathbf{r}_3 \rightarrow \mathbf{r}_3 \end{array} \qquad\qquad (-1)\mathbf{r}_2 \rightarrow \mathbf{r}_2$$

$$\rightarrow \begin{bmatrix} 1 & 2 & 3 & \Big| & 0 \\ 0 & 1 & 2 & \Big| & 0 \\ 0 & -3 & -6 & \Big| & 0 \end{bmatrix} \rightarrow \begin{bmatrix} 1 & 0 & -1 & \Big| & 0 \\ 0 & 1 & 2 & \Big| & 0 \\ 0 & 0 & 0 & \Big| & 0 \end{bmatrix}$$

$(-2)\mathbf{r}_2 + \mathbf{r}_1 \rightarrow \mathbf{r}_1$ \qquad reduced row echelon form
$3\mathbf{r}_2 + \mathbf{r}_3 \rightarrow \mathbf{r}_3$

Thus we have the equivalent linear system

$$\begin{aligned} x \quad - \quad z &= 0 \\ y + 2z &= 0. \end{aligned}$$

Hence $x = z$ and $y = -2z$. It follows that we can choose $z = r$, any real number, and all solutions are of the form $x = r$, $y = -2r$, $z = r$.

(d) We row reduce the augmented matrix to obtain:

$$\begin{bmatrix} 1 & 2 & 3 & \Big| & 0 \\ 1 & 2 & 1 & \Big| & 0 \end{bmatrix} \rightarrow \begin{bmatrix} 1 & 2 & 3 & \Big| & 0 \\ 0 & 0 & -2 & \Big| & 0 \end{bmatrix} \rightarrow \begin{bmatrix} 1 & 2 & 3 & \Big| & 0 \\ 0 & 0 & 1 & \Big| & 0 \end{bmatrix} \rightarrow \begin{bmatrix} 1 & 2 & 0 & \Big| & 0 \\ 0 & 0 & 1 & \Big| & 0 \end{bmatrix}$$

$(-1)\mathbf{r}_1 + \mathbf{r}_2 \rightarrow \mathbf{r}_2$ \qquad $\left(-\tfrac{1}{2}\right)\mathbf{r}_2 \rightarrow \mathbf{r}_2$ \qquad $(-3)\mathbf{r}_2 + \mathbf{r}_1 \rightarrow \mathbf{r}_1$ \qquad reduced row echelon form

Thus we have the equivalent linear systems

$$\begin{aligned} x + 2y \quad &= 0 \\ z &= 0. \end{aligned}$$

Hence $x = -2y$ and $z = 0$. It follows that we can choose $y = r$, any real number, and all solutions have the form $x = -2r$, $y = r$, $z = 0$.

9. (a) We row reduce the augmented matrix to obtain:

$$\begin{bmatrix} 1 & 2 & 1 & \Big| & 7 \\ 2 & 0 & 1 & \Big| & 4 \\ 1 & 0 & 2 & \Big| & 5 \\ 1 & 2 & 3 & \Big| & 11 \\ 2 & 1 & 4 & \Big| & 12 \end{bmatrix} \rightarrow \begin{bmatrix} 1 & 2 & 1 & \Big| & 7 \\ 0 & -4 & -1 & \Big| & -10 \\ 0 & -2 & 1 & \Big| & -2 \\ 0 & 0 & 2 & \Big| & 4 \\ 0 & -3 & 2 & \Big| & -2 \end{bmatrix} \rightarrow \begin{bmatrix} 1 & 2 & 1 & \Big| & 7 \\ 0 & 1 & 3 & \Big| & 8 \\ 0 & -2 & 1 & \Big| & -2 \\ 0 & 0 & 2 & \Big| & 4 \\ 0 & -3 & 2 & \Big| & -2 \end{bmatrix}$$

$(-2)\mathbf{r}_1 + \mathbf{r}_2 \rightarrow \mathbf{r}_2$ \qquad $(-1)\mathbf{r}_2 \rightarrow \mathbf{r}_2$ \qquad $(-2)\mathbf{r}_2 + \mathbf{r}_1 \rightarrow \mathbf{r}_1$
$(-1)\mathbf{r}_1 + \mathbf{r}_3 \rightarrow \mathbf{r}_3$ \qquad $1\mathbf{r}_5 + \mathbf{r}_2 \rightarrow \mathbf{r}_2$ \qquad $2\mathbf{r}_2 + \mathbf{r}_3 \rightarrow \mathbf{r}_3$
$(-1)\mathbf{r}_1 + \mathbf{r}_4 \rightarrow \mathbf{r}_4$ \qquad\qquad\qquad $3\mathbf{r}_2 + \mathbf{r}_5 \rightarrow \mathbf{r}_5$
$(-2)\mathbf{r}_1 + \mathbf{r}_5 \rightarrow \mathbf{r}_5$

$$\rightarrow \begin{bmatrix} 1 & 0 & -5 & \Big| & -9 \\ 0 & 1 & 3 & \Big| & 8 \\ 0 & 0 & 7 & \Big| & 14 \\ 0 & 0 & 2 & \Big| & 4 \\ 0 & 0 & 11 & \Big| & 22 \end{bmatrix} \rightarrow \begin{bmatrix} 1 & 0 & -5 & \Big| & -9 \\ 0 & 1 & 3 & \Big| & 8 \\ 0 & 0 & 1 & \Big| & 2 \\ 0 & 0 & 2 & \Big| & 4 \\ 0 & 0 & 11 & \Big| & 22 \end{bmatrix} \rightarrow \begin{bmatrix} 1 & 0 & 0 & \Big| & 1 \\ 0 & 1 & 0 & \Big| & 2 \\ 0 & 0 & 1 & \Big| & 2 \\ 0 & 0 & 0 & \Big| & 0 \\ 0 & 0 & 0 & \Big| & 0 \end{bmatrix}$$

$\tfrac{1}{7}\mathbf{r}_3 \rightarrow \mathbf{r}_3$ \qquad $5\mathbf{r}_3 + \mathbf{r}_1 \rightarrow \mathbf{r}_1$ \qquad reduced row echelon form
$(-3)\mathbf{r}_3 + \mathbf{r}_2 \rightarrow \mathbf{r}_2$
$(-2)\mathbf{r}_3 + \mathbf{r}_4 \rightarrow \mathbf{r}_4$
$(-11)\mathbf{r}_3 + \mathbf{r}_5 \rightarrow \mathbf{r}_5$

Thus there is a unique solution $x = 1$, $y = 2$, $z = 2$.

(b) We row reduce the augmented matrix to obtain:

$$\begin{bmatrix} 1 & 2 & 1 & | & 0 \\ 2 & 3 & 0 & | & 0 \\ 0 & 1 & 2 & | & 0 \\ 2 & 1 & 4 & | & 0 \end{bmatrix} \rightarrow \begin{bmatrix} 1 & 2 & 1 & | & 0 \\ 0 & -1 & -2 & | & 0 \\ 0 & 1 & 2 & | & 0 \\ 0 & -3 & 2 & | & 0 \end{bmatrix} \rightarrow \begin{bmatrix} 1 & 2 & 1 & | & 0 \\ 0 & 1 & 2 & | & 0 \\ 0 & 1 & 2 & | & 0 \\ 0 & -3 & 2 & | & 0 \end{bmatrix}$$

$$(-2)\mathbf{r}_1 + \mathbf{r}_2 \rightarrow \mathbf{r}_2 \qquad\qquad (-1)\mathbf{r}_2 \rightarrow \mathbf{r}_2 \qquad\qquad (-2)\mathbf{r}_2 + \mathbf{r}_1 \rightarrow \mathbf{r}_1$$
$$(-2)\mathbf{r}_1 + \mathbf{r}_4 \rightarrow \mathbf{r}_4 \qquad\qquad\qquad\qquad\qquad\qquad (-1)\mathbf{r}_2 + \mathbf{r}_3 \rightarrow \mathbf{r}_3$$
$$\qquad\qquad\qquad\qquad\qquad\qquad\qquad\qquad\qquad\qquad\qquad 3\mathbf{r}_2 + \mathbf{r}_4 \rightarrow \mathbf{r}_4$$

$$\rightarrow \begin{bmatrix} 1 & 0 & -3 & | & 0 \\ 0 & 1 & 2 & | & 0 \\ 0 & 0 & 0 & | & 0 \\ 0 & 0 & 8 & | & 0 \end{bmatrix} \rightarrow \begin{bmatrix} 1 & 0 & -3 & | & 0 \\ 0 & 1 & 2 & | & 0 \\ 0 & 0 & 1 & | & 0 \\ 0 & 0 & 0 & | & 0 \end{bmatrix} \rightarrow \begin{bmatrix} 1 & 0 & 0 & | & 0 \\ 0 & 1 & 0 & | & 0 \\ 0 & 0 & 1 & | & 0 \\ 0 & 0 & 0 & | & 0 \end{bmatrix}$$

$$\mathbf{r}_4 \leftrightarrow \mathbf{r}_3 \qquad\qquad (-2)\mathbf{r}_3 + \mathbf{r}_2 \rightarrow \mathbf{r}_2 \qquad \text{reduced row echelon form}$$
$$\tfrac{1}{8}\mathbf{r}_3 \rightarrow \mathbf{r}_3 \qquad\qquad 3\mathbf{r}_3 + \mathbf{r}_1 \rightarrow \mathbf{r}_1$$

Thus there is a unique solution $x = 0$, $y = 0$, $z = 0$.

11. Start with the matrix equation $A\mathbf{x} = 3\mathbf{x}$ and rearrange it as follows:

$$A\mathbf{x} = 3\mathbf{x} \quad\Longrightarrow\quad A\mathbf{x} - 3\mathbf{x} = \mathbf{0} \quad\Longrightarrow\quad A\mathbf{x} - 3I_2\mathbf{x} = \mathbf{0} \quad\Longrightarrow\quad (A - 3I_2)\mathbf{x} = \mathbf{0}.$$

Then

$$(A - 3I_2)\mathbf{x} = \left(\begin{bmatrix} 2 & 1 \\ 1 & 2 \end{bmatrix} - 3\begin{bmatrix} 1 & 0 \\ 0 & 1 \end{bmatrix} \right) \begin{bmatrix} x_1 \\ x_2 \end{bmatrix} = \begin{bmatrix} -1 & 1 \\ 1 & -1 \end{bmatrix} \begin{bmatrix} x_1 \\ x_2 \end{bmatrix} = \begin{bmatrix} 0 \\ 0 \end{bmatrix}.$$

Form the augmented matrix and find its reduced row echelon form:

$$\begin{bmatrix} -1 & 1 & | & 0 \\ 1 & -1 & | & 0 \end{bmatrix} \rightarrow \begin{bmatrix} -1 & 1 & | & 0 \\ 0 & 0 & | & 0 \end{bmatrix} \rightarrow \begin{bmatrix} 1 & -1 & | & 0 \\ 0 & 0 & | & 0 \end{bmatrix}$$

$$\mathbf{r}_1 + \mathbf{r}_2 \rightarrow \mathbf{r}_2 \qquad\qquad (-1)\mathbf{r}_1 \rightarrow \mathbf{r}_1$$

The equivalent linear system is $x_1 - x_2 = 0$, or $x_1 = x_2$. Hence we have the solution $x_1 = r$, $x_2 = r$, where r is any real number. In matrix form,

$$\mathbf{x} = \begin{bmatrix} r \\ r \end{bmatrix}, \quad r \neq 0.$$

For example, setting $r = 7$ then $\mathbf{x} = \begin{bmatrix} 7 \\ 7 \end{bmatrix}$ is a nonzero 2×1 matrix such that $A\mathbf{x} = 3\mathbf{x}$.

13. Start with the matrix equation $A\mathbf{x} = 1\mathbf{x}$ and rearrange it as follows:

$$A\mathbf{x} = 1\mathbf{x} \quad\Longrightarrow\quad A\mathbf{x} - 1\mathbf{x} = \mathbf{0} \quad\Longrightarrow\quad A\mathbf{x} - 1I_3\mathbf{x} = \mathbf{0} \quad\Longrightarrow\quad (A - 1I_3)\mathbf{x} = \mathbf{0}.$$

Then

$$(A - 1I_3)\mathbf{x} = \left(\begin{bmatrix} 1 & 2 & -1 \\ 1 & 0 & 0 \\ 4 & -4 & 5 \end{bmatrix} - \begin{bmatrix} 1 & 0 & 0 \\ 0 & 1 & 0 \\ 0 & 0 & 1 \end{bmatrix} \right) \begin{bmatrix} x_1 \\ x_2 \\ x_3 \end{bmatrix} = \begin{bmatrix} 0 & 2 & -1 \\ 1 & -1 & 1 \\ 4 & -4 & 4 \end{bmatrix} \begin{bmatrix} x_1 \\ x_2 \\ x_3 \end{bmatrix} = \begin{bmatrix} 0 \\ 0 \\ 0 \end{bmatrix}.$$

Form the augmented matrix and find the reduced row echelon form:

$$\begin{bmatrix} 0 & 2 & -1 & | & 0 \\ 1 & -1 & 1 & | & 0 \\ 4 & -4 & 4 & | & 0 \end{bmatrix} \rightarrow \begin{bmatrix} 1 & -1 & 1 & | & 0 \\ 0 & 2 & -1 & | & 0 \\ 4 & -4 & 4 & | & 0 \end{bmatrix} \rightarrow \begin{bmatrix} 1 & -1 & 1 & | & 0 \\ 0 & 2 & -1 & | & 0 \\ 0 & 0 & 0 & | & 0 \end{bmatrix}$$

$$\mathbf{r}_2 \leftrightarrow \mathbf{r}_1 \qquad\qquad (-4)\mathbf{r}_1 + \mathbf{r}_3 \rightarrow \mathbf{r}_3 \qquad\qquad \tfrac{1}{2}\mathbf{r}_2 \rightarrow \mathbf{r}_2$$

$$\rightarrow \begin{bmatrix} 1 & -1 & 1 & \Big| & 0 \\ 0 & 1 & -\frac{1}{2} & \Big| & 0 \\ 0 & 0 & 0 & \Big| & 0 \end{bmatrix} \rightarrow \begin{bmatrix} 1 & 0 & \frac{1}{2} & \Big| & 0 \\ 0 & 1 & -\frac{1}{2} & \Big| & 0 \\ 0 & 0 & 0 & \Big| & 0 \end{bmatrix}$$

$$\mathbf{r}_1 \rightarrow \mathbf{r}_2 + \mathbf{r}_1 \qquad\qquad \text{reduced row echelon form}$$

The equivalent linear system is

$$\begin{aligned} x_1 \quad\; + \tfrac{1}{2}x_3 &= 0 \\ x_2 - \tfrac{1}{2}x_3 &= 0. \end{aligned}$$

Hence we have the solution $x_1 = -\frac{1}{2}r$, $x_2 = \frac{1}{2}r$, $x_3 = r$, where r is any real number. In matrix form,

$$\mathbf{x} = \begin{bmatrix} -\frac{1}{2}r \\ \frac{1}{2}r \\ r \end{bmatrix}, \quad r \neq 0.$$

For example, setting $r = 2$, $\mathbf{x} = \begin{bmatrix} -1 \\ 1 \\ 2 \end{bmatrix}$ is a nonzero 2×1 matrix such that $A\mathbf{x} = 1\mathbf{x}$.

15. Form the augmented matrix and row reduce it:

$$\begin{bmatrix} 1 & 1 & 1 & \Big| & 2 \\ 2 & 3 & 2 & \Big| & 5 \\ 2 & 3 & a^2 - 1 & \Big| & a + 1 \end{bmatrix} \rightarrow \begin{bmatrix} 1 & 1 & 1 & \Big| & 2 \\ 0 & 1 & 0 & \Big| & 1 \\ 0 & 1 & a^2 - 3 & \Big| & a - 3 \end{bmatrix} \rightarrow \begin{bmatrix} 1 & 0 & 1 & \Big| & 1 \\ 0 & 1 & 0 & \Big| & 1 \\ 0 & 0 & a^2 - 3 & \Big| & a - 4 \end{bmatrix}$$

$$\begin{aligned} (-2)\mathbf{r}_1 + \mathbf{r}_2 &\rightarrow \mathbf{r}_2 \\ (-2)\mathbf{r}_1 + \mathbf{r}_3 &\rightarrow \mathbf{r}_3 \end{aligned} \qquad \begin{aligned} (-1)\mathbf{r}_2 + \mathbf{r}_1 &\rightarrow \mathbf{r}_1 \\ (-1)\mathbf{r}_2 + \mathbf{r}_3 &\rightarrow \mathbf{r}_3 \end{aligned}$$

(a) There will be no solution if the system is inconsistent. That is, if there is a row of the form $\begin{bmatrix} 0 & 0 & 0 & | & * \end{bmatrix}$, where $*$ represents a nonzero number. For this to occur we require

$$a^2 - 3 = 0 \quad\Longrightarrow\quad a = \pm\sqrt{3}$$

in which case $a - 4 \neq 0$. So there is no solution when $a = \pm\sqrt{3}$.

(b) There will be a unique solution provided the reduced row echelon form of the coefficient matrix is I_3. This will occur when the last row has $a^2 - 3 \neq 0$ since in this case we can multiply row 3 by $1/(a^2 - 3)$, then add -1 times it to row 1 and thus obtain I_3. We have

$$a^2 - 3 \neq 0 \quad\Longrightarrow\quad a^2 \neq 3 \quad\Longrightarrow\quad a \neq \pm\sqrt{3}.$$

For $a \neq \pm\sqrt{3}$ there is a unique solution. The numerical values of such a solution depend upon the value of a.

(c) For infinitely many solutions we need the third row to be a zero row. This case only happen if both $a^2 - 3$ and $a - 4$ are zero. There is no value of a when this is true. Thus there is no value of a such that this system has infinitely many solutions.

17. Form the augmented matrix and row reduce it:

$$\begin{bmatrix} 1 & 1 & \Big| & 3 \\ 1 & a^2 - 8 & \Big| & a \end{bmatrix} \rightarrow \begin{bmatrix} 1 & 1 & \Big| & 3 \\ 0 & a^2 - 9 & \Big| & a - 3 \end{bmatrix}$$

$$(-1)\mathbf{r}_1 + \mathbf{r}_2 \rightarrow \mathbf{r}_2$$

(a) There will be no solution if the system is inconsistent. That is, if there is a row of the form $\begin{bmatrix} 0 & 0 & | & * \end{bmatrix}$, where $*$ represents a nonzero number. For this to occur we require

$$a^2 - 9 = 0 \quad\Longrightarrow\quad a = \pm 3.$$

If $a = -3$, then $a - 3 \neq 0$. So there is no solution when $a = -3$.

(b) There will be a unique solution provided the reduced row echelon form of the coefficient matrix is I_2. This will occur when the last row has $a^2 - 9 \neq 0$ since in this case we can multiply row 2 by $1/(a^2 - 9)$, then add -1 times it to row 1 and thus obtain I_2. We have

$$a^2 - 9 \neq 0 \quad \Longrightarrow \quad a^2 \neq 9 \quad \Longrightarrow \quad a \neq \pm 3.$$

For $a \neq \pm 3$ there is a unique solution. The numerical values of such a solution depend upon the value of a.

(c) For infinitely many solutions we need the second row to be a zero row. This can only happen if both $a^2 - 9$ and $a - 3$ are zero. If $a = 3$, then both expressions are zero and thus the system has infinitely many solutions.

19. We are asked to prove an "if and only if" statement. There are two things that must be proved.

(a) If A is row equivalent to I_2, then $ad - bc \neq 0$.

Proof: We proceed to show this statement by verifying its contrapositive; that is, we will show that if $ad - bc = 0$, then A is not row equivalent to I_2. Suppose that $ad - bc = 0$. Then $ad = bc$. We show that the two rows of

$$A = \begin{bmatrix} a & b \\ c & d \end{bmatrix}$$

are multiples of one another:

$$c \begin{bmatrix} a & b \end{bmatrix} = \begin{bmatrix} ac & bc \end{bmatrix} = \begin{bmatrix} ac & ad \end{bmatrix} = a \begin{bmatrix} c & d \end{bmatrix}.$$

Thus any elementary row operations applied to A will produce a matrix with rows that are multiples of one another. In particular, elementary row operations cannot produce I_2 since its rows are not multiples of one another. Hence A is not row equivalent to I_2. Since we have verified the contrapositive of what we wanted to show the original statement has been proved.

(b) If $ad - bc \neq 0$, then A is row equivalent to I_2.

Proof: We proceed directly. Since $ad - bc \neq 0$, then a and c are not both zero. Suppose $a \neq 0$. (A corresponding proof can be given if we assume $c \neq 0$.) Then we can multiply row 1 by $1/a$ to obtain

$$\begin{bmatrix} 1 & \dfrac{b}{a} \\ c & d \end{bmatrix}.$$

Next we add $-c$ times row 1 to row 2 to obtain:

$$\begin{bmatrix} 1 & \dfrac{b}{a} \\ 0 & d - \dfrac{bc}{a} \end{bmatrix} = \begin{bmatrix} 1 & \dfrac{b}{a} \\ 0 & \dfrac{ad - bc}{a} \end{bmatrix}.$$

Next

$$\text{multiply row 2 by } \dfrac{a}{ad - bc}$$

to create a 1 in entry $(2, 2)$, then add $-\dfrac{b}{a}$ times row 2 to row 1 to get the identity matrix I_2. Hence, A is row equivalent to I_2.

21. The equation

$$f\left(\begin{bmatrix} x \\ y \\ z \end{bmatrix} \right) = \begin{bmatrix} 1 & 2 & 3 \\ -3 & -2 & -1 \\ -2 & 0 & 2 \end{bmatrix} \begin{bmatrix} x \\ y \\ z \end{bmatrix} = \begin{bmatrix} x + 2y + 3z \\ -3x - 2y - z \\ -2x + 2z \end{bmatrix} = \begin{bmatrix} 2 \\ 2 \\ 4 \end{bmatrix}$$

leads to the linear system

$$
\begin{aligned}
x + 2y + 3z &= 2 \\
-3x - 2y - z &= 2 \\
-2x + 2z &= 4
\end{aligned}
$$

whose augmented matrix is

$$
B = \left[\begin{array}{ccc|c}
1 & 2 & 3 & 2 \\
-3 & -2 & -1 & 2 \\
-2 & 0 & 2 & 4
\end{array} \right].
$$

The reduced row echelon form of B is

$$
\left[\begin{array}{ccc|c}
1 & 0 & -1 & -2 \\
0 & 1 & 2 & 2 \\
0 & 0 & 0 & 0
\end{array} \right].
$$

This means there are infinitely many solutions:

$$
\begin{aligned}
x &= -2 - r \\
y &= 2 - 2r \\
z &= r
\end{aligned}
$$

where r is any real number. For example, if $r = 0$, then

$$
\left[\begin{array}{c} -2 \\ 2 \\ 0 \end{array} \right] \quad \text{is such that} \quad f\left(\left[\begin{array}{c} -2 \\ 2 \\ 0 \end{array} \right] \right) = \left[\begin{array}{c} 2 \\ 2 \\ 4 \end{array} \right].
$$

23. This is the same linear transformation as the one discussed in Exercise 21. Thus, we have the augmented matrix

$$
\left[\begin{array}{ccc|c}
1 & 2 & 3 & a \\
-3 & -2 & -1 & b \\
-2 & 0 & 2 & c
\end{array} \right].
$$

whose reduced row echelon form is

$$
\left[\begin{array}{ccc|c}
1 & 0 & -1 & a - \frac{1}{2}(3a + b) \\
0 & 1 & 2 & \frac{1}{4}(3a + b) \\
0 & 0 & 0 & -a - b + c
\end{array} \right].
$$

In order for this linear system to have a solution we must have $-a - b + c = 0$.

25. Using Exercise 24(b) we can assume that every $m \times n$ matrix A is column equivalent to a matrix in column echelon form. That is, A is column equivalent to a matrix B that satisfies the following:

   (a) All columns consisting entirely of zeros, if any, are at the right side of the matrix.

   (b) The first nonzero entry in each column that is not all zeros is a 1, called the leading entry of the column.

   (c) If the columns j and $j + 1$ are two successive columns that are not all zeros, then the leading entry of column $j + 1$ is below the leading entry of column j.

   We start with matrix B and show that it is possible to find a matrix C that is column equivalent to B that satisfies

   (d) If a row contains a leading entry of some column then all other entries in that row are zero.

If column j of B contains a nonzero element, then its first (counting top to bottom) nonzero element is a 1. Suppose the 1 appears in row r_j. We can perform column operations of the form $ac_j + c_k$ for each of the nonzero columns c_k of B such that the resulting matrix has row r_j with a 1 in the (r_j, j) entry and zeros everywhere else. This can be done for each column that contains a nonzero entry hence we can produce a matrix C satisfying (d). It follows that C is the unique matrix in reduced column echelon form and column equivalent to the original matrix A.

27. The augmented matrix of the linear system is

$$A = \left[\begin{array}{ccc|c} 2 & 2 & 3 & a \\ 3 & -1 & 5 & b \\ 1 & -3 & 2 & c \end{array}\right].$$

We find that

$$A_{\substack{\mathbf{r}_1 - 2\mathbf{r}_3 \to \mathbf{r}_1 \\ \mathbf{r}_2 - 3\mathbf{r}_3 \to \mathbf{r}_2 \\ \mathbf{r}_1 - \mathbf{r}_2 \to \mathbf{r}_1}} = \left[\begin{array}{ccc|c} 0 & 0 & 0 & -a+b-c \\ 0 & 8 & -1 & b-3c \\ 1 & -3 & 2 & c \end{array}\right].$$

Therefore, to be a consistent linear system, we must have $-a + b - c = 0$.

29. (a) $A(\mathbf{x}_p + \mathbf{x}_h) = A\mathbf{x}_p + A\mathbf{x}_h = \mathbf{b} + \mathbf{0} = \mathbf{b}$.

 (b) Let $\mathbf{x}_p$ be a particular solution to $A\mathbf{x} = \mathbf{b}$ and let $\mathbf{x}$ be any solution to $A\mathbf{x} = \mathbf{b}$. Let $\mathbf{x}_h = \mathbf{x} - \mathbf{x}_p$. Then $\mathbf{x} = \mathbf{x}_p + \mathbf{x}_h = \mathbf{x}_p + (\mathbf{x} - \mathbf{x}_p)$ and $A\mathbf{x}_h = A(\mathbf{x} - \mathbf{x}_p) = A\mathbf{x} - A\mathbf{x}_p = \mathbf{b} - \mathbf{b} = \mathbf{0}$. Thus $\mathbf{x}_h$ is in fact a solution to $A\mathbf{x} = \mathbf{0}$.

31. First compute $f'(x)$ and $f''(x)$:

$$\begin{aligned} f(x) = e^{2x} &\implies f(0) = 1 \\ f'(x) = 2e^{2x} &\implies f'(0) = 2 \\ f''(x) = 4e^{2x} &\implies f''(0) = 4. \end{aligned}$$

Hence

$$\begin{aligned} p(0) = c = f(0) = 1 \\ p'(0) = b = f'(0) = 2 \\ p''(0) = 2a = f''(0) = 4. \end{aligned}$$

Therefore $a = 2$, $b = 2$, $c = 1$, and hence $p(x) = 2x^2 + 2x + 1$.

33. The temperature at each of the four nodes T_1, T_2, T_3, T_4, is the average of the temperature at the four surrounding nodes. Thus:

$$\begin{aligned} T_1 &= \frac{50 + 30 + T_2 + T_3}{4} &\text{or} \quad T_1 - 4T_2 + T_4 = -80 \\ T_2 &= \frac{30 + 50 + T_4 + T_1}{4} &\text{or} \quad T_1 - 4T_2 + T_4 = -80 \\ T_3 &= \frac{0 + 50 + T_1 + T_4}{4} &\text{or} \quad T_1 - 4T_3 + T_4 = -50 \\ T_4 &= \frac{50 + 0 + T_3 + T_2}{4} &\text{or} \quad T_2 + T_3 - 4T_4 = -50 \end{aligned}$$

Therefore we have the linear system

$$\begin{aligned} 4T_1 - T_2 - T_3 \qquad &= 80 \\ T_1 - 4T_2 \qquad + T_4 &= -80 \\ T_1 \qquad - 4T_3 + T_4 &= -50 \\ T_2 + T_3 - 4T_4 &= -50. \end{aligned}$$

The augmented matrix of this linear system is

$$\begin{bmatrix} 4 & -1 & -1 & 0 & 80 \\ 1 & -4 & 0 & 1 & -80 \\ 1 & 0 & -4 & 1 & -50 \\ 0 & 1 & 1 & -4 & -50 \end{bmatrix}$$ whose reduced echelon form is $$\begin{bmatrix} 1 & 0 & 0 & 0 & 36.25 \\ 0 & 1 & 0 & 0 & 36.25 \\ 0 & 0 & 1 & 0 & 28.75 \\ 0 & 0 & 0 & 1 & 28.75 \end{bmatrix}$$

Therefore, $T_1 = 36.25°$, $T_2 = 36.25°$, $T_3 = 28.75°$, and $T_4 = 28.75°$.

35. Here $(a_1, b_1) = (-16, 38)$, $(a_2, b_2) = (7, -57)$, $(a_3, b_3) = (32, 80)$, and $r_2 = 61$, $r_3 = 85$. Using Equation (9), we find that:

$$\begin{bmatrix} 2(-23) & 2(95) \\ 2(25) & 2(137) \end{bmatrix} \begin{bmatrix} x \\ y \end{bmatrix} = \begin{bmatrix} (61^2 - r_1^2) + 207 + (-1805) \\ -3504 + 975 + 3151 \end{bmatrix}$$

or

$$\begin{bmatrix} -46 & 190 \\ 50 & 274 \end{bmatrix} \begin{bmatrix} x \\ y \end{bmatrix} = \begin{bmatrix} 2123 - r_2^2 \\ 622 \end{bmatrix}.$$

If $(-4, 3)$ is the location of the GPS receiver, then $x = -4$, $y = 3$ must satisfy this equation:

$$\begin{bmatrix} 2(-23) & 2(95) \\ 2(25) & 2(137) \end{bmatrix} \begin{bmatrix} -4 \\ 3 \end{bmatrix} = \begin{bmatrix} (61^2 - r_1^2) + 207 + (-1805) \\ -3504 + 975 + 3151 \end{bmatrix}$$

or

$$\begin{bmatrix} -46 & 190 \\ 50 & 274 \end{bmatrix} \begin{bmatrix} -4 \\ 3 \end{bmatrix} = \begin{bmatrix} 2123 - r_2^2 \\ 622 \end{bmatrix}$$

or

$$\begin{bmatrix} 730 \\ 622 \end{bmatrix} = \begin{bmatrix} 2123 - r_1^2 \\ 622 \end{bmatrix}.$$

Equating the $(1, 1)$ entries, we obtain

$$2123 - r_1^2 = 730 \quad \Longrightarrow \quad r_1 \approx 37.32.$$

Thus, r_1 is approximately 37 miles.

37. The location will be at one of the intersection points of the two circles, since the location is the intersection point of the 3 circles.

39. Write the equation to be balanced as follows:

$$xC_2H_6 + yO_2 \rightarrow zCO_2 + wH_2O.$$

We must determine x, y, z, and w. Equating the values of the three elements, we find:

$$\text{C: } 2x = z$$
$$\text{H: } 6x = 2w$$
$$\text{O: } 2y = 2z + w$$

From the second equation, $x = \frac{1}{3}w$. From the first equation, $z = 2x = \frac{2}{3}w$. From the third equation, $y = z + \frac{1}{2}w = \frac{2}{3}w + \frac{1}{2}w = \frac{7}{6}w$. So the solution to the linear system is

$$x = \frac{1}{3}w, \quad y = \frac{7}{6}w, \quad z = \frac{2}{3}w, \quad w = t = \text{any positive real number.}$$

Setting $w = 6$, we get the solution $x = 2$, $y = 7$, $z = 4$, and $w = 6$ which gives the balanced equation:

$$2C_2H_6 + 7O_2 \rightarrow 4CO_2 + 6H_2O.$$

41. The augmented matrix is

$$A = \begin{bmatrix} 1 & 1 & 0 & 3-i \\ i & 1 & 1 & 3 \\ 0 & 1 & i & 3 \end{bmatrix}.$$

Performing elementary row operations, we obtain the reduced echelon form of this matrix:

$$A_{\substack{\mathbf{r}_2 - i\mathbf{r}_1 \to \mathbf{r}_2 \\ \mathbf{r}_2 \to \mathbf{r}_3 \\ \mathbf{r}_3 - (1-i)\mathbf{r}_2 \to \mathbf{r}_3 \\ \mathbf{r}_1 - vecr_2 \to \mathbf{r}_1 \\ \mathbf{r}_2 - i\mathbf{r}_3 \to \mathbf{r}_2 \\ \mathbf{r}_1 + \mathbf{r}_3 \to \mathbf{r}_1}} = \begin{bmatrix} 1 & 0 & 0 & 1-i \\ 0 & 1 & 0 & 2 \\ 0 & 0 & 1 & -i \end{bmatrix}.$$

The solution is $\begin{bmatrix} x \\ y \\ z \end{bmatrix} = \begin{bmatrix} 1-i \\ 2 \\ -i \end{bmatrix}.$

43. $\begin{bmatrix} 1 & i & -i & -2+2i \\ 2i & -i & 2 & -2 \\ 1 & 2 & 3i & 2i \end{bmatrix} \underset{\substack{\mathbf{r}_2 - 2i\mathbf{r}_1 \to \mathbf{r}_2 \\ \mathbf{r}_3 - \mathbf{r}_2 \to \mathbf{r}_3 \\ \frac{1}{2-i}\mathbf{r}_2 \to \mathbf{r}_2 \\ \mathbf{r}_1 - i\mathbf{r}_2 \to \mathbf{r}_1 \\ \frac{1}{4i}\mathbf{r}_3 \to \mathbf{r}_3 \\ \mathbf{r}_1 + i\mathbf{r}_3 \to \mathbf{r}_1}}{=} \begin{bmatrix} 1 & 0 & 0 & i \\ 0 & 1 & 0 & 2i \\ 0 & 0 & 1 & -1 \end{bmatrix}.$ The solution is $\begin{bmatrix} i \\ 2i \\ -1 \end{bmatrix}.$

Section 2.3, p. 124

1. The elementary matrix E which results from I_m by a type I interchange of the ith and jth rows differs from I_m by having 1's in the (i,j) and (j,i) positions and 0's in the (i,i) and (j,j) positions. For that E, EA has as its ith row the jth row of A and for its jth row the ith row of A.

The elementary matrix E which results from I_m by a type II operation differs from I_m by having $c \neq 0$ in the (i,i) position. Then EA has as its ith row c times the ith row of A.

The elementary matrix E which results from I_m by a type III operation differs from I_m by having c in the (j,i) position. Then EA has as its jth row the sum of the jth row of A and c times the ith row of A.

3. Let A be a 3×4 matrix and let I_4 be the 4×4 identity matrix.

(a) $F = (I_4)_{\mathbf{c}_2 - 4\mathbf{c}_1 \to \mathbf{c}_2} = \begin{bmatrix} 1 & -4 & 0 & 0 \\ 0 & 1 & 0 & 0 \\ 0 & 0 & 1 & 0 \\ 0 & 0 & 0 & 1 \end{bmatrix}$

(b) $F = (I_4)_{\mathbf{c}_3 \leftrightarrow \mathbf{c}_2} = \begin{bmatrix} 1 & 0 & 0 & 0 \\ 0 & 0 & 1 & 0 \\ 0 & 1 & 0 & 0 \\ 0 & 0 & 0 & 1 \end{bmatrix}$

(c) $F = (I_4)_{4\mathbf{c}_3 \rightarrow \mathbf{c}_3} = \begin{bmatrix} 1 & 0 & 0 & 0 \\ 0 & 1 & 0 & 0 \\ 0 & 0 & 4 & 0 \\ 0 & 0 & 0 & 1 \end{bmatrix}$

5. (a) $\begin{bmatrix} 1 & 0 & 1 \\ 2 & 1 & 0 \\ 0 & -1 & 1 \end{bmatrix} \rightarrow \begin{bmatrix} 1 & 0 & 1 \\ 0 & 1 & -2 \\ 0 & -1 & 1 \end{bmatrix} \rightarrow \begin{bmatrix} 1 & 0 & 1 \\ 0 & 1 & -2 \\ 0 & 0 & -1 \end{bmatrix} \rightarrow \begin{bmatrix} 1 & 0 & 0 \\ 0 & 1 & -2 \\ 0 & 0 & -1 \end{bmatrix} \rightarrow$

      $\mathbf{r}_2 - 2\mathbf{r}_1 \rightarrow \mathbf{r}_2$ $\mathbf{r}_2 + \mathbf{r}_3 \rightarrow \mathbf{r}_3$ $\mathbf{r}_1 + \mathbf{r}_3 \rightarrow \mathbf{r}_1$ $\mathbf{r}_2 - 2\mathbf{r}_3 \rightarrow \mathbf{r}_2$

      $\begin{bmatrix} 1 & 0 & 0 \\ 0 & 1 & 0 \\ 0 & 0 & -1 \end{bmatrix} \rightarrow \begin{bmatrix} 1 & 0 & 0 \\ 0 & 1 & 0 \\ 0 & 0 & 1 \end{bmatrix} = C$

      $-\mathbf{r}_3 \rightarrow \mathbf{r}_3$

    (b) $\begin{bmatrix} 1 & 0 & 0 \\ 0 & 1 & 0 \\ 0 & 0 & 1 \end{bmatrix} \rightarrow \begin{bmatrix} 1 & 0 & 0 \\ -2 & 1 & 0 \\ 0 & 0 & 1 \end{bmatrix} \rightarrow \begin{bmatrix} 1 & 0 & 0 \\ -2 & 1 & 0 \\ -2 & 1 & 1 \end{bmatrix} \rightarrow \begin{bmatrix} -1 & 1 & 1 \\ -2 & 1 & 0 \\ -2 & 1 & 1 \end{bmatrix} \rightarrow$

      $\begin{bmatrix} -1 & 1 & 1 \\ 2 & -1 & -2 \\ -2 & 1 & 1 \end{bmatrix} \rightarrow \begin{bmatrix} -1 & 1 & 1 \\ 2 & -1 & -2 \\ 2 & -1 & -1 \end{bmatrix} = B$

    (c) $AB = \begin{bmatrix} 1 & 0 & 1 \\ 2 & 1 & 0 \\ 0 & -1 & 1 \end{bmatrix} \begin{bmatrix} -1 & 1 & 1 \\ 2 & -1 & -2 \\ 2 & -1 & -1 \end{bmatrix} = \begin{bmatrix} 1 & 0 & 0 \\ 0 & 1 & 0 \\ 0 & 0 & 1 \end{bmatrix}.$

      $BA = \begin{bmatrix} -1 & 1 & 1 \\ 2 & -1 & -2 \\ 2 & -1 & -1 \end{bmatrix} \begin{bmatrix} 1 & 0 & 1 \\ 2 & 1 & 0 \\ 0 & -1 & 1 \end{bmatrix} = \begin{bmatrix} 1 & 0 & 0 \\ 0 & 1 & 0 \\ 0 & 0 & 1 \end{bmatrix}.$

    Therefore A and B are inverses of each other.

7. $\begin{bmatrix} 1 & 3 & | & 1 & 0 \\ 2 & 4 & | & 0 & 1 \end{bmatrix} \rightarrow \begin{bmatrix} 1 & 3 & | & 1 & 0 \\ 0 & -2 & | & -2 & 1 \end{bmatrix} \rightarrow \begin{bmatrix} 1 & 3 & | & 1 & 0 \\ 0 & 1 & | & 1 & -\frac{1}{2} \end{bmatrix} \rightarrow \begin{bmatrix} 1 & 0 & | & -2 & \frac{3}{2} \\ 0 & 1 & | & 1 & -\frac{1}{2} \end{bmatrix}$

  $\mathbf{r}_2 - 2\mathbf{r}_1 \rightarrow \mathbf{r}_2$ $-\frac{1}{2}\mathbf{r}_2 \rightarrow \mathbf{r}_2$ $\mathbf{r}_1 - 3\mathbf{r}_2 \rightarrow \mathbf{r}_1$ reduced row
                                                                     echelon form

    Thus $A^{-1} = \begin{bmatrix} -2 & \frac{3}{2} \\ 1 & -\frac{1}{2} \end{bmatrix}$.

9. (a) $\begin{bmatrix} 1 & 3 & | & 1 & 0 \\ 2 & 6 & | & 0 & 1 \end{bmatrix} \rightarrow \begin{bmatrix} 1 & 3 & | & 1 & 0 \\ 0 & 0 & | & -2 & 1 \end{bmatrix}$

     $\mathbf{r}_2 - 2\mathbf{r}_1 \rightarrow \mathbf{r}_2$ reduced row
                             echelon form

    Thus the matrix $A = \begin{bmatrix} 1 & 3 \\ 2 & 6 \end{bmatrix}$ is row equivalent to a matrix with a row of zeros. By Theorem
    1.18, A is singular.

    (b) $\begin{bmatrix} 1 & 3 & | & 1 & 0 \\ -2 & 6 & | & 0 & 1 \end{bmatrix} \rightarrow \begin{bmatrix} 1 & 3 & | & 1 & 0 \\ 0 & 12 & | & 2 & 1 \end{bmatrix} \rightarrow \begin{bmatrix} 1 & 3 & | & 1 & 0 \\ 0 & 1 & | & \frac{1}{6} & \frac{1}{12} \end{bmatrix} \rightarrow \begin{bmatrix} 1 & 0 & | & \frac{1}{2} & -\frac{1}{4} \\ 0 & 1 & | & \frac{1}{6} & \frac{1}{12} \end{bmatrix}$

     $2\mathbf{r}_1 + \mathbf{r}_2 \rightarrow \mathbf{r}_2$ $\frac{1}{12}\mathbf{r}_2 \rightarrow \mathbf{r}_2$ $\mathbf{r}_1 - 3\mathbf{r}_2 \rightarrow \mathbf{r}_1$ reduced row
                                                                 echelon form

Thus $A^{-1} = \begin{bmatrix} \frac{1}{2} & -\frac{1}{4} \\ \frac{1}{6} & \frac{1}{12} \end{bmatrix}$.

(c) $\begin{bmatrix} 1 & 2 & 3 & | & 1 & 0 & 0 \\ 1 & 1 & 2 & | & 0 & 1 & 0 \\ 0 & 1 & 2 & | & 0 & 0 & 1 \end{bmatrix} \rightarrow \begin{bmatrix} 1 & 2 & 3 & | & 1 & 0 & 0 \\ 0 & -1 & -1 & | & -1 & 1 & 0 \\ 0 & 1 & 2 & | & 0 & 0 & 1 \end{bmatrix} \rightarrow \begin{bmatrix} 1 & 2 & 3 & | & 1 & 0 & 0 \\ 0 & 1 & 1 & | & 1 & -1 & 0 \\ 0 & 1 & 2 & | & 0 & 0 & 1 \end{bmatrix} \rightarrow$

$\mathbf{r}_2 - \mathbf{r}_1 \rightarrow \mathbf{r}_2$ \qquad\qquad $-\mathbf{r}_2 \rightarrow \mathbf{r}_2$ \qquad\qquad $\mathbf{r}_1 - 2\mathbf{r}_2 \rightarrow \mathbf{r}_1$
$\mathbf{r}_3 - \mathbf{r}_2 \rightarrow \mathbf{r}_3$

$\begin{bmatrix} 1 & 0 & 1 & | & -1 & 2 & 0 \\ 0 & 1 & 1 & | & 1 & -1 & 0 \\ 0 & 0 & 1 & | & -1 & 1 & 1 \end{bmatrix} \rightarrow \begin{bmatrix} 1 & 0 & 0 & | & 0 & 1 & -1 \\ 0 & 1 & 0 & | & 2 & -2 & -1 \\ 0 & 0 & 1 & | & -1 & 1 & 1 \end{bmatrix}$

$\mathbf{r}_2 - \mathbf{r}_3 \rightarrow \mathbf{r}_2$ \qquad\qquad reduced row echelon form
$\mathbf{r}_1 - \mathbf{r}_3 \rightarrow \mathbf{r}_1$

Thus $A^{-1} = \begin{bmatrix} 0 & 1 & -1 \\ 2 & -2 & -1 \\ -1 & 1 & 1 \end{bmatrix}$.

(d) $\begin{bmatrix} 1 & 2 & 3 & | & 1 & 0 & 0 \\ 1 & 1 & 2 & | & 0 & 1 & 0 \\ 0 & 1 & 2 & | & 0 & 0 & 1 \end{bmatrix} \rightarrow \begin{bmatrix} 1 & 2 & 3 & | & 1 & 0 & 0 \\ 0 & -1 & -1 & | & -1 & 1 & 0 \\ 0 & 1 & 2 & | & 0 & 0 & 1 \end{bmatrix} \rightarrow \begin{bmatrix} 1 & 2 & 3 & | & 1 & 0 & 0 \\ 0 & 1 & 1 & | & 1 & -1 & 0 \\ 0 & 1 & 2 & | & 0 & 0 & 1 \end{bmatrix} \rightarrow$

$\mathbf{r}_2 - \mathbf{r}_1 \rightarrow \mathbf{r}_2$ \qquad\qquad $-\mathbf{r}_2 \rightarrow \mathbf{r}_2$ \qquad\qquad $\mathbf{r}_1 - 2\mathbf{r}_2 \rightarrow \mathbf{r}_1$
$\mathbf{r}_3 - \mathbf{r}_2 \rightarrow \mathbf{r}_3$

$\begin{bmatrix} 1 & 0 & 1 & | & -1 & 2 & 0 \\ 0 & 1 & 1 & | & 1 & -1 & 0 \\ 0 & 0 & 0 & | & -1 & 1 & 1 \end{bmatrix}$

reduced row echelon form

Thus the matrix $A = \begin{bmatrix} 1 & 2 & 3 \\ 1 & 1 & 2 \\ 0 & 1 & 1 \end{bmatrix}$ is row equivalent to a matrix with a row of zeros. By Theorem 2.10, A is singular.

11. (a) $\begin{bmatrix} 1 & 1 & 1 & | & 1 & 0 & 0 \\ 1 & 2 & 3 & | & 0 & 1 & 0 \\ 0 & 1 & 1 & | & 0 & 0 & 1 \end{bmatrix} \rightarrow \begin{bmatrix} 1 & 1 & 1 & | & 1 & 0 & 0 \\ 0 & 1 & 2 & | & -1 & 1 & 0 \\ 0 & 1 & 1 & | & 0 & 0 & 1 \end{bmatrix} \rightarrow \begin{bmatrix} 1 & 0 & -1 & | & 2 & -1 & 0 \\ 0 & 1 & 2 & | & -1 & 1 & 0 \\ 0 & 0 & -1 & | & 1 & -1 & 1 \end{bmatrix} \rightarrow$

$\mathbf{r}_2 - \mathbf{r}_1 \rightarrow \mathbf{r}_2$ \qquad\qquad $\mathbf{r}_1 - \mathbf{r}_2 \rightarrow \mathbf{r}_1$ \qquad\qquad $-\mathbf{r}_3 \rightarrow \mathbf{r}_3$
\qquad\qquad\qquad\qquad $\mathbf{r}_3 - \mathbf{r}_2 \rightarrow \mathbf{r}_3$

$\begin{bmatrix} 1 & 0 & -1 & | & 2 & -1 & 0 \\ 0 & 1 & 2 & | & -1 & 1 & 0 \\ 0 & 0 & 1 & | & -1 & 1 & -1 \end{bmatrix} \rightarrow \begin{bmatrix} 1 & 0 & 0 & | & 1 & 0 & -1 \\ 0 & 1 & 0 & | & 1 & -1 & 2 \\ 0 & 0 & 1 & | & -1 & 1 & -1 \end{bmatrix}$

$\mathbf{r}_2 - 2\mathbf{r}_3 \rightarrow \mathbf{r}_2$ \qquad\qquad reduced row echelon form
$\mathbf{r}_1 + \mathbf{r}_3 \rightarrow \mathbf{r}_1$

Thus $A^{-1} = \begin{bmatrix} 1 & 0 & -1 \\ 1 & -1 & 2 \\ -1 & 1 & -1 \end{bmatrix}$.

(b) $\begin{bmatrix} 1 & 1 & 1 & 1 & | & 1 & 0 & 0 & 0 \\ 1 & 2 & -1 & 2 & | & 0 & 1 & 0 & 0 \\ 1 & -1 & 2 & 1 & | & 0 & 0 & 1 & 0 \\ 1 & 3 & 3 & 2 & | & 0 & 0 & 0 & 1 \end{bmatrix} \rightarrow \begin{bmatrix} 1 & 1 & 1 & 1 & | & 1 & 0 & 0 & 0 \\ 0 & 1 & -2 & 1 & | & -1 & 1 & 0 & 0 \\ 0 & -2 & 1 & 0 & | & -1 & 0 & 1 & 0 \\ 0 & 2 & 2 & 1 & | & -1 & 0 & 0 & 1 \end{bmatrix} \rightarrow \begin{bmatrix} 1 & 0 & 3 & 0 & | & 2 & -1 & 0 & 0 \\ 0 & 1 & -2 & 1 & | & -1 & 1 & 0 & 0 \\ 0 & 0 & -3 & 2 & | & -3 & 2 & 1 & 0 \\ 0 & 0 & 6 & -1 & | & 1 & -2 & 0 & 1 \end{bmatrix} \rightarrow$

$\mathbf{r}_2 - \mathbf{r}_1 \rightarrow \mathbf{r}_2$ \qquad\qquad $\mathbf{r}_1 - \mathbf{r}_2 \rightarrow \mathbf{r}_1$ \qquad\qquad $-\frac{1}{3}\mathbf{r}_3 \rightarrow \mathbf{r}_3$
$\mathbf{r}_3 - \mathbf{r}_1 \rightarrow \mathbf{r}_3$ \qquad\qquad $\mathbf{r}_3 + 2\mathbf{r}_2 \rightarrow \mathbf{r}_3$
$\mathbf{r}_4 - \mathbf{r}_1 \rightarrow \mathbf{r}_4$ \qquad\qquad $\mathbf{r}_4 - 2\mathbf{r}_2 \rightarrow \mathbf{r}_4$

$$
\begin{bmatrix}
1 & 0 & 3 & 0 & 2 & -1 & 0 & 0 \\
0 & 1 & -2 & 1 & -1 & 1 & 0 & 0 \\
0 & 0 & 1 & -\frac{2}{3} & 1 & -\frac{2}{3} & -\frac{1}{3} & 0 \\
0 & 0 & 6 & -1 & 1 & -2 & 0 & 1
\end{bmatrix}
\rightarrow
\begin{bmatrix}
1 & 0 & 0 & 2 & -1 & 1 & 1 & 0 \\
0 & 1 & 0 & -\frac{1}{3} & -1 & -\frac{1}{3} & -\frac{2}{3} & 0 \\
0 & 0 & 1 & -\frac{2}{3} & 1 & -\frac{2}{3} & -\frac{1}{3} & 0 \\
0 & 0 & 0 & 3 & -5 & 2 & 2 & 1
\end{bmatrix}
\rightarrow
\begin{bmatrix}
1 & 0 & 0 & 0 & \frac{7}{3} & -\frac{1}{3} & -\frac{1}{3} & -\frac{2}{3} \\
0 & 1 & 0 & 0 & \frac{4}{9} & -\frac{1}{9} & -\frac{4}{9} & \frac{1}{9} \\
0 & 0 & 1 & 0 & -\frac{1}{9} & -\frac{2}{9} & \frac{1}{9} & \frac{2}{9} \\
0 & 0 & 0 & 1 & -\frac{5}{3} & \frac{2}{3} & \frac{2}{3} & \frac{1}{3}
\end{bmatrix}
$$

$$
\begin{array}{l}
\mathbf{r}_1 \rightarrow \mathbf{r}_1 - 3\mathbf{r}_3 \\
\mathbf{r}_2 + 2\mathbf{r}_3 \rightarrow \mathbf{r}_2 \\
\mathbf{r}_4 - 6\mathbf{r}_3 \rightarrow \mathbf{r}_4
\end{array}
\qquad
\begin{array}{l}
\mathbf{r}_4 \rightarrow \frac{1}{3}\mathbf{r}_4 \\
\mathbf{r}_1 - 2\mathbf{r}_4 \rightarrow \mathbf{r}_1 \\
\mathbf{r}_2 + \frac{1}{3}\mathbf{r}_4 \rightarrow \mathbf{r}_2 \\
\mathbf{r}_3 + \frac{2}{3}\mathbf{r}_4 \rightarrow \mathbf{r}_3
\end{array}
\qquad
\text{reduced row echelon form}
$$

Thus $A^{-1} = \begin{bmatrix} \frac{7}{3} & -\frac{1}{3} & -\frac{1}{3} & -\frac{2}{3} \\ \frac{4}{9} & -\frac{1}{9} & -\frac{4}{9} & \frac{1}{9} \\ -\frac{1}{9} & -\frac{2}{9} & \frac{1}{9} & \frac{2}{9} \\ -\frac{5}{3} & \frac{2}{3} & \frac{2}{3} & \frac{1}{3} \end{bmatrix}.$

(c) $\begin{bmatrix}
1 & 1 & 1 & 1 & 1 & 0 & 0 & 0 \\
1 & 3 & 1 & 2 & 0 & 1 & 0 & 0 \\
1 & 2 & -1 & 1 & 0 & 0 & 1 & 0 \\
5 & 9 & 1 & 6 & 0 & 0 & 0 & 1
\end{bmatrix}
\rightarrow
\begin{bmatrix}
1 & 1 & 1 & 1 & 1 & 0 & 0 & 0 \\
0 & 2 & 0 & 1 & -1 & 1 & 0 & 0 \\
0 & 1 & -2 & 0 & -1 & 0 & 1 & 0 \\
0 & 4 & -4 & 1 & -5 & 0 & 0 & 1
\end{bmatrix}
\rightarrow
\begin{bmatrix}
1 & 1 & 1 & 1 & 1 & 0 & 0 & 0 \\
0 & 1 & -2 & 0 & -1 & 0 & 1 & 0 \\
0 & 2 & 0 & 1 & -1 & 1 & 0 & 0 \\
0 & 4 & -4 & 1 & -5 & 0 & 0 & 1
\end{bmatrix}
\rightarrow$

$$
\begin{array}{l}
\mathbf{r}_2 - \mathbf{r}_1 \rightarrow \mathbf{r}_2 \\
\mathbf{r}_3 - \mathbf{r}_1 \rightarrow \mathbf{r}_3 \\
\mathbf{r}_4 + 5\mathbf{r}_1 \rightarrow \mathbf{r}_4
\end{array}
\qquad
\begin{array}{l}
\mathbf{r}_3 \leftrightarrow \mathbf{r}_2
\end{array}
\qquad
\begin{array}{l}
\mathbf{r}_1 - \mathbf{r}_2 \rightarrow \mathbf{r}_1 \\
\mathbf{r}_3 - 2\mathbf{r}_2 \rightarrow \mathbf{r}_3 \\
\mathbf{r}_4 - 4\mathbf{r}_2 \rightarrow \mathbf{r}_4
\end{array}
$$

$$
\begin{bmatrix}
1 & 0 & 3 & 1 & 2 & 0 & -1 & 0 \\
0 & 1 & -2 & 0 & -1 & 0 & 1 & 0 \\
0 & 0 & 4 & 1 & 1 & 1 & -2 & 0 \\
0 & 0 & 4 & 1 & -1 & 0 & -4 & 1
\end{bmatrix}
\rightarrow
\begin{bmatrix}
1 & 0 & 3 & 1 & 2 & 0 & -1 & 0 \\
0 & 1 & -2 & 0 & -1 & 0 & 1 & 0 \\
0 & 0 & 4 & 1 & 1 & 1 & -2 & 0 \\
0 & 0 & 0 & 0 & -2 & -1 & -2 & 1
\end{bmatrix}
$$

$$
\mathbf{r}_4 - \mathbf{r}_3 \rightarrow \mathbf{r}_4
$$

Thus the matrix $A = \begin{bmatrix} 1 & 1 & 1 & 1 \\ 1 & 3 & 1 & 2 \\ 1 & 2 & -1 & 1 \\ 5 & 9 & 1 & 6 \end{bmatrix}$ is row equivalent to a matrix with a row of zeros. By Theorem 2.10, A is singular.

(d) $\begin{bmatrix}
1 & 2 & 1 & 1 & 0 & 0 \\
1 & 3 & 2 & 0 & 1 & 0 \\
1 & 0 & 1 & 0 & 0 & 1
\end{bmatrix}
\rightarrow
\begin{bmatrix}
1 & 2 & 1 & 1 & 0 & 0 \\
0 & 1 & 1 & -1 & 1 & 0 \\
0 & -2 & 0 & -1 & 0 & 1
\end{bmatrix}
\rightarrow
\begin{bmatrix}
1 & 0 & -1 & 3 & -2 & 0 \\
0 & 1 & 1 & -1 & 1 & 0 \\
0 & 0 & 2 & -3 & 2 & 1
\end{bmatrix}
\rightarrow$

$$
\begin{array}{l}
\mathbf{r}_2 - \mathbf{r}_1 \rightarrow \mathbf{r}_2 \\
\mathbf{r}_3 - \mathbf{r}_1 \rightarrow \mathbf{r}_3
\end{array}
\qquad
\begin{array}{l}
\mathbf{r}_1 - 2\mathbf{r}_2 \rightarrow \mathbf{r}_1 \\
\mathbf{r}_3 + 2\mathbf{r}_2 \rightarrow \mathbf{r}_3
\end{array}
\qquad
\begin{array}{l}
\frac{1}{2}\mathbf{r}_3 \rightarrow \mathbf{r}_3
\end{array}
$$

$$
\begin{bmatrix}
1 & 0 & -1 & 3 & -2 & 0 \\
0 & 1 & 1 & -1 & 1 & 0 \\
0 & 0 & 1 & -\frac{3}{2} & 1 & \frac{1}{2}
\end{bmatrix}
\rightarrow
\begin{bmatrix}
1 & 0 & 0 & \frac{3}{2} & -1 & \frac{1}{2} \\
0 & 1 & 0 & \frac{1}{2} & 0 & -\frac{1}{2} \\
0 & 0 & 1 & -\frac{3}{2} & 1 & \frac{1}{2}
\end{bmatrix}
$$

$$
\begin{array}{l}
\mathbf{r}_2 - \mathbf{r}_3 \rightarrow \mathbf{r}_2 \\
\mathbf{r}_1 + \mathbf{r}_3 \rightarrow \mathbf{r}_1
\end{array}
\qquad
\text{reduced row echelon form}
$$

Thus $A^{-1} = \begin{bmatrix} \frac{3}{2} & -1 & \frac{1}{2} \\ \frac{1}{2} & 0 & -\frac{1}{2} \\ -\frac{3}{2} & 1 & \frac{1}{2} \end{bmatrix}.$

(e) $\begin{bmatrix}
1 & 2 & 2 & 1 & 0 & 0 \\
1 & 3 & 1 & 0 & 1 & 0 \\
1 & 1 & 3 & 0 & 0 & 1
\end{bmatrix}
\rightarrow
\begin{bmatrix}
1 & 2 & 2 & 1 & 0 & 0 \\
0 & 1 & -1 & -1 & 1 & 0 \\
0 & -1 & 1 & -1 & 0 & 1
\end{bmatrix}
\rightarrow
\begin{bmatrix}
1 & 2 & 2 & 1 & 0 & 0 \\
0 & 1 & -1 & -1 & 1 & 0 \\
0 & 0 & 0 & -2 & 1 & 1
\end{bmatrix}$

$$
\begin{array}{l}
\mathbf{r}_2 - \mathbf{r}_1 \rightarrow \mathbf{r}_2 \\
\mathbf{r}_3 - \mathbf{r}_1 \rightarrow \mathbf{r}_3
\end{array}
\qquad
\begin{array}{l}
\mathbf{r}_3 + \mathbf{r}_2 \rightarrow \mathbf{r}_3
\end{array}
$$

Thus the matrix $A = \begin{bmatrix} 1 & 2 & 2 \\ 1 & 3 & 1 \\ 1 & 1 & 3 \end{bmatrix}$ is row equivalent to a matrix with a row of zeros. By Theorem 2.10, A is singular.

13. Let $A = \begin{bmatrix} 1 & 2 \\ 3 & 4 \end{bmatrix}$. Following Theorem 2.8 we row reduce A to I_2 forming the corresponding elementary matrices for the row operations employed in the reduction.

Step 1. Apply $\mathbf{r}_2 - 3\mathbf{r}_1 \to \mathbf{r}_2$:

$$\begin{bmatrix} 1 & 2 \\ 3 & 4 \end{bmatrix} \quad \to \quad \begin{bmatrix} 1 & 2 \\ 0 & -2 \end{bmatrix} = E_1 A, \quad \text{where } E_1 = \begin{bmatrix} 1 & 0 \\ -3 & 1 \end{bmatrix}.$$

Step 2. Apply $-\frac{1}{2}|\mathbf{r}_2 \to \mathbf{r}_2$:

$$\begin{bmatrix} 1 & 2 \\ 0 & -2 \end{bmatrix} \quad \to \quad \begin{bmatrix} 1 & 2 \\ 0 & 1 \end{bmatrix} = E_2(E_1 A), \quad \text{where } E_2 = \begin{bmatrix} 1 & 0 \\ 0 & -\frac{1}{2} \end{bmatrix}.$$

Step 3. Apply $\mathbf{r}_1 - 2\mathbf{r}_2 \to \mathbf{r}_1$:

$$\begin{bmatrix} 1 & 2 \\ 0 & 1 \end{bmatrix} \quad \to \quad \begin{bmatrix} 1 & 0 \\ 0 & 1 \end{bmatrix} = E_3(E_2 E_1 A) = I_2, \quad \text{where } E_3 = \begin{bmatrix} 1 & -2 \\ 0 & 1 \end{bmatrix}.$$

Thus $E_3 E_2 E_1 A = I_2$ and it follows that $A = E_1^{-1} E_2^{-1} E_3^{-1} I_2$. It is not hard to show that

$$E_1^{-1} = \begin{bmatrix} 1 & 0 \\ 3 & 1 \end{bmatrix}, \quad E_2^{-1} = \begin{bmatrix} 1 & 0 \\ 0 & -2 \end{bmatrix}, \quad E_3^{-1} = \begin{bmatrix} 1 & 2 \\ 0 & 1 \end{bmatrix}$$

so that A is the product of the elementary matrices

$$A = \begin{bmatrix} 1 & 0 \\ 3 & 1 \end{bmatrix} \begin{bmatrix} 1 & 0 \\ 0 & -2 \end{bmatrix} \begin{bmatrix} 1 & 2 \\ 0 & 1 \end{bmatrix}.$$

15. Let $A^{-1} = \begin{bmatrix} 4 & 2 \\ 1 & 1 \end{bmatrix}$. Then to find A we determine the reduced row echelon form of the augmented matrix $\left[A^{-1} \mid I_2 \right]$

$$\begin{bmatrix} 4 & 2 & 1 & 0 \\ 1 & 1 & 0 & 1 \end{bmatrix} \to \begin{bmatrix} 1 & 1 & 0 & 1 \\ 4 & 2 & 1 & 0 \end{bmatrix} \to \begin{bmatrix} 1 & 1 & 0 & 1 \\ 0 & -2 & 1 & -4 \end{bmatrix} \to \begin{bmatrix} 1 & 1 & 0 & 1 \\ 0 & 1 & -\frac{1}{2} & 2 \end{bmatrix} \to \begin{bmatrix} 1 & 0 & \frac{1}{2} & -1 \\ 0 & 1 & -\frac{1}{2} & 2 \end{bmatrix}$$

$$\quad \mathbf{r}_2 \leftrightarrow \mathbf{r}_1 \qquad \mathbf{r}_2 - 4\mathbf{r}_1 \to \mathbf{r}_2 \qquad -\frac{1}{2}\mathbf{r}_2 \to \mathbf{r}_2 \qquad \mathbf{r}_1 - \mathbf{r}_2 \to \mathbf{r}_1 \qquad \text{reduced row} \\ \text{echelon form}$$

Thus $A = \begin{bmatrix} \frac{1}{2} & -1 \\ -\frac{1}{2} & 2 \end{bmatrix}$.

17. Following Theorem 2.9, we check to see if the coefficient matrix A is singular. Row reduce A. If the result is not I_3, then A is singular.

(a) $A = \begin{bmatrix} 1 & 2 & 3 \\ 0 & 2 & 2 \\ 1 & 2 & 3 \end{bmatrix} \to \begin{bmatrix} 1 & 2 & 3 \\ 0 & 2 & 2 \\ 0 & 0 & 0 \end{bmatrix}$

$\quad \mathbf{r}_3 - \mathbf{r}_1 \to \mathbf{r}_3$

Thus matrix A is row equivalent to a matrix with a zero row and hence cannot be row equivalent to I_3. We conclude that A is singular and that the homogeneous system has a nontrivial solution.

(b) $A = \begin{bmatrix} 2 & 1 & -1 \\ 1 & -2 & -3 \\ -3 & -1 & 2 \end{bmatrix} \rightarrow \begin{bmatrix} 1 & -2 & -3 \\ 2 & 1 & -1 \\ -3 & -1 & 2 \end{bmatrix} \rightarrow \begin{bmatrix} 1 & -2 & -3 \\ 0 & 5 & 5 \\ 0 & -7 & -7 \end{bmatrix} \rightarrow$

$\quad\quad \mathbf{r}_2 \leftrightarrow \mathbf{r}_1 \quad\quad\quad\quad \mathbf{r}_2 - 2\mathbf{r}_1 \rightarrow \mathbf{r}_2 \quad\quad \frac{1}{5}\mathbf{r}_2 \rightarrow \mathbf{r}_2$

$\begin{bmatrix} 1 & -2 & -3 \\ 0 & 1 & 1 \\ 0 & -7 & -7 \end{bmatrix} \rightarrow \begin{bmatrix} 1 & -2 & -3 \\ 0 & 1 & 1 \\ 0 & 0 & 0 \end{bmatrix}$

$\mathbf{r}_3 + 7\mathbf{r}_2 \rightarrow \mathbf{r}_3$

Thus matrix A is row equivalent to a matrix with a zero row and hence cannot be row equivalent to I_3. We conclude that A is singular and that the homogeneous system has a nontrivial solution.

(c) $A = \begin{bmatrix} 3 & 1 & 3 \\ -2 & 2 & -4 \\ 2 & -3 & 5 \end{bmatrix} \rightarrow \begin{bmatrix} 1 & 3 & -1 \\ -2 & 2 & -4 \\ 2 & -3 & 5 \end{bmatrix} \rightarrow \begin{bmatrix} 1 & 3 & -1 \\ 0 & 8 & -6 \\ 0 & -9 & 7 \end{bmatrix} \rightarrow$

$\quad\quad \mathbf{r}_1 + \mathbf{r}_2 \rightarrow \mathbf{r}_1 \quad\quad\quad \mathbf{r}_2 + 2\mathbf{r}_1 \rightarrow \mathbf{r}_2 \quad\quad \frac{1}{8}\mathbf{r}_2 \rightarrow \mathbf{r}_2$
$\quad\quad\quad\quad\quad\quad\quad\quad\quad\quad \mathbf{r}_3 - 2\mathbf{r}_1 \rightarrow \mathbf{r}_3$

$\begin{bmatrix} 1 & 3 & -1 \\ 0 & 1 & -\frac{6}{8} \\ 0 & -9 & 7 \end{bmatrix} \rightarrow \begin{bmatrix} 1 & 3 & -1 \\ 0 & 1 & -\frac{6}{8} \\ 0 & 0 & \frac{2}{8} \end{bmatrix}$

$\mathbf{r}_3 + 9\mathbf{r}_2 \rightarrow \mathbf{r}_3$

It follows that multiplying the third row by 4 gives a 1 in the $(3,3)$ entry and that we can continue to use row operations to obtain I_3. Thus this homogeneous system has only the trivial solution.

19. Let $A = \begin{bmatrix} 1 & 1 & 0 \\ 1 & 0 & 0 \\ 1 & 2 & a \end{bmatrix}$. We form the partitioned matrix $\begin{bmatrix} A \mid I_3 \end{bmatrix}$ and row reduce it in order to determine the values of a for which we obtain the form $\begin{bmatrix} I_3 \mid B \end{bmatrix}$. Then $B = A^{-1}$.

$\begin{bmatrix} 1 & 1 & 0 & 1 & 0 & 0 \\ 1 & 0 & 0 & 0 & 1 & 0 \\ 1 & 2 & a & 0 & 0 & 1 \end{bmatrix} \rightarrow \begin{bmatrix} 1 & 0 & 0 & 0 & 1 & 0 \\ 1 & 1 & 0 & 1 & 0 & 0 \\ 1 & 2 & a & 0 & 0 & 1 \end{bmatrix} \rightarrow \begin{bmatrix} 1 & 0 & 0 & 0 & 1 & 0 \\ 0 & 1 & 0 & 1 & -1 & 0 \\ 0 & 2 & a & 0 & -1 & 1 \end{bmatrix} \rightarrow \begin{bmatrix} 1 & 0 & 0 & 0 & 1 & 0 \\ 0 & 1 & 0 & 1 & -1 & 0 \\ 0 & 0 & a & -2 & 1 & 1 \end{bmatrix}$

$\quad\quad \mathbf{r}_2 \rightarrow \mathbf{r}_1 \quad\quad\quad\quad \mathbf{r}_2 - \mathbf{r}_1 \rightarrow \mathbf{r}_2 \quad\quad\quad \mathbf{r}_3 - 2\mathbf{r}_2 \rightarrow \mathbf{r}_3$
$\quad\quad\quad\quad\quad\quad\quad\quad \mathbf{r}_3 - \mathbf{r}_1 \rightarrow \mathbf{r}_3$

We see that for $a \neq 0$, A will be row equivalent to I_3, hence nonsingular. Thus for $a \neq 0$ we can multiply row 3 by $\frac{1}{a}$ to obtain

$$A^{-1} = \begin{bmatrix} 0 & 1 & 0 \\ 1 & -1 & 0 \\ -\dfrac{2}{a} & \dfrac{1}{a} & \dfrac{1}{a} \end{bmatrix}.$$

21. This follows directly from Exercise 19 of Section 2.2 and Corollary 2.2. To show that

$$A^{-1} = \frac{1}{ad - bc} \begin{bmatrix} d & -b \\ -c & a \end{bmatrix}$$

we proceed as follows:

$$\frac{1}{ad - bc} \begin{bmatrix} d & -b \\ -c & a \end{bmatrix} \begin{bmatrix} a & b \\ c & d \end{bmatrix} = \frac{1}{ad - bc} \begin{bmatrix} ad - bc & db - bd \\ -ca + ac & -bc + ad \end{bmatrix} = \begin{bmatrix} 1 & 0 \\ 0 & 1 \end{bmatrix}.$$

23. We are asked to prove an "if and only if" statement. Hence there are two things that must be proved.

(a) If A and B are row equivalent then there exists a nonsingular matrix P such that $B = PA$.

Proof: From Theorem 2.6 there exist elementary matrices E_1, E_2, ..., E_k such that $B = E_k \cdots E_2 E_1 A$. By Theorem 2.7 each elementary matrix is nonsingular and the product of nonsingular matrices is nonsingular by Theorem 2.8. Thus let $P = E_k \cdots E_2 E_1$ and we have a nonsingular matrix P such that $B = PA$.

(b) If there exists a nonsingular matrix P such that $B = PA$ then A and B are row equivalent.

Proof: By Theorem 2.8 the nonsingular matrix P can be expressed as a product of elementary matrices. Hence by Theorem 2.6 it follows that A and B are row equivalent.

25. (a) The matrix B is nonsingular.

Proof: We develop a proof by contradiction. First assume that B is singular. We will show that this is not possible since AB is nonsingular. By Theorem 2.9, if B is singular there is an $n \times 1$ matrix $\mathbf{x} \neq \mathbf{0}$ such that $B\mathbf{x} = \mathbf{0}$. Then multiplying both sides on the left by A gives the homogeneous system $AB\mathbf{x} = \mathbf{0}$. But this system has a nontrivial solution and hence by Theorem 2.9 the coefficient matrix AB must be singular. However this is a contradiction of the hypothesis that AB is nonsingular. Thus we have a contradiction. It follows then that our assumption that B is singular is incorrect. Therefore B must be nonsingular.

(b) The matrix A is nonsingular.

Proof: As in part (a), we develop a proof by contradiction. Assume that A is singular. We will show that this is not possible since AB is nonsingular. By Theorem 2.9, if A is singular there is an $n \times 1$ matrix $\mathbf{y} \neq \mathbf{0}$ such that $A\mathbf{y} = \mathbf{0}$. Since B is nonsingular, there exists an $\mathbf{x} \neq \mathbf{0}$ such that $B\mathbf{x} = \mathbf{y}$, namely $\mathbf{x} = B^{-1}\mathbf{y}$. Multiplying both sides of $B\mathbf{x} = \mathbf{y}$ by A on the left gives $AB\mathbf{x} = A\mathbf{y} = \mathbf{0}$. Thus the homogeneous system $AB\mathbf{x} = \mathbf{0}$ has a nontrivial solution. But by the same argument given in the first case this is a contradiction of the hypothesis that AB is nonsingular. It follows then that our assumption that A is singular is incorrect. Therefore A must be nonsingular.

27. We are asked to prove an "if and only if" statement. Hence there are two things that must be proved.

(a) If A is row equivalent to B, then A^T is column equivalent to B^T.

Proof: By Exercise 23 matrix A is row equivalent to B if and only if there exists a nonsingular matrix P such that $B = PA$. It follows that P^T is also nonsingular and by properties of transposes that $B^T = A^T P^T$. Since P^T is a product of elementary matrices which multiply A^T on the right to give B^T we have that A^T is column equivalent to B^T.

(b) If A^T is column equivalent to B^T, then A is row equivalent to B.

Proof: Since A^T is column equivalent to B^T there exists a nonsingular matrix Q such that $B^T = A^T Q$. We parallel the argument given above in part (a) to conclude that A is row equivalent to B.

29. (a) No; it is not true that $(A + B)^{-1} = A^{-1} + B^{-1}$. We verify this by example. Let

$$A = \begin{bmatrix} 1 & 0 \\ 0 & 0 \end{bmatrix} \quad \text{and} \quad B = \begin{bmatrix} 0 & 0 \\ 0 & 1 \end{bmatrix}.$$

Then $A + B = I_2$ and $(A + B)^{-1} = I_2$, but neither A nor B is nonsingular.

(b) Yes. If A is nonsingular and $c \neq 0$, then $(cA)\left(\frac{1}{c}A^{-1}\right) = c\left(\frac{1}{c}\right)AA^{-1} = 1 \cdot I_n = I_n$ and therefore $(cA)^{-1} = \frac{1}{c}A^{-1}$.

31. We consider the case that A is nonsingular and upper triangular. A similar argument can be given for A lower triangular. By Theorem 2.8 A is a product of elementary matrices which are the inverses of the elementary matrices that "reduce" A to I_n. That is, $A = E_1^{-1} \cdots E_k^{-1}$. The elementary matrix E_i will be upper triangular since it is used to introduce zeros into the upper triangular part of A in the reduction process. The inverse of E_i is an elementary matrix of the same type and also an upper triangular matrix. Since the product of upper triangular matrices is upper triangular and since $A^{-1} = E_k \cdots E_1$ we conclude that A^{-1} is upper triangular.

Section 2.4, p. 129

1. (a) A is equivalent to itself: the sequence of operations is the empty sequence.

   (b) Each elementary row and column operation of types I, II or III has a corresponding inverse operation of the same type which "undoes" the effect of the original operation. For example, the inverse of the operation "add d times row r of A to row s of A" is "subtract d times row r of A from row s of A." Since B is assumed equivalent to A, there is a sequence of elementary row and column operations which gets from A to B. Take those operations in the reverse order, and for each operation do its inverse, and that takes B to A. Thus A is equivalent to B.

   (c) Follow the operations which take A to B with those which take B to C.

3. We apply appropriate row or column operations to obtain the form described in Theorem 2.12. In parts (a) and (b) we obtain the reduced row echelon form before applying column operations.

(a) $\begin{bmatrix} 1 & 2 & 3 & -1 \\ 1 & 0 & 2 & 3 \\ 3 & 4 & 8 & 1 \end{bmatrix} \rightarrow \begin{bmatrix} 1 & 2 & 3 & -1 \\ 0 & -2 & -1 & 4 \\ 0 & -2 & -1 & 4 \end{bmatrix} \rightarrow \begin{bmatrix} 1 & 0 & 2 & 3 \\ 0 & 1 & \frac{1}{2} & 2 \\ 0 & 0 & 0 & 0 \end{bmatrix} \rightarrow \begin{bmatrix} 1 & 0 & 0 & 0 \\ 0 & 1 & 0 & 0 \\ 0 & 0 & 0 & 0 \end{bmatrix}$

$\begin{array}{ll} \mathbf{r}_2 - \mathbf{r}_1 \rightarrow \mathbf{r}_2 \\ \mathbf{r}_3 - 3\mathbf{r}_1 \rightarrow \mathbf{r}_3 \end{array}$ $\quad \begin{array}{l} \mathbf{r}_3 - \mathbf{r}_2 \rightarrow \mathbf{r}_3 \\ -\frac{1}{2}\mathbf{r}_2 \rightarrow \mathbf{r}_2 \\ \mathbf{r}_1 - 2\mathbf{r}_2 \rightarrow \mathbf{r}_1 \end{array}$ $\quad \begin{array}{l} \mathbf{c}_3 - 2\mathbf{c}_1 \rightarrow \mathbf{c}_3 \\ \mathbf{c}_4 - 3\mathbf{c}_1 \rightarrow \mathbf{c}_4 \\ \mathbf{c}_3 - \frac{1}{2}\mathbf{c}_2 \rightarrow \mathbf{c}_3 \\ \mathbf{c}_4 - 2\mathbf{c}_2 \rightarrow \mathbf{c}_4 \end{array}$

(b) $\begin{bmatrix} 3 & 4 & 1 \\ 1 & 2 & -2 \\ 5 & 6 & 4 \\ 5 & 8 & -1 \end{bmatrix} \rightarrow \begin{bmatrix} 1 & 2 & -2 \\ 3 & 4 & 1 \\ 5 & 6 & 4 \\ 5 & 8 & -1 \end{bmatrix} \rightarrow \begin{bmatrix} 1 & 2 & -2 \\ 0 & -2 & 7 \\ 0 & -4 & 14 \\ 0 & -2 & 9 \end{bmatrix} \rightarrow$

$\mathbf{r}_2 \leftrightarrow \mathbf{r}_1$ $\quad \begin{array}{l} \mathbf{r}_2 - 3\mathbf{r}_1 \rightarrow \mathbf{r}_2 \\ \mathbf{r}_3 - 5\mathbf{r}_3 \rightarrow \mathbf{r}_3 \\ \mathbf{r}_4 - 5\mathbf{r}_1 \rightarrow \mathbf{r}_4 \end{array}$ $\quad \begin{array}{l} \mathbf{r}_1 + \mathbf{r}_2 \rightarrow \mathbf{r}_1 \\ \mathbf{r}_3 - 2\mathbf{r}_2 \rightarrow \mathbf{r}_3 \\ \mathbf{r}_4 - \mathbf{r}_2 \rightarrow \mathbf{r}_4 \end{array}$

$\begin{bmatrix} 1 & 0 & 5 \\ 0 & -2 & 7 \\ 0 & 0 & 0 \\ 0 & 0 & 2 \end{bmatrix} \rightarrow \begin{bmatrix} 1 & 0 & 5 \\ 0 & 1 & -\frac{7}{2} \\ 0 & 0 & 1 \\ 0 & 0 & 0 \end{bmatrix} \rightarrow \begin{bmatrix} 1 & 0 & 0 \\ 0 & 1 & 0 \\ 0 & 0 & 1 \\ 0 & 0 & 0 \end{bmatrix}$

$\begin{array}{l} -\frac{1}{2}\mathbf{r}_2 \rightarrow \mathbf{r}_2 \\ \frac{1}{2}\mathbf{r}_4 \rightarrow \mathbf{r}_4 \\ \mathbf{r}_4 \rightarrow \mathbf{r}_3 \end{array}$ $\quad \begin{array}{l} \mathbf{r}_2 + \frac{7}{2}\mathbf{r}_3 \rightarrow \mathbf{r}_2 \\ \mathbf{r}_1 - 5\mathbf{r}_3 \rightarrow \mathbf{r}_1 \end{array}$

(c) $\begin{bmatrix} 2 & 3 & 4 & -1 \\ 1 & 2 & 1 & -1 \\ 2 & -1 & 1 & 1 \\ 4 & 2 & 5 & 0 \\ 4 & 3 & 3 & -1 \end{bmatrix} \rightarrow \begin{bmatrix} 1 & 2 & 1 & -1 \\ 2 & 3 & 4 & -1 \\ 2 & -1 & 1 & 1 \\ 4 & 2 & 5 & 0 \\ 4 & 3 & 3 & -1 \end{bmatrix} \rightarrow \begin{bmatrix} 1 & 2 & 1 & -1 \\ 0 & -1 & 2 & 1 \\ 0 & -5 & -1 & 3 \\ 0 & -6 & 1 & 4 \\ 0 & -5 & -1 & 3 \end{bmatrix} \rightarrow$

$\mathbf{r}_2 \leftrightarrow \mathbf{r}_1$ $\quad \begin{array}{l} \mathbf{r}_2 - 2\mathbf{r}_1 \rightarrow \mathbf{r}_2 \\ \mathbf{r}_3 - 2\mathbf{r}_1 \rightarrow \mathbf{r}_3 \\ \mathbf{r}_4 - 4\mathbf{r}_1 \rightarrow \mathbf{r}_4 \\ \mathbf{r}_5 - 4\mathbf{r}_1 \rightarrow \mathbf{r}_5 \end{array}$ $\quad \begin{array}{l} \mathbf{r}_1 + 2\mathbf{r}_2 \rightarrow \mathbf{r}_1 \\ \mathbf{r}_3 - 5\mathbf{r}_2 \rightarrow \mathbf{r}_4 \\ \mathbf{r}_4 - 6\mathbf{r}_2 \rightarrow \mathbf{r}_4 \\ \mathbf{r}_5 - 5\mathbf{r}_2 \rightarrow \mathbf{r}_5 \end{array}$

$$
\begin{bmatrix} 1 & 0 & 5 & 1 \\ 0 & 1 & -2 & -1 \\ 0 & 0 & -11 & -2 \\ 0 & 0 & -11 & -2 \\ 0 & 0 & -11 & -2 \end{bmatrix} \rightarrow
\begin{bmatrix} 1 & 0 & 0 & 0 \\ 0 & 1 & 0 & 0 \\ 0 & 0 & -11 & -2 \\ 0 & 0 & -11 & -2 \\ 0 & 0 & -11 & -2 \end{bmatrix} \rightarrow
\begin{bmatrix} 1 & 0 & 0 & 0 \\ 0 & 1 & 0 & 0 \\ 0 & 0 & 1 & -\frac{2}{11} \\ 0 & 0 & 0 & 0 \\ 0 & 0 & 0 & 0 \end{bmatrix} \rightarrow
\begin{bmatrix} 1 & 0 & 0 & 0 \\ 0 & 1 & 0 & 0 \\ 0 & 0 & 1 & 0 \\ 0 & 0 & 0 & 0 \\ 0 & 0 & 0 & 0 \end{bmatrix}
$$

$$
\begin{array}{c}
\mathbf{c}_3 - 5\mathbf{c}_1 \to \mathbf{c}_3 \\
\mathbf{c}_4 - \mathbf{c}_1 \to \mathbf{c}_4 \\
\mathbf{c}_3 + 2\mathbf{c}_2 \to \mathbf{c}_3 \\
\mathbf{c}_4 + \mathbf{c}_2 \to \mathbf{c}_4
\end{array}
\qquad
\begin{array}{c}
\mathbf{r}_4 - \mathbf{r}_3 \to \mathbf{r}_4 \\
\mathbf{r}_5 - \mathbf{r}_3 \to \mathbf{r}_5 \\
-\frac{1}{11}\mathbf{r}_3 \to \mathbf{r}_3
\end{array}
\qquad
\mathbf{c}_4 + \tfrac{2}{11}\mathbf{c}_3 \to \mathbf{c}_4
$$

(d)
$$
\begin{bmatrix} 1 & 2 & 3 \\ 1 & -1 & 0 \\ 0 & 1 & 2 \end{bmatrix} \rightarrow
\begin{bmatrix} 1 & 2 & 3 \\ 0 & -3 & -3 \\ 0 & 1 & 2 \end{bmatrix} \rightarrow
\begin{bmatrix} 1 & 2 & 3 \\ 0 & 1 & 2 \\ 0 & -3 & -3 \end{bmatrix} \rightarrow
\begin{bmatrix} 1 & 0 & -1 \\ 0 & 1 & 2 \\ 0 & 0 & 3 \end{bmatrix} \rightarrow
\begin{bmatrix} 1 & 0 & 0 \\ 0 & 1 & 0 \\ 0 & 0 & 1 \end{bmatrix}
$$

$$
\mathbf{r}_2 - \mathbf{r}_1 \to \mathbf{r}_2
\qquad
\mathbf{r}_3 \leftrightarrow \mathbf{r}_2
\qquad
\begin{array}{c}
\mathbf{r}_1 - 2\mathbf{r}_2 \to \mathbf{r}_1 \\
\mathbf{r}_3 + 3\mathbf{r}_2 \to \mathbf{r}_3
\end{array}
\qquad
\begin{array}{c}
\mathbf{c}_3 + \mathbf{c}_1 \to \mathbf{c}_3 \\
\mathbf{c}_3 - 2\mathbf{c}_2 \to \mathbf{c}_3 \\
\frac{1}{3}\mathbf{r}_3 \to \mathbf{r}_3
\end{array}
$$

5. We are asked to prove an "if and only if" statement. Hence two things must be proved.

   (a) If A and B are equivalent then $B = PAQ$ for some nonsingular matrices P and Q.

   Proof: From Definition 2.5, since A and B are equivalent there exist a sequence of elementary row operations denoted $E_1, \ldots, E_k$ or elementary column operations denoted $F_1, \ldots, F_j$ (or both) such that $B = E_1 \cdots E_k A F_1 \cdots F_j$. Each elementary matrix is nonsingular so $P = E_1 \cdots E_k$ and $Q = F_1 \cdots F_j$ are nonsingular matrices such that $B = PAQ$.

   (b) If $B = PAQ$ for some nonsingular matrices P and Q then A and B are equivalent.

   Proof: From Theorem 2.8, both P and Q can be expressed as a product of elementary matrices. Thus it is possible to obtain B from A by a sequence of elementary row or elementary column operations. Hence A and B are equivalent.

7. Let

$$
A = \begin{bmatrix} 1 & -1 & 2 & 3 \\ 2 & -1 & 3 & 1 \\ 4 & -3 & 7 & 7 \\ 0 & -1 & 1 & 5 \end{bmatrix}.
$$

We use elementary row and column operations to transform A to the form in Theorem 2.12 and keep track of the corresponding elementary matrices.

$$
\begin{array}{c}
\mathbf{r}_2 - 2\mathbf{r}_1 \to \mathbf{r}_2: \\
\mathbf{r}_3 - 4\mathbf{r}_1 \to \mathbf{r}_3:
\end{array}
\begin{bmatrix} 1 & -1 & 2 & 3 \\ 2 & -1 & 3 & 1 \\ 4 & -3 & 7 & 7 \\ 0 & -1 & 1 & 5 \end{bmatrix} \longrightarrow
\begin{bmatrix} 1 & -1 & 2 & 3 \\ 0 & 1 & -1 & -5 \\ 0 & 1 & -1 & -5 \\ 0 & -1 & 1 & 5 \end{bmatrix} = E_2 E_1 A,
$$

where

$$
E_1 \text{ is obtained from } I_4 \text{ by applying } \mathbf{r}_2 - 2\mathbf{r}_1 \to \mathbf{r}_2;
$$
$$
E_2 \text{ is obtained from } I_4 \text{ by applying } \mathbf{r}_3 - 4\mathbf{r}_1 \to \mathbf{r}_3.
$$

$$
\begin{array}{c}
\mathbf{r}_1 + \mathbf{r}_2 \to \mathbf{r}_1: \\
\mathbf{r}_3 - \mathbf{r}_2 \to \mathbf{r}_3: \\
\mathbf{r}_4 + \mathbf{r}_2 \to \mathbf{r}_4:
\end{array}
\begin{bmatrix} 1 & -1 & 2 & 3 \\ 0 & 1 & -1 & -5 \\ 0 & 1 & -1 & -5 \\ 0 & -1 & 1 & 5 \end{bmatrix} \longrightarrow
\begin{bmatrix} 1 & 0 & 1 & -2 \\ 0 & 1 & -1 & -5 \\ 0 & 0 & 0 & 0 \\ 0 & 0 & 0 & 0 \end{bmatrix} = E_5 E_4 E_3 (E_2 E_1 A),
$$

where

E_3 is obtained from I_4 by applying $\mathbf{r}_1 + \mathbf{r}_2 \to \mathbf{r}_1$;

E_4 is obtained from I_4 by applying $\mathbf{r}_3 - \mathbf{r}_2 \to \mathbf{r}_3$;

E_5 is obtained from I_4 by applying $\mathbf{r}_4 + \mathbf{r}_2 \to \mathbf{r}_4$.

$$
\begin{matrix}
\mathbf{c}_3 - \mathbf{c}_1 \to \mathbf{c}_3: \\
\mathbf{c}_4 + 2\mathbf{c}_1 \to \mathbf{c}_4: \\
\mathbf{c}_3 + \mathbf{c}_2 \to \mathbf{c}_3: \\
\mathbf{c}_4 + 5\mathbf{c}_2 \to \mathbf{c}_4:
\end{matrix}
\begin{bmatrix}
1 & 0 & 1 & -2 \\
0 & 1 & -1 & -5 \\
0 & 0 & 0 & 0 \\
0 & 0 & 0 & 0
\end{bmatrix}
\longrightarrow
\begin{bmatrix}
1 & 0 & 0 & 0 \\
0 & 1 & 0 & 0 \\
0 & 0 & 0 & 0 \\
0 & 0 & 0 & 0
\end{bmatrix}
= PAQ = B,
$$

where $P = E_5 E_4 E_3 E_2 E_1$, $Q = F_1 F_2 F_3 F_4$, and

F_1 is obtained from I_4 by applying $\mathbf{c}_3 + \mathbf{c}_1 \to \mathbf{c}_3$;

F_2 is obtained from I_4 by applying $\mathbf{c}_4 + 2\mathbf{c}_1 \to \mathbf{c}_4$;

F_3 is obtained from I_4 by applying $\mathbf{c}_3 + \mathbf{c}_2 \mathbf{c}_3 \to$;

F_4 is obtained from I_4 by applying $\mathbf{c}_4 + 5\mathbf{c}_2 \to \mathbf{c}_4$.

Then after forming E_i, $i = 1, \ldots, 5$, we have

$$
P = E_5 E_4 E_3 E_2 E_1 =
\begin{bmatrix}
-1 & 1 & 0 & 0 \\
2 & -1 & 0 & 0 \\
-4 & 0 & 1 & 0 \\
0 & 0 & 0 & 1
\end{bmatrix}
$$

and after forming F_i, $i = 1, \ldots, 4$, we have

$$
Q = F_1 F_2 F_3 F_4 =
\begin{bmatrix}
1 & 0 & -1 & 2 \\
0 & 1 & 1 & 5 \\
0 & 0 & 1 & 0 \\
0 & 0 & 0 & 1
\end{bmatrix}.
$$

The matrices P and Q are not unique.

9. We are asked to prove an "if and only if" statement. Hence two things must be proved.

   (a) If A is equivalent to B, then A^T is equivalent to B^T.

   Proof: If A is equivalent to B, then Theorem 2.13 guarantees that there exist nonsingular matrices P and Q such that $B = PAQ$. Using properties of transposes we have $B^T = (PAQ)^T = Q^T A^T P^T$. It follows that Q^T and P^T are nonsingular since Q and P are nonsingular. Hence Theorem 2.13 implies that A^T and B^T are equivalent.

   (b) If A^T and B^T are equivalent, then A and B are equivalent.

   Proof: If A^T is equivalent to B^T, then, by part (a), $(A^T)^T$ is equivalent to $(B^T)^T$. But, by Theorem 1.4(a), $(A^T)^T = A$ and $(B^T)^T = B$. Therefore A is equivalent to B.

11. We are asked to prove an "if and only if" statement. Hence two things must be proved.

   (a) If A is nonsingular then B is nonsingular.

   Proof: From Theorem 2.13 there exist nonsingular matrices P and Q such that $B = PAQ$. Hence B is the product of nonsingular matrices and it follows that B is nonsingular.

   (b) If B is nonsingular then A is nonsingular.

   Proof: Use the preceding proof with A and B interchanged.

Section 2.5, p. 136

1. Solve $L\mathbf{z} = \begin{bmatrix} 2 & 0 & 0 \\ 2 & -3 & 0 \\ 1 & -1 & 4 \end{bmatrix} \begin{bmatrix} z_1 \\ z_2 \\ z_3 \end{bmatrix} = \mathbf{b} = \begin{bmatrix} 18 \\ 3 \\ 12 \end{bmatrix}$ by forward substitution:

$$z_1 = \frac{18}{2} = 9$$
$$z_2 = \frac{3 - 2z_1}{-3} = \frac{-15}{-3} = 5$$
$$z_3 = \frac{12 + z_2 + z_1}{4} = \frac{8}{4} = 2.$$

Solve $U\mathbf{x} = \begin{bmatrix} 1 & 4 & 0 \\ 0 & 2 & 1 \\ 0 & 0 & 2 \end{bmatrix} \begin{bmatrix} x_1 \\ x_2 \\ x_3 \end{bmatrix} = \mathbf{z} = \begin{bmatrix} 9 \\ 5 \\ 2 \end{bmatrix}$ by back substitution:

$$x_3 = \frac{2}{2} = 1$$
$$x_2 = \frac{5 - x_3}{2} = \frac{4}{2} = 2$$
$$x_1 = \frac{9 - 0x_3 - 4x_2}{1} = \frac{1}{1} = 1$$

Thus the solution is $\mathbf{x} = \begin{bmatrix} 1 \\ 2 \\ 1 \end{bmatrix}$.

3. Solve $L\mathbf{z} = \begin{bmatrix} 1 & 0 & 0 & 0 \\ 2 & 1 & 0 & 0 \\ -1 & 3 & 1 & 0 \\ 4 & 3 & 2 & 1 \end{bmatrix} \begin{bmatrix} z_1 \\ z_2 \\ z_3 \\ z_4 \end{bmatrix} = \mathbf{b} = \begin{bmatrix} -2 \\ -2 \\ -16 \\ -66 \end{bmatrix}$ by forward substitution:

$$z_1 = \frac{-2}{1} = -2$$
$$z_2 = \frac{-2 - 2z_1}{1} = \frac{2}{1} = 2$$
$$z_3 = \frac{-16 - 3z_2 + z_1}{1} = \frac{-24}{1} = -24$$
$$z_4 = \frac{-66 - 2z_3 - 3z_2 - 4z_1}{1} = \frac{-16}{1} = -16$$

Solve $U\mathbf{x} = \begin{bmatrix} 2 & 3 & 0 & 1 \\ 0 & -1 & 3 & 1 \\ 0 & 0 & -2 & 5 \\ 0 & 0 & 0 & 4 \end{bmatrix} \begin{bmatrix} x_1 \\ x_2 \\ x_3 \\ x_4 \end{bmatrix} = \mathbf{z} = \begin{bmatrix} -2 \\ 2 \\ -24 \\ -16 \end{bmatrix}$ by back substitution:

$$x_4 = \frac{-16}{4} = -4$$
$$x_3 = \frac{-24 - 5x_4}{-2} = \frac{-4}{-2} = 2$$
$$x_2 = \frac{2 - x_4 - 3x_3}{-1} = \frac{0}{-1} = 0$$
$$x_1 = \frac{-2 - x_4 + 0x_3 - 3x_2}{2} = \frac{2}{2} = 1$$

Thus the solution is $\mathbf{x} = \begin{bmatrix} 1 \\ 0 \\ 2 \\ -4 \end{bmatrix}$.

5. To find an *LU*-factorization of $A = \begin{bmatrix} 2 & 3 & 4 \\ 4 & 5 & 10 \\ 4 & 8 & 2 \end{bmatrix}$ we follow the procedure used in Example 3.

Step 1. "Zero out" below the first diagonal entry of A. Add -2 times the first row of A to the second row of A. Add -2 times the first row of A to the third row of A. Call the new matrix U_1.

$$U_1 = \begin{bmatrix} 2 & 3 & 4 \\ 0 & -1 & 2 \\ 0 & 2 & -6 \end{bmatrix}$$

We begin building a lower triangular matrix, with 1's on the main diagonal, to record the row operations. Enter the *negatives of the multipliers* used in the row operations in the first column of L_1, below the first diagonal entry of L_1.

$$L_1 = \begin{bmatrix} 1 & 0 & 0 \\ 2 & 1 & 0 \\ 2 & * & 1 \end{bmatrix}$$

Step 2. "Zero out" below the second diagonal entry of U_1. Add 2 times the second row of U_1 to the third row of U_1. Call the new matrix U_2.

$$U_2 = \begin{bmatrix} 2 & 3 & 4 \\ 0 & -1 & 2 \\ 0 & 0 & -2 \end{bmatrix}$$

Enter the *negatives of the multipliers* from the row operations below the second diagonal entry of L_1. Call the new matrix L_2.

$$L_2 = \begin{bmatrix} 1 & 0 & 0 \\ 2 & 1 & 0 \\ 2 & -2 & 1 \end{bmatrix}$$

Let $L = L_2$ and $U = U_2$. Solve

$$L\mathbf{z} = \begin{bmatrix} 1 & 0 & 0 \\ 2 & 1 & 0 \\ 2 & -2 & 1 \end{bmatrix} \begin{bmatrix} z_1 \\ z_2 \\ z_3 \end{bmatrix} = \mathbf{b} = \begin{bmatrix} 6 \\ 16 \\ 2 \end{bmatrix}$$

by forward substitution:

$$z_1 = 6$$
$$z_2 = 16 - 2z_1 = 4$$
$$z_3 = 2 + 2z_2 - 2z_1 = -2.$$

Solve

$$U\mathbf{x} = \begin{bmatrix} 2 & 3 & 4 \\ 0 & -1 & 2 \\ 0 & 0 & -2 \end{bmatrix} \begin{bmatrix} x_1 \\ x_2 \\ x_3 \end{bmatrix} = \mathbf{z} = \begin{bmatrix} 6 \\ 4 \\ -2 \end{bmatrix}$$

by back substitution:

$$x_3 = \frac{-2}{-2} = 1$$
$$x_2 = \frac{4 - 2x_3}{-1} = -2$$
$$x_1 = \frac{6 - 4x_3 - 3x_2}{2} = 4$$

Thus the solution is $\mathbf{x} = \begin{bmatrix} 4 \\ -2 \\ 1 \end{bmatrix}$.

7. To find an LU-factorization of $A = \begin{bmatrix} 4 & 2 & 3 \\ 2 & 0 & 5 \\ 1 & 2 & 1 \end{bmatrix}$ we follow the procedure used in Example 3.

Step 1. "Zero out" below the first diagonal entry of A. Add $-\frac{1}{2}$ times the first row of A to the second row of A. Add $-\frac{1}{4}$ times the first row of A to the third row of A. Call the new matrix U_1.

$$U_1 = \begin{bmatrix} 4 & 2 & 3 \\ 0 & -1 & \frac{7}{2} \\ 0 & \frac{3}{2} & \frac{1}{4} \end{bmatrix}$$

We begin building a lower triangular matrix, with 1's on the main diagonal, to record the row operations. Enter the *negatives of the multipliers* used in the row operations in the first column of L_1, below the first diagonal entry of L_1.

$$L_1 = \begin{bmatrix} 1 & 0 & 0 \\ \frac{1}{2} & 1 & 0 \\ \frac{1}{4} & * & 1 \end{bmatrix}$$

Step 2. "Zero out" below the second diagonal entry of U_1. Add $\frac{3}{2}$ times the second row of U_1 to the third row of U_1. Call the new matrix U_2.

$$U_2 = \begin{bmatrix} 4 & 2 & 3 \\ 0 & -1 & \frac{7}{2} \\ 0 & 0 & \frac{11}{2} \end{bmatrix}$$

Enter the *negatives of the multipliers* from the row operations below the second diagonal entry of L_1. Call the new matrix L_2.

$$L_2 = \begin{bmatrix} 1 & 0 & 0 \\ \frac{1}{2} & 1 & 0 \\ \frac{1}{4} & -\frac{3}{2} & 1 \end{bmatrix}$$

Let $L = L_2$ and $U = U_2$. Solve

$$L\mathbf{z} = \begin{bmatrix} 1 & 0 & 0 \\ \frac{1}{2} & 1 & 0 \\ \frac{1}{4} & -\frac{3}{2} & 1 \end{bmatrix} \begin{bmatrix} z_1 \\ z_2 \\ z_3 \end{bmatrix} = \mathbf{b} = \begin{bmatrix} 1 \\ -1 \\ -3 \end{bmatrix}$$

by using forward substitution :

$$z_1 = 1$$
$$z_2 = -1 - \frac{1}{2}z_1 = -\frac{3}{2}$$
$$z_3 = -3 + \frac{3}{2}z_2 - \frac{1}{4}z_1 = -\frac{11}{2}.$$

Solve

$$U\mathbf{x} = \begin{bmatrix} 4 & 2 & 3 \\ 0 & -1 & \frac{7}{2} \\ 0 & 0 & \frac{11}{2} \end{bmatrix} \begin{bmatrix} x_1 \\ x_2 \\ x_3 \end{bmatrix} = \mathbf{z} = \begin{bmatrix} 1 \\ -\frac{3}{2} \\ -\frac{11}{2} \end{bmatrix}$$

by back substitution:

$$x_3 = -1$$
$$x_2 = \frac{-\frac{3}{2} - \frac{7}{2}x_3}{-1} = -2$$
$$x_1 = \frac{1 - 3x_3 - 2x_2}{4} = 2.$$

Thus the solution is $\mathbf{x} = \begin{bmatrix} 2 \\ -2 \\ -1 \end{bmatrix}$.

9. To find an LU-factorization of $A = \begin{bmatrix} 2 & 1 & 0 & -4 \\ 1 & 0 & .25 & -1 \\ -2 & -1.1 & .25 & 6.2 \\ 4 & 2.2 & .30 & -2.4 \end{bmatrix}$ we follow the procedure in Example 3.

Step 1. "Zero out" below the first diagonal entry of A. Add $-.5$ times the first row of A to the second row of A. Add 1 times the first row of A to the third row of A. Add -2 times the first row of A to the fourth row of A. Call the new matrix U_1.

$$U_1 = \begin{bmatrix} 2 & 1 & 0 & -4 \\ 0 & -.5 & .25 & 1 \\ 0 & -.1 & .25 & 2.2 \\ 0 & .2 & .30 & 5.6 \end{bmatrix}$$

We begin building a lower triangular matrix, with 1's on the main diagonal, to record the row operations. Enter the *negatives of the multipliers* used in the row operations in the first column of L_1, below the first diagonal entry of L_1.

$$L_1 = \begin{bmatrix} 1 & 0 & 0 & 0 \\ .5 & 1 & 0 & 0 \\ -1 & * & 1 & 0 \\ 2 & * & * & 1 \end{bmatrix}$$

Step 2. "Zero out" below the second diagonal entry of U_1. Add $-.2$ times the second row of U_1 to the third row of U_1. Add $.4$ times the second row of U_1 to the fourth row of U_1. Call the new matrix U_2.

$$U_2 = \begin{bmatrix} 2 & 1 & 0 & -4 \\ 0 & -.5 & .25 & 1 \\ 0 & 0 & .20 & 2 \\ 0 & 0 & .40 & 6 \end{bmatrix}$$

Enter the *negatives of the multipliers* from the row operations below the second diagonal entry of L_1. Call the new matrix L_2.

$$L_2 = \begin{bmatrix} 1 & 0 & 0 & 0 \\ .5 & 1 & 0 & 0 \\ -1 & .2 & 1 & 0 \\ 2 & -.4 & * & 1 \end{bmatrix}$$

Step 3. "Zero out" below the third diagonal entry of U_2. Add -2 times the third row of U_2 to the fourth row of U_2. Call the new matrix U_3.

$$U_3 = \begin{bmatrix} 2 & 1 & 0 & -4 \\ 0 & -.5 & .25 & 1 \\ 0 & 0 & .20 & 2 \\ 0 & 0 & 0 & 2 \end{bmatrix}$$

Enter the *negatives of the multipliers* from the row operations below the third diagonal entry of L_2. Call the new matrix L_3.

$$L_3 = \begin{bmatrix} 1 & 0 & 0 & 0 \\ .5 & 1 & 0 & 0 \\ -1 & .2 & 1 & 0 \\ 2 & -.4 & 2 & 1 \end{bmatrix}$$

Let $L = L_3$ and $U = U_3$. Solve

$$L\mathbf{z} = \begin{bmatrix} 1 & 0 & 0 & 0 \\ .5 & 1 & 0 & 0 \\ -1 & .2 & 1 & 0 \\ 2 & -.4 & 2 & 1 \end{bmatrix} \begin{bmatrix} z_1 \\ z_2 \\ z_3 \end{bmatrix} = \mathbf{b} = \begin{bmatrix} -3 \\ -1.5 \\ 5.6 \\ 2.2 \end{bmatrix}$$

by using forward substitution :

$$z_1 = -3$$
$$z_2 = -1.5 - .5z_1 = 0$$
$$z_3 = 5.6 - .2z_2 + z_1 = 2.6$$
$$z_4 = 2.2 - 2z_3 + .4z_2 - 2z_1 = 3$$

Solve

$$U\mathbf{x} = \begin{bmatrix} 2 & 1 & 0 & -4 \\ 0 & -.5 & .25 & 1 \\ 0 & 0 & .20 & 2 \\ 0 & 0 & 0 & 2 \end{bmatrix} \begin{bmatrix} x_1 \\ x_2 \\ x_3 \\ x_4 \end{bmatrix} = \mathbf{z} = \begin{bmatrix} -3 \\ 0 \\ 2.6 \\ 3 \end{bmatrix}$$

by back substitution:

$$x_4 = 1.5$$
$$x_3 = \frac{2.6 - 2x_4}{.2} = -2$$
$$x_2 = \frac{0 - x_4 - .25x_3}{-.5} = 2$$
$$x_1 = \frac{-3 + 4x_4 - 0x_3 - x_2}{2} = 0.5.$$

Thus the solution is $\mathbf{x} = \begin{bmatrix} 0.5 \\ 2 \\ -2 \\ 1.5 \end{bmatrix}$.

Supplementary Exercises for Chapter 2, p. 137

1. First add 4 times row 2 to row 1:

$$\begin{bmatrix} 2 & 0 & 12 \\ 0 & 1 & 3 \\ 0 & 0 & 6 \end{bmatrix}.$$

Add $\left(-\frac{1}{2}\right)$ times row 3 to row 2 and (-2) times row 3 to row 1:

$$\begin{bmatrix} 2 & 0 & 0 \\ 0 & 1 & 0 \\ 0 & 0 & 6 \end{bmatrix}.$$

Finally, multiply row 1 by $\frac{1}{2}$ and row 3 by $\frac{1}{6}$ to obtain

$$B = \begin{bmatrix} 1 & 0 & 0 \\ 0 & 1 & 0 \\ 0 & 0 & 1 \end{bmatrix}.$$

3. The coefficient matrix of the homogeneous system is

$$\begin{bmatrix} 1-a & 0 & 1 \\ 0 & -a & 1 \\ 0 & 1 & 1 \end{bmatrix}.$$

Add a times row 3 to row 2: $\begin{bmatrix} 1-a & 0 & 1 \\ 0 & 0 & 1+a \\ 0 & 1 & 1 \end{bmatrix}$. Interchange rows 2 and 3: $\begin{bmatrix} 1-a & 0 & 1 \\ 0 & 1 & 1 \\ 0 & 0 & 1+a \end{bmatrix}$.

In order for the corresponding homogeneous system to have a nontrivial solution we must have $a = 1$ or $a = -1$.

5. (a) Multiply the jth row of B by $\frac{1}{k}$.

   (b) Interchange the ith and jth rows of B.

   (c) Add $-k$ times the jth row of B to its ith row.

7. We first adjoin the 3×3 identity matrix:

$$\left[\begin{array}{ccc|ccc} 1 & 0 & 1 & 1 & 0 & 0 \\ 1 & 1 & 0 & 0 & 1 & 0 \\ 0 & 1 & 1 & 0 & 0 & 1 \end{array}\right]$$

Add -1 times row 1 to row 2:

$$\left[\begin{array}{ccc|ccc} 1 & 0 & 1 & 1 & 0 & 0 \\ 0 & 1 & -1 & -1 & 1 & 0 \\ 0 & 1 & 1 & 0 & 0 & 1 \end{array}\right]$$

Add -1 times row 2 to row 3:

$$\left[\begin{array}{ccc|ccc} 1 & 0 & 1 & 1 & 0 & 0 \\ 0 & 1 & -1 & -1 & 1 & 0 \\ 0 & 0 & 2 & 1 & -1 & 1 \end{array}\right]$$

Multiply row 3 by $\frac{1}{2}$:

$$\left[\begin{array}{ccc|ccc} 1 & 0 & 1 & 1 & 0 & 0 \\ 0 & 1 & -1 & -1 & 1 & 0 \\ 0 & 0 & 1 & \frac{1}{2} & -\frac{1}{2} & \frac{1}{2} \end{array}\right]$$

Add (-1) times row 3 to row 1 and 1 times row 3 to row 2:

$$\left[\begin{array}{ccc|ccc} 1 & 0 & 0 & \frac{1}{2} & \frac{1}{2} & -\frac{1}{2} \\ 0 & 1 & 0 & -\frac{1}{2} & \frac{1}{2} & \frac{1}{2} \\ 0 & 0 & 1 & \frac{1}{2} & -\frac{1}{2} & \frac{1}{2} \end{array}\right]$$

Therefore

$$A^{-1} = \left[\begin{array}{ccc} \frac{1}{2} & \frac{1}{2} & -\frac{1}{2} \\ -\frac{1}{2} & \frac{1}{2} & \frac{1}{2} \\ \frac{1}{2} & -\frac{1}{2} & \frac{1}{2} \end{array}\right] = \frac{1}{2}\left[\begin{array}{ccc} 1 & 1 & -1 \\ -1 & 1 & 1 \\ 1 & -1 & 1 \end{array}\right].$$

9. (a) The results must be identical, since an inverse is unique.

   (b) Compute AA_1 and AA_2. If the result is I_{10}, then the answers submitted by the students are correct.

11. First interchange rows 1 and 3: $\left[\begin{array}{ccc} 1 & s & 2 \\ 2 & 1 & 1 \\ 0 & 1 & 2 \end{array}\right]$.

Add -2 times row 1 to row 2: $\left[\begin{array}{ccc} 1 & s & 2 \\ 0 & 1-2s & -3 \\ 0 & 1 & 2 \end{array}\right]$.

Add $-s$ times row 3 to row 1 and $-1+2s$ times row 3 to row 2:

$$\left[\begin{array}{ccc} 1 & 0 & 2-2s \\ 0 & 0 & -5+4s \\ 0 & 1 & 2 \end{array}\right]$$

Interchange rows 2 and 3:

$$\begin{bmatrix} 1 & 0 & 2-2s \\ 0 & 1 & 2 \\ 0 & 0 & -5+4s \end{bmatrix}$$

This matrix can be reduced to the 3×3 identity matrix, and hence A is nonsingular, provided $-5+4s \neq 0$, or, $s \neq \frac{5}{4}$.

13. For any angle θ, $\cos\theta$ and $\sin\theta$ are never simultaneously zero. Thus at least one element in column 1 is not zero. Assume $\cos\theta \neq 0$. (If $\cos\theta = 0$, then interchange rows 1 and 2 and proceed in a similar manner to that described below.) To show that the matrix is nonsingular and determine its inverse, we put

$$\left[\begin{array}{cc|cc} \cos\theta & \sin\theta & 1 & 0 \\ -\sin\theta & \cos\theta & 0 & 1 \end{array}\right]$$

into reduced row echelon form. Apply row operations $\frac{1}{\cos\theta}$ times row 1 and $\sin\theta$ times row 1 added to row 2 to obtain

$$\left[\begin{array}{cc|cc} 1 & \dfrac{\sin\theta}{\cos\theta} & \dfrac{1}{\cos\theta} & 0 \\[2ex] 0 & \dfrac{\sin^2\theta}{\cos\theta}+\cos\theta & \dfrac{\sin\theta}{\cos\theta} & 1 \end{array}\right].$$

Since

$$\frac{\sin^2\theta}{\cos\theta}+\cos\theta = \frac{\sin^2\theta+\cos^2\theta}{\cos\theta} = \frac{1}{\cos\theta},$$

the $(2,2)$-element is not zero. Applying row operations $\cos\theta$ times row 2 and $\left(-\frac{\sin\theta}{\cos\theta}\right)$ times row 2 added to row 1 we obtain

$$\left[\begin{array}{cc|cc} 1 & 0 & \cos\theta & -\sin\theta \\ 0 & 1 & \sin\theta & \cos\theta \end{array}\right].$$

It follows that the matrix is nonsingular and its inverse is

$$\begin{bmatrix} \cos\theta & -\sin\theta \\ \sin\theta & \cos\theta \end{bmatrix}.$$

15. If $A\mathbf{u} = \mathbf{b}$ and $A\mathbf{v} = \mathbf{b}$, then $A(\mathbf{u} - \mathbf{v}) = A\mathbf{u} - A\mathbf{v} = \mathbf{b} - \mathbf{b} = \mathbf{0}$.

17. Let $\mathbf{u}$ be one solution to $A\mathbf{x} = \mathbf{b}$. Since A is singular, the homogeneous system $A\mathbf{x} = \mathbf{0}$ has a nontrivial solution $\mathbf{u}_0$. Then for any real number r, $\mathbf{v} = r\mathbf{u}_0$ is also a solution to the homogeneous system. Finally, by Exercise 29, Sec. 2.2, for each of the infinitely many matrices $\mathbf{v}$, the matrix $\mathbf{w} = \mathbf{u} + \mathbf{v}$ is a solution to the nonhomogeneous system $A\mathbf{x} = \mathbf{b}$.

19. $LU = \begin{bmatrix} 2 & 0 & 0 \\ t & s & 0 \\ 1 & 0 & -1 \end{bmatrix}\begin{bmatrix} r & 1 & 4 \\ 0 & 2 & -1 \\ 0 & 0 & p \end{bmatrix} = \begin{bmatrix} 2r & 2 & 8 \\ tr & t+2s & 4t-s \\ r & 1 & 4-p \end{bmatrix} = \begin{bmatrix} 6 & 2 & 8 \\ 9 & 5 & 11 \\ 3 & 1 & 6 \end{bmatrix}$

Therefore $2r = 6$, $tr = 9$, $t + 2s = 5$, $4t - s = 11$, and $4 - p = 6$. So $r = 3$, $t = 3$, $s = 1$, and $p = -2$.

21. The outer product of X and Y can be written in the form

$$XY^T = \begin{bmatrix} x_1\begin{bmatrix} y_1 & y_2 & \cdots & y_n \end{bmatrix} \\ x_2\begin{bmatrix} y_1 & y_2 & \cdots & y_n \end{bmatrix} \\ \vdots \\ x_n\begin{bmatrix} y_1 & y_2 & \cdots & y_n \end{bmatrix} \end{bmatrix}.$$

If either $\overset{.}{X} = O$ or $Y = O$, then $XY^T = O$. Thus assume that there is at least one nonzero component in X, say x_i, and at least one nonzero component in Y, say y_j. Then $\left(\frac{1}{x_i}\right)\text{Row}_i(XY^T)$ makes the ith row exactly Y^T. Since all the other rows are multiples of Y^T, row operations of the form $-x_k R_i + R_p$, for $p \neq i$, can be performed to zero out everything but the ith row. It follows that either XY^T is row equivalent to O or to a matrix with $n - 1$ zero rows.

Chapter Review for Chapter 2, p. 138

True or False
1. False.   2. True.   3. False.   4. True.   5. True.
6. True.   7. True.   8. True.   9. True.   10. False.

Quiz

1. $A = \begin{bmatrix} 1 & 1 & 5 \\ 2 & -2 & -2 \\ -3 & 1 & -3 \end{bmatrix} \underset{\substack{-2\mathbf{r}_1 + \mathbf{r}_2 \to \mathbf{r}_2 \\ 3\mathbf{r}_1 + \mathbf{r}_3 \to \mathbf{r}_3}}{\to} \begin{bmatrix} 1 & 1 & 5 \\ 0 & -4 & -12 \\ 0 & 4 & 12 \end{bmatrix} \underset{\substack{-\frac{1}{4}\mathbf{r}_2 \\ \mathbf{r}_2 + \mathbf{r}_3 \to \mathbf{r}_3}}{\to} \begin{bmatrix} 1 & 1 & 5 \\ 0 & 1 & 3 \\ 0 & 0 & 0 \end{bmatrix} \underset{-1\mathbf{r}_2 + \mathbf{r}_1 \to \mathbf{r}_1}{\to}$

$\to \begin{bmatrix} 1 & 0 & 2 \\ 0 & 1 & 3 \\ 0 & 0 & 0 \end{bmatrix}$

2. (a) No. The $(1,3)$ entry is not zero.
   (b) Infinitely many.
   (c) No. C is not the identity matrix I_4.
   (d) The solution is $x_1 = 2x_2 - 4x_3 - 5x_4 - 6$, $x_3 = -3x_4$, $x_2 = r =$ any number, $x_4 = s =$ any number. Therefore $x_1 = 2x_2 + 12x_4 - 5x_4 - 6 = -6 + 2x_2 + 7x_4 - 6 = -6 + 2r + 7s$, $x_2 = r$, $x_3 = -3s$, $x_4 = s$, or

$$\mathbf{x} = \begin{bmatrix} -6 + 2r + 7s \\ r \\ -3s \\ s \end{bmatrix}.$$

3. $\begin{bmatrix} 2 & 1 & 4 \\ 1 & -2 & 1 \\ 2 & 6 & k \end{bmatrix} \to \begin{bmatrix} 1 & -2 & 1 \\ 0 & 5 & 2 \\ 0 & 10 & k-2 \end{bmatrix}$. For A to be singular, we must have $k - 2 = 4$, or $k = 6$.

4. $\begin{bmatrix} 2 & 0 & 1 \\ 1 & 2 & 1 \\ 1 & 2 & 3 \\ 1 & 0 & 2 \\ 2 & 1 & 4 \end{bmatrix} \to \begin{bmatrix} 1 & 0 & 2 \\ 0 & 0 & -3 \\ 0 & 2 & -1 \\ 0 & 2 & 1 \\ 0 & 1 & 0 \end{bmatrix} \to \begin{bmatrix} 1 & 0 & 2 \\ 0 & 1 & 0 \\ 0 & 0 & -1 \\ 0 & 0 & 1 \\ 0 & 0 & -3 \end{bmatrix} \to x_1 = x_2 = x_3 = 0$, or $\mathbf{x} = \begin{bmatrix} 0 \\ 0 \\ 0 \end{bmatrix}$.

5. $\left[\begin{array}{ccc|ccc} 1 & 2 & 1 & 1 & 0 & 0 \\ 1 & 1 & 1 & 0 & 1 & 0 \\ 2 & 1 & 0 & 0 & 0 & 1 \end{array}\right] \to \left[\begin{array}{ccc|ccc} 1 & 2 & 1 & 1 & 0 & 0 \\ 0 & -1 & 0 & -1 & 1 & 0 \\ 0 & -3 & -2 & -2 & 0 & 1 \end{array}\right] \to \left[\begin{array}{ccc|ccc} 1 & 0 & 1 & -1 & 2 & 0 \\ 0 & 1 & 0 & 1 & -1 & 0 \\ 0 & 0 & -2 & 1 & -3 & 1 \end{array}\right]$

$\to \left[\begin{array}{ccc|ccc} 1 & 0 & 1 & -1 & 2 & 0 \\ 0 & 1 & 0 & 1 & -1 & 0 \\ 0 & 0 & 1 & -\frac{1}{2} & \frac{3}{2} & -\frac{1}{2} \end{array}\right] \to \left[\begin{array}{ccc|ccc} 1 & 0 & 0 & -\frac{1}{2} & \frac{1}{2} & \frac{1}{2} \\ 0 & 1 & 0 & 1 & -1 & 0 \\ 0 & 0 & 1 & -\frac{1}{2} & \frac{3}{2} & -\frac{1}{2} \end{array}\right]$

So $A^{-1} = \begin{bmatrix} -\frac{1}{2} & \frac{1}{2} & \frac{1}{2} \\ 1 & -1 & 0 \\ -\frac{1}{2} & \frac{3}{2} & -\frac{1}{2} \end{bmatrix}$.

6. Let $P = A^{-1}$, $Q = B$. Then $PAQ = A^{-1}AB = I_nB = B$.

7. Diagonal, zero, or symmetric. In each case, $A^T = A$.

Chapter 3

Determinants

Section 3.1, p. 145

1. (a) Permutation 52134 has 5 inversions: 5 precedes 2, 1, 3, and 4; 2 precedes 1.

    (b) Permutation 45213 has 7 inversions: 4 precedes 2, 1, and 3; 5 precedes 2, 1, and 3; 2 precedes 1.

    (c) Permutations 42135 has 4 inversions: 4 precedes 2, 1, and 3; 2 precedes 1

3. (a) Permutation 4213 is even since it has 4 inversions: 4 precedes 2, 1, and 3; 2 precedes 1.

    (b) Permutation 1243 is odd since it has 1 inversion: 4 precedes 3.

    (c) Permutation 1234 is even since it has 0 inversions.

5. We determine whether the permutation is even or odd. If it is even, a plus sign is associated with it; otherwise, a minus sign is associated.

    (a) Permutation 25431 is odd since it has 7 inversions: 2 precedes 1; 5 precedes 4, 3, and 1; 4 precedes 3 and 1; 3 precedes 1. Thus we associate a minus sign with the permutation.

    (b) Permutation 31245 is even since it has 2 inversions: 3 precedes 1 and 2. Thus we associate a plus sign.

    (c) Permutation 21345 is odd since it has 1 inversion: 2 precedes 1. Thus we associate a minus sign.

7. (a) There are 9 inversions in 436215: 4 precedes 3, 2, and 1; 3 precedes 2 and 1; 6 precedes 2, 1, and 5; 2 precedes 1.

    (b) There are 6 inversions in 416235: 4 precedes 1, 2, and 3; 6 precedes 2, 3, and 5. Thus the difference between the number of inversions in this permutation and that in (a) is 3 which is odd.

9. Use the procedure in Example 7.

    (a) $\begin{vmatrix} 1 & 2 \\ 2 & 4 \end{vmatrix} = (1)(4) - (2)(2) = 4 - 4 = 0$

    (b) $\begin{vmatrix} 3 & 1 \\ -3 & -1 \end{vmatrix} = (3)(-1) - (-3)(1) = -3 + 3 = 0$

11. We use the schematic procedure in Example 8 for (a) and (b).

    (a) $\begin{vmatrix} 2 & 1 & 3 \\ 3 & 2 & 1 \\ 0 & 1 & 2 \end{vmatrix} \begin{matrix} 2 & 1 \\ 3 & 2 \\ 0 & 1 \end{matrix} = (2)(2)(2) + (1)(1)(0) + (3)(3)(1) - (0)(2)(3) - (1)(1)(2) - (2)(3)(1) = 9$

    (b) $\begin{vmatrix} 2 & 1 & 3 \\ -3 & 2 & 1 \\ -1 & 3 & 4 \end{vmatrix} \begin{matrix} 2 & 1 \\ -3 & 2 \\ -1 & 3 \end{matrix} = (2)(2)(4) + (1)(1)(-1) + (3)(-3)(3) - (-1)(2)(3) - (3)(1)(2) - (4)(-3)(1) = 0$

(c) Here we use Definition 3.2. Observe that there is only one nonzero entry in each row and that each of these entries lies in a different column. Hence there will be only one permutation of the column subscripts such that each of the four numbers in the product $a_{1j_1} a_{2j_2} a_{3j_3} a_{4j_4}$ is different from zero. Hence

$$|A| = (\text{sign of permutation } 4321)\,(3)(4)(2)(6) = 144.$$

13. (a) $\begin{vmatrix} t-1 & 2 \\ 3 & t-2 \end{vmatrix} = (t-1)(t-2) - (3)(2) = t^2 - 3t - 4$

(b) $\begin{vmatrix} t-1 & -1 & -2 \\ 0 & t & 2 \\ 0 & 0 & t-3 \end{vmatrix} = \begin{matrix} (t-1)(t)(t-3) + (-1)(2)(0) + (-2)(0)(0) \\ -(0)(t)(-2) - (0)(2)(t-1) - (t-3)(-1)(0) \end{matrix} = t^3 - 4t^2 + 3t$

15. (a) $\begin{vmatrix} t-1 & 2 \\ 3 & t-2 \end{vmatrix} = t^2 - 3t - 4 = (t-4)(t+1) = 0$, provided $t = 4$ or $t = -1$.

(b) $\begin{vmatrix} t-1 & -2 & -2 \\ 0 & t & 2 \\ 0 & 0 & t-3 \end{vmatrix} = t^3 - 4t^2 + 3t = t(t-3)(t-1) = 0$, provided $t = 0$, $t = 3$, or $t = 1$.

Section 3.2, p. 154

1. (a) Use Theorem 3.7: $\begin{vmatrix} 3 & 0 \\ 2 & 1 \end{vmatrix} = (3)(1) = 3.$

(b) Use the method in Example 8 and Theorem 3.7:

$$\begin{vmatrix} 2 & 1 \\ 4 & 3 \end{vmatrix} = \begin{vmatrix} 2 & 1 \\ 0 & 1 \end{vmatrix} = (2)(1) = 2$$
$$-2\mathbf{r}_1 + \mathbf{r}_2$$

(c) Use Theorem 3.7: $\begin{vmatrix} 4 & 0 & 0 \\ 0 & 2 & 0 \\ 0 & 0 & 3 \end{vmatrix} = (4)(2)(3) = 24$

(d) Use the method in Example 8 and Theorem 3.7:

$$\begin{vmatrix} 4 & 1 & 3 \\ 2 & 3 & 0 \\ 1 & 3 & 2 \end{vmatrix} = (-1)\begin{vmatrix} 1 & 3 & 2 \\ 2 & 3 & 0 \\ 4 & 1 & 3 \end{vmatrix} = (-1)\begin{vmatrix} 1 & 3 & 2 \\ 0 & -3 & -4 \\ 0 & -11 & -5 \end{vmatrix} = (-1)(-3)\begin{vmatrix} 1 & 3 & 2 \\ 0 & 1 & \frac{4}{3} \\ 0 & -11 & -5 \end{vmatrix}$$
$$\mathbf{r}_1 \leftrightarrow \mathbf{r}_3 \qquad \begin{matrix} -2\ \mathbf{r}_1 + \mathbf{r}_2 \\ -2\ \mathbf{r}_1 + \mathbf{r}_3 \end{matrix} \qquad -\tfrac{4}{3}\ \mathbf{r}_2 \qquad 11\ \mathbf{r}_2 + \mathbf{r}_3$$

$$= (3)\begin{vmatrix} 1 & 3 & 2 \\ 0 & 1 & \frac{4}{3} \\ 0 & 0 & \frac{29}{3} \end{vmatrix} = (3)(1)(1)\left(\frac{29}{3}\right) = 29$$

(e) Use the method in Example 8 and Theorem 3.7:

$$\begin{vmatrix} 4 & 2 & 2 & 0 \\ 2 & 0 & 0 & 0 \\ 3 & 0 & 0 & 1 \\ 0 & 0 & 1 & 0 \end{vmatrix} = \begin{vmatrix} 4 & 2 & 2 & 0 \\ 0 & -1 & -1 & 0 \\ 0 & -\frac{3}{2} & -\frac{3}{2} & 1 \\ 0 & 0 & 1 & 0 \end{vmatrix} = \begin{vmatrix} 4 & 2 & 2 & 0 \\ 0 & -1 & -1 & 0 \\ 0 & 0 & 0 & 1 \\ 0 & 0 & 1 & 0 \end{vmatrix} = (-1)\begin{vmatrix} 4 & 2 & 2 & 0 \\ 0 & -1 & -1 & 0 \\ 0 & 0 & 1 & 0 \\ 0 & 0 & 0 & 1 \end{vmatrix}$$
$$\begin{matrix} -\frac{1}{2}\ \mathbf{r}_1 + \mathbf{r}_2 \\ -\frac{3}{4}\ \mathbf{r}_1 + \mathbf{r}_3 \end{matrix} \qquad -\tfrac{3}{2}\ \mathbf{r}_2 + \mathbf{r}_3 \qquad \mathbf{r}_3 \leftrightarrow \mathbf{r}_4$$

$$= (-1)(4)(-1)(1)(1) = 4$$

(f) Use the method of Example 8 and Theorem 3.7:

$$
\begin{vmatrix} 4 & 2 & 3 & -4 \\ 3 & -2 & 1 & 5 \\ -2 & 0 & 1 & -3 \\ 8 & -2 & 6 & 4 \end{vmatrix}
=
\begin{vmatrix} 1 & 4 & 2 & -9 \\ 3 & -2 & 1 & 5 \\ -2 & 0 & 1 & -3 \\ 8 & -2 & 6 & 4 \end{vmatrix}
=
\begin{vmatrix} 1 & 4 & 2 & -9 \\ 0 & -14 & -5 & 32 \\ 0 & 8 & 5 & -21 \\ 0 & -34 & -10 & 76 \end{vmatrix}
$$

$$-1 \;\; \mathbf{r}_2 + \mathbf{r}_1 \qquad\qquad \begin{matrix} -3 \;\; \mathbf{r}_1 + \mathbf{r}_2 \\ 2 \;\; \mathbf{r}_1 + \mathbf{r}_3 \\ -8 \;\; \mathbf{r}_1 + \mathbf{r}_4 \end{matrix} \qquad\qquad -\tfrac{1}{14} \;\; \mathbf{r}_2$$

$$
= (-14)\begin{vmatrix} 1 & 4 & 2 & -9 \\ 0 & 1 & \frac{5}{14} & -\frac{16}{7} \\ 0 & 8 & 5 & -21 \\ 0 & -34 & -10 & 76 \end{vmatrix}
= (-14)\begin{vmatrix} 1 & 4 & 2 & -9 \\ 0 & 1 & -\frac{5}{14} & \frac{32}{14} \\ 0 & 0 & \frac{15}{7} & -\frac{19}{7} \\ 0 & 0 & \frac{15}{7} & -\frac{12}{7} \end{vmatrix}
= (-14)\begin{vmatrix} 1 & 4 & 2 & -9 \\ 0 & 1 & -\frac{5}{14} & \frac{32}{14} \\ 0 & 0 & \frac{15}{7} & -\frac{19}{7} \\ 0 & 0 & 0 & 1 \end{vmatrix}
$$

$$-1 \;\; \mathbf{r}_2 + \mathbf{r}_1 \qquad\qquad \begin{matrix} -3 \;\; \mathbf{r}_1 + \mathbf{r}_2 \\ 2 \;\; \mathbf{r}_1 + \mathbf{r}_3 \\ -8 \;\; \mathbf{r}_1 + \mathbf{r}_4 \end{matrix} \qquad\qquad -\tfrac{1}{14} \;\; \mathbf{r}_2$$

$$= (-14)(1)(1)(\tfrac{15}{7})(1) = -30$$

3. Note that the final matrix is obtained from the original matrix by using the row operations $2\mathbf{r}_2 + \mathbf{r}_1$ and $-3\mathbf{r}_3 + \mathbf{r}_1$. Since such row operations do not change the value of the determinant, the determinant of the final matrix is also 3.

5. Note that the final matrix is obtained by using the column operations $4\mathbf{c}_3$, $\mathbf{c}_3 - 2\mathbf{c}_1$, and then row operation $\frac{1}{2}\mathbf{r}_3$. Therefore the final value of the determinant is $(4)(\frac{1}{2})(\text{original value}) = (4)(\frac{1}{2})(4) = 8$.

7. (a) Both row and column operations may be used in the fashion of Example 8. Thus

$$
\begin{vmatrix} -4 & 2 & 0 & 0 \\ 2 & 3 & 1 & 0 \\ 3 & 1 & 0 & 2 \\ 1 & 3 & 0 & 3 \end{vmatrix}
= (-1)\begin{vmatrix} 0 & 2 & -4 & 0 \\ 1 & 3 & 2 & 0 \\ 0 & 1 & 3 & 2 \\ 0 & 3 & 1 & 3 \end{vmatrix}
= \begin{vmatrix} 1 & 3 & 2 & 0 \\ 0 & 2 & -4 & 0 \\ 0 & 1 & 3 & 2 \\ 0 & 3 & 1 & 3 \end{vmatrix}
$$

$$\mathbf{c}_1 \leftrightarrow \mathbf{c}_3 \qquad\qquad \mathbf{r}_1 \leftrightarrow \mathbf{r}_2 \qquad\qquad \mathbf{r}_2 \leftrightarrow \mathbf{r}_3$$

$$
= (-1)\begin{vmatrix} 1 & 3 & 2 & 0 \\ 0 & 1 & 3 & 2 \\ 0 & 2 & -4 & 0 \\ 0 & 3 & 1 & 3 \end{vmatrix}
= (-1)\begin{vmatrix} 1 & 3 & 2 & 0 \\ 0 & 1 & 3 & 2 \\ 0 & 0 & -10 & -4 \\ 0 & 0 & -8 & -3 \end{vmatrix}
= (-1)\begin{vmatrix} 1 & 3 & 2 & 0 \\ 0 & 1 & 3 & 2 \\ 0 & 0 & -10 & -4 \\ 0 & 0 & 0 & \frac{1}{5} \end{vmatrix} = 2
$$

$$\begin{matrix} -2 \;\; \mathbf{r}_2 + \mathbf{r}_3 \\ -3 \;\; \mathbf{r}_2 + \mathbf{r}_4 \end{matrix} \qquad\qquad -\tfrac{4}{5} \;\; \mathbf{r}_3 + \mathbf{r}_4$$

(b) Apply Theorem 3.7. The determinant is -120.

(c) Apply Theorem 3.7. The determinant is $(t-1)(t-2)(t-3) = t^3 - 6t^2 + 11t - 6$.

(d) Using Example 7 from Section 3.1 we have

$$\begin{vmatrix} t+1 & 4 \\ 2 & t-3 \end{vmatrix} = (t+1)(t-3) - 8 = t^2 - 2t - 11.$$

9. Yes. Using Theorem 3.9, $\det(AB) = \det(A)\det(B)$. If $\det(AB) = 0$, then the product of the real numbers $\det(A)$ and $\det(B)$ is zero. It follows that one of the numbers $\det(A)$ and $\det(B)$ is zero.

11. Since A is skew symmetric, $A^T = -A$. Therefore

$$\begin{aligned}
\det(A) &= \det(A^T) &&\text{by Theorem 3.1}\\
&= \det(-A) &&\text{since } A \text{ is skew symmetric}\\
&= (-1)^n \det(A) &&\text{by Exercise 10}\\
&= -\det(A) &&\text{since } n \text{ is odd}
\end{aligned}$$

The only number equal to its negative is zero, so $\det(A) = 0$.

13. By Theorem 3.9, $\det(AB^{-1}) = \det(A)\det(B^{-1})$. Hence, by Corollary 3.3,

$$\det(AB^{-1}) = \det(A)\det(B^{-1}) = \det(A)\frac{1}{\det(B)} = \frac{\det(A)}{\det(B)}.$$

15. (a) By Corollary 3.3, $\det(A^{-1}) = 1/\det(A)$. Since $A = A^{-1}$, we have

$$\det(A) = \frac{1}{\det(A)} \quad \Longrightarrow \quad (\det(A))^2 = 1.$$

Hence $\det(A) = \pm 1$.

(b) If $A^T = A^{-1}$, then $\det(A^T) = \det(A^{-1})$. But

$$\det(A) = \det(A^T) \quad \text{and} \quad \det(A^{-1}) = \frac{1}{\det(A)}$$

hence we have

$$\det(A) = \frac{1}{\det(A)} \quad \Longrightarrow \quad (\det(A))^2 = 1 \quad \Longrightarrow \quad \det(A) = \pm 1.$$

17. If A is nonsingular and $A^2 = A$, then

$$A^{-1}(A^2) = A^{-1}A \quad \Longrightarrow \quad A = I_n \quad \Longrightarrow \quad \det(A) = 1.$$

19. From Definition 3.2, the only time we get terms which do not contain a zero factor is when the terms involved come from A and B alone. Each one of the column permutations of terms from A can be associated with every one of the column permutations of B. Hence by factoring we have

$$\begin{vmatrix} A & O \\ C & B \end{vmatrix} = \sum (\text{terms from } A \text{ for any column permutations}) \det B$$

$$= \det B \sum \text{terms from } A \text{ for any column permutation})$$

$$= \det B \det A$$

21. Using row operations $-3\mathbf{r}_1 + \mathbf{r}_2$ and $\frac{3}{2}\mathbf{r}_3 + \mathbf{r}_4$, we obtain

$$\begin{vmatrix} 1 & 2 & 0 & 0 \\ 3 & 4 & 0 & 0 \\ 0 & 0 & 2 & 1 \\ 0 & 0 & -3 & 2 \end{vmatrix} = \begin{vmatrix} 1 & 2 & 0 & 0 \\ 0 & -2 & 0 & 0 \\ 0 & 0 & 2 & 1 \\ 0 & 0 & 0 & \frac{7}{2} \end{vmatrix} = -14.$$

On the other hand,

$$\begin{vmatrix} 1 & 2 \\ 3 & 4 \end{vmatrix} = -2 \quad \text{and} \quad \begin{vmatrix} 2 & 1 \\ -3 & 2 \end{vmatrix} = 7 \quad \Longrightarrow \quad \begin{vmatrix} 1 & 2 \\ 3 & 4 \end{vmatrix}\begin{vmatrix} 2 & 1 \\ -3 & 2 \end{vmatrix} = -14.$$

23. If $\det(A) = 2$, then $\det(A^5) = \det(AAAAA) = (\det(A))^5 = 32$.

25. (a) Using row operations we obtain

$$\begin{vmatrix} 1 & 3 & 2 \\ 2 & 1 & 4 \\ 1 & -7 & 2 \end{vmatrix} = \begin{vmatrix} 1 & 3 & 2 \\ 0 & -5 & 0 \\ 0 & -10 & 0 \end{vmatrix} = \begin{vmatrix} 1 & 3 & 2 \\ 0 & -5 & 0 \\ 0 & 0 & 0 \end{vmatrix} = 0.$$

Thus the matrix is singular.

(b) Using row operations we obtain

$$\begin{vmatrix} 1 & 2 & 0 & 5 \\ 3 & 4 & 1 & 7 \\ -2 & 5 & 2 & 0 \\ 0 & 1 & 2 & -7 \end{vmatrix} = \begin{vmatrix} 1 & 2 & 0 & 5 \\ 0 & 1 & 2 & -7 \\ 0 & 9 & 2 & 10 \\ 0 & -2 & 1 & -8 \end{vmatrix} = \begin{vmatrix} 1 & 2 & 0 & 5 \\ 0 & 1 & 2 & -7 \\ 0 & 0 & -16 & 73 \\ 0 & 0 & 0 & \frac{29}{16} \end{vmatrix} = -29 \neq 0.$$

Thus the matrix is nonsingular.

27. Compute the determinant of the coefficient matrix. We have

$$\begin{vmatrix} 1 & -2 & 1 \\ 2 & 3 & 1 \\ 3 & 1 & 2 \end{vmatrix} = 0,$$

thus the system has a nontrivial solution.

29. Since A is upper triangular, $\det(A) = a_{11}a_{22}\cdots a_{nn}$. From Theorem 3.8 we have that A is nonsingular if and only if $a_{11}a_{22}\cdots a_{nn} \neq 0$. Therefore A is nonsingular if and only if $a_{ii} \neq 0$ for $i = 1, 2, \ldots, n$.

31. (a) Since $\det(A) = 0$, A is a singular matrix by Theorem 3.8. Therefore the reduced row echelon form of A has at least one row of zeros.

(b) A row of zeros in the reduced row echelon form of A means that the solutions of $A\mathbf{x} = \mathbf{0}$ involves at least one real number that may assume any value. Hence there must be infinitely many solultions.

33. If A and B are similar, then there exists a nonsingular matrix P such that $B = P^{-1}AP$. Then

$$\det(B) = \det(P^{-1}BP) = \det(P^{-1})\det(A)\det(P) = \frac{1}{\det(P)}\det(P)\det(A) = \det(A).$$

Section 3.3, p. 164

1. Let $A = \begin{bmatrix} 1 & 0 & -2 \\ 3 & 1 & 4 \\ 5 & 2 & -3 \end{bmatrix}$.

(a) $\det(M_{13}) = \begin{vmatrix} 3 & 1 \\ 5 & 2 \end{vmatrix} = 1$ (b) $\det(M_{22}) = \begin{vmatrix} 1 & -2 \\ 5 & -3 \end{vmatrix} = 7$

(c) $\det(M_{31}) = \begin{vmatrix} 0 & -2 \\ 1 & 4 \end{vmatrix} = 2$ (d) $\det(M_{32}) = \begin{vmatrix} 1 & -2 \\ 3 & 4 \end{vmatrix} = 10.$

3. Let $A = \begin{bmatrix} -1 & 2 & 3 \\ -2 & 5 & 4 \\ 0 & 1 & -3 \end{bmatrix}$.

(a) $A_{13} = (-1)^{1+3}|M_{13}| = (-1)^4 \begin{vmatrix} -2 & 5 \\ 0 & 1 \end{vmatrix} = -2$

(b) $A_{21} = (-1)^{2+1}|M_{21}| = (-1)\begin{array}{c} 2+1 = 3 \\ \boxed{2} \end{array}\begin{vmatrix} 2 & 3 \\ 1 & -3 \end{vmatrix} = 9$

(c) $A_{32} = (-1)^{3+2}|M_{32}| = (-1)^5 \begin{vmatrix} -1 & 3 \\ -2 & 4 \end{vmatrix} = -2$

(d) $A_{33} = (-1)^{3+3}|M_{33}| = (-1)^6 \begin{vmatrix} -1 & 2 \\ -2 & 5 \end{vmatrix} = -1$

5. Referring to Exercise 1 in Section 3.2 we have the following.

   (a) Use cofactor expansion along the first row:

   $$\begin{vmatrix} 3 & 0 \\ 2 & 1 \end{vmatrix} = a_{11}A_{11} + a_{12}A_{12} = (3)(-1)^{1+1}|M_{11}| + (0)A_{12} = 3|1| = 3.$$

   (d) Use cofactor expansion along the third column to take advantage of the zero in the $(2,3)$ element:

   $$\begin{vmatrix} 4 & 1 & 3 \\ 2 & 3 & 0 \\ 1 & 3 & 2 \end{vmatrix} = a_{13}A_{13} + a_{23}A_{23} + a_{33}A_{33}$$

   $$= 3(-1)^{1+4}|M_{13}| + (0)A_{23} + 2(-1)^{3+3}|M_{33}|$$

   $$= 3\begin{vmatrix} 2 & 3 \\ 1 & 3 \end{vmatrix} + 2\begin{vmatrix} 4 & 1 \\ 2 & 3 \end{vmatrix}$$

   $$= 3(3) + 2(10) = 29.$$

   (e) Use cofactor expansion along the fourth column to take advantage of the three zeros. (The second row could also be used.)

   $$\begin{vmatrix} 4 & 2 & 2 & 0 \\ 2 & 0 & 0 & 0 \\ 3 & 0 & 0 & 1 \\ 0 & 0 & 1 & 0 \end{vmatrix} = a_{14}A_{14} + a_{24}A_{24} + a_{34}A_{34} + a_{44}A_{44}$$

   $$= (0)A_{14} + (0)A_{24} + (1)(-1)^{3+4}|M_{34}| + (0)A_{44}$$

   $$= (-1)\begin{vmatrix} 4 & 2 & 2 \\ 2 & 0 & 0 \\ 0 & 0 & 1 \end{vmatrix} \qquad \text{(expand along row 3)}$$

   $$= (-1)\left[(0)A_{31} + (0)A_{32} + (1)(-1)^{3+3}|M_{33}|\right]$$

   $$= (-1)\begin{vmatrix} 4 & 2 \\ 2 & 0 \end{vmatrix} = 4.$$

7. Referring to Exercise 2 in Section 3.2 we have the following.

   (a) Use cofactor expansion along the first row:

   $$\begin{vmatrix} 2 & -2 \\ 3 & -1 \end{vmatrix} = a_{11}A_{11} + a_{12}A_{12} = (2)(-1)^{1+1}|M_{11}| + (-2)(-1)^{1+2}|M_{12}| = (2)(-1) + 2(3) = 4.$$

   (c) Use cofactor expansion along the third column to take advantage of the two zeros in column 3:

   $$\begin{vmatrix} 3 & 4 & 2 \\ 2 & 5 & 0 \\ 3 & 0 & 0 \end{vmatrix} = a_{13}A_{13} + a_{23}A_{23} + a_{33}A_{33}$$

   $$= 2(-1)^{1+3}|M_{13}| + (0)A_{23} + (0)A_{33}$$

   $$= 2\begin{vmatrix} 2 & 5 \\ 3 & 0 \end{vmatrix} = (2)(-15) = -30.$$

(f) Use cofactor expansion along the first row to take advantage of the zero. (The third row could also be used.)

$$\begin{vmatrix} 2 & 0 & 1 & 4 \\ 3 & 2 & -4 & -2 \\ 2 & 3 & -1 & 0 \\ 11 & 8 & -4 & 6 \end{vmatrix} = a_{11}A_{11} + a_{12}A_{12} + a_{13}A_{13} + a_{14}A_{14}$$

$$= 2(-1)^{1+1}|M_{11}| + (0)A_{24} + (1)(-1)^{1+3}|M_{13}| + 4(-1)^{1+4}|M_{14}|$$

$$= 2\begin{vmatrix} 2 & -4 & -2 \\ 3 & -1 & 0 \\ 8 & -4 & 6 \end{vmatrix} + \begin{vmatrix} 3 & 2 & -2 \\ 2 & 3 & 0 \\ 11 & 8 & 6 \end{vmatrix} - 4\begin{vmatrix} 3 & 2 & -4 \\ 2 & 3 & -1 \\ 11 & 8 & -4 \end{vmatrix}$$

expand along column 3   expand along column 3   expand along row 1

$$= 2\left[-2(-1)^{1+3}\begin{vmatrix} 3 & -1 \\ 8 & -4 \end{vmatrix} + 6(-1)^{3+3}\begin{vmatrix} 2 & -4 \\ 3 & -1 \end{vmatrix} \right]$$

$$+ \left[-2(-1)^{1+3}\begin{vmatrix} 2 & 3 \\ 11 & 8 \end{vmatrix} + 6(-1)^{3+3}\begin{vmatrix} 3 & 2 \\ 2 & 3 \end{vmatrix} \right]$$

$$- 4\left[3(-1)^{1+1}\begin{vmatrix} 3 & -1 \\ 8 & -4 \end{vmatrix} + 2(-1)^{1+2}\begin{vmatrix} 2 & -1 \\ 11 & -4 \end{vmatrix} + (-4)(-1)^{1+3}\begin{vmatrix} 2 & 3 \\ 11 & 8 \end{vmatrix} \right]$$

$$= 2[(-2)(-4) + (6)(10)] + [(-2)(-17) + (6)(5)]$$
$$- 4[(3)(-4) - (2)(3) - (4)(-17)]$$
$$= 2(68) + (64) - 4(50) = 0$$

9. We proceed by successive expansions along first columns:

$$|A| = a_{11}\begin{vmatrix} a_{22} & a_{23} & \cdots & a_{2n} \\ 0 & a_{33} & \cdots & a_{3n} \\ \vdots & \vdots & \vdots & \vdots \\ 0 & 0 & \cdots & a_{nn} \end{vmatrix} = a_{11}a_{22}\begin{vmatrix} a_{33} & a_{34} & \cdots & a_{3n} \\ 0 & a_{44} & \cdots & a_{4n} \\ \vdots & \vdots & \vdots & \vdots \\ 0 & 0 & \cdots & a_{nn} \end{vmatrix} = \cdots = a_{11}a_{22}\cdots a_{nn}.$$

11. (a) $\begin{vmatrix} t-2 & 2 \\ 3 & t-3 \end{vmatrix} = (t-2)(t-3) - (2)(3) = t^2 - 5t = t(t-5) = 0$ only if $t = 0$ or $t = 5$.

(b) $\begin{vmatrix} t-1 & -4 \\ 0 & t-4 \end{vmatrix} = (t-1)(t-4) = 0$ only if $t = 1$ or $t = 4$.

13. (a) From Definition 3.2 each term in the expansion of the determinant of an $n \times n$ matrix is a product of n entries of the matrix. Each of these products contains exactly one entry from each row and exactly one entry from each column. Thus each such product from $\det(tI_n - A)$ contains at most n terms of the form $t - a_{ii}$. Hence each of these products is at most a polynomial of degree n. Since one of the products has the form $(t - a_{11})(t - a_{22})\cdots(t - a_{nn})$ it follows that the sum of the products is a polynomial of degree n in t.

(b) The coefficient of t^n is 1 since it only appears in the term $(t - a_{11})(t - a_{22})\cdots(t - a_{nn})$ which we discussed in part (a). (The permutation of the column indices is even here so a plus sign is associated with this term.)

(c) Using part (a), suppose that

$$\det(tI_n - A) = t^n + c_1 t^{n-1} + c_2 t^{n-2} + \cdots + c_{n-1}t + c_n.$$

Set $t = 0$ and we have $\det(-A) = c_n$ which implies that $c_n = (-1)^n \det(A)$. (See Exercise 10 in Section 3.2.)

15. (a) Using Equation (3), the area of the triangle is

$$\frac{1}{2}\left|\det\left(\begin{bmatrix} 3 & 3 & 1 \\ -1 & -1 & 1 \\ 4 & 1 & 1 \end{bmatrix}\right)\right| = \frac{1}{2}\left|(1)\begin{vmatrix} -1 & -1 \\ 4 & 1 \end{vmatrix} - (1)\begin{vmatrix} 3 & 3 \\ 4 & 1 \end{vmatrix} + (1)\begin{vmatrix} 3 & 3 \\ -1 & -1 \end{vmatrix}\right| = 6$$

(b) Here $L : R^2 \to R^2$ by $L(\mathbf{v}) = A\mathbf{v}$ for $\mathbf{v}$ in R^2. The image of the vertices under L are:

$$L\left(\begin{bmatrix} 3 \\ 3 \end{bmatrix}\right) = \begin{bmatrix} 4 & -3 \\ -4 & 2 \end{bmatrix}\begin{bmatrix} 3 \\ 3 \end{bmatrix} = \begin{bmatrix} 3 \\ -6 \end{bmatrix}$$

$$L\left(\begin{bmatrix} -1 \\ -1 \end{bmatrix}\right) = \begin{bmatrix} 4 & -3 \\ -4 & 2 \end{bmatrix}\begin{bmatrix} -1 \\ -1 \end{bmatrix} = \begin{bmatrix} -1 \\ 2 \end{bmatrix}$$

$$L\left(\begin{bmatrix} 4 \\ 1 \end{bmatrix}\right) = \begin{bmatrix} 4 & -3 \\ -4 & 2 \end{bmatrix}\begin{bmatrix} 4 \\ 1 \end{bmatrix} = \begin{bmatrix} 13 \\ -14 \end{bmatrix}$$

Thus the vertices of the image of the triangle T under L are $(3, -6)$, $(-1, 2)$, and $(13, -14)$.

(c) The area of the image triangle is

$$\frac{1}{2}\left|\det\left(\begin{bmatrix} 3 & -6 & 1 \\ -1 & 2 & 1 \\ 13 & -14 & 1 \end{bmatrix}\right)\right| = \frac{1}{2}\left|(1)\begin{vmatrix} -1 & 2 \\ 13 & -14 \end{vmatrix} - (1)\begin{vmatrix} 3 & -6 \\ 13 & -14 \end{vmatrix} + (1)\begin{vmatrix} 3 & -6 \\ -1 & 2 \end{vmatrix}\right| = 24.$$

17. We divide the quadrilateral into two triangles, one with vertices $(-2, 3)$, $(1, 4)$, and $(3, 0)$, the other with vertices $(-2, 3)$, $(3, 0)$, and $(-1, -3)$, find the area of each using Equation (2), then add the two areas together.

$$\text{area of triangle with vertices } (-2, 3), (1, 4), \text{ and } (3, 0) = \frac{1}{2}\left|\det\left(\begin{bmatrix} -2 & 3 & 1 \\ 1 & 4 & 1 \\ 3 & 0 & 1 \end{bmatrix}\right)\right| = 7$$

$$\text{area of triangle with vertices } (-2, 3), (3, 0), \text{ and } (-1, -3) = \frac{1}{2}\left|\det\left(\begin{bmatrix} -2 & 3 & 1 \\ 3 & 0 & 1 \\ -1 & -3 & 1 \end{bmatrix}\right)\right| = \frac{27}{2}$$

Therefore the area of quadrilateral Q is $7 + \frac{27}{2} = \frac{41}{2}$.

19. Let T be the triangle with vertices (x_1, y_1), (x_2, y_2), and (x_3, y_3). Let

$$A = \begin{bmatrix} a & b \\ c & d \end{bmatrix}$$

and define the linear operator $f : R^2 \to R^2$ by $f(\mathbf{v}) = A\mathbf{v}$ for $\mathbf{v}$ in R^2. The vertices of $f(T)$ are

$$(ax_1 + by_1, cx_1 + dy_1), \quad (ax_2 + by_2, cx_2 + dy_2), \quad \text{and} \quad (ax_3 + by_3, cx_3 + dy_3).$$

Then by Equation (3),

$$\text{Area of } T = \frac{1}{2}\left|x_1 y_2 - x_1 y_3 - x_2 y_1 + x_2 y_3 + x_3 y_1 - x_3 y_2\right|$$

and

$$\text{Area of } f(T) = \frac{1}{2}|ax_1 dy_2 - ax_1 dy_3 - ax_2 dy_1 + ax_2 dy_3 + ax_3 dy_1 - ax_3 dy_2$$
$$- bcx_1 y_2 + bcx_1 y_3 + bcx_2 y_1 - bcx_2 y_3 - bcx_3 y_1 + bcx_3 y_2|$$

Now,

$$|\det(A)| \cdot \text{Area of } T = |ad - bc| \frac{1}{2} |x_1y_2 - x_1y_3 - x_2y_1 + x_2y_3 + x_3y_1 - x_3y_2|$$

$$= \frac{1}{2} |ax_1dy_2 - ax_1dy_3 - ax_2dy_1 + ax_2dy_3 + ax_3dy_1 - ax_3dy_2$$

$$- bcx_1y_2 + bcx_1y_3 + bcx_2y_1 - bcx_2y_3 - bcx_3y_1 + bcx_3y_2|$$

$$= \text{Area of } f(T)$$

Section 3.4, p. 169

1. Let $A = \begin{bmatrix} -2 & 3 & 0 \\ 4 & 1 & -3 \\ 2 & 0 & 1 \end{bmatrix}$. Then

$$a_{11}A_{12} + a_{21}A_{22} + a_{31}A_{32} = (-2)(-1)^{1+2}\begin{vmatrix} 4 & -3 \\ 2 & 1 \end{vmatrix} + (4)(-1)^{2+2}\begin{vmatrix} -2 & 0 \\ 2 & 1 \end{vmatrix} + (2)(-1)^{3+2}\begin{vmatrix} -2 & 0 \\ 4 & -3 \end{vmatrix}$$

$$= (2)(10) + (4)(-2) - (2)(6) = 0$$

3. Let $A = \begin{bmatrix} 6 & 2 & 8 \\ -3 & 4 & 1 \\ 4 & -4 & 5 \end{bmatrix}$.

(a) $\text{adj } A = \begin{bmatrix} A_{11} & A_{21} & A_{31} \\ A_{12} & A_{22} & A_{32} \\ A_{13} & A_{23} & A_{33} \end{bmatrix} = \begin{bmatrix} 24 & -42 & -30 \\ 19 & -2 & -30 \\ -4 & 32 & 30 \end{bmatrix}$ where

$$A_{11} = (-1)^{1+1}\begin{vmatrix} 4 & 1 \\ -4 & 5 \end{vmatrix} = 24, \quad A_{21} = (-1)^{2+1}\begin{vmatrix} 2 & 8 \\ -4 & 5 \end{vmatrix} = -42, \quad A_{31} = (-1)^{3+1}\begin{vmatrix} 2 & 8 \\ 4 & 1 \end{vmatrix} = -30,$$

$$A_{12} = (-1)^{1+2}\begin{vmatrix} -3 & 1 \\ 4 & 5 \end{vmatrix} = 19, \quad A_{22} = (-1)^{2+2}\begin{vmatrix} 6 & 8 \\ 4 & 5 \end{vmatrix} = -2, \quad A_{32} = (-1)^{3+2}\begin{vmatrix} 6 & 8 \\ -3 & 1 \end{vmatrix} = -30,$$

$$A_{13} = (-1)^{1+3}\begin{vmatrix} -3 & 4 \\ 4 & -4 \end{vmatrix} = -4, \quad A_{23} = (-1)^{2+3}\begin{vmatrix} 6 & 2 \\ 4 & -4 \end{vmatrix} = 32, \quad A_{33} = (-1)^{3+3}\begin{vmatrix} 6 & 2 \\ -3 & 4 \end{vmatrix} = 30$$

(b) $\det(A) = a_{11}A_{11} + a_{12}A_{12} + a_{13}A_{13} = (6)(24) + (2)(19) + (8)(-4) = 150$

(c) $A(\text{adj } A) = \begin{bmatrix} 6 & 2 & 8 \\ -3 & 4 & 1 \\ 4 & -4 & 5 \end{bmatrix}\begin{bmatrix} 24 & -42 & -30 \\ 19 & -2 & -30 \\ -4 & 32 & 30 \end{bmatrix} = 150I_3 = \det(A)I_3$

$(\text{adj } A)A = \begin{bmatrix} 24 & -42 & -30 \\ 19 & -2 & -30 \\ -4 & 32 & 30 \end{bmatrix}\begin{bmatrix} 6 & 2 & 8 \\ -3 & 4 & 1 \\ 4 & -4 & 5 \end{bmatrix} = 150I_3 = \det(A)I_3$

5. (a) Let

$$A = \begin{bmatrix} 1 & 1 & 1 \\ 1 & 2 & 3 \\ 0 & 1 & 1 \end{bmatrix}.$$

Using the technique of Example 8 in Section 3.1 we have $\det(A) = -1$. Computing the cofactors we have $A_{11} = -1$, $A_{12} = -1$, $A_{13} = 1$, $A_{21} = 0$, $A_{22} = 1$, $A_{23} = -1$, $A_{31} = 1$, $A_{32} = -2$, $A_{33} = 1$. Hence

$$\text{adj } A = \begin{bmatrix} -1 & 0 & 1 \\ -1 & 1 & -2 \\ 1 & -1 & 1 \end{bmatrix}.$$

Thus it follows from Corollary 3.4 that

$$A^{-1} = \frac{1}{\det(A)}(\text{adj } A) = (-1)(\text{adj } A) = \begin{bmatrix} 1 & 0 & -1 \\ 1 & -1 & 2 \\ -1 & 1 & -1 \end{bmatrix}.$$

(b) Let

$$A = \begin{bmatrix} 1 & 1 & 1 & 1 \\ 1 & 2 & -1 & 2 \\ 1 & -1 & 2 & 1 \\ 1 & 3 & 3 & 2 \end{bmatrix}.$$

Using the technique of Example 8 in Section 3.2 we find that $\det(A) = -9$. Computing the cofactors of the 16 terms we have

$$\text{adj } A = \begin{bmatrix} -21 & 3 & 3 & 6 \\ -4 & 1 & 4 & -1 \\ 1 & 2 & -1 & -2 \\ 15 & -6 & -6 & -3 \end{bmatrix}.$$

Thus

$$A^{-1} = \frac{1}{\det(A)}(\text{adj } A) = -\frac{1}{9}(\text{adj } A) = \begin{bmatrix} \frac{7}{3} & -\frac{1}{3} & -\frac{1}{3} & -\frac{2}{3} \\ \frac{4}{9} & -\frac{1}{9} & -\frac{4}{9} & \frac{1}{9} \\ -\frac{1}{9} & -\frac{2}{9} & \frac{1}{9} & \frac{2}{9} \\ -\frac{5}{3} & \frac{2}{3} & \frac{2}{3} & \frac{1}{3} \end{bmatrix}.$$

(c) Let

$$A = \begin{bmatrix} 1 & 1 & 1 & 1 \\ 1 & 3 & 1 & 2 \\ 1 & 2 & -1 & 1 \\ 5 & 9 & 1 & 6 \end{bmatrix}.$$

Using the technique of Example 8 in Section 3.2 we find that $\det(A) = 0$. Hence A^{-1} does not exist.

(d) Let

$$A = \begin{bmatrix} 1 & 2 & 1 \\ 1 & 3 & 2 \\ 1 & 0 & 1 \end{bmatrix}.$$

Using the technique of Example 8 in Section 3.1 we have $\det(A) = 2$. Computing the cofactors we have $A_{11} = 3$, $A_{12} = 1$, $A_{13} = -3$, $A_{21} = -2$, $A_{22} = 0$, $A_{23} = 2$, $A_{31} = 1$, $A_{32} = -1$, $A_{33} = 1$. Hence

$$\text{adj } A = \begin{bmatrix} 3 & -2 & 1 \\ 1 & 0 & -1 \\ -3 & 2 & 1 \end{bmatrix}.$$

Thus it follows from Corollary 3.4 that

$$A^{-1} = \frac{1}{\det(A)}(\text{adj } A) = \frac{1}{2}(\text{adj } A) = \begin{bmatrix} \frac{3}{2} & -1 & \frac{1}{2} \\ \frac{1}{2} & 0 & -\frac{1}{2} \\ -\frac{3}{2} & 1 & \frac{1}{2} \end{bmatrix}.$$

(e) Let

$$A = \begin{bmatrix} 1 & 2 & 2 \\ 1 & 3 & 1 \\ 1 & 1 & 3 \end{bmatrix}.$$

Using the technique of Example 8 in Section 3.1 we have that $\det(A) = 0$. Thus A^{-1} does not exist.

7. (a) Let
$$A = \begin{bmatrix} 0 & 2 & 1 & 3 \\ 2 & -1 & 3 & 4 \\ -2 & 1 & 5 & 2 \\ 0 & 1 & 0 & 2 \end{bmatrix}.$$

Using the technique of Example 8 in Section 3.2 we find that $\det(A) = -28$. Computing the cofactors of the 16 terms we have

$$\text{adj } A = \begin{bmatrix} -30 & -5 & 9 & 46 \\ -32 & 4 & 4 & 36 \\ -12 & -2 & -2 & 24 \\ 16 & -2 & -2 & -32 \end{bmatrix}.$$

Thus

$$A^{-1} = \frac{1}{\det(A)}(\text{adj } A) = -\frac{1}{28}(\text{adj } A) = \begin{bmatrix} \frac{15}{14} & \frac{5}{28} & -\frac{9}{28} & -\frac{23}{14} \\ \frac{8}{7} & -\frac{1}{7} & -\frac{1}{7} & -\frac{9}{7} \\ \frac{3}{7} & \frac{1}{14} & \frac{1}{14} & -\frac{6}{7} \\ -\frac{4}{7} & \frac{1}{14} & \frac{1}{14} & \frac{8}{7} \end{bmatrix}.$$

(b) Let
$$A = \begin{bmatrix} 4 & 2 & 2 \\ 0 & 1 & 2 \\ 1 & 0 & 3 \end{bmatrix}.$$

Using the technique of Example 8 in Section 3.1 we have $\det(A) = 14$. Computing the cofactors we have $A_{11} = 3$, $A_{12} = 2$, $A_{13} = -1$, $A_{21} = -6$, $A_{22} = 10$, $A_{23} = 2$, $A_{31} = 2$, $A_{32} = -8$, $A_{33} = 4$. Hence

$$\text{adj } A = \begin{bmatrix} 3 & -6 & 2 \\ 2 & 10 & -8 \\ -1 & 2 & 4 \end{bmatrix}.$$

Thus it follows from Corollary 3.4 that

$$A^{-1} = \frac{1}{\det(A)}(\text{adj } A) = \frac{1}{14}(\text{adj } A) = \begin{bmatrix} \frac{3}{14} & -\frac{3}{7} & \frac{1}{7} \\ \frac{1}{7} & \frac{5}{7} & -\frac{4}{7} \\ -\frac{1}{14} & \frac{1}{7} & \frac{2}{7} \end{bmatrix}.$$

(c) Let $A = \begin{bmatrix} 3 & 2 \\ -3 & 4 \end{bmatrix}$. Then $\det(A) = 18$ and $\text{adj } A = \begin{bmatrix} 4 & -2 \\ 3 & 3 \end{bmatrix}$. Hence

$$A^{-1} = \frac{1}{\det(A)}(\text{adj } A) = \begin{bmatrix} \frac{2}{9} & -\frac{1}{9} \\ \frac{1}{6} & \frac{1}{6} \end{bmatrix}.$$

9. Let $A = \begin{bmatrix} a & b \\ c & d \end{bmatrix}$. Then $\det(A) = ad - bc$ which we are told is not zero. It follows that

$$\text{adj } A = \begin{bmatrix} d & -b \\ -c & a \end{bmatrix} \quad \text{and hence} \quad A^{-1} = \frac{1}{ad - bc}\begin{bmatrix} d & -b \\ -c & a \end{bmatrix}.$$

11. Let $A = \begin{bmatrix} 4 & 0 & 0 \\ 0 & -3 & 0 \\ 0 & 0 & 2 \end{bmatrix}$. Then $\det(A) = -24$ and we find that

$$\text{adj } A = \begin{bmatrix} -6 & 0 & 0 \\ 0 & 8 & 0 \\ 0 & 0 & -12 \end{bmatrix}.$$

Hence it follows that

$$A^{-1} = \begin{bmatrix} \frac{1}{4} & 0 & 0 \\ 0 & -\frac{1}{3} & 0 \\ 0 & 0 & \frac{1}{2} \end{bmatrix}.$$

13. We follow the hint. If A is singular then $\det(A) = 0$. Hence $A(\text{adj } A) = \det(A)\, I_n = 0 I_n = O$. If adj A were nonsingular, $(\text{adj } A)^{-1}$ exists. Then we have

$$A(\text{adj } A)(\text{adj } A)^{-1} = A = O(\text{adj } A)^{-1} = O,$$

that is, $A = O$. But the adjoint of the zero matrix must be a matrix of all zeros. Thus adj $A = O$ so adj A is singular. This is a contradiction. Hence it follows that adj A is singular.

Section 3.5, p. 172

1. Follow the steps used in Example 1. Let $A = \begin{bmatrix} 2 & 4 & 6 \\ 1 & 0 & 2 \\ 2 & 3 & -1 \end{bmatrix}$. We have

$$|A| = \begin{vmatrix} 2 & 4 & 6 \\ 1 & 0 & 2 \\ 2 & 3 & -1 \end{vmatrix} = 26.$$

Then

$$x_1 = \frac{\begin{vmatrix} 2 & 4 & 6 \\ 1 & 0 & 2 \\ 2 & 3 & -1 \end{vmatrix}}{|A|} = \frac{-52}{26} = -2, \quad x_2 = \frac{\begin{vmatrix} 2 & 2 & 6 \\ 1 & 0 & 2 \\ 2 & -5 & -1 \end{vmatrix}}{|A|} = \frac{0}{26} = 0, \quad x_3 = \frac{\begin{vmatrix} 2 & 4 & 2 \\ 1 & 0 & 0 \\ 2 & 3 & -5 \end{vmatrix}}{|A|} = \frac{26}{26} = 1.$$

3. $x_3 = \dfrac{\begin{vmatrix} 2 & 1 & 6 \\ 3 & 2 & -2 \\ 1 & 1 & -4 \end{vmatrix}}{\begin{vmatrix} 2 & 1 & 1 \\ 3 & 2 & -2 \\ 1 & 1 & 2 \end{vmatrix}} = \dfrac{4}{5}.$

5. Follow the steps used in Example 1. Let $A = \begin{bmatrix} 2 & -1 & 3 \\ 1 & 2 & -3 \\ 4 & 2 & 1 \end{bmatrix}$. We have

$$|A| = \begin{vmatrix} 2 & -1 & 3 \\ 1 & 2 & -3 \\ 4 & 2 & 1 \end{vmatrix} = 11.$$

Then

$$x_1 = \frac{\begin{vmatrix} 0 & -1 & 3 \\ 0 & 2 & -3 \\ 0 & 2 & 1 \end{vmatrix}}{|A|} = \frac{0}{11} = 0, \quad x_2 = \frac{\begin{vmatrix} 2 & 0 & 3 \\ 1 & 0 & -3 \\ 4 & 0 & 1 \end{vmatrix}}{|A|} = \frac{0}{11} = 0, \quad x_3 = \frac{\begin{vmatrix} 2 & -1 & 0 \\ 1 & 2 & 0 \\ 4 & 2 & 0 \end{vmatrix}}{|A|} = \frac{0}{13} = 0.$$

Note that since $|A| \neq 0$, this homogeneous system has a nonsingular coefficient matrix. Hence, by previous work, we are guaranteed that the only solution is the trivial solution.

7. Let A be the coefficient matrix of the linear system. Then

$$\det(A) = \begin{vmatrix} 2 & 3 & 7 \\ -2 & 0 & -4 \\ 1 & 2 & 4 \end{vmatrix} = 0.$$

Since the determinant is 0, Cramer's rule cannot be used to solve the linear system.

Supplementary Exercises for Chapter 3, p. 174

1. (a) Use the "trick" for evaluating 3×3 determinants in Example 8 of Section 3.1.

$$\begin{vmatrix} 2 & 3 & 4 \\ 1 & 2 & 4 \\ 4 & 3 & 1 \end{vmatrix} = 5$$

(b) Use the "trick" for evaluating 3×3 determinants in Example 8 of Section 3.1.

$$\begin{vmatrix} 2 & 1 & 0 \\ 1 & 2 & 1 \\ 0 & 1 & 2 \end{vmatrix} = 4$$

(c) Use the method that employs row and/or column operations that appears in Example 8 of Section 3.2.

$$\begin{vmatrix} 2 & 1 & -1 & 2 \\ 2 & -3 & -1 & 4 \\ 1 & 3 & 2 & -3 \\ 1 & -2 & -1 & 1 \end{vmatrix} = (-1) \begin{vmatrix} 1 & 3 & 2 & -3 \\ 2 & -3 & -1 & 4 \\ 2 & 1 & -1 & 2 \\ 1 & -2 & -1 & 1 \end{vmatrix} = (-1) \begin{vmatrix} 1 & 3 & 2 & -3 \\ 0 & -9 & -5 & 10 \\ 0 & -5 & -5 & 8 \\ 0 & -5 & -3 & 4 \end{vmatrix}$$

$$\mathbf{r}_1 \leftrightarrow \mathbf{r}_3 \qquad\qquad -2\ \mathbf{r}_1 + \mathbf{r}_2 \qquad\qquad -\tfrac{1}{5}\mathbf{r}_3$$
$$-2\ \mathbf{r}_1 + \mathbf{r}_3 \qquad\qquad \mathbf{r}_2 \leftrightarrow \mathbf{r}_3$$
$$-1\ \mathbf{r}_1 + \mathbf{r}_4$$

$$= (-1)(-5)(-1) \begin{vmatrix} 1 & 3 & 2 & -3 \\ 0 & 1 & 1 & -\tfrac{8}{5} \\ 0 & -9 & -5 & 10 \\ 0 & -5 & -3 & 4 \end{vmatrix} = (-5) \begin{vmatrix} 1 & 3 & 2 & -3 \\ 0 & 1 & 1 & -\tfrac{8}{5} \\ 0 & 0 & 4 & -\tfrac{22}{5} \\ 0 & 0 & 2 & -4 \end{vmatrix} = (-5) \begin{vmatrix} 1 & 3 & 2 & -3 \\ 0 & 1 & 1 & -\tfrac{8}{5} \\ 0 & 0 & 4 & -\tfrac{22}{5} \\ 0 & 0 & 0 & -\tfrac{9}{5} \end{vmatrix}$$

$$9\ \mathbf{r}_2 + \mathbf{r}_3 \qquad\qquad -\tfrac{1}{2}\ \mathbf{r}_3 + \mathbf{r}_4$$
$$5\ \mathbf{r}_2 + \mathbf{r}_4$$

$$= (-5)(1)(1)(4)(-\tfrac{9}{5}) = 36$$

(d) Proceeding as in part (c), we have

$$\begin{vmatrix} 2 & 1 & 0 & 0 \\ 1 & 2 & 1 & 0 \\ 0 & 1 & 2 & 1 \\ 0 & 0 & 1 & 2 \end{vmatrix} = \begin{vmatrix} 2 & 1 & 0 & 0 \\ 0 & \tfrac{3}{2} & 1 & 0 \\ 0 & 1 & 2 & 1 \\ 0 & 0 & 1 & 2 \end{vmatrix} = \begin{vmatrix} 2 & 1 & 0 & 0 \\ 0 & \tfrac{3}{2} & 1 & 0 \\ 0 & 0 & \tfrac{4}{3} & 1 \\ 0 & 0 & 1 & 2 \end{vmatrix} = \begin{vmatrix} 2 & 1 & 0 & 0 \\ 0 & \tfrac{3}{2} & 1 & 0 \\ 0 & 0 & \tfrac{4}{3} & 1 \\ 0 & 0 & 0 & \tfrac{5}{4} \end{vmatrix} = (2)(\tfrac{3}{2})(\tfrac{4}{3})(\tfrac{5}{4}) = 5$$

$$-\tfrac{1}{2}\ \mathbf{r}_1 + \mathbf{r}_2 \qquad -\tfrac{2}{3}\ \mathbf{r}_2 + \mathbf{r}_3 \qquad -\tfrac{3}{4}\ \mathbf{r}_3 + \mathbf{r}_4$$

3. If $A^n = O$ for some positive integer n, then

$$0 = \det(O) = \det(A^n) = \det \underbrace{(A\,A \cdots A)}_{n \text{ times}} = \underbrace{\det(A)\det(A) \cdots \det(A)}_{n \text{ times}} = (\det(A))^n.$$

It follows that $\det(A) = 0$.

5. If A is an $n \times n$ matrix then

$$\det(AA^T) = \det(A)\det(A^T) = \det(A)\det(A) = (\det(A))^2.$$

(Here we used Theorems 3.9 and 3.1.) Since the square of any real number is ≥ 0 we have $\det(AA^T) \geq 0$.

7. Since A is nonsingular, Corollary 3.4 implies that

$$A^{-1} = \frac{1}{\det(A)}(\text{adj } A).$$

Multiplying both sides on the left by A gives

$$AA^{-1} = I_n = \frac{1}{\det(A)}A(\text{adj } A).$$

Hence we have that

$$(\text{adj } A)^{-1} = \frac{1}{\det(A)}A.$$

From Corollary 3.4 it follows that for any nonsingular matrix B, adj $B = \det(B)B^{-1}$. Let $B = A^{-1}$ and we have

$$\text{adj } (A^{-1}) = \det(A^{-1})(A^{-1})^{-1} = \frac{1}{\det(A)}A = (\text{adj } A)^{-1}.$$

9. Matrix Q is $n \times n$ with each entry equal to 1. Then, adding row j to row 1 for $j = 2, 3, \ldots, n$, we have

$$\det(Q - nI_n) = \begin{vmatrix} 1-n & 1 & 1 & \cdots & 1 \\ 1 & 1-n & 1 & \cdots & 1 \\ \vdots & \vdots & \vdots & \cdots & \vdots \\ 1 & 1 & 1 & \cdots & 1-n \end{vmatrix} = \begin{vmatrix} 0 & 0 & 0 & \cdots & 0 \\ 1 & 1-n & 1 & \cdots & 1 \\ \vdots & \vdots & \vdots & \cdots & \vdots \\ 1 & 1 & 1 & \cdots & 1-n \end{vmatrix} = 0$$

by Theorem 3.4.

11. If A and $\mathbf{b}$ have integer entries and $\det(A) = \pm 1$, then using Cramer's rule to solve $A\mathbf{x} = \mathbf{b}$, we find that the numerator in the fraction giving x_i is an integer and the denominator is ± 1, so x_i is an integer for $i = 1, 2, \ldots, n$.

Chapter Review for Chapter 3, p. 174

True or False

1. False.	2. True.	3. False.	4. True.	5. True.	6. False.
7. False.	8. True.	9. True.	10. False.	11. True.	12. False.

Quiz

1. $\det(6A^T BC^{-1}) = 6^2 \det(A)\det(B)\dfrac{1}{\det(C)} = (36)(3)(-2)\left(\frac{1}{4}\right) = -54.$

2. False. Let $A = \begin{bmatrix} 1 & 2 \\ 3 & 4 \end{bmatrix}$. Then

$$B = \begin{bmatrix} 6 & 7 \\ 8 & 9 \end{bmatrix} \implies \det(B) = 54 - 56 = -2$$

but $5 + \det(A) = 5 + (-2) = 3$.

3. $\det(B) = (2)(-1)\det(A) = 2 \implies \det(A) = -1.$

4. First add 2 times row 1 to row 3 and (-3) times row 1 to row 4:

$$\begin{bmatrix} 1 & 2 & 1 & 0 \\ 0 & 1 & 2 & 3 \\ 0 & 4 & 1 & -1 \\ 0 & -6 & -3 & -1 \end{bmatrix}.$$

Next, add (-4) times row 2 to row 3 and 6 times row 2 to row 4:

$$\begin{bmatrix} 1 & 2 & 1 & 0 \\ 0 & 1 & 2 & 3 \\ 0 & 0 & -7 & -13 \\ 0 & 0 & 9 & 17 \end{bmatrix}.$$

Next, multiply row 4 by 7:

$$\begin{bmatrix} 1 & 2 & 1 & 0 \\ 0 & 1 & 2 & 3 \\ 0 & 0 & -7 & -13 \\ 0 & 0 & 63 & 119 \end{bmatrix}.$$

Now add 6 times row 3 to row 4:

$$\begin{bmatrix} 1 & 2 & 1 & 0 \\ 0 & 1 & 2 & 3 \\ 0 & 0 & -7 & -13 \\ 0 & 0 & 0 & 2 \end{bmatrix}.$$

Therefore, $\det(A) = \frac{1}{7}(1)(1)(-7)(2) = -2$.

5. Let the diagonal entries of A be $d_{11}, \ldots, d_{nn}$. Then $\det(A) = d_{11} \cdots d_{nn}$. Since A is singular if and only if $\det(A) = 0$, A is singular if and only if some diagonal entry d_{ii} is zero.

6. Expand along the 4th row:

$$\det(A) = (-1)\det\left(\begin{bmatrix} 1 & 2 & 0 \\ 3 & 4 & 5 \\ 0 & -1 & 1 \end{bmatrix}\right) + (-1)\det\left(\begin{bmatrix} 0 & 2 & 0 \\ 1 & 4 & 5 \\ -2 & -1 & 1 \end{bmatrix}\right) = (-1)(3) + (-1)(-22) = 19.$$

7. $\det(A) = 2$. Now

$$\begin{array}{lll} A_{11} = -1 & A_{21} = 5 & A_{31} = 2 \\ A_{12} = 2 & A_{22} = -6 & A_{32} = -2 \\ A_{13} = -2 & A_{23} = 8 & A_{33} = 2 \end{array}$$

Therefore,

$$A^{-1} = \frac{1}{2}\begin{bmatrix} -1 & 5 & 2 \\ 2 & -6 & -2 \\ -2 & 8 & 2 \end{bmatrix} = \begin{bmatrix} -\frac{1}{2} & \frac{5}{2} & 1 \\ 1 & -3 & -1 \\ -1 & 4 & 1 \end{bmatrix}.$$

8. $\det(A) = 14$. Therefore

$$x_1 = \frac{\det\left(\begin{bmatrix} 1 & 4 & 2 \\ -1 & 2 & 2 \\ 4 & 0 & 1 \end{bmatrix}\right)}{14} = \frac{11}{7}, \quad x_2 = \frac{\det\left(\begin{bmatrix} 3 & 1 & 2 \\ 1 & -1 & 2 \\ 3 & 4 & 1 \end{bmatrix}\right)}{14} = -\frac{4}{7},$$

$$x_3 = \frac{\det\left(\begin{bmatrix} 3 & 4 & 1 \\ 1 & 2 & -1 \\ 3 & 0 & 4 \end{bmatrix}\right)}{14} = -\frac{5}{7}.$$

Chapter 4

Real Vector Spaces

Section 4.1, p. 187

1. (a) $\mathbf{u} = \begin{bmatrix} -2 \\ 3 \end{bmatrix}$ (b) $\mathbf{v} = \begin{bmatrix} 3 \\ 4 \end{bmatrix}$ (c) $\mathbf{w} = \begin{bmatrix} -3 \\ -3 \end{bmatrix}$ (d) $\mathbf{z} = \begin{bmatrix} 0 \\ -3 \end{bmatrix}$

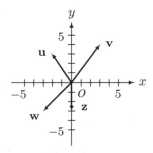

Exercise 1

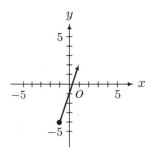

Exercise 3

3. Let the tail of the vector $\begin{bmatrix} 2 \\ 6 \end{bmatrix}$ be denoted by $P(x, y)$. Since the head is given as $Q(1, 2)$, we have that

$$\begin{bmatrix} 2 \\ 6 \end{bmatrix} = \begin{bmatrix} 1 - x \\ 2 - y \end{bmatrix} \implies \begin{array}{ccc} 2 & = & 1 - x \\ 6 & = & 2 - y \end{array} \implies \begin{array}{ccc} x & = & -1 \\ y & = & -4 \end{array}$$

5. The two vectors are equal provided corresponding components are equal. That is,

$$\begin{bmatrix} a - b \\ 2 \end{bmatrix} = \begin{bmatrix} 4 \\ a + b \end{bmatrix} \iff a - b = 4 \quad \text{and} \quad a + b = 2.$$

Solve the corresponding linear system to obtain $a = 3$, $b = -1$.

7. (a) The vector from P to Q for $P(1, 2)$ and $Q(3, 5)$ is given by

$$\begin{bmatrix} 3 - 1 \\ 5 - 2 \end{bmatrix} = \begin{bmatrix} 2 \\ 3 \end{bmatrix}.$$

(b) The vector from P to Q for $P(-2, 2, 3)$ and $Q(-3, 5, 2)$ is given by

$$\begin{bmatrix} -3 - (-2) \\ 5 - 2 \\ 2 - 3 \end{bmatrix} = \begin{bmatrix} -1 \\ 3 \\ -1 \end{bmatrix}.$$

9. We are to find a vector with tail at the origin representing the vector from P to Q.

    (a) For $P(-1, 2)$ and $Q(3, 5)$, the vector from P to Q is given by

$$\mathbf{v} = \begin{bmatrix} 3 - (-1) \\ 5 - 2 \end{bmatrix} = \begin{bmatrix} 4 \\ 3 \end{bmatrix}.$$

    Then $\mathbf{v}$ can be considered as a vector with tail at $(0,0)$ and head at $(4,3)$.

    (b) For $P(1, 1, -2)$ and $Q(3, 4, 5)$, the vector from P to Q is given by

$$\mathbf{v} = \begin{bmatrix} 3 - 1 \\ 4 - 1 \\ 5 - (-2) \end{bmatrix} = \begin{bmatrix} 2 \\ 3 \\ 7 \end{bmatrix}.$$

    Then $\mathbf{v}$ can be considered as a vector with tail at $(0,0,0)$ and head at $(2,3,7)$.

11.   (a) For $\mathbf{u} = \begin{bmatrix} 2 \\ 3 \end{bmatrix}$ and $\mathbf{v} = \begin{bmatrix} -2 \\ 5 \end{bmatrix}$, we find that

$$\mathbf{u} + \mathbf{v} = \begin{bmatrix} 2 + (-2) \\ 3 + 5 \end{bmatrix} = \begin{bmatrix} 0 \\ 8 \end{bmatrix} \qquad \mathbf{u} - \mathbf{v} = \begin{bmatrix} 2 - (-2) \\ 3 - 5 \end{bmatrix} = \begin{bmatrix} 4 \\ -2 \end{bmatrix}$$

$$2\mathbf{u} = 2 \begin{bmatrix} 2 \\ 3 \end{bmatrix} = \begin{bmatrix} 4 \\ 6 \end{bmatrix} \qquad 3\mathbf{u} - 2\mathbf{v} = \begin{bmatrix} 3(2) - 2(-2) \\ 3(3) - 2(5) \end{bmatrix} = \begin{bmatrix} 10 \\ -1 \end{bmatrix}.$$

    (b) For $\mathbf{u} = \begin{bmatrix} 0 \\ 3 \end{bmatrix}$ and $\mathbf{v} = \begin{bmatrix} 3 \\ 2 \end{bmatrix}$, we find that

$$\mathbf{u} + \mathbf{v} = \begin{bmatrix} 0 + 3 \\ 3 + 2 \end{bmatrix} = \begin{bmatrix} 3 \\ 5 \end{bmatrix} \qquad \mathbf{u} - \mathbf{v} = \begin{bmatrix} 0 - 3 \\ 3 - 2 \end{bmatrix} = \begin{bmatrix} -3 \\ 1 \end{bmatrix}$$

$$2\mathbf{u} = 2 \begin{bmatrix} 0 \\ 3 \end{bmatrix} = \begin{bmatrix} 0 \\ 6 \end{bmatrix} \qquad 3\mathbf{u} - 2\mathbf{v} = \begin{bmatrix} 3(0) - 2(3) \\ 3(3) - 2(2) \end{bmatrix} = \begin{bmatrix} -6 \\ 5 \end{bmatrix}.$$

    (c) For $\mathbf{u} = \begin{bmatrix} 2 \\ 6 \end{bmatrix}$ and $\mathbf{v} = \begin{bmatrix} 3 \\ 2 \end{bmatrix}$, we find that

$$\mathbf{u} + \mathbf{v} = \begin{bmatrix} 2 + 3 \\ 6 + 2 \end{bmatrix} = \begin{bmatrix} 5 \\ 8 \end{bmatrix} \qquad \mathbf{u} - \mathbf{v} = \begin{bmatrix} 2 - 3 \\ 6 - 2 \end{bmatrix} = \begin{bmatrix} -1 \\ 4 \end{bmatrix}$$

$$2\mathbf{u} = 2 \begin{bmatrix} 2 \\ 6 \end{bmatrix} = \begin{bmatrix} 4 \\ 12 \end{bmatrix} \qquad 3\mathbf{u} - 2\mathbf{v} = \begin{bmatrix} 3(2) - 2(3) \\ 3(6) - 2(2) \end{bmatrix} = \begin{bmatrix} 0 \\ 14 \end{bmatrix}$$

13. For $\mathbf{u} = \begin{bmatrix} 2 \\ 3 \\ -1 \end{bmatrix}$, $\mathbf{v} = \begin{bmatrix} -1 \\ 2 \\ 4 \end{bmatrix}$, $\mathbf{w} = \begin{bmatrix} 0 \\ 1 \\ -1 \end{bmatrix}$, $c = -2$, and $d = 3$, we have:

    (a) $\mathbf{u} + \mathbf{v} = \begin{bmatrix} 1 \\ 5 \\ 3 \end{bmatrix}$ (b) $c\mathbf{u} + d\mathbf{w} = \begin{bmatrix} -4 \\ -6 \\ 2 \end{bmatrix} + \begin{bmatrix} 0 \\ 3 \\ -3 \end{bmatrix} = \begin{bmatrix} -4 \\ -3 \\ -1 \end{bmatrix}$

    (c) $\mathbf{u} + \mathbf{v} + \mathbf{w} = \begin{bmatrix} 1 \\ 6 \\ 2 \end{bmatrix}$ (d) $c\mathbf{u} + d\mathbf{v} + \mathbf{w} = \begin{bmatrix} -4 \\ -6 \\ 2 \end{bmatrix} + \begin{bmatrix} -3 \\ 6 \\ 12 \end{bmatrix} + \begin{bmatrix} 0 \\ 1 \\ -1 \end{bmatrix} = \begin{bmatrix} -7 \\ 1 \\ 13 \end{bmatrix}$

15. Let

$$\mathbf{x} = \begin{bmatrix} 1 \\ -2 \\ 3 \end{bmatrix}, \quad \mathbf{y} = \begin{bmatrix} -3 \\ 1 \\ 3 \end{bmatrix}, \quad \mathbf{z} = \begin{bmatrix} r \\ -1 \\ s \end{bmatrix}, \quad \mathbf{u} = \begin{bmatrix} 3 \\ t \\ 2 \end{bmatrix}.$$

To determine r, s, and t in each of the following cases we equate corresponding components of the matrices involved.

(a) $\mathbf{z} = \frac{1}{2}\mathbf{x} \implies \begin{bmatrix} r \\ -1 \\ s \end{bmatrix} = \begin{bmatrix} \frac{1}{2} \\ -1 \\ \frac{3}{2} \end{bmatrix} \implies \begin{array}{rcl} r &=& \frac{1}{2} \\ s &=& \frac{3}{2} \end{array}$

(b) $\mathbf{z} + \mathbf{u} = \mathbf{x} \implies \begin{bmatrix} r+3 \\ -1+t \\ s+2 \end{bmatrix} = \begin{bmatrix} 1 \\ -2 \\ 3 \end{bmatrix} \implies \begin{array}{rcl} r+3 &=& 1 \\ -1+t &=& -2 \\ s+2 &=& 3 \end{array} \implies \begin{array}{rcl} r &=& -2 \\ t &=& -1 \\ s &=& 1 \end{array}$

(c) $\mathbf{z} - \mathbf{x} = \mathbf{y} \implies \begin{bmatrix} r-1 \\ -1-(-2) \\ s-3 \end{bmatrix} = \begin{bmatrix} -3 \\ 1 \\ 3 \end{bmatrix} \implies \begin{array}{rcl} r-1 &=& -3 \\ 1 &=& 1 \\ s-3 &=& 3 \end{array} \implies \begin{array}{rcl} r &=& -2 \\ s &=& 6 \end{array}$

17. Performing the indicated operations on the vectors we have

$$c_1 \begin{bmatrix} 1 \\ 2 \\ -3 \end{bmatrix} + c_2 \begin{bmatrix} -1 \\ 1 \\ 1 \end{bmatrix} + c_3 \begin{bmatrix} -1 \\ 4 \\ -1 \end{bmatrix} = \begin{bmatrix} c_1 - c_2 - c_3 \\ 2c_1 + c_2 + 4c_3 \\ -3c_1 + c_2 - c_3 \end{bmatrix} = \begin{bmatrix} 2 \\ -2 \\ 3 \end{bmatrix}.$$

Equating corresponding components gives the linear system

$$\begin{array}{rcrcrcr} c_1 & - & c_2 & - & c_3 & = & 2 \\ 2c_1 & + & c_2 & + & 4c_3 & = & -2 \\ -3c_1 & + & c_2 & - & c_3 & = & 3 \end{array}$$

Next form the augmented matrix and use row operations to attempt to solve this system:

$$\left[\begin{array}{rrr|r} 1 & -1 & -1 & 2 \\ 2 & 1 & 4 & -2 \\ -3 & 1 & -1 & 3 \end{array}\right] \longrightarrow \left[\begin{array}{rrr|r} 1 & -1 & -1 & 2 \\ 0 & 3 & 6 & -6 \\ 0 & -2 & -4 & 9 \end{array}\right] \longrightarrow \left[\begin{array}{rrr|r} 1 & -1 & -1 & 2 \\ 0 & 3 & 6 & -6 \\ 0 & 0 & 0 & 5 \end{array}\right]$$

add -2 (row 1) to row 2   add $\frac{2}{3}$ (row 2) to row 3
add 3 (row 1) to row 3

The system is inconsistent, hence no such scalars c_1, c_2, c_3 exist.

19. Performing the indicated operations on the vectors we have

$$c_1 \begin{bmatrix} 1 \\ 2 \\ -1 \end{bmatrix} + c_2 \begin{bmatrix} 1 \\ 3 \\ -2 \end{bmatrix} + c_3 \begin{bmatrix} 3 \\ 7 \\ -4 \end{bmatrix} = \begin{bmatrix} c_1 + c_2 + 3c_3 \\ 2c_1 + 3c_2 + 7c_3 \\ -c_1 - 2c_2 - 4c_3 \end{bmatrix} = \begin{bmatrix} 0 \\ 0 \\ 0 \end{bmatrix}.$$

Equating corresponding components gives the linear system

$$\begin{array}{rcrcrcr} c_1 & + & c_2 & + & 3c_3 & = & 0 \\ 2c_1 & + & 3c_2 & + & 7c_3 & = & 0 \\ -c_1 & - & 2c_2 & - & 4c_3 & = & 0 \end{array}$$

Next form the augmented matrix and use row operations to attempt to solve this system:

$$\left[\begin{array}{rrr|r} 1 & 1 & 3 & 0 \\ 2 & 3 & 7 & 0 \\ -1 & -2 & -4 & 0 \end{array}\right] \longrightarrow \left[\begin{array}{rrr|r} 1 & 1 & 3 & 0 \\ 0 & 1 & 1 & 0 \\ 0 & -1 & -1 & 0 \end{array}\right] \longrightarrow \left[\begin{array}{rrr|r} 1 & 0 & 2 & 0 \\ 0 & 1 & 1 & 0 \\ 0 & 0 & 0 & 0 \end{array}\right]$$

add -2 (row 1) to row 2   add -1 (row 2) to row 1
add row 1 to row 3   add row 2 to row 3

The augmented matrix represents the linear system

$$\begin{array}{rcrcrcr} c_1 & + & & + & 2c_3 & = & 0 \\ & & c_2 & + & c_3 & = & 0 \end{array}$$

It follows that $c_1 = -2c_3$ and $c_2 = -c_3$. Hence the value for c_3 can be chosen to be any real number. Let $c_3 = r$, then we have the solution $c_1 = -2r$, $c_2 = -r$, $c_3 = r$. One solution that is not all zero is, for $r = 1$, $c_1 = -2$, $c_2 = -1$, $c_3 = 1$.

21. Let $\mathbf{u} = \begin{bmatrix} x \\ y \end{bmatrix}$ be a vector in R^2. Then

$$\mathbf{u} + \mathbf{0} = \begin{bmatrix} x \\ y \end{bmatrix} + \begin{bmatrix} 0 \\ 0 \end{bmatrix} = \begin{bmatrix} x + 0 \\ y + 0 \end{bmatrix} = \begin{bmatrix} x \\ y \end{bmatrix} = \mathbf{u}.$$

A similar proof can be given for $\mathbf{u}$ in R^3.

23. Parts (b) and (d) through (h) require that we show equality of certain vectors. Since the vectors are column matrices, this is equivalent to showing that corresponding entries of the matrices involved are equal. Hence, instead of displaying the matrices we need only work with the matrix entries. Suppose that $\mathbf{u}$, $\mathbf{v}$, and $\mathbf{w}$ are in R^3 with c and d real scalars. It follows that all the components of matrices involved will be real numbers, hence when appropriate we will use properties of real numbers.

(b) $(\mathbf{u} + (\mathbf{v} + \mathbf{w}))_i = u_i + (v_i + w_i)$; $((\mathbf{u} + \mathbf{v}) + \mathbf{w})_i = (u_i + v_i) + w_i$

Since real numbers $u_i + (v_i + w_i)$ and $(u_i + v_i) + w_i$ are equal for $i = 1, 2, 3$ we have that $\mathbf{u} + (\mathbf{v} + \mathbf{w}) = (\mathbf{u} + \mathbf{v}) + \mathbf{w}$.

(d) $(\mathbf{u} + (-\mathbf{u}))_i = u_i + (-u_i)$; $(\mathbf{0})_i = 0$

Since real numbers $u_i + (-u_i)$ and 0 are equal for $i = 1, 2, 3$ we have that $\mathbf{u} + (-\mathbf{u}) = \mathbf{0}$.

(e) $c(\mathbf{u} + \mathbf{v}))_i = c(u_i + v_i)$; $(c\mathbf{u} + c\mathbf{v})_i = cu_i + cv_i$

Since real numbers $c(u_i + v_i)$ and $cu_i + cv_i$ are equal for $i = 1, 2, 3$ we have that $c(\mathbf{u} + \mathbf{v}) = c\mathbf{u} + c\mathbf{v}$.

(f) $((c + d)\mathbf{u})_i = (c + d)u_i$; $(c\mathbf{u} + d\mathbf{u})_i = cu_i + du_i$

Since real numbers $(c + d)u_i$ and $cu_i + du_i$ are equal for $i = 1, 2, 3$ we have that $(c + d)\mathbf{u} = c\mathbf{u} + d\mathbf{u}$.

(g) $(c(d\mathbf{u}))_i = c(du_i)$; $((cd)\mathbf{u})_i = (cd)u_i$

Since real numbers $c(du_i)$ and $(cd)u_i$ are equal for $i = 1, 2, 3$ we have that $c(d\mathbf{u}) = (cd)\mathbf{u}$.

(h) $(1\mathbf{u})_i = 1u_i$; $(\mathbf{u})_i = u_i$

Since real numbers $1u_i$ and u_i are equal for $i = 1, 2, 3$ we have that $1\mathbf{u} = \mathbf{u}$.

The proof for vectors in R^2 is obtained by changing the values for i to 1 and 2 only.

Section 4.2, p. 196

1. (a) The polynomials $t^2 + t$ and $-t^2 - 1$ are in P_2, but their sum $(t^2 + t) + (-t^2 - 1) = t - 1$ is not in P_2. Therefore, V is not closed under addition.

(b) No, since $0(t^2 + 1) = 0$ is not in P_2.

3. (a) Yes. To show this, let

$$A = \begin{bmatrix} a_1 & b_1 \\ 2b_1 & d_1 \end{bmatrix} \quad \text{and} \quad B = \begin{bmatrix} a_2 & b_2 \\ 2b_2 & d_2 \end{bmatrix}$$

be in V. Then

$$A + B = \begin{bmatrix} a_1 + a_2 & b_1 + b_2 \\ 2(b_1 + b_2) & d_1 + d_2 \end{bmatrix}$$

is in V, so V is closed under addition.

(b) Yes. If $A = \begin{bmatrix} a & b \\ 2b & d \end{bmatrix}$ is in V and c is any scalar, then

$$cA = \begin{bmatrix} ca & cb \\ 2(cb) & cd \end{bmatrix}$$

is in V, so V is closed under scalar multiplication.

(c) $O = \begin{bmatrix} 0 & 0 \\ 0 & 0 \end{bmatrix}$ is the zero vector.

(d) Yes. If $A = \begin{bmatrix} a & b \\ 2b & d \end{bmatrix}$ is in V, then

$$-A = \begin{bmatrix} -a & -b \\ -2b & -d \end{bmatrix} = \begin{bmatrix} -a & -b \\ 2(-b) & -d \end{bmatrix}$$

is the negative of A since

$$A + (-A) = \begin{bmatrix} 0 & 0 \\ 0 & 0 \end{bmatrix} = O.$$

(e) Yes. It satisfies all the properties of Definition 4.4

5. The proof that R^n is a vector space is very similar to the proof of Theorem 4.1 in Section 4.1. (See Exercise 23 in Section 4.1.) Properties (1) through (8) are shown just as before, but now the subscript i ranges over the values from 1 to n. Thus here we verify only properties (a) and (b) of Definition 4.4.

(a) Let $\mathbf{u} = \begin{bmatrix} u_1 \\ u_2 \\ \vdots \\ u_n \end{bmatrix}$ and $\mathbf{v} = \begin{bmatrix} v_1 \\ v_2 \\ \vdots \\ v_n \end{bmatrix}$. Then $\mathbf{u} + \mathbf{v} = \begin{bmatrix} u_1 + v_1 \\ u_2 + v_2 \\ \vdots \\ u_n + v_n \end{bmatrix}$, which is an element of R^n.

(b) Let $\mathbf{u}$ be as in (a) and let c be any real scalar. Then $c\mathbf{u} = \begin{bmatrix} cu_1 \\ cu_2 \\ \vdots \\ cu_n \end{bmatrix}$, which is an element of R^n.

7. Let V be the set of positive real numbers. The operations on V are addition and multiplication of real numbers. Let $\mathbf{u}$, $\mathbf{v}$, and $\mathbf{w}$ be in V and c and d any real scalars.

(a) The sum of two positive numbers is positive, so $\mathbf{u} \oplus \mathbf{v}$ is in V. Thus property (a) holds.

(1) $\mathbf{u} \oplus \mathbf{v} = \mathbf{v} \oplus \mathbf{u}$ since we are adding real numbers. Thus property (1) holds.

(2) $\mathbf{u} \oplus (\mathbf{v} \oplus \mathbf{w}) = (\mathbf{u} \oplus \mathbf{v}) \oplus \mathbf{w}$ since we are adding real numbers. Thus property (2) holds.

(3) Since 0 is not positive, there does not exist an element in V to satisfy property (3). Thus property (3) fails to hold.

(4) If $\mathbf{u}$ is in V then $-\mathbf{u}$ is not in V, so there does not exist an element of V to satisfy property (4). Thus property (4) fails to hold.

(b) If scalar c is zero or negative, then $c\mathbf{u}$ is not in V. Thus property (b) fails to hold.

Since property (b) fails, any property involving arbitrary scalars will also fail to hold in all possible cases. Hence properties (5), (6), and (7) fail to hold. Since (8) involves the positive scalar 1, it must be checked individually.

(8) Since $1\mathbf{u}$ is just a product of reals, it follows that $1\mathbf{u} = \mathbf{u}$ for each element of V. Thus property (8) holds.

9. Let V be the set of ordered triples of real numbers. The operations on V are $(x, y, z) \oplus (x', y', z') = (x + x', y + y', z + z')$, which is ordinary addition in R_3, and $r \odot (x, y, z) = (x, 1, z)$. From Example 2, M_{13} is a vector space with the usual operations of matrix addition and scalar multiplication. Since R_3 is identical to M_{13}, it follows that V with $\oplus$ satisfies properties (a) and (1)–(4). Hence we will investigate properties (b) and (5)–(8).

(b) From the definition of $\odot$ we see that $c \odot (x, y, z)$ is indeed an ordered triple, hence in V. Thus property (b) holds.

(5) $c \odot ((x, y, z) \oplus (x', y', z')) = c \odot (x + x', y + y', z + z') = (x + x', 1, z + z')$
$c \odot (x, y, z) \oplus c \odot (x', y', z') = (x, 1, z) \oplus (x', 1, z') = (x + x', 2, z + z')$
It follows that property (5) fails to hold.

(6) $(c + d) \odot (x, y, z) = (x, 1, z)$
$c \odot (x, y, z) \oplus d \odot (x, y, z) = (x, 1, z) \oplus (x, 1, z) = (2x, 2, 2z)$
It follows that property (6) fails to hold.

(7) $c \odot (d \odot (x, y, z)) = c \odot (x, 1, z) = (x, 1, z)$
$(cd) \odot (x, y, z) = (x, 1, z)$
Thus property (7) holds.

(8) $1 \odot (x, y, z) = (x, 1, z) \neq (x, y, z)$ unless $y = 1$. Thus property (8) fails to hold.

11. Let V be the set of ordered pairs of reals with $(x, y) \oplus (x', y') = (x + x', y + y')$ and $r \odot (x, y) = (0, 0)$. V is identical to R_2 and since operation $\oplus$ is just ordinary addition in R_2, it follows that property (a) and properties (1)–(4) are satisfied. We investigate properties (b) and (5)–(8) individually.

(b) Since operation $\odot$ yields an ordered pair, property (b) holds.

(5) $c \odot ((x, y) \oplus (x', y')) = c \odot (x + x', y + y') = (0, 0)$
$c \odot (x, y) \oplus d \odot (x', y') = (0, 0) \oplus (0, 0) = (0, 0)$
Thus property (5) holds.

(6) $(c + d) \odot (x, y) = (0, 0)$
$c \odot (x, y) \oplus d \odot (x, y) = (0, 0) \oplus (0, 0) = (0, 0)$
Thus property (6) holds.

(7) $c \odot (d \odot (x, y)) = c \odot (0, 0) = (0, 0)$
$(cd) \odot (x, y) = (0, 0)$
Thus property (7) holds.

(8) $1 \odot (x, y) = (0, 0) \neq (x, y)$ for arbitrary x and y. Thus property (8) fails to hold.

13. Let V be the set of real-valued continuous functions with operations $(f \oplus g)(t) = f(t) + g(t)$ and $(c \odot f)(t) = cf(t)$.

(a) From Calculus, the sum of two continuous functions is continuous, thus property (a) holds.

(1) $(f \oplus g)(t) = f(t) + g(t); (g \oplus f)(t) = g(t) + f(t)$
Since $f(t)$ and $g(t)$ are real numbers, $f(t) + g(t) = g(t) + f(t)$. Thus property (1) holds.

(2) $(f \oplus (g \oplus h))(t) = f(t) + (g \oplus h)(t) = f(t) + (g(t) + h(t))$
$((f \oplus g) \oplus h)(t) = (f \oplus g)(t) + h(t) = (f(t) + g(t)) + h(t)$
Since $f(t)$, $g(t)$, and $h(t)$ are real numbers $f(t) + (g(t) + h(t)) = (f(t) + g(t)) + h(t)$. Thus property (2) holds.

(3) Let $z(t) = 0$ for all t. Since $z(t)$ is a constant function it is continuous and hence in V. We have $(f \oplus z)(t) = f(t) + z(t) = f(t) + 0 = f(t)$ and $(z \oplus f)(t) = z(t) + f(t) = 0 + f(t) = f(t)$. Thus, $(f \oplus z)(t) = (z \oplus f)(t) = f(t)$ and property (3) holds.

(4) If f is in V, then so is $-f$ where $-f(t) = (-1)f(t)$. It follows that $(f \oplus -f)(t) = (-f \oplus f)(t) = z(t)$, where $z(t)$ is defined in (3). Thus property (4) holds.

(b) From Calculus, a constant multiple of a continuous function is continuous. Hence property (b) holds.

(5) $(c \odot (f \oplus g))(t) = c(f(t) + g(t))$
$((c \odot f) \oplus (c \odot g))(t) = (c \odot f)(t) + (c \odot g)(t) = cf(t) + cg(t)$
By properties of real numbers $c(f(t) + g(t)) = cf(t) + cg(t)$, thus property (5) holds.

(6) $((c + d) \odot f)(t) = (c + d)f(t)$
$((c \odot f) \oplus (d \odot f))(t) = (c \odot f)(t) + (d \odot f)(t) = cf(t) + df(t)$
By properties of real numbers $(c + d)f(t) = cf(t) + df(t)$, thus property (6) holds.

(7) $(c \odot (d \odot f))(t) = c(d \odot f(t)) = c(df(t))$
$((cd) \odot f)(t) = (cd)f(t)$
By properties of real numbers $c(df(t)) = (cd)f(t)$. Thus property (7) holds.

(8) $(1 \odot f)(t) = 1f(t) = f(t)$. Thus property (8) holds.

It follows that Definition 4.4 holds, hence V is a vector space.

15. Let V be the set of real-valued functions that satisfy the differential equation $y'' - y' + 2y = 0$. That is, $f(t)$ is in V provided $f''(t) - f'(t) + 2f(t) = 0$ for all real values t. The operations on members of V are defined in Exercise 13. We will use properties of differentiation from Calculus, namely

$$(f(t) + g(t))'' = f''(t) + g''(t) \qquad (f(t) + g(t))' = f'(t) + g'(t)$$
$$(cf(t))'' = cf''(t) \qquad\qquad (cf(t))' = cf'(t)$$

Let f, g, and h be in V and let c and d be real scalars.

(a) For f and g in V to show $f \oplus g$ is in V we show that $f(t) + g(t)$ is a solution of the differential equation:

$$\begin{aligned}
(f \oplus g)''(t) - (f \oplus g)'(t) + 2(f \oplus g)(t) &= (f(t) + g(t))'' - (f(t) + g(t))' + 2(f(t) + g(t)) \\
&= f''(t) + g''(t) - (f'(t) + g'(t)) + 2(f(t) + g(t)) \\
&= (f''(t) - f'(t) + 2f(t)) + (g''(t) - g'(t) + 2g(t)) \\
&\qquad \text{(since both } f \text{ and } g \text{ are in } V) \\
&= 0 + 0 = 0
\end{aligned}$$

(1) If f and g are in V, then they are real-valued functions. From Exercise 13, $f \oplus g = g \oplus f$.

(2) If f, g, and h are in V, then they are real-valued functions. From Exercise 13, $(f \oplus g) \oplus h = f \oplus (g \oplus h)$.

(3) If we let $z(t) = 0$ for all t, then we must have that $z(t)$ is in V. This follows immediately since $0'' - 0' + 2(0) = 0$. Hence from Exercise 13 we have $f \oplus z = z \oplus f = f$.

(4) If $f(t)$ is in V, then $-f(t)$ is in V since

$$(-f(t))'' - (-f(t))' + 2(-f(t)) = -(f''(t) - f'(t) + 2f(t)) = -(0) = 0.$$

Hence from Exercise 13 we have $f \oplus -f = -f \oplus f = z$.

(b) For f in V to show $c \odot f$ is in V we show that $cf(t)$ is a solution of the differential equation:

$$\begin{aligned}
(c \odot f)''(t) - (c \odot f)'(t) + 2(c \odot f)(t) &= (cf(t))'' - (cf(t))' + 2(cf(t)) \\
&= c(f''(t) - f'(t) + 2f(t)) \\
&= c(0) = 0
\end{aligned}$$

(5) If f and g are in V they are real-valued functions. From Exercise 13, $c \odot (f + g) = c \odot f \oplus c \odot g$.

(6) If f is in V then it is a real-valued function. From Exercise 13, $(c + d) \odot f = c \odot f \oplus d \odot f$.

(7) If f is in V then it is a real-valued function. From Exercise 13, $c \odot (d \odot f) = (cd) \odot f$.

(8) If f is in V then it is a real-valued function. From Exercise 13, $1 \odot f = f$.

17. No. The zero element for $\oplus$ would have to be the real number 1, but then $\mathbf{u} = 0$ has no "negative" $\mathbf{v}$ such that $\mathbf{u} \oplus \mathbf{v} = 0 \cdot \mathbf{v} = 1$. Thus (4) fails to hold. Condition (5) fails since $c \odot (\mathbf{u} \oplus \mathbf{v}) = c + (\mathbf{uv}) \neq (c + \mathbf{u})(c + \mathbf{v}) = c \odot \mathbf{u} \oplus c \odot \mathbf{v}$.

19. Let $\mathbf{0}_1$ and $\mathbf{0}_2$ be zero vectors for a vector space. Then $\mathbf{0}_1 + \mathbf{0}_2 = \mathbf{0}_1$ and $\mathbf{0}_1 + \mathbf{0}_2 = \mathbf{0}_2$. So $\mathbf{0}_1 = \mathbf{0}_2$.

21. Theorem 4.2 (b): $c \odot \mathbf{0} = \mathbf{0}$ for any scalar c.

Proof: We have $c \odot \mathbf{0} = c \odot (\mathbf{0} + \mathbf{0}) = c \odot \mathbf{0} + c \odot \mathbf{0}$. Since $c \odot \mathbf{0}$ is in V it has a negative. Add its negative to both sides of the preceding equality to get $\mathbf{0} = c \odot \mathbf{0}$.

Theorem 4.2 (c): If $c \odot \mathbf{u} = \mathbf{0}$, then either $c = 0$ or $\mathbf{u} = \mathbf{0}$.

Proof: If $c = 0$, then we are done. If $c \neq 0$, then we can multiply both sides of the equation $c \odot \mathbf{u} = \mathbf{0}$ by $\frac{1}{c}$ to give

$$\frac{1}{c} \odot (c \odot \mathbf{u}) = \left(\frac{1}{c} c\right) \odot \mathbf{u} = 1 \odot \mathbf{u} = \mathbf{u}$$

which must equal $\frac{1}{c} \odot \mathbf{0} = \mathbf{0}$. Thus $\mathbf{u} = \mathbf{0}$.

23. $\mathbf{v} \oplus (-\mathbf{v}) = \mathbf{0}$, so $-(-\mathbf{v}) = \mathbf{v}$.

25. If $a \odot \mathbf{u} = b \odot \mathbf{u}$, then $(a - b) \odot \mathbf{u} = a \odot \mathbf{u} - b \odot \mathbf{u} = \mathbf{0}$. It follows from Theorem 4.2 (c) and the fact that $\mathbf{u} \neq \mathbf{0}$ that $a - b = 0$, or $a = b$.

Section 4.3, p. 205

1. Yes; requirements (a) and (b) of Theorem 4.3 are satisfied.

3. No; for example, $(1, 0)$ is in W but $2(1, 0) = (2, 0)$ is not.

5. We are in vector space R^3. Let W stand for the given subset.

(a) W is not a subspace. For if

$$\mathbf{u} = \begin{bmatrix} a_1 \\ b_1 \\ 1 \end{bmatrix} \quad \text{and} \quad \mathbf{v} = \begin{bmatrix} a_2 \\ b_2 \\ 1 \end{bmatrix}$$

are vectors in W, then

$$\mathbf{u} + \mathbf{v} = \begin{bmatrix} a_1 + a_2 \\ b_1 + b_2 \\ 2 \end{bmatrix}$$

which is not in W since the last component is not 1. By Theorem 4.3 (a), W is not a subspace.

(b) W is a subspace. For if

$$\mathbf{u} = \begin{bmatrix} a_1 \\ b_1 \\ a_1 + 2b_1 \end{bmatrix} \quad \text{and} \quad \mathbf{v} = \begin{bmatrix} a_2 \\ b_2 \\ a_2 + 2b_2 \end{bmatrix}$$

are vectors in W, then

$$\mathbf{u} + \mathbf{v} = \begin{bmatrix} a_1 \\ b_1 \\ a_1 + 2b_1 \end{bmatrix} + \begin{bmatrix} a_2 \\ b_2 \\ a_2 + 2b_2 \end{bmatrix} = \begin{bmatrix} a_1 + a_2 \\ b_1 + b_2 \\ a_1 + a_2 + 2(b_1 + b_2) \end{bmatrix}.$$

Since the third entry of $\mathbf{u} + \mathbf{v}$ is the correct combination of the first two entries, $\mathbf{u} + \mathbf{v}$ is in W. Next, for any real number r,

$$r\mathbf{u} = \begin{bmatrix} ra_1 \\ rb_1 \\ r(a_1 + 2b_1) \end{bmatrix} = \begin{bmatrix} ra_1 \\ rb_1 \\ ra_1 + 2(rb_1) \end{bmatrix}.$$

The third entry is the correct combination of the first two entries so that $r\mathbf{u}$ belongs to W. Thus by Theorem 4.3 W is a subspace of R^3.

(c) W is a subspace. For if

$$\mathbf{u} = \begin{bmatrix} a_1 \\ 0 \\ 0 \end{bmatrix} \quad \text{and} \quad \mathbf{v} = \begin{bmatrix} a_2 \\ 0 \\ 0 \end{bmatrix}$$

then

$$\mathbf{u} + \mathbf{v} = \begin{bmatrix} a_1 + a_2 \\ 0 \\ 0 \end{bmatrix} \quad \text{and} \quad r\mathbf{u} = \begin{bmatrix} ra_1 \\ 0 \\ 0 \end{bmatrix}.$$

Since the second and third entries are zero, both $\mathbf{u} + \mathbf{v}$ and $r\mathbf{u}$ are in W. Thus by Theorem 4.3 W is a subspace of R^3.

(d) W is a subspace. To see this, let

$$\mathbf{u} = \begin{bmatrix} a_1 \\ b_1 \\ c_1 \end{bmatrix} \quad \text{and} \quad \mathbf{v} = \begin{bmatrix} a_2 \\ b_2 \\ c_2 \end{bmatrix}$$

where $a_1 + 2b_1 - c_1 = 0$ and $a_2 + 2b_2 - c_2 = 0$. Then

$$\mathbf{u} + \mathbf{v} = \begin{bmatrix} a_1 + a_2 \\ b_1 + b_2 \\ c_1 + c_2 \end{bmatrix}.$$

Since

$$(a_1 + a_2) + 2(b_1 + b_2) - (c_1 + c_2) = (a_1 + 2b_1 - c_1) + (a_2 + 2b_2 - c_2) = 0 + 0 = 0,$$

it follows that $\mathbf{u} + \mathbf{v}$ is in W. In the same way we find that

$$r\mathbf{u} = \begin{bmatrix} ra_1 \\ rb_1 \\ rc_1 \end{bmatrix}$$

and $ra_1 + 2(rb_1) - rc_1 = r(a_1 + 2b_1 - c_1) = r(0) = 0$. Therefore $r\mathbf{u}$ is in W. Hence by Theorem 4.3 W is a subspace of R^3.

7. We are in vector space R_4. Let W be the given subset and let

$$\mathbf{u} = \begin{bmatrix} a_1 & b_1 & c_1 & d_1 \end{bmatrix} \quad \text{and} \quad \mathbf{v} = \begin{bmatrix} a_2 & b_2 & c_2 & d_2 \end{bmatrix}$$

be vectors in W. In each case below the components of $\mathbf{u}$ and $\mathbf{v}$ satisfy certain conditions. To determine if $\mathbf{u} + \mathbf{v}$ and $r\mathbf{u}$ are in W, we first find

$$\mathbf{u} + \mathbf{v} = \begin{bmatrix} a_1 + a_2 & b_1 + b_2 & c_1 + c_2 & d_1 + d_2 \end{bmatrix} \quad \text{and} \quad r\mathbf{u} = \begin{bmatrix} ra_1 & rb_1 & rc_1 & rd_1 \end{bmatrix},$$

and then determine if the components of $\mathbf{u} + \mathbf{v}$ and $r\mathbf{u}$ satisfy the same conditions.

(a) In this case the components of $\mathbf{u}$ and $\mathbf{v}$ satisfy the equations $a_1 - b_1 = 2$ and $a_2 - b_2 = 2$. Then for the vector $\mathbf{u} + \mathbf{v}$ we find

$$(a_1 + a_2) - (b_1 + b_2) = (a_1 - b_1) + (a_2 - b_2) = 2 - 2 = 0.$$

Since $(a_1 + a_2) - (b_1 + b_2) \neq 2$, $\mathbf{u} + \mathbf{v}$ is not in W and hence W is not a subspace.

(b) In this case the components of $\mathbf{u}$ and $\mathbf{v}$ satisfy the equations

$$c_1 = a_1 + 2b_1, \quad d_1 = a_1 - 3b_1$$
$$c_2 = a_2 + 2b_2, \quad d_2 = a_2 - 3b_2$$

Then for the vector $\mathbf{u} + \mathbf{v}$ we find that

$$c_1 + c_2 = (a_1 + 2b_1) + (a_2 + 2b_2) = (a_1 + a_2) + 2(b_1 + b_2)$$
$$d_1 + d_2 = (a_1 - 3b_1) + (a_2 - 3b_2) = (a_1 + a_2) - 3(b_1 + b_2)$$

hence $\mathbf{u} + \mathbf{v}$ is in W, and for $r\mathbf{u}$ we find

$$rc_1 = r(a_1 + 2b_1) = ra_1 + 2(rb_1) \quad \text{and} \quad rd_1 = r(a_1 - 3b_1) = ra_1 - 3(rb_1)$$

so that $r\mathbf{u}$ is in W. Therefore W is a subspace.

(c) In this case the components of $\mathbf{u}$ and $\mathbf{v}$ satisfy the equations $a_1 = 0$, $b_1 = -d_1$, and $a_2 = 0$, $b_2 = -d_2$. The for the vector $\mathbf{u} + \mathbf{v}$ we find that

$$a_1 + a_2 = 0 + 0 = 0 \quad \text{and} \quad b_1 + b_2 = -d_1 - d_2 = -(d_1 + d_2).$$

Therefore $\mathbf{u} + \mathbf{v}$ is in W. For the vector $r\mathbf{u}$ we find that

$$ra_1 = r(0) = 0 \quad \text{and} \quad rb_1 = r(-d_1) = -(rd_1).$$

Therefore $r\mathbf{u}$ is in W. Hence W is a subspace.

9. We are in vector space M_{23}. Following the method used in Exercise 7, we let W be the given subset and determine if the components of $\mathbf{u} + \mathbf{v}$ and $r\mathbf{u}$ satisfy the required conditions for $\mathbf{u}$, $\mathbf{v}$ in W.

(a) In this case the components of $\mathbf{u}$ and $\mathbf{v}$ satisfy the equations $b_1 = a_1 + c_1$ and $b_2 = a_2 + c_2$. Then for $\mathbf{u} + \mathbf{v}$ and $r\mathbf{u}$ we find that

$$b_1 + b_2 = (a_1 + c_1) + (a_2 + c_2) = (a_1 + a_2) + (c_1 + c_2) \quad \text{and} \quad rb_1 = r(a_1 + c_1) = ra_1 + rc_1.$$

Thus $\mathbf{u} + \mathbf{v}$ and $r\mathbf{u}$ are in W so W is a subspace.

(b) In this case the components of $\mathbf{u}$ and $\mathbf{v}$ satisfy the conditions $c_1 > 0$ and $c_2 > 0$. Then for the vector $\mathbf{u} + \mathbf{v}$ we find that $c_1 + c_2 > 0$. But for $r\mathbf{u}$ the component rc_1 need not always be positive. Hence $r\mathbf{u}$ need not be in W. Therefore W is not a subspace.

(c) In this case the components of $\mathbf{u}$ and $\mathbf{v}$ satisfy the conditions

$$a_1 = -2c_1, \quad f_1 = 2e_1 + d_1$$
$$a_2 = -2c_2, \quad f_2 = 2e_2 + d_2.$$

Then for the vectors $\mathbf{u} + \mathbf{v}$ and $r\mathbf{u}$ we find that

$$a_1 + a_2 = -2c_1 - 2c_2 = -2(c_1 + c_2), \quad ra_1 = r(-2c_1) = -2(rc_1)$$
$$f_1 + f_2 = -2e_1 - 2e_2 = -2(e_1 + e_2), \quad rf_1 = r(-2e_1) = -2(re_1).$$

Therefore $\mathbf{u} + \mathbf{v}$ and $r\mathbf{u}$ are in W and hence W is a subspace.

11. Let A and B be 3×3 matrices in W. Then $\text{Tr}(A) = \text{Tr}(B) = 0$. Therefore $\text{Tr}(A+B) = \text{Tr}(A) + \text{Tr}(B) = 0$ and hence $A + B$ is in W. Moreover, if k is a scalar, then $\text{Tr}(kA) = k\,\text{Tr}(A) = k(0) = 0$ and hence kA is in W. Therefore W is closed under addition and scalar multiplication and hence is a subspace of M_{33}.

13. Yes. If

$$A = \begin{bmatrix} a_1 & b_1 \\ c_1 & d_1 \end{bmatrix} \quad \text{and} \quad B = \begin{bmatrix} a_2 & b_2 \\ c_2 & d_2 \end{bmatrix}$$

are matrices in W, then $a_1 + b_1 + c_1 + d_1 = a_2 + b_2 + c_2 + d_2 = 0$. Therefore

$$A + B = \begin{bmatrix} a_1 + a_2 & b_1 + b_2 \\ c_1 + c_2 & d_1 + d_2 \end{bmatrix}$$

and

$$(a_1 + a_2) + (b_1 + b_2) + (c_1 + c_2) + (d_1 + d_2) = (a_1 + b_1 + c_1 + d_1) + (a_2 + b_2 + c_2 + d_2) = 0,$$

so $A + B$ is in W. Moreover, if k is a scalar, then

$$kA = \begin{bmatrix} ka_1 & kb_1 \\ kc_1 & kd_1 \end{bmatrix}$$

and

$$ka_1 + kb_1 + kc_1 + kd_1 = k(a_1 + b_1 + c_1 + d_1) = k(0) = 0,$$

so kA is in W. Therefore W is a subspace of M_{22}.

15. We are in vector space P_2. Let W be the given subset of polynomials.

   (a) W is a subspace of P_2. To show this, let $\mathbf{u} = a_2 t^2 + a_1 t + a_0$ and $\mathbf{v} = b_2 t^2 + b_1 t + b_0$ be in W. Then $a_0 = 0$ and $b_0 = 0$. Now,

   $$\mathbf{u} + \mathbf{v} = (a_2 + b_2)t^2 + (a_1 + b_1)t + (a_0 + b_0)$$
   $$r\mathbf{u} = (ra_2)t^2 + (ra_1)t + (ra_0)$$

   Since $a_0 + b_0 = 0 + 0 = 0$ and $ra_0 = r(0) = 0$, $\mathbf{u} + \mathbf{v}$ and $r\mathbf{u}$ are in W. Therefore W is a subspace.

   (b) W is not a subspace of P_2. To show this, let $\mathbf{u} = a_2 t^2 + a_1 t + a_0$ and $\mathbf{v} = b_2 t^2 + b_1 t + b_0$ be in W. Then $a_1 = 2$ and $b_1 = 2$. Now,

   $$\mathbf{u} + \mathbf{v} = (a_2 + b_2)t^2 + (a_1 + b_1)t + (a_0 + b_0).$$

   Since $a_0 + b_0 = 2 + 2 = 4 \neq 2$, $\mathbf{u} + \mathbf{v}$ is not in W. Therefore W is not a subspace of P_2.

   (c) W is a subspace of P_2. To show this, let $\mathbf{u} = a_2 t^2 + a_1 t + a_0$ and $\mathbf{v} = b_2 t^2 + b_1 t + b_0$ be in W. Then $a_2 + a_1 = a_0$ and $b_2 + b_1 = b_0$. Now

   $$\mathbf{u} + \mathbf{v} = (a_2 + b_2)t^2 + (a_1 + b_1)t + (a_0 + b_0)$$
   $$r\mathbf{u} = (ta_2)t^2 + (ra_1) + (ra_0).$$

   Since $(a_2 + b_2) + (a_1 + b_1) = (a_2 + a_1) + (b_2 + b_1) = a_0 + b_0$ and $ra_2 + ra_1 = (r(a_2 + a_1) = ra_0$, $\mathbf{u} + \mathbf{v}$ and $r\mathbf{u}$ are in W. Therefore W is a subspace.

17. We are in vector space M_{nn}. Let W be the given subset of matrices.

   (a) W is a subspace of M_{nn}. To show this, let $\mathbf{u}$ and $\mathbf{v}$ be symmetric matrices. Then $\mathbf{u}^T = \mathbf{u}$, $\mathbf{v}^T = \mathbf{v}$. It follows that $(\mathbf{u} + \mathbf{v})^T = \mathbf{u}^T + \mathbf{v}^T = \mathbf{u} + \mathbf{v}$ and $(r\mathbf{u})^T = r\mathbf{u}^T = r\mathbf{u}$. Therefore $\mathbf{u} + \mathbf{v}$ and $r\mathbf{u}$ are symmetric and hence in W. Thus W is a subspace of M_{nn}.

   (b) W is a subspace of M_{nn}. To show this, let $\mathbf{u} = \begin{bmatrix} u_{ij} \end{bmatrix}$ and $\mathbf{v} = \begin{bmatrix} v_{ij} \end{bmatrix}$ be diagonal matrices. Then $u_{ij} = 0$ and $v_{ij} = 0$ for $i \neq j$. Now, $\mathbf{u} + \mathbf{v} = \begin{bmatrix} u_{ij} + v_{ij} \end{bmatrix}$ and $r\mathbf{u} = \begin{bmatrix} ru_{ij} \end{bmatrix}$. Since $u_{ij} + v_{ij} = 0$ and $ru_{ij} = 0$ for $i \neq j$, $\mathbf{u} + \mathbf{v}$ and $r\mathbf{u}$ are diagonal matrices and hence in W. Thus W is a subspace of M_{nn}.

   (c) W is not a subspace of M_{nn}. For example, let

   $$\mathbf{u} = \begin{bmatrix} 1 & 0 \\ 0 & 1 \end{bmatrix} \quad \text{and} \quad \mathbf{v} = \begin{bmatrix} -1 & 0 \\ 0 & -1 \end{bmatrix}.$$

   Then $\mathbf{u}$ and $\mathbf{v}$ are in W since they are nonsingular, but $\mathbf{u} + \mathbf{v} = \begin{bmatrix} 0 & 0 \\ 0 & 0 \end{bmatrix}$ is singular and hence not in W. So W is not a subspace of M_{nn}.

19. We are in vector space $C(-\infty, \infty)$. Let W be the given subset.

    (a) W is not a subspace. For if we let $f(x) = 1$ for all x, then f is nonnegative and hence in W, but if $r = -1$, then rf is the function $(rf)(x) = -1$ for all x, so rf is not in W since it is not nonnegative. Therefore W is not a subspace.

    (b) W is a subspace. To show this, let $f(t) = k_1$ and $g(t) = k_2$ be in W. Then $(f \oplus g)(t) = f(t) + g(t) = k_1 + k_2$, which is constant for all t, and $(c \odot f)(t) = cf(t) = ck_1$, which is constant for all t. Thus, both $f \oplus g$ and $c \odot f$ are in W. Therefore W is a subspace of $C(-\infty, \infty)$.

    (c) W is a subspace. To show this, let f and g be functions in W. Then $f(0) = g(0) = 0$. Therefore $(f \oplus g)(0) = f(0) + g(0) = 0 + 0 = 0$ and $(c \odot f)(0) = cf(0) = c(0) = 0$. Hence $f \oplus g$ and $c \odot f$ are in W and therefore W is a subspace of $C(-\infty, \infty)$.

    (d) W is not a subspace. For if f and g are in W, then $f(0) = g(0) = 5$, but $(f \oplus g)(0) = f(0) + g(0) = 5 + 5 \neq 5$. Therefore $f \oplus g$ is not in W and hence W is not a subspace of $C(-\infty, \infty)$.

    (e) W is a subspace. To show this, let f and g be in W. Then f and g are differentiable. It follows from Calculus that the function $(f \oplus g)((t) = f(t) + g(t)$ is differentiable (since it is the sum of two differentiable functions) and the function $(c \odot f)(t) = cf(t)$ is differentiable (since it is a constant multiple of a differentiable function). Therefore $f \oplus g$ and $c \odot f$ are in W and hence W is a subspace of $C(-\infty, \infty)$.

21. We first note that P, the set of all polynomials, is in fact a subset of $C(-\infty, \infty)$ since a polynomial is a continuous function on $(-\infty, \infty)$. Now, if f and g are in P, then $(f \oplus g)(t) = f(t) + g(t)$ is a polynomial, and $(c \odot f)(t) = cf(t)$ is a polynomial. Therefore $f \oplus g$ and $c \odot f$ are in P. Hence P is a subspace of $C(-\infty, \infty)$.

23. Let W be the set of solutions of the linear system $A\mathbf{x} = \mathbf{b}$, $\mathbf{b} \neq \mathbf{0}$. Let $\mathbf{u}$ and $\mathbf{v}$ be in W. Then $A\mathbf{u} = \mathbf{b}$ and $A\mathbf{v} = \mathbf{b}$. It follows that $A(\mathbf{u} + \mathbf{v}) = A\mathbf{u} + A\mathbf{v} = \mathbf{b} + \mathbf{b} \neq \mathbf{b}$ since $\mathbf{b} \neq \mathbf{0}$. Therefore $\mathbf{u} + \mathbf{v}$ is not in W and hence W is not a subspace of R^n.

25. Let t be any real number. Then

$$A \begin{bmatrix} t \\ -t \\ t \end{bmatrix} = \begin{bmatrix} 1 & 2 & 1 \\ -1 & 0 & 1 \\ 2 & 6 & 4 \end{bmatrix} \begin{bmatrix} t \\ -t \\ t \end{bmatrix} = \begin{bmatrix} 0 \\ 0 \\ 0 \end{bmatrix}.$$

Therefore, $\begin{bmatrix} t \\ -t \\ t \end{bmatrix}$ is in the null space of A.

27. No, it is not a subspace. To show this, let $\mathbf{x}$ and $\mathbf{y}$ be in W, so that $A\mathbf{x} \neq \mathbf{0}$ and $A\mathbf{y} \neq \mathbf{0}$. Then $A(\mathbf{x} + \mathbf{y}) = A\mathbf{x} + A\mathbf{y}$, and $A\mathbf{x} + A\mathbf{y}$ may be equal to $\mathbf{0}$. So $\mathbf{x} + \mathbf{y}$ is not necessarily in W. Thus, W is not a subspace of R^n.

29. Certainly $\{\mathbf{0}\}$ and R^2 are subspaces of R^2. If $\mathbf{u}$ is any nonzero vector then span $\{\mathbf{u}\}$ is a subspace of R^2. To show this, observe that span $\{\mathbf{u}\}$ consists of all vectors in R^2 that are scalar multiples of $\mathbf{u}$. Let $\mathbf{v} = c\mathbf{u}$ and $\mathbf{w} = d\mathbf{u}$ be in span $\{\mathbf{u}\}$ where c and d are any real numbers. Then $\mathbf{v} + \mathbf{w} = c\mathbf{u} + d\mathbf{u} = (c+d)\mathbf{u}$ is in span $\{\mathbf{u}\}$ and if k is any real number, then $k\mathbf{v} = k(c\mathbf{u}) = (kc)\mathbf{u}$ is in span $\{\mathbf{u}\}$. Then by Theorem 4.3, span $\{\mathbf{u}\}$ is a subspace of R^2.

To show that these are the only subspaces of R^2 we proceed as follows. Let W be any subspace of R^2. Since W is a vector space in its own right, it contains the zero vector $\mathbf{0}$. If $W \neq \{\mathbf{0}\}$, then W contains a nonzero vector $\mathbf{u}$. But then by property (b) of Definition 4.4, W must contain every scalar multiple of $\mathbf{u}$. If every vector in W is a scalar multiple of $\mathbf{u}$ then W is span $\{\mathbf{u}\}$. Otherwise, W contains span $\{\mathbf{u}\}$ and another vector which is not a multiple of $\mathbf{u}$. Call this other vector $\mathbf{v}$. It follows that W contains span $\{\mathbf{u}, \mathbf{v}\}$. But in fact span $\{\mathbf{u}, \mathbf{v}\} = R^2$. To show this, let $\mathbf{y}$ be any vector in R^2 and let

$$\mathbf{u} = \begin{bmatrix} u_1 \\ u_2 \end{bmatrix}, \quad \mathbf{v} = \begin{bmatrix} v_1 \\ v_2 \end{bmatrix}, \quad \text{and} \quad \mathbf{y} = \begin{bmatrix} y_1 \\ y_2 \end{bmatrix}.$$

We must show there are scalars c_1 and c_2 such that $c_1\mathbf{u} + c_2\mathbf{v} = \mathbf{y}$. This equation leads to the linear system

$$\begin{bmatrix} u_1 & v_1 \\ u_2 & v_2 \end{bmatrix}\begin{bmatrix} c_1 \\ c_2 \end{bmatrix} = \begin{bmatrix} y_1 \\ y_2 \end{bmatrix}.$$

Consider the transpose of the coefficient matrix:

$$\begin{bmatrix} u_1 & v_1 \\ u_2 & v_2 \end{bmatrix}^T = \begin{bmatrix} u_1 & u_2 \\ v_1 & v_2 \end{bmatrix}.$$

This matrix is row equivalent to I_2 since its rows are not multiples of each other. Therefore the matrix is nonsingular. It follows that the coefficient matrix is nonsingular and hence the linear system has a solution. Therefore span $\{\mathbf{u}, \mathbf{v}\} = R^2$, as required, and hence the only subspaces of R^2 are $\{\mathbf{0}\}$, R^2, or scalar multiples of a single nonzero vector.

31. Let $\mathbf{w} = \begin{bmatrix} a & b & c \\ a & 0 & 0 \end{bmatrix}$ be a vector in W. Then $c = a + b$, so

$$\mathbf{w} = \begin{bmatrix} a & b & c \\ a & 0 & 0 \end{bmatrix} = \begin{bmatrix} a & b & a+b \\ a & 0 & 0 \end{bmatrix} = a\begin{bmatrix} 1 & 0 & 1 \\ 1 & 0 & 0 \end{bmatrix} + b\begin{bmatrix} 0 & 1 & 1 \\ 0 & 0 & 0 \end{bmatrix} = a\mathbf{w}_1 + b\mathbf{w}_2.$$

Therefore, every vector in W is a linear combination of $\mathbf{w}_1$ and $\mathbf{w}_2$.

33. To determine if a vector $\mathbf{w}$ is a linear combination of $\mathbf{v}_1$, $\mathbf{v}_2$, $\mathbf{v}_3$, we construct a linear system from the equation $a_1\mathbf{v}_1 + a_2\mathbf{v}_2 + a_3\mathbf{v}_3 = \mathbf{w}$. Let $\mathbf{w}$ have components w_1, w_2, and w_3. Then we have

$$a_1\begin{bmatrix} 4 \\ 2 \\ -3 \end{bmatrix} + a_2\begin{bmatrix} 2 \\ 1 \\ -2 \end{bmatrix} + a_3\begin{bmatrix} -2 \\ -1 \\ 0 \end{bmatrix} = \begin{bmatrix} w_1 \\ w_2 \\ w_3 \end{bmatrix}.$$

Combining the matrices on the left and equating corresponding components gives the linear system

$$\begin{array}{rcrcrcl} 4a_1 & + & 2a_2 & - & 2a_3 & = & w_1 \\ 2a_1 & + & a_2 & - & a_3 & = & w_2 \\ -3a_1 & - & 2a_2 & & & = & w_3 \end{array}$$

Then $\mathbf{w}$ is a linear combination of $\mathbf{v}_1, \mathbf{v}_2$, $\mathbf{v}_3$ provided the preceding system has a solution. This is determined by using elementary row operations to reduce the augmented matrix.

(a) $\mathbf{w} = \begin{bmatrix} 1 & 1 & 1 \end{bmatrix}^T$ and the corresponding augmented matrix is

$$\left[\begin{array}{ccc|c} 4 & 2 & -2 & 1 \\ 2 & 1 & -1 & 1 \\ -3 & -2 & 0 & 1 \end{array}\right] \longrightarrow \left[\begin{array}{ccc|c} 1 & \frac{1}{2} & -\frac{1}{2} & \frac{1}{4} \\ 2 & 1 & -1 & 1 \\ -3 & -2 & 0 & 1 \end{array}\right] \longrightarrow \left[\begin{array}{ccc|c} 1 & \frac{1}{2} & -\frac{1}{2} & \frac{1}{4} \\ 0 & 0 & 0 & \frac{1}{2} \\ 0 & -\frac{1}{2} & -\frac{3}{2} & \frac{7}{4} \end{array}\right]$$

$\frac{1}{4}$ (row 1)   add -2 (row 1) to row 2
         add 3 (row 1) to row 3

From the second row we see that the system is inconsistent and hence $\mathbf{w}$ is not a linear combination of $\mathbf{v}_1$, $\mathbf{v}_2$, $\mathbf{v}_3$.

(b) $\mathbf{w} = \begin{bmatrix} 4 & 2 & -6 \end{bmatrix}^T$ and the corresponding augmented matrix is

$$\left[\begin{array}{ccc|c} 4 & 2 & -2 & 4 \\ 2 & 1 & -1 & 2 \\ -3 & -2 & 0 & -6 \end{array}\right] \longrightarrow \left[\begin{array}{ccc|c} 1 & \frac{1}{2} & -\frac{1}{2} & 1 \\ 2 & 1 & -1 & 2 \\ -3 & -2 & 0 & -6 \end{array}\right] \longrightarrow \left[\begin{array}{ccc|c} 1 & \frac{1}{2} & -\frac{1}{2} & 1 \\ 0 & 0 & 0 & 0 \\ 0 & -\frac{1}{2} & -\frac{3}{2} & -3 \end{array}\right]$$

$\frac{1}{4}$ (row 1)   add -2 (row 1) to row 2   interchange
         add 3 (row 1) to row 3

$$\longrightarrow \begin{bmatrix} 1 & \frac{1}{2} & -\frac{1}{2} & \Big| & 1 \\ 0 & -\frac{1}{2} & -\frac{3}{2} & \Big| & -3 \\ 0 & 0 & 0 & \Big| & 0 \end{bmatrix}$$

If we now multiply row 2 by -2, row 2 then has a leading 1 and we have obtained row echelon form. Since there are no rows of the form $\begin{bmatrix} 0 & \cdots & 0 & | & * \end{bmatrix}$, where $*$ is some nonzero number, the system is consistent. Thus $\mathbf{w}$ is a linear combination of $\mathbf{v}_1$, $\mathbf{v}_2$, and $\mathbf{v}_3$.

(c) $\mathbf{w} = \begin{bmatrix} -2 & -1 & 1 \end{bmatrix}^T$ and the corresponding augmented matrix is

$$\begin{bmatrix} 4 & 2 & -2 & \Big| & -2 \\ 2 & 1 & -1 & \Big| & -1 \\ -3 & -2 & 0 & \Big| & 1 \end{bmatrix} \longrightarrow \begin{bmatrix} 1 & \frac{1}{2} & -\frac{1}{2} & \Big| & -\frac{1}{2} \\ 2 & 1 & -1 & \Big| & -1 \\ -3 & -2 & 0 & \Big| & 1 \end{bmatrix} \longrightarrow \begin{bmatrix} 1 & \frac{1}{2} & -\frac{1}{2} & \Big| & 1 \\ 0 & 0 & 0 & \Big| & 0 \\ 0 & -\frac{1}{2} & -\frac{3}{2} & \Big| & -\frac{1}{2} \end{bmatrix}$$
$$\quad\frac{1}{4}\text{ (row 1)} \qquad\qquad\qquad \text{add } -2 \text{ (row 1) to row 2} \qquad\qquad \text{interchange}$$
$$\qquad\qquad\qquad\qquad\qquad \text{add } 3 \text{ (row 1) to row 3} \qquad\qquad\qquad \text{rows 2 and 3}$$

$$\longrightarrow \begin{bmatrix} 1 & \frac{1}{2} & -\frac{1}{2} & \Big| & 1 \\ 0 & -\frac{1}{2} & -\frac{3}{2} & \Big| & -\frac{1}{2} \\ 0 & 0 & 0 & \Big| & 0 \end{bmatrix}$$

If we multiply row 2 by -2, then row 2 has a leading 1 and we have obtained row echelon form. Since there are no rows of the form $\begin{bmatrix} 0 & \cdots & 0 & | & * \end{bmatrix}$, where $*$ is some nonzero number, the system is consistent. Thus $\mathbf{w}$ is a linear combination of $\mathbf{v}_1$, $\mathbf{v}_2$, and $\mathbf{v}_3$.

(d) $\mathbf{w} = \begin{bmatrix} -1 & 2 & 3 \end{bmatrix}^T$ and the corresponding augmented matrix is

$$\begin{bmatrix} 4 & 2 & -2 & \Big| & -1 \\ 2 & 1 & -1 & \Big| & 2 \\ -3 & -2 & 0 & \Big| & 3 \end{bmatrix} \longrightarrow \begin{bmatrix} 1 & \frac{1}{2} & -\frac{1}{2} & \Big| & -\frac{1}{4} \\ 2 & 1 & -1 & \Big| & 2 \\ -3 & -2 & 0 & \Big| & 3 \end{bmatrix} \longrightarrow \begin{bmatrix} 1 & \frac{1}{2} & -\frac{1}{2} & \Big| & -\frac{1}{4} \\ 0 & 0 & 0 & \Big| & \frac{5}{2} \\ 0 & -\frac{1}{2} & -\frac{3}{2} & \Big| & -3 \end{bmatrix}$$
$$\quad\frac{1}{4}\text{ (row 1)} \qquad\qquad\qquad \text{add } -2 \text{ (row 1) to row 2}$$
$$\qquad\qquad\qquad\qquad\qquad \text{add } 3 \text{ (row 1) to row 3}$$

From the second row we see that the system is inconsistent and hence $\mathbf{w}$ is not a linear combination of $\mathbf{v}_1$, $\mathbf{v}_2$, and $\mathbf{v}_3$.

35. (a) The line ℓ_0 consists of all vectors of the form

$$\begin{bmatrix} x \\ y \\ z \end{bmatrix} = t \begin{bmatrix} u \\ v \\ w \end{bmatrix}.$$

It follows from Theorem 4.3 that ℓ_0 is a subspace of R^3.

(b) The line ℓ through the point $P_0(x_0, y_0, z_0)$ consists of all vectors of the form

$$\begin{bmatrix} x \\ y \\ z \end{bmatrix} = \begin{bmatrix} x_0 \\ y_0 \\ z_0 \end{bmatrix} + t \begin{bmatrix} u \\ v \\ w \end{bmatrix}.$$

If P_0 is not the origin, the conditions of Theorem 4.3 are not satisfied.

37. (a) Yes. Let $t = 2$; then $x = 4 - 2(2) = 0$, $y = -3 + 2(2) = 1$, $z = 4 - 5(2) = -6$.

(b) No. If $x = 4 - 2t = 1$, then $t = \frac{3}{2}$, but $y = -3 + 2\left(\frac{1}{2}\right) = -2 \neq 2$.

(c) Yes. Let $t = 0$; then $x = 4 - 2(0) = 4$, $y = -3 + 2(0) = -3$, $z = 4 - 5(0) = 4$.

(d) No. If $x = 4 - 2t = 0$, then $t = 2$, but $z = 4 - 5(2) = -6 \neq -1$.

39. (a) The line through $(2, -3, 1)$ and $(4, 2, 5)$ is parallel to vector

$$\mathbf{v} = \begin{bmatrix} 4 - 2 \\ 2 - (-3) \\ 5 - 1 \end{bmatrix} = \begin{bmatrix} 2 \\ 5 \\ 4 \end{bmatrix}.$$

Thus we can write the line in parametric form as

$$x = 2 + 2t, \quad y = -3 + 5t, \quad z = 1 + 4t, \quad -\infty < t < \infty.$$

(b) The line through $(-3, -2, -2)$ and $(5, 5, 4)$ is parallel to vector

$$\mathbf{v} = \begin{bmatrix} 5 - (-3) \\ 5 - (-2) \\ 4 - (-2) \end{bmatrix} = \begin{bmatrix} 8 \\ 7 \\ 6 \end{bmatrix}.$$

Thus we can write the line in parametric form as

$$x = -3 + 8t, \quad y = -2 + 7t, \quad z = -2 + 6t, \quad -\infty < t < \infty.$$

Section 4.4, p. 215

1. (a) Possible answers: $\left\{ \begin{bmatrix} 1 \\ 0 \\ 0 \end{bmatrix}, \begin{bmatrix} 1 \\ 1 \\ 0 \end{bmatrix}, \begin{bmatrix} 1 \\ 1 \\ 1 \end{bmatrix} \right\}, \left\{ \begin{bmatrix} 0 \\ 0 \\ 1 \end{bmatrix}, \begin{bmatrix} 0 \\ 1 \\ 1 \end{bmatrix}, \begin{bmatrix} 1 \\ 1 \\ 1 \end{bmatrix} \right\}.$

(b) Possible answers: $\left\{ \begin{bmatrix} 1 & 0 \\ 0 & 0 \end{bmatrix}, \begin{bmatrix} 0 & 1 \\ 0 & 0 \end{bmatrix}, \begin{bmatrix} 0 & 0 \\ 1 & 0 \end{bmatrix}, \begin{bmatrix} 0 & 0 \\ 0 & 1 \end{bmatrix} \right\}, \left\{ \begin{bmatrix} 1 & 0 \\ 0 & 0 \end{bmatrix}, \begin{bmatrix} 1 & 1 \\ 0 & 0 \end{bmatrix}, \right.$
$\left. \begin{bmatrix} 1 & 1 \\ 1 & 0 \end{bmatrix}, \begin{bmatrix} 1 & 1 \\ 1 & 1 \end{bmatrix} \right\}.$

(c) Possible answers: $\{t^2, t + 1, t - 1\}, \{t^2 + t, t^2 - t, t + 1\}.$

3. To determine if a vector $p(t)$ belongs to the span of vectors

$$p_1(t) = t^2 + 2t + 1, \quad p_2(t) = t^2 + 3, \quad p_3(t) = t - 1$$

we construct a linear system from the equation

$$c_1 p_1(t) + c_2 p_2(t) + c_3 p_3(t) = p(t) = at^2 + bt + c$$

or

$$c_1(t^2 + 2t + 1) + c_2(t^2 + 3) + c_3(t - 1) = at^2 + bt + c.$$

Equating coefficients of like powers of t gives

$$\begin{array}{rcrcrcl} c_1 & + & c_2 & & & = & a \\ 2c_1 & & & + & c_3 & = & b \\ c_1 & + & 3c_2 & - & c_3 & = & c \end{array}$$

Vector $p(t)$ is a linear combination of $p_1(t)$, $p_2(t)$, and $p_3(t)$ provided the preceding system has a solution. This is determined by using elementary row operations to reduce the augmented matrix.

(a) For $p(t) = t^2 + t + 2$ the corresponding augmented matrix is

$$\left[\begin{array}{ccc|c} 1 & 1 & 0 & 1 \\ 2 & 0 & 1 & 1 \\ 1 & 3 & -1 & 2 \end{array} \right] \longrightarrow \left[\begin{array}{ccc|c} 1 & 1 & 0 & 1 \\ 0 & -2 & 1 & -1 \\ 0 & 2 & -1 & 1 \end{array} \right] \longrightarrow \left[\begin{array}{ccc|c} 1 & 1 & 0 & 1 \\ 0 & -2 & 1 & -1 \\ 0 & 0 & 0 & 0 \end{array} \right]$$

add -2 (row 1) to row 2
add -1 (row 1) to row 3
    add row 2 to row 3

If we multiply row 2 by $-\frac{1}{2}$, then row 2 has a leading 1 and we have obtained row echelon form. Since there are no rows of the form $\begin{bmatrix} 0 & \cdots & 0 & | & * \end{bmatrix}$, where $*$ is some nonzero number, the system is consistent. Thus $p(t)$ is in the span of $p_1(t)$, $p_2(t)$, $p_3(t)$.

(b) For $p(t) = 2t^2 + 2t + 3$ the corresponding augmented matrix is

$$\begin{bmatrix} 1 & 1 & 0 & | & 2 \\ 2 & 0 & 1 & | & 2 \\ 1 & 3 & -1 & | & 3 \end{bmatrix} \longrightarrow \begin{bmatrix} 1 & 1 & 0 & | & 2 \\ 0 & -2 & 1 & | & -2 \\ 0 & 2 & -1 & | & 1 \end{bmatrix} \longrightarrow \begin{bmatrix} 1 & 1 & 0 & | & 2 \\ 0 & -2 & 1 & | & -2 \\ 0 & 0 & 0 & | & -1 \end{bmatrix}$$

add -2 (row 1) to row 2   add row 2 to row 3
add -1 (row 1) to row 3

From the third row we see that the system is inconsistent and hence $p(t)$ is not in the span of $p_1(t)$, $p_2(t)$, and $p_3(t)$.

(c) For $p(t) = -t^2 + t - 4$ the corresponding augmented matrix is

$$\begin{bmatrix} 1 & 1 & 0 & | & -1 \\ 2 & 0 & 1 & | & 1 \\ 1 & 3 & -1 & | & -4 \end{bmatrix} \longrightarrow \begin{bmatrix} 1 & 1 & 0 & | & -1 \\ 0 & -2 & 1 & | & 3 \\ 0 & 2 & -1 & | & -3 \end{bmatrix} \longrightarrow \begin{bmatrix} 1 & 1 & 0 & | & -1 \\ 0 & -2 & 1 & | & 3 \\ 0 & 0 & 0 & | & 0 \end{bmatrix}$$

add -2 (row 1) to row 2   add row 2 to row 3
add -1 (row 1) to row 3

If we multiply row 2 by $-\frac{1}{2}$, then row 2 has a leading 1 and we have obtained row echelon form. Since there are no rows of the form $\begin{bmatrix} 0 & \cdots & 0 & | & * \end{bmatrix}$, where $*$ is some nonzero number, the system is consistent. Thus $p(t)$ is in the span of $p_1(t)$, $p_2(t)$, $p_3(t)$.

(d) For $p(t) = -2t^2 + 3t + 1$ the corresponding augmented matrix is

$$\begin{bmatrix} 1 & 1 & 0 & | & -2 \\ 2 & 0 & 1 & | & 3 \\ 1 & 3 & -1 & | & 1 \end{bmatrix} \longrightarrow \begin{bmatrix} 1 & 1 & 0 & | & -2 \\ 0 & -2 & 1 & | & 7 \\ 0 & 2 & -1 & | & 3 \end{bmatrix} \longrightarrow \begin{bmatrix} 1 & 1 & 0 & | & -2 \\ 0 & -2 & 1 & | & 7 \\ 0 & 0 & 0 & | & 10 \end{bmatrix}$$

add -2 (row 1) to row 2   add row 2 to row 3
add -1 (row 1) to row 3

From the third row we see that the system is inconsistent and hence $p(t)$ is not in the span of $p_1(t)$, $p_2(t)$, and $p_3(t)$.

5. For a set to span R_2 we must be able to write every element of R_2 as a linear combination of vectors in the set. We form a linear combination of the vectors of the set with unknown coefficients and set it equal to an arbitrary vector in R_2. Next construct a linear system of equations from this relation to solve for the unknown coefficients. If this system is consistent, then the set spans R_2. If the system is inconsistent for certain vectors in R_2, then the set does not span R_2. Let $\begin{bmatrix} a & b \end{bmatrix}$ be an arbitrary vector in R_2.

(a) $S = \left\{ \begin{bmatrix} 1 & 2 \end{bmatrix}, \begin{bmatrix} -1 & 1 \end{bmatrix} \right\}$; $c_1 \begin{bmatrix} 1 & 2 \end{bmatrix} + c_2 \begin{bmatrix} -1 & 1 \end{bmatrix} = \begin{bmatrix} a & b \end{bmatrix}$

Combining the terms on the left side and equating corresponding entries gives the linear system

$$\begin{aligned} c_1 - c_2 &= a \\ 2c_1 + c_2 &= b \end{aligned}$$

We form the augmented matrix and row reduce it.

$$\begin{bmatrix} 1 & -1 & | & a \\ 2 & 1 & | & b \end{bmatrix} \longrightarrow \begin{bmatrix} 1 & -1 & | & a \\ 0 & 3 & | & b - 2a \end{bmatrix}$$

add -2 (row 1) to row 2

If we now multiply row 2 by $\frac{1}{3}$ we get a leading 1 in row 2. The resulting matrix is in row echelon form and has a leading 1 in each row. Hence the system is consistent for all vectors $\begin{bmatrix} a & b \end{bmatrix}$. Thus span S is R_2.

(b) $S = \{[\ 0\ \ 0\ ], [\ 1\ \ 1\ ], [\ -2\ \ -2\ ]\}$; $c_1 [\ 0\ \ 0\ ] + c_2 [\ 1\ \ 1\ ] + c_3 [\ -2\ \ -2\ ] = [\ a\ \ b\ ]$

Combining the terms on the left side and equating corresponding entries gives the linear system

$$\begin{array}{rrrrr} 0c_1 & + & c_2 & - & 2c_3 & = & a \\ 0c_1 & + & c_2 & - & 2c_3 & = & b \end{array}$$

We form the augmented matrix and row reduce it:

$$\left[\begin{array}{ccc|c} 0 & 1 & -2 & a \\ 0 & 1 & -2 & b \end{array}\right] \longrightarrow \left[\begin{array}{ccc|c} 0 & 1 & -2 & a \\ 0 & 0 & 0 & b-a \end{array}\right]$$

add -1 (row 1) to row 2

The system is consistent only when $b - a = 0$, hence not every vector in R_2 can be written as a linear combination of vectors in S. Thus S does not span R_2.

(c) $S = \{[\ 1\ \ 3\ ], [\ 2\ \ -3\ ], [\ 0\ \ 2\ ]\}$; $c_1 [\ 1\ \ 3\ ] + c_2 [\ 2\ \ -3\ ] + c_3 [\ 0\ \ 2\ ] = [\ a\ \ b\ ]$

Combining the terms on the left side and equating corresponding entries gives the linear system

$$\begin{array}{rrrrr} c_1 & + & 2c_2 & & & = & a \\ 3c_1 & - & 3c_2 & + & 2c_3 & = & b \end{array}$$

We form the augmented matrix and row reduce it:

$$\left[\begin{array}{ccc|c} 1 & 2 & 0 & a \\ 3 & -3 & 2 & b \end{array}\right] \longrightarrow \left[\begin{array}{ccc|c} 1 & 2 & 0 & a \\ 0 & -9 & 2 & b-3a \end{array}\right]$$

add -3 (row 1) to row 2

Multiplying row 2 by $-\frac{1}{9}$ gives a leading 1 in row 2. The resulting matrix is in row echelon form and has a leading 1 in each row. Hence the system is consistent for all vectors $[\ a\ \ b\ ]$. Thus span S is R_2.

(d) $S = \{[\ 2\ \ 4\ ], [\ -1\ \ 2\ ]\}$; $c_1 [\ 2\ \ 4\ ] + c_2 [\ -1\ \ 2\ ] = [\ a\ \ b\ ]$

Combining the terms on the left side and equating corresponding entries gives the linear system

$$\begin{array}{rrrr} 2c_1 & - & c_2 & = & a \\ 4c_1 & + & 2c_2 & = & b \end{array}$$

We form the augmented matrix and row reduce it:

$$\left[\begin{array}{cc|c} 2 & -1 & a \\ 4 & 2 & b \end{array}\right] \longrightarrow \left[\begin{array}{cc|c} 2 & -1 & a \\ 0 & 4 & b-2a \end{array}\right]$$

add -2 (row 1) to row 2

Multiplying row 1 by $\frac{1}{2}$ and row 2 by $\frac{1}{4}$ gives a leading 1 in both rows. The resulting matrix is in row echelon form. Hence the system is consistent for all vectors $[\ a\ \ b\ ]$. Thus span S is R_2.

7. For a set to span R_4 we must be able to write every vector in R_4 as a linear combination of vectors in the set. We form a linear combination of the vectors of the set with unknown coefficients and set it equal to an arbitrary vector of R_4. Next construct a linear system of equations from this relation to solve for the unknown coefficients. If this system is consistent, then the set spans R_4. If the system is inconsistent for certain vectors in R_4, then the set does not span R_4. Let $[\ a\ \ b\ \ c\ \ d\ ]$ be an arbitrary vector in R_4.

(a) $S = \{[\ 1\ \ 0\ \ 0\ \ 1\ ], [\ 0\ \ 1\ \ 0\ \ 0\ ], [\ 1\ \ 1\ \ 1\ \ 1\ ], [\ 1\ \ 1\ \ 1\ \ 0\ ]\}$

$c_1 [\ 1\ \ 0\ \ 0\ \ 1\ ] + c_2 [\ 0\ \ 1\ \ 0\ \ 0\ ] + c_3 [\ 1\ \ 1\ \ 1\ \ 1\ ] + c_4 [\ 1\ \ 1\ \ 1\ \ 0\ ] = [\ a\ \ b\ \ c\ \ d\ ]$

Combining the terms on the left side and equating corresponding entries gives the linear system

$$
\begin{aligned}
c_1 \quad\quad\; + c_3 + c_4 &= a \\
c_2 + c_3 + c_4 &= b \\
c_3 + c_4 &= c \\
c_1 \quad\quad\; + c_3 \quad\quad\; &= d
\end{aligned}
$$

We form the augmented matrix and row reduce it.

$$
\left[\begin{array}{cccc|c}
1 & 0 & 1 & 1 & a \\
0 & 1 & 1 & 1 & b \\
0 & 0 & 1 & 1 & c \\
1 & 0 & 1 & 0 & d
\end{array}\right]
\longrightarrow
\left[\begin{array}{cccc|c}
1 & 0 & 1 & 1 & a \\
0 & 1 & 1 & 1 & b \\
0 & 0 & 1 & 1 & c \\
0 & 0 & 0 & -1 & d-a
\end{array}\right]
$$
$$\text{add } -1 \text{ (row 1) to row 4}$$

Multiplying row 1 by -1 gives us a leading 1 in each row. The resulting matrix is in row echelon form. Hence the system is consistent for all vectors $\begin{bmatrix} a & b & c & d \end{bmatrix}$. Thus span S is R_4.

(b) $S = \left\{ \begin{bmatrix} 1 & 2 & 1 & 0 \end{bmatrix}, \begin{bmatrix} 1 & 1 & -1 & 0 \end{bmatrix}, \begin{bmatrix} 0 & 0 & 0 & 1 \end{bmatrix} \right\}$

$c_1 \begin{bmatrix} 1 & 2 & 1 & 0 \end{bmatrix} + c_2 \begin{bmatrix} 1 & 1 & -1 & 0 \end{bmatrix} + c_3 \begin{bmatrix} 0 & 0 & 0 & 1 \end{bmatrix} = \begin{bmatrix} a & b & c & d \end{bmatrix}$

Combining the terms on the left side and equating corresponding entries gives the linear system

$$
\begin{aligned}
c_1 + c_2 \quad\quad &= a \\
2c_1 + c_2 \quad\quad &= b \\
c_1 - c_2 \quad\quad &= c \\
c_3 &= d
\end{aligned}
$$

We form the augmented matrix and row reduce it.

$$
\left[\begin{array}{ccc|c}
1 & 1 & 0 & a \\
2 & 1 & 0 & b \\
1 & -1 & 0 & c \\
0 & 0 & 1 & d
\end{array}\right]
\longrightarrow
\left[\begin{array}{ccc|c}
1 & 1 & 0 & a \\
0 & -1 & 0 & b-2a \\
0 & -2 & 0 & c-a \\
0 & 0 & 1 & d
\end{array}\right]
\longrightarrow
\left[\begin{array}{ccc|c}
1 & 1 & 0 & a \\
0 & 1 & 0 & 2a-b \\
0 & 0 & 0 & 3a-2b+c \\
0 & 0 & 1 & d
\end{array}\right]
$$
$$\text{add } -2 \text{ (row 1) to row 2} \quad\quad\quad -1 \text{ (row 2)}$$
$$\text{add } -1 \text{ (row 1) to row 3} \quad\quad \text{add } 2 \text{ (row 2) to row 3}$$

The system is consistent only when $3a - 2b + c = 0$, hence not every vector of R_4 can be written as a linear combination of vectors in S. Thus S does not span R_4.

(c) $S = \left\{ \begin{bmatrix} 6 & 4 & -2 & 4 \end{bmatrix}, \begin{bmatrix} 2 & 0 & 0 & 1 \end{bmatrix}, \begin{bmatrix} 3 & 2 & -1 & 2 \end{bmatrix}, \begin{bmatrix} 5 & 6 & -3 & 2 \end{bmatrix}, \begin{bmatrix} 0 & 4 & -2 & -1 \end{bmatrix} \right\}$

$c_1 \begin{bmatrix} 6 & 4 & -2 & 4 \end{bmatrix} + c_2 \begin{bmatrix} 2 & 0 & 0 & 1 \end{bmatrix} + c_3 \begin{bmatrix} 3 & 2 & -1 & 2 \end{bmatrix}$
$\quad\quad + c_4 \begin{bmatrix} 5 & 6 & -3 & 2 \end{bmatrix} + c_5 \begin{bmatrix} 0 & 4 & -2 & -1 \end{bmatrix} = \begin{bmatrix} a & b & c & d \end{bmatrix}$

Combining the terms on the left side and equating corresponding entries gives the linear system

$$
\begin{aligned}
6c_1 + 2c_2 + 3c_3 + 5c_4 \quad\quad\quad &= a \\
4c_1 \quad\quad\; + 2c_3 + 6c_4 + 4c_5 &= b \\
-2c_1 \quad\quad\; - c_3 - 3c_4 - 2c_5 &= c \\
4c_1 + c_2 + 2c_3 + 2c_4 - c_5 &= d
\end{aligned}
$$

We form the augmented matrix and row reduce it.

$$\left[\begin{array}{ccccc|c} 6 & 2 & 3 & 5 & 0 & a \\ 4 & 0 & 2 & 6 & 4 & b \\ -2 & 0 & -1 & -3 & -2 & c \\ 4 & 1 & 2 & 2 & -1 & d \end{array}\right] \longrightarrow \left[\begin{array}{ccccc|c} 1 & \frac{1}{3} & \frac{1}{2} & \frac{5}{6} & 0 & \frac{1}{6}a \\ 4 & 0 & 2 & 6 & 4 & b \\ -2 & 0 & -1 & -3 & -2 & c \\ 4 & 1 & 2 & 2 & -1 & d \end{array}\right] \longrightarrow \left[\begin{array}{ccccc|c} 1 & \frac{1}{3} & \frac{1}{2} & \frac{5}{6} & 0 & \frac{1}{6}a \\ 0 & -\frac{4}{3} & 0 & \frac{8}{3} & 4 & b - \frac{2}{3}a \\ 0 & \frac{2}{3} & 0 & -\frac{4}{3} & -2 & c - \frac{1}{3}a \\ 0 & -\frac{1}{3} & 0 & -\frac{4}{3} & -1 & d - \frac{2}{3}a \end{array}\right] \longrightarrow$$

$$\frac{1}{6}\ (\text{row } 1) \qquad\qquad\qquad \begin{array}{l}\text{add } -4\ (\text{row } 1) \text{ to row } 2 \\ \text{add } 2\ (\text{row } 1) \text{ to row } 3 \\ \text{add } -4\ (\text{row } 1) \text{ to row } 4\end{array} \qquad\qquad \text{add } \tfrac{1}{2}\ (\text{row } 2) \text{ to row } 3$$

$$\left[\begin{array}{ccccc|c} 1 & \frac{1}{3} & \frac{1}{2} & \frac{5}{6} & 0 & \frac{1}{6}a \\ 0 & -\frac{4}{3} & 0 & \frac{8}{3} & 4 & b - \frac{2}{3}a \\ 0 & 0 & 0 & 0 & 0 & * \\ 0 & -\frac{1}{3} & 0 & -\frac{4}{3} & -1 & d - \frac{2}{3}a \end{array}\right]$$

where $*$ is a combination of constants a, b, and c. It follows that the system is inconsistent, hence not every member of R_4 can be written as a linear combination of members of S. Thus S does not span R_4.

(d) $S = \left\{ \begin{bmatrix} 1 & 1 & 0 & 0 \end{bmatrix}, \begin{bmatrix} 1 & 2 & -1 & 1 \end{bmatrix}, \begin{bmatrix} 0 & 0 & 1 & 1 \end{bmatrix}, \begin{bmatrix} 2 & 1 & 2 & 1 \end{bmatrix} \right\}$

$c_1 \begin{bmatrix} 1 & 1 & 0 & 0 \end{bmatrix} + c_2 \begin{bmatrix} 1 & 2 & -1 & 1 \end{bmatrix} + c_3 \begin{bmatrix} 0 & 0 & 1 & 1 \end{bmatrix} + c_4 \begin{bmatrix} 2 & 1 & 2 & 1 \end{bmatrix} = \begin{bmatrix} a & b & c & d \end{bmatrix}$

Combining the terms on the left side and equating corresponding entries gives the linear system

$$\begin{array}{rcl} c_1 + c_2 \phantom{{} + c_3} + 2c_4 &=& a \\ c_1 + 2c_2 \phantom{{} + c_3} + c_4 &=& b \\ -c_2 + c_3 + 2c_4 &=& c \\ c_2 + c_3 + c_4 &=& d \end{array}$$

We form the augmented matrix and row reduce it.

$$\left[\begin{array}{cccc|c} 1 & 1 & 0 & 2 & a \\ 1 & 2 & 0 & 1 & b \\ 0 & -1 & 1 & 2 & c \\ 0 & 1 & 1 & 1 & d \end{array}\right] \longrightarrow \left[\begin{array}{cccc|c} 1 & 1 & 0 & 2 & a \\ 0 & 1 & 0 & -1 & b - a \\ 0 & -1 & 1 & 2 & c \\ 0 & 1 & 1 & 1 & d \end{array}\right] \longrightarrow \left[\begin{array}{cccc|c} 1 & 1 & 0 & 2 & a \\ 0 & 1 & 0 & -1 & b - a \\ 0 & 0 & 1 & 1 & c + b - a \\ 0 & 0 & 1 & 2 & d - b + a \end{array}\right] \longrightarrow$$

$$\text{add } -1\ (\text{row } 1) \text{ to row } 2 \qquad\quad \begin{array}{l}\text{add row } 2 \text{ to row } 3 \\ \text{add } -1\ (\text{row } 2) \text{ to row } 4\end{array} \qquad\quad \text{add } -1\ (\text{row } 3) \text{ to row } 4$$

$$\left[\begin{array}{cccc|c} 1 & 1 & 0 & 2 & a \\ 0 & 1 & 0 & -1 & b - a \\ 0 & 0 & 1 & 1 & c + b - a \\ 0 & 0 & 0 & 1 & * \end{array}\right]$$

Thus we have a leading 1 in each row. Hence the system is consistent for all vectors $\begin{bmatrix} a & b & c & d \end{bmatrix}$. Thus span S is R_4.

9. For a set to span P_3 we must be able to write every element of P_3 as a linear combination of members of the set. We form a linear combination of the members of the set with unknown coefficients and set it equal to an arbitrary member of P_3. Next construct a linear system of equations from this relation to solve for the unknown coefficients. If this system is consistent, then the set spans P_3. If the system is inconsistent for certain vectors in P_3, then the set does not span P_3. Let $at^3 + bt^2 + ct + d$ be an arbitrary vector in P_3. Then set

$$c_1(t^3 + 2t + 1) + c_2(t^2 - t + 2) + c_3(t^3 + 2) + c_4(-t^3 + t^2 - 5t + 2) = at^3 + bt^2 + ct + d.$$

Combining like terms on the left side and equating corresponding coefficients gives the linear system

$$
\begin{array}{rcrcrcrcl}
c_1 & & & + & c_3 & - & c_4 & = & a \\
& & c_2 & & & + & c_4 & = & b \\
2c_1 & - & c_2 & & & - & 5c_4 & = & c \\
c_1 & + & 2c_2 & + & 2c_3 & + & 2c_4 & = & d
\end{array}
$$

We form the augmented matrix and row reduce it.

$$
\begin{bmatrix}
1 & 0 & 1 & -1 & a \\
0 & 1 & 0 & 1 & b \\
2 & -1 & 0 & -5 & c \\
1 & 2 & 2 & 2 & d
\end{bmatrix}
\longrightarrow
\begin{bmatrix}
1 & 0 & 1 & -1 & a \\
0 & 1 & 0 & 1 & b \\
0 & -1 & -2 & -3 & c-2a \\
1 & 2 & 2 & 2 & d-a
\end{bmatrix}
\longrightarrow
\begin{bmatrix}
1 & 0 & 1 & -1 & a \\
0 & 1 & 0 & 1 & b \\
0 & 0 & -2 & -2 & c-2a+b \\
0 & 0 & 1 & 1 & d-a-2b
\end{bmatrix}
\longrightarrow
$$

add -2 (row 1) to row 3　　　add row 2 to row 3　　　add $-\frac{1}{2}$ (row 3) to row 4

add -1 (row 1) to row 4　　　add -2 (row 2) to row 4

$$
\begin{bmatrix}
1 & 0 & 1 & -1 & a \\
0 & 1 & 0 & 1 & b \\
0 & 0 & -2 & -2 & c-2a+b \\
0 & 0 & 0 & 0 & *
\end{bmatrix}
$$

Thus the system is inconsistent and the specified polynomials do not span P_3.

11. Following the method used in Example 3, we find the reduced row echelon form of the augmented matrix $\begin{bmatrix} A & | & \mathbf{0} \end{bmatrix}$ and determined the general solution of the homogeneous system.

$$
\begin{bmatrix}
1 & 0 & 1 & 0 & 0 \\
1 & 2 & 3 & 1 & 0 \\
2 & 1 & 3 & 1 & 0 \\
1 & 1 & 2 & 1 & 0
\end{bmatrix}
\longrightarrow
\begin{bmatrix}
1 & 0 & 1 & 0 & 0 \\
0 & 2 & 2 & 1 & 0 \\
0 & 1 & 1 & 1 & 0 \\
0 & 1 & 1 & 1 & 0
\end{bmatrix}
\longrightarrow
\begin{bmatrix}
1 & 0 & 1 & 0 & 0 \\
0 & 1 & 1 & 1 & 0 \\
0 & 2 & 2 & 1 & 0 \\
0 & 0 & 0 & 0 & 0
\end{bmatrix}
\longrightarrow
$$

add -1 (row 1) to row 2　　　add -1 (row 3) to row 4　　　add -2 (row 2) to row 3

add -2 (row 1) to row 3　　　interchange rows 3 and 2

add -1 (row 1) to row 4

$$
\begin{bmatrix}
1 & 0 & 1 & 0 & 0 \\
0 & 1 & 1 & 1 & 0 \\
0 & 0 & 0 & -1 & 0 \\
0 & 0 & 0 & 0 & 0
\end{bmatrix}
\longrightarrow
\begin{bmatrix}
1 & 0 & 1 & 0 & 0 \\
0 & 1 & 1 & 0 & 0 \\
0 & 0 & 0 & 1 & 0 \\
0 & 0 & 0 & 0 & 0
\end{bmatrix}
$$

add row 3 to row 2

-1 (row 3)

Thus the general solution is given by $x_4 = 0$, $x_2 = -x_3$, $x_1 = -x_3$. Hence we can choose x_3 arbitrarily. Letting $x_3 = r$ we can write the general solution in the form

$$
\mathbf{x} =
\begin{bmatrix}
-r \\
-r \\
r \\
0
\end{bmatrix}
= r
\begin{bmatrix}
-1 \\
-1 \\
1 \\
0
\end{bmatrix}.
$$

Hence the vector

$$
\begin{bmatrix}
-1 \\
-1 \\
1 \\
0
\end{bmatrix}
$$

spans the solution of $A\mathbf{x} = \mathbf{0}$.

13. Every vector A in W is of the form $A = \begin{bmatrix} a & b \\ c & -a \end{bmatrix}$, where a, b, and c are any real numbers. We have

$$\begin{bmatrix} a & b \\ c & -a \end{bmatrix} = a \begin{bmatrix} 1 & 0 \\ 0 & -1 \end{bmatrix} + b \begin{bmatrix} 0 & 1 \\ 0 & 0 \end{bmatrix} + c \begin{bmatrix} 0 & 0 \\ 1 & 0 \end{bmatrix},$$

so A is in span S. Thus, every vector in W is in span S. Hence, span $S = W$.

15. Observe that

$$\begin{bmatrix} a & 0 & b \\ 0 & c & 0 \\ d & 0 & e \end{bmatrix} = a \begin{bmatrix} 1 & 0 & 0 \\ 0 & 0 & 0 \\ 0 & 0 & 0 \end{bmatrix} + b \begin{bmatrix} 0 & 0 & 1 \\ 0 & 0 & 0 \\ 0 & 0 & 0 \end{bmatrix} + c \begin{bmatrix} 0 & 0 & 0 \\ 0 & 1 & 0 \\ 0 & 0 & 0 \end{bmatrix} + d \begin{bmatrix} 0 & 0 & 0 \\ 0 & 0 & 0 \\ 1 & 0 & 0 \end{bmatrix} + e \begin{bmatrix} 0 & 0 & 0 \\ 0 & 0 & 0 \\ 0 & 0 & 1 \end{bmatrix}.$$

Therefore, if

$$S = \left\{ \begin{bmatrix} 1 & 0 & 0 \\ 0 & 0 & 0 \\ 0 & 0 & 0 \end{bmatrix}, \begin{bmatrix} 0 & 0 & 1 \\ 0 & 0 & 0 \\ 0 & 0 & 0 \end{bmatrix}, \begin{bmatrix} 0 & 0 & 0 \\ 0 & 1 & 0 \\ 0 & 0 & 0 \end{bmatrix}, \begin{bmatrix} 0 & 0 & 0 \\ 0 & 0 & 0 \\ 1 & 0 & 0 \end{bmatrix}, \begin{bmatrix} 0 & 0 & 0 \\ 0 & 0 & 0 \\ 0 & 0 & 1 \end{bmatrix} \right\},$$

then span $S = W$.

Section 4.5, p. 226

1. We form Equation (1):

$$c_1 \begin{bmatrix} 2 \\ 1 \\ 3 \end{bmatrix} + c_2 \begin{bmatrix} 3 \\ -1 \\ 2 \end{bmatrix} + c_3 \begin{bmatrix} 10 \\ 0 \\ 10 \end{bmatrix} = \begin{bmatrix} 0 \\ 0 \\ 0 \end{bmatrix}.$$

By observation, this equation has nontrivial solutions; for example, $c_1 = 2$, $c_2 = 2$, $c_3 = -1$. Therefore, S is linearly dependent.

3. We form Equation (1):

$$c_1 \begin{bmatrix} 2 \\ 1 \\ 3 \end{bmatrix} + c_2 \begin{bmatrix} 3 \\ -1 \\ 2 \end{bmatrix} + c_3 \begin{bmatrix} 10 \\ 0 \\ 10 \end{bmatrix} = \begin{bmatrix} 0 \\ 0 \\ 0 \end{bmatrix}.$$

The coefficient matrix of this linear system is

$$\begin{bmatrix} 1 & 4 & 2 \\ 2 & 3 & 0 \\ 1 & 1 & 1 \\ -1 & 0 & 3 \end{bmatrix}$$

whose reduced row echelon form is

$$\begin{bmatrix} 1 & 0 & 0 \\ 0 & 1 & 0 \\ 0 & 0 & 1 \\ 0 & 0 & 0 \end{bmatrix}.$$

Thus the only sollution is the trivial solution $c_1 = c_2 = c_3 = 0$ and hence the set S is linearly independent.

5. The reduced row echelon form of the augmented matrix is

$$\left[\begin{array}{cccc|c} 1 & 0 & 0 & 0 & 0 \\ 0 & 1 & 0 & 0 & 0 \\ 0 & 0 & 1 & 0 & 0 \\ 0 & 0 & 0 & 1 & 0 \end{array} \right].$$

Yes, the set S is linearly independent.

7. The reduced row echelon form of the augmented matrix is

$$\left[\begin{array}{ccc|c} 1 & 0 & 0 & 0 \\ 0 & 1 & 0 & 0 \\ 0 & 0 & 1 & 0 \end{array}\right].$$

Yes, the set S is linearly independent.

9. Suppose that $c_1\mathbf{x}_1 + c_2\mathbf{x}_2 + c_3\mathbf{x}_3 = \mathbf{0}$ for some scalars c_1, c_2, and c_3. Then

$$c_1\left[\begin{array}{c} 2 \\ -1 \\ 1 \end{array}\right] + c_2\left[\begin{array}{c} 4 \\ -7 \\ -1 \end{array}\right] + c_3\left[\begin{array}{c} 1 \\ 2 \\ 2 \end{array}\right] = \left[\begin{array}{c} 0 \\ 0 \\ 0 \end{array}\right]$$

whcih gives the homogeneous system

$$\begin{array}{rrrrrrr} 2c_1 & + & 4c_2 & + & c_3 & = & 0 \\ -c_1 & - & 7c_2 & + & 2c_3 & = & 0 \\ c_1 & - & c_2 & + & 2c_3 & = & 0 \end{array}$$

Row reducing the augmented matrix of this system we obtain

$$\left[\begin{array}{ccc|c} 2 & 4 & 1 & 0 \\ -1 & -7 & 2 & 0 \\ 1 & -1 & 2 & 0 \end{array}\right] \longrightarrow \left[\begin{array}{ccc|c} 1 & -1 & 2 & 0 \\ -1 & -7 & 2 & 0 \\ 2 & 4 & 1 & 0 \end{array}\right] \longrightarrow \left[\begin{array}{ccc|c} 1 & -1 & 2 & 0 \\ 0 & -8 & 4 & 0 \\ 0 & 6 & -3 & 0 \end{array}\right] \longrightarrow$$

$$\begin{array}{ccc} \text{interchange} & \text{add row 1 to row 2} & \text{add } \frac{3}{4}(\text{row 2}) \text{ to row 3} \\ \text{row 1 and 3} & \text{add } -2(\text{row 1}) \text{ to row 3} & -\frac{1}{8}(\text{row 2}) \end{array}$$

$$\left[\begin{array}{ccc|c} 1 & -1 & 2 & 0 \\ 0 & 1 & -\frac{1}{2} & 0 \\ 0 & 0 & 0 & 0 \end{array}\right] \longrightarrow \left[\begin{array}{ccc|c} 1 & 0 & \frac{3}{2} & 0 \\ 0 & 1 & -\frac{1}{2} & 0 \\ 0 & 0 & 0 & 0 \end{array}\right]$$

add row 2 to row 1

It follows from the reduced echelon form that the system has many nontrivial solutions. Therefore the set $\{\mathbf{x}_1, \mathbf{x}_2, \mathbf{x}_3\}$ is not linearly independent.

11. To determine which sets are linearly dependent we form a linear combination of the vectors with unknown coefficients and set it equal to the zero vector. Next perform the matrix operations and equate corresponding entries to obtain a homogeneous linear system. If the homogeneous system has any nontrivial solutions, then the vectors are linearly dependent.

(a) $S = \{[\,1\ \ 1\ \ 0\,], [\,0\ \ 2\ \ 3\,], [\,1\ \ 2\ \ 3\,], [\,3\ \ 6\ \ 6\,]\}$

$c_1[\,1\ \ 1\ \ 0\,] + c_2[\,0\ \ 2\ \ 3\,] + c_3[\,1\ \ 2\ \ 3\,] + c_4[\,3\ \ 6\ \ 6\,] = [\,0\ \ 0\ \ 0\,]$

$$\left[\begin{array}{cccc|c} 1 & 0 & 1 & 3 & 0 \\ 1 & 2 & 2 & 6 & 0 \\ 0 & 3 & 3 & 6 & 0 \end{array}\right] \longrightarrow \left[\begin{array}{cccc|c} 1 & 0 & 1 & 3 & 0 \\ 0 & 2 & 1 & 3 & 0 \\ 0 & 3 & 3 & 6 & 0 \end{array}\right] \longrightarrow \left[\begin{array}{cccc|c} 1 & 0 & 1 & 3 & 0 \\ 0 & 1 & \frac{1}{2} & \frac{3}{2} & 0 \\ 0 & 3 & 3 & 6 & 0 \end{array}\right] \longrightarrow$$

$$\begin{array}{ccc} \text{add } -1 \ (\text{row 1}) \text{ to row 2} & \frac{1}{2} \ (\text{row 1}) & \text{add } -3 \ (\text{row 2}) \text{ to row 3} \end{array}$$

$$\left[\begin{array}{cccc|c} 1 & 0 & 1 & 3 & 0 \\ 0 & 1 & \frac{1}{2} & \frac{3}{2} & 0 \\ 0 & 0 & \frac{3}{2} & \frac{3}{2} & 0 \end{array}\right] \longrightarrow \left[\begin{array}{cccc|c} 1 & 0 & 0 & 2 & 0 \\ 0 & 1 & 0 & 1 & 0 \\ 0 & 0 & 1 & 1 & 0 \end{array}\right]$$

$$\begin{array}{c} \frac{2}{3} \ (\text{row 3}) \\ \text{add } -\frac{1}{2} \ (\text{row 3}) \text{ to row 2} \\ \text{add } -1 \ (\text{row 3}) \text{ to row 1} \end{array}$$

Hence the general solution is $c_1 = -2r$, $c_2 = -r$, $c_3 = -r$, $c_4 = r$, where r is any real number. Thus we have many nontrivial solutions, hence the vectors are linearly dependent. Note that

$$[\,3\ \ 6\ \ 6\,] = 2[\,1\ \ 1\ \ 0\,] + [\,0\ \ 2\ \ 3\,] + [\,1\ \ 2\ \ 3\,].$$

(b) $S = \{[\,1\ \ 1\ \ 0\,], [\,3\ \ 4\ \ 2\,]\}$

$c_1[\,1\ \ 1\ \ 0\,] + c_2[\,3\ \ 4\ \ 2\,] = [\,0\ \ 0\ \ 0\,]$

$$
\left[\begin{array}{cc|c} 1 & 3 & 0 \\ 1 & 4 & 0 \\ 0 & 2 & 0 \end{array}\right]
\longrightarrow
\left[\begin{array}{cc|c} 1 & 3 & 0 \\ 0 & 1 & 0 \\ 0 & 2 & 0 \end{array}\right]
\longrightarrow
\left[\begin{array}{cc|c} 1 & 0 & 0 \\ 0 & 1 & 0 \\ 0 & 0 & 0 \end{array}\right]
$$

add -1 (row 1) to row 2   add -3 (row 2) to row 1

                                    add -2 (row 2) to row 3

Thus the only solution is $c_1 = c_2 = 0$ and hence the vectors are linearly independent.

(c) $S = \{[\,1\ \ 1\ \ 0\,], [\,0\ \ 2\ \ 3\,], [\,1\ \ 2\ \ 3\,], [\,0\ \ 0\ \ 0\,]\}$

Since this set of vectors contains the zero vector, it is linearly dependent. For example,

$$[\,0\ \ 0\ \ 0\,] = 0[\,1\ \ 1\ \ 0\,] + 0[\,0\ \ 2\ \ 3\,] + 0[\,1\ \ 2\ \ 3\,].$$

13. To determine which sets are linearly dependent we form a linear combination of the polynomials with unknown coefficients and set it equal to the zero polynomial. Next perform the operations, group together like terms, and equate corresponding coefficients from the polynomials on each side of the equal sign to obtain a homogeneous linear system. If the homogeneous system has any nontrivial solutions, then the polynomials are linearly dependent.

(a) $c_1(t^2 + 1) + c_2(t - 2) + c_3(t + 3) = 0t^2 + 0t + 0$

Equating coefficients of like powers of t, we obtain the system

$$
\begin{array}{rcrcrcl}
c_1 & & & & & = & 0 \\
& & c_2 & + & c_3 & = & 0 \\
c_1 & - & 2c_2 & + & 3c_3 & = & 0
\end{array}
$$

Next, row reduce the augmented matrix of the system:

$$
\left[\begin{array}{ccc|c} 1 & 0 & 0 & 0 \\ 0 & 1 & 1 & 0 \\ 1 & -2 & 3 & 0 \end{array}\right]
\longrightarrow
\left[\begin{array}{ccc|c} 1 & 0 & 0 & 0 \\ 0 & 1 & 1 & 0 \\ 0 & -2 & 3 & 0 \end{array}\right]
\longrightarrow
\left[\begin{array}{ccc|c} 1 & 0 & 0 & 0 \\ 0 & 1 & 1 & 0 \\ 0 & 0 & 3 & 0 \end{array}\right]
\longrightarrow
$$

add -1 (row 1) to row 3   add 2 (row 2) to row 3 $\frac{1}{3}$ (row 3)

                                              add -1 (row 3) to row 2

$$
\left[\begin{array}{ccc|c} 1 & 0 & 0 & 0 \\ 0 & 1 & 0 & 0 \\ 0 & 0 & 1 & 0 \end{array}\right]
$$

Hence $c_1 = c_2 = c_3 = 0$ and the polynomials are linearly independent.

(b) $c_1(2t^2 + t) + c_2(t^2 + 3) + c_3 t - 0t^2 + 0t + 0$

Equating coefficients of like powers of t, we obtain the system

$$
\begin{array}{rcrcrcl}
2c_1 & + & c_2 & & & = & 0 \\
c_1 & & & + & c_3 & = & 0 \\
& & 3c_2 & & & = & 0
\end{array}
$$

Next, row reduce the augmented matrix of the system:

$$
\left[\begin{array}{ccc|c} 2 & 1 & 0 & 0 \\ 1 & 0 & 1 & 0 \\ 0 & 3 & 0 & 0 \end{array}\right]
\longrightarrow
\left[\begin{array}{ccc|c} 1 & 0 & 1 & 0 \\ 2 & 1 & 0 & 0 \\ 0 & 3 & 0 & 0 \end{array}\right]
\longrightarrow
\left[\begin{array}{ccc|c} 1 & 0 & 1 & 0 \\ 0 & 1 & -2 & 0 \\ 0 & 3 & 0 & 0 \end{array}\right]
\longrightarrow
$$

interchange rows 1 and 2   add -2 (row 1) to row 2   add -3 (row 2) to row 3

$$\begin{bmatrix} 1 & 0 & 1 & | & 0 \\ 0 & 1 & -2 & | & 0 \\ 0 & 0 & 6 & | & 0 \end{bmatrix} \longrightarrow \begin{bmatrix} 1 & 0 & 0 & | & 0 \\ 0 & 1 & 0 & | & 0 \\ 0 & 0 & 1 & | & 0 \end{bmatrix}$$

$$\frac{1}{6} \text{ (row 3)}$$
add 2 (row 3) to row 2
add -1 (row 3) to row 1

Thus $c_1 = c_2 = c_3 = 0$ and the polynomials are linearly independent.

(c) $c_1(2t^2 + t + 1) + c_2(3t^2 + t - 5) + c_3(t + 13) = 0t^2 + 0t + 0$

Equating coefficients of like powers of t, we obtain the system

$$\begin{array}{rcrcrcl} 2c_1 & + & 3c_2 & & & = & 0 \\ c_1 & + & c_2 & + & c_3 & = & 0 \\ c_1 & - & 5c_2 & + & 13c_3 & = & 0 \end{array}$$

Next, row reduce the augmented matrix of the system:

$$\begin{bmatrix} 2 & 3 & 0 & | & 0 \\ 1 & 1 & 1 & | & 0 \\ 1 & -5 & 13 & | & 0 \end{bmatrix} \longrightarrow \begin{bmatrix} 1 & 1 & 1 & | & 0 \\ 2 & 3 & 0 & | & 0 \\ 1 & -5 & 13 & | & 0 \end{bmatrix} \longrightarrow \begin{bmatrix} 1 & 1 & 1 & | & 0 \\ 0 & 1 & -2 & | & 0 \\ 0 & -6 & 12 & | & 0 \end{bmatrix} \longrightarrow$$

interchange rows 1 and 2 $\qquad$ add -2 (row 1) to row 2 $\qquad$ add -1 (row 2) to row 1
$\qquad\qquad\qquad\qquad\qquad$ add -1 (row 1) to row 3 $\qquad$ add 6 (row 2) to row 3

$$\begin{bmatrix} 1 & 0 & 3 & | & 0 \\ 0 & 1 & -2 & | & 0 \\ 0 & 0 & 0 & | & 0 \end{bmatrix}$$

We have $c_1 = -3c_3$, $c_2 = 2c_3$, $c_3 = r$, where r is any real number. Thus the polynomials are linearly dependent and, with $r = 1$, we obtain

$$t + 13 = 3(2t^2 + t + 1) - 2(3t^2 + t - 5).$$

15. To determine which sets are linearly dependent we form a linear combination of the vectors with unknown coefficients and set it equal to the zero polynomial. Next perform the matrix operations and equate corresponding entries to obtain a homogeneous linear system. If the homogeneous system has any nontrivial solutions, then the vectors are linearly dependent.

(a) $c_1 \begin{bmatrix} 1 \\ 0 \\ 0 \end{bmatrix} + c_2 \begin{bmatrix} 0 \\ 1 \\ 1 \end{bmatrix} + c_3 \begin{bmatrix} 1 \\ 2 \\ -1 \end{bmatrix} = \begin{bmatrix} 0 \\ 0 \\ 0 \end{bmatrix}$

Equating corresponding entries, we obtain the system

$$\begin{array}{rcrcrcl} c_1 & & & + & c_3 & = & 0 \\ & & c_2 & + & 2c_3 & = & 0 \\ & & c_2 & - & c_3 & = & 0 \end{array}$$

Next, row reduce the augmented matrix of the system:

$$\begin{bmatrix} 1 & 0 & 1 & | & 0 \\ 0 & 1 & 2 & | & 0 \\ 0 & 1 & -1 & | & 0 \end{bmatrix} \longrightarrow \begin{bmatrix} 1 & 0 & 1 & | & 0 \\ 0 & 1 & 2 & | & 0 \\ 0 & 0 & -3 & | & 0 \end{bmatrix} \longrightarrow \begin{bmatrix} 1 & 0 & 0 & | & 0 \\ 0 & 1 & 0 & | & 0 \\ 0 & 0 & 1 & | & 0 \end{bmatrix}$$

add -1 (row 2) to row 3 $\qquad\qquad -\frac{1}{3}$ (row 3)
$\qquad\qquad\qquad\qquad\qquad$ add -2 (row 3) to row 2
$\qquad\qquad\qquad\qquad\qquad$ add -1 (row 3) to row 1

Hence $c_1 = c_2 = c_3 = 0$ and the vectors are linearly independent.

(b) $c_1 \begin{bmatrix} 1 \\ 1 \\ -1 \end{bmatrix} + c_2 \begin{bmatrix} 0 \\ 1 \\ 1 \end{bmatrix} + c_3 \begin{bmatrix} 1 \\ 1 \\ 1 \end{bmatrix} + c_4 \begin{bmatrix} 1 \\ 2 \\ -2 \end{bmatrix} = \begin{bmatrix} 0 \\ 0 \\ 0 \end{bmatrix}$

Equating corresponding entries, we obtain the system

$$\begin{array}{rcrcrcrcl} c_1 & & & + & c_3 & + & c_4 & = & 0 \\ c_1 & + & c_2 & + & c_3 & + & 2c_4 & = & 0 \\ -c_1 & + & c_2 & + & c_3 & - & 2c_4 & = & 0 \end{array}$$

Next, row reduce the augmented matrix of the system:

$$\left[\begin{array}{cccc|c} 1 & 0 & 1 & 1 & 0 \\ 1 & 1 & 1 & 2 & 0 \\ -1 & 1 & 1 & -2 & 0 \end{array}\right] \longrightarrow \left[\begin{array}{cccc|c} 1 & 0 & 1 & 1 & 0 \\ 0 & 1 & 0 & 1 & 0 \\ 0 & 1 & 2 & -1 & 0 \end{array}\right] \longrightarrow \left[\begin{array}{cccc|c} 1 & 0 & 1 & 1 & 0 \\ 0 & 1 & 0 & 1 & 0 \\ 0 & 0 & 2 & -2 & 0 \end{array}\right] \longrightarrow$$

add -1 (row 1) to row 2   add -1 (row 2) to row 3 $\frac{1}{2}$ (row 3)
add row 1 to row 3

$$\left[\begin{array}{cccc|c} 1 & 0 & 1 & 1 & 0 \\ 0 & 1 & 0 & 1 & 0 \\ 0 & 0 & 1 & -1 & 0 \end{array}\right] \longrightarrow \left[\begin{array}{cccc|c} 1 & 0 & 0 & 2 & 0 \\ 0 & 1 & 0 & 1 & 0 \\ 0 & 0 & 1 & -1 & 0 \end{array}\right]$$

add -1 (row 3) to row 1

Thus $c_1 = -2c_4$, $c_2 = -c_4$, $c_3 = c_4$, $c_4 = r$, where r is any real number. Thus the vectors are linearly dependent and, with $r = 1$, we have that

$$\begin{bmatrix} 1 \\ 2 \\ -2 \end{bmatrix} = 2 \begin{bmatrix} 1 \\ 1 \\ -1 \end{bmatrix} + \begin{bmatrix} 0 \\ 1 \\ 1 \end{bmatrix} - \begin{bmatrix} 1 \\ 1 \\ 1 \end{bmatrix}.$$

(c) $c_1 \begin{bmatrix} 1 \\ 0 \\ 0 \end{bmatrix} + c_2 \begin{bmatrix} 2 \\ 1 \\ 1 \end{bmatrix} + c_3 \begin{bmatrix} -1 \\ 2 \\ 1 \end{bmatrix} = \begin{bmatrix} 0 \\ 0 \\ 0 \end{bmatrix}$

Equating corresponding entries, we obtain the system

$$\begin{array}{rcrcrcl} c_1 & + & 2c_2 & - & c_3 & = & 0 \\ & & c_2 & + & 2c_3 & = & 0 \\ & & c_2 & + & c_3 & = & 0 \end{array}$$

Next, row reduce the augmented matrix of the system:

$$\left[\begin{array}{ccc|c} 1 & 2 & -1 & 0 \\ 0 & 1 & 2 & 0 \\ 0 & 1 & 1 & 0 \end{array}\right] \longrightarrow \left[\begin{array}{ccc|c} 1 & 0 & -5 & 0 \\ 0 & 1 & 2 & 0 \\ 0 & 0 & -1 & 0 \end{array}\right] \longrightarrow \left[\begin{array}{ccc|c} 1 & 0 & 0 & 0 \\ 0 & 1 & 0 & 0 \\ 0 & 0 & 1 & 0 \end{array}\right]$$

add -2 (row 2) to row 1 -1 (row 3)
add -1 (row 2) to row 3   add -2 (row 3) to row 2
add -5 (row 3) to row 1

Thus $c_1 = c_2 = c_3 = 0$ and the vectors are linearly independent.

17. Suppose that

$$c_1(t + 3) + c_2(2t + c^2 + 2) = 0$$

for some scalars c_1, c_2. Combining terms and equating coefficients of powers of t to zero, we obtain the linear system

$$\begin{array}{rcrcl} c_1 & + & 2c_2 & = & 0 \\ 3c_1 & + & (c^2 + 2)c_2 & = & 0 \end{array}$$

Row reducing the augmented matrix of this system, we have

$$\left[\begin{array}{cc|c} 1 & 2 & 0 \\ 3 & c^2+2 & 0 \end{array}\right] \longrightarrow \left[\begin{array}{cc|c} 1 & 2 & 0 \\ 0 & c^2-4 & 0 \end{array}\right]$$

add -3 (row 1) to row 2

If $c^2 - 4 \neq 0$ we can multiply row 2 by $1/(c^2 - 4)$ to obtain a leading 1. The reduced echelon form of the matrix is then $\left[\begin{array}{c|c} I_2 & 0 \end{array}\right]$ and hence the solution is $c_1 = c_2 = 0$. In this case the polynomials are linearly independent. If $c^2 - 4 = 0$, however, the system has nontrivial solutions, in which case the polynomials are linearly dependent. Thus, the polynomials are linearly independent if and only if $c \neq \pm 2$.

19. We are asked to prove an if and only if statement. Hence two things must be proved.

   (a) If S is linearly independent, then one of the vectors in S is a linear combination of all the other vectors in S.

   Proof: If S is linear independent then $a_1\mathbf{v}_1 + a_2\mathbf{v}_2 + \cdots + a_k\mathbf{v}_k = \mathbf{0}$, where at least one of the coefficients $a_1, a_2, \ldots, a_k$ is not zero. Suppose that $a_j \neq 0$, then we can rewrite the preceding expression as follows:

$$a_j\mathbf{v}_j = -(a_1\mathbf{v}_1 + a_2\mathbf{v}_2 + \cdots + a_{j-1}\mathbf{v}_{j-1} + a_{j+1}\mathbf{v}_{j+1} + \cdots + a_k\mathbf{v}_k)$$
$$\mathbf{v}_j = -\frac{a_1}{a_j}\mathbf{v}_1 - \frac{a_2}{a_j}\mathbf{v}_2 - \cdots - \frac{a_{j-1}}{a_j}\mathbf{v}_{j-1} - \frac{a_{j+1}}{a_j}\mathbf{v}_{j+1} - \cdots - \frac{a_k}{a_j}\mathbf{v}_k$$

   Thus one of the vectors in S is a linear combination of all the other vectors.

   (b) If one of the vectors in S is a linear combination of all the other vectors, then S is linearly independent.

   Proof: Suppose $\mathbf{v}_j$ is a linear combination of the other vectors in S. Then there exist coefficients $c_1, c_2, \ldots, c_{j-1}, c_{j+1}, \ldots, c_k$ such that

$$\mathbf{v}_j = c_1\mathbf{v}_1 + c_2\mathbf{v}_2 + \cdots + c_{j-1}\mathbf{v}_{j-1} + c_{j+1}\mathbf{v}_{j+1} + \cdots + c_k\mathbf{v}_k.$$

   Then subtracting $\mathbf{v}_j$ from both sides we have

$$c_1\mathbf{v}_1 + c_2\mathbf{v}_2 + \cdots + c_{j-1}\mathbf{v}_{j-1} + (-1)\mathbf{v}_j + c_{j+1}\mathbf{v}_{j+1} + \cdots + c_k\mathbf{v}_k = \mathbf{0}.$$

   Hence we have a linear combination of the members of S which give the zero vector and not all the coefficients are zero. Thus S is linearly dependent.

21. Form the linear combination $c_1\mathbf{w}_1 + c_2\mathbf{w}_2 + c_3\mathbf{w}_3 = \mathbf{0}$. Then

$$c_1(\mathbf{v}_1 + \mathbf{v}_2) + c_2(\mathbf{v}_1 + \mathbf{v}_3) + c_3(\mathbf{v}_2 + \mathbf{v}_3) = (c_1 + c_2)\mathbf{v}_1 + (c_1 + c_3)\mathbf{v}_2 + (c_2 + c_3)\mathbf{v}_3 = \mathbf{0}.$$

Since S is linearly independent, we obtain the linear system

$$\begin{array}{rcl} c_1 + c_2 & = & 0 \\ c_1 \quad + c_3 & = & 0 \\ c_2 + c_3 & = & 0 \end{array}$$

The augmented matrix of this system and its reduced row echelon form are

$$\left[\begin{array}{ccc|c} 1 & 1 & 0 & 0 \\ 1 & 0 & 1 & 0 \\ 0 & 1 & 1 & 0 \end{array}\right] \longrightarrow \quad \cdots \quad \longrightarrow \left[\begin{array}{ccc|c} 1 & 0 & 0 & 0 \\ 0 & 1 & 0 & 0 \\ 0 & 0 & 1 & 0 \end{array}\right]$$

$$\begin{array}{cc} \text{steps} & \text{reduced row} \\ \text{omitted} & \text{echelon form} \end{array}$$

Therefore $c_1 = c_2 = c_3 = 0$ which implies that $\{\mathbf{w}_1, \mathbf{w}_2, \mathbf{w}_3\}$ is linearly independent.

23. Suppose the set $\{\mathbf{v}_1, \mathbf{v}_2, \mathbf{v}_3\}$ is linearly dependent. Then one of the $\mathbf{v}_j$'s is a linear combination of the preceding vectors in the list. It must be $\mathbf{v}_3$ since $\{\mathbf{v}_1, \mathbf{v}_2\}$ is linearly independent. Thus $\mathbf{v}_3$ belongs to span $\{\mathbf{v}_1, \mathbf{v}_2\}$. This is a contradiction, so the set $\{\mathbf{v}_1, \mathbf{v}_2, \mathbf{v}_3\}$ is linearly independent.

25. Suppose A has k nonzero rows and denote them as $\mathbf{v}_1, \mathbf{v}_2, \ldots, \mathbf{v}_k$. Let the numbers of the columns in which the leading 1's appear be denoted by $c_1, c_2, \ldots, c_k$. Thus

$$\mathbf{v}_i = \begin{bmatrix} 0 & \cdots & 0 & 1 & a_{ic_i+1} & \cdots & a_{in} \end{bmatrix}.$$

Suppose

$$a_1\mathbf{v}_1 + a_2\mathbf{v}_2 + \cdots + a_k\mathbf{v}_k = \mathbf{0} = \begin{bmatrix} 0 & 0 & \cdots & 0 \end{bmatrix}.$$

Then the c_1 entry on the left is a_1, so that $a_1 = 0$. Similarly, examining the c_2 entry on the left yields $a_2 = 0$, and so forth. Therefore the set $\{\mathbf{v}_1, \mathbf{v}_2, \ldots, \mathbf{v}_k\}$ is linearly independent.

27. In R^1 let $S_1 = \{1\}$ and $S_2 = \{1, 0\}$. Then S_1 is linearly independent and S_2 is linearly dependent.

Section 4.6, p. 242

1. Let $V = R^2$

   (a) Let

   $$S = \left\{ \begin{bmatrix} 1 \\ 3 \end{bmatrix}, \begin{bmatrix} 1 \\ -1 \end{bmatrix} \right\}$$

   and let $\begin{bmatrix} a \\ b \end{bmatrix}$ be an arbitrary vector in V. To check if S spans V we form the linear combination

   $$a_1 \begin{bmatrix} 1 \\ 3 \end{bmatrix} + a_2 \begin{bmatrix} 1 \\ -1 \end{bmatrix} = \begin{bmatrix} a \\ b \end{bmatrix}.$$

   Performing the matrix algebra on the left and equating corresponding entries leads to the linear system

   $$\begin{array}{rcl} a_1 + a_2 & = & a \\ 3a_1 - a_2 & = & b \end{array}$$

   We form the augmented matrix and row reduce it:

   $$\begin{bmatrix} 1 & 1 & | & a \\ 3 & -1 & | & b \end{bmatrix} \longrightarrow \begin{bmatrix} 1 & 1 & | & a \\ 0 & -4 & | & b-3a \end{bmatrix} \longrightarrow \begin{bmatrix} 1 & 1 & | & a \\ 0 & 1 & | & \frac{1}{4}(-b+3a) \end{bmatrix} \longrightarrow$$
   $$\text{add } -3 \text{ (row 1) to row 2} \qquad -\tfrac{1}{4} \text{ (row 2)} \qquad \text{add } -1 \text{ (row 2) to row 1}$$

   $$\begin{bmatrix} 1 & 0 & | & \frac{1}{4}(a+b) \\ 0 & 1 & | & \frac{1}{4}(-b+3a) \end{bmatrix}$$

   Thus we see that the system is consistent and hence S spans V. To check for linear independence we set $a = b = 0$ and ask if the final equivalent system determined above has only the zero solution $a_1 = a_2 = 0$. The final system is $\begin{bmatrix} I_2 & | & 0 \end{bmatrix}$, which does indeed imply that $a_1 = a_2 = 0$ so S is linearly independent. Hence S is a basis for V.

   (b) Let

   $$S = \left\{ \begin{bmatrix} 0 \\ 0 \end{bmatrix}, \begin{bmatrix} 1 \\ 2 \end{bmatrix}, \begin{bmatrix} 2 \\ 4 \end{bmatrix} \right\}.$$

   We can proceed as in part (a). However in this case we note that S contains the zero vector of R^2. Any set containing the zero vector is linearly dependent hence S is not a basis for R^2.

(c) Let

$$S = \left\{ \begin{bmatrix} 1 \\ 2 \end{bmatrix}, \begin{bmatrix} 2 \\ -3 \end{bmatrix}, \begin{bmatrix} 3 \\ 2 \end{bmatrix} \right\}$$

and let $\begin{bmatrix} a \\ b \end{bmatrix}$ be an arbitrary vector in V. To check if S spans V we form the linear combination

$$a_1 \begin{bmatrix} 1 \\ 2 \end{bmatrix} + a_2 \begin{bmatrix} 2 \\ -3 \end{bmatrix} + a_3 \begin{bmatrix} 3 \\ 2 \end{bmatrix} = \begin{bmatrix} a \\ b \end{bmatrix}.$$

Performing the matrix algebra on the left and equating corresponding entries leads to the linear system

$$\begin{array}{rcrcrcl} a_1 & + & 2a_2 & + & 3a_3 & = & a \\ 2a_1 & - & 3a_2 & + & 2a_3 & = & b \end{array}$$

We form the augmented matrix and row reduce it:

$$\begin{bmatrix} 1 & 2 & 3 & a \\ 2 & -3 & 2 & b \end{bmatrix} \longrightarrow \begin{bmatrix} 1 & 2 & 3 & a \\ 0 & -7 & -4 & b - 2a \end{bmatrix} \longrightarrow \begin{bmatrix} 1 & 2 & 3 & a \\ 0 & 1 & \frac{4}{7} & \frac{1}{7}(-b + 2a) \end{bmatrix} \longrightarrow$$

add -2 (row 1) to row 2 $\qquad -\frac{1}{7}$ (row 2) $\qquad$ add -2 (row 2) to row 1

$$\begin{bmatrix} 1 & 0 & \frac{13}{7} & \frac{1}{7}(3a + 2b) \\ 0 & 1 & \frac{4}{7} & \frac{1}{7}(-b + 2a) \end{bmatrix}$$

Since there is no row of the form $\begin{bmatrix} 0 & 0 & 0 & | & * \end{bmatrix}$, where $*$ is nonzero, the system is consistent and hence S spans V. To check for linear independence we set $a = b = 0$ and ask if the final equivalent system determined above has only the zero solution $a_1 = a_2 = 0$. The general solution of this homogeneous system is

$$a_1 = -\frac{13}{7}a_3, \quad a_2 = -\frac{4}{7}a_3, \quad a_3 = r$$

where r is any real number. Thus there are nontrivial solutions so the set S is linearly dependent and not a basis for V.

(d) Let

$$S = \left\{ \begin{bmatrix} 1 \\ 3 \end{bmatrix}, \begin{bmatrix} -2 \\ 6 \end{bmatrix} \right\}$$

and let $\begin{bmatrix} a \\ b \end{bmatrix}$ be an arbitrary vector in V. To check if S spans V we form the linear combination

$$a_1 \begin{bmatrix} 1 \\ 3 \end{bmatrix} + a_2 \begin{bmatrix} -2 \\ 6 \end{bmatrix} = \begin{bmatrix} a \\ b \end{bmatrix}.$$

Performing the matrix algebra on the left and equating corresponding entries leads to the linear system

$$\begin{array}{rcrcl} a_1 & - & 2a_2 & = & a \\ 3a_1 & + & 6a_2 & = & b \end{array}$$

We form the augmented matrix and row reduce it:

$$\begin{bmatrix} 1 & -2 & a \\ 3 & 6 & b \end{bmatrix} \longrightarrow \begin{bmatrix} 1 & -2 & a \\ 0 & 12 & b - 3a \end{bmatrix} \longrightarrow \begin{bmatrix} 1 & -2 & a \\ 0 & 1 & \frac{1}{12}(b - 3a) \end{bmatrix} \longrightarrow$$

add -3 (row 1) to row 2 $\qquad \frac{1}{12}$ (row 2) $\qquad$ add 2 (row 2) to row 1

$$\begin{bmatrix} 1 & 0 & \frac{1}{6}(3a + b) \\ 0 & 1 & \frac{1}{12}(b - 3a) \end{bmatrix}$$

Thus we see that the system is consistent and hence S spans V. To check for linear independence we set $a = b = 0$ and ask if the final equivalent system determined above has only the trivial solution $a_1 = a_2 = 0$. The final system is $\begin{bmatrix} I_2 & | & 0 \end{bmatrix}$, which does indeed imply that $a_1 = a_2 = 0$, so S is linearly independent. Hence S is a basis for V.

3. Let $V = R_4$.

(a) Let

$$S = \{\begin{bmatrix} 1 & 0 & 0 & 1 \end{bmatrix}, \begin{bmatrix} 0 & 1 & 0 & 0 \end{bmatrix}, \begin{bmatrix} 1 & 1 & 1 & 1 \end{bmatrix}, \begin{bmatrix} 0 & 1 & 1 & 1 \end{bmatrix}\}$$

and let $\begin{bmatrix} a & b & c & d \end{bmatrix}$ be an arbitrary vector in V. To check if S spans V we form the linear combination

$$a_1 \begin{bmatrix} 1 & 0 & 0 & 1 \end{bmatrix} + a_2 \begin{bmatrix} 0 & 1 & 0 & 0 \end{bmatrix} + a_3 \begin{bmatrix} 1 & 1 & 1 & 1 \end{bmatrix} + a_4 \begin{bmatrix} 0 & 1 & 1 & 1 \end{bmatrix} = \begin{bmatrix} a & b & c & d \end{bmatrix}.$$

Performing the matrix algebra on the left and equating corresponding entries leads to the linear system

$$
\begin{array}{rcl}
a_1 \quad\quad + a_3 \quad\quad &=& a \\
a_2 + a_3 + a_4 &=& b \\
a_3 + a_4 &=& c \\
a_1 \quad\quad + a_3 + a_4 &=& d
\end{array}
$$

We form the augmented matrix and row reduce it:

$$
\left[\begin{array}{cccc|c}
1 & 0 & 1 & 0 & a \\
0 & 1 & 1 & 1 & b \\
0 & 0 & 1 & 1 & c \\
1 & 0 & 1 & 1 & d
\end{array}\right]
\longrightarrow
\left[\begin{array}{cccc|c}
1 & 0 & 1 & 0 & a \\
0 & 1 & 1 & 1 & b \\
0 & 0 & 1 & 1 & c \\
0 & 0 & 0 & 1 & d-a
\end{array}\right]
\longrightarrow
\left[\begin{array}{cccc|c}
1 & 0 & 0 & -1 & a-c \\
0 & 1 & 0 & 0 & b-c \\
0 & 0 & 1 & 1 & c \\
0 & 0 & 0 & 1 & d-a
\end{array}\right]
\longrightarrow
$$

add -1 (row 1) to row 4   add -1 (row 3) to row 1   add -1 (row 4) to row 3
add -1 (row 3) to row 2   add row 4 to row 1

$$
\left[\begin{array}{cccc|c}
1 & 0 & 0 & 0 & d-c \\
0 & 1 & 0 & 0 & b-c \\
0 & 0 & 1 & 0 & c-d+a \\
0 & 0 & 0 & 1 & d-a
\end{array}\right]
$$

Thus we see that the system is consistent and hence S spans V. To check for linear independence we set $a = b = c = d = 0$ and ask if the final equivalent system determined above has only the zero solution $a_1 = a_2 = a_3 = a_4 = 0$. The final system is $\begin{bmatrix} I_4 & | & 0 \end{bmatrix}$, which does indeed imply that $a_1 = a_2 = a_3 = a_4 = 0$ so S is linearly independent. Hence S is a basis for V.

(b) Let

$$S = \{\begin{bmatrix} 1 & -1 & 0 & 2 \end{bmatrix}, \begin{bmatrix} 3 & -1 & 2 & 1 \end{bmatrix}, \begin{bmatrix} 1 & 0 & 0 & 1 \end{bmatrix}\}$$

and let $\begin{bmatrix} a & b & c & d \end{bmatrix}$ be an arbitrary vector in V. To check to see if S spans V we form the linear combination

$$a_1 \begin{bmatrix} 1 & -1 & 0 & 2 \end{bmatrix} + a_2 \begin{bmatrix} 3 & -1 & 2 & 1 \end{bmatrix} + a_3 \begin{bmatrix} 1 & 0 & 0 & 1 \end{bmatrix} = \begin{bmatrix} a & b & c & d \end{bmatrix}.$$

Performing the matrix algebra on the left and equating corresponding entries leads to the linear system

$$
\begin{array}{rcl}
a_1 + 3a_2 + a_3 &=& a \\
-a_1 - a_2 \quad\quad &=& b \\
2a_2 \quad\quad &=& c \\
2a_1 + a_2 + a_3 &=& d
\end{array}
$$

We form the augmented matrix and row reduce it:

$$\begin{bmatrix} 1 & 3 & 1 & a \\ -1 & -1 & 0 & b \\ 0 & 2 & 0 & c \\ 2 & 1 & 1 & d \end{bmatrix} \longrightarrow \begin{bmatrix} 1 & 3 & 1 & a \\ 0 & 2 & 1 & a+b \\ 0 & 2 & 0 & c \\ 0 & -5 & -1 & d-2a \end{bmatrix} \longrightarrow \begin{bmatrix} 1 & 3 & 1 & a \\ 0 & 1 & 0 & \frac{1}{2}c \\ 0 & 2 & 1 & a+b \\ 0 & -5 & -1 & d-2a \end{bmatrix} \longrightarrow$$

$$\begin{array}{c} \text{add row 1 to row 2} \\ \text{add } -2 \text{ (row 1) to row 4} \end{array} \qquad \begin{array}{c} \text{interchange rows 3 and 2} \\ \frac{1}{2} \text{ (row 2)} \end{array} \qquad \begin{array}{c} \text{add } -3 \text{ (row 2) to row 1} \\ \text{add } -2 \text{ (row 2) to row 3} \\ \text{add } 5 \text{ (row 2) to row 4} \end{array}$$

$$\begin{bmatrix} 1 & 0 & 1 & a-\frac{3}{2}c \\ 0 & 1 & 0 & \frac{1}{2}c \\ 0 & 0 & 1 & a+b-c \\ 0 & 0 & -1 & d-2a+\frac{5}{2}c \end{bmatrix} \longrightarrow \begin{bmatrix} 1 & 0 & 1 & a-\frac{3}{2}c \\ 0 & 1 & 0 & \frac{1}{2}c \\ 0 & 0 & 1 & a+b-c \\ 0 & 0 & 0 & * \end{bmatrix}$$

$$\text{add row 3 to row 4}$$

where $* \neq 0$. Thus we see that the system is inconsistent for some vectors $\begin{bmatrix} a & b & c & d \end{bmatrix}$ and so S does not span V. Hence S is not a basis for V.

Alternative analysis: Here S contains 3 vectors from R_4. From the discussion following Example 1 we can infer that $\mathbf{e}_1^T$, $\mathbf{e}_2^T$, $\mathbf{e}_3^T$, $\mathbf{e}_4^T$ is a basis for R_4. Then from Corollary 4.1 it follows that every basis of R_4 must contain 4 vectors. Hence S cannot be a basis for R_4.

(c) Let

$$S = \left\{ \begin{bmatrix} -2 & 4 & 6 & 4 \end{bmatrix}, \begin{bmatrix} 0 & 1 & 2 & 0 \end{bmatrix}, \begin{bmatrix} -1 & 2 & 3 & 2 \end{bmatrix}, \begin{bmatrix} -3 & 2 & 5 & 6 \end{bmatrix}, \begin{bmatrix} -2 & -1 & 0 & 4 \end{bmatrix} \right\}$$

and let $\begin{bmatrix} a & b & c & d \end{bmatrix}$ be an arbitrary vector in V. To check if S spans V we form the linear combination

$$a_1 \begin{bmatrix} -2 & 4 & 6 & 4 \end{bmatrix} + a_2 \begin{bmatrix} 0 & 1 & 2 & 0 \end{bmatrix} + a_3 \begin{bmatrix} -1 & 2 & 3 & 2 \end{bmatrix}$$
$$+ a_4 \begin{bmatrix} -3 & 2 & 5 & 6 \end{bmatrix} + a_5 \begin{bmatrix} -2 & -1 & 0 & 4 \end{bmatrix} = \begin{bmatrix} a & b & c & d \end{bmatrix}.$$

Performing the matrix algebra on the left and equating corresponding entries leads to the linear system

$$\begin{array}{rcrcrcrcrcl} -2a_1 & & & - & a_3 & - & 3a_4 & - & 2a_5 & = & a \\ 4a_1 & + & a_2 & + & 2a_3 & + & 2a_4 & - & a_5 & = & b \\ 6a_1 & + & 2a_2 & + & 3a_3 & + & 5a_4 & & & = & c \\ 4a_1 & & & + & 2a_3 & + & 6a_4 & + & 4a_5 & = & d \end{array}$$

We form the augmented matrix and row reduce it:

$$\begin{bmatrix} -2 & 0 & -1 & -3 & -2 & a \\ 4 & 1 & 2 & 2 & -1 & b \\ 6 & 2 & 3 & 5 & 0 & c \\ 4 & 0 & 2 & 6 & 4 & d \end{bmatrix}$$

Inspecting this matrix we see that the first and fourth rows are closely related (except in the augmented column). In fact adding 2 times row 1 to row 4 gives an equivalent system whose augmented matrix has its fourth row in the form $\begin{bmatrix} 0 & 0 & 0 & 0 & | & * \end{bmatrix}$, where $* \neq 0$. Thus the system is inconsistent and S does not span V. Hence S is not a basis for V.

Alternative analysis: As in part (b), S contains 5 vectors and hence cannot be a basis for R_4.

(d) Let

$$S = \left\{ \begin{bmatrix} 0 & 0 & 1 & 1 \end{bmatrix}, \begin{bmatrix} -1 & 1 & 1 & 2 \end{bmatrix}, \begin{bmatrix} 1 & 1 & 0 & 0 \end{bmatrix}, \begin{bmatrix} 2 & 1 & 2 & 1 \end{bmatrix} \right\}$$

and let $\begin{bmatrix} a & b & c & d \end{bmatrix}$ be an arbitrary vector in V. To check if S spans V we form the linear combination

$$a_1 \begin{bmatrix} 0 & 0 & 1 & 1 \end{bmatrix} + a_2 \begin{bmatrix} -1 & 1 & 1 & 2 \end{bmatrix} + a_3 \begin{bmatrix} 1 & 1 & 0 & 0 \end{bmatrix} + a_4 \begin{bmatrix} 2 & 1 & 2 & 1 \end{bmatrix} = \begin{bmatrix} a & b & c & d \end{bmatrix}.$$

Performing the matrix algebra on the left and equating corresponding entries leads to the linear system

$$
\begin{array}{rrrrr}
 & -a_2 & + a_3 & + 2a_4 & = a \\
 & a_2 & + a_3 & + a_4 & = b \\
a_1 & + a_2 & & + 2a_4 & = c \\
a_1 & + 2a_2 & & + a_4 & = d
\end{array}
$$

We form the augmented matrix and row reduce it:

$$
\left[\begin{array}{cccc|c}
0 & -1 & 1 & 2 & a \\
0 & 1 & 1 & 1 & b \\
1 & 1 & 0 & 2 & c \\
1 & 2 & 0 & 1 & d
\end{array}\right]
\longrightarrow
\left[\begin{array}{cccc|c}
1 & 1 & 0 & 2 & c \\
0 & 1 & 1 & 1 & b \\
0 & -1 & 1 & 2 & a \\
0 & 1 & 0 & -1 & d-c
\end{array}\right]
\longrightarrow
\left[\begin{array}{cccc|c}
1 & 0 & -1 & 1 & c-b \\
0 & 1 & 1 & 1 & b \\
0 & 0 & 2 & 3 & a+b \\
0 & 0 & -1 & -2 & d-c-b
\end{array}\right]
\longrightarrow
$$

add -1 (row 3) to row 4   add -1 (row 2) to row 1 -1 (row 4)
interchange rows 3 and 1   add row 2 to row 3   interchange rows 3 and 4
            add -1 (row 2) to row 4

$$
\left[\begin{array}{cccc|c}
1 & 0 & -1 & 1 & c-b \\
0 & 1 & 1 & 1 & b \\
0 & 0 & 1 & 2 & d-c-b \\
0 & 0 & 2 & 3 & a+b
\end{array}\right]
\longrightarrow
\left[\begin{array}{cccc|c}
1 & 0 & -1 & 1 & c-b \\
0 & 1 & 1 & 1 & b \\
0 & 0 & 1 & 2 & d-c-b \\
0 & 0 & 0 & 1 & *
\end{array}\right]
$$

add -2 (row 3) to row 4
-1 (row 4)

Since there is a leading 1 in each row the system will be equivalent to $\left[\begin{array}{c|c} I_4 & * \end{array}\right]$, where $*$ is a column whose entries involve a, b, c d. Thus we see that the system is consistent for all vectors and so S spans V. To check for linear independence we set $a = b = c = d = 0$ and ask if the final equivalent system determined above has only the zero solution $a_1 = a_2 = a_3 = a_4 = 0$. The final system is $\left[\begin{array}{c|c} I_4 & 0 \end{array}\right]$, which does indeed imply $a_1 = a_2 = a_3 = a_4 = 0$ so S is linearly independent. Hence S is a basis for V.

5. Let $V = P_3$.

(a) Let

$$
S = \left\{ t^3 + 2t^2 + 3t, 2t^3 + 1, 6t^3 + 8t^2 + 6t + 4, t^3 + 2t^2 + t + 1 \right\}
$$

and let $at^3 + bt^2 + ct + d$ be an arbitrary vector in V. To check if S spans V we form the linear combination

$$
a_1(t^3 + 2t^2 + 3t) + a_2(2t^3 + 1) + a_3(6t^3 + 8t^2 + 6t + 4)
$$
$$
+ a_4(t^3 + 2t^2 + t + 1) = at^3 + bt^2 + ct + d.
$$

Performing the algebra on the left side and equating like power terms leads to the linear system

$$
\begin{array}{rrrrr}
a_1 & + 2a_2 & + 6a_3 & + a_4 & = a \\
2a_1 & & + 8a_3 & + 2a_4 & = b \\
3a_1 & & + 6a_3 & + a_4 & = c \\
 & a_2 & + 4a_3 & + a_4 & = d
\end{array}
$$

We form the augmented matrix and row reduce it:

$$
\left[\begin{array}{cccc|c}
1 & 2 & 6 & 1 & a \\
2 & 0 & 8 & 2 & b \\
3 & 0 & 6 & 1 & c \\
0 & 1 & 4 & 1 & d
\end{array}\right]
\longrightarrow
\left[\begin{array}{cccc|c}
1 & 2 & 6 & 1 & a \\
0 & -4 & -4 & 0 & b-2a \\
0 & -6 & -12 & -2 & c-3a \\
0 & 1 & 4 & 1 & d
\end{array}\right]
\longrightarrow
\left[\begin{array}{cccc|c}
1 & 2 & 6 & 1 & a \\
0 & 1 & 4 & 1 & d \\
0 & -6 & -12 & -2 & c-3a \\
0 & -4 & -4 & 0 & b-2a
\end{array}\right]
\longrightarrow
$$

add -2 (row 1) to row 2   interchange rows 2 and 4   add 6 (row 2) to row 3
add -3 (row 1) to row 3   add 4 (row 2) to row 4

$$\begin{bmatrix} 1 & 2 & 6 & 1 & | & a \\ 0 & 1 & 4 & 1 & | & d \\ 0 & 0 & 12 & 4 & | & c - 3a + 6d \\ 0 & 0 & 12 & 4 & | & b - 2a + 4d \end{bmatrix} \longrightarrow \begin{bmatrix} 1 & 2 & 6 & 1 & | & a \\ 0 & 1 & 4 & 1 & | & d \\ 0 & 0 & 12 & 4 & | & c - 3a + 6d \\ 0 & 0 & 0 & 0 & | & * \end{bmatrix}$$
add -1 (row 3) to row 4

where $* \neq 0$. Thus there are polynomials in P_3 for which this system is inconsistent. Hence S does not span V and is then not a basis for V.

(b) Let

$$S = \{t^3 + t^2 + 1, t^3 - 1, t^3 + t^2 + t\}$$

and let $at^3 + bt^2 + ct + d$ be an arbitrary vector in V. To check if S spans V we form the linear combination

$$a_1(t^3 + t^2 + 1) + a_2(t^3 - 1) + a_3(t^3 + t^2 + t) = at^3 + bt^2 + ct + d.$$

Performing the algebra on the left and equating like powered terms to the linear system

$$\begin{array}{rrrrr} a_1 & + & a_2 & + & a_3 & = & a \\ a_1 & & & + & a_3 & = & b \\ & & & & a_3 & = & c \\ a_1 & - & a_2 & & & = & d \end{array}$$

We form the augmented matrix and row reduce it:

$$\begin{bmatrix} 1 & 1 & 1 & | & a \\ 1 & 0 & 1 & | & b \\ 0 & 0 & 1 & | & c \\ 1 & -1 & 0 & | & d \end{bmatrix} \longrightarrow \begin{bmatrix} 1 & 1 & 1 & | & a \\ 0 & -1 & 0 & | & b - a \\ 0 & 0 & 1 & | & c \\ 0 & -2 & -1 & | & d - a \end{bmatrix} \longrightarrow \begin{bmatrix} 1 & 1 & 1 & | & a \\ 0 & 1 & 0 & | & a - b \\ 0 & 0 & 1 & | & c \\ 0 & 0 & -1 & | & a - 2b + d \end{bmatrix} \longrightarrow$$
add -1 (row 1) to row 2 -1 (row 2)   add row 3 to row 4
add -1 (row 1) to row 4   add 2 (row 2) to row 4

$$\begin{bmatrix} 1 & 1 & 1 & | & a \\ 0 & 1 & 0 & | & a - b \\ 0 & 0 & 1 & | & c \\ 0 & 0 & 0 & | & * \end{bmatrix}$$

where $* \neq 0$. Thus there are polynomials in P_3 for which this system is inconsistent. Hence S does not span V and is then not a basis for V.

Alternative analysis: It is easily shown that a natural basis for P_3 is $\{t^3, t^2, t, 1\}$. By Corollary 4.1 any other basis for P_3 must have 4 vectors. Since S has only 3 vectors, it cannot be a basis for P_3.

(c) Let

$$S = \{t^3 + t^2 + t + 1, t^3 + 2t^2 + t + 3, 2t^3 + t^2 + 3t + 2, t^3 + t^2 + 2t + 2\}$$

and let $at^3 + bt^2 + ct + d$ be an arbitrary vector in V. To check if S spans V we form the linear combination

$$a_1(t^3 + t^2 + t + 1) + a_2(t^3 + 2t^2 + t + 3) + a_3(2t^3 + t^2 + 3t + 2)$$
$$+ a_4(t^3 + t^2 + 2t + 2) = at^3 + bt^2 + ct + d.$$

Performing the algebra on the left side and equating like power terms leads to the linear system

$$\begin{array}{rrrrrrrr} a_1 & + & a_2 & + & 2a_3 & + & a_4 & = & a \\ a_1 & + & 2a_2 & + & a_3 & + & a_4 & = & b \\ a_1 & + & a_2 & + & 3a_3 & + & 2a_4 & = & c \\ a_1 & + & 3a_2 & + & 2a_3 & + & 2a_4 & = & d \end{array}$$

We form the augmented matrix and row reduce it:

$$\left[\begin{array}{cccc|c} 1 & 1 & 2 & 1 & a \\ 1 & 2 & 1 & 1 & b \\ 1 & 1 & 3 & 2 & c \\ 1 & 3 & 2 & 2 & d \end{array}\right] \longrightarrow \left[\begin{array}{cccc|c} 1 & 1 & 2 & 1 & a \\ 0 & 1 & -1 & 0 & b-a \\ 0 & 0 & 1 & 1 & c-a \\ 0 & 2 & 0 & 1 & d-a \end{array}\right] \longrightarrow \left[\begin{array}{cccc|c} 1 & 0 & 3 & 1 & 2a-b \\ 0 & 1 & -1 & 0 & b-a \\ 0 & 0 & 1 & 1 & c-a \\ 0 & 0 & 2 & 1 & a-2b+d \end{array}\right] \longrightarrow$$

$$\begin{array}{c} \text{add } -1 \text{ (row 1) to} \\ \text{rows 2, 3, 4} \end{array} \qquad \begin{array}{c} \text{add } -1 \text{ (row 2) to row 1} \\ \text{add } -2 \text{ (row 2) to row 4} \end{array} \qquad \begin{array}{c} \text{add } -2 \text{ (row 3) to row 4} \end{array}$$

$$\left[\begin{array}{cccc|c} 1 & 0 & 3 & 1 & 2a-b \\ 0 & 1 & -1 & 0 & b-a \\ 0 & 0 & 1 & 1 & c-a \\ 0 & 0 & 0 & -1 & 3a-2b-2c+d \end{array}\right]$$

If we next multiply row 4 by -1, we see that each row has a leading 1 which implies that the system is consistent for all polynomials in P_3. Hence S spans V. Furthermore, row operations can be applied to show that this augmented matrix is equivalent to $\left[\ I_4\ |\ *\ \right]$, where the components of $*$ are linear combinations of a, b, c, and d. It follows that if $a = b = c = d = 0$, then the only solution is the trivial solution which implies that S is a linearly independent set. Thus S is a basis for P_3.

(d) Applying the alternative analysis given in part (b), we can immediately say that $S = \{t^3 - t, t^3 + t^2 + 1, t - 1\}$ is not a basis for V

7. Let $V = R^3$.

(a) Let

$$S = \left\{ \begin{bmatrix} 1 \\ 1 \\ 1 \end{bmatrix}, \begin{bmatrix} 1 \\ 2 \\ 3 \end{bmatrix}, \begin{bmatrix} 0 \\ 1 \\ 0 \end{bmatrix} \right\}.$$

Form a linear combination of the three vectors in S and set it equal to an arbitrary vector in V:

$$a_1 \begin{bmatrix} 1 \\ 1 \\ 1 \end{bmatrix} + a_2 \begin{bmatrix} 1 \\ 2 \\ 3 \end{bmatrix} + a_3 \begin{bmatrix} 0 \\ 1 \\ 0 \end{bmatrix} = \begin{bmatrix} a \\ b \\ c \end{bmatrix}.$$

Now perform the matrix algebra on the left, equate corresponding entries, and row reduce the resulting matrix:

$$\left[\begin{array}{ccc|c} 1 & 1 & 0 & a \\ 1 & 2 & 1 & b \\ 1 & 3 & 0 & c \end{array}\right] \longrightarrow \left[\begin{array}{ccc|c} 1 & 1 & 0 & a \\ 0 & 1 & 1 & b-a \\ 0 & 2 & 0 & c-a \end{array}\right] \longrightarrow \left[\begin{array}{ccc|c} 1 & 0 & -1 & 2a-b \\ 0 & 1 & 1 & b-a \\ 0 & 0 & -2 & a-2b+c \end{array}\right]$$

$$\begin{array}{c} \text{add } -1 \text{ (row 1) to row 2} \\ \text{add } -1 \text{ (row 1) to row 3} \end{array} \qquad \begin{array}{c} \text{add } -1 \text{ (row 2) to row 1} \\ \text{add } -2 \text{ (row 2) to row 3} \end{array}$$

Multiplying row 3 by $-\frac{1}{2}$ gives a leading 1 in each row. Further row operations can be used to obtain an equivalent augmented matrix of the form $\left[\ I_3\ |\ *\ \right]$, where the components of $*$ are linear combinations of a, b, and c. Hence S spans V. If $a = b = c = 0$, it follows that the only solution is the trivial solution so S is linearly independent. Thus S is a basis for V.

To write the vector $\begin{bmatrix} 2 & 1 & 3 \end{bmatrix}^T$ in terms of the basis we set $a = 2$, $b = 1$ and $c = 3$ to get the matrix

$$\left[\begin{array}{ccc|c} 1 & 0 & -1 & 3 \\ 0 & 1 & 1 & -1 \\ 0 & 0 & -2 & 3 \end{array}\right]$$

Back substitution then gives

$$a_3 = -\frac{3}{2}$$

$$a_2 = -1 - a_3 = -1 + \frac{3}{2} = \frac{1}{2}$$

$$a_1 = 3 + a_3 = 3 - \frac{3}{2} = \frac{3}{2}$$

Thus

$$\begin{bmatrix} 2 \\ 1 \\ 3 \end{bmatrix} = \frac{3}{2}\begin{bmatrix} 1 \\ 1 \\ 1 \end{bmatrix} + \frac{1}{2}\begin{bmatrix} 1 \\ 2 \\ 3 \end{bmatrix} - \frac{3}{2}\begin{bmatrix} 0 \\ 1 \\ 0 \end{bmatrix}.$$

(b) Let

$$S = \left\{ \begin{bmatrix} 1 \\ 2 \\ 3 \end{bmatrix}, \begin{bmatrix} 2 \\ 1 \\ 3 \end{bmatrix}, \begin{bmatrix} 0 \\ 0 \\ 0 \end{bmatrix} \right\}.$$

From Section 4.5, any set containing the zero vector is linearly dependent, so S is dependent and hence cannot be a basis for V.

9. Let $V = P_2$.

(a) Let $S = \{t^2 + t, t - 1, t + 1\}$. Form a linear combination of the vectors in S and set it equal to an arbitrary vector in V:

$$a_1(t^2 + t) + a_2(t - 1) + a_3(t + 1) = at^2 + bt + c.$$

Now combine like terms on the left, equate corresponding coefficients, and row reduce the resulting augmented matrix:

$$\begin{bmatrix} 1 & 0 & 0 & a \\ 1 & 1 & 1 & b \\ 0 & -1 & 1 & c \end{bmatrix} \longrightarrow \begin{bmatrix} 1 & 0 & 0 & a \\ 0 & 1 & 1 & b - a \\ 0 & -1 & 1 & c \end{bmatrix} \longrightarrow \begin{bmatrix} 1 & 0 & 0 & a \\ 0 & 1 & 1 & b - a \\ 0 & 0 & 2 & c + b - a \end{bmatrix}$$

add -1 (row 1) to row 2   add row 2 to row 3

Multiplying row 3 by $\frac{1}{2}$ gives a leading 1 in each row. Further row operations can be used to obtain an equivalent augmented matrix of the form $\begin{bmatrix} I_3 & | & * \end{bmatrix}$, where the components of $*$ are linear combinations of a, b, and c. Hence S spans V. If $a = b = c = 0$, it follows that the only solution is the trivial solution so S is linearly independent. Thus S is a basis for P_2. To write $5t^2 - 3t + 8$ in terms of the basis we set $a = 5$, $b = -3$, and $c = 8$ to get the matrix

$$\begin{bmatrix} 1 & 0 & 0 & 5 \\ 0 & 1 & 1 & -8 \\ 0 & 0 & 2 & 0 \end{bmatrix}$$

Back substitution then gives $a_3 = 0$, $a_2 = -8 - a_3 = -8$, $a_1 = 5$. Thus

$$5t^2 - 3t + 8 = 5(t^2 + t) - 8(t - 1).$$

(b) Let $S = \{t^2 + 1, t - 1\}$. Since a natural basis for V is $\{t^2, t, 1\}$, any basis must contain 3 vectors. Since S has only 2 elements, it cannot be a basis for V.

11. Let

$$S = \left\{ \begin{bmatrix} 1 \\ 2 \\ 2 \end{bmatrix}, \begin{bmatrix} 3 \\ 2 \\ 1 \end{bmatrix}, \begin{bmatrix} 11 \\ 10 \\ 7 \end{bmatrix}, \begin{bmatrix} 7 \\ 6 \\ 4 \end{bmatrix} \right\}$$

and let $W = \text{span } S$. Then S is linearly dependent since it is set of 4 vectors in R^3 and $\dim R^3 = 3$. To find a subset of S that is linearly independent, we set an arbitrary element in W equal to the zero vector:

$$a_1 \begin{bmatrix} 1 \\ 2 \\ 2 \end{bmatrix} + a_2 \begin{bmatrix} 3 \\ 2 \\ 1 \end{bmatrix} + a_3 \begin{bmatrix} 11 \\ 10 \\ 7 \end{bmatrix} + a_4 \begin{bmatrix} 7 \\ 6 \\ 4 \end{bmatrix} = \begin{bmatrix} 0 \\ 0 \\ 0 \end{bmatrix}.$$

The corresponding augmented matrix of the resulting linear system is

$$\left[\begin{array}{cccc|c} 1 & 3 & 11 & 7 & 0 \\ 2 & 2 & 10 & 6 & 0 \\ 2 & 1 & 7 & 4 & 0 \end{array} \right]$$

We now row reduce this matrix:

$$\left[\begin{array}{cccc|c} 1 & 3 & 11 & 7 & 0 \\ 2 & 2 & 10 & 6 & 0 \\ 2 & 1 & 7 & 4 & 0 \end{array} \right] \longrightarrow \left[\begin{array}{cccc|c} 1 & 3 & 11 & 7 & 0 \\ 0 & -4 & -12 & -8 & 0 \\ 0 & -5 & -15 & -10 & 0 \end{array} \right] \longrightarrow \left[\begin{array}{cccc|c} 1 & 3 & 11 & 7 & 0 \\ 0 & 1 & 3 & 2 & 0 \\ 0 & 0 & 0 & 0 & 0 \end{array} \right]$$

$$\begin{array}{c} \text{add } -2 \text{ (row 1) to row 2} \\ \text{add } -2 \text{ (row 1) to row 3} \end{array} \qquad \begin{array}{c} -\frac{1}{4} \text{ (row 2)} \\ \text{add } 5 \text{ (row 2) to row 3} \end{array}$$

The leading 1's in column 1 and column 2 imply that the first and second vectors in S are a basis for W. Thus $\dim W = 2$.

13. Let

$$S = \{t^3 + t^2 - 2t + 1, t^2 + 1, t^3 - 2t, 2t^3 + 3t^2 - 4t + 3\}$$

and let $W = \text{span } S$. To determine possible dependence relations between the vectors in S, we set an arbitrary vector in W equal to the zero vector:

$$a_1(t^3 + t^2 - 2t + 1) + a_2(t^2 + 1) + a_3(t^3 - 2t) + a_4(2t^3 + 3t^2 - 4t + 3) = 0t^3 + 0t^2 + 0t + 0.$$

Combining terms on the left and equating coefficients of like powered terms gives the linear system

$$\begin{array}{rcrcrcrcl} a_1 & & & + & a_3 & + & 2a_4 & = & 0 \\ a_1 & + & a_2 & & & + & 3a_4 & = & 0 \\ -2a_1 & & & - & 2a_3 & - & 4a_4 & = & 0 \\ a_1 & + & a_2 & & & + & 3a_4 & = & 0 \end{array}$$

Row reducing the augmented matrix, we obtain:

$$\left[\begin{array}{cccc|c} 1 & 0 & 1 & 2 & 0 \\ 1 & 1 & 0 & 3 & 0 \\ -2 & 0 & -2 & -4 & 0 \\ 1 & 1 & 0 & 3 & 0 \end{array} \right] \longrightarrow \left[\begin{array}{cccc|c} 1 & 0 & 1 & 2 & 0 \\ 0 & 1 & -1 & 1 & 0 \\ 0 & 0 & 0 & 0 & 0 \\ 0 & 1 & -1 & 1 & 0 \end{array} \right] \longrightarrow \left[\begin{array}{cccc|c} 1 & 0 & 1 & 2 & 0 \\ 0 & 1 & -1 & 1 & 0 \\ 0 & 0 & 0 & 0 & 0 \\ 0 & 0 & 0 & 0 & 0 \end{array} \right]$$

$$\begin{array}{c} \text{add } -1 \text{ (row 1) to row 2} \\ \text{add } 2 \text{ (row 1) to row 3} \\ \text{add } -1 \text{ (row 1) to row 4} \end{array} \qquad \text{add } -1 \text{ (row 2) to row 4}$$

Since there are leading 1's in the first and second columns, it follows that the first and second vectors in S are a basis for W. Thus $\dim W = 2$.

15. Let

$$S = \{ \begin{bmatrix} a^2 & 0 & 1 \end{bmatrix}, \begin{bmatrix} 0 & a & 2 \end{bmatrix}, \begin{bmatrix} 1 & 0 & 1 \end{bmatrix} \}.$$

Since there are 3 vectors in S and $\dim R_3 = 3$, S is a basis for R_3 when the vectors in S are linearly independent. To determine when this is the case, we choose an arbitrary vector in span S and set it equal to the zero vector:

$$a_1 \begin{bmatrix} a^2 & 0 & 1 \end{bmatrix} + a_2 \begin{bmatrix} 0 & a & 2 \end{bmatrix} + a_3 \begin{bmatrix} 1 & 0 & 1 \end{bmatrix} = \begin{bmatrix} 0 & 0 & 0 \end{bmatrix}.$$

Equating corresponding entries leads to the linear system

$$
\begin{array}{rcrcrcl}
a^2 a_1 & & & + & a_3 & = & 0 \\
& & a a_2 & & & = & 0 \\
a_1 & + & 2a_2 & + & a_3 & = & 0
\end{array}
$$

Observe that if $a = 0$, then the first and second vectors in S are $\begin{bmatrix} 0 & 0 & 1 \end{bmatrix}$ and $\begin{bmatrix} 0 & 0 & 2 \end{bmatrix}$, which are linearly dependent. In this case S is not a basis for R_3. Thus we assume that $a \neq 0$. Next, row reduce the augmented matrix of this system:

$$
\left[\begin{array}{ccc|c} a^2 & 0 & 1 & 0 \\ 0 & a & 0 & 0 \\ 1 & 2 & 1 & 0 \end{array}\right]
\longrightarrow
\left[\begin{array}{ccc|c} a^2 & 0 & 1 & 0 \\ 0 & 1 & 0 & 0 \\ 1 & 0 & 1 & 0 \end{array}\right]
\longrightarrow
\left[\begin{array}{ccc|c} 1 & 0 & 1 & 0 \\ 0 & 1 & 0 & 0 \\ 0 & 0 & 1-a^2 & 0 \end{array}\right]
$$

$$
\begin{array}{cc}
\frac{1}{a} \text{ (row 2)} & \text{add } -a^2 \text{ (row 3) to row 1} \\
\text{add } -2 \text{ (row 2) to row 3} & \text{interchange rows 1 and 3}
\end{array}
$$

If $a \neq \pm 1$, then we can multiply row 3 by $1/(1-a^2)$ to get a leading 1 and use it to zero out the column. Thus we obtain the reduced row echelon form $\begin{bmatrix} I_3 & | & 0 \end{bmatrix}$, which implies that the three vectors are linearly independent and hence form a basis for R_3. Thus, S is a basis for R_3 when $a \neq 0, \pm 1$.

17. Let D be the subspace of M_{33} consisting of all diagonal matrices and let

$$
\begin{bmatrix} a & 0 & 0 \\ 0 & b & 0 \\ 0 & 0 & c \end{bmatrix}
$$

be an arbitrary element in D. Then

$$
\begin{bmatrix} a & 0 & 0 \\ 0 & b & 0 \\ 0 & 0 & c \end{bmatrix} = a \begin{bmatrix} 1 & 0 & 0 \\ 0 & 0 & 0 \\ 0 & 0 & 0 \end{bmatrix} + b \begin{bmatrix} 0 & 0 & 0 \\ 0 & 1 & 0 \\ 0 & 0 & 0 \end{bmatrix} + c \begin{bmatrix} 0 & 0 & 0 \\ 0 & 0 & 0 \\ 0 & 0 & 1 \end{bmatrix}.
$$

It follows that the set

$$
S = \left\{ \begin{bmatrix} 1 & 0 & 0 \\ 0 & 0 & 0 \\ 0 & 0 & 0 \end{bmatrix}, \begin{bmatrix} 0 & 0 & 0 \\ 0 & 1 & 0 \\ 0 & 0 & 0 \end{bmatrix}, \begin{bmatrix} 0 & 0 & 0 \\ 0 & 0 & 0 \\ 0 & 0 & 1 \end{bmatrix} \right\}
$$

spans D. This set is linearly independent. Therefore S is a basis for D.

19. Let W be the given subspace.

(a) Let $\begin{bmatrix} a \\ b \\ c \end{bmatrix}$ be an arbitrary element in W, where $b = a + c$. Then

$$
\begin{bmatrix} a \\ b \\ c \end{bmatrix} = \begin{bmatrix} a \\ a+c \\ c \end{bmatrix} = a \begin{bmatrix} 1 \\ 1 \\ 0 \end{bmatrix} + c \begin{bmatrix} 0 \\ 1 \\ 1 \end{bmatrix}.
$$

It follows that the set

$$
S = \left\{ \begin{bmatrix} 1 \\ 1 \\ 0 \end{bmatrix}, \begin{bmatrix} 0 \\ 1 \\ 1 \end{bmatrix} \right\}
$$

spans the subspace W. This set is linearly independent; for if $a = b = c = 0$ on the left side of the above equation, then $a = c = 0$. Thus S is linearly independent and spans W and is therefore a basis for W.

(b) Let $\begin{bmatrix} a \\ b \\ c \end{bmatrix}$ be an arbitrary element in W, where $b = a$. Then

$$\begin{bmatrix} a \\ b \\ c \end{bmatrix} = \begin{bmatrix} a \\ a \\ c \end{bmatrix} = a \begin{bmatrix} 1 \\ 1 \\ 0 \end{bmatrix} + c \begin{bmatrix} 0 \\ 0 \\ 1 \end{bmatrix}.$$

It follows that the set

$$S = \left\{ \begin{bmatrix} 1 \\ 1 \\ 0 \end{bmatrix}, \begin{bmatrix} 0 \\ 0 \\ 1 \end{bmatrix} \right\}$$

spans the subspace W. This set is linearly independent; for if $a = b = c = 0$ on the left side of the above equation, then $a = c = 0$. Thus S is linearly independent and spans W and is therefore a basis for W.

(c) Let $\begin{bmatrix} a \\ b \\ c \end{bmatrix}$ be an arbitrary element in W, where $2a + b - c = 0$. Then $c = 2a + b$ and hence

$$\begin{bmatrix} a \\ b \\ c \end{bmatrix} = \begin{bmatrix} a \\ b \\ 2a + b \end{bmatrix} = a \begin{bmatrix} 1 \\ 0 \\ 2 \end{bmatrix} + b \begin{bmatrix} 0 \\ 1 \\ 1 \end{bmatrix}.$$

It follows that the set

$$S = \left\{ \begin{bmatrix} 1 \\ 0 \\ 2 \end{bmatrix}, \begin{bmatrix} 0 \\ 1 \\ 1 \end{bmatrix} \right\}$$

spans the subspace W. This set is linearly independent; for if $a = b = c = 0$ on the left of the above equation, then $a = b = 0$. Thus S is linearly independent and spans W and is therefore a basis for W.

21. Let W be the given subspace and let $at^2 + bt + c$ be an arbitrary element in W. Then $c = 2a - 3b$ so

$$at^2 + bt + c = at^2 + bt + (2a - 3b) = a(t^2 + 2) + b(t - 3).$$

It follows that the set $S = \{t^2 + 2, t - 3\}$ spans the subspace W. This set is linearly independent; for if

$$a(t^2 + 2) + b(t - 3) = 0$$

for some scalars a, b, then $at^2 + bt + (2a - 3b) = 0$ so $a = 0$, $b = 0$. Thus S is linearly independent and spans W and is therefore a basis for W.

23. We find a basis for each subspace W of R_4.

(a) W consists of all vectors of the form $\begin{bmatrix} a & b & c & a+b \end{bmatrix}$. Now,

$$\begin{bmatrix} a & b & c & a+b \end{bmatrix} = a \begin{bmatrix} 1 & 0 & 0 & 1 \end{bmatrix} + b \begin{bmatrix} 0 & 1 & 0 & 1 \end{bmatrix} + c \begin{bmatrix} 0 & 0 & 1 & 0 \end{bmatrix}.$$

Let

$$S = \left\{ \begin{bmatrix} 1 & 0 & 0 & 1 \end{bmatrix}, \begin{bmatrix} 0 & 1 & 0 & 1 \end{bmatrix}, \begin{bmatrix} 0 & 0 & 1 & 0 \end{bmatrix} \right\}.$$

Then the preceding relation implies that S spans W. This set is linearly independent; for if

$$a \begin{bmatrix} 1 & 0 & 0 & 1 \end{bmatrix} + b \begin{bmatrix} 0 & 1 & 0 & 1 \end{bmatrix} + c \begin{bmatrix} 0 & 0 & 1 & 0 \end{bmatrix} = \begin{bmatrix} a & b & c & a+b \end{bmatrix} = \begin{bmatrix} 0 & 0 & 0 & 0 \end{bmatrix},$$

then $a = b = c = 0$. Thus S is a basis for W and therefore $\dim W = 3$.

(b) W consists of all vectors of the form $[\ a\ \ b\ \ a-b\ \ a+b\ ]$. Now,

$$[\ a\ \ b\ \ a-b\ \ a+b\ ] = a[\ 1\ \ 0\ \ 1\ \ 1\ ] + b[\ 0\ \ 1\ \ -1\ \ 1\ ].$$

Let

$$S = \{[\ 1\ \ 0\ \ 1\ \ 1\ ], [\ 0\ \ 1\ \ -1\ \ 1\ ]\}.$$

Then the preceding relation implies that S spans W. This set is linearly independent; for if

$$a[\ 1\ \ 0\ \ 1\ \ 1\ ] + b[\ 0\ \ 1\ \ -1\ \ 1\ ] = [\ a\ \ b\ \ a-b\ \ a+b\ ] = [\ 0\ \ 0\ \ 0\ \ 0\ ],$$

then $a = b = 0$. Thus S spans W and is linearly independent. Thus S is a basis for W and therefore dim $W = 2$.

25. (a) From Exercise 1(a), $S = \left\{ \begin{bmatrix} 1 \\ 3 \end{bmatrix}, \begin{bmatrix} 1 \\ -1 \end{bmatrix} \right\}$ is a basis for R^2. Therefore dim span $S = $ dim $R^2 = 2$.

(b) From Exercise 1,

$$S = \left\{ \begin{bmatrix} 0 \\ 0 \end{bmatrix}, \begin{bmatrix} 1 \\ 2 \end{bmatrix}, \begin{bmatrix} 2 \\ 4 \end{bmatrix} \right\}$$

is linearly dependent. We see that each vector in S is a multiple of $\begin{bmatrix} 1 \\ 2 \end{bmatrix}$. Hence span S is a subspace consisting of all scalar multiples of this (nonzero) vector and hence dim span $S = 1$.

(c) From Exercise 1,

$$S = \left\{ \begin{bmatrix} 1 \\ 2 \end{bmatrix}, \begin{bmatrix} 2 \\ -3 \end{bmatrix}, \begin{bmatrix} 3 \\ 2 \end{bmatrix} \right\}$$

spans R^2 but is linearly dependent. Hence span $S = R^2$ and hence dim span $S = $ dim $R^2 = 2$.

(d) From Exercise 1,

$$S = \left\{ \begin{bmatrix} 1 \\ 3 \end{bmatrix}, \begin{bmatrix} -2 \\ 6 \end{bmatrix} \right\}$$

is a basis for R^2, thus dim span $S = $ dim $R^2 = 2$.

27. (a) From Exercise 3,

$$S = \{[\ 1\ \ 0\ \ 0\ \ 1\ ], [\ 0\ \ 1\ \ 0\ \ 0\ ], [\ 1\ \ 1\ \ 1\ \ 1\ ], [\ 0\ \ 1\ \ 1\ \ 1\ ]\}$$

is a basis for R_4. Hence span $S = R_4$ and dim span $S = $ dim $R_4 = 4$.

(b) Let

$$S = \{[\ 1\ \ -1\ \ 0\ \ 2\ ], [\ 3\ \ -1\ \ 2\ \ 1\ ], [\ 1\ \ 0\ \ 0\ \ 1\ ]\}.$$

From Exercise 3 when we form a linear combination of these vectors and set it equal to zero, the row echelon form of the corresponding augmented matrix has three leading 1's. Thus S is linearly independent. Hence span S has a basis consisting of 3 vectors which implies dim span $S = 3$.

(c) Let

$$S = \{[\ -2\ \ 4\ \ 6\ \ 4\ ], [\ 0\ \ 1\ \ 2\ \ 0\ ], [\ -1\ \ 2\ \ 3\ \ 2\ ], [\ -3\ \ 2\ \ 5\ \ 6\ ], [\ -2\ \ -1\ \ 0\ \ 4\ ]\}.$$

Since span S is a subspace of R_4, dim span $S \leq$ dim $R_4 = 4$. To determine dim span S we find the number of linearly independent vectors in S. We use the procedure suggested by the alternate proof of Theorem 4.9. From Exercise 3(c) we have the following augmented matrix where a, b, c, and d have been set equal to zero:

$$\begin{bmatrix} -2 & 0 & -1 & -3 & -2 & | & 0 \\ 4 & 1 & 2 & 2 & -1 & | & 0 \\ 6 & 2 & 3 & 5 & 0 & | & 0 \\ 4 & 0 & 2 & 6 & 4 & | & 0 \end{bmatrix} \longrightarrow \begin{bmatrix} -2 & 0 & -1 & -3 & -2 & | & 0 \\ 0 & 1 & 0 & -4 & -5 & | & 0 \\ 0 & 2 & 0 & -4 & -6 & | & 0 \\ 0 & 0 & 0 & 0 & 0 & | & 0 \end{bmatrix} \longrightarrow \begin{bmatrix} -2 & 0 & -1 & -3 & -2 & | & 0 \\ 0 & 1 & 0 & -4 & -5 & | & 0 \\ 0 & 0 & 0 & 4 & 4 & | & 0 \\ 0 & 0 & 0 & 0 & 0 & | & 0 \end{bmatrix}$$

add 2 (row 1) to row 2   add -2 (row 2) to row 3
add 3 (row 1) to row 3
add 2 (row 1) to row 4

We see that by using three more row operations we can obtain an equivalent augmented matrix with 3 leading 1's. Hence S contains 3 linearly independent vectors so $\dim \operatorname{span} S = 3$.

(d) From Exercise 3(d),

$$S = \{[\,0 \ \ 0 \ \ 1 \ \ 1\,], [\,-1 \ \ 1 \ \ 1 \ \ 2\,], [\,1 \ \ 1 \ \ 0 \ \ 0\,], [\,2 \ \ 1 \ \ 2 \ \ 1\,]\}$$

is a basis for R_4, thus $\dim \operatorname{span} S = \dim R_4 = 4$.

29. Let $S = \{t^3 + t, t^2 - t\}$. Following the technique used in the proof of Theorem 4.11, we append a basis to set S and proceed to determine which vectors are dependent on preceding vectors. Here we use the natural basis for P_3, so let $T = \{t^3 + t, t^2 - t, t^3, t^2, t, 1\}$. Next we use the technique developed in the alternate proof of Theorem 4.9. Set

$$a_1(t^3 + t) + a_2(t^2 - t) + a_3 t^3 + a_4 t^2 + a_5 t + a_6 = 0t^3 + 0t^2 + 0t + 0.$$

From the corresponding system of equations, we find the augmented matrix of the system and row reduce it:

$$\begin{bmatrix} 1 & 0 & 1 & 0 & 0 & 0 & | & 0 \\ 0 & 1 & 0 & 1 & 0 & 0 & | & 0 \\ 1 & -1 & 0 & 0 & 1 & 0 & | & 0 \\ 0 & 0 & 0 & 0 & 0 & 1 & | & 0 \end{bmatrix} \longrightarrow \begin{bmatrix} 1 & 0 & 1 & 0 & 0 & 0 & | & 0 \\ 0 & 1 & 0 & 1 & 0 & 0 & | & 0 \\ 0 & 0 & -1 & 1 & 1 & 0 & | & 0 \\ 0 & 0 & 0 & 0 & 0 & 1 & | & 0 \end{bmatrix}$$

add -1 (row 1) to row 3
add row 2 to row 3

It follows that we can produce leading 1's in columns 1, 2, 3, and 6. Therefore a basis for P_3 is $\{t^3 + t, t^2 - t, t^3, 1\}$ and this basis contains S.

31. Let W be the given subspace. We first find a basis for W. Let $at^2 + bt + c$ be an arbitrary element in W. Then $c = b - 2a$, so

$$at^2 + bt + c = at^2 + bt + (b - 2a) = a(t^2 - 2) + b(t + 1).$$

It follows that the set $S = \{t^2 - 2, t + 1\}$ spans the subspace W. This set is linearly independent; for if

$$a(t^2 - 2) + b(t + 1) = 0$$

for some scalars a, b, then

$$at^2 + bt + (b - 2a) = 0.$$

Therefore $a = 0$, $b = 0$. Thus S is linearly independent and spans W, and therefore is a basis for W. Since S contains two vectors, it follows that the dimension of W is 2.

33. Let

$$S = \left\{ \begin{bmatrix} 1 \\ 0 \\ 0 \\ 0 \end{bmatrix}, \begin{bmatrix} 0 \\ 1 \\ 0 \\ 0 \end{bmatrix} \right\}.$$

Then the vectors in S are linearly independent and hence span a two-dimensional subspace of R^4.

35. Let $S = \{\mathbf{v}_1, \mathbf{v}_2, \ldots, \mathbf{v}_k\}$ and $T = \{c\mathbf{v}_1, \mathbf{v}_2, \ldots, \mathbf{v}_k\}$. Since S is a basis for V, $\dim V = k$. Hence a maximal linearly independent set in V contains k vectors. We show that T spans V; then by Theorem 4.12(b) T is a basis for V. Let $\mathbf{v}$ be any vector in V. Then $\mathbf{v}$ can be expressed as a linear combination of the members of S; that is, there are scalars a_i, $i = 1, 2, \ldots, k$ such that

$$\mathbf{v} = a_1 \mathbf{v}_1 + a_2 \mathbf{v}_2 + \cdots + a_k \mathbf{v}_k.$$

Since $c \neq 0$, we have

$$\mathbf{v} = \frac{a_1}{c}(c\mathbf{v}_1) + a_2 \mathbf{v}_2 + \cdots + a_k \mathbf{v}_k.$$

Hence $\mathbf{v}$ is a linear combination of the vectors of T. Since $\mathbf{v}$ was an arbitrary vector in V, T spans V.

37. Since $\dim V = n$, Theorem 4.10 guarantees that any linearly independent subset of V has at most n vectors. Hence any subset with $n + 1$ vectors must be linearly dependent.

39. Since $\dim V = n$, Theorem 4.10 guarantees that any linearly independent subset of V has at most n vectors. Hence any subset with $n + 1$ vectors must be linearly dependent.

41. Let $\dim V = n$. First note that any set of vectors in W that is linearly independent in W is linearly independent in V. If $W = \{\mathbf{0}\}$, then $\dim W = 0$ and we are done. Suppose now that W is a nonzero subspace of V. Then W contains a nonzero vector $\mathbf{v}_1$, so $\{\mathbf{v}_1\}$ is linearly independent in W and therefore in V. If span $\{\mathbf{v}_1\} = W$, then $\dim W = 1$ and we are done. If span $\{\mathbf{v}_1\} \neq W$, then there is a vector $\mathbf{v}_2$ in W which is not in span $\{\mathbf{v}_1\}$. Then $\{\mathbf{v}_1, \mathbf{v}_2\}$ is linearly independent in W and hence in V. Since $\dim V = n$, no linearly independent set in V can have more than n vectors. Hence, no linearly independent set in W can have more than n vectors. Continuing the above process we find a basis for W containing at most n vectors. Hence W is finite-dimensional and $\dim W \leq \dim V$.

43. Let $V = R^3$. The trivial subspaces of any vector space are $\{\mathbf{0}\}$ and V. Hence $\{\mathbf{0}\}$ and R^3 are subspaces of R^3. In Exercise 35 of Section 4.3 we showed that any line ℓ through the origin is a subspace of R^3. Thus we need only show that any plane π passing through the origin is a subspace of R^3. A plane π through the origin in R^3 has an equation of the form $ax + by + cz = 0$. Sums and scalar multiples of any point on π will also satisfy this equation, hence π is a subspace of R^3.

To show that $\{\mathbf{0}\}$, V, lines, and planes are the only subspaces of R^3 we argue in a manner similar to that given in Exercise 29 of Section 4.3 which considered a similar problem in R^2. Let W be any subspace of R^3. Hence W contains the zero vector $\mathbf{0}$. If $W \neq \{\mathbf{0}\}$, then it contains a nonzero vector $\mathbf{v} = \begin{bmatrix} a & b & c \end{bmatrix}^T$ where at least one of a, b, or c is not zero. Since W is a subspace it contains span $\{\mathbf{v}\}$. If $W = $ span $\{\mathbf{v}\}$ then W is a line in R^3 through the origin. Otherwise, there is a vector $\mathbf{u}$ in W which is not in span $\{\mathbf{v}\}$. Hence $\{\mathbf{v}, \mathbf{u}\}$ is a linearly independent set. But then W contains span $\{\mathbf{u}, \mathbf{v}\}$. If $W = $ span $\{\mathbf{v}, \mathbf{u}\}$ then W is a plane through the origin. Otherwise there is a vector $\mathbf{x}$ in W that is not in span $\{\mathbf{v}, \mathbf{u}\}$. Hence $\{\mathbf{v}, \mathbf{u}, \mathbf{x}\}$ is a linearly independent set in W and W contains span $\{\mathbf{v}, \mathbf{u}, \mathbf{x}\}$. But $\{\mathbf{v}, \mathbf{u}, \mathbf{x}\}$ is a maximal linearly independent set in R^3, hence a basis for R^3. It follows in this case that $W = R^3$.

45. Let V be an n-dimensional vector space.

   (a) If $S = \{\mathbf{v}_1, \mathbf{v}_2, \ldots, \mathbf{v}_n\}$ is a linearly independent set of vectors, then S is a basis for V.

   Proof: If span $S \neq V$, then there is a vector $\mathbf{v}$ in V that is not in span S. Vector $\mathbf{v}$ cannot be the zero vector since the zero vector is in every subspace and hence in span S. Hence $S_1 = \{\mathbf{v}_1, \mathbf{v}_2, \ldots, \mathbf{v}_n, \mathbf{v}\}$ is a linearly independent set. This follows since $\mathbf{v}_i$, $i = 1, \ldots, n$ are linearly independent and $\mathbf{v}$ is not a linear combination of the $\mathbf{v}_i$. But this contradicts Corollary 3.4. Hence our assumption that span $S \neq V$ is incorrect. Thus span $S = V$. Since S is linearly independent and spans V, it is a basis for V.

   (b) If $S = \{\mathbf{v}_1, \mathbf{v}_2, \ldots, \mathbf{v}_n\}$ spans V, then S is a basis for V.

   Proof: We want to show that S is linearly independent. Suppose S is linearly dependent. Then there is a subset of S consisting of at most $n - 1$ vectors which is a basis for V. (This follows from Theorem 4.9.) But this contradicts $\dim V = n$. Hence our assumption is false and S is linearly independent. Since S spans V and is linearly independent, it is a basis for V.

47. If A is nonsingular, then the linear system $A\mathbf{x} = \mathbf{0}$ has only the trivial solution $\mathbf{x} = \mathbf{0}$. Let

$$c_1 A\mathbf{v}_1 + c_2 A\mathbf{v}_2 + \cdots + c_n A\mathbf{v}_n = \mathbf{0}.$$

Then $A(c_1\mathbf{v}_1 + c_2\mathbf{v}_2 + \cdots + c_n\mathbf{v}_n) = \mathbf{0}$ and, since A is nonsingular, it follows that $c_1\mathbf{v}_1 + c_2\mathbf{v}_2 + \cdots + c_n\mathbf{v}_n = \mathbf{0}$. Since $\{\mathbf{v}_1, \mathbf{v}_2, \ldots, \mathbf{v}_n\}$ is linearly independent, it follows that $c_1 = \cdots = c_n = 0$. Therefore $\{A\mathbf{v}_1, A\mathbf{v}_2, \ldots, A\mathbf{v}_n\}$ is linearly independent and hence is a basis for R^n.

49. Let

$$A = \begin{bmatrix} a_1 & b_1 & c_1 \\ a_2 & b_2 & c_2 \\ a_3 & b_3 & c_3 \end{bmatrix}$$

be in W. Then $\mathrm{Tr}(A) = a_1 + b_2 + c_3 = 0 \Longrightarrow c_3 = -a_1 - b_2$. Thus, a typical matrix in W has the form

$$\begin{bmatrix} a_1 & b_1 & c_1 \\ a_2 & b_2 & c_2 \\ a_3 & b_3 & -a_1-b_2 \end{bmatrix} = a_1 \begin{bmatrix} 1 & 0 & 0 \\ 0 & 0 & 0 \\ 0 & 0 & -1 \end{bmatrix} + b_1 \begin{bmatrix} 0 & 1 & 0 \\ 0 & 0 & 0 \\ 0 & 0 & 0 \end{bmatrix} + c_1 \begin{bmatrix} 0 & 0 & 1 \\ 0 & 0 & 0 \\ 0 & 0 & 0 \end{bmatrix} + a_2 \begin{bmatrix} 0 & 0 & 0 \\ 1 & 0 & 0 \\ 0 & 0 & 0 \end{bmatrix}$$

$$+ b_2 \begin{bmatrix} 0 & 0 & 0 \\ 0 & 1 & 0 \\ 0 & 0 & -1 \end{bmatrix} + c_2 \begin{bmatrix} 0 & 0 & 0 \\ 0 & 0 & 1 \\ 0 & 0 & 0 \end{bmatrix} + a_3 \begin{bmatrix} 0 & 0 & 0 \\ 0 & 0 & 0 \\ 1 & 0 & 0 \end{bmatrix} + b_3 \begin{bmatrix} 0 & 0 & 0 \\ 0 & 0 & 0 \\ 0 & 1 & 0 \end{bmatrix}$$

Clearly, if this expression is equal to the zero matrix, $\begin{bmatrix} 0 & 0 & 0 \\ 0 & 0 & 0 \\ 0 & 0 & 0 \end{bmatrix}$, then all coefficients must be equal

to zero. Therefore the set

$$\left\{ \begin{bmatrix} 1 & 0 & 0 \\ 0 & 0 & 0 \\ 0 & 0 & -1 \end{bmatrix}, \begin{bmatrix} 0 & 1 & 0 \\ 0 & 0 & 0 \\ 0 & 0 & 0 \end{bmatrix}, \begin{bmatrix} 0 & 0 & 1 \\ 0 & 0 & 0 \\ 0 & 0 & 0 \end{bmatrix}, \begin{bmatrix} 0 & 0 & 0 \\ 1 & 0 & 0 \\ 0 & 0 & 0 \end{bmatrix}, \right.$$

$$\left. \begin{bmatrix} 0 & 0 & 0 \\ 0 & 1 & 0 \\ 0 & 0 & -1 \end{bmatrix}, \begin{bmatrix} 0 & 0 & 0 \\ 0 & 0 & 1 \\ 0 & 0 & 0 \end{bmatrix}, \begin{bmatrix} 0 & 0 & 0 \\ 0 & 0 & 0 \\ 1 & 0 & 0 \end{bmatrix}, \begin{bmatrix} 0 & 0 & 0 \\ 0 & 0 & 0 \\ 0 & 1 & 0 \end{bmatrix} \right\}$$

is a basis for W. Hence $\dim W = 8$.

Section 4.7, p. 251

1. Let W be the solution space to $A\mathbf{x} = \mathbf{0}$.

    (a) To find W, we row reduce the augmented matrix of A:

$$\begin{bmatrix} 2 & -1 & -2 & \big| & 0 \\ -4 & 2 & 4 & \big| & 0 \\ -8 & 4 & 8 & \big| & 0 \end{bmatrix} \rightarrow \begin{bmatrix} 2 & -1 & -2 & \big| & 0 \\ 0 & 0 & 0 & \big| & 0 \\ 0 & 0 & 0 & \big| & 0 \end{bmatrix} \rightarrow \begin{bmatrix} 2 & -1 & -2 & \big| & 0 \\ 0 & 0 & 0 & \big| & 0 \\ 0 & 0 & 0 & \big| & 0 \end{bmatrix}$$

        add 2(row 1) to row 2   multiply row 1 by $\frac{1}{2}$
        add 4(row 1) to row 3

The solutions to the system are therefore $x_1 = \frac{1}{2}r + s$, $x_2 = r$, $x_3 = s$, where r, s are arbitrary real numbers.

    (b) Every solution has the form

$$\mathbf{x} = \begin{bmatrix} x_1 \\ x_2 \\ x_3 \end{bmatrix} = \begin{bmatrix} \frac{1}{2}r + s \\ r \\ s \end{bmatrix} = r \begin{bmatrix} \frac{1}{2} \\ 1 \\ 0 \end{bmatrix} + s \begin{bmatrix} 1 \\ 0 \\ 1 \end{bmatrix}.$$

Thus, every solution to $A\mathbf{x} = \mathbf{0}$ may be expressed as a linear combination of the two vectors

$$\mathbf{x}_1 = \begin{bmatrix} \frac{1}{2} \\ 1 \\ 0 \end{bmatrix} \quad \text{and} \quad \mathbf{x}_2 = \begin{bmatrix} 1 \\ 0 \\ 1 \end{bmatrix}.$$

(c)

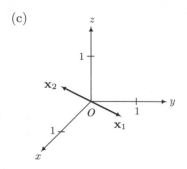

3. Let W be the solution space of the linear system. To find W, we row reduce the augmented matrix:

$$\left[\begin{array}{cccc|c} 1 & 1 & 1 & 1 & 0 \\ 2 & 1 & -1 & 1 & 0 \end{array}\right] \longrightarrow \left[\begin{array}{cccc|c} 1 & 1 & 1 & 1 & 0 \\ 0 & -1 & -3 & -1 & 0 \end{array}\right] \longrightarrow \left[\begin{array}{cccc|c} 1 & 0 & -2 & 0 & 0 \\ 0 & 1 & 3 & 1 & 0 \end{array}\right]$$

add -2 (row 1) to row 2 -1 (row 2)

The solutions of the system are therefore $x_1 = 2r$, $x_2 = -3r - s$, $x_3 = r$, $x_4 = s$, where r and s are any real numbers. Thus every solution has the form

$$\mathbf{x} = \left[\begin{array}{c} 2r \\ -3r - s \\ r \\ s \end{array}\right] = r\left[\begin{array}{c} 2 \\ -3 \\ 1 \\ 0 \end{array}\right] + s\left[\begin{array}{c} 0 \\ -1 \\ 0 \\ 1 \end{array}\right]$$

where r, s are any real numbers. Let

$$\mathbf{x}_1 = \left[\begin{array}{c} 2 \\ -3 \\ 1 \\ 0 \end{array}\right] \quad \text{and} \quad \mathbf{x}_2 = \left[\begin{array}{c} 0 \\ -1 \\ 0 \\ 1 \end{array}\right].$$

Then $\{\mathbf{x}_1, \mathbf{x}_2\}$ is a basis for the solution space W and $\dim W = 2$.

5. Let W be the solution space of the linear system. To find W, we row reduce the augmented matrix:

$$\left[\begin{array}{cccc|c} 1 & 2 & -1 & 3 & 0 \\ 2 & 2 & -1 & 2 & 0 \\ 1 & 0 & 3 & 3 & 0 \end{array}\right] \longrightarrow \left[\begin{array}{cccc|c} 1 & 2 & -1 & 3 & 0 \\ 0 & -2 & 1 & -4 & 0 \\ 0 & -2 & 4 & 0 & 0 \end{array}\right] \longrightarrow \left[\begin{array}{cccc|c} 1 & 0 & 3 & 3 & 0 \\ 0 & 0 & -3 & -4 & 0 \\ 0 & 1 & -2 & 0 & 0 \end{array}\right] \longrightarrow$$

add -2 (row 1) to row 2 $-\frac{1}{2}$ (row 3)   interchange rows 2 and 3
add -1 (row 1) to row 3   add -2 (row 3) to row 1 $-\frac{1}{3}$ (row 3)
                                          add 2 (row 3) to row 2

$$\left[\begin{array}{cccc|c} 1 & 0 & 3 & 3 & 0 \\ 0 & 1 & -2 & 0 & 0 \\ 0 & 0 & 1 & \frac{4}{3} & 0 \end{array}\right] \longrightarrow \left[\begin{array}{cccc|c} 1 & 0 & 0 & -1 & 0 \\ 0 & 1 & 0 & \frac{8}{3} & 0 \\ 0 & 0 & 1 & \frac{4}{3} & 0 \end{array}\right]$$

add -3 (row 3) to row 1
add 2 (row 3) to row 2

The solutions of the system are therefore $x_1 = r$, $x_2 = -\frac{8}{3}r$, $x_3 = -\frac{4}{3}r$, $x_4 = r$, where r is any real number. Thus every solution has the form

$$\mathbf{x} = \left[\begin{array}{c} r \\ -\frac{8}{3}r \\ -\frac{4}{3}r \\ r \end{array}\right] = r\left[\begin{array}{c} 1 \\ -\frac{8}{3} \\ -\frac{4}{3} \\ 1 \end{array}\right]$$

where r is any real numbers. Let

$$\mathbf{x}_1 = \begin{bmatrix} 1 \\ -\frac{8}{3} \\ -\frac{4}{3} \\ 1 \end{bmatrix}.$$

Then $\{\mathbf{x}_1\}$ is a basis for the solution space W and $\dim W = 1$.

7. Let W be the solution space of the linear system. To find W, we row reduce the augmented matrix:

$$\begin{bmatrix} 1 & 2 & 1 & 2 & 1 & | & 0 \\ 1 & 2 & 2 & 1 & 2 & | & 0 \\ 2 & 4 & 3 & 3 & 3 & | & 0 \\ 0 & 0 & 1 & -1 & -1 & | & 0 \end{bmatrix} \longrightarrow \begin{bmatrix} 1 & 2 & 1 & 2 & 1 & | & 0 \\ 0 & 0 & 1 & -1 & 1 & | & 0 \\ 0 & 0 & 1 & 1 & 1 & | & 0 \\ 0 & 0 & 1 & -1 & -1 & | & 0 \end{bmatrix} \longrightarrow \begin{bmatrix} 1 & 2 & 0 & 3 & 0 & | & 0 \\ 0 & 0 & 1 & -1 & 1 & | & 0 \\ 0 & 0 & 0 & 0 & 0 & | & 0 \\ 0 & 0 & 0 & 0 & -2 & | & 0 \end{bmatrix} \longrightarrow$$

add -1 (row 1) to row 2   add -1 (row 2) to row 1 $-\frac{1}{2}$ (row 4)

add -2 (row 1) to row 3   add -1 (row 2) to row 3   add -1 (row 4) to row 2

add -1 (row 2) to row 4   interchange rows 3 and 4

$$\begin{bmatrix} 1 & 2 & 0 & 3 & 0 & | & 0 \\ 0 & 0 & 1 & -1 & 0 & | & 0 \\ 0 & 0 & 0 & 0 & 1 & | & 0 \\ 0 & 0 & 0 & 0 & 0 & | & 0 \end{bmatrix}$$

The solutions of the system are therefore $x_1 = -2r - 3s$, $x_2 = r$, $x_3 = s$, $x_4 = s$, $x_5 = 0$, where r is any real number. Thus every solution has the form

$$\mathbf{x} = \begin{bmatrix} -2r - 3s \\ r \\ s \\ s \\ 0 \end{bmatrix} = r \begin{bmatrix} -2 \\ 1 \\ 0 \\ 0 \\ 0 \end{bmatrix} + s \begin{bmatrix} -3 \\ 0 \\ 1 \\ 1 \\ 0 \end{bmatrix}$$

where r is any real numbers. Let

$$\mathbf{x}_1 = \begin{bmatrix} -2 \\ 1 \\ 0 \\ 0 \\ 0 \end{bmatrix} \quad \text{and} \quad \mathbf{x}_2 = \begin{bmatrix} -3 \\ 0 \\ 1 \\ 1 \\ 0 \end{bmatrix}.$$

Then $\{\mathbf{x}_1, \mathbf{x}_2\}$ is a basis for the solution space W and $\dim W = 2$.

9. Let W be the solution space of the linear system. To find W, we row reduce the augmented matrix:

$$\begin{bmatrix} 1 & 2 & 2 & -1 & 1 & | & 0 \\ 0 & 2 & 2 & -2 & -1 & | & 0 \\ 2 & 6 & 2 & -4 & 1 & | & 0 \\ 1 & 4 & 0 & -3 & 0 & | & 0 \end{bmatrix} \longrightarrow \begin{bmatrix} 1 & 2 & 2 & -1 & 1 & | & 0 \\ 0 & 2 & 2 & -2 & -1 & | & 0 \\ 0 & 2 & -2 & -2 & -1 & | & 0 \\ 0 & 2 & -2 & -2 & -1 & | & 0 \end{bmatrix} \longrightarrow \begin{bmatrix} 1 & 0 & 0 & 1 & 2 & | & 0 \\ 0 & 1 & 1 & -1 & -\frac{1}{2} & | & 0 \\ 0 & 0 & -4 & 0 & 0 & | & 0 \\ 0 & 0 & 0 & 0 & 0 & | & 0 \end{bmatrix} \longrightarrow$$

add -2 (row 1) to row 3   add -1 (row 3) to row 4 $-\frac{1}{4}$ (row 3)

add -1 (row 1) to row 4   add -1 (row 2) to row 3   add -1 (row 3) to row 2

                        add -1 (row 2) to row 1

                        $\frac{1}{2}$ (row 2)

$$\begin{bmatrix} 1 & 0 & 0 & 1 & 2 & | & 0 \\ 0 & 1 & 0 & -1 & -\frac{1}{2} & | & 0 \\ 0 & 0 & 1 & 0 & 0 & | & 0 \\ 0 & 0 & 0 & 0 & 0 & | & 0 \end{bmatrix}$$

The solutions of the system are therefore $x_1 = -r - 2s$, $x_2 = r + \frac{1}{2}s$, $x_3 = 0$, $x_4 = r$, $x_5 = s$, where r and s are any real numbers. Thus every solution has the form

$$\mathbf{x} = \begin{bmatrix} -r - 2s \\ r + \frac{1}{2}s \\ 0 \\ r \\ s \end{bmatrix} = r \begin{bmatrix} -1 \\ 1 \\ 0 \\ 1 \\ 0 \end{bmatrix} + s \begin{bmatrix} -2 \\ \frac{1}{2} \\ 0 \\ 0 \\ 1 \end{bmatrix}$$

where r and s are any real numbers. Let

$$\mathbf{x}_1 = \begin{bmatrix} -1 \\ 1 \\ 0 \\ 1 \\ 0 \end{bmatrix} \quad \text{and} \quad \mathbf{x}_2 = \begin{bmatrix} -2 \\ \frac{1}{2} \\ 0 \\ 0 \\ 1 \end{bmatrix}.$$

Then $\{\mathbf{x}_1, \mathbf{x}_2\}$ is a basis for the solution space W and $\dim W = 2$.

11. The null space of A is the solution space of the linear system $A\mathbf{x} = \mathbf{0}$. To find a basis for it, we row reduce the augmented coefficient matrix:

$$\begin{bmatrix} 1 & 2 & 3 & -1 & | & 0 \\ 2 & 3 & 2 & 0 & | & 0 \\ 3 & 4 & 1 & 1 & | & 0 \\ 1 & 1 & -1 & 1 & | & 0 \end{bmatrix} \longrightarrow \begin{bmatrix} 1 & 2 & 3 & -1 & | & 0 \\ 0 & -1 & -4 & 2 & | & 0 \\ 0 & -2 & -8 & 4 & | & 0 \\ 0 & -1 & -4 & 2 & | & 0 \end{bmatrix} \longrightarrow \begin{bmatrix} 1 & 0 & -5 & 3 & | & 0 \\ 0 & 1 & 4 & -2 & | & 0 \\ 0 & 0 & 0 & 0 & | & 0 \\ 0 & 0 & 0 & 0 & | & 0 \end{bmatrix}$$

add -2 (row 1) to row 2   add 2 (row 2) to row 1
add -3 (row 1) to row 3   add -2 (row 2) to row 3
add -1 (row 1) to row 4   add -1 (row 2) to row 4
         -1 (row 2)

The solutions of the system are therefore $x_1 = 5r - 3s$, $x_2 = -4r + 2s$, $x_3 = r$, $x_4 = s$, where r and s are any real numbers. Thus every solution has the form

$$\mathbf{x} = \begin{bmatrix} 5r - 3s \\ -4r + 2s \\ r \\ s \end{bmatrix} = r \begin{bmatrix} 5 \\ -4 \\ 1 \\ 0 \end{bmatrix} + s \begin{bmatrix} -3 \\ 2 \\ 0 \\ 1 \end{bmatrix}$$

where r and s are any real numbers. Let

$$\mathbf{x}_1 = \begin{bmatrix} 5 \\ -4 \\ 1 \\ 0 \end{bmatrix} \quad \text{and} \quad \mathbf{x}_2 = \begin{bmatrix} -3 \\ 2 \\ 0 \\ 1 \end{bmatrix}.$$

Then $\{\mathbf{x}_1, \mathbf{x}_2\}$ is a basis for the null space of A.

13. For the given matrix A and λ, we find

$$\lambda I_2 - A = \begin{bmatrix} 1 & 0 \\ 0 & 1 \end{bmatrix} - \begin{bmatrix} 3 & 2 \\ 1 & 2 \end{bmatrix} = \begin{bmatrix} -2 & -2 \\ -1 & -1 \end{bmatrix}.$$

We now form the augmented matrix for $(I_2 - A)\mathbf{x} = \mathbf{0}$ and row reduce it:

$$\begin{bmatrix} -2 & -2 & | & 0 \\ -1 & -1 & | & 0 \end{bmatrix} \longrightarrow \begin{bmatrix} 1 & 1 & | & 0 \\ 0 & 0 & | & 0 \end{bmatrix}$$

add -2 (row 2) to row 1
interchange rows 1 and 2

The solutions of this system are therefore $x_1 = -r$, $x_2 = r$, where r is any real number. Thus every solution has the form

$$\mathbf{x} = \begin{bmatrix} -r \\ r \end{bmatrix} = r \begin{bmatrix} -1 \\ 1 \end{bmatrix}.$$

Let

$$\mathbf{x_1} = \begin{bmatrix} -1 \\ 1 \end{bmatrix}.$$

Then $\{\mathbf{x_1}\}$ is a basis for the solution space of $(I_2 - A)\mathbf{x} = \mathbf{0}$.

15. For the given matrix A and λ, we find

$$\lambda I_3 - A = \begin{bmatrix} 1 & 0 & 0 \\ 0 & 1 & 0 \\ 0 & 0 & 1 \end{bmatrix} - \begin{bmatrix} 0 & 0 & 1 \\ 1 & 0 & -3 \\ 0 & 1 & 3 \end{bmatrix} = \begin{bmatrix} 1 & 0 & -1 \\ -1 & 1 & 3 \\ 0 & -1 & -2 \end{bmatrix}.$$

We now form the augmented matrix for $(I_2 - A)\mathbf{x} = \mathbf{0}$ and row reduce it:

$$\begin{bmatrix} 1 & 0 & -1 & 0 \\ -1 & 1 & 3 & 0 \\ 0 & -1 & -2 & 0 \end{bmatrix} \longrightarrow \begin{bmatrix} 1 & 0 & -1 & 0 \\ 0 & 1 & 2 & 0 \\ 0 & -1 & -2 & 0 \end{bmatrix} \longrightarrow \begin{bmatrix} 1 & 0 & -1 & 0 \\ 0 & 1 & 2 & 0 \\ 0 & 0 & 0 & 0 \end{bmatrix}$$
$$\text{add row 1 to row 2} \qquad \text{add row 2 to row 3}$$

The solutions of this system are therefore $x_1 = r$, $x_2 = -2r$, $x_3 = r$ where r is any real number. Thus every solution has the form

$$\mathbf{x} = \begin{bmatrix} r \\ -2r \\ r \end{bmatrix} = r \begin{bmatrix} 1 \\ -2 \\ 1 \end{bmatrix}.$$

Let

$$\mathbf{x_1} = \begin{bmatrix} 1 \\ -2 \\ 1 \end{bmatrix}.$$

Then $\{\mathbf{x_1}\}$ is a basis for the solution space of $(I_3 - A)\mathbf{x} = \mathbf{0}$.

17. We have

$$\lambda I_2 - A = \lambda \begin{bmatrix} 1 & 0 \\ 0 & 1 \end{bmatrix} - \begin{bmatrix} 2 & 3 \\ 2 & -3 \end{bmatrix} = \begin{bmatrix} \lambda - 2 & -3 \\ -2 & \lambda + 3 \end{bmatrix}.$$

By Exercise 18 in Section 2.2, the corresponding homogeneous system has a nontrivial solution if and only if

$$(\lambda - 2)(\lambda + 3) - 6 = 0 \quad \Longleftrightarrow \quad \lambda^2 + \lambda - 12 = 0 \quad \Longleftrightarrow \quad \lambda = -4 \quad \text{or} \quad \lambda = 3.$$

Thus, when $\lambda = -4$ or $\lambda = 3$ the homogeneous system $(\lambda I_2 - A)\mathbf{x} = \mathbf{0}$ has a nontrivial solution.

19. We have

$$\lambda I_3 - A = \lambda \begin{bmatrix} 1 & 0 & 0 \\ 0 & 1 & 0 \\ 0 & 0 & 1 \end{bmatrix} - \begin{bmatrix} 0 & 0 & 0 \\ 0 & 1 & -1 \\ 1 & 0 & 0 \end{bmatrix} = \begin{bmatrix} \lambda & 0 & 0 \\ 0 & \lambda - 1 & 1 \\ -1 & 0 & \lambda \end{bmatrix}.$$

To determine when the system $(\lambda I_3 - A)\mathbf{x} = \mathbf{0}$ has a nontrivial solution, we form the augmented matrix and row reduce it:

$$\begin{bmatrix} \lambda & 0 & 0 & 0 \\ 0 & \lambda - 1 & 1 & 0 \\ -1 & 0 & \lambda & 0 \end{bmatrix} \longrightarrow \begin{bmatrix} 1 & 0 & -\lambda & 0 \\ 0 & \lambda - 1 & 1 & 0 \\ 0 & 0 & \lambda^2 & 0 \end{bmatrix}$$
$$\text{add } \lambda \text{ (row 3) to row 1}$$
$$\text{interchange rows 1 and 3}$$

If $\lambda = 0$, the matrix has the form

$$\left[\begin{array}{ccc|c} 1 & 0 & 0 & 0 \\ 0 & -1 & 1 & 0 \\ -1 & 0 & 0 & 0 \end{array}\right]$$

which has the nontrivial solutions $x_1 = 0$, $x_2 = r$, $x_3 = r$, where r is any real number. If $\lambda \neq 0$, multiply row 3 by $1/\lambda^2$ to get a leading 1, then use it to zero out the other entries in column 3. If $\lambda \neq 1$, multiplying row 2 by $1/(\lambda - 1)$ gives the matrix $[\ I_3 \ | \ 0\ ]$, which shows that the system has only the trivial solution. If $\lambda = 1$ the matrix has the form

$$\left[\begin{array}{ccc|c} 1 & 0 & -1 & 0 \\ 0 & 0 & 1 & 0 \\ 0 & 0 & 1 & 0 \end{array}\right] \longrightarrow \left[\begin{array}{ccc|c} 1 & 0 & 0 & 0 \\ 0 & 0 & 1 & 0 \\ 0 & 0 & 0 & 0 \end{array}\right]$$

which has nontrivial solutions $x_1 = 0$, $x_2 = 0$, $x_3 = r$, where r is any real number. Thus, the homogeneous system $(\lambda I_3 - A)\mathbf{x} = \mathbf{0}$ has nontrivial solutions when $\lambda = 0$ or $\lambda = 1$.

21. The reduced row echelon form of the augmented matrix

$$\left[\begin{array}{cc|c} 1 & 2 & -2 \\ 2 & 4 & -4 \end{array}\right] \quad \text{is} \quad \left[\begin{array}{cc|c} 1 & 2 & -2 \\ 0 & 0 & 0 \end{array}\right].$$

Therefore, the solution to the linear system $A\mathbf{x} = \mathbf{b}$ is $x = -2 - 2r$, $y = r = 0 + 1r$, where r is any number. If we let $\mathbf{x}_p = \begin{bmatrix} -2 \\ 0 \end{bmatrix}$ and $\mathbf{x}_h = \begin{bmatrix} -2 \\ 1 \end{bmatrix}$, then the solution may be written in the form $\mathbf{x} = \mathbf{x}_p + \mathbf{x}_h$.

23. Let $\mathbf{v} = a_1\mathbf{x}_1 + a_2\mathbf{x}_2 + \cdots + a_k\mathbf{x}_k$ be a vector in span S. Since each of the vectors $\mathbf{x}_i$ is a solution of the system $A\mathbf{x} = \mathbf{0}$, we have $A\mathbf{x}_i = \mathbf{0}$ for $i = 1, 2, \ldots, k$ and hence

$$\begin{aligned} A\mathbf{v} &= A(a_1\mathbf{x}_1 + a_2\mathbf{x}_2 + \cdots + a_k\mathbf{x}_k) \\ &= a_1 A\mathbf{x}_1 + a_2 A\mathbf{x}_2 + \cdots + a_k A\mathbf{x}_k \\ &= a_1(\mathbf{0}) + a_2(\mathbf{0}) + \cdots + a_k(\mathbf{0}) = \mathbf{0} \end{aligned}$$

Thus any vector in span S is a solution to the system $A\mathbf{x} = \mathbf{0}$.

25. (a) Set $A = [\ a_{ij}\ ]$. Since the dimension of the null space of A is 3, the null space of A is R^3. Then the natural basis $\{\mathbf{e}_1, \mathbf{e}_2, \mathbf{e}_3\}$ is a basis for the null space of A. Forming $A\mathbf{e}_1 = \mathbf{0}$, $A\mathbf{e}_2 = \mathbf{0}$, $A\mathbf{e}_3 = \mathbf{0}$, we find that all the columns of A must be zero. Hence, $A = \mathbf{0}$.

   (b) Since $A\mathbf{x} = \mathbf{0}$ has a nontrivial solution, the null space of A contains a nonzero vector, so the dimension of the null space of A is not zero. If this dimension is 3, then by part (a), $A = \mathbf{0}$, a contradiction. Hence, the dimension is either 1 or 2.

Section 4.8, p. 267

1. Since S is the natural basis for R^2, $[\ \mathbf{v}\ ]_S = \begin{bmatrix} 3 \\ -2 \end{bmatrix}$.

3. Set $\mathbf{v} = t + 4 = a_1(t + 1) + a_2(t - 2)$. Equating like powers of t leads to the linear system

$$\begin{aligned} a_1 + a_2 &= 1 \\ a_1 - 2a_2 &= 4 \end{aligned}$$

Subtracting these two equations gives $3a_2 = -3$, and hence $a_2 = -1$ and $a_1 = 2$. Therefore

$$\mathbf{v} = 2(t + 1) + (-1)(t - 2) \quad \Longrightarrow \quad [\ \mathbf{v}\ ]_S = \begin{bmatrix} 2 \\ -1 \end{bmatrix}.$$

5. Since

$$\mathbf{v} = \begin{bmatrix} 1 & 0 \\ -1 & 2 \end{bmatrix} = (1)\begin{bmatrix} 1 & 0 \\ 0 & 0 \end{bmatrix} + (-1)\begin{bmatrix} 0 & 0 \\ 1 & 0 \end{bmatrix} + (0)\begin{bmatrix} 0 & 1 \\ 0 & 0 \end{bmatrix} + (2)\begin{bmatrix} 0 & 0 \\ 0 & 1 \end{bmatrix}$$

it follows that $\begin{bmatrix} \mathbf{v} \end{bmatrix}_S = \begin{bmatrix} 1 \\ -1 \\ 0 \\ 2 \end{bmatrix}.$

7. $\mathbf{v} = (1)\begin{bmatrix} 2 \\ 1 \end{bmatrix} + (2)\begin{bmatrix} -1 \\ 1 \end{bmatrix} = \begin{bmatrix} 0 \\ 3 \end{bmatrix}.$

9. $\mathbf{v} = (-2)t + (3)(2t - 1) = 4t - 3.$

11. $\mathbf{v} = (2)\begin{bmatrix} -1 & 0 \\ 1 & 0 \end{bmatrix} + (1)\begin{bmatrix} 2 & 2 \\ 0 & 1 \end{bmatrix} + (-1)\begin{bmatrix} 1 & 2 \\ -1 & 3 \end{bmatrix} + (3)\begin{bmatrix} 0 & 0 \\ 2 & 3 \end{bmatrix} = \begin{bmatrix} -1 & 0 \\ 9 & 7 \end{bmatrix}.$

13. (a) The reduced row echelon form of the matrix

$$\begin{bmatrix} \mathbf{v}_1 & \mathbf{v}_2 \end{bmatrix} = \begin{bmatrix} 1 & 1 \\ -1 & -2 \end{bmatrix} \quad \text{is} \quad \begin{bmatrix} 1 & 0 \\ 0 & 1 \end{bmatrix}.$$

Therefore S is a basis for R^2.

(b) Let $\mathbf{v} = a\mathbf{v}_1 + b\mathbf{v}_2$. Then

$$\begin{bmatrix} 2 \\ 6 \end{bmatrix} = a\begin{bmatrix} 1 \\ -1 \end{bmatrix} + b\begin{bmatrix} 1 \\ -2 \end{bmatrix} = \begin{bmatrix} a + b \\ -1 - 2b \end{bmatrix}.$$

The reduced row echelon form of the augmented matrix

$$\left[\begin{array}{cc|c} 1 & 1 & 2 \\ -1 & -2 & 6 \end{array}\right] \quad \text{is} \quad \left[\begin{array}{cc|c} 1 & 0 & 10 \\ 0 & 1 & -8 \end{array}\right].$$

Therefore $\begin{bmatrix} \mathbf{v} \end{bmatrix}_S = \begin{bmatrix} 10 \\ -8 \end{bmatrix}.$

(c) $A\mathbf{v}_1 = \lambda_1\mathbf{v}_1 \implies \begin{bmatrix} -0.85 & -0.55 \\ 1.10 & 0.80 \end{bmatrix}\begin{bmatrix} 1 \\ -1 \end{bmatrix} = \lambda_1\begin{bmatrix} 1 \\ -1 \end{bmatrix} \implies \begin{bmatrix} -0.3 \\ 0.3 \end{bmatrix} = \begin{bmatrix} \lambda_1 \\ -\lambda_1 \end{bmatrix} \implies \lambda_1 = -0.3.$

(d) $A\mathbf{v}_2 = \lambda_2\mathbf{v}_2 \implies \begin{bmatrix} -0.85 & -0.55 \\ 1.10 & 0.80 \end{bmatrix}\begin{bmatrix} 1 \\ -2 \end{bmatrix} = \lambda_2\begin{bmatrix} 1 \\ -2 \end{bmatrix} \implies \begin{bmatrix} 0.25 \\ -0.5 \end{bmatrix} = \begin{bmatrix} \lambda_2 \\ -\lambda_2 \end{bmatrix} \implies \lambda_2 = 0.25.$

(e) $A^n\mathbf{v} = A^n(10\mathbf{v}_1 - 8\mathbf{v}_2) = 10A^n\mathbf{v}_1 - 8A^n\mathbf{v}_2 = 10\lambda_1^n\mathbf{v}_1 - 8\lambda_2^n\mathbf{v}_2.$

(f) Since $\lim_{n\to\infty}\lambda_1^n = \lim_{n\to\infty}\lambda_2^n = 0$, $\lim_{n\to\infty} = 10\left(\lim_{n\to\infty}\lambda_1^n\right)\mathbf{v}_1 - 8\left(\lim_{n\to\infty}\lambda_2^n\right)\mathbf{v}_2 = 0.$

15. (a) We first express $\mathbf{v}$ and $\mathbf{w}$ in terms of the vectors in the T-basis by solving the equations

$$a_1\begin{bmatrix} 1 \\ 1 \end{bmatrix} + a_2\begin{bmatrix} 2 \\ 3 \end{bmatrix} = \mathbf{v} = \begin{bmatrix} 1 \\ 5 \end{bmatrix} \quad \text{and} \quad b_1\begin{bmatrix} 1 \\ 1 \end{bmatrix} + b_2\begin{bmatrix} 2 \\ 3 \end{bmatrix} = \mathbf{w} = \begin{bmatrix} 5 \\ 4 \end{bmatrix}.$$

Equating corresponding entries leads to two linear systems whose augmented matrices may be combined into one partitioned matrix and row reduced:

$$\left[\begin{array}{cc|c|c} 1 & 2 & 1 & 5 \\ 1 & 3 & 5 & 4 \end{array}\right] \longrightarrow \left[\begin{array}{cc|c|c} 1 & 2 & 1 & 5 \\ 0 & 1 & 4 & -1 \end{array}\right] \longrightarrow \left[\begin{array}{cc|c|c} 1 & 0 & -7 & 7 \\ 0 & 1 & 4 & -1 \end{array}\right]$$

$\qquad\qquad$ add -1 (row 1) to row 2 $\qquad$ add -2 (row 2) to row 1

Therefore $a_1 = -7$, $a_2 = 4$, and $b_1 = 7$, $b_2 = -1$. Hence

$$\mathbf{v} = \begin{bmatrix} 1 \\ 5 \end{bmatrix} = (-7) \begin{bmatrix} 1 \\ 1 \end{bmatrix} + (4) \begin{bmatrix} 2 \\ 3 \end{bmatrix} \implies [\,\mathbf{v}\,]_T = \begin{bmatrix} -7 \\ 4 \end{bmatrix}$$

$$\mathbf{w} = \begin{bmatrix} 5 \\ 4 \end{bmatrix} = (7) \begin{bmatrix} 1 \\ 1 \end{bmatrix} + (-1) \begin{bmatrix} 2 \\ 3 \end{bmatrix} \implies [\,\mathbf{w}\,]_T = \begin{bmatrix} 7 \\ -1 \end{bmatrix}$$

(b) We first find the coordinates of the vectors in the T-basis with respect to the S-basis. Following the method used in part (a), we must solve the equations

$$a_1 \begin{bmatrix} 1 \\ 2 \end{bmatrix} + a_2 \begin{bmatrix} 0 \\ 1 \end{bmatrix} = \begin{bmatrix} 1 \\ 1 \end{bmatrix} \quad \text{and} \quad b_1 \begin{bmatrix} 1 \\ 2 \end{bmatrix} + b_2 \begin{bmatrix} 0 \\ 1 \end{bmatrix} = \begin{bmatrix} 2 \\ 3 \end{bmatrix}.$$

Form the partitioned matrix of these linear systems and row reduce it:

$$\left[\begin{array}{cc|c|c} 1 & 0 & 1 & 2 \\ 2 & 1 & 1 & 3 \end{array} \right] \longrightarrow \left[\begin{array}{cc|c|c} 1 & 0 & 1 & 2 \\ 0 & 1 & -1 & -1 \end{array} \right]$$

add -2 (row 1) to row 2

Thus the transition matrix from the T-basis to the S-basis is $P_{S \leftarrow T} = \begin{bmatrix} 1 & 2 \\ -1 & -1 \end{bmatrix}$.

(c) We find the coordinates of $\mathbf{v}$ and $\mathbf{w}$ relative to the S-basis by multiplying their coordinates relative to the T-basis by the transition matrix. Using the results from parts (a) and (b) we have

$$[\,\mathbf{v}\,]_S = P_{S \leftarrow T}[\,\mathbf{v}\,]_T = \begin{bmatrix} 1 & 2 \\ -1 & -1 \end{bmatrix} \begin{bmatrix} -7 \\ 4 \end{bmatrix} = \begin{bmatrix} 1 \\ 3 \end{bmatrix}$$

$$[\,\mathbf{w}\,]_S = P_{S \leftarrow T}[\,\mathbf{w}\,]_T = \begin{bmatrix} 1 & 2 \\ -1 & -1 \end{bmatrix} \begin{bmatrix} 7 \\ -1 \end{bmatrix} = \begin{bmatrix} 5 \\ -6 \end{bmatrix}.$$

(d) To find the coordinates of $\mathbf{v}$ and $\mathbf{w}$ with respect to the S-basis directly we proceed as in part (a). Here we find the partitioned matrix

$$\left[\begin{array}{cc|c|c} 1 & 0 & 1 & 5 \\ 2 & 1 & 5 & 4 \end{array} \right] \longrightarrow \left[\begin{array}{cc|c|c} 1 & 0 & 1 & 5 \\ 0 & 1 & 3 & -6 \end{array} \right]$$

add -2 (row 1) to row 2

Hence $[\,\mathbf{v}\,]_S = \begin{bmatrix} 1 \\ 3 \end{bmatrix}$ and $[\,\mathbf{w}\,]_S = \begin{bmatrix} 5 \\ -6 \end{bmatrix}$.

(e) We first find the coordinates of the vectors in the S-basis relative to the T-basis. Hence we must solve the equations

$$a_1 \begin{bmatrix} 1 \\ 1 \end{bmatrix} + a_2 \begin{bmatrix} 2 \\ 3 \end{bmatrix} = \begin{bmatrix} 1 \\ 2 \end{bmatrix} \quad \text{and} \quad b_1 \begin{bmatrix} 1 \\ 1 \end{bmatrix} + b_2 \begin{bmatrix} 2 \\ 3 \end{bmatrix} = \begin{bmatrix} 0 \\ 1 \end{bmatrix}.$$

Combining these systems into the partitioned matrix form as above, we find:

$$\left[\begin{array}{cc|c|c} 1 & 2 & 1 & 0 \\ 1 & 3 & 2 & 1 \end{array} \right] \longrightarrow \left[\begin{array}{cc|c|c} 1 & 2 & 1 & 0 \\ 0 & 1 & 1 & 1 \end{array} \right] \longrightarrow \left[\begin{array}{cc|c|c} 1 & 0 & -1 & -2 \\ 0 & 1 & 1 & 1 \end{array} \right]$$

add -1 (row 1) to row 2   add -2 (row 2) to row 1

Hence $Q_{T \leftarrow S} = \begin{bmatrix} -1 & -2 \\ 1 & 1 \end{bmatrix}$.

(f) From part (d) we have $\left[\, \mathbf{v}\, \right]_S = \begin{bmatrix} 1 \\ 3 \end{bmatrix}$ and $\left[\, \mathbf{w}\, \right]_S = \begin{bmatrix} 5 \\ -6 \end{bmatrix}$. Then,

$$\left[\, \mathbf{v}\, \right]_T = Q_{T \leftarrow S}\left[\, \mathbf{v}\, \right]_S = \begin{bmatrix} -1 & -2 \\ 1 & 1 \end{bmatrix}\begin{bmatrix} 1 \\ 3 \end{bmatrix} = \begin{bmatrix} -7 \\ 4 \end{bmatrix}$$

$$\left[\, \mathbf{w}\, \right]_T = Q_{T \leftarrow S}\left[\, \mathbf{w}\, \right]_S = \begin{bmatrix} -1 & -2 \\ 1 & 1 \end{bmatrix}\begin{bmatrix} 5 \\ -6 \end{bmatrix} = \begin{bmatrix} 7 \\ -1 \end{bmatrix}$$

which agrees with the results obtained in part (a).

17. (a) We first express $\mathbf{v}$ and $\mathbf{w}$ in terms of the vectors in the T-basis by solving the equations

$$a_1(2t^2 + t) + a_2(t^2 + 3) + a_3(t) = 8t^2 - 4t + 6$$
$$b_1(2t^2 + t) + b_2(t^2 + 3) + b_3(t) = 7t^2 - t + 9$$

Combining terms and equating like powers of t, we are lead to a system of linear equations. We then form the partitioned matrix of these systems and row reduce it as follows:

$$\left[\begin{array}{ccc|c|c} 2 & 1 & 0 & 8 & 7 \\ 1 & 0 & 1 & -4 & -1 \\ 0 & 3 & 0 & 6 & 9 \end{array}\right] \longrightarrow \left[\begin{array}{ccc|c|c} 1 & 0 & 1 & -4 & -1 \\ 0 & 1 & -2 & 16 & 9 \\ 0 & 3 & 0 & 6 & 9 \end{array}\right] \longrightarrow \left[\begin{array}{ccc|c|c} 1 & 0 & 1 & -4 & -1 \\ 0 & 1 & -2 & 16 & 9 \\ 0 & 0 & 6 & -42 & -18 \end{array}\right] \longrightarrow$$

interchange rows 1 and 2   add -3 (row 2) to row 3 $\frac{1}{6}$ (row 3)
add -2 (row 1) to row 2   add 2 (row 3) to row 2
                                                                   add -1 (row 3) to row 1

$$\left[\begin{array}{ccc|c|c} 1 & 0 & 0 & 3 & 2 \\ 0 & 1 & 0 & 2 & 3 \\ 0 & 0 & 1 & -7 & -3 \end{array}\right]$$

Thus
$$\left[\, \mathbf{v}\, \right]_T = \begin{bmatrix} 3 \\ 2 \\ -7 \end{bmatrix} \quad \text{and} \quad \left[\, \mathbf{w}\, \right]_T = \begin{bmatrix} 2 \\ 3 \\ -3 \end{bmatrix}.$$

(b) We first find the coordinates of the vectors in the T-basis with respect to the S-basis. Hence we must solve three equations:

$$a_1(t^2 + 1) + a_2(t - 2) + a_3(t + 3) = 2t^2 + t$$
$$b_1(t^2 + 1) + b_2(t - 2) + b_3(t + 3) = t^2 + 3$$
$$c_1(t^2 + 1) + c_2(t - 2) + c_3(t + 3) = t$$

Combine terms on the left and equate like powers of t. This leads to three systems of linear equations. We row reduce the partitioned augmented matrix:

$$\left[\begin{array}{ccc|c|c|c} 1 & 0 & 0 & 2 & 1 & 0 \\ 0 & 1 & 1 & 1 & 0 & 1 \\ 1 & -2 & 3 & 0 & 3 & 0 \end{array}\right] \longrightarrow \left[\begin{array}{ccc|c|c|c} 1 & 0 & 0 & 2 & 1 & 0 \\ 0 & 1 & 1 & 1 & 0 & 1 \\ 0 & -2 & 3 & -2 & 2 & 0 \end{array}\right] \longrightarrow \left[\begin{array}{ccc|c|c|c} 1 & 0 & 0 & 2 & 1 & 0 \\ 0 & 1 & 1 & 1 & 0 & 1 \\ 0 & 0 & 5 & 0 & 2 & 2 \end{array}\right] \longrightarrow$$

add -1 (row 1) to row 3   add 2 (row 2) to row 3 $\frac{1}{5}$ (row 3)
                                                                 add -1 (row 3) to row 2

$$\left[\begin{array}{ccc|c|c|c} 1 & 0 & 0 & 2 & 1 & 0 \\ 0 & 1 & 0 & 1 & -\frac{2}{5} & \frac{3}{5} \\ 0 & 0 & 1 & 0 & \frac{2}{5} & \frac{2}{5} \end{array}\right]$$

Thus
$$P_{S \leftarrow T} = \begin{bmatrix} 2 & 1 & 0 \\ 1 & -\frac{2}{5} & \frac{3}{5} \\ 0 & \frac{2}{5} & \frac{2}{5} \end{bmatrix}.$$

(c) $[\mathbf{v}]_S = P_{S \leftarrow T}[\mathbf{v}]_T = \begin{bmatrix} 2 & 1 & 0 \\ 1 & -\frac{2}{5} & \frac{3}{5} \\ 0 & \frac{2}{5} & \frac{2}{5} \end{bmatrix} \begin{bmatrix} 3 \\ 2 \\ -7 \end{bmatrix} = \begin{bmatrix} 8 \\ -2 \\ -2 \end{bmatrix};$

$[\mathbf{w}]_S = P_{S \leftarrow T}[\mathbf{w}]_T = \begin{bmatrix} 2 & 1 & 0 \\ 1 & -\frac{2}{5} & \frac{3}{5} \\ 0 & \frac{2}{5} & \frac{2}{5} \end{bmatrix} \begin{bmatrix} 2 \\ 3 \\ -3 \end{bmatrix} = \begin{bmatrix} 7 \\ -1 \\ 0 \end{bmatrix}$

(d) We find the coordinates using the partitioned matrix approach.

$$\begin{bmatrix} 1 & 0 & 0 & 8 & 7 \\ 0 & 1 & 1 & -4 & -1 \\ 1 & -2 & 3 & 6 & 9 \end{bmatrix} \longrightarrow \begin{bmatrix} 1 & 0 & 0 & 8 & 7 \\ 0 & 1 & 1 & -4 & -1 \\ 0 & -2 & 3 & -2 & 2 \end{bmatrix} \longrightarrow \begin{bmatrix} 1 & 0 & 0 & 8 & 7 \\ 0 & 1 & 1 & -4 & -1 \\ 0 & 0 & 5 & -10 & 0 \end{bmatrix} \longrightarrow$$

add -1 (row 1) to row 3 $\qquad$ add 2 (row 2) to row 3 $\qquad$ $\frac{1}{5}$ (row 3)

add -1 (row 3) to row 2

$$\begin{bmatrix} 1 & 0 & 0 & 8 & 7 \\ 0 & 1 & 0 & -2 & -1 \\ 0 & 0 & 1 & -2 & 0 \end{bmatrix}$$

Thus $[\mathbf{v}]_S = \begin{bmatrix} 8 \\ -2 \\ -2 \end{bmatrix}$ and $[\mathbf{w}]_S = \begin{bmatrix} 7 \\ -1 \\ 0 \end{bmatrix}$.

(e) We find the coordinates of the vectors in the S-basis relative to the T-basis. This leads to the equations

$$a_1(2t^2 + t) + a_2(t^2 + 3) + a_3(t) = t^2 + 1$$
$$b_1(2t^2 + t) + b_2(t^2 + 3) + b_3(t) = t - 2$$
$$c_1(2t^2 + t) + c_2(t^2 + 3) + c_3(t) = t + 3$$

Combining terms and equating like powers of t, we are lead to three systems of linear equations. We form the partitioned matrix of these systems and row reduce it as follows:

$$\begin{bmatrix} 2 & 1 & 0 & 1 & 0 & 0 \\ 1 & 0 & 1 & 0 & 1 & 1 \\ 0 & 3 & 0 & 1 & -2 & 3 \end{bmatrix} \longrightarrow \begin{bmatrix} 1 & 0 & 1 & 0 & 1 & 1 \\ 0 & 1 & -2 & 1 & -2 & -2 \\ 0 & 3 & 0 & 1 & -2 & 3 \end{bmatrix} \longrightarrow \begin{bmatrix} 1 & 0 & 1 & 0 & 1 & 1 \\ 0 & 1 & -2 & 1 & -2 & -2 \\ 0 & 0 & 6 & -2 & 4 & 9 \end{bmatrix} \longrightarrow$$

interchange rows 1 and 2 $\qquad$ add -3 (row 2) to row 3 $\qquad$ $\frac{1}{6}$ (row 3)

add -2 (row 1) to row 2 $\qquad\qquad\qquad\qquad\qquad\qquad\qquad$ add 2 (row 3) to row 2

add -1 (row 3) to row 1

$$\begin{bmatrix} 1 & 0 & 0 & \frac{1}{3} & \frac{1}{3} & -\frac{1}{2} \\ 0 & 1 & 0 & \frac{1}{3} & -\frac{2}{3} & 1 \\ 0 & 0 & 1 & -\frac{1}{3} & \frac{2}{3} & \frac{3}{2} \end{bmatrix}$$

Thus $Q_{T \leftarrow S} = \begin{bmatrix} \frac{1}{3} & \frac{1}{3} & -\frac{1}{2} \\ \frac{1}{3} & -\frac{2}{3} & 1 \\ -\frac{1}{3} & \frac{2}{3} & \frac{3}{2} \end{bmatrix}$.

(f) $[\mathbf{v}]_T = Q_{T \leftarrow S}[\mathbf{v}]_S = \begin{bmatrix} \frac{1}{3} & \frac{1}{3} & -\frac{1}{2} \\ \frac{1}{3} & -\frac{2}{3} & 1 \\ -\frac{1}{3} & \frac{2}{3} & \frac{3}{2} \end{bmatrix} \begin{bmatrix} 8 \\ -2 \\ -2 \end{bmatrix} = \begin{bmatrix} 3 \\ 2 \\ -7 \end{bmatrix}$

$[\mathbf{w}]_T = Q_{T \leftarrow S}[\mathbf{w}]_S = \begin{bmatrix} \frac{1}{3} & \frac{1}{3} & -\frac{1}{2} \\ \frac{1}{3} & -\frac{2}{3} & 1 \\ -\frac{1}{3} & \frac{2}{3} & \frac{3}{2} \end{bmatrix} \begin{bmatrix} 7 \\ -1 \\ 0 \end{bmatrix} = \begin{bmatrix} 2 \\ 3 \\ -3 \end{bmatrix}$

19. Let

$$S = \{s_1, s_2, s_3, s_4\} = \left\{ \begin{bmatrix} 1 & 0 \\ 0 & 0 \end{bmatrix}, \begin{bmatrix} 0 & 1 \\ 1 & 0 \end{bmatrix}, \begin{bmatrix} 0 & 2 \\ 0 & 1 \end{bmatrix}, \begin{bmatrix} 0 & 0 \\ 1 & 1 \end{bmatrix} \right\}$$

$$T = \{t_1, t_2, t_3, t_4\} = \left\{ \begin{bmatrix} 1 & 1 \\ 0 & 0 \end{bmatrix}, \begin{bmatrix} 0 & 0 \\ 1 & 0 \end{bmatrix}, \begin{bmatrix} 0 & 0 \\ 0 & 1 \end{bmatrix}, \begin{bmatrix} 1 & 0 \\ 0 & 0 \end{bmatrix} \right\}$$

(a) We express v and w in terms of the vectors in the T-basis. The expression

$$a_1 t_1 + a_2 t_2 + a_3 t_3 + a_4 t_4 = v$$

leads to the matrix equation

$$\begin{bmatrix} a_1 + a_4 & a_1 \\ a_2 & a_3 \end{bmatrix} = \begin{bmatrix} 1 & 1 \\ 1 & 1 \end{bmatrix}.$$

Equating corresponding entries gives a linear system. Row reducing the augmented matrix of this system, we obtain

$$\begin{bmatrix} 1 & 0 & 0 & 1 & | & 1 \\ 1 & 0 & 0 & 0 & | & 1 \\ 0 & 1 & 0 & 0 & | & 1 \\ 0 & 0 & 1 & 0 & | & 1 \end{bmatrix} \longrightarrow \begin{bmatrix} 1 & 0 & 0 & 1 & | & 1 \\ 0 & 0 & 0 & -1 & | & 0 \\ 0 & 1 & 0 & 0 & | & 1 \\ 0 & 0 & 1 & 0 & | & 1 \end{bmatrix}$$

add -1 (row 1) to row 2

It follows that $a_1 = a_2 = a_3 = 1$ and $a_4 = 0$. Thus

$$[\, v \,]_T = \begin{bmatrix} 1 \\ 1 \\ 1 \\ 0 \end{bmatrix}.$$

To determine $[\, w \,]_T$ we replace v in the previous steps by w. This leads to the matrix equation

$$\begin{bmatrix} a_1 + a_4 & a_1 \\ a_2 & a_3 \end{bmatrix} = \begin{bmatrix} 1 & 2 \\ -2 & 1 \end{bmatrix}.$$

Equating corresponding entries gives the augmented matrix

$$\begin{bmatrix} 1 & 0 & 0 & 1 & | & 1 \\ 1 & 0 & 0 & 0 & | & 2 \\ 0 & 1 & 0 & 0 & | & -2 \\ 0 & 0 & 1 & 0 & | & 1 \end{bmatrix}$$

Row reducing this matrix gives $a_1 = 2$, $a_2 = -2$, $a_3 = 1$, $a_4 = -1$. Thus

$$[\, w \,]_T = \begin{bmatrix} 2 \\ -2 \\ 1 \\ -1 \end{bmatrix}.$$

(b) We first find the coordinates of the vectors in the T-basis with respect to the S-basis. Hence we must solve the equations

$$a_1 s_1 + a_2 s_2 + a_3 s_3 + a_4 s_4 = t_1 \mid t_2 \mid t_3 \mid t_4$$

That is, there are really four equations to be solved. The right-hand side is the only thing that changes in each of the corresponding systems. Using the ideas from Exercise 13, we substitute in the vectors from M_{22}, equate corresponding components and obtain a set of linear systems with the same coefficient matrix but with different right-hand sides. We may express this as in Exercise 13 using a partitioned matrix. We obtain the following matrix which we row reduce:

$$
\left[\begin{array}{cccc|c|c|c|c}
1 & 0 & 0 & 0 & 1 & 0 & 0 & 1 \\
0 & 1 & 2 & 0 & 1 & 0 & 0 & 0 \\
0 & 1 & 0 & 1 & 0 & 1 & 0 & 0 \\
0 & 0 & 1 & 1 & 0 & 0 & 1 & 0
\end{array}\right]
\rightarrow
\left[\begin{array}{cccc|c|c|c|c}
1 & 0 & 0 & 0 & 1 & 0 & 0 & 1 \\
0 & 1 & 2 & 0 & 1 & 0 & 0 & 0 \\
0 & 0 & -2 & 1 & -1 & 1 & 0 & 0 \\
0 & 0 & 1 & 1 & 0 & 0 & 1 & 0
\end{array}\right]
\rightarrow
\left[\begin{array}{cccc|c|c|c|c}
1 & 0 & 0 & 0 & 1 & 0 & 0 & 1 \\
0 & 1 & 2 & 0 & 1 & 0 & 0 & 0 \\
0 & 0 & 1 & 1 & 0 & 0 & 1 & 0 \\
0 & 0 & -2 & 1 & -1 & 1 & 0 & 0
\end{array}\right]
\rightarrow
$$

add -1 (row 2) to row 3   interchange rows 3 and 4   add -2 (row 3) to row 2

                                                                 add 2 (row 3) to row 4

$$
\left[\begin{array}{cccc|c|c|c|c}
1 & 0 & 0 & 0 & 1 & 0 & 0 & 1 \\
0 & 1 & 0 & -2 & 1 & 0 & -2 & 0 \\
0 & 0 & 1 & 1 & 0 & 0 & 1 & 0 \\
0 & 0 & 0 & 3 & -1 & 1 & 2 & 0
\end{array}\right]
\rightarrow
\left[\begin{array}{cccc|c|c|c|c}
1 & 0 & 0 & 0 & 1 & 0 & 0 & 1 \\
0 & 1 & 0 & 0 & \frac{1}{3} & \frac{2}{3} & -\frac{2}{3} & 0 \\
0 & 0 & 1 & 1 & \frac{1}{3} & -\frac{1}{3} & \frac{1}{3} & 0 \\
0 & 0 & 0 & 1 & -\frac{1}{3} & \frac{1}{3} & \frac{2}{3} & 0
\end{array}\right]
$$

add $-\frac{1}{3}$ (row 4) to row 3

add $\frac{2}{3}$ (row 4) to row 2

$\frac{1}{3}$ (row 4)

The coordinates of $\mathbf{t}_j$, $j = 1, 2, 3, 4$ are respectively the last four columns of the preceding matrix. Hence the transition matrix $P_{S \leftarrow T}$ from the T-basis to the S-basis is

$$
P_{S \leftarrow T} = \left[\begin{array}{cccc}
1 & 0 & 0 & 1 \\
\frac{1}{3} & \frac{2}{3} & -\frac{2}{3} & 0 \\
\frac{1}{3} & -\frac{1}{3} & \frac{1}{3} & 0 \\
-\frac{1}{3} & \frac{1}{3} & \frac{2}{3} & 0
\end{array}\right].
$$

(c) We find the coordinates of $\mathbf{v}$ and $\mathbf{w}$ relative to the S-basis by multiplying their coordinates relative to the T-basis by the transition matrix. Using the results from parts (a) and (b) we have

$$
[\mathbf{v}]_S = P_{S \leftarrow T} [\mathbf{v}]_T = \left[\begin{array}{cccc}
1 & 0 & 0 & 1 \\
\frac{1}{3} & \frac{2}{3} & -\frac{2}{3} & 0 \\
\frac{1}{3} & -\frac{1}{3} & \frac{1}{3} & 0 \\
-\frac{1}{3} & \frac{1}{3} & \frac{2}{3} & 0
\end{array}\right]
\left[\begin{array}{c} 1 \\ 1 \\ 1 \\ 0 \end{array}\right]
= \left[\begin{array}{c} 1 \\ \frac{1}{3} \\ \frac{1}{3} \\ \frac{2}{3} \end{array}\right]
$$

$$
[\mathbf{w}]_S = P_{S \leftarrow T} [\mathbf{w}]_T = \left[\begin{array}{cccc}
1 & 0 & 0 & 1 \\
\frac{1}{3} & \frac{2}{3} & -\frac{2}{3} & 0 \\
\frac{1}{3} & -\frac{1}{3} & \frac{1}{3} & 0 \\
-\frac{1}{3} & \frac{1}{3} & \frac{2}{3} & 0
\end{array}\right]
\left[\begin{array}{c} 2 \\ -2 \\ 1 \\ -1 \end{array}\right]
= \left[\begin{array}{c} 1 \\ -\frac{4}{3} \\ \frac{5}{3} \\ -\frac{2}{3} \end{array}\right]
$$

(d) To find the coordinate vectors of $\mathbf{v}$ and $\mathbf{w}$ with respect to the S-basis directly we proceed as in part (a). Here we find the coordinates using the partitioned matrix approach. From the equation

$$
a_1 \mathbf{s}_1 + a_2 \mathbf{s}_2 + a_3 \mathbf{s}_3 + a_4 \mathbf{s}_4 = \mathbf{v} \mid \mathbf{w}
$$

we obtain linear systems whose augmented matrices written in partitioned form are represented by

$$
\left[\begin{array}{cccc|c|c}
1 & 0 & 0 & 0 & 1 & 1 \\
0 & 1 & 2 & 0 & 1 & 2 \\
0 & 1 & 0 & 1 & 1 & -2 \\
0 & 0 & 1 & 1 & 1 & 1
\end{array}\right]
$$

Performing the same row operations as in part (b) we obtain the reduced echelon form

$$
\left[\begin{array}{cccc|c|c}
1 & 0 & 0 & 0 & 1 & 1 \\
0 & 1 & 0 & 0 & \frac{1}{3} & -\frac{4}{3} \\
0 & 0 & 1 & 0 & \frac{1}{3} & \frac{5}{3} \\
0 & 0 & 0 & 1 & \frac{2}{3} & -\frac{2}{3}
\end{array}\right]
$$

Thus $[\mathbf{v}]_S = \begin{bmatrix} 1 \\ \frac{1}{3} \\ \frac{1}{3} \\ \frac{2}{3} \\ \frac{2}{3} \end{bmatrix}$ and $[\mathbf{w}]_S = \begin{bmatrix} 1 \\ -\frac{4}{3} \\ \frac{5}{3} \\ -\frac{2}{3} \end{bmatrix}$.

(e) We could proceed directly as in part (b) reversing the roles of the S and T bases. However, if we call the transition matrix from the S-basis to the T-basis $Q_{T \leftarrow S}$, then we have that $Q_{T \leftarrow S} = P_{S \leftarrow T}^{-1}$. We form the partitioned matrix $[\, P_{S \leftarrow T} \ \mid \ I_4 \,]$ and obtain its reduced echelon form, which is $[\, I_4 \ \mid \ P_{S \leftarrow T}^{-1} \,]$. The result is

$$Q_{T \leftarrow S} = P_{S \leftarrow T}^{-1} = \begin{bmatrix} 0 & 1 & 2 & 0 \\ 0 & 1 & 0 & 1 \\ 0 & 0 & 1 & 1 \\ 1 & -1 & -2 & 0 \end{bmatrix}.$$

(f) From parts (d) and (e) we have that

$$[\mathbf{v}]_T = Q_{T \leftarrow S}[\mathbf{v}]_S = \begin{bmatrix} 1 \\ 1 \\ 1 \\ 0 \end{bmatrix} \quad \text{and} \quad [\mathbf{w}]_T = Q_{T \leftarrow S}[\mathbf{w}]_S = \begin{bmatrix} 2 \\ -2 \\ 1 \\ -1 \end{bmatrix}.$$

21. Since $[\mathbf{v}]_T = \begin{bmatrix} -1 \\ 3 \end{bmatrix}$, it follows that

$$\mathbf{v} = (-1)(t-5) + (3)(t-2) = 2t - 1 = (t+1) + (t-2).$$

Therefore $[\mathbf{v}]_S = \begin{bmatrix} 1 \\ 1 \end{bmatrix}$.

23. Since $[\mathbf{v}]_T = \begin{bmatrix} 1 \\ 2 \\ 3 \end{bmatrix}$, it follows that

$$\mathbf{v} = (1)(t^2) + (2)(t-1) + (3)(1) = t^2 + 2t + 1 = (1)(t^2 + t + 1) + (1)(t+1) + (-1)(1).$$

Therefore $[\mathbf{v}]_S = \begin{bmatrix} 1 \\ 1 \\ -1 \end{bmatrix}$.

25. For any vector $\mathbf{v}$, $[\mathbf{v}]_T = P_{T \leftarrow S}[\mathbf{v}]_S$. Therefore

$$[\mathbf{v}_1]_T = P_{S \leftarrow T}[\mathbf{v}_1]_S = \begin{bmatrix} 2 & 3 \\ -1 & 2 \end{bmatrix}\begin{bmatrix} 1 \\ 0 \end{bmatrix} = \begin{bmatrix} 2 \\ -1 \end{bmatrix}$$
$$\Longrightarrow \mathbf{v}_1 = 2\mathbf{w}_1 - \mathbf{w}_2 = 2(t) - (t-1) = t+1$$

$$[\mathbf{v}_2]_T = P_{S \leftarrow T}[\mathbf{v}_2]_S = \begin{bmatrix} 2 & 3 \\ -1 & 2 \end{bmatrix}\begin{bmatrix} 0 \\ 1 \end{bmatrix} = \begin{bmatrix} 3 \\ 2 \end{bmatrix}$$
$$\Longrightarrow \mathbf{v}_2 = 3\mathbf{w}_1 + 2\mathbf{w}_2 = 3(t) + 2(t-1) = 5t - 2.$$

Hence $S = \{\mathbf{v}_1, \mathbf{v}_2\} = \{t+1, 5t-2\}$.

27. For any vector $\mathbf{v}$, $[\mathbf{v}]_S = P_{S \leftarrow T}[\mathbf{v}]_T$. Therefore $[\mathbf{v}]_T = P_{S \leftarrow T}^{-1}[\mathbf{v}]_S$. To find $P_{S \leftarrow T}^{-1}$ we form the augmented matrix $[\, P_{S \leftarrow T} \ \mid \ I_2 \,]$ and row reduce it to obtain

$$\begin{bmatrix} 1 & 0 & | & -3 & 2 \\ 0 & 1 & | & 2 & -1 \end{bmatrix} \implies P_{S \leftarrow T}^{-1} = \begin{bmatrix} -3 & 2 \\ 2 & -1 \end{bmatrix}.$$

Thus

$$\left[\,\mathbf{v}_1\,\right]_T = P_{S \leftarrow T}^{-1}\left[\,\mathbf{v}_1\,\right]_S = \begin{bmatrix} -3 & 2 \\ 2 & -1 \end{bmatrix}\begin{bmatrix} 1 \\ 0 \end{bmatrix} = \begin{bmatrix} -3 \\ 2 \end{bmatrix}$$

$$\implies \mathbf{v}_1 = -3\mathbf{w}_1 + 2\mathbf{w}_2 = -3(t-1) + 2(t+1) = -t + 5$$

$$\left[\,\mathbf{v}_2\,\right]_T = P_{S \leftarrow T}^{-1}\left[\,\mathbf{v}_2\,\right]_S = \begin{bmatrix} -3 & 2 \\ 2 & -1 \end{bmatrix}\begin{bmatrix} 0 \\ 1 \end{bmatrix} = \begin{bmatrix} 2 \\ -1 \end{bmatrix}$$

$$\implies \mathbf{v}_2 = 2\mathbf{w}_1 + (-1)\mathbf{w}_2 = 2(t-1) - (t+1) = t - 3$$

Hence $S = \{\mathbf{v}_1, \mathbf{v}_2\} = \{-t + 5, t - 3\}$.

29. (a) Since $\mathbf{0}_V = \mathbf{0}_V + \mathbf{0}_V$, we have that $L(\mathbf{0}_V) = L(\mathbf{0}_V + \mathbf{0}_V) = L(\mathbf{0}_V) + L(\mathbf{0}_V)$. Thus $L(\mathbf{0}_V) = \mathbf{0}_W$.

(b) $L(\mathbf{v} - \mathbf{w}) = L(\mathbf{v} + (-1)\mathbf{w}) = L(\mathbf{v}) + L((-1)\mathbf{w}) = L(\mathbf{v}) + (-1)L(\mathbf{w}) = L(\mathbf{v}) - L(\mathbf{w})$.

(c) We use induction to prove this statement. If $k = 1$, then $L(a_1\mathbf{v}_1) = a_1 L(\mathbf{v}_1)$. Now assume the statement is true for $k = n$; that is, assume that

$$L(a_1\mathbf{v}_1 + \cdots + a_n\mathbf{v}_n) = a_1 L(\mathbf{v}_1) + \cdots + a_n L(\mathbf{v}_n).$$

Then for $k = n + 1$ we have

$$L(a_1\mathbf{v}_1 + \cdots + a_{n+1}\mathbf{v}_{n+1}) = L((a_1\mathbf{v}_1 + \cdots + a_n\mathbf{v}_n) + a_{n+1}\mathbf{v}_{n+1})$$
$$= L(a_1\mathbf{v}_1 + \cdots + a_n\mathbf{v}_n) + L(a_{n+1}\mathbf{v}_{n+1})$$
$$= a_1 L(\mathbf{v}_1) + \cdots + a_n L(\mathbf{v}_n) + a_{n+1}L(\mathbf{v}_n)$$

Therefore the statement is true for all positive integers k by mathematical induction.

31. Define $L : R_n \to R^n$ by letting $L(\mathbf{v}) = \mathbf{v}^T$ for all vectors $\mathbf{v}$ in R_n. Then L is one-to-one since

$$L(\mathbf{v}) = L(\mathbf{w}) \implies \mathbf{v}^T = \mathbf{w}^T \implies (\mathbf{v}^T)^T = (\mathbf{w}^T)^T \implies \mathbf{v} = \mathbf{w}.$$

L is an onto mapping since, for any vector $\mathbf{w}$ in R^n, $L(\mathbf{w}^T) = (\mathbf{w}^T)^T = \mathbf{w}$. Finally, for any vectors $\mathbf{v}$, $\mathbf{w}$, and any scalar c, we have

$$L(\mathbf{v} + \mathbf{w}) = (\mathbf{v} + \mathbf{w})^T = \mathbf{v}^T + \mathbf{w}^T = L(\mathbf{v}) + L(\mathbf{w})$$
$$L(c\mathbf{v}) = (c\mathbf{v})^T = c\mathbf{v}^T = L(\mathbf{v})$$

Therefore L is an isomorphism from R_n to R^n.

33. (a) Define the function $L : M_{22} \to R^4$ by

$$L\left(\begin{bmatrix} a & b \\ c & d \end{bmatrix}\right) = \begin{bmatrix} a \\ b \\ c \\ d \end{bmatrix} \quad \text{for any matrix } \begin{bmatrix} a & b \\ c & d \end{bmatrix} \text{ in } M_{22}.$$

To show that L is one-to-one, onto, and satisfies the conditions of Definition 4.13, let

$$\mathbf{v} = \begin{bmatrix} a & b \\ c & d \end{bmatrix} \quad \text{and} \quad \mathbf{w} = \begin{bmatrix} a' & b' \\ c' & d' \end{bmatrix}$$

Then L is one-to-one since

$$L(\mathbf{v}) = L(\mathbf{w}) \implies \begin{bmatrix} a \\ b \\ c \\ d \end{bmatrix} = \begin{bmatrix} a' \\ b' \\ c' \\ d' \end{bmatrix} \implies \begin{matrix} a = a' \\ b = b' \\ c = c' \\ d = d' \end{matrix} \implies \mathbf{v} = \mathbf{w}.$$

L is onto, for if $\begin{bmatrix} a & b & c & d \end{bmatrix}^T$ is any matrix in R^4, then

$$L(\mathbf{x}) = \begin{bmatrix} a \\ b \\ c \\ d \end{bmatrix}, \quad \text{where } \mathbf{x} = \begin{bmatrix} a & b \\ c & d \end{bmatrix}.$$

Finally,

$$L(\mathbf{v} + \mathbf{w}) = L\left(\begin{bmatrix} a + a' & b + b' \\ c + c' & d + d' \end{bmatrix}\right) = \begin{bmatrix} a + a' \\ b + b' \\ c + c' \\ d + d' \end{bmatrix} = \begin{bmatrix} a \\ b \\ c \\ d \end{bmatrix} + \begin{bmatrix} a' \\ b' \\ c' \\ d' \end{bmatrix} = L(\mathbf{v}) + L(\mathbf{w})$$

$$L(k\mathbf{v}) = L\left(\begin{bmatrix} ka & kb \\ kc & kd \end{bmatrix}\right) = \begin{bmatrix} ka \\ kb \\ kc \\ kd \end{bmatrix} = k\begin{bmatrix} a \\ b \\ c \\ d \end{bmatrix} = kL(\mathbf{v}), \quad \text{for any scalar } k.$$

Therefore L is an isomorphism.

(b) By Corollary 3.6, $\dim M_{22} = \dim R^4 = 4$.

35. From Exercise 18 in Section 4.6, $V = \text{span } S$ has a basis $\{\cos^2 t, \sin^2 t\}$ and hence $\dim V = 2$. It follows from Theorem 4.14 that V is isomorphic to R_2.

37. Let $\mathbf{v} = \mathbf{w}$. The coordinates of a vector relative to basis S are the coefficients used to expressed the vector in terms of the members of S. A vector has a unique expression in terms of the vectors of a basis, hence it follows that $\begin{bmatrix} \mathbf{v} \end{bmatrix}_S$ must equal $\begin{bmatrix} \mathbf{w} \end{bmatrix}_S$. Conversely, suppose that

$$\begin{bmatrix} \mathbf{v} \end{bmatrix}_S = \begin{bmatrix} \mathbf{w} \end{bmatrix}_S = \begin{bmatrix} a_1 \\ a_2 \\ \vdots \\ a_n \end{bmatrix}.$$

Then

$$\mathbf{v} = a_1\mathbf{v}_1 + a_2\mathbf{v}_2 + \cdots + a_n\mathbf{v}_n$$
$$\mathbf{w} = a_1\mathbf{v}_1 + a_2\mathbf{v}_2 + \cdots + a_n\mathbf{v}_n$$

Hence $\mathbf{v} = \mathbf{w}$.

39. Consider the homogeneous system

$$M_S\mathbf{x} = \mathbf{0}, \quad \text{where } \mathbf{x} = \begin{bmatrix} a_1 \\ a_2 \\ \vdots \\ a_n \end{bmatrix}.$$

This system can be written in terms of the columns of M_S as

$$a_1\mathbf{v}_1 + a_2\mathbf{v}_2 + \cdots + a_n\mathbf{v}_n = \mathbf{0},$$

where $\mathbf{v}_j$ is the jth column of M_S. Since $\mathbf{v}_1, \mathbf{v}_2, \ldots, \mathbf{v}_n$ are linearly independent, we have $a_1 = a_2 = \cdots = a_n = 0$. Thus, $\mathbf{x} = \mathbf{0}$ is the only solution to $M_S\mathbf{x} = \mathbf{0}$, so by Theorem 2.9 we conclude that M_S is nonsingular. In the same way it follows that M_T is nonsingular.

41. (a) From Exercise 40 we have $M_S\begin{bmatrix} \mathbf{v} \end{bmatrix}_S = M_T\begin{bmatrix} \mathbf{v} \end{bmatrix}_T$. From Exercise 39 we know that M_S is nonsingular so $\begin{bmatrix} \mathbf{v} \end{bmatrix}_S = M_S^{-1}M_T\begin{bmatrix} \mathbf{v} \end{bmatrix}_T$. By Equation (3), $\begin{bmatrix} \mathbf{v} \end{bmatrix}_S = P_{S\leftarrow T}\begin{bmatrix} \mathbf{v} \end{bmatrix}_T$. Therefore $P_{S\leftarrow T} = M_S^{-1}M_T$.

(b) $P_{S\leftarrow T}$ is nonsingular since it is a product of two nonsingular matrices.

(c) From Example 4 we have

$$M_S = \begin{bmatrix} 2 & 1 & 1 \\ 0 & 2 & 1 \\ 1 & 0 & 1 \end{bmatrix} \quad \text{and} \quad M_T = \begin{bmatrix} 6 & 4 & 5 \\ 3 & -1 & 5 \\ 3 & 3 & 2 \end{bmatrix}.$$

To find M_S^{-1}, form the partitioned matrix $\begin{bmatrix} M_S & | & I_3 \end{bmatrix}$ and row reduce it to get $\begin{bmatrix} I_3 & | & M_S^{-1} \end{bmatrix}$. We find

$$M_S^{-1} = \begin{bmatrix} \frac{2}{3} & -\frac{1}{3} & -\frac{1}{3} \\ \frac{1}{3} & \frac{1}{3} & -\frac{2}{3} \\ -\frac{2}{3} & \frac{1}{3} & \frac{4}{3} \end{bmatrix} = \frac{1}{3}\begin{bmatrix} 2 & -1 & -1 \\ 1 & 1 & -2 \\ -2 & 1 & 4 \end{bmatrix}.$$

Then

$$M_S^{-1}M_T = \frac{1}{3}\begin{bmatrix} 2 & -1 & -1 \\ 1 & 1 & -2 \\ -2 & 1 & 4 \end{bmatrix}\begin{bmatrix} 6 & 4 & 5 \\ 3 & -1 & 5 \\ 3 & 3 & 2 \end{bmatrix} = \begin{bmatrix} 2 & 2 & 1 \\ 1 & -1 & 2 \\ 1 & 1 & 1 \end{bmatrix} = P_{S\leftarrow T}.$$

43. From Exercise 42 we know that $T = \{\begin{bmatrix} \mathbf{v}_1 \end{bmatrix}_S, \begin{bmatrix} \mathbf{v}_2 \end{bmatrix}_S, \dots, \begin{bmatrix} \mathbf{v}_n \end{bmatrix}_S\}$ is a linearly independent set of vectors in R^n. By Theorem 4.12, T spans R^n and is thus a basis for R^n.

Section 4.9, p. 282

1. The subspace spanned by S is the column space of the matrix

$$A = \begin{bmatrix} 1 & 2 & -1 & 0 & 1 \\ 2 & 1 & -1 & 1 & 1 \\ 3 & 4 & 2 & 2 & 1 \end{bmatrix}.$$

This subspace is isomorphic to the row space of A^T. Thus we first row reduce A^T:

$$A^T = \begin{bmatrix} 1 & 2 & 3 \\ 2 & 1 & 4 \\ -1 & -1 & 2 \\ 0 & 1 & 2 \\ 1 & 1 & 1 \end{bmatrix} \longrightarrow \begin{bmatrix} 1 & 2 & 3 \\ 0 & -3 & -2 \\ 0 & 1 & 5 \\ 0 & 1 & 2 \\ 0 & -1 & -2 \end{bmatrix} \longrightarrow \begin{bmatrix} 1 & 2 & 3 \\ 0 & 1 & 5 \\ 0 & -3 & -2 \\ 0 & 1 & 2 \\ 0 & 0 & 0 \end{bmatrix} \longrightarrow$$

add -2 (row 1) to row 2   add row 4 to row 5   add -2 (row 2) to row 1

add row 1 to row 3   interchange rows 3 and 2   add 3 (row 2) to row 3

add -1 (row 1) to row 5   add -1 (row 2) to row 4

$$\begin{bmatrix} 1 & 0 & -7 \\ 0 & 1 & 5 \\ 0 & 0 & 13 \\ 0 & 0 & -3 \\ 0 & 0 & 0 \end{bmatrix} \longrightarrow \begin{bmatrix} 1 & 0 & -7 \\ 0 & 1 & 5 \\ 0 & 0 & 1 \\ 0 & 0 & 0 \\ 0 & 0 & 0 \end{bmatrix} \longrightarrow \begin{bmatrix} 1 & 0 & 0 \\ 0 & 1 & 0 \\ 0 & 0 & 1 \\ 0 & 0 & 0 \\ 0 & 0 & 0 \end{bmatrix}$$

$\frac{1}{13}$ (row 3)   add 7 (row 3) to row 1

add 3 (row 3) to row 4   add -5 (row 3) to row 1

The transpose of the first three rows of the last matrix gives a basis for V. This is the natural basis $\{\mathbf{v}_1, \mathbf{v}_2, \mathbf{v}_3\}$ for R^3.

(a) $[\ 3\ \ 4\ \ 12\ ]^T = 3\mathbf{v}_1 + 4\mathbf{v}_2 + 12\mathbf{v}_3$

(b) $[\ 3\ \ 2\ \ 2\ ]^T = 3\mathbf{v}_1 + 2\mathbf{v}_2 + 2\mathbf{v}_3$

(c) $[\ 1\ \ 2\ \ 6\ ]^T = \mathbf{v}_1 + 2\mathbf{v}_2 + 6\mathbf{v}_3$

3. To use the ideas associated with row spaces, we use the isomorphism $L : M_{22} \to R_4$ defined by

$$L\left(\begin{bmatrix} a & b \\ c & d \end{bmatrix}\right) = [\ a\ \ b\ \ c\ \ d\ ].$$

Let S' be the image of the matrices in S under L:

$$S' = \{[\ 1\ \ 2\ \ 1\ \ 1\ ], [\ 2\ \ 1\ \ 3\ \ 1\ ], [\ 0\ \ 2\ \ 1\ \ 2\ ], [\ 3\ \ 2\ \ 1\ \ 4\ ], [\ 5\ \ 0\ \ 0\ \ -1\ ]\}.$$

Next, we find a basis for span S'. Let A be the matrix whose rows are the vectors in S'.

$$A = \begin{bmatrix} 1 & 2 & 1 & 1 \\ 2 & 1 & 3 & 1 \\ 0 & 2 & 1 & 2 \\ 3 & 2 & 1 & 4 \\ 5 & 0 & 0 & -1 \end{bmatrix} \longrightarrow \begin{bmatrix} 1 & 2 & 1 & 1 \\ 0 & -3 & 1 & -1 \\ 0 & 2 & 1 & 2 \\ 0 & -4 & -2 & 1 \\ 0 & -10 & -5 & -6 \end{bmatrix} \longrightarrow \begin{bmatrix} 1 & 0 & 0 & -1 \\ 0 & -3 & 1 & -1 \\ 0 & 2 & 1 & 2 \\ 0 & 0 & 0 & 5 \\ 0 & 0 & 0 & 4 \end{bmatrix} \longrightarrow$$

add -2 (row 1) to row 2   add 2 (row 3) to row 4 $\frac{1}{5}$ (row 4)

add -3 (row 1) to row 4   add 5 (row 3) to row 5   add -4 (row 4) to row 5

add -5 (row 1) to row 5   add row 4 to row 1

add -2 (row 4) to row 3

add row 4 to row 2

$$\begin{bmatrix} 1 & 0 & 0 & 0 \\ 0 & -3 & 1 & 0 \\ 0 & 2 & 1 & 0 \\ 0 & 0 & 0 & 1 \\ 0 & 0 & 0 & 0 \end{bmatrix} \longrightarrow \begin{bmatrix} 1 & 0 & 0 & 0 \\ 0 & 1 & -2 & 0 \\ 0 & 2 & 1 & 0 \\ 0 & 0 & 0 & 1 \\ 0 & 0 & 0 & 0 \end{bmatrix} \longrightarrow \begin{bmatrix} 1 & 0 & 0 & 0 \\ 0 & 1 & -2 & 0 \\ 0 & 0 & 1 & 0 \\ 0 & 0 & 0 & 1 \\ 0 & 0 & 0 & 0 \end{bmatrix}$$

add row 3 to row 2   add -2 (row 2) to row 3

-1 (row 2) $\frac{1}{5}$ (row 3)

Adding 2 times row 3 of the last matrix to row 2, we obtain the reduced echelon form

$$\begin{bmatrix} 1 & 0 & 0 & 0 \\ 0 & 1 & 0 & 0 \\ 0 & 0 & 1 & 0 \\ 0 & 0 & 0 & 1 \\ 0 & 0 & 0 & 0 \end{bmatrix}.$$

Thus a basis for span S' is the natural basis for R_4 and hence a basis for span S is the natural basis for M_{22} which is

$$\left\{\begin{bmatrix} 1 & 0 \\ 0 & 0 \end{bmatrix}, \begin{bmatrix} 0 & 1 \\ 0 & 0 \end{bmatrix}, \begin{bmatrix} 0 & 0 \\ 1 & 0 \end{bmatrix}, \begin{bmatrix} 0 & 0 \\ 0 & 1 \end{bmatrix}\right\}.$$

5. We row reduce the matrix A. First use the leading 1 in row 1 to zero out the first column.

$$A = \begin{bmatrix} 1 & 2 & -1 \\ 1 & 9 & -1 \\ -3 & 8 & 3 \\ -2 & 3 & 2 \end{bmatrix} \longrightarrow \begin{bmatrix} 1 & 2 & -1 \\ 0 & 7 & 0 \\ 0 & 14 & 0 \\ 0 & 7 & 0 \end{bmatrix} \longrightarrow \begin{bmatrix} 1 & 0 & -1 \\ 0 & 1 & 0 \\ 0 & 0 & 0 \\ 0 & 0 & 0 \end{bmatrix}$$

Next, multiply the second row by $\frac{1}{7}$, then use the leading 1 in column 2 to zero out the second column.

(a) It follows from the last matrix that

$$\{[\,1 \quad 0 \quad -1\,],[\,0 \quad 1 \quad 0\,]\}$$

is a basis for the row space of A consisting of vectors that are not row vectors of A.

(b) The reduced echelon form shows that the first two rows of A

$$\{[\,1 \quad 2 \quad -1\,],[\,1 \quad 9 \quad -1\,]\}$$

is a basis for the row space of A consisting of vectors that are row vectors of A.

7. We form the transpose A^T and row reduce it:

$$\begin{bmatrix} 1 & 1 & 3 & 2 \\ -2 & -1 & 2 & 1 \\ 7 & 4 & -3 & -1 \\ 0 & 0 & 5 & 3 \end{bmatrix} \longrightarrow \begin{bmatrix} 1 & 1 & 3 & 2 \\ 0 & 1 & 8 & 5 \\ 0 & -3 & -24 & -15 \\ 0 & 0 & 5 & 3 \end{bmatrix} \longrightarrow \begin{bmatrix} 1 & 0 & -5 & -3 \\ 0 & 1 & 8 & 5 \\ 0 & 0 & 0 & 0 \\ 0 & 0 & 5 & 3 \end{bmatrix}$$

add 2 (row 1) to row 2   add -1 (row 2) to row 1   add row 4 to row 1

add -7 (row 1) to row 3   add 3 (row 2) to row 3 $\frac{1}{5}$ (row 4)

add -8 (row 4) to row 2

Multiplying row 4 of the last matrix by $\frac{1}{5}$ to create a leading 1, using it to zero out column 3, and then interchanging rows 3 and 4, we obtain the reduced row echelon form

$$\begin{bmatrix} 1 & 0 & 0 & 0 \\ 0 & 1 & 0 & \frac{1}{5} \\ 0 & 0 & 1 & \frac{3}{5} \\ 0 & 0 & 0 & 0 \end{bmatrix}.$$

(a) Thus the vectors $[\,1 \quad 0 \quad 0 \quad 0\,]$, $[\,0 \quad 1 \quad 0 \quad \frac{1}{5}\,]$, and $[\,0 \quad 0 \quad 1 \quad \frac{3}{5}\,]$ form a basis for the row space of A^T. Therefore

$$\left\{ \begin{bmatrix} 1 \\ 0 \\ 0 \\ 0 \end{bmatrix}, \begin{bmatrix} 0 \\ 1 \\ 0 \\ \frac{1}{5} \end{bmatrix}, \begin{bmatrix} 0 \\ 0 \\ 1 \\ \frac{3}{5} \end{bmatrix} \right\}$$

is a basis for the column space of A consisting of vectors that are not column vectors of A.

(b) The first, second and fourth rows of A^T form a basis for the row space of A^T. Hence their transposes

$$\left\{ \begin{bmatrix} 1 \\ 1 \\ 3 \\ 2 \end{bmatrix}, \begin{bmatrix} -2 \\ -1 \\ 2 \\ 1 \end{bmatrix}, \begin{bmatrix} 0 \\ 0 \\ 5 \\ 3 \end{bmatrix} \right\}$$

is a basis for the column space of A consisting of vectors that are column vectors of A.

9. Since row rank = column rank, we determine the number of nonzero rows in the row reduced echelon form of the matrix. In some cases we need not go as far as the reduced row echelon form.

(a) $$\begin{bmatrix} 1 & 2 & 3 & 2 & 1 \\ 3 & 1 & -5 & -2 & 1 \\ 7 & 8 & -1 & 2 & 5 \end{bmatrix} \longrightarrow \begin{bmatrix} 1 & 2 & 3 & 2 & 1 \\ 0 & -5 & -14 & -8 & -2 \\ 0 & -6 & -22 & -12 & -2 \end{bmatrix} \longrightarrow \begin{bmatrix} 1 & 2 & 3 & 2 & 1 \\ 0 & 1 & 8 & 4 & 0 \\ 0 & -6 & -22 & -12 & -2 \end{bmatrix} \longrightarrow$$

add -3 (row 1 to row 2)   add -1 (row 3) to row 2   add 6 (row 2) to row 3

add -7 (row 1) to row 3

$$\begin{bmatrix} 1 & 2 & 3 & 2 & 1 \\ 0 & 1 & 8 & 4 & 0 \\ 0 & 0 & 26 & 12 & -2 \end{bmatrix}$$

Finally using $\frac{1}{26}$ (row 3) gives us a leading 1 in each row, so row rank = 3 = column rank.

(b)
$$\begin{bmatrix} 1 & 3 & 2 & 0 & 0 & 1 \\ 2 & 1 & -5 & 1 & 2 & 0 \\ 3 & 2 & 5 & 1 & -2 & 1 \\ 5 & 8 & 9 & 1 & -2 & 2 \\ 9 & 9 & 4 & 2 & 0 & 2 \end{bmatrix} \longrightarrow \begin{bmatrix} 1 & 3 & 2 & 0 & 0 & 1 \\ 0 & -5 & -9 & 1 & 2 & -2 \\ 0 & -7 & -1 & 1 & -2 & -2 \\ 0 & -7 & -1 & 1 & -2 & -3 \\ 0 & -18 & -14 & 2 & 0 & -7 \end{bmatrix} \longrightarrow \begin{bmatrix} 1 & 3 & 2 & 0 & 0 & 1 \\ 0 & 2 & -8 & 0 & 4 & 0 \\ 0 & -7 & -1 & 1 & -2 & -2 \\ 0 & 0 & 0 & 0 & 0 & -1 \\ 0 & -18 & -14 & 2 & 0 & -7 \end{bmatrix} \longrightarrow$$

add −2 (row 1) to row 2   add −1 (row 3) to row 4   −1 (row 4)
add −3 (row 1) to row 3   add −1 (row 3) to row 2   add 9 (row 2) to row 5
add −5 (row 1) to row 4
add −9 (row 1) to row 5

$$\begin{bmatrix} 1 & 3 & 2 & 0 & 0 & 1 \\ 0 & 2 & -8 & 0 & 4 & 0 \\ 0 & -7 & -1 & 1 & -2 & -2 \\ 0 & 0 & 0 & 0 & 0 & 1 \\ 0 & 0 & -86 & 2 & 36 & -7 \end{bmatrix} \longrightarrow \begin{bmatrix} 1 & 3 & 2 & 0 & 0 & 1 \\ 0 & 1 & -4 & 0 & 2 & 0 \\ 0 & 0 & -57 & 1 & 26 & -2 \\ 0 & 0 & 0 & 0 & 0 & 1 \\ 0 & 0 & -86 & 2 & 36 & -7 \end{bmatrix}$$

$\frac{1}{2}$ (row 2)
add 7 (row 2) to row 3

If we now add $-\frac{86}{57}$ times row 3 to row 5, we obtain in the fifth row a nonzero entry in the fourth column. Further row operations can be used to show that we can obtain a leading 1 in each row. Hence row rank = 5 = column rank.

11. The nonzero rows of A have leading 1's that appear in different columns. Such a set of row vectors is linearly independent. Since these rows span the row space of A, they are a basis for the row space. Hence their number, the number of nonzero rows of A, is the dimension of the row space. Thus, row rank = the number of nonzero rows of A.

13. We row reduce the given matrix to find its rank, then use Theorem 2.18 to find the nullity.

(a)
$$\begin{bmatrix} 1 & -1 & 2 & 3 \\ 2 & 6 & -8 & 1 \\ 5 & 3 & -2 & 10 \end{bmatrix} \longrightarrow \begin{bmatrix} 1 & -1 & 2 & 3 \\ 0 & 8 & -12 & -5 \\ 0 & 8 & -12 & -5 \end{bmatrix} \longrightarrow \begin{bmatrix} 1 & -1 & 2 & 3 \\ 0 & 1 & -\frac{3}{2} & -\frac{5}{8} \\ 0 & 0 & 0 & 0 \end{bmatrix}$$

add −2 (row 1) to row 2   add −1 (row 2) to row 3
add −5 (row 1) to row 3 $\frac{1}{8}$ (row 2)

Thus rank = 2 and therefore nullity = 4 − rank = 2.

(b)
$$\begin{bmatrix} 1 & 2 & 0 & 3 \\ 3 & 2 & -1 & 0 \\ 2 & -1 & 0 & 1 \end{bmatrix} \longrightarrow \begin{bmatrix} 1 & 2 & 0 & 3 \\ 0 & -4 & -1 & -9 \\ 0 & -5 & 0 & -5 \end{bmatrix} \longrightarrow \begin{bmatrix} 1 & 2 & 0 & 3 \\ 0 & 1 & \frac{1}{4} & \frac{9}{4} \\ 0 & 0 & \frac{5}{4} & \frac{25}{4} \end{bmatrix}$$

add −3 (row 1) to row 2 $-\frac{1}{4}$ (row 2)
add −2 (row 1) to row 3   add 5 (row 2) to row 3

Further row operations give a leading 1 in each row. Thus rank = 3 and nullity = 4 − rank = 1.

15. We summarize the echelon form of each matrix. Equivalent matrices will have the same reduced echelon form, meaning both row reduced and column reduced. To obtain this form, we apply row and column operations.

$$\text{reduced ef of } A = \begin{bmatrix} 1 & 0 & 0 & 0 \\ 0 & 1 & 0 & 0 \\ 0 & 0 & 0 & 0 \\ 0 & 0 & 0 & 0 \end{bmatrix}; \quad \text{reduced ef of } B = I_4; \quad \text{reduced ef of } C = I_4;$$

$$\text{reduced ef of } D = \begin{bmatrix} 1 & 0 & 0 & 0 \\ 0 & 1 & 0 & 0 \\ 0 & 0 & 0 & 0 \\ 0 & 0 & 0 & 0 \end{bmatrix}; \quad \text{reduced ef of } E = \begin{bmatrix} 1 & 0 & 0 & 0 \\ 0 & 1 & 0 & 0 \\ 0 & 0 & 0 & 0 \\ 0 & 0 & 0 & 0 \end{bmatrix}$$

Thus B and C are equivalent, and A, D and E are equivalent.

17. Let the coefficient matrix be denoted by A and the right-hand side by $\mathbf{b}$. Then the augmented matrix is $\begin{bmatrix} A & | & \mathbf{b} \end{bmatrix}$. Row reducing $\begin{bmatrix} A & | & \mathbf{b} \end{bmatrix}$ we can determine rank A and rank $\begin{bmatrix} A & | & \mathbf{b} \end{bmatrix}$ at the same time. If these two ranks are the same, the system is consistent.

(a) $\begin{bmatrix} A & | & \mathbf{b} \end{bmatrix} = \left[\begin{array}{cccc|c} 1 & -2 & -3 & 4 & 1 \\ 4 & -1 & -5 & 6 & 2 \\ 2 & 3 & 1 & -2 & 2 \end{array}\right] \longrightarrow \left[\begin{array}{cccc|c} 1 & -2 & -3 & 4 & 1 \\ 0 & 7 & 7 & -10 & -2 \\ 0 & 7 & 7 & -10 & 0 \end{array}\right] \longrightarrow \left[\begin{array}{cccc|c} 1 & -2 & -3 & 4 & 1 \\ 0 & 7 & 7 & -10 & -2 \\ 0 & 0 & 0 & 0 & 2 \end{array}\right]$

add -4 (row 1) to row 2   add -1 (row 2) to row 3
add -2 (row 1) to row 3

It follows that rank $A = 2$ while rank $\begin{bmatrix} A & | & \mathbf{b} \end{bmatrix} = 3$. Thus the system is inconsistent.

(b) $\begin{bmatrix} A & | & \mathbf{b} \end{bmatrix} = \left[\begin{array}{ccc|c} 1 & 1 & 1 & 6 \\ 1 & -1 & 1 & 2 \\ 5 & 1 & 5 & 5 \end{array}\right] \longrightarrow \left[\begin{array}{ccc|c} 1 & 1 & 1 & 6 \\ 0 & -2 & 0 & -4 \\ 0 & -4 & 0 & -25 \end{array}\right] \longrightarrow \left[\begin{array}{ccc|c} 1 & 1 & 1 & 6 \\ 0 & -2 & 0 & -4 \\ 0 & 0 & 0 & -17 \end{array}\right]$

add -1 (row 1) to row 2   add -2 (row 2) to row 3
add -5 (row 2) to row 3

It follows that rank $A = 2$ while rank $\begin{bmatrix} A & | & \mathbf{b} \end{bmatrix} = 3$. Thus the system is inconsistent.

19. We determine the rank of the matrices.

(a) We use row operations to get the matrix to the form

$$\begin{bmatrix} 1 & 1 & 2 \\ 0 & 4 & 6 \\ 0 & 0 & 0 \end{bmatrix}$$

Therefore the rank of the matrix is 2, hence it is singular.

(b) We use row operations to get the matrix to the form

$$\begin{bmatrix} 1 & 1 & 4 & -1 \\ 0 & 1 & -1 & 3 \\ 0 & 0 & 10 & -12 \\ 0 & 0 & 0 & 1 \end{bmatrix}$$

Therefore the rank is 4 and hence the matrix is nonsingular.

21. (a) Yes; $\det(A) = \begin{vmatrix} 1 & 0 & 1 \\ 1 & 1 & 0 \\ 2 & 1 & 0 \end{vmatrix} = -1 \neq 0.$

(b) No; $\det(A) = \begin{vmatrix} 2 & 3 & -1 \\ 1 & 2 & -2 \\ -1 & -3 & 5 \end{vmatrix} = 0.$

23. (a) We use row operations to get the matrix to the form

$$\begin{bmatrix} 1 & 2 & -1 \\ 0 & -5 & 5 \\ 0 & 0 & -6 \end{bmatrix}$$

Therefore the rank of the matrix is 3, hence the homogeneous system has only the trivial solution.

(b) We use row operations to get the matrix to the form

$$\begin{bmatrix} 1 & 0 & 2 \\ 0 & 1 & \frac{1}{2} \\ 0 & 0 & 0 \end{bmatrix}$$

Therefore the rank of the matrix is 2, hence the homogeneous system has a nontrivial solution.

25. (a)
$$\begin{bmatrix} 1 & 1 & -2 \\ 1 & 2 & 3 \\ 3 & 4 & -1 \end{bmatrix} \longrightarrow \begin{bmatrix} 1 & 1 & -2 \\ 0 & 1 & 5 \\ 0 & 1 & 5 \end{bmatrix} \longrightarrow \begin{bmatrix} 1 & 0 & -7 \\ 0 & 1 & 5 \\ 0 & 0 & 0 \end{bmatrix} \longrightarrow$$

add −1 (row 1) to row 2　　add −1 (row 2) to row 3　　add 7 (col 1) top col 3
add −3 (row 1) to row 3　　add −1 (row 2) to row 1　　add −5 (col 2) to col 3

$$\begin{bmatrix} 1 & 0 & 0 \\ 0 & 1 & 0 \\ 0 & 0 & 0 \end{bmatrix} \longrightarrow \begin{bmatrix} I_2 & 0 \\ 0 & 0 \end{bmatrix}$$

Hence the rank of the matrix is 2.

(b)
$$\begin{bmatrix} 1 & -1 & 2 & 3 \\ 2 & 2 & 0 & 1 \\ 1 & -5 & 6 & 8 \\ 4 & 0 & 4 & 6 \end{bmatrix} \longrightarrow \begin{bmatrix} 1 & -1 & 2 & 3 \\ 0 & 4 & -4 & -5 \\ 0 & -4 & 4 & 5 \\ 0 & 4 & -4 & -6 \end{bmatrix} \longrightarrow \begin{bmatrix} 1 & -1 & 2 & 3 \\ 0 & 4 & -4 & -5 \\ 0 & 0 & 0 & 0 \\ 0 & 0 & 0 & -1 \end{bmatrix} \longrightarrow$$

add −2 (row 1) to row 2　　add row 2 to row 3　　interchange rows 4 and 3
add −1 (row 1) to row 3　　add −1 (row 2) to row 4　　−1 (row 3
add −4 (row 1) to row 4　　　　　　　　　　　　$\frac{1}{4}$ (row 2)

$$\begin{bmatrix} 1 & -1 & 2 & 3 \\ 0 & 1 & -1 & -\frac{5}{4} \\ 0 & 0 & 0 & 1 \\ 0 & 0 & 0 & 0 \end{bmatrix} \longrightarrow \begin{bmatrix} 1 & 0 & 0 & 0 \\ 0 & 1 & -1 & -\frac{5}{4} \\ 0 & 0 & 0 & 1 \\ 0 & 0 & 0 & 0 \end{bmatrix} \longrightarrow \begin{bmatrix} 1 & 0 & 0 & 0 \\ 0 & 1 & 0 & 0 \\ 0 & 0 & 0 & 1 \\ 0 & 0 & 0 & 0 \end{bmatrix} \longrightarrow$$

add col 1 to col 2　　add col 2 to col 3　　interchange cols 3 and 4
add −2 (col 1) to col 3　　add $\frac{5}{4}$ (col 2) to col 4
add −3 (col 1) to col 4

$$\begin{bmatrix} 1 & 0 & 0 & 0 \\ 0 & 1 & 0 & 0 \\ 0 & 0 & 1 & 0 \\ 0 & 0 & 0 & 0 \end{bmatrix} \longrightarrow \begin{bmatrix} I_3 & 0 \\ 0 & 0 \end{bmatrix}$$

Hence the rank of the matrix is 3.

27. We find the reduced row echelon form of the matrix A to be I_3, which has rank 3. Therefore the system $A\mathbf{x} = \mathbf{b}$ has a unique solution for every 3×1 matrix $\mathbf{b}$.

29. We form the matrix whose rows are the three vectors and row reduce it:

$$\begin{bmatrix} 4 & 1 & 2 \\ 2 & 5 & -5 \\ 2 & -1 & 3 \end{bmatrix} \longrightarrow \quad \cdots \quad \longrightarrow \begin{bmatrix} 1 & 0 & \frac{5}{6} \\ 0 & 1 & -\frac{4}{3} \\ 0 & 0 & 0 \end{bmatrix}$$

steps
omitted

It follows that the rank of the matrix is 2 and hence the vectors are not linearly independent.

31. We use Corollary 4.8. To determine the coordinates of the vectors in S relative to some basis for P_3. The easiest basis to use is the natural basis $\{t^3, t^2, t, 1\}$. The matrix whose columns are the coordinates of the elements of S relative to this basis is

$$A = \begin{bmatrix} 1 & 2 & 0 & 2 \\ 0 & 0 & 0 & -2 \\ 1 & 0 & 1 & 0 \\ 1 & 3 & -1 & 0 \end{bmatrix}.$$

We have $|A| = -2 \neq 0$, hence rank $A = 3$ and therefore S is linearly independent.

33. We use Corollary 4.8. To determine the coordinates of the vectors in $S = \{t + 3, 2t + c^2 + 2\}$ relative to some basis for P_1. The easiest basis to use is the natural basis $\{t, 1\}$. The matrix whose columns are the coordinates of the elements of S relative to this basis is

$$A = \begin{bmatrix} 1 & 2 \\ 3 & c^2 + 2 \end{bmatrix}.$$

We have $|A| = c^2 + 2 - 6 = c^2 - 4$. It follows that rank $A = 3$ if and only if $c \neq \pm 2$. Hence S is linearly independent for $c \neq \pm 2$.

35. (a) The 7 rows of A span a row space of dimension 3. Thus the rows are linearly dependent.

   (b) The 3 columns of A span a column space of dimension 3. Thus the columns are linearly independent.

37. S is linearly independent if and only if the n rows of A are linearly independent, which is the case if and only if rank $A = n$.

39. If the system $A\mathbf{x} = \mathbf{0}$ has a nontrivial solution then A is singular, in which case rank $A < n$. Therefore the columns of A are linearly dependent. Conversely, if the columns of A are linearly dependent, then rank $A < n$. In this case the matrix A is singular and therefore $A\mathbf{x} = \mathbf{0}$ has a nontrivial solution.

41. The rows of A are linearly independent if and only if rank $A = n$, which is the case if and only if the columns span R^n.

43. Let rank $A = n$. Then Corollary 4.7 implies that A is nonsingular, so $\mathbf{x} = A^{-1}\mathbf{b}$ is a solution of the system $A\mathbf{x} = \mathbf{b}$. If $\mathbf{x}_1$ and $\mathbf{x}_2$ are solutions, then $A\mathbf{x}_1 = A\mathbf{x}_2$ and multiplying both sides by A^{-1} we have $\mathbf{x}_1 = \mathbf{x}_2$. Thus, the system $A\mathbf{x} = \mathbf{b}$ has a unique solution. Conversely, suppose that $A\mathbf{x} = \mathbf{b}$ has a unique solution for every $n \times 1$ matrix $\mathbf{b}$. Then the n linear systems $A\mathbf{x}_1 = \mathbf{e}_1$, $A\mathbf{x}_2 = \mathbf{e}_2$, ..., $A\mathbf{x}_n = \mathbf{e}_n$, where $\mathbf{e}_1$, $\mathbf{e}_2$, ..., $\mathbf{e}_n$ are the columns of I_n, have solutions $\mathbf{x}_1$, $\mathbf{x}_2$, ..., $\mathbf{x}_n$. Let B be the matrix whose jth column is $\mathbf{x}_j$. Then the n linear systems above can be written as $AB = I_n$. Hence $B = A^{-1}$, so A is nonsingular and Corollary 4.7 implies rank $A = n$.

45. Since the rank of a matrix is the same as its row rank and column rank, the number of linearly independent rows of a matrix is the same as the number of linearly independent columns. It follows that the largest the rank can be is the minimum of m and n. Since $m \neq n$, it must be that either the rows or columns are linearly dependent.

47. The solution space is a vector space of dimension d, where $2 \leq d \leq 7$.

49. Suppose that $S = \{\mathbf{v}_1, \mathbf{v}_2, \ldots, \mathbf{v}_n\}$ spans R^n (R_n). Then by Theorem 4.12(b), S is linearly independent and hence the dimension of the column space of A is n. Thus, rank $A = n$. Conversely, if rank $A = n$, then the set S consisting of the columns (rows) of A is linearly independent. By Theorem 4.12(a), S spans R^n.

Supplementary Exercises for Chapter 4, p. 285

1. (a) The verification of Definition 4.4 follows from the properties of continuous functions and real numbers. In particular, in calculus it is shown that the sum of continuous functions is continuous and a real number times a continuous function is again continuous. This verifies (a) and (b) of Definition 4.4. To verify the remaining portions, let f, g and h belong to $C[a,b]$ and c and d be real scalars. It is important to remember that for t in $[a,b]$, $f(t)$, $g(t)$ and $h(t)$ are just real numbers. Thus when working with these quantities we can use properties of real numbers.

(1) $(f \oplus g)(t) = f(t) + g(t) = g(t) + f(t) = (g \oplus f)(t)$

(2) $(f \oplus (g \oplus h))(t) = f(t) + (g(t) + h(t)) = (f(t) + g(t)) + h(t) = ((f \oplus g) \oplus h)(t)$

(3) Let $z(t) = 0$ for all t in $[a,b]$. Then

$$(f \oplus z)(t) = f(t) + z(t) = f(t) + 0 = 0 + f(t) = z(t) + f(t) = (z \oplus f)(t).$$

Uniqueness of z follows from the uniqueness of the additive identity, 0, for real numbers.

(4) For f in $C[a,b]$, $-f$ is defined by $(-f)(t) = -f(t)$. From calculus, if f is continuous, so is $-f$. Furthermore,

$$(f \oplus -f)(t) = f(t) + (-f(t)) = -f(t) + f(t) = ((-f) \oplus f)(t).$$

Uniqueness of $-f$ follows from the uniqueness of the additive inverse for real numbers.

(5) $(c \odot (f \oplus g))(t) = c(f(t) + g(t)) = cf(t) + cg(t) = (c \odot f)(t) + (c \odot g)(t) = (c \odot f) \oplus (c \odot g)(t)$

(6) $((c+d) \odot f)(t) = (c+d)f(t) = cf(t) + df(t) = (c \odot f)(t) + (d \odot f)(t) = (c \odot f \oplus d \odot f)(t)$

(7) $(c \odot (d \odot f))(t) = c(df(t)) = (cd)f(f) = ((cd) \odot f)(t)$

(8) $(1 \odot f)(t) = 1f(t) = f(t)$

(b) To determine when $W(k)$ is a subspace of $C[a,b]$ we use Theorem 4.3. Let f and g be in $W(k)$ and let c be a scalar. Then $f(a) = k$ and $g(a) = k$. Therefore

$$(f \oplus g)(a) = f(a) + g(a) = k + k = 2k.$$

In order for $f \oplus g$ to be in $W(k)$, we must have its value at $t = a$ be k. This is true if and only if $k = 0$. We now check the second condition when $k = 0$. In this case

$$(c \odot f)(a) = cf(a) = ck = c0 = 0 = k.$$

Thus the second condition is satisfied when $k = 0$. Thus, $W(k)$ is a subspace if and only if $k = 0$.

(c) Again we use Theorem 4.3. Suppose that f and g have roots at t_i, $i = 1, 2, \ldots, n$. Then

$$(f \oplus g)(t_i) = f(t_i) + g(t_i) = 0 + 0 = 0$$
$$(c \odot f)(t_i) = cf(t_i) = c(0) = 0.$$

Since both conditions of Theorem 4.3 are satisfied, the subset of all functions in $C[a,b]$ which have roots at t_i, $i = 1, 2, \ldots, n$ is a subspace.

3. For each vector $\mathbf{v}$ we construct a linear system from the expression

$$a_1\mathbf{v}_1 + a_2\mathbf{v}_2 + a_3\mathbf{v}_3 = \mathbf{v}.$$

If the system is consistent, then $\mathbf{v}$ is in span $\{\mathbf{v}_1, \mathbf{v}_2, \mathbf{v}_3\}$, otherwise it is not. For each vector $\mathbf{v}$ the linear system constructed has the same coefficient matrix but the right-hand side is different. We use the partitioned matrix approach developed in the solution of Exercise 15 in Section 4.8. Substituting into the previous expression and equating corresponding components from the left and right sides of the expression we obtain the following partitioned linear system for the three cases:

$$
\left[\begin{array}{rrr|r|r|r}
1 & 1 & 3 & 0 & 5 & 3 \\
1 & 5 & 0 & 7 & 6 & -8 \\
-2 & 2 & 2 & 4 & 2 & -6 \\
1 & -1 & 1 & 0 & 1 & 5
\end{array}\right] \rightarrow
\left[\begin{array}{rrr|r|r|r}
1 & 1 & 3 & 0 & 5 & 3 \\
0 & 4 & -3 & 7 & 1 & -11 \\
0 & 4 & 8 & 4 & 12 & 0 \\
0 & -2 & -2 & 0 & -4 & 2
\end{array}\right] \rightarrow
\left[\begin{array}{rrr|r|r|r}
1 & 1 & 3 & 0 & 5 & 3 \\
0 & 1 & 1 & 0 & 2 & -1 \\
0 & 4 & 8 & 4 & 12 & 0 \\
0 & 4 & -3 & 7 & 1 & -11
\end{array}\right] \rightarrow
$$

add -1 (row 1) to row 2 $\qquad$ $-\frac{1}{2}$ (row 4) $\qquad$ add -1 (row 2) to row 1

add 2 (row 1) to row 3 $\qquad$ interchange rows 4 and 2 $\qquad$ add -4 (row 2) to row 3

add -1 (row 1) to row 4 $\qquad\qquad\qquad\qquad\qquad\qquad$ add -4 (row 2) to row 4

$$
\left[\begin{array}{rrr|r|r|r}
1 & 0 & 2 & 0 & 3 & 4 \\
0 & 1 & 1 & 0 & 2 & -1 \\
0 & 0 & 4 & 4 & 4 & 4 \\
0 & 0 & -7 & 7 & -7 & -7
\end{array}\right] \rightarrow
\left[\begin{array}{rrr|r|r|r}
1 & 0 & 0 & -2 & 1 & 2 \\
0 & 1 & 0 & -1 & 1 & -2 \\
0 & 0 & 1 & 1 & 1 & 1 \\
0 & 0 & 0 & 14 & 0 & 0
\end{array}\right]
$$

$\frac{1}{4}$ (row 3)

add -2 (row 3) to row 1

add -1 (row 3) to row 2

add 7 (row 3) to row 4

(a) Inspecting the last row of the preceding matrix we see that the linear system obtained by using the vector $\mathbf{v} = \begin{bmatrix} 0 & 7 & 4 & 0 \end{bmatrix}^T$ is not consistent. Hence the vector $\mathbf{v}$ is not in span $\{\mathbf{v}_1, \mathbf{v}_2, \mathbf{v}_3\}$.

(b) It follows from column 5 of the preceding matrix that the system is consistent for the vector $\mathbf{v} = \begin{bmatrix} 5 & 6 & 2 & 1 \end{bmatrix}^T$. Hence this vector is in span $\{\mathbf{v}_1, \mathbf{v}_2, \mathbf{v}_3\}$.

(c) It follows from column 6 of the preceding matrix that the system is consistent for the vector $\mathbf{v} = \begin{bmatrix} 3 & -8 & -6 & 5 \end{bmatrix}^T$. Hence this vector is in span $\{\mathbf{v}_1, \mathbf{v}_2, \mathbf{v}_3\}$.

5. (a) To show that $W \cup U$ is not necessarily a subspace of V we construct an example. Let $V = R_2$, $W = \text{span} \{\begin{bmatrix} 1 & 0 \end{bmatrix}\}$, and $U = \text{span} \{\begin{bmatrix} 0 & 1 \end{bmatrix}\}$. If $W \cup U$ is to be a subspace, it must contain the sum of any two vectors that it contains. Certainly $\begin{bmatrix} 1 & 0 \end{bmatrix}$ and $\begin{bmatrix} 0 & 1 \end{bmatrix}$ are in $W \cup U$, but

$$\begin{bmatrix} 1 & 0 \end{bmatrix} + \begin{bmatrix} 0 & 1 \end{bmatrix} = \begin{bmatrix} 1 & 1 \end{bmatrix}$$

is not since it is neither in W nor in U. Hence $W \cup U$ is not a subspace of V.

(b) Let $\mathbf{v}$ and $\mathbf{w}$ be in $W \cap U$ and let c be a scalar. Since vectors $\mathbf{v}$ and $\mathbf{w}$ are in both W and U so is $\mathbf{v} + \mathbf{w}$, since both W and U are subspace. Hence $\mathbf{v} + \mathbf{w}$ is in $W \cap U$. Similarly, $c\mathbf{v}$ is in both W and U, so it is in $W \cap U$. By Theorem 4.3 $W \cap U$ is a subspace of V.

7. (a) Yes. To show this, let $\mathbf{b}_1$ and $\mathbf{b}_2$ be in W. Then there are vectors $\mathbf{x}_1$ and $\mathbf{x}_2$ such that $A\mathbf{x}_1 = \mathbf{b}_1$ and $A\mathbf{x}_2 = \mathbf{b}_2$. Therefore

$$A(\mathbf{x}_1 + \mathbf{x}_2) = A\mathbf{x}_1 + A\mathbf{x}_2 = \mathbf{b}_1 + \mathbf{b}_2$$

and hence $\mathbf{b}_1 + \mathbf{b}_2$ is in W. Similarly, for any scalar c we have

$$A(c\mathbf{x}_1) = cA\mathbf{x}_1 = c\mathbf{b}_1,$$

so that $c\mathbf{b}_1$ is in W. Hence W is a subspace of R^m.

(b) The column space of A is spanned by the columns of A so if $\mathbf{b}$ is in the column space of A, it is a linear combination of the columns of A. Hence the column space of A is contained in W. If $\mathbf{b}$ is in W, then for some $\mathbf{x}$, $A\mathbf{x} = \mathbf{b}$, that is, $\mathbf{b}$ is a linear combination of the columns of A. Hence W is contained in the column space of A. It follows that W is the same as the column space of A.

9. Suppose that W is a subspace of V. Let $\mathbf{u}$ and $\mathbf{v}$ be in W and let r and s be scalars. Then $r\mathbf{u}$ and $s\mathbf{u}$ are in W, so $r\mathbf{u} + s\mathbf{v}$ is in W. Conversely, if $r\mathbf{u} + s\mathbf{v}$ is in W for any $\mathbf{u}$ and $\mathbf{v}$ in W and any scalars r and s, then for $r = s = 1$ we have $\mathbf{u} + \mathbf{v}$ is in W. Also, for $s = 0$ we have $r\mathbf{u}$ is in W. Hence, W is a subspace of V.

11. We must determine for what values of a there are scalars c_1, c_2, c_3 such that

$$c_1 \begin{bmatrix} 1 \\ 2 \\ 3 \end{bmatrix} + c_2 \begin{bmatrix} 0 \\ 1 \\ 1 \end{bmatrix} + c_3 \begin{bmatrix} 1 \\ 3 \\ 4 \end{bmatrix} = \begin{bmatrix} a^2 \\ -3a \\ -2 \end{bmatrix}.$$

We equate corresponding entries of the matrices and row reduce the augmented matrix:

$$\begin{bmatrix} 1 & 0 & 1 & | & a^2 \\ 2 & 1 & 3 & | & -3a \\ 3 & 1 & 4 & | & -2 \end{bmatrix} \longrightarrow \begin{bmatrix} 1 & 0 & 1 & | & a^2 \\ 0 & 1 & 1 & | & -2a^2 - 3a \\ 0 & 1 & 1 & | & -3a^2 - 2 \end{bmatrix} \longrightarrow \begin{bmatrix} 1 & 0 & 1 & | & a^2 \\ 0 & 1 & 1 & | & -2a^2 - 3a \\ 0 & 0 & 0 & | & -a^2 + 3a - 2 \end{bmatrix}$$

add -2 (row 1) to row 2   add -1 (row 2) to row 3
add -3 (row 1) to row 3

This system is consistent provided $-a^2 + 3a - 2 = 0$. That is, if $a = 1$ or $a = 2$. Thus the given vector is in the span of the three vectors if and only if $a = 1$ or $a = 2$.

13. Since there are 6 vectors in S, they will be a basis for the 6-dimensional vector space R^6 provided they are linearly independent. (See Theorem 4.12.) Hence we form a linear combination of the vectors in S, set it equal to the zero vector, and construct the corresponding homogeneous system. The coefficient matrix in this case has columns which are exactly the vectors in S. The homogeneous system has only the zero solution, that is, the vectors are linearly independent, provided the coefficient matrix is row equivalent to I_6. We perform the row reduction and determine the values of k so that the coefficient matrix reduces to I_6. (The order in which the vectors are used is immaterial. Thus we arrange them so that after eliminating in the first column the $(2, 2)$ element would be a 1. This choice simplifies the next step of the reduction.)

$$\begin{bmatrix} 1 & 0 & 0 & 0 & 0 & 0 \\ 2 & 1 & 2 & 5 & 2 & 3 \\ -2 & -2 & 0 & k & 1 & 1 \\ 1 & 4 & 0 & 1 & k & 1 \\ 1 & -3 & 0 & 0 & 0 & 1 \\ 1 & 1 & 0 & 0 & 0 & 0 \end{bmatrix} \rightarrow \begin{bmatrix} 1 & 0 & 0 & 0 & 0 & 0 \\ 0 & 1 & 2 & 5 & 2 & 3 \\ 0 & -2 & 0 & k & 1 & 1 \\ 0 & 4 & 0 & 1 & k & 1 \\ 0 & -3 & 0 & 0 & 0 & 1 \\ 0 & 1 & 0 & 0 & 0 & 0 \end{bmatrix} \rightarrow \begin{bmatrix} 1 & 0 & 0 & 0 & 0 & 0 \\ 0 & 1 & 0 & 0 & 0 & 0 \\ 0 & 0 & 0 & k & 1 & 1 \\ 0 & 0 & 0 & 1 & k & 1 \\ 0 & 0 & 0 & 0 & 0 & 1 \\ 0 & 0 & 2 & 5 & 2 & 3 \end{bmatrix} \rightarrow$$

add -2 (row 1) to row 2   interchange rows 2 and 6   interchange rows 3 and 6
add 2 (row 1) to row 3   add 2 (row 2) to row 3   interchange rows 5 and 6
add -1 (row 1) to row 4   add -4 (row 2) to row 4
add -1 (row 1) to row 5   add 3 (row 2) to row 5
add -1 (row 1) to row 6   add -1 (row 2) to row 6

$$\begin{bmatrix} 1 & 0 & 0 & 0 & 0 & 0 \\ 0 & 1 & 0 & 0 & 0 & 0 \\ 0 & 0 & 2 & 5 & 2 & 3 \\ 0 & 0 & 0 & 1 & k & 1 \\ 0 & 0 & 0 & k & 1 & 1 \\ 0 & 0 & 0 & 0 & 0 & 1 \end{bmatrix} \rightarrow \begin{bmatrix} 1 & 0 & 0 & 0 & 0 & 0 \\ 0 & 1 & 0 & 0 & 0 & 0 \\ 0 & 0 & 2 & 5 & 2 & 3 \\ 0 & 0 & 0 & 1 & k & 1 \\ 0 & 0 & 0 & 0 & 1 - k^2 & 1 - k \\ 0 & 0 & 0 & 0 & 0 & 1 \end{bmatrix}$$

add $-k$ (row 4) to row 5

In order for this matrix to be row equivalent to I_6 we must choose k so that $1 - k^2 \neq 0$. Hence S will be a basis for R^6 for all values of k other than $k = 1$ or $k = -1$.

15. Since S is a basis for a subspace W of V, S is a linearly independent set in V. The result follows directly from Theorem 4.11.

17. The vector $\mathbf{b}$ must be in the column space of A in order for the linear system $A\mathbf{x} = \mathbf{b}$ to be consistent. The system is consistent if and only if the row reduced echelon form of $\begin{bmatrix} A & | & \mathbf{b} \end{bmatrix}$ contains no row of the form $\begin{bmatrix} 0 & \cdots & 0 & | & * \end{bmatrix}$, where $*$ is nonzero. Let $\mathbf{b} = \begin{bmatrix} a & b & c \end{bmatrix}^T$

(a) $[\,A\ \mid\ \mathbf{b}\,] = \begin{bmatrix} 1 & -2 & 1 & 0 & a \\ 2 & 1 & 1 & 2 & b \\ 1 & -7 & 2 & -2 & c \end{bmatrix} \longrightarrow \begin{bmatrix} 1 & -2 & 1 & 0 & a \\ 0 & 5 & -1 & 2 & b-2a \\ 0 & -5 & 1 & -2 & c-a \end{bmatrix} \longrightarrow \begin{bmatrix} 1 & -2 & 1 & 0 & a \\ 0 & 5 & -1 & 2 & b-2a \\ 0 & 0 & 0 & 0 & b+c-3a \end{bmatrix}$

add -2 (row 1) to row 2   add row 2 to row 3
add -1 (row 1) to row 3

Thus the system is consistent provided $b+c-3a=0$

(b) $[\,A\ \mid\ \mathbf{b}\,] = \begin{bmatrix} 1 & 2 & 1 & a \\ 1 & 3 & 1 & b \\ 2 & 4 & 3 & c \end{bmatrix} \longrightarrow \begin{bmatrix} 1 & 2 & 1 & a \\ 0 & 1 & 0 & b-a \\ 0 & 0 & 1 & c-2a \end{bmatrix}$

add -1 (row 1) to row 2
add -2 (row 1) to row 3

Since each row has a leading 1 we can show that the coefficient matrix A is row equivalent to I_3. Hence A is nonsingular and thus the linear system is consistent for all vectors $\mathbf{b}$.

19. rank A^T = row rank A^T = column rank A = rank A.

21. (a) From the definition of matrix multiplication, the rows of AB are linear combinations of the rows of B. Hence, the row space of AB is a subspace of the row space of B. Thus we have

$$\dim \text{row space of } AB \le \dim \text{row space of } B$$

which implies that rank $AB \le$ rank B.

(b) Let $A = [\,1\ \ 0\,]$ and $B = \begin{bmatrix} 0 \\ 1 \end{bmatrix}$. Then

$$AB = [\,1\ \ 0\,] \begin{bmatrix} 0 \\ 1 \end{bmatrix} = [\,0\,] \quad \Longrightarrow \quad \text{rank } (AB) = 0$$

but $\min\{\text{rank } A, \text{rank } B\} = 1$.

(c) Since $A = (AB)B^{-1}$, part (a) implies that

$$\text{rank } A \le \min\{\text{rank } (AB), \text{rank } B^{-1}\}$$

so certainly rank $A \le$ rank (AB). But part (a) also implies that rank $(AB) \le$ rank A. Hence rank (AB) = rank A.

(d) Since $B = A^{-1}(AB)$, part (a) implies that

$$\text{rank } B \le \min\{\text{rank } A^{-1}, \text{rank } (AB)\}$$

so certainly rank $B \le$ rank (AB). But part (a) also implies that rank $(AB) \le$ rank B. Hence rank (AB) = rank B.

(e) If P and Q are nonsingular, then rank (PAQ) = rank (PA) by part (c) and rank (PA) = rank A by part (d). So rank (PAQ) = rank A.

23. From Exercise 22, NS(BA) is the set of all vectors $\mathbf{x}$ such that $BA\mathbf{x} = \mathbf{0}$. If $\mathbf{x}$ is in NS(BA), then $BA\mathbf{x} = \mathbf{0}$ and hence $B^{-1}(BA\mathbf{x}) = B^{-1}\mathbf{0} = \mathbf{0}$, so $A\mathbf{x} = \mathbf{0}$. Therefore $\mathbf{x}$ is in NS(A). Hence NS(BA) $\subseteq$ NS(A). Conversely, if $\mathbf{x}$ is in NS(A), then $A\mathbf{x} = \mathbf{0}$, so $BA\mathbf{x} = B\mathbf{0} = \mathbf{0}$, hence $\mathbf{x}$ is in NS(BA). Thus NS(A) $\subseteq$ NS(BA). It follows that NS(BA) = NS(A).

25. (a) Since P is the transition matrix from the T-basis to the S-basis, the columns of P are the coordinates of the vectors in the T-basis relative to the S-basis.

$$\left[\,\mathbf{w}_1\,\right]_S = \begin{bmatrix} 1 \\ 0 \\ 1 \end{bmatrix} \implies \mathbf{w}_1 = (1)\mathbf{v}_1 + (0)\mathbf{v}_2 + (1)\mathbf{v}_3 = \begin{bmatrix} 2 \\ 2 \\ 1 \end{bmatrix}$$

$$\left[\,\mathbf{w}_2\,\right]_S = \begin{bmatrix} 1 \\ 2 \\ 0 \end{bmatrix} \implies \mathbf{w}_2 = (1)\mathbf{v}_1 + (2)\mathbf{v}_2 + (0)\mathbf{v}_3 = \begin{bmatrix} 3 \\ 1 \\ 2 \end{bmatrix}$$

$$\left[\,\mathbf{w}_3\,\right]_S = \begin{bmatrix} 1 \\ 1 \\ 3 \end{bmatrix} \implies \mathbf{w}_3 = (1)\mathbf{v}_1 + (1)\mathbf{v}_2 + (3)\mathbf{v}_3 = \begin{bmatrix} 5 \\ 4 \\ 4 \end{bmatrix}$$

(b) In order to find the S-basis, we use the preceding strategy with P^{-1} in place of P. Computing P^{-1}, we find

$$P^{-1} = \frac{1}{5}\begin{bmatrix} 6 & -3 & -1 \\ 1 & 2 & -1 \\ -2 & 1 & 2 \end{bmatrix}.$$

Then:

$$\left[\,\mathbf{v}_1\,\right]_S = \frac{1}{5}\begin{bmatrix} 6 \\ 1 \\ -2 \end{bmatrix} \implies \mathbf{v}_1 = \left(\frac{6}{5}\right)\mathbf{w}_1 + \left(\frac{1}{5}\right)\mathbf{w}_2 + \left(-\frac{2}{5}\right)\mathbf{w}_3 = \begin{bmatrix} \frac{6}{5} \\ -\frac{2}{5} \\ \frac{13}{5} \end{bmatrix}$$

$$\left[\,\mathbf{v}_2\,\right]_S = \frac{1}{5}\begin{bmatrix} -3 \\ 2 \\ 1 \end{bmatrix} \implies \mathbf{v}_2 = \left(-\frac{3}{5}\right)\mathbf{w}_1 + \left(\frac{2}{5}\right)\mathbf{w}_2 + \left(\frac{1}{5}\right)\mathbf{w}_3 = \begin{bmatrix} \frac{2}{5} \\ \frac{1}{5} \\ -\frac{4}{5} \end{bmatrix}$$

$$\left[\,\mathbf{v}_3\,\right]_S = \frac{1}{5}\begin{bmatrix} -1 \\ -1 \\ 2 \end{bmatrix} \implies \mathbf{v}_3 = \left(-\frac{1}{5}\right)\mathbf{w}_1 + \left(-\frac{1}{5}\right)\mathbf{w}_2 + \left(\frac{2}{5}\right)\mathbf{w}_3 = \begin{bmatrix} -\frac{1}{5} \\ \frac{2}{5} \\ -\frac{3}{5} \end{bmatrix}$$

27. Let $\mathbf{x}$ be nonzero. Then $A\mathbf{x} \neq \mathbf{x}$ and hence $A\mathbf{x} - \mathbf{x} = (A - I_n)\mathbf{x} \neq \mathbf{0}$. That is, there is no nonzero solution to the homogeneous system with square coefficient matrix $A - I_n$. Hence the only solution to the homogeneous system with coefficient matrix $A - I_n$ is the zero solution which implies $A - I_n$ is nonsingular.

29. (a) False. Let

$$A = \begin{bmatrix} 1 & 0 \\ 0 & 0 \end{bmatrix} \quad \text{and} \quad B = \begin{bmatrix} 0 & 0 \\ 0 & 1 \end{bmatrix}.$$

Both A and B have rank 1 but $A + B = I_2$ and hence $\text{rank}(A + B) = 2$.

(b) False. Let

$$A = \begin{bmatrix} 1 & -9 \\ 7 & 2 \end{bmatrix} \quad \text{and} \quad B = -A.$$

Both A and B have rank 2 but $A + B = \mathbf{0}$, which has rank 0.

(c) False. Use the same matrices as in part (b). Both A and B have rank 2, so the sum of their ranks is 4, but $\text{rank}(A + B) = 0$.

31. Suppose that the linear system $A\mathbf{x} = \mathbf{b}$ has at most one solution for every $m \times 1$ matrix $\mathbf{b}$. Since $A\mathbf{x} = \mathbf{0}$ always has the trivial solution, then $A\mathbf{x} = \mathbf{0}$ has only the trivial solution. Conversely, suppose that $A\mathbf{x} = \mathbf{0}$ has only the trivial solution. Then nullity $A = 0$, so by Theorem 4.19, rank $A = n$. Thus, dim column space $A = n$, so the n columns of A, which span its column space, form a basis for the column space. If $\mathbf{b}$ is an $m \times 1$ matrix then $\mathbf{b}$ is a vector in R^m. If $\mathbf{b}$ is in the column space of A,

then **b** can be written as a linear combination of the columns of A in one and only one way. That is, $A\mathbf{x} = \mathbf{b}$ has exactly one solution. If **b** is not in the column space of A, then $A\mathbf{x} = \mathbf{b}$ has no solution. Thus, $A\mathbf{x} = \mathbf{b}$ has at most one solution.

33. Let A be an $m \times n$ matrix whose rank is k. Then the dimension of the solution space of the associated homogeneous system $A\mathbf{x} = \mathbf{0}$ is $n - k$, so the general solution to the homogeneous system has $n - k$ arbitrary parameters. As we noted at the end of Section 4.7, every solution **x** to the nonhomogeneous system $A\mathbf{x} = \mathbf{b}$ can be written as $\mathbf{x}_p + \mathbf{x}_h$, where $\mathbf{x}_p$ is a particular solution to the nonhomogeneous system and $\mathbf{x}_h$ is a solution to the associated homogeneous system $A\mathbf{x} = \mathbf{0}$. Hence, the general solution to the given nonhomogeneous system has $n - k$ arbitrary parameters.

35. Since $V = W_1 + W_2$, every vector **v** in W can be written as $\mathbf{w}_1 + \mathbf{w}_2$, $\mathbf{w}_1$ in W_1 and $\mathbf{w}_2$ in W_2. Suppose now that

$$\mathbf{v} = \mathbf{w}_1 + \mathbf{w}_2 \quad \text{and} \quad \mathbf{v} = \mathbf{w}_1' + \mathbf{w}_2'.$$

Then $\mathbf{w}_1 + \mathbf{w}_2 = \mathbf{w}_1' + \mathbf{w}_2'$ so that

$$\mathbf{w}_1 - \mathbf{w}_1' = \mathbf{w}_2' - \mathbf{w}_2$$

Since $\mathbf{w}_1 - \mathbf{w}_1'$ is in W_1 and $\mathbf{w}_2' - \mathbf{w}_2$ is in W_2, $\mathbf{w}_1 - \mathbf{w}_1'$ is in $W_1 \cap W_2 = \{\mathbf{0}\}$. Hence $\mathbf{w}_1 = \mathbf{w}_1'$. Similarly; we conclude that $\mathbf{w}_2 = \mathbf{w}_2'$. Thus, every vector in V can be written uniquely in the form $\mathbf{w}_1 + \mathbf{w}_2$, where $\mathbf{w}_1$ is in W_1 and $\mathbf{w}_2$ is in W_2.

Chapter Review for Chapter 4, p. 217

True or False
1. True.	2. True.	3. False.	4. False.	5. True.	6. False.
7. True.	8. True.	9. True.	10. False.	11. False.	12. True.
13. False.	14. True.	15. True.	16. True.	17. True.	18. True.
19. False.	20. False.	21. True.			

Quiz

1. No. Property 1 in Definition 4.4 is not satisfied.

2. No. Properties 5–8 in Definition 4.4 are not satisfied.

3. Yes. If $\mathbf{x} = \begin{bmatrix} a_1 \\ b_1 \\ -a_1 \end{bmatrix}$, $\mathbf{y} = \begin{bmatrix} a_2 \\ b_2 \\ -a_2 \end{bmatrix}$ are in the set and c is any scalar, then

$$\mathbf{x} + \mathbf{y} = \begin{bmatrix} a_1 + a_2 \\ b_1 + b_2 \\ -a_1 - a_2 \end{bmatrix} = \begin{bmatrix} a_1 + a_2 \\ b_1 + b_2 \\ -(a_1 + a_2) \end{bmatrix} \quad \text{and} \quad c\mathbf{x} = \begin{bmatrix} ca_1 \\ cb_1 \\ -ca_1 \end{bmatrix}$$

are in the set. By Theorem 4.3, the set is a subspace. Let $\begin{bmatrix} a \\ b \\ -a \end{bmatrix}$ be a typical vector in the set. Then

$$\begin{bmatrix} a \\ b \\ -a \end{bmatrix} = a \begin{bmatrix} 1 \\ 0 \\ -1 \end{bmatrix} + b \begin{bmatrix} 0 \\ 1 \\ 0 \end{bmatrix}.$$

Therefore the set

$$\left\{ \begin{bmatrix} 1 \\ 0 \\ -1 \end{bmatrix}, \begin{bmatrix} 0 \\ 1 \\ 0 \end{bmatrix} \right\}$$

spans the subspace. Since the vectors $\begin{bmatrix} 1 \\ 0 \\ -1 \end{bmatrix}$ and $\begin{bmatrix} 0 \\ 1 \\ 0 \end{bmatrix}$ are linearly independent, this set is a basis for the subspace.

4. No. Property (b) in Theorem 4.3 is not satisfied. For example, if

$$\mathbf{x} = \begin{bmatrix} a \\ b \\ c \end{bmatrix}, \quad \text{where } a + b + c > 0,$$

then

$$(-1)\mathbf{x} = \begin{bmatrix} -a \\ -b \\ -c \end{bmatrix}, \quad \text{but } -a - b - c < 0.$$

5. If $p(t)$ and $q(t)$ are in W and c is any scalar, then

$$(p+q)(0) = p(0) + q(0) = 0 + 0 = 0$$
$$(cp)(0) = cp(0) = c0 = 0.$$

Hence $p+q$ and cp are in W. Therefore, W is a subspace of P_2. If $p(t) = at^2 + bt + c$ is any polynomial in P_2, then $p(t)$ is in W when $c = 0$. Therefore $p(t) = at^2 + bt$ and hence a basis for W is the set $\{t^2, t\}$.

6. No. S is linearly dependent since the reduced row echelon form of the matrix

$$\begin{bmatrix} 2 & 3 & -2 \\ 1 & -1 & 4 \\ -1 & 2 & -6 \end{bmatrix} \quad \text{is} \quad \begin{bmatrix} 1 & 0 & 2 \\ 0 & 1 & -2 \\ 0 & 0 & 0 \end{bmatrix}.$$

7. The reduced row echelon form of the matrix

$$\begin{bmatrix} 1 & 1 & 2 & 3 & 4 \\ 2 & 0 & 8 & 0 & 0 \\ 0 & 1 & -2 & 0 & 1 \end{bmatrix} \quad \text{is} \quad \begin{bmatrix} 1 & 0 & 4 & 0 & -1 \\ 0 & 1 & -2 & 0 & 1 \\ 0 & 0 & 0 & 1 & 1 \end{bmatrix}.$$

The three rows are linearly independent, so S spans R^3. A basis consists of the first, second, and fourth vectors:

$$\left\{ \begin{bmatrix} 1 \\ 2 \\ 0 \end{bmatrix}, \begin{bmatrix} 1 \\ 0 \\ 1 \end{bmatrix}, \begin{bmatrix} 3 \\ 0 \\ 0 \end{bmatrix} \right\}.$$

8. The reduced row echelon form of A is

$$\begin{bmatrix} 1 & 0 & 1 & -1 \\ 0 & 1 & 0 & 1 \\ 0 & 0 & 0 & 0 \\ 0 & 0 & 0 & 0 \end{bmatrix}.$$

Hence $x_1 = -r + s$, $x_2 = -s$, $x_3 = r$, $x_4 = s$, where r and s are any numbers. So

$$\mathbf{x} = \begin{bmatrix} -r+s \\ -s \\ r \\ s \end{bmatrix} = r \begin{bmatrix} -1 \\ 0 \\ 1 \\ 0 \end{bmatrix} + s \begin{bmatrix} 1 \\ -1 \\ 0 \\ 1 \end{bmatrix}.$$

Hence a basis for the null space is

$$\left\{ \begin{bmatrix} -1 \\ 0 \\ 1 \\ 0 \end{bmatrix}, \begin{bmatrix} 1 \\ -1 \\ 0 \\ 1 \end{bmatrix} \right\}.$$

9. Row reducing the matrix we obtain the reduced row echelon form

$$\begin{bmatrix} 1 & 0 & 2 \\ 0 & 1 & -2 \\ 0 & 0 & 0 \end{bmatrix}.$$

Hence a basis for the row space is $\{ \begin{bmatrix} 1 & 0 & 2 \end{bmatrix}, \begin{bmatrix} 0 & 1 & -2 \end{bmatrix} \}$.

10. Dimension of null space $= n - \operatorname{rank} A = 3 - 2 = 1$.

11. Row reducing the augmented matrix we obtain the reduced row echelon form

$$\left[\begin{array}{cccc|c} 1 & 0 & 0 & 1 & -\frac{1}{3} \\ 0 & 1 & 0 & \frac{1}{4} & \frac{7}{6} \\ 0 & 0 & 1 & -\frac{3}{2} & 0 \\ 0 & 0 & 0 & 0 & 0 \end{array}\right].$$

Therefore $x_1 = -x_4 - \frac{1}{3}$, $x_2 = -\frac{1}{4}x_4 + \frac{7}{6}$, $x_3 = \frac{3}{2}x_4$, $x_4 = r$, where r is any number. Hence

$$\mathbf{x} = \begin{bmatrix} -r - \frac{1}{3} \\ -\frac{1}{4}r + \frac{7}{6} \\ \frac{3}{2}r \\ r \end{bmatrix} = \begin{bmatrix} -\frac{1}{3} \\ \frac{7}{6} \\ 0 \\ 0 \end{bmatrix} + r \begin{bmatrix} -1 \\ -\frac{1}{4} \\ \frac{3}{2} \\ 1 \end{bmatrix}$$

and hence

$$\mathbf{x}_p = \begin{bmatrix} -\frac{1}{3} \\ \frac{7}{6} \\ 0 \\ 0 \end{bmatrix} \quad \text{and} \quad \mathbf{x}_h = r \begin{bmatrix} -1 \\ -\frac{1}{4} \\ \frac{3}{2} \\ 1 \end{bmatrix}, \quad r \text{ any number.}$$

12. Let c_1 and c_2 be scalars so that

$$c_1(t + 3) + c_2(2t + c^2 + 2) = 0.$$

Then

$$(c_1 + 2c_2)t + [3c_1 + (c^2 + 2)c_2] = 0$$

and hecne

$$c_1 + 2c_2 = 0$$
$$3c_1 + (c^2 + 2)c_2 = 0.$$

The coefficient matrix of this linear system is

$$\begin{bmatrix} 1 & 2 \\ 3 & c^2 + 2 \end{bmatrix}.$$

In order that the set $\{t + 3, 2t + c^2 + 2\}$ be linearly independent, the reduced row echelon form of this coefficient matrix must be I_2. Now, adding (-3) times row 1 to row 2, we obtain

$$\begin{bmatrix} 1 & 2 \\ 0 & c^2 - 4 \end{bmatrix}.$$

Thus, if $c^2 - 4 \neq 0$, or $c \neq \pm 2$, then $\{t + 3, 2t + c^2 + 2\}$ is linearly independent.

Chapter 5

Inner Product Spaces

Section 5.1, p. 297

1. (a) For $\mathbf{v} = \begin{bmatrix} 1 \\ 0 \end{bmatrix}$, $\|\mathbf{v}\| = \sqrt{1^2 + 0^2} = 1$.

    (b) For $\mathbf{v} = \begin{bmatrix} 0 \\ 0 \end{bmatrix}$, $\|\mathbf{v}\| = \sqrt{0^2 + 0^2} = 0$.

    (c) For $\mathbf{v} = \begin{bmatrix} 1 \\ 2 \end{bmatrix}$, $\|\mathbf{v}\| = \sqrt{1^2 + 2^2} = \sqrt{5}$.

3. (a) For $\mathbf{u} = \begin{bmatrix} 1 \\ 0 \end{bmatrix}$ and $\mathbf{v} = \begin{bmatrix} 1 \\ 1 \end{bmatrix}$, we have $\|\mathbf{u} - \mathbf{v}\| = \sqrt{(1-1)^2 + (0-1)^2} = 1$.

    (b) For $\mathbf{u} = \begin{bmatrix} 0 \\ 0 \end{bmatrix}$ and $\mathbf{v} = \begin{bmatrix} 1 \\ -1 \end{bmatrix}$, we have $\|\mathbf{u} - \mathbf{v}\| = \sqrt{(0-1)^2 + (0+1)^2} = \sqrt{2}$.

5. (a) $\|\mathbf{v} - \mathbf{u}\| = \sqrt{((-4)-1)^2 + ((-5)-2)^2} = \sqrt{74}$.
    (b) $\|\mathbf{v} - \mathbf{u}\| = \sqrt{(4-1)^2 + ((-5)-2)^2} = \sqrt{58}$.

7. We have
$$\|\mathbf{u}\| = \sqrt{4 + c^2} = 3 \quad \Longrightarrow \quad 4 + c^2 = 9 \quad \Longrightarrow \quad c = \pm\sqrt{5}.$$

9. We use $\cos\theta = \dfrac{\mathbf{u} \cdot \mathbf{v}}{\|\mathbf{u}\| \, \|\mathbf{v}\|}$. For $\mathbf{u} = \begin{bmatrix} 1 \\ 2 \end{bmatrix}$ and $\mathbf{v} = \begin{bmatrix} -4 \\ -5 \end{bmatrix}$, we have
$$\cos\theta = \frac{(1)(-4) + (2)(-5)}{\sqrt{1^2 + 2^2}\,\sqrt{(-4)^2 + (-5)^2}} = \frac{-14}{\sqrt{5}\,\sqrt{41}}.$$

    For $\mathbf{u} = \begin{bmatrix} 1 \\ 2 \end{bmatrix}$ and $\mathbf{v} = \begin{bmatrix} 4 \\ -5 \end{bmatrix}$, we have
$$\cos\theta = \frac{(1)(4) + (2)(-5)}{\sqrt{1^2 + 2^2}\,\sqrt{4^2 + (-5)^2}} = \frac{-6}{\sqrt{5}\,\sqrt{41}}.$$

11. Represent the direction of the positive x-axis, the direction of the positive y-axis, and the direction of the positive z-axis by
$$\mathbf{e}_1 = \begin{bmatrix} 1 \\ 0 \\ 0 \end{bmatrix}, \quad \mathbf{e}_2 = \begin{bmatrix} 0 \\ 1 \\ 0 \end{bmatrix}, \quad \text{and} \quad \mathbf{e}_3 = \begin{bmatrix} 0 \\ 0 \\ 1 \end{bmatrix},$$
    respectively.

(a) For $\mathbf{v} = \begin{bmatrix} 1 \\ 0 \\ 0 \end{bmatrix}$, we have $\cos\theta_1 = \dfrac{\mathbf{v}\cdot\mathbf{e}_1}{\|\mathbf{v}\|\,\|\mathbf{e}_1\|} = 1$; $\cos\theta_2 = \dfrac{\mathbf{v}\cdot\mathbf{e}_2}{\|\mathbf{v}\|\,\|\mathbf{e}_2\|} = 0$; $\cos\theta_3 = \dfrac{\mathbf{v}\cdot\mathbf{e}_3}{\|\mathbf{v}\|\,\|\mathbf{e}_3\|} = 0$.

(b) For $\mathbf{v} = \begin{bmatrix} 1 \\ 3 \\ 2 \end{bmatrix}$, we have

$$\cos\theta_1 = \frac{\mathbf{v}\cdot\mathbf{e}_1}{\|\mathbf{v}\|\,\|\mathbf{e}_1\|} = \frac{1}{\sqrt{14}}; \quad \cos\theta_2 = \frac{\mathbf{v}\cdot\mathbf{e}_2}{\|\mathbf{v}\|\,\|\mathbf{e}_2\|} = \frac{3}{\sqrt{14}}; \quad \cos\theta_3 = \frac{\mathbf{v}\cdot\mathbf{e}_3}{\|\mathbf{v}\|\,\|\mathbf{e}_3\|} = \frac{2}{\sqrt{14}}.$$

(c) For $\mathbf{v} = \begin{bmatrix} -1 \\ -2 \\ -3 \end{bmatrix}$, we have

$$\cos\theta_1 = \frac{\mathbf{v}\cdot\mathbf{e}_1}{\|\mathbf{v}\|\,\|\mathbf{e}_1\|} = \frac{-1}{\sqrt{14}}; \quad \cos\theta_2 = \frac{\mathbf{v}\cdot\mathbf{e}_2}{\|\mathbf{v}\|\,\|\mathbf{e}_2\|} = \frac{-2}{\sqrt{14}}; \quad \cos\theta_3 = \frac{\mathbf{v}\cdot\mathbf{e}_3}{\|\mathbf{v}\|\,\|\mathbf{e}_3\|} = \frac{-3}{\sqrt{14}}.$$

(d) For $\mathbf{v} = \begin{bmatrix} 4 \\ -3 \\ 2 \end{bmatrix}$, we have

$$\cos\theta_1 = \frac{\mathbf{v}\cdot\mathbf{e}_1}{\|\mathbf{v}\|\,\|\mathbf{e}_1\|} = \frac{4}{\sqrt{29}}; \quad \cos\theta_2 = \frac{\mathbf{v}\cdot\mathbf{e}_2}{\|\mathbf{v}\|\,\|\mathbf{e}_2\|} = \frac{-3}{\sqrt{29}}; \quad \cos\theta_3 = \frac{\mathbf{v}\cdot\mathbf{e}_3}{\|\mathbf{v}\|\,\|\mathbf{e}_3\|} = \frac{2}{\sqrt{29}}.$$

13. We prove Theorem 5.1 for vectors in R^2. The proof for vectors in R^3 is similar. Let

$$\mathbf{u} = \begin{bmatrix} u_1 \\ u_2 \end{bmatrix}, \quad \mathbf{v} = \begin{bmatrix} v_1 \\ v_2 \end{bmatrix}, \quad \text{and} \quad \mathbf{w} = \begin{bmatrix} w_1 \\ w_2 \end{bmatrix}.$$

(a) $\mathbf{u}\cdot\mathbf{u} \geq 0$ if $\mathbf{u} \neq \mathbf{0}$.

Proof: If $\mathbf{u} \neq \mathbf{0}$, then at least one of the components u_1, u_2 is nonzero. Hence it follows that $\mathbf{u}\cdot\mathbf{u} = u_1^2 + u_2^2 > 0$.

$\mathbf{u}\cdot\mathbf{u} = 0$ if and only if $\mathbf{u} = \mathbf{0}$.

Proof: $\mathbf{u}\cdot\mathbf{u} = u_1^2 + u_2^2 = 0$ if and only if $u_1 = u_2 = 0$ if and only if $\mathbf{u} = \mathbf{0}$.

(b) $\mathbf{u}\cdot\mathbf{v} = \mathbf{v}\cdot\mathbf{u}$

Proof: $\mathbf{u}\cdot\mathbf{v} = u_1v_1 + u_2v_2 = v_1u_1 + v_2u_2 = \mathbf{v}\cdot\mathbf{u}$.

(c) $(\mathbf{u}+\mathbf{v})\cdot\mathbf{w} = \mathbf{u}\cdot\mathbf{w} + \mathbf{v}\cdot\mathbf{w}$

Proof: $\begin{aligned}[t] (\mathbf{u}+\mathbf{v})\cdot\mathbf{w} &= (u_1+v_1)w_1 + (u_2+v_2)w_2 = u_1w_1 + v_1w_1 + u_2w_2 + v_2w_2 \\ &= (u_1w_1 + u_2w_2) + (v_1w_1 + v_2w_2) = \mathbf{u}\cdot\mathbf{w} + \mathbf{v}\cdot\mathbf{w} \end{aligned}$

(d) $c\mathbf{u}\cdot\mathbf{v} = c(\mathbf{u}\cdot\mathbf{v})$

Proof: $c\mathbf{u}\cdot\mathbf{v} = (cu_1)v_1 + (cu_2)v_2 = c(u_1v_1 + u_2v_2) = c(\mathbf{u}\cdot\mathbf{v})$.

15. (a) $\mathbf{i}\cdot\mathbf{j} = (1)(1) + (0)(0) = 1$ and $\mathbf{j}\cdot\mathbf{j} = (0)(0) + (1)(1) = 1$.

(b) $\mathbf{i}\cdot\mathbf{j} = (1)(0) + (0)(1) = 0$.

17. (a) To determine which pairs of vectors are orthogonal we compute the inner products of all possible

pairs

$$\mathbf{v}_1 \cdot \mathbf{v}_2 = 2;$$
$$\mathbf{v}_1 \cdot \mathbf{v}_3 = -10;$$
$$\mathbf{v}_1 \cdot \mathbf{v}_4 = 0; \qquad \text{hence } \mathbf{v}_1 \text{ and } \mathbf{v}_4 \text{ are orthogonal}$$

$$\mathbf{v}_1 \cdot \mathbf{v}_5 = 10;$$
$$\mathbf{v}_1 \cdot \mathbf{v}_6 = 0; \qquad \text{hence } \mathbf{v}_1 \text{ and } \mathbf{v}_6 \text{ are orthogonal}$$

$$\mathbf{v}_2 \cdot \mathbf{v}_3 = -4;$$
$$\mathbf{v}_2 \cdot \mathbf{v}_4 = 1;$$
$$\mathbf{v}_2 \cdot \mathbf{v}_5 = 4;$$
$$\mathbf{v}_2 \cdot \mathbf{v}_6 = 3;$$

$$\mathbf{v}_3 \cdot \mathbf{v}_4 = 0; \qquad \text{hence } \mathbf{v}_3 \text{ and } \mathbf{v}_4 \text{ are orthogonal}$$
$$\mathbf{v}_3 \cdot \mathbf{v}_5 = -20;$$
$$\mathbf{v}_3 \cdot \mathbf{v}_6 = 0; \qquad \text{hence } \mathbf{v}_3 \text{ and } \mathbf{v}_6 \text{ are orthgonal}$$

$$\mathbf{v}_4 \cdot \mathbf{v}_5 = 0; \qquad \text{hence } \mathbf{v}_4 \text{ and } \mathbf{v}_5 \text{ are orthogonal}$$
$$\mathbf{v}_4 \cdot \mathbf{v}_6 = 15;$$

$$\mathbf{v}_5 \cdot \mathbf{v}_6 = 0; \qquad \text{hence } \mathbf{v}_5 \text{ and } \mathbf{v}_6 \text{ are orthogonal}$$

(b) A pair of vectors are in the same direction if one is a positive scalar multiple of the other. Inspecting the vectors we see that $\mathbf{v}_5 = 2\mathbf{v}_1$ and $\mathbf{v}_6 = 3\mathbf{v}_4$. Hence vectors $\mathbf{v}_1$ and $\mathbf{v}_5$ are in the same direction and vectors $\mathbf{v}_4$ and $\mathbf{v}_6$ are in the same direction.

(c) A pair of vectors are in the opposite direction if one is a negative scalar multiple of the other. Inspecting the vectors we see that $\mathbf{v}_3 = -2\mathbf{v}_1$ and $\mathbf{v}_5 = -1\mathbf{v}_3$. Hence vectors $\mathbf{v}_1$ and $\mathbf{v}_3$ are in opposite directions and vectors $\mathbf{v}_3$ and $\mathbf{v}_5$ are in opposite directions.

19. Using the material on lines in R^3 discussed in Section 4.3, we first write the line in vector form in order to determine its direction. It follows that a pair of lines are perpendicular provided that the vectors in the direction of the lines are orthogonal.

(a) In vector form the lines are

$$L_1 : \begin{bmatrix} 2 \\ -3 \\ 4 \end{bmatrix} + t \begin{bmatrix} 2 \\ -3 \\ 4 \end{bmatrix} \qquad \text{with direction } \mathbf{v} = \begin{bmatrix} 2 \\ -3 \\ 4 \end{bmatrix}$$

$$L_2 : \begin{bmatrix} 2 \\ 4 \\ 5 \end{bmatrix} + t \begin{bmatrix} 1 \\ -1 \\ 1 \end{bmatrix} \qquad \text{with direction } \mathbf{w} = \begin{bmatrix} 1 \\ -1 \\ 1 \end{bmatrix}.$$

We have $\mathbf{v} \cdot \mathbf{w} = 9$, hence the lines are not perpendicular.

(b) In vector form the lines are

$$L_1 : \begin{bmatrix} 3 \\ 4 \\ 2 \end{bmatrix} + t \begin{bmatrix} -1 \\ 1 \\ 2 \end{bmatrix} \qquad \text{with direction } \mathbf{v} = \begin{bmatrix} -1 \\ 1 \\ 2 \end{bmatrix}$$

$$L_2 : \begin{bmatrix} 0 \\ 3 \\ 4 \end{bmatrix} + t \begin{bmatrix} 2 \\ -2 \\ 2 \end{bmatrix} \qquad \text{with direction } \mathbf{w} = \begin{bmatrix} 2 \\ -2 \\ 2 \end{bmatrix}$$

We have $\mathbf{v} \cdot \mathbf{w} = 0$, hence the lines are perpendicular.

21. Following the ideas in Examples 9 and 10, we have the following sketch.

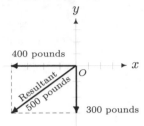

The resultant force is the sum of $300(-\mathbf{j})$ and $400(-\mathbf{i})$ which gives the resultant $(-400, -300)$ and $\|(-400, -300)\| = 500$.

23. To show that $ABCD$ is a parallelogram, we show that opposite sides are parallel.

side AB is in the direction of $B - A = (-3, 1, 2)$
side CD is in the direction of $C - D = (-3, 1, 2)$ $\implies$ $AB \parallel CD$

side BC is in the direction of $C - B = (2, 0, 1)$
side AD is in the direction of $D - A = (2, 0, 1)$ $\implies$ $BC \parallel AD$

25. $\mathbf{v} \cdot \mathbf{w} = 2 - 2c + 3 = 0$ $\iff$ $5 - 2c = 0$ $\iff$ $c = \frac{5}{2}$.

27. Setting the inner products equal to zero, we have the equations

$$\mathbf{v} \cdot \mathbf{w} = 2a + b + 2 = 0$$
$$\mathbf{v} \cdot \mathbf{x} = a + 2 = 0$$

It follows that $a = -2$ and $b = 2$. Thus $\mathbf{v} = \begin{bmatrix} -2 \\ 2 \\ 2 \end{bmatrix}$.

29. If $\mathbf{u}$ and $\mathbf{v}$ are parallel, then $\mathbf{v} = k\mathbf{u}$ for some scalar k and hence

$$\cos\theta = \frac{\mathbf{u} \cdot \mathbf{v}}{\|\mathbf{u}\| \, \|\mathbf{v}\|} = \frac{\mathbf{u} \cdot k\mathbf{u}}{\|\mathbf{u}\| \, \|k\mathbf{u}\|} = \frac{k \, \|\mathbf{u}\|^2}{|k| \, \|\mathbf{u}\|^2} = \pm 1.$$

31. Suppose that $\mathbf{v}$ is orthogonal to both $\mathbf{w}$ and $\mathbf{x}$. Then $\mathbf{v} \cdot \mathbf{w} = \mathbf{v} \cdot \mathbf{x} = 0$. Any vector in span $\{\mathbf{w}, \mathbf{x}\}$ is of the form $a\mathbf{w} + b\mathbf{x}$ for some scalars a, b. Hence

$$\mathbf{v} \cdot (a\mathbf{w} + b\mathbf{x}) = \mathbf{v} \cdot (a\mathbf{w}) + \mathbf{v} \cdot (b\mathbf{x}) = a(\mathbf{v} \cdot \mathbf{w}) + b(\mathbf{v} \cdot \mathbf{x}) = a(0) + b(0) = 0.$$

Thus $\mathbf{v}$ is orthogonal to every vector in span $\{\mathbf{w}, \mathbf{x}\}$.

33. We prove the result for $\mathbf{v}$ in R^2. Let $\mathbf{v} = (v_1, v_2)$. Then

$$\|c\mathbf{v}\| = \sqrt{(cv_1)^2 + (cv_2)^2} = \sqrt{c^2(v_1^2 + v_2^2)} = |c|\sqrt{v_1^2 + v_2^2} = |c| \, \|\mathbf{v}\|.$$

35. Suppose that $c_1\mathbf{v}_1 + c_2\mathbf{v}_2 + c_3\mathbf{v}_3 = \mathbf{0}$ for some scalars c_1, c_2, c_3. Taking the inner product of both sides of this equation with the vector $\mathbf{v}_1$, we obtain

$$\mathbf{v}_1 \cdot (c_1\mathbf{v}_1 + c_2\mathbf{v}_2 + c_3\mathbf{v}_3) = c_1(\mathbf{v}_1 \cdot \mathbf{v}_1) + c_2(\mathbf{v}_1 \cdot \mathbf{v}_2) + c_3(\mathbf{v}_1 \cdot \mathbf{v}_3)$$
$$= c_1 \, \|\mathbf{v}_1\|^2 + c_2(0) + c_3(0)$$
$$= c_1 \, \|\mathbf{v}_1\|^2 = 0$$

Since $\mathbf{v}_1 \neq \mathbf{0}$, $\|\mathbf{v}_1\| \neq 0$ and hence $c_1 = 0$. Repeating this argument with $\mathbf{v}_2$ and $\mathbf{v}_3$ we find $c_2 = c_3 = 0$. Thus S is linearly independent.

37. (a) $(\mathbf{u} + c\mathbf{v}) \cdot \mathbf{w} = \mathbf{u} \cdot \mathbf{w} + (c\mathbf{v} \cdot \mathbf{w}) = \mathbf{u} \cdot \mathbf{w} + c(\mathbf{v} \cdot \mathbf{w})$

   (b) $\mathbf{u} \cdot (c\mathbf{v}) = (c\mathbf{v}) \cdot \mathbf{u} = c(\mathbf{v} \cdot \mathbf{u}) = c(\mathbf{u} \cdot \mathbf{v})$

   (c) $(\mathbf{u} + \mathbf{v}) \cdot c\mathbf{w} = \mathbf{u} \cdot c\mathbf{w} + \mathbf{v} \cdot c\mathbf{w} = c(\mathbf{u} \cdot \mathbf{w}) + c(\mathbf{v} \cdot \mathbf{w})$

39. Let the vertices of an isosceles triangle be denoted by A, B and C. We show that the cosine of the angles between sides CA, AB and AC, CB are the same.

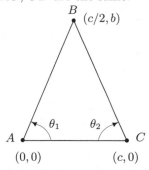

To simplify the expressions involved, let $A(0,0)$, $B(c/2, b)$ and $C(c, 0)$. (Note that the perpendicular bisector from B to AC bisects AC so we have the form of a general isosceles triangle.) Let

$$\mathbf{v} = \text{vector from } A \text{ to } B = \begin{bmatrix} c/2 \\ b \end{bmatrix}$$

$$\mathbf{w} = \text{vector from } A \text{ to } C = \begin{bmatrix} c \\ 0 \end{bmatrix}$$

$$\mathbf{u} = \text{vector from } C \text{ to } B = \begin{bmatrix} -c/2 \\ b \end{bmatrix}$$

Let θ_1 be the angle between $\mathbf{v}$ and $\mathbf{w}$. Then

$$\cos \theta_1 = \frac{\mathbf{v} \cdot \mathbf{w}}{\|\mathbf{v}\| \, \|\mathbf{w}\|} = \frac{c^2/2}{\sqrt{(c^2/4) + b^2} \, \sqrt{c^2}}.$$

Let θ_2 be the angle between $-\mathbf{w}$ and $\mathbf{u}$. Then

$$\cos \theta_2 = \frac{-\mathbf{w} \cdot \mathbf{u}}{\|\mathbf{w}\| \, \|\mathbf{u}\|} = \frac{c^2/2}{\sqrt{(c^2/4) + b^2} \, \sqrt{c^2}}.$$

Hence $\cos \theta_1 = \cos \theta_2$, which implies that $\theta_1 = \theta_2$ since an angle θ between vectors is between 0 and π radians.

Section 5.2, p. 306

1. We use the vector form for the cross product given in Equation (2) in the text.

   (a) $\mathbf{u} \times \mathbf{v} = ((3)(-1) - (4)(3))\mathbf{i} + ((4)(-1) - (2)(-1))\mathbf{j} + ((2)(3) - (3)(-1))\mathbf{k} = -15\mathbf{i} - 2\mathbf{j} + 9\mathbf{k}$

   (b) $\mathbf{u} \times \mathbf{v} = ((0)(-1) - (1)(3))\mathbf{i} + ((1)(2) - (1)(3))\mathbf{j} + ((1)(3) - (0)(2))\mathbf{k} = -3\mathbf{i} + 3\mathbf{j} + 3\mathbf{k}$

   (c) $\mathbf{u} \times \mathbf{v} = ((-1)(1) - (2)(-4))\mathbf{i} + ((2)(3) - (1)(1))\mathbf{j} + ((1)(-4) - (-1)(3))\mathbf{k} = 7\mathbf{i} + 5\mathbf{j} - \mathbf{k}$

   (d) $\mathbf{u} \times \mathbf{v} = \mathbf{u} \times (-2\mathbf{u}) = -((-2\mathbf{u}) \times \mathbf{u}) = 2(\mathbf{u} \times \mathbf{u}) = 2(\mathbf{0}) = \mathbf{0} = 0\mathbf{i} + 0\mathbf{j} + 0\mathbf{k}$

   Here we have used cross product properties (a), (d), and (e) which are listed preceding Example 2.

3. (a) $\mathbf{u} \times \mathbf{v} = ((2)(1) - (-3)(3))\mathbf{i} + ((-3)(2) - (1)(1))\mathbf{j} + ((1)(3) - (2)(2))\mathbf{k} = 11\mathbf{i} - 7\mathbf{j} - \mathbf{k}$

$-(\mathbf{v} \times \mathbf{u}) = -[((3)(-3) - (1)(2))\mathbf{i} + ((1)(1) - (2)(-3))\mathbf{j} + ((2)(2) - (3)(1))\mathbf{k}]$

$\qquad\qquad = -[-11\mathbf{i} + 7\mathbf{j} + \mathbf{k}] = 11\mathbf{i} - 7\mathbf{j} - \mathbf{k}$

Thus $\mathbf{u} \times \mathbf{v} = -(\mathbf{v} \times \mathbf{u})$.

(b) $\mathbf{u} \times (\mathbf{v} + \mathbf{w}) = (\mathbf{i} + 2\mathbf{j} - 3\mathbf{k}) \times (4\mathbf{i} + 2\mathbf{j} + 3\mathbf{k})$

$\qquad\qquad = ((2)(3) - (-3)(2))\mathbf{i} + ((-3)(4) - (1)(3))\mathbf{j} + ((1)(2) - (2)(4))\mathbf{k}$

$\qquad\qquad = 12\mathbf{i} - 15\mathbf{j} - 6\mathbf{k}$

$\mathbf{u} \times \mathbf{v} + \mathbf{u} \times \mathbf{w} = 11\mathbf{i} - 7\mathbf{j} - \mathbf{k}$

$\qquad\qquad\quad + ((2)(2) - (-3)(-1))\mathbf{i} + ((-3)(2) - (1)(2))\mathbf{j} + ((1)(-1) - (2)(2))\mathbf{k}$

$\qquad\qquad = 11\mathbf{i} - 7\mathbf{j} - \mathbf{k} + \mathbf{i} - 8\mathbf{j} - 5\mathbf{k}$

$\qquad\qquad = 12\mathbf{i} - 15\mathbf{j} - 6\mathbf{k}$

Thus $\mathbf{u} \times (\mathbf{v} + \mathbf{w}) = \mathbf{u} \times \mathbf{v} + \mathbf{u} \times \mathbf{w}$.

(c) $(\mathbf{u} + \mathbf{v}) \times \mathbf{w} = (3\mathbf{i} + 5\mathbf{j} - 2\mathbf{k}) \times (2\mathbf{i} - \mathbf{j} + 2\mathbf{k})$

$\qquad\qquad = ((5)(2) - (-2)(-1))\mathbf{i} + ((-2)(2) - (3)(2))\mathbf{j} + ((3)(-1) - (5)(2))\mathbf{k}$

$\qquad\qquad = 8\mathbf{i} - 10\mathbf{j} - 13\mathbf{k}$

$\mathbf{u} \times \mathbf{w} + \mathbf{v} \times \mathbf{w} = \mathbf{i} - 8\mathbf{j} - 5\mathbf{k}$

$\qquad\qquad\quad + ((3)(2) - (1)(-1))\mathbf{i} + ((1)(2) - (2)(2))\mathbf{j} + ((2)(-1) - (3)(2))\mathbf{k}$

$\qquad\qquad = \mathbf{i} - 8\mathbf{j} - 5\mathbf{k} + 7\mathbf{i} - 2\mathbf{j} - 8\mathbf{k}$

$\qquad\qquad = 8\mathbf{i} - 10\mathbf{j} - 13\mathbf{k}$

Thus $(\mathbf{u} + \mathbf{v}) \times \mathbf{w} = \mathbf{u} \times \mathbf{w} + \mathbf{v} \times \mathbf{w}$.

(d) $c(\mathbf{u} \times \mathbf{v}) = -3(11\mathbf{i} - 7\mathbf{j} - \mathbf{k}) = -33\mathbf{i} + 21\mathbf{j} + 3\mathbf{k}$

$(c\mathbf{u}) \times \mathbf{v} = (-3\mathbf{i} - 6\mathbf{j} + 9\mathbf{k}) \times (2\mathbf{i} + 3\mathbf{j} + \mathbf{k})$

$\qquad\qquad = ((-6)(1) - (9)(3))\mathbf{i} + ((9)(2) - (-3)(1))\mathbf{j} + ((-3)(3) - (-6)(2))\mathbf{k}$

$\qquad\qquad = -33\mathbf{i} + 21\mathbf{j} + 3\mathbf{k}$

$\mathbf{u} \times (c\mathbf{v}) = (\mathbf{i} + 2\mathbf{j} - 3\mathbf{k}) \times (-6\mathbf{i} - 9\mathbf{j} - 3\mathbf{k})$

$\qquad\qquad = ((2)(-3) - (-3)(-9))\mathbf{i} + ((-3)(-6) - (1)(-3))\mathbf{j} + ((1)(-9) - (2)(-6))\mathbf{k}$

$\qquad\qquad = -33\mathbf{i} + 21\mathbf{j} + 3\mathbf{k}$

Thus $c(\mathbf{u} \times \mathbf{v}) = (c\mathbf{u}) \times \mathbf{v} = \mathbf{u} \times (c\mathbf{v})$.

(e) $\mathbf{u} \times \mathbf{u} = ((2)(-3) - (-3)(2))\mathbf{i} + ((-3)(1) - (1)(-3))\mathbf{j} + ((1)(2) - (2)(1))\mathbf{k} = \mathbf{0}$

Thus $\mathbf{u} \times \mathbf{u} = \mathbf{0}$.

(f) $\mathbf{0} \times \mathbf{u} = (0\mathbf{i} + 0\mathbf{j} + 0\mathbf{k}) \times (\mathbf{i} + 2\mathbf{j} - 3\mathbf{k})$

$\qquad\qquad = ((0)(-3) - (0)(1))\mathbf{i} + ((0)(1) - (0)(-3))\mathbf{j} + ((0)(2) - (0)(1))\mathbf{k} = \mathbf{0}$

$\mathbf{u} \times \mathbf{0} = -(\mathbf{0} \times \mathbf{u}) = -\mathbf{0} = \mathbf{0}$

Thus $\mathbf{0} \times \mathbf{u} = \mathbf{u} \times \mathbf{0} = \mathbf{0}$.

(g) $\mathbf{u} \times (\mathbf{v} \times \mathbf{w}) = (\mathbf{i} + 2\mathbf{j} - 3\mathbf{k}) \times (7\mathbf{i} - 2\mathbf{j} - 8\mathbf{k})$

$\qquad\qquad = ((2)(-8) - (-3)(-2))\mathbf{i} + ((-3)(7) - (1)(-8))\mathbf{j} + ((1)(-2) - (2)(7))\mathbf{k}$

$\qquad\qquad = -22\mathbf{i} - 13\mathbf{j} - 16\mathbf{k}$

$(\mathbf{u} \cdot \mathbf{w})\mathbf{v} - (\mathbf{u} \cdot \mathbf{v})\mathbf{w} = ((1)(2) + (2)(-1) + (-3)(2))\mathbf{v} - ((1)(2) + (2)(3) + (-3)(1))\mathbf{w}$

$\qquad\qquad = -6\mathbf{v} - 5\mathbf{w} = -22\mathbf{i} - 13\mathbf{j} - 16\mathbf{k}$

Thus $\mathbf{u} \times (\mathbf{v} \times \mathbf{w}) = (\mathbf{u} \cdot \mathbf{w})\mathbf{v} - (\mathbf{u} \cdot \mathbf{v})\mathbf{w}$.

(h) $(\mathbf{u} \times \mathbf{v}) \times \mathbf{w} = (11\mathbf{i} - 7\mathbf{j} - \mathbf{k}) \times (2\mathbf{i} - \mathbf{j} + 2\mathbf{k})$

$\qquad\qquad = ((-7)(2) - (-1)(-1))\mathbf{i} + ((-1)(2) - (11)(2))\mathbf{j} + ((11)(-1) - (-7)(2))\mathbf{k}$

$\qquad\qquad = -6\mathbf{v} - 3\mathbf{u} = -15\mathbf{i} - 24\mathbf{j} + 3\mathbf{k}$

$(\mathbf{w} \cdot \mathbf{u})\mathbf{v} - (\mathbf{w} \cdot \mathbf{v})\mathbf{u} = ((2)(1) - (-1)(2) + (2)(-3))\mathbf{v} - ((2)(2) + (-1)(3) + (2)(1))\mathbf{u}$

Thus $(\mathbf{u} \times \mathbf{v}) \times \mathbf{w} = (\mathbf{w} \cdot \mathbf{u})\mathbf{v} - (\mathbf{w} \cdot \mathbf{v})\mathbf{u}$.

5. $(\mathbf{u} \times \mathbf{v}) \cdot \mathbf{w} = [((-2)(-1) - (3)(1))\mathbf{i} + ((3)(3) - (2)(-1))\mathbf{j} + ((2)(1) - (-2)(3))\mathbf{k}] \cdot \mathbf{w}$

$\qquad\qquad = (-\mathbf{i} + 11\mathbf{j} + 8\mathbf{k}) \cdot (3\mathbf{i} + \mathbf{j} + 2\mathbf{k}) = 24$

$\mathbf{u} \cdot (\mathbf{v} \times \mathbf{w}) = \mathbf{u} \cdot [((1)(2) - (-1)(1))\mathbf{i} + ((-1)(3) - (3)(2))\mathbf{j} + ((3)(1) - (1)(3))\mathbf{k}]$

$\qquad\qquad = (2\mathbf{i} - 2\mathbf{j} + 3\mathbf{k}) \cdot (3\mathbf{i} - 9\mathbf{j}) = 24$

Thus $(\mathbf{u} \times \mathbf{v}) \cdot \mathbf{w} = \mathbf{u} \cdot (\mathbf{v} \times \mathbf{w})$.

7. Let $\mathbf{u} = u_1\mathbf{i} + u_2\mathbf{j} + u_3\mathbf{k}$, $\mathbf{v} = v_1\mathbf{i} + v_2\mathbf{j} + v_3\mathbf{k}$, and $\mathbf{w} = w_1\mathbf{i} + w_2\mathbf{j} + w_3\mathbf{k}$.

$$
\begin{aligned}
(\mathbf{u} \times \mathbf{v}) \cdot \mathbf{w} &= [(u_2v_3 - u_3v_2)\mathbf{i} + (u_3v_1 - u_1v_3)\mathbf{j} + (u_1v_2 - u_2v_1)\mathbf{k}] \cdot \mathbf{w} \\
&= (u_2v_3 - u_3v_2)w_1 + (u_3v_1 - u_1v_3)w_2 + (u_1v_2 - u_2v_1)w_3 \\
&= u_1(v_2w_3 - v_3w_2) + u_2(v_3w_1 - v_1w_3) + u_3(v_1w_2 - v_2w_1) \\
&= \mathbf{u} \cdot (\mathbf{v} \times \mathbf{w})
\end{aligned}
$$

9. If $\mathbf{u}$ and $\mathbf{v}$ are parallel, then the angle θ between them is either 0 or π radians. Hence $\sin\theta = 0$ in either case. Then from Equation (6) we have $\|\mathbf{u} \times \mathbf{v}\| = 0$. But the only vector of length zero is the zero vector. Thus $\mathbf{u} \times \mathbf{v} = \mathbf{0}$. Conversely, if $\mathbf{u} \times \mathbf{v} = \mathbf{0}$, then from Equation (6) $\|\mathbf{u}\|\,\|\mathbf{v}\|\sin\theta = 0$. It follows that either $\sin\theta = 0$ or one (or both) of $\mathbf{u}$ and $\mathbf{v}$ is the zero vector. If $\sin\theta = 0$, then the angle between them is 0 or π radians so the vectors are parallel. If one of the vectors is the zero vector, then they are also parallel because the zero vector is parallel to every vector.

11. Use three applications of property (h) for cross products and simplify:

$$
\begin{aligned}
(\mathbf{u} \times \mathbf{v}) \times \mathbf{w} + (\mathbf{v} \times \mathbf{w}) \times \mathbf{u} + (\mathbf{w} \times \mathbf{u}) \times \mathbf{v} &= (\mathbf{w} \cdot \mathbf{u})\mathbf{v} - (\mathbf{w} \cdot \mathbf{v})\mathbf{u} + (\mathbf{u} \cdot \mathbf{v})\mathbf{w} - (\mathbf{u} \cdot \mathbf{w})\mathbf{v} \\
&\quad + (\mathbf{v} \cdot \mathbf{w})\mathbf{u} - (\mathbf{v} \cdot \mathbf{u})\mathbf{w} \\
&= [(\mathbf{w} \cdot \mathbf{u}) - (\mathbf{u} \cdot \mathbf{w})]\mathbf{v} + [(\mathbf{v} \cdot \mathbf{w}) - (\mathbf{w} \cdot \mathbf{v})]\mathbf{u} \\
&\quad + [(\mathbf{u} \cdot \mathbf{v}) - (\mathbf{v} \cdot \mathbf{u})]\mathbf{w} \\
&= 0\mathbf{v} + 0\mathbf{u} + 0\mathbf{w} = \mathbf{0}
\end{aligned}
$$

13. $\text{Area of triangle} = \frac{1}{2}\|(\mathbf{u}_2 - \mathbf{u}_1) \times (\mathbf{u}_3 - \mathbf{u}_1)\| = \frac{1}{2}\|(2\mathbf{i} + 3\mathbf{j} - \mathbf{k}) \times (\mathbf{i} + 2\mathbf{j} + 2\mathbf{k})\|$

$$
= \frac{1}{2}\|((3)(2) - (-1)(2))\mathbf{i} + ((-1)(1) - (2)(2))\mathbf{j} + ((2)(2) - (3)(1))\mathbf{k}\|
$$

$$
= \frac{1}{2}\|8\mathbf{i} - 5\mathbf{j} + \mathbf{k}\| = \frac{1}{2}\sqrt{64 + 25 + 1} = \frac{3}{2}\sqrt{10}
$$

15. Volume of parallelepiped $= |\mathbf{u} \cdot (\mathbf{v} \times \mathbf{w})| = |(2\mathbf{i} - \mathbf{j}) \cdot (-4\mathbf{i} - 7\mathbf{j} + 5\mathbf{k})| = |-8 + 7| = 1$

17. Determine if the coordinates of the points satisfy the equation of the plane.

   (a) $3(0 - 2) + 2(-2 + 3) - 4(3 - 4) = 0$. Thus the point is on the plane.

   (b) $3(1 - 2) + 2(-2 + 3) - 4(3 - 4) = 3 \neq 0$. Thus the point is not on the plane

19. We use the coordinates of the specified points to find vectors which lie in the plane determined by the points. Then we find a normal to the plane and apply Equation (9).

   (a) Let the points be denoted $A(0, 1, 2)$, $B(3, -2, 5)$ and $C(2, 3, 4)$. Let $\mathbf{v}$ be the vector from A to B:

   $$
   \mathbf{v} = (3 - 0)\mathbf{i} + (-2 - 1)\mathbf{j} + (5 - 2)\mathbf{k} = 3\mathbf{i} - 3\mathbf{j} + 3\mathbf{k}.
   $$

   Let $\mathbf{w}$ be the vector from A to C:

   $$
   \mathbf{w} = (2 - 0)\mathbf{i} + (3 - 1)\mathbf{j} + (4 - 2)\mathbf{k} = 2\mathbf{i} + 2\mathbf{j} + 2\mathbf{k}.
   $$

   The normal $\mathbf{n}$ to the plane determined by A, B, and C is $\mathbf{n} = \mathbf{v} \times \mathbf{w} = -12\mathbf{i} + 12\mathbf{k}$. The equation of the plane where we use point A is

   $$
   \mathbf{n} \cdot [(x - 0)\mathbf{i} + (y - 1)\mathbf{j} + (z - 2)\mathbf{k}] = -12x + 12(z - 2) = -12x + 12z - 24 = 0
   $$

   or equivalently $x - z + 2 = 0$.

(b) Let the points be denoted $A(2,3,4)$, $B(-1,-2,3)$ and $C(-5,-4,2)$. Let $\mathbf{v}$ be the vector from A to B:

$$\mathbf{v} = (-1-2)\mathbf{i} + (-2-3)\mathbf{j} + (3-4)\mathbf{k} = -3\mathbf{i} - 5\mathbf{j} - \mathbf{k}.$$

Let $\mathbf{w}$ be the vector from A to C:

$$\mathbf{w} = (-5-2)\mathbf{i} + (-4-3)\mathbf{j} + (2-4)\mathbf{k} = -7\mathbf{i} - 7\mathbf{j} - 2\mathbf{k}.$$

The normal $\mathbf{n}$ to the plane determined by A, B, and C is $\mathbf{n} = \mathbf{v} \times \mathbf{w} = 3\mathbf{i} + \mathbf{j} - 14\mathbf{k}$. The equation of the plane where we use point A is

$$\mathbf{n} \cdot [(x-2)\mathbf{i} + (y-3)\mathbf{j} + (z-4)\mathbf{k}] = 3(x-2) + (y-3) - 14(z-4) = 3x + y - 14z + 47 = 0.$$

21. The plane we seek is perpendicular to the line through the points $(4,-2,5)$ and $(0,2,4)$. Hence it is normal to the vector $\mathbf{v}$ determined by the points: $\mathbf{v} = 4\mathbf{i} - 4\mathbf{j} + \mathbf{k}$. Using the point $(-2,3,4)$, which is to lie in the plane, in Equation (9), the equation of the plane is given by

$$4(x+2) - 4(y-3) + (z-4) = 4x - 4y + z + 16 = 0.$$

23. The vector $\mathbf{n} = 2\mathbf{i} - 3\mathbf{j} + 4\mathbf{k}$ is normal to the plane. (See the discussion preceding Equation (9).) Vector $\mathbf{n}$ is to be parallel to the line we seek and this line is to go through the point $(-2,5,-3)$. Using the material on lines in R^3 from Section 4.3, the parametric equations of the line are:

$$x = -2 + 2t, \quad y = 5 - 3t, \quad z = -3 + 4t.$$

25. The subspace given by the plane $-3x + 2y + 5z = 0$ consists of all vectors $\begin{bmatrix} a \\ b \\ c \end{bmatrix}$ such that

$$\begin{bmatrix} a \\ b \\ c \end{bmatrix} \cdot \begin{bmatrix} -3 \\ 2 \\ 5 \end{bmatrix} = -3a + 2b + 5c = 0.$$

Therefore $a = \frac{2}{3}b + \frac{5}{3}c$ and hence

$$\begin{bmatrix} a \\ b \\ c \end{bmatrix} = \begin{bmatrix} \frac{2}{3}b + \frac{5}{3}c \\ b \\ c \end{bmatrix} = b \begin{bmatrix} \frac{2}{3} \\ 1 \\ 0 \end{bmatrix} + c \begin{bmatrix} \frac{5}{3} \\ 0 \\ 1 \end{bmatrix}.$$

Since the vectors $\begin{bmatrix} \frac{2}{3} \\ 1 \\ 0 \end{bmatrix}$ and $\begin{bmatrix} \frac{5}{3} \\ 0 \\ 1 \end{bmatrix}$ are linearly independent, it follows that

$$\left\{ \begin{bmatrix} \frac{2}{3} \\ 1 \\ 0 \end{bmatrix}, \begin{bmatrix} \frac{5}{3} \\ 0 \\ 1 \end{bmatrix} \right\}$$

is a basis for the subspace given by $-3x + 2y + 5z = 0$.

27. Following Example 4 we form the matrix corresponding to C for each of the pairs of vectors.

(a) $\mathbf{u} \times \mathbf{v} = \begin{vmatrix} \mathbf{i} & \mathbf{j} & \mathbf{k} \\ 2 & 3 & 4 \\ -1 & 3 & -1 \end{vmatrix} = -15\mathbf{i} - 2\mathbf{j} + 9\mathbf{k}$

(b) $\mathbf{u} \times \mathbf{v} = \begin{vmatrix} \mathbf{i} & \mathbf{j} & \mathbf{k} \\ 1 & 0 & 1 \\ 2 & 3 & -1 \end{vmatrix} = -3\mathbf{i} + 3\mathbf{j} + 3\mathbf{k}$

(c) $\mathbf{u} \times \mathbf{v} = \begin{vmatrix} \mathbf{i} & \mathbf{j} & \mathbf{k} \\ 1 & -1 & 2 \\ 3 & -4 & 1 \end{vmatrix} = 7\mathbf{i} + 5\mathbf{j} - \mathbf{k}$

(d) $\mathbf{u} \times \mathbf{v} = \begin{vmatrix} \mathbf{i} & \mathbf{j} & \mathbf{k} \\ 2 & 1 & -2 \\ 1 & 0 & 3 \end{vmatrix} = 3\mathbf{i} - 8\mathbf{j} - \mathbf{k}.$

29. Using the row operations $-\mathbf{r}_1 + \mathbf{r}_2 \to \mathbf{r}_2$, $-\mathbf{r}_1 + \mathbf{r}_3 \to \mathbf{r}_3$, and $-\mathbf{r}_1 + \mathbf{r}_4 \to \mathbf{r}_4$ we have

$$0 = \begin{vmatrix} x & y & z & 1 \\ x_1 & y_1 & z_1 & 1 \\ x_2 & y_2 & z_2 & 1 \\ x_3 & y_3 & z_3 & 1 \end{vmatrix} = \begin{vmatrix} x & y & z & 1 \\ x_1 - x & y_1 - y & z_1 - z & 1 \\ x_2 - x & y_2 - y & z_2 - z & 1 \\ x_3 - x & y_3 - y & z_3 - z & 1 \end{vmatrix} = (-1) \begin{vmatrix} x_1 - x & y_1 - y & z_1 - z \\ x_2 - x & y_2 - y & z_2 - z \\ x_3 - x & y_3 - y & z_3 - z \end{vmatrix}.$$

Using the row operations $-\mathbf{r}_1 \to \mathbf{r}_1$, $\mathbf{r}_1 + \mathbf{r}_2 \to \mathbf{r}_2$, and $\mathbf{r}_1 + \mathbf{r}_3 \to \mathbf{r}_3$, we have

$$
\begin{aligned}
0 &= \begin{vmatrix} x - x_1 & y - y_1 & z - z_1 \\ x_2 - x_1 & y_2 - y_1 & z_2 - z_1 \\ x_3 - x_1 & y_3 - y_1 & z_3 - z_1 \end{vmatrix} \\
&= (x - x_1)[y_2 - y_1 + z_3 - z_1 - y_3 + y_1 - z_2 + z_1] \\
&\quad + (y - y_1)[z_2 - z_1 + x_3 - x_1 - z_3 + z_1 - x_2 + x_1] \\
&\quad + (z - z_1)[x_2 - x_1 + y_3 - y_1 - x_3 + x_1 - y_2 + y_1] \\
&= (x - x_1)[y_2 - y_3 + z_3 - z_2] \\
&\quad + (y - y_1)[z_2 - z_3 + x_3 - x_2] \\
&\quad + (z - z_1)[x_2 - x_3 + y_3 - y_2]
\end{aligned}
$$

This is a linear equation of the form $Ax + By + Cz + D = 0$ and hence represents a plane. If we replace (x, y, z) in the original expression by (x_i, y_i, z_i), $i = 1, 2$

Section 5.3, p. 317

1. Let

$$\mathbf{u} = \begin{bmatrix} u_1 \\ u_2 \\ \vdots \\ u_n \end{bmatrix}, \quad \mathbf{v} = \begin{bmatrix} v_1 \\ v_2 \\ \vdots \\ v_n \end{bmatrix}, \quad \text{and} \quad \mathbf{w} = \begin{bmatrix} w_1 \\ w_2 \\ \vdots \\ w_n \end{bmatrix}$$

be vectors in R^n and let c be a real scalar.

(a) Suppose $\mathbf{u}$ is not the zero vector. Then some component u_j of $\mathbf{u}$ is not zero. Hence $(\mathbf{u}, \mathbf{u}) = u_1^2 + u_2^2 + \cdots + u_n^2 > 0$. If $(\mathbf{u}, \mathbf{u}) = 0$, then $u_1^2 + u_2^2 + \cdots + u_n^2 = 0$ which implies that $u_1 = u_2 = \cdots = u_n = 0$ hence $\mathbf{u} = \mathbf{0}$. If $\mathbf{u} = \mathbf{0}$, then $u_1 = u_2 = \cdots = u_n = 0$ and $(\mathbf{u}, \mathbf{u}) = u_1^2 + u_2^2 + \cdots + u_n^2 = 0$.

(b) $(\mathbf{v}, \mathbf{u}) = v_1 u_1 + v_2 u_2 + \cdots + v_n u_n = u_1 v_1 + u_2 v_2 + \cdots + u_n v_n = (\mathbf{u}, \mathbf{v})$

(c) $\begin{aligned}[t] (\mathbf{u} + \mathbf{v}, \mathbf{w}) &= (u_1 + v_1)w_1 + (u_2 + v_2)w_2 + \cdots + (u_n + v_n)w_n \\ &= u_1 w_1 + v_1 w_1 + u_2 w_2 + v_2 w_2 + \cdots + u_n w_n + v_n w_n \\ &= (u_1 w_1 + u_2 w_2 + \cdots + u_n w_n) + (v_1 w_1 + v_2 w_2 + \cdots + v_n w_n) \\ &= (\mathbf{u}, \mathbf{w}) + (\mathbf{v}, \mathbf{w}) \end{aligned}$

(d) $(c\mathbf{u}, \mathbf{v}) = (cu_1)v_1 + (cu_2)v_2 + \cdots + (cu_n)v_n = c(u_1 v_1 + u_2 v_2 + \cdots + u_n v_n) = c(\mathbf{u}, \mathbf{v})$

3. Let $A = [\, a_{ij} \,]$, $B = [\, b_{ij} \,]$, and $C = [\, c_{ij} \,]$ be $n \times n$ matrices with k a real scalar. We verify the requirements for inner product using properties of the trace function from Exercise 43, Section 1.3.

(a) Suppose A is not the zero matrix. Then some element a_{ij} of A is not zero. Hence

$$(A, A) = \text{Tr}(A^T A) = \sum_{i=1}^{n} \sum_{j=1}^{n} (a_{ij})^2 > 0.$$

If $(A, A) = 0$, then

$$\sum_{i=1}^{n} \sum_{j=1}^{n} (a_{ij})^2 = 0$$

which implies that each element of matrix A is zero, hence $A = O$. If $A = O$, the zero matrix, then every element of A is zero and hence

$$(A, A) = \sum_{i=1}^{n} \sum_{j=1}^{n} (a_{ij})^2 = 0.$$

(b) $(A, B) = \text{Tr}(B^T A) = \text{Tr}\left((B^T A)^T\right) = \text{Tr}(A^T B) = (B, A)$

(c) $(A + B, C) = \text{Tr}\left(C^T (A + B)\right) = \text{Tr}(C^T A + C^T B) = \text{Tr}(C^T A) + \text{Tr}(C^T B) = (A, C) + (B, C)$

(d) $(kA, B) = \text{Tr}\left(B^T (kA)\right) = \text{Tr}(kB^T A) = k\,\text{Tr}(B^T A) = k(A, B)$

5. Let $\mathbf{u} = \begin{bmatrix} u_1 & u_2 \end{bmatrix}$, $\mathbf{v} = \begin{bmatrix} v_1 & v_2 \end{bmatrix}$, and $\mathbf{w} = \begin{bmatrix} w_1 & w_2 \end{bmatrix}$ be vectors in R_2 and let c be a scalar. We define $(\mathbf{u}, \mathbf{v}) = u_1 v_1 - u_2 v_1 - u_1 v_2 + 5 u_2 v_2$.

(a) Suppose $\mathbf{u}$ is not the zero vector. Then one of u_1 and u_2 is not zero. Hence

$$(\mathbf{u}, \mathbf{u}) = u_1 u_1 - u_2 u_1 - u_1 u_2 + 5 u_2 u_2 = (u_1 - u_2)^2 + 4(u_2)^2 > 0.$$

If $(\mathbf{u}, \mathbf{u}) = 0$, then

$$u_1 u_1 - u_2 u_1 - u_1 u_2 + 5 u_2 u_2 = (u_1 - u_2)^2 + 4(u_2)^2 = 0$$

which implies that $u_1 = u_2 = 0$ hence $\mathbf{u} = \mathbf{0}$. If $\mathbf{u} = \mathbf{0}$, then $u_1 = u_2 = 0$ and

$$(\mathbf{u}, \mathbf{u}) = u_1 u_1 - u_2 u_1 - u_1 u_2 + 5 u_2 u_2 = 0.$$

(b) $(\mathbf{u}, \mathbf{v}) = u_1 v_1 - u_2 v_1 - u_1 v_2 + 5 u_2 v_2 = v_1 u_1 - v_2 u_1 - v_1 u_2 + 5 v_2 u_2 = (\mathbf{v}, \mathbf{u})$

(c) $\begin{aligned}(\mathbf{u} + \mathbf{v}, \mathbf{w}) &= (u_1 + v_1) w_1 - (u_2 + v_2) w_2 - (u_1 + v_1) w_2 + 5(u_2 + v_2) w_2 \\ &= u_1 w_1 + v_1 w_1 - u_2 w_2 - v_2 w_2 - u_1 w_2 - v_1 w_2 + 5 u_2 w_2 + 5 v_2 w_2 \\ &= (u_1 w_1 - u_2 w_2 - u_1 w_2 + 5 u_2 w_2) + (v_1 w_1 - v_2 w_2 - v_1 w_2 + 5 v_2 w_2) \\ &= (\mathbf{u}, \mathbf{w}) + (\mathbf{v}, \mathbf{w})\end{aligned}$

(d) $(c\mathbf{u}, \mathbf{v}) = (cu_1) v_1 - (cu_2) v_1 - (cu_1) v_2 + 5(cu_2) v_2 = c(u_1 v_1 - u_2 v_1 - u_1 v_2 + 5 u_2 v_2) = c(\mathbf{u}, \mathbf{v})$

7. (a) $\|\mathbf{0}\| = \sqrt{(\mathbf{0}, \mathbf{0})} = \sqrt{0} = 0$

(b) For any vector $\mathbf{u}$, $(\mathbf{0}, \mathbf{u}) = (\mathbf{0} + \mathbf{0}, \mathbf{u}) = (\mathbf{0}, \mathbf{u}) + (\mathbf{0}, \mathbf{u}) = 2(\mathbf{0}, \mathbf{u})$. Since the value of an inner product is a real number, it follows that $(\mathbf{0}, \mathbf{u}) = 0$.

(c) Let $\mathbf{v} = \mathbf{u}$. Then it follows from the assumption that $(\mathbf{u}, \mathbf{u}) = 0$. Therefore $\mathbf{u} = \mathbf{0}$.

(d) Since $(\mathbf{u}, \mathbf{w}) = (\mathbf{v}, \mathbf{w})$ for all vectors $\mathbf{w}$, it follows that $(\mathbf{u}, \mathbf{w}) - (\mathbf{v}, \mathbf{w}) = (\mathbf{u} - \mathbf{v}, \mathbf{w}) = 0$. This is true for every vector $\mathbf{w}$ in V. Hence by part (c), $\mathbf{u} - \mathbf{v} = \mathbf{0}$ which implies $\mathbf{u} = \mathbf{v}$.

(e) Since $(\mathbf{w}, \mathbf{u}) = (\mathbf{u}, \mathbf{w})$ and $(\mathbf{w}, \mathbf{v}) = (\mathbf{v}, \mathbf{w})$, it follows that $(\mathbf{u}, \mathbf{w}) = (\mathbf{v}, \mathbf{w})$ for all vectors $\mathbf{w}$. Hence, by part (d), $\mathbf{u} = \mathbf{v}$.

9. (a) $(\mathbf{u}, \mathbf{v}) = (1)(-1) + (2)(0) + (3)(-1) + (4)(-1) = -8$

(b) $(\mathbf{u}, \mathbf{v}) = (0)(2) + (-1)(0) + (1)(-8) + (4)(2) = 0$

(c) $(\mathbf{u}, \mathbf{v}) = (0)(2) + (0)(3) + (-1)(-1) + (2)(0) = 1$

11. (a) $(f, g) = \int_0^1 (3t)(2t^2)\, dt = \int_0^1 6t^3\, dt = \frac{3}{2}t^2\big|_0^1 = \frac{3}{2}$

    (b) $(f, g) = \int_0^1 (t)(e^t)\, dt = \int_0^1 te^t\, dt = (te^t - e^t)\big|_0^1 = (e - e) - (0 - 1) = 1$

    (c) $(f, g) = \int_0^1 (\sin t)(\cos t)\, dt = \frac{1}{2}\sin^2 t\big|_0^1 = \frac{1}{2}\sin^2 1$

13. (a) $\left\|\begin{bmatrix} 0 \\ -2 \end{bmatrix}\right\| = \sqrt{0^2 + (-2)^2} = 2$

    (b) $\left\|\begin{bmatrix} -2 \\ -4 \end{bmatrix}\right\| = \sqrt{(-2)^2 + (-4)^2} = \sqrt{20} = 2\sqrt{5}$

    (c) $\left\|\begin{bmatrix} 2 \\ 2 \end{bmatrix}\right\| = \sqrt{2^2 + 2^2} = \sqrt{8} = 2\sqrt{2}$

15. (a) For $p(t) = 1$ and $q(t) = 1$ we have $(p(t), q(t)) = \int_0^1 (1)(1)\, dt = t\big|_0^1 = 1$. Therefore

$$\cos\theta = \frac{(p(t), q(t))}{\|p(t)\|\, \|q(t)\|} = \frac{1}{1} = 1.$$

    (b) For $p(t) = t^2$ and $q(t) = 2t^3 - \frac{4}{3}t$ we have $(p(t), q(t)) = \int_0^1 (t^2)(2t^3 - \frac{4}{3}t)\, dt = \frac{1}{3}(t^6 - t^4)\big|_0^1 = 0$. Therefore

$$\cos\theta = \frac{(p(t), q(t))}{\|p(t)\|\, \|q(t)\|} = \frac{0}{\|p(t)\|\, \|q(t)\|} = 0.$$

    (c) For $p(t) = \sin t$ and $q(t) = \cos t$ we have

$$(p(t), q(t)) = \int_0^1 (\sin t)(\cos t)\, dt = \frac{1}{2}\sin^2 t\,\bigg|_0^1 = \frac{1}{2}\sin^2 1$$

$$\|p(t)\|^2 = (p(t), p(t)) = \int_0^1 \sin^2 t\, dt = \int_0^1 \frac{1 - \cos 2t}{2}\, dt = \left[\frac{1}{2}t - \frac{1}{4}\sin 2t\right]\bigg|_0^2 = \frac{1}{2} - \frac{1}{4}\sin 2$$

$$\|q(t)\|^2 = (q(t), q(t)) = \int_0^1 \cos^2 t\, dt = \int_0^1 \frac{1 + \cos 2t}{2}\, dt = \left[\frac{1}{2}t + \frac{1}{4}\sin 2t\right]\bigg|_0^1 = \frac{1}{2} + \frac{1}{4}\sin 2$$

    Therefore

$$\cos\theta = \frac{(p(t), q(t))}{\|p(t)\|\, \|q(t)\|} = \frac{\frac{1}{2}\sin^2 1}{\sqrt{\frac{1}{4}[2 - \sin 2]}\, \sqrt{\frac{1}{4}[2 + \sin 2]}} = \frac{2\sin^2 1}{\sqrt{4 - \sin^2 2}}.$$

17. $\|c\mathbf{u}\| = \sqrt{(c\mathbf{u}, c\mathbf{u})} = \sqrt{c(\mathbf{u}, c\mathbf{u})} = \sqrt{c(c\mathbf{u}, \mathbf{u})} = \sqrt{c^2(\mathbf{u}, \mathbf{u})} = |c|\sqrt{(\mathbf{u}, \mathbf{u})} = |c|\, \|\mathbf{u}\|$

19. Suppose that $\mathbf{u}$ and $\mathbf{v}$ are vectors in V. Then

$$\|\mathbf{u} + \mathbf{v}\|^2 = (\mathbf{u} + \mathbf{v}, \mathbf{u} + \mathbf{v}) = (\mathbf{u}, \mathbf{u} + \mathbf{v}) + (\mathbf{v}, \mathbf{u} + \mathbf{v}) = (\mathbf{u}, \mathbf{u}) + (\mathbf{u}, \mathbf{v}) + (\mathbf{v}, \mathbf{u}) + (\mathbf{v}, \mathbf{v})$$

$$= \|\mathbf{u}\|^2 + (\mathbf{u}, \mathbf{v}) + (\mathbf{u}, \mathbf{v}) + \|\mathbf{v}\|^2 = \|\mathbf{u}\|^2 + 2(\mathbf{u}, \mathbf{v}) + \|\mathbf{v}\|^2$$

It follows that $\|\mathbf{u} + \mathbf{v}\|^2 = \|\mathbf{u}\|^2 + \|\mathbf{v}\|^2$ if and only if $2(\mathbf{u}, \mathbf{v}) = 0$, or equivalently, $(\mathbf{u}, \mathbf{v}) = 0$.

21. Let $\mathbf{u}$ and $\mathbf{v}$ be vectors in V. Then

$$\frac{1}{4}\|\mathbf{u} + \mathbf{v}\|^2 - \frac{1}{4}\|\mathbf{u} - \mathbf{v}\|^2 = \frac{1}{4}(\mathbf{u} + \mathbf{v}, \mathbf{u} + \mathbf{v}) - \frac{1}{4}(\mathbf{u} - \mathbf{v}, \mathbf{u} - \mathbf{v})$$

$$= \frac{1}{4}[(\mathbf{u}, \mathbf{u}) + 2(\mathbf{u}, \mathbf{v}) + (\mathbf{v}, \mathbf{v})] - \frac{1}{4}[(\mathbf{v}, \mathbf{v}) - 2(\mathbf{u}, \mathbf{v}) + (\mathbf{v}, \mathbf{v})]$$

$$= (\mathbf{u}, \mathbf{v})$$

23. Let S be the set of all vectors in V orthogonal to $\mathbf{u}$. We show that S is closed under vector addition and scalar multiplication. Let $\mathbf{v}$ and $\mathbf{w}$ be in S. Then $(\mathbf{u}, \mathbf{v}) = (\mathbf{u}, \mathbf{w}) = 0$ and hence

$$(\mathbf{u}, \mathbf{v} + \mathbf{w}) = (\mathbf{u}, \mathbf{v}) + (\mathbf{u}, \mathbf{w}) = 0 + 0 = 0 \quad \text{and} \quad (\mathbf{u}, c\mathbf{v}) = c(\mathbf{u}, \mathbf{v}) = c(0) = 0.$$

Hence $\mathbf{v} + \mathbf{w}$ and $c\mathbf{v}$ are in S for all scalars c and therefore S is a subspace of V.

25. From Theorem 5.2, in order for C to be the matrix of an inner product with respect to the natural ordered basis $S = \{\mathbf{e}_1, \mathbf{e}_2\}$ we must have $(\mathbf{e}_1, \mathbf{e}_1) = 3$, $(\mathbf{e}_1, \mathbf{e}_2) = (\mathbf{e}_2, \mathbf{e}_1) = -2$, and $(\mathbf{e}_2, \mathbf{e}_2) = 3$. Thus for $\mathbf{u} = u_1\mathbf{e}_1 + u_2\mathbf{e}_2$ and $\mathbf{v} = v_1\mathbf{e}_1 + v_2\mathbf{e}_2$, we define

$$\begin{aligned}(\mathbf{u}, \mathbf{v}) &= (u_1\mathbf{e}_1 + u_2\mathbf{e}_2, v_1\mathbf{e}_1 + v_2\mathbf{e}_2) \\ &= u_1v_1(\mathbf{e}_1, \mathbf{e}_1) + u_1v_2(\mathbf{e}_1, \mathbf{e}_2) + u_2v_1(\mathbf{e}_2, \mathbf{e}_1) + u_2v_2(\mathbf{e}_2, \mathbf{e}_2) \\ &= 3u_1v_1 - 2u_1v_2 - 2u_2v_1 + 3u_2v_2\end{aligned}$$

27. We use Definition 5.4.

(a) $\begin{aligned}d(\sin t, \cos t) &= \|\sin t - \cos t\| = \left[\int_0^1 (\sin t - \cos t)^2\, dt\right]^{1/2} = \left[\int_0^1 [\sin^2 t - 2\sin t \cos t + \cos^2 t]\, dt\right]^{1/2} \\ &= \left[\int_0^1 [1 - 2\sin t \cos t]\, dt\right]^{1/2} = \left[(t - \sin^2 t)\big|_0^1\right]^{1/2} = \sqrt{1 - \sin^2 1}\end{aligned}$

(b) $\begin{aligned}d(t, t^2) &= \|t - t^2\| = \left[\int_0^1 (t - t^2)^2\, dt\right]^{1/2} = \left[\int_0^1 (t^2 - 2t^3 + t^4)\, dt\right]^{1/2} \\ &= \left[\left(\tfrac{1}{3}t^3 - \tfrac{1}{2}t^4 + \tfrac{1}{5}t^5\right)\big|_0^1\right]^{1/2} = \left[\tfrac{1}{3} - \tfrac{1}{2} + \tfrac{1}{5}\right]^{1/2} = \sqrt{\tfrac{1}{30}}\end{aligned}$

29. (a) Let

$$\mathbf{v}_1 = \begin{bmatrix} \frac{1}{\sqrt{2}} \\ 0 \\ \frac{1}{\sqrt{2}} \end{bmatrix}, \quad \mathbf{v}_2 = \begin{bmatrix} -\frac{1}{\sqrt{2}} \\ 0 \\ \frac{1}{\sqrt{2}} \end{bmatrix}, \quad \mathbf{v}_3 = \begin{bmatrix} 0 \\ 1 \\ 0 \end{bmatrix}.$$

Then $(\mathbf{v}_1, \mathbf{v}_2) = 0$, $(\mathbf{v}_1, \mathbf{v}_3) = 0$, $(\mathbf{v}_2, \mathbf{v}_3) = 0$, $\|\mathbf{v}_1\| = 1$, $\|\mathbf{v}_2\| = 1$, $\|\mathbf{v}_3\| = 1$. Thus the vectors form an orthonormal set.

(b) Let

$$\mathbf{v}_1 = \begin{bmatrix} 1 \\ 1 \\ 0 \end{bmatrix}, \quad \mathbf{v}_2 = \begin{bmatrix} 0 \\ 0 \\ 1 \end{bmatrix}, \quad \mathbf{v}_3 = \begin{bmatrix} 0 \\ 1 \\ 0 \end{bmatrix}.$$

Then $(\mathbf{v}_1, \mathbf{v}_2) = 0$, $(\mathbf{v}_1, \mathbf{v}_3) = 1$, $(\mathbf{v}_2, \mathbf{v}_3) = 0$. The vectors are not orthogonal so they cannot be orthonormal. Hence this set is neither orthogonal nor orthonormal.

(c) Let

$$\mathbf{v}_1 = \begin{bmatrix} 1 \\ -1 \\ 0 \end{bmatrix}, \quad \mathbf{v}_2 = \begin{bmatrix} 0 \\ 1 \\ 1 \end{bmatrix}, \quad \mathbf{v}_3 = \begin{bmatrix} 0 \\ 0 \\ 1 \end{bmatrix}.$$

Then $(\mathbf{v}_1, \mathbf{v}_2) = -1$, $(\mathbf{v}_1, \mathbf{v}_3) = 0$, $(\mathbf{v}_2, \mathbf{v}_3) = 1$. The vectors are not orthogonal so they cannot be orthonormal. Hence this set is neither orthogonal nor orthonormal.

31. The polynomials are orthogonal provided $(p(t), q(t)) = \int_0^1 (3t + 1)(at)\, dt = 0$. We have

$$\int_0^1 (3t + 1)(at)\, dt = \int_0^1 (3at^2 + at)\, dt = \left[at^3 + \frac{a}{2}t^2\right]\Big|_0^1 = \frac{3}{2}a = 0$$

only if $a = 0$.

33. $(\mathbf{u}, \mathbf{v}) = a - 5 = 0$ if and only if $a = 5$.

35. Let
$$B = \begin{bmatrix} b_{11} & b_{12} \\ b_{21} & b_{22} \end{bmatrix}.$$

Then
$$(A, B) = \text{Tr}\left(B^T A\right) = \text{Tr}\left(\begin{bmatrix} b_{11} & b_{21} \\ b_{12} & b_{22} \end{bmatrix} \begin{bmatrix} 1 & 2 \\ 3 & 4 \end{bmatrix}\right) = \text{Tr}\left(\begin{bmatrix} b_{11} + 3b_{21} & 2b_{11} + 4b_{21} \\ b_{12} + 3b_{22} & 2b_{12} + 4b_{22} \end{bmatrix}\right)$$
$$= b_{11} + 3b_{21} + 2b_{12} + 4b_{22} = 0$$

This is one equation in four unknowns, the elements of B. Hence three of the elements of B can be chosen arbitrarily as long as at least one of them is not zero. For example, if we let $b_{21} = 1$, $b_{12} = 1$, and $b_{22} = 3$, then from the preceding relation we have that $b_{11} = -17$. Certainly there is more than one matrix B which is orthogonal to A.

37. We must verify Definition 5.2 for
$$(\mathbf{u}, \mathbf{v}) = \sum_{i=1}^{n} \sum_{j=1}^{n} a_i c_{ij} b_j = \begin{bmatrix} \mathbf{v} \end{bmatrix}_S^T C \begin{bmatrix} \mathbf{w} \end{bmatrix}_S$$

We choose to use the matrix formulation of this inner product which appears in Equation (2) since we can then use matrix algebra to verify the parts of Definition 5.2

(a) $(\mathbf{v}, \mathbf{v}) = \begin{bmatrix} \mathbf{v} \end{bmatrix}_S^T C \begin{bmatrix} \mathbf{v} \end{bmatrix}_S > 0$ whenever $\begin{bmatrix} \mathbf{v} \end{bmatrix}_S \neq \mathbf{0}$ since C is positive definite.

   $(\mathbf{v}, \mathbf{v}) = 0$ if and only if $\begin{bmatrix} \mathbf{v} \end{bmatrix}_S = \mathbf{0}$ since C is positive definite. But $\begin{bmatrix} \mathbf{v} \end{bmatrix}_S = \mathbf{0}$ is true if and only if $\mathbf{v} = \mathbf{0}$.

(b) $(\mathbf{v}, \mathbf{w}) = \begin{bmatrix} \mathbf{v} \end{bmatrix}_S^T C \begin{bmatrix} \mathbf{w} \end{bmatrix}_S$ is a real number so it is equal to its transpose. Therefore
$$(\mathbf{v}, \mathbf{w}) = \begin{bmatrix} \mathbf{v} \end{bmatrix}_S^T C \begin{bmatrix} \mathbf{w} \end{bmatrix}_S = \left[\begin{bmatrix} \mathbf{v} \end{bmatrix}_S^T C \begin{bmatrix} \mathbf{w} \end{bmatrix}_S\right]^T$$
$$= \begin{bmatrix} \mathbf{w} \end{bmatrix}_S^T C^T \begin{bmatrix} \mathbf{v} \end{bmatrix}_S = \begin{bmatrix} \mathbf{w} \end{bmatrix}_S^T C \begin{bmatrix} \mathbf{v} \end{bmatrix}_S = (\mathbf{w}, \mathbf{v})$$

(c) $(\mathbf{u} + \mathbf{v}, \mathbf{w}) = \left[\begin{bmatrix} \mathbf{u} \end{bmatrix}_S + \begin{bmatrix} \mathbf{v} \end{bmatrix}_S\right]^T C \begin{bmatrix} \mathbf{w} \end{bmatrix}_S = \left[\begin{bmatrix} \mathbf{u} \end{bmatrix}_S^T + \begin{bmatrix} \mathbf{v} \end{bmatrix}_S^T\right] C \begin{bmatrix} \mathbf{w} \end{bmatrix}_S$
$$= \begin{bmatrix} \mathbf{u} \end{bmatrix}_S^T C \begin{bmatrix} \mathbf{w} \end{bmatrix}_S + \begin{bmatrix} \mathbf{v} \end{bmatrix}_S^T C \begin{bmatrix} \mathbf{w} \end{bmatrix}_S = (\mathbf{u}, \mathbf{w}) + (\mathbf{v}, \mathbf{w})$$

(d) $(k\mathbf{v}, \mathbf{w}) = \begin{bmatrix} k\mathbf{v} \end{bmatrix}_S^T C \begin{bmatrix} \mathbf{w} \end{bmatrix}_S = k \begin{bmatrix} \mathbf{v} \end{bmatrix}_S^T C \begin{bmatrix} \mathbf{w} \end{bmatrix}_S = k(\mathbf{v}, \mathbf{w})$

39. Let $\mathbf{u} = \begin{bmatrix} u_1 \\ u_2 \\ \vdots \\ u_n \end{bmatrix}$ and $\mathbf{v} = \begin{bmatrix} v_1 \\ v_2 \\ \vdots \\ v_n \end{bmatrix}$ be in R^n. Then

$$\mathbf{u}^T \mathbf{v} = \begin{bmatrix} u_1 & u_2 & \cdots & u_n \end{bmatrix} \begin{bmatrix} v_1 \\ v_2 \\ \vdots \\ v_n \end{bmatrix} = u_1 v_1 + u_2 v_2 + \cdots + u_n v_n$$

Since the standard inner product on R^n is defined by $(\mathbf{u}, \mathbf{v}) = u_1 v_1 + u_2 v_2 + \cdots + u_n v_n$, it follows that $(\mathbf{u}, \mathbf{v}) = \mathbf{u}^T \mathbf{v}$.

41. Let $S = \{\mathbf{w}_1, \mathbf{w}_2, \ldots \mathbf{w}_k\}$. By assumption $\mathbf{v}$ is orthogonal to every vector in S so $(\mathbf{v}, \mathbf{w}_i) = 0$ for $i = 1$, $\ldots$, k. Let $\mathbf{u}$ be in span S. Then $\mathbf{u} = c_1 \mathbf{w}_1 + c_2 \mathbf{w}_2 + \cdots + c_k \mathbf{w}_k$ for some scalars $c_1, c_2, \ldots, c_k$. Hence
$$(\mathbf{v}, \mathbf{u}) = (\mathbf{v}, c_1 \mathbf{w}_1 + c_2 \mathbf{w}_2 + \cdots + c_k \mathbf{w}_k)$$
$$= c_1(\mathbf{v}, \mathbf{w}_1) + c_2(\mathbf{v}, \mathbf{w}_2) + \cdots + c_k(\mathbf{v}, \mathbf{w}_k)$$
$$= c_1(0) + c_2(0) + \cdots + c_k(0) = 0$$

Thus $\mathbf{u}$ is orthogonal to every vector in span S.

43. In this exercise it is important to note that the set of vectors is assumed to be orthogonal, but there is no restriction that the vectors must be nonzero. Hence Theorem 5.4 does not apply. That is, we cannot conclude that the vectors must be linearly independent. Indeed, if we choose the zero vector as one of the vectors in the given set, then the matrix has a column of all zeros and hence is singular.

45. Since C is positive definite, we have $\mathbf{x}^T C \mathbf{x} > 0$. Multiply both sides of $C\mathbf{x} = k\mathbf{x}$ on the left by $\mathbf{x}^T$ to obtain $\mathbf{x}^T C \mathbf{x} = k\mathbf{x}^T \mathbf{x} > 0$. But $\mathbf{x}^T \mathbf{x} > 0$ since $\mathbf{x} \neq \mathbf{0}$. Hence k must be positive.

47. Let C be positive definite and let r be any scalar. For the matrix rC to be positive definite, we must show that for any vector $\mathbf{x} \neq \mathbf{0}$, $\mathbf{x}^T(rC)\mathbf{x} > 0$. But $\mathbf{x}^T(rC)\mathbf{x} = r(\mathbf{x}^T C\mathbf{x})$ using properties of matrix algebra. Since C is positive definite $\mathbf{x}^T C\mathbf{x} > 0$. However, since r can be any scalar there is no guarantee that $r(\mathbf{x}^T C\mathbf{x})$ is positive. For example, if $r = -5$, then $r(\mathbf{x}^T C\mathbf{x}) < 0$. Hence we have shown that rC is not necessarily positive definite.

49. By Exercise 48, S is closed under addition; but, by Exercise 47 it is not closed under scalar multiplication. Thus S is not a subspace of M_{nn}.

Section 5.4, p. 329

1. (a) $\mathbf{v}_1 = \begin{bmatrix} 1 \\ 2 \end{bmatrix}$, $\mathbf{v}_2 = \begin{bmatrix} -3 \\ 4 \end{bmatrix} - \dfrac{5}{5}\begin{bmatrix} 1 \\ 2 \end{bmatrix} = \begin{bmatrix} -4 \\ 2 \end{bmatrix}$

   (b) $\mathbf{w}_1 = \dfrac{\mathbf{v}_1}{\|\mathbf{v}_1\|} = \dfrac{\mathbf{v}_1}{\sqrt{5}} = \dfrac{1}{\sqrt{5}}\begin{bmatrix} 1 \\ 2 \end{bmatrix}$, $\mathbf{w}_2 = \dfrac{\mathbf{v}_2}{\|\mathbf{v}_2\|} = \dfrac{\mathbf{v}_1}{\sqrt{20}} = \dfrac{1}{\sqrt{5}}\begin{bmatrix} -2 \\ 1 \end{bmatrix}$

3. Let $S = \{\mathbf{u}_1, \mathbf{u}_2\} = \{\begin{bmatrix} 1 & 1 & -1 & 0 \end{bmatrix}, \begin{bmatrix} 0 & 2 & 0 & 1 \end{bmatrix}\}$. Using the Gram–Schmidt process we have:

$$\mathbf{v}_1 = \mathbf{u}_1 = \begin{bmatrix} 1 & 1 & -1 & 0 \end{bmatrix}$$

$$\mathbf{v}_2 = \mathbf{u}_2 - \dfrac{(\mathbf{u}_2, \mathbf{v}_1)}{(\mathbf{v}_1, \mathbf{v}_1)}\mathbf{v}_1 = \begin{bmatrix} 0 & 2 & 0 & 1 \end{bmatrix} - \dfrac{2}{3}\begin{bmatrix} 1 & 1 & -1 & 0 \end{bmatrix} = \dfrac{1}{3}\begin{bmatrix} -2 & 4 & 2 & 3 \end{bmatrix}$$

The set $\{\mathbf{v}_1, \mathbf{v}_2\}$ is an orthogonal basis for W. To obtain an orthonormal basis for W we determine unit vectors in the same direction as the $\mathbf{v}_i$, $i = 1, 2$. Let

$$\mathbf{w}_1 = \dfrac{\mathbf{v}_1}{\|\mathbf{v}_1\|} = \dfrac{1}{\sqrt{3}}\begin{bmatrix} 1 & 1 & -1 & 0 \end{bmatrix}$$

$$\mathbf{w}_2 = \dfrac{\mathbf{v}_2}{\|\mathbf{v}_2\|} = \dfrac{1}{\sqrt{33}}\begin{bmatrix} -2 & 4 & 2 & 3 \end{bmatrix}$$

Then $\{\mathbf{w}_1, \mathbf{w}_2\}$ is an orthonormal basis for W.

5. Let $S = \{\mathbf{u}_1, \mathbf{u}_2\} = \{t, 1\}$. Using the Gram–Schmidt process we have:

$$\mathbf{v}_1 = \mathbf{u}_1 = t$$

$$\mathbf{v}_2 = \mathbf{u}_2 - \dfrac{(\mathbf{u}_2, \mathbf{v}_1)}{(\mathbf{v}_1, \mathbf{v}_1)}\mathbf{v}_1 = 1 - \dfrac{\int_0^1 (1)(t)\,dt}{\int_0^1 t^2\,dt}t = 1 - \dfrac{\frac{1}{2}}{\frac{1}{3}}t = 1 - \dfrac{3}{2}t$$

The set $\{\mathbf{v}_1, \mathbf{v}_2\}$ is an orthogonal basis for W. To obtain an orthonormal basis for W we determine unit vectors in the same direction as the $\mathbf{v}_i$, $i = 1, 2$. Let

$$\mathbf{w}_1 = \dfrac{\mathbf{v}_1}{\|\mathbf{v}_1\|} = \dfrac{t}{\sqrt{\int_0^1 t^2\,dt}} = \dfrac{t}{\sqrt{\frac{1}{3}}} = \sqrt{3}\,t$$

$$\mathbf{w}_2 = \dfrac{\mathbf{v}_2}{\|\mathbf{v}_2\|} = \dfrac{1 - \frac{3}{2}t}{\sqrt{\int_0^1 \left[1 - \frac{3}{2}t\right]^2\,dt}} = \dfrac{1 - \frac{3}{2}t}{\sqrt{\frac{1}{4}}} = 2 - 3t$$

Then $\{\mathbf{w}_1, \mathbf{w}_2\}$ is an orthonormal basis for W.

7. Let $S = \{\mathbf{u}_1, \mathbf{u}_2\} = \{t, \sin 2\pi t\}$. Using the Gram–Schmidt process we have:

$$\mathbf{v}_1 = \mathbf{u}_1 = t$$

$$\mathbf{v}_2 = \mathbf{u}_2 - \frac{(\mathbf{u}_2, \mathbf{v}_1)}{(\mathbf{v}_1, \mathbf{v}_1)} \mathbf{v}_1 = \sin 2\pi t - \frac{\int_0^1 t \sin 2\pi t \, dt}{\int_0^1 t^2 \, dt} t = \sin 2\pi t - \frac{-\frac{1}{2\pi}}{\frac{1}{3}} t = \sin 2\pi t + \frac{3}{2\pi} t$$

The integral in the numerator above was found using integration by parts; the denominator was computed in Exercise 5. The set $\{\mathbf{v}_1, \mathbf{v}_2\}$ is an orthogonal basis for W. To obtain an orthonormal basis for W we determine unit vectors in the same direction as the $\mathbf{v}_i$, $i = 1, 2$. Let

$$\mathbf{w}_1 = \frac{\mathbf{v}_1}{\|\mathbf{v}_1\|} = \frac{t}{\sqrt{\int_0^1 t^2 \, dt}} = \frac{t}{\sqrt{\frac{1}{3}}} = \sqrt{3}\, t$$

$$\mathbf{w}_2 = \frac{\mathbf{v}_2}{\|\mathbf{v}_2\|} = \frac{\sin 2\pi t + \frac{3}{2\pi} t}{\sqrt{\int_0^1 \left[\sin 2\pi t + \frac{3}{2\pi} t \right]^2 dt}} = \frac{\sin 2\pi t + \frac{3}{2\pi} t}{\sqrt{\int_0^1 \left[\sin^2 2\pi t + \frac{3}{\pi} t \sin 2\pi t + \frac{9}{4\pi^2} t^2 \right] dt}}$$

We integrate the terms in the preceding expression:

$$\int_0^1 \sin^2 2\pi t \, dt = \left[\frac{1}{2} t - \frac{1}{8\pi} \sin 4\pi t \right]\Bigg|_0^1 = \frac{1}{2}$$

$$\int_0^1 \frac{3}{\pi} t \sin 2\pi t \, dt = \frac{3}{\pi} \int_0^1 t \sin 2\pi t \, dt = \left(\frac{3}{\pi} \right) \left(-\frac{1}{2\pi} \right) = -\frac{3}{2\pi^2}$$

$$\int_0^1 \frac{9}{4\pi^2} t^2 \, dt = \frac{9}{4\pi^2} \int_0^1 t^2 \, dt = \left(\frac{9}{4\pi^2} \right) \left(\frac{1}{3} \right) = \frac{3}{4\pi^2}$$

Then

$$\mathbf{w}_2 = \frac{\sin 2\pi t + \frac{3}{2\pi} t}{\sqrt{\frac{1}{2} - \frac{3}{2\pi^2} + \frac{3}{4\pi^2}}} = \frac{\sin 2\pi t + \frac{3}{2\pi} t}{\sqrt{\frac{1}{2} - \frac{3}{4\pi^2}}}$$

Therefore $\{\mathbf{w}_1, \mathbf{w}_2\}$ is an orthonormal basis for W.

9. We must first find a basis for R^3 that contains

$$\mathbf{u}_1 = \begin{bmatrix} \frac{2}{3} \\ -\frac{2}{3} \\ \frac{1}{3} \end{bmatrix} \quad \text{and} \quad \mathbf{u}_2 = \begin{bmatrix} \frac{2}{3} \\ \frac{1}{3} \\ -\frac{2}{3} \end{bmatrix}.$$

Note that $\mathbf{u}_1$ and $\mathbf{u}_2$ are linearly independent since they are not scalar multiples of one another. Using techniques from Section 4.4 we form the set $T = \{\mathbf{u}_1, \mathbf{u}_2, \mathbf{e}_1, \mathbf{e}_2, \mathbf{e}_3\}$, where the $\mathbf{e}_j$ are the standard basis vectors of R^3. Certainly this set spans R^3, and we may use the techniques developed in the alternate proof of Theorem 4.9 to determine a basis that contains $\mathbf{u}_1$ and $\mathbf{u}_2$. We form the matrix whose columns are the vectors in T and row reduce it. The reduced echelon form of this matrix is

$$\begin{bmatrix} 1 & 0 & 0 & -2 & -1 \\ 0 & 1 & 0 & -1 & -2 \\ 0 & 0 & 1 & 2 & 2 \end{bmatrix}.$$

The vectors of T corresponding to columns with leading 1's form a basis for R^3. Let $S = \{\mathbf{u}_1, \mathbf{u}_2, \mathbf{u}_3\}$, where $\mathbf{u}_3 = \mathbf{e}_1$. Then S is a basis for R^3 to which we apply the Gram–Schmidt procedure. Let

$$\mathbf{v}_1 = \mathbf{u}_1 = \begin{bmatrix} \frac{2}{3} \\ -\frac{2}{3} \\ \frac{1}{3} \end{bmatrix}$$

$$\mathbf{v}_2 = \mathbf{u}_2 - \frac{(\mathbf{u}_2, \mathbf{v}_1)}{(\mathbf{v}_1, \mathbf{v}_1)} \mathbf{v}_1 = \begin{bmatrix} \frac{2}{3} \\ \frac{1}{3} \\ -\frac{2}{3} \end{bmatrix} - \frac{0}{1} \mathbf{v}_1 = \mathbf{u}_2$$

$$\mathbf{v}_3 = \mathbf{u}_3 - \frac{(\mathbf{u}_3, \mathbf{v}_1)}{(\mathbf{v}_1, \mathbf{v}_1)} \mathbf{v}_1 - \frac{(\mathbf{u}_3, \mathbf{v}_2)}{(\mathbf{v}_2, \mathbf{v}_2)} \mathbf{v}_2 = \mathbf{u}_3 - \frac{\frac{2}{3}}{1} \mathbf{v}_1 - \frac{\frac{2}{3}}{1} \mathbf{v}_2 = \begin{bmatrix} \frac{1}{9} \\ \frac{2}{9} \\ \frac{2}{9} \end{bmatrix}$$

The set $\{\mathbf{v}_1, \mathbf{v}_2, \mathbf{v}_3\}$ is an orthogonal basis for R^3. To obtain an orthonormal basis we determine unit vectors in the same direction as the $\mathbf{v}_i$, $i = 1, 2, 3$. Let

$$\mathbf{w}_1 = \frac{\mathbf{v}_1}{\|\mathbf{v}_1\|} = \mathbf{v}_1$$

$$\mathbf{w}_2 = \frac{\mathbf{v}_2}{\|\mathbf{v}_2\|} = \mathbf{v}_2$$

$$\mathbf{w}_3 = \frac{\mathbf{v}_3}{\|\mathbf{v}_3\|} = \frac{\mathbf{v}_3}{\sqrt{\frac{1}{9}}} = 3\mathbf{v}_3 = \begin{bmatrix} \frac{1}{3} \\ \frac{2}{3} \\ \frac{2}{3} \end{bmatrix}$$

Then $\{\mathbf{w}_1, \mathbf{w}_2, \mathbf{w}_3\}$ is an orthonormal basis for R^3 that contains the vectors $\mathbf{u}_1$, $\mathbf{u}_2$.

11. Note that

$$\begin{bmatrix} 2 \\ 2 \\ 2 \end{bmatrix} = 2 \begin{bmatrix} 1 \\ 1 \\ 1 \end{bmatrix}$$

and so the set is linearly dependent. Omit $\begin{bmatrix} 2 & 2 & 2 \end{bmatrix}^T$ from the set and let

$$S = \{\mathbf{u}_1, \mathbf{u}_2, \mathbf{u}_3\} = \left\{ \begin{bmatrix} 1 \\ 1 \\ 1 \end{bmatrix}, \begin{bmatrix} 0 \\ 0 \\ 1 \end{bmatrix}, \begin{bmatrix} 1 \\ 2 \\ 3 \end{bmatrix} \right\}.$$

S is linearly independent since the reduced row echelon form of

$$\begin{bmatrix} 1 & 0 & 1 \\ 1 & 0 & 2 \\ 1 & 1 & 3 \end{bmatrix}$$

is I_3. Thus by Theorem 4.12(a) S is a basis for R^3. Applying the Gram–Schmidt process to the vectors in S we have:

$$\mathbf{v}_1 = \mathbf{u}_1 = \begin{bmatrix} 1 \\ 1 \\ 1 \end{bmatrix}$$

$$\mathbf{v}_2 = \mathbf{u}_2 - \frac{(\mathbf{u}_2, \mathbf{v}_1)}{(\mathbf{v}_1, \mathbf{v}_1)} \mathbf{v}_1 = \begin{bmatrix} 0 \\ 0 \\ 1 \end{bmatrix} - \frac{1}{3} \begin{bmatrix} 1 \\ 1 \\ 1 \end{bmatrix} = \begin{bmatrix} -\frac{1}{3} \\ -\frac{1}{3} \\ \frac{2}{3} \end{bmatrix}$$

$$\mathbf{v}_3 = \mathbf{u}_3 - \frac{(\mathbf{u}_3, \mathbf{v}_1)}{(\mathbf{v}_1, \mathbf{v}_1)} \mathbf{v}_1 - \frac{(\mathbf{u}_3, \mathbf{v}_2)}{(\mathbf{v}_2, \mathbf{v}_2)} \mathbf{v}_2 = \mathbf{u}_3 - \frac{6}{3} \mathbf{v}_1 - \frac{1}{\frac{6}{9}} \mathbf{v}_2$$

$$= \begin{bmatrix} 1 \\ 2 \\ 3 \end{bmatrix} - 2 \begin{bmatrix} 1 \\ 1 \\ 1 \end{bmatrix} - \frac{3}{2} \begin{bmatrix} -\frac{1}{3} \\ -\frac{1}{3} \\ \frac{2}{3} \end{bmatrix} = \begin{bmatrix} -\frac{1}{2} \\ \frac{1}{2} \\ 0 \end{bmatrix}$$

The set $\{\mathbf{v}_1, \mathbf{v}_2, \mathbf{v}_3\}$ is an orthogonal basis for R^3. To obtain an orthonormal basis we determine unit vectors in the same direction as the $\mathbf{v}_i$, $i = 1, 2, 3$. Let

$$\mathbf{w}_1 = \frac{\mathbf{v}_1}{\|\mathbf{v}_1\|} = \frac{\mathbf{v}_1}{\sqrt{3}} = \frac{1}{\sqrt{3}} \begin{bmatrix} 1 \\ 1 \\ 1 \end{bmatrix}$$

$$\mathbf{w}_2 = \frac{\mathbf{v}_2}{\|\mathbf{v}_2\|} = \frac{\mathbf{v}_2}{\sqrt{\frac{6}{9}}} = \frac{1}{\sqrt{6}} \begin{bmatrix} -1 \\ -1 \\ 2 \end{bmatrix}$$

$$\mathbf{w}_3 = \frac{\mathbf{v}_3}{\|\mathbf{v}_3\|} = \frac{\mathbf{v}_3}{\sqrt{\frac{1}{2}}} = \frac{1}{\sqrt{2}} \begin{bmatrix} -1 \\ 1 \\ 0 \end{bmatrix}$$

Then $\{\mathbf{w}_1, \mathbf{w}_2, \mathbf{w}_3\}$ is an orthonormal basis for R^3.

13. Let W be the given subspace. Since

$$\begin{bmatrix} a \\ a+b \\ b \end{bmatrix} = a \begin{bmatrix} 1 \\ 1 \\ 0 \end{bmatrix} + b \begin{bmatrix} 0 \\ 1 \\ 1 \end{bmatrix}$$

the vectors

$$\mathbf{u}_1 = \begin{bmatrix} 1 \\ 1 \\ 0 \end{bmatrix}, \quad \mathbf{u}_2 = \begin{bmatrix} 0 \\ 1 \\ 1 \end{bmatrix}$$

span the subspace W and form a basis for W. Using the Gram-Schmidt process we have:

$$\mathbf{v}_1 = \mathbf{u}_1 = \begin{bmatrix} 1 \\ 1 \\ 0 \end{bmatrix}$$

$$\mathbf{v}_2 = \mathbf{u}_2 - \frac{(\mathbf{u}_2, \mathbf{v}_1)}{(\mathbf{v}_1, \mathbf{v}_1)}\mathbf{v}_1 = \begin{bmatrix} 0 \\ 1 \\ 1 \end{bmatrix} - \frac{1}{2}\begin{bmatrix} 1 \\ 1 \\ 0 \end{bmatrix} = \begin{bmatrix} -\frac{1}{2} \\ \frac{1}{2} \\ 1 \end{bmatrix}$$

The set $\{\mathbf{v}_1, \mathbf{v}_2\}$ is an orthogonal basis for W. To obtain an orthonormal basis for W we determine unit vectors in the same direction as the $\mathbf{v}_i$, $i = 1, 2$. Let

$$\mathbf{w}_1 = \frac{\mathbf{v}_1}{\|\mathbf{v}_1\|} = \frac{\mathbf{v}_1}{\sqrt{2}} = \frac{1}{\sqrt{2}} \begin{bmatrix} 1 \\ 1 \\ 0 \end{bmatrix}$$

$$\mathbf{w}_2 = \frac{\mathbf{v}_2}{\|\mathbf{v}_2\|} = \frac{\mathbf{v}_2}{\sqrt{\frac{3}{2}}} = \frac{2}{\sqrt{6}} \begin{bmatrix} -\frac{1}{2} \\ \frac{1}{2} \\ 1 \end{bmatrix} = \frac{1}{\sqrt{6}} \begin{bmatrix} -1 \\ 1 \\ 2 \end{bmatrix}$$

Therefore $\{\mathbf{w}_1, \mathbf{w}_2\}$ is an orthonormal basis for W.

15. Let W be the given subspace. Since

$$\begin{bmatrix} a \\ b \\ c \end{bmatrix} = \begin{bmatrix} a \\ b \\ -a-b \end{bmatrix} = a \begin{bmatrix} 1 \\ 0 \\ -1 \end{bmatrix} + b \begin{bmatrix} 0 \\ 1 \\ -1 \end{bmatrix}$$

the vectors

$$\mathbf{u}_1 = \begin{bmatrix} 1 \\ 0 \\ -1 \end{bmatrix}, \quad \mathbf{u}_2 = \begin{bmatrix} 0 \\ 1 \\ -1 \end{bmatrix}$$

span W and form a basis for W. Using the Gram-Schmidt process on the set $S = \{\mathbf{u}_1, \mathbf{u}_2\}$ we have:

$$\mathbf{v}_1 = \mathbf{u}_1 = \begin{bmatrix} 1 \\ 0 \\ -1 \end{bmatrix}$$

$$\mathbf{v}_2 = \mathbf{u}_2 - \frac{(\mathbf{u}_2, \mathbf{v}_1)}{(\mathbf{v}_1, \mathbf{v}_1)}\mathbf{v}_1 = \begin{bmatrix} 0 \\ 1 \\ -1 \end{bmatrix} - \frac{1}{2}\begin{bmatrix} 1 \\ 0 \\ -1 \end{bmatrix} = \begin{bmatrix} -\frac{1}{2} \\ 1 \\ -\frac{1}{2} \end{bmatrix}$$

The set $\{\mathbf{v}_1, \mathbf{v}_2\}$ is an orthogonal basis for W. To obtain an orthonormal basis for W we determine unit vectors in the same direction as the $\mathbf{v}_i$, $i = 1, 2$. Let

$$\mathbf{w}_1 = \frac{\mathbf{v}_1}{\|\mathbf{v}_1\|} = \frac{\mathbf{v}_1}{\sqrt{2}} = \frac{1}{\sqrt{2}}\begin{bmatrix} 1 \\ 0 \\ -1 \end{bmatrix}$$

$$\mathbf{w}_2 = \frac{\mathbf{v}_2}{\|\mathbf{v}_2\|} = \frac{\mathbf{v}_2}{\sqrt{\frac{3}{2}}} = \frac{2}{\sqrt{6}}\begin{bmatrix} -\frac{1}{2} \\ 1 \\ -\frac{1}{2} \end{bmatrix} = \frac{1}{\sqrt{6}}\begin{bmatrix} -1 \\ 2 \\ -1 \end{bmatrix}$$

Therefore the set $\{\mathbf{w}_1, \mathbf{w}_2\}$ is an orthonormal basis for W.

17. We first solve the homogeneous system by forming the coefficient matrix and finding its row reduced echelon form. We obtain

$$\begin{bmatrix} 1 & 0 & 3 \\ 0 & 1 & -4 \end{bmatrix}.$$

The solutions to the system are therefore $x_1 = -3r$, $x_2 = 4r$, $x_3 = r$, where r is any real number. The general solution is therefore

$$\begin{bmatrix} x_1 \\ x_2 \\ x_3 \end{bmatrix} = \begin{bmatrix} -3r \\ 4r \\ r \end{bmatrix} = r\begin{bmatrix} -3 \\ 4 \\ 1 \end{bmatrix}.$$

Let $\mathbf{u}_1 = \begin{bmatrix} -3 & 4 & 1 \end{bmatrix}^T$ and let $\mathbf{v}_1 = \mathbf{u}_1$. Then $\{\mathbf{v}_1\}$ is a basis for the solution space. To determine a unit vector in the direction of $\mathbf{v}_1$, we let

$$\mathbf{w}_1 = \frac{\mathbf{v}_1}{\|\mathbf{v}_1\|} = \frac{\mathbf{v}_1}{\sqrt{26}} = \frac{1}{\sqrt{26}}\begin{bmatrix} -3 \\ 4 \\ 1 \end{bmatrix}.$$

Then the set $\{\mathbf{w}_1\}$ is an orthonormal basis for the solution space of the system.

19. Since S is a basis for V we can express $\mathbf{v}$ as a linear combination of the vectors in S. Let

$$\mathbf{v} = c_1\mathbf{u}_1 + c_2\mathbf{u}_2 + \cdots + c_n\mathbf{u}_n.$$

Next we take the inner product of $\mathbf{v}$ with each of the basis vectors and use the fact that the vectors in S are an orthonormal set; that is,

$$(\mathbf{u}_i, \mathbf{u}_j) = \begin{cases} 0, & \text{if } i \neq j \\ 1, & \text{if } i = j \end{cases}$$

We have

$$(\mathbf{v}, \mathbf{u}_1) = c_1(\mathbf{u}_1, \mathbf{u}_1) + c_2(\mathbf{u}_2, \mathbf{u}_1) + \cdots + c_n(\mathbf{u}_n, \mathbf{u}_1) = c_1$$
$$(\mathbf{v}, \mathbf{u}_2) = c_1(\mathbf{u}_1, \mathbf{u}_2) + c_2(\mathbf{u}_2, \mathbf{u}_2) + \cdots + c_n(\mathbf{u}_n, \mathbf{u}_2) = c_2$$
$$\vdots$$
$$(\mathbf{v}, \mathbf{u}_n) = c_1(\mathbf{u}_1, \mathbf{u}_n) + c_2(\mathbf{u}_2, \mathbf{u}_n) + \cdots + c_n(\mathbf{u}_n, \mathbf{u}_n) = c_n$$

Thus we have $c_i = (\mathbf{v}, \mathbf{u}_i)$, for $i = 1, 2, \ldots, n$.

21. Since $\begin{bmatrix} \mathbf{v} \end{bmatrix}_T$ is given we know that $\mathbf{v} = a_1\mathbf{u}_1 + a_2\mathbf{u}_2 + \cdots + a_n\mathbf{u}_n$. Then

$$(\mathbf{v}, \mathbf{v}) = a_1(\mathbf{u}_1, \mathbf{v}) + a_2(\mathbf{u}_2, \mathbf{v}) + \cdots + a_n(\mathbf{u}_n, \mathbf{v}).$$

Since T is an orthonormal basis

$$(\mathbf{u}_i, \mathbf{u}_j) = \begin{cases} 0, & \text{if } i \neq j \\ 1, & \text{if } i = j \end{cases}$$

hence $(\mathbf{u}_j, \mathbf{v}) = a_j$ as shown in Theorem 3.5. Thus it follows that

$$(\mathbf{v}, \mathbf{v}) = a_1^2 + a_2^2 + \cdots + a_n^2 \quad \text{and} \quad \|\mathbf{v}\| = \sqrt{a_1^2 + a_2^2 + \cdots + a_n^2}.$$

23. Let $S = \{\mathbf{u}_1, \mathbf{u}_2, \mathbf{u}_3\} = \left\{ \begin{bmatrix} \frac{1}{3} \\ \frac{2}{3} \\ \frac{2}{3} \end{bmatrix}, \begin{bmatrix} \frac{2}{3} \\ \frac{1}{3} \\ -\frac{2}{3} \end{bmatrix}, \begin{bmatrix} \frac{2}{3} \\ -\frac{2}{3} \\ \frac{1}{3} \end{bmatrix} \right\}.$

(a) This follows directly from Exercise 9.

(b) Let $\mathbf{v} = \begin{bmatrix} 15 \\ 3 \\ 3 \end{bmatrix}$. Then $c_1 = (\mathbf{v}, \mathbf{u}_1) = 9$, $c_2 = (\mathbf{v}, \mathbf{u}_2) = 9$, and $c_3 = (\mathbf{v}, \mathbf{u}_3) = 9$. Thus

$$\begin{bmatrix} \mathbf{v} \end{bmatrix}_S = \begin{bmatrix} 9 \\ 9 \\ 9 \end{bmatrix}.$$

(c) $\|\mathbf{v}\| = \sqrt{(15)^2 + (3)^2 + (3)^2} = \sqrt{243} = 9\sqrt{3}$. Since S is an orthonormal basis,

$$\|\mathbf{v}\| = \|\begin{bmatrix} \mathbf{v} \end{bmatrix}_S\| = \sqrt{9^2 + 9^2 + 9^2} = \sqrt{243} = 9\sqrt{3}$$

25. Let $S = \{\mathbf{u}_1, \mathbf{u}_2, \mathbf{u}_3, \mathbf{u}_4\} = \left\{ \begin{bmatrix} 1 & 0 \\ 0 & 0 \end{bmatrix}, \begin{bmatrix} 0 & 1 \\ 0 & 0 \end{bmatrix}, \begin{bmatrix} 0 & 0 \\ 1 & 0 \end{bmatrix}, \begin{bmatrix} 0 & 0 \\ 0 & 1 \end{bmatrix} \right\}.$

(a) S is the natural basis for V. To show that S is an orthonormal basis, we show that

$$(\mathbf{u}_i, \mathbf{u}_j) = \begin{cases} 0, & \text{if } i \neq j \\ 1, & \text{if } i = j \end{cases}$$

where the inner product is given by $(A, B) = \text{Tr}(B^T A)$.

$$(\mathbf{u}_1, \mathbf{u}_1) = \text{Tr}(\mathbf{u}_1^T \mathbf{u}_1) = \text{Tr}\left(\begin{bmatrix} 1 & 0 \\ 0 & 0 \end{bmatrix} \right) = 1, \quad (\mathbf{u}_3, \mathbf{u}_3) = \text{Tr}(\mathbf{u}_3^T \mathbf{u}_3) = \text{Tr}\left(\begin{bmatrix} 1 & 0 \\ 0 & 0 \end{bmatrix} \right) = 1$$

$$(\mathbf{u}_2, \mathbf{u}_2) = \text{Tr}(\mathbf{u}_2^T \mathbf{u}_2) = \text{Tr}\left(\begin{bmatrix} 0 & 0 \\ 0 & 1 \end{bmatrix} \right) = 1, \quad (\mathbf{u}_4, \mathbf{u}_4) = \text{Tr}(\mathbf{u}_4^T \mathbf{u}_4) = \text{Tr}\left(\begin{bmatrix} 0 & 0 \\ 0 & 1 \end{bmatrix} \right) = 1$$

$$(\mathbf{u}_1, \mathbf{u}_2) = \text{Tr}(\mathbf{u}_2^T \mathbf{u}_1) = \text{Tr}(O) = 0, \qquad (\mathbf{u}_2, \mathbf{u}_3) = \text{Tr}(\mathbf{u}_3^T \mathbf{u}_2) = \text{Tr}(O) = 0$$

$$(\mathbf{u}_1, \mathbf{u}_3) = \text{Tr}(\mathbf{u}_3^T \mathbf{u}_1) = \text{Tr}(O) = 0, \qquad (\mathbf{u}_2, \mathbf{u}_4) = \text{Tr}(\mathbf{u}_4^T \mathbf{u}_2) = \text{Tr}(O) = 0$$

$$(\mathbf{u}_1, \mathbf{u}_4) = \text{Tr}(\mathbf{u}_4^T \mathbf{u}_1) = \text{Tr}(O) = 0, \qquad (\mathbf{u}_3, \mathbf{u}_4) = \text{Tr}(\mathbf{u}_4^T \mathbf{u}_3) = \text{Tr}(O) = 0$$

Therefore S is an orthonormal basis for M_{22}.

(b) Let $\mathbf{v} = c_1\mathbf{u}_1 + c_2\mathbf{u}_2 + c_3\mathbf{u}_3 + c_4\mathbf{u}_4$. By Theorem 5.5, $c_i = (\mathbf{v}, \mathbf{u}_i)$ for $i = 1, 2, 3, 4$. We find:

$$c_1 = (\mathbf{v}, \mathbf{u}_1) = \text{Tr}(\mathbf{u}_1^T \mathbf{v}) = \text{Tr}\left(\begin{bmatrix} 1 & 2 \\ 0 & 0 \end{bmatrix} \right) = 1 \quad c_2 = (\mathbf{v}, \mathbf{u}_2) = \text{Tr}(\mathbf{u}_2^T \mathbf{v}) = \text{Tr}\left(\begin{bmatrix} 0 & 0 \\ 1 & 2 \end{bmatrix} \right) = 2$$

$$c_3 = (\mathbf{v}, \mathbf{u}_3) = \text{Tr}(\mathbf{u}_3^T \mathbf{v}) = \text{Tr}\left(\begin{bmatrix} 3 & 4 \\ 0 & 0 \end{bmatrix} \right) = 3 \quad c_4 = (\mathbf{v}, \mathbf{u}_4) = \text{Tr}(\mathbf{u}_4^T \mathbf{v}) = \text{Tr}\left(\begin{bmatrix} 0 & 0 \\ 3 & 4 \end{bmatrix} \right) = 4$$

Therefore

$$\mathbf{v} = \begin{bmatrix} 1 & 2 \\ 3 & 4 \end{bmatrix} = (1)\mathbf{u}_1 + (2)\mathbf{u}_2 + (3)\mathbf{u}_3 + (4)\mathbf{u}_4 \implies [\,\mathbf{v}\,]_S = \begin{bmatrix} 1 \\ 2 \\ 3 \\ 4 \end{bmatrix}.$$

27. Let W be the subspace spanned by S and let

$$\mathbf{u}_1 = \begin{bmatrix} 1 & 0 \\ 0 & 0 \end{bmatrix}, \quad \mathbf{u}_2 = \begin{bmatrix} 0 & 1 \\ 1 & 0 \end{bmatrix}, \quad \mathbf{u}_3 = \begin{bmatrix} 1 & -1 \\ 0 & 1 \end{bmatrix}.$$

Since $\mathbf{u}_1$ is not a scalar multiple of $\mathbf{u}_2$, $\{\mathbf{u}_1, \mathbf{u}_2\}$ is a linearly independent set of vectors. We first apply the Gram-Schmidt process to find an orthogonal basis for $\{\mathbf{u}_1, \mathbf{u}_2\}$. Let

$$\mathbf{v}_1 = \mathbf{u}_1 = \begin{bmatrix} 1 & 0 \\ 0 & 0 \end{bmatrix}$$

$$\mathbf{v}_2 = \mathbf{u}_2 - \frac{(\mathbf{u}_2, \mathbf{v}_1)}{(\mathbf{v}_1, \mathbf{v}_1)}\mathbf{v}_1$$

$$\mathbf{v}_3 = \mathbf{u}_3 - \frac{(\mathbf{u}_3, \mathbf{v}_1)}{(\mathbf{v}_1, \mathbf{v}_1)}\mathbf{v}_1 - \frac{(\mathbf{u}_3, \mathbf{v}_2)}{(\mathbf{v}_2, \mathbf{v}_2)}\mathbf{v}_2.$$

Using the inner product defined on the Euclidean space of all 2×2 matrices in Exercise 25, we find:

$$(\mathbf{u}_2, \mathbf{v}_1) = \mathrm{Tr}(\mathbf{v}_1^T \mathbf{u}_2) = \mathrm{Tr}\left(\begin{bmatrix} 1 & 0 \\ 0 & 1 \end{bmatrix}\begin{bmatrix} 0 & 1 \\ 1 & 0 \end{bmatrix}\right) = 0$$

$$(\mathbf{v}_1, \mathbf{v}_1) = \mathrm{Tr}(\mathbf{v}_1^T \mathbf{v}_1) = \mathrm{Tr}\left(\begin{bmatrix} 1 & 0 \\ 0 & 0 \end{bmatrix}\begin{bmatrix} 1 & 0 \\ 0 & 0 \end{bmatrix}\right) = 1$$

$$(\mathbf{u}_3, \mathbf{v}_1) = \mathrm{Tr}(\mathbf{v}_1^T \mathbf{u}_3) = \mathrm{Tr}\left(\begin{bmatrix} 1 & 0 \\ 0 & 0 \end{bmatrix}\begin{bmatrix} 1 & -1 \\ 0 & 1 \end{bmatrix}\right) = 1.$$

Therefore

$$\mathbf{v}_2 = \begin{bmatrix} 0 & 1 \\ 1 & 0 \end{bmatrix} - \frac{0}{3}\begin{bmatrix} 1 & 0 \\ 0 & 1 \end{bmatrix} = \begin{bmatrix} 0 & 1 \\ 1 & 0 \end{bmatrix}.$$

Now,

$$(\mathbf{u}_3, \mathbf{v}_2) = \mathrm{Tr}(\mathbf{v}_2^T \mathbf{u}_3) = \mathrm{Tr}\left(\begin{bmatrix} 0 & 1 \\ 1 & 0 \end{bmatrix}\begin{bmatrix} 1 & -1 \\ 0 & 1 \end{bmatrix}\right) = -1$$

$$(\mathbf{v}_2, \mathbf{v}_2) = \mathrm{Tr}(\mathbf{v}_2^T \mathbf{v}_2) = \mathrm{Tr}\left(\begin{bmatrix} 0 & 1 \\ 1 & 0 \end{bmatrix}\begin{bmatrix} 0 & 1 \\ 1 & 0 \end{bmatrix}\right) = 2.$$

Thus

$$\mathbf{v}_3 = \begin{bmatrix} 1 & -1 \\ 0 & 1 \end{bmatrix} - \frac{1}{1}\begin{bmatrix} 1 & 0 \\ 0 & 0 \end{bmatrix} - \frac{-1}{2}\begin{bmatrix} 0 & 1 \\ 1 & 0 \end{bmatrix} = \frac{1}{2}\begin{bmatrix} 0 & -1 \\ 1 & 2 \end{bmatrix}.$$

The set $\{\mathbf{v}_1, \mathbf{v}_2, \mathbf{v}_3\}$ is an orthogonal basis for W. To find an orthonormal basis, we determine unit vectors in the same direction as $\mathbf{v}_1$, $\mathbf{v}_2$, and $\mathbf{v}_3$. Let

$$\mathbf{w}_1 = \frac{\mathbf{v}_1}{\|\mathbf{v}_1\|} = \begin{bmatrix} 1 & 0 \\ 0 & 0 \end{bmatrix}$$

$$\mathbf{w}_2 = \frac{\mathbf{v}_2}{\|\mathbf{v}_2\|} = \frac{1}{\sqrt{2}}\begin{bmatrix} 0 & 1 \\ 1 & 0 \end{bmatrix}.$$

Since

$$(\mathbf{v}_3, \mathbf{v}_3) = \mathrm{Tr}(\mathbf{v}_3^T \mathbf{v}_3) = \mathrm{Tr}\left(\frac{1}{2}\begin{bmatrix} 0 & 1 \\ 1 & 0 \end{bmatrix}\frac{1}{2}\begin{bmatrix} 0 & -1 \\ 1 & 2 \end{bmatrix}\right) = \frac{3}{2},$$

we find that

$$\mathbf{w}_3 = \frac{\mathbf{v}_3}{\|\mathbf{v}_3\|} = \frac{1}{\sqrt{3/2}}\frac{1}{2}\begin{bmatrix} 0 & -1 \\ 1 & 2 \end{bmatrix} = \frac{1}{\sqrt{6}}\begin{bmatrix} 0 & -1 \\ 1 & 2 \end{bmatrix}.$$

The set $\{\mathbf{w}_1, \mathbf{w}_2, \mathbf{w}_3\}$ is an orthonormal basis for W.

29. Let A be the given matrix and let W stand for the column space of A.

(a) Let $\mathbf{u}_1$ and $\mathbf{u}_2$ be the column vectors of A:

$$\mathbf{u}_1 = \begin{bmatrix} 1 \\ -1 \end{bmatrix}, \quad \mathbf{u}_2 = \begin{bmatrix} 2 \\ 3 \end{bmatrix}.$$

The vectors $\mathbf{u}_1, \mathbf{u}_2$ are linearly independent since neither is a multiple of the other. Hence $\{\mathbf{u}_1, \mathbf{u}_2\}$ is a basis for the column space W. We apply the Gram–Schmidt process to find an orthogonal basis. Let

$$\mathbf{v}_1 = \mathbf{u}_1 = \begin{bmatrix} 1 \\ -1 \end{bmatrix}$$

$$\mathbf{v}_2 = \mathbf{u}_2 - \frac{(\mathbf{u}_2, \mathbf{v}_1)}{(\mathbf{v}_1, \mathbf{v}_1)}\mathbf{v}_1 = \begin{bmatrix} 2 \\ 3 \end{bmatrix} - \frac{-1}{2}\begin{bmatrix} 1 \\ -1 \end{bmatrix} = \frac{5}{2}\begin{bmatrix} 1 \\ 1 \end{bmatrix}$$

The set $\{\mathbf{v}_1, \mathbf{v}_2\}$ is an orthogonal basis for W. To find an orthonormal basis, we determine unit vectors in the same direction as $\mathbf{v}_1$ and $\mathbf{v}_2$. Let

$$\mathbf{w}_1 = \frac{\mathbf{v}_1}{\|\mathbf{v}_1\|} = \frac{\mathbf{v}_1}{\sqrt{2}} = \frac{1}{\sqrt{2}}\begin{bmatrix} 1 \\ -1 \end{bmatrix}$$

$$\mathbf{w}_2 = \frac{\mathbf{v}_2}{\|\mathbf{v}_2\|} = \frac{\mathbf{v}_2}{\frac{5}{\sqrt{2}}} = \frac{\sqrt{2}}{5}\frac{5}{2}\begin{bmatrix} 1 \\ 1 \end{bmatrix} = \frac{1}{\sqrt{2}}\begin{bmatrix} 1 \\ 1 \end{bmatrix}$$

Then $\{\mathbf{w}_1, \mathbf{w}_2\}$ is an orthonormal basis for W. Therefore

$$Q = \begin{bmatrix} \mathbf{w}_1 & \mathbf{w}_2 \end{bmatrix} = \begin{bmatrix} \frac{1}{\sqrt{2}} & \frac{1}{\sqrt{2}} \\ -\frac{1}{\sqrt{2}} & \frac{1}{\sqrt{2}} \end{bmatrix} \approx \begin{bmatrix} 0.7071 & 0.7071 \\ -0.7071 & 0.7071 \end{bmatrix}.$$

To find the matrix $R = \begin{bmatrix} r_{ij} \end{bmatrix}$ we use $r_{ji} = (\mathbf{u}_i, \mathbf{w}_j)$ to get:

$$r_{11} = (\mathbf{u}_1, \mathbf{w}_1) = \frac{2}{\sqrt{2}} = \sqrt{2}, \qquad r_{12} = (\mathbf{u}_2, \mathbf{w}_1) = -\frac{1}{\sqrt{2}}$$

$$r_{21} = (\mathbf{u}_1, \mathbf{w}_2) = 0, \qquad r_{22} = (\mathbf{u}_2, \mathbf{w}_2) = \frac{5\sqrt{2}}{2} = \frac{5}{\sqrt{2}}$$

Therefore

$$R = \begin{bmatrix} \sqrt{2} & -\frac{1}{\sqrt{2}} \\ 0 & \frac{5}{\sqrt{2}} \end{bmatrix} \approx \begin{bmatrix} 1.4142 & -0.7071 \\ 0 & 3.5355 \end{bmatrix}.$$

(b) Let $\mathbf{u}_1$ and $\mathbf{u}_2$ be the columns of A:

$$\mathbf{u}_1 = \begin{bmatrix} 1 \\ -1 \\ 1 \end{bmatrix}, \quad \mathbf{u}_2 = \begin{bmatrix} 2 \\ -2 \\ 1 \end{bmatrix}.$$

The vectors $\mathbf{u}_1$ and $\mathbf{u}_2$ are linearly independent since neither is a multiple of the other. Hence $\{\mathbf{u}_1, \mathbf{u}_2\}$ is a basis for the column space W. We apply the Gram–Schmidt process to find an orthogonal basis for W. Let

$$\mathbf{v}_1 = \mathbf{u}_1 = \begin{bmatrix} 1 \\ -1 \\ 1 \end{bmatrix}$$

$$\mathbf{v}_2 = \mathbf{u}_2 - \frac{(\mathbf{u}_2, \mathbf{v}_1)}{(\mathbf{v}_1, \mathbf{v}_1)}\mathbf{v}_1 = \begin{bmatrix} 2 \\ -2 \\ 1 \end{bmatrix} - \frac{5}{3}\begin{bmatrix} 1 \\ -1 \\ 1 \end{bmatrix} = \begin{bmatrix} \frac{1}{3} \\ -\frac{1}{3} \\ -\frac{2}{3} \end{bmatrix}$$

The set $\{\mathbf{v}_1, \mathbf{v}_2\}$ is an orthogonal basis for W. To find an orthonormal basis, we determine unit vectors in the same direction as $\mathbf{v}_1$ and $\mathbf{v}_2$. Let

$$\mathbf{w}_1 = \frac{\mathbf{v}_1}{\|\mathbf{v}_1\|} = \frac{\mathbf{v}_1}{\sqrt{3}} = \frac{1}{\sqrt{3}}\begin{bmatrix} 1 \\ -1 \\ 1 \end{bmatrix}$$

$$\mathbf{w}_2 = \frac{\mathbf{v}_2}{\|\mathbf{v}_2\|} = \frac{\mathbf{v}_2}{\sqrt{\frac{2}{3}}} = \sqrt{\frac{3}{2}}\begin{bmatrix} \frac{1}{3} \\ -\frac{1}{3} \\ -\frac{2}{3} \end{bmatrix} = \frac{1}{\sqrt{6}}\begin{bmatrix} 1 \\ -1 \\ -2 \end{bmatrix}$$

Then $\{\mathbf{w}_1, \mathbf{w}_2\}$ is an orthonormal basis for W. Therefore

$$Q = \begin{bmatrix} \mathbf{w}_1 & \mathbf{w}_2 \end{bmatrix} = \begin{bmatrix} \frac{1}{\sqrt{3}} & \frac{1}{\sqrt{6}} \\ -\frac{1}{\sqrt{3}} & -\frac{1}{\sqrt{6}} \\ \frac{1}{\sqrt{3}} & -\frac{2}{\sqrt{6}} \end{bmatrix} \approx \begin{bmatrix} 0.5774 & 0.4082 \\ -0.5774 & -0.4082 \\ 0.5774 & -0.8165 \end{bmatrix}$$

To find the matrix $R = \begin{bmatrix} r_{ij} \end{bmatrix}$ we use $r_{ji} = (\mathbf{u}_i, \mathbf{w}_j)$ to get:

$$r_{11} = (\mathbf{u}_1, \mathbf{w}_1) = \frac{3}{\sqrt{3}} = \sqrt{3}, \qquad r_{12} = (\mathbf{u}_2, \mathbf{w}_1) = \frac{5}{\sqrt{3}}$$

$$r_{21} = (\mathbf{u}_1, \mathbf{w}_2) = 0, \qquad r_{22} = (\mathbf{u}_2, \mathbf{w}_2) = \frac{2}{\sqrt{6}}$$

Therefore

$$R = \begin{bmatrix} \sqrt{3} & \frac{5}{\sqrt{3}} \\ 0 & \frac{2}{\sqrt{6}} \end{bmatrix} \approx \begin{bmatrix} 1.7321 & 2.8868 \\ 0 & 0.8165 \end{bmatrix}$$

(c) Let $\mathbf{u}_1$, $\mathbf{u}_2$, and $\mathbf{u}_3$ be the columns of A:

$$\mathbf{u}_1 = \begin{bmatrix} 1 \\ 2 \\ -1 \end{bmatrix}, \qquad \mathbf{u}_2 = \begin{bmatrix} 0 \\ -3 \\ 2 \end{bmatrix}, \qquad \mathbf{u}_3 = \begin{bmatrix} -1 \\ 3 \\ 4 \end{bmatrix}.$$

Row reducing the matrix A gives I_3. Therefore the vectors $\mathbf{u}_1$, $\mathbf{u}_2$, $\mathbf{u}_3$ are linearly independent and hence form a basis for the column space W. We apply the Gram–Schmidt process to find an orthonormal basis. Let

$$\mathbf{v}_1 = \mathbf{u}_1 = \begin{bmatrix} 1 \\ 2 \\ -1 \end{bmatrix}$$

$$\mathbf{v}_2 = \mathbf{u}_2 - \frac{(\mathbf{u}_2, \mathbf{v}_1)}{(\mathbf{v}_1, \mathbf{v}_1)}\mathbf{v}_1 = \begin{bmatrix} 0 \\ -3 \\ 2 \end{bmatrix} - \frac{-8}{6}\begin{bmatrix} 1 \\ 2 \\ -1 \end{bmatrix} = \begin{bmatrix} \frac{4}{3} \\ -\frac{1}{3} \\ \frac{2}{3} \end{bmatrix}$$

$$\mathbf{v}_3 = \mathbf{u}_3 - \frac{(\mathbf{u}_3, \mathbf{v}_1)}{(\mathbf{v}_1, \mathbf{v}_1)}\mathbf{v}_1 - \frac{(\mathbf{u}_3, \mathbf{v}_2)}{(\mathbf{v}_2, \mathbf{v}_2)}\mathbf{v}_2 = \begin{bmatrix} -1 \\ 3 \\ 4 \end{bmatrix} - \frac{1}{6}\begin{bmatrix} 1 \\ 2 \\ -1 \end{bmatrix} - \frac{7}{3}\begin{bmatrix} \frac{4}{3} \\ -\frac{1}{3} \\ \frac{2}{3} \end{bmatrix} = \begin{bmatrix} -\frac{19}{14} \\ \frac{38}{14} \\ \frac{57}{14} \end{bmatrix}$$

The set $\{\mathbf{v}_1, \mathbf{v}_2, \mathbf{v}_3\}$ is an orthogonal basis for W. To find an orthonormal basis, we determine unit vectors in the same direction as $\mathbf{v}_1$, $\mathbf{v}_2$, and $\mathbf{v}_3$. Let

$$\mathbf{w}_1 = \frac{\mathbf{v}_1}{\|\mathbf{v}_1\|} = \frac{1}{\sqrt{6}}\begin{bmatrix} 1 \\ 2 \\ -1 \end{bmatrix}$$

$$\mathbf{w}_2 = \frac{\mathbf{v}_2}{\|\mathbf{v}_2\|} = \frac{1}{\sqrt{\frac{21}{9}}}\frac{1}{3}\begin{bmatrix} 4 \\ -1 \\ 2 \end{bmatrix} = \frac{1}{\sqrt{21}}\begin{bmatrix} 4 \\ -1 \\ 2 \end{bmatrix}$$

$$\mathbf{w}_3 = \frac{\mathbf{v}_3}{\|\mathbf{v}_3\|} = \frac{1}{\frac{19}{\sqrt{14}}}\frac{19}{14}\begin{bmatrix} -1 \\ 2 \\ 3 \end{bmatrix} = \frac{1}{\sqrt{14}}\begin{bmatrix} -1 \\ 2 \\ 3 \end{bmatrix}$$

Therefore

$$Q = \begin{bmatrix} \mathbf{w}_1 & \mathbf{w}_2 & \mathbf{w}_3 \end{bmatrix} = \begin{bmatrix} \frac{1}{\sqrt{6}} & \frac{4}{\sqrt{21}} & -\frac{1}{\sqrt{14}} \\ \frac{2}{\sqrt{6}} & -\frac{1}{\sqrt{21}} & \frac{2}{\sqrt{14}} \\ -\frac{1}{\sqrt{6}} & \frac{2}{\sqrt{21}} & \frac{3}{\sqrt{14}} \end{bmatrix} \approx \begin{bmatrix} 0.4082 & 0.8729 & -0.2673 \\ 0.8165 & -0.2182 & 0.5345 \\ -0.4082 & 0.4364 & 0.8018 \end{bmatrix}$$

To find the matrix $R = \begin{bmatrix} r_{ij} \end{bmatrix}$ we use $r_{ji} = (\mathbf{u}_i, \mathbf{w}_j)$ to get:

$$r_{11} = (\mathbf{u}_1, \mathbf{w}_1) = \frac{6}{\sqrt{6}}, \qquad r_{12} = (\mathbf{u}_2, \mathbf{w}_1) = -\frac{8}{\sqrt{6}}, \qquad r_{13} = (\mathbf{u}_3, \mathbf{w}_1) = \frac{1}{\sqrt{6}}$$

$$r_{21} = (\mathbf{u}_1, \mathbf{w}_2) = 0, \qquad r_{22} = (\mathbf{u}_2, \mathbf{w}_2) = \frac{7}{\sqrt{21}}, \qquad r_{23} = (\mathbf{u}_3, \mathbf{w}_2) = \frac{1}{\sqrt{21}}$$

$$r_{31} = (\mathbf{u}_1, \mathbf{w}_3) = 0, \qquad r_{32} = (\mathbf{u}_2, \mathbf{w}_3) = 0, \qquad r_{33} = (\mathbf{u}_3, \mathbf{w}_3) = \frac{19}{\sqrt{14}}$$

Therefore

$$R = \begin{bmatrix} \frac{6}{\sqrt{6}} & -\frac{8}{\sqrt{6}} & \frac{1}{\sqrt{6}} \\ 0 & \frac{7}{\sqrt{21}} & \frac{1}{\sqrt{21}} \\ 0 & 0 & \frac{19}{\sqrt{14}} \end{bmatrix} \approx \begin{bmatrix} 2.4475 & -3.2660 & 0.4082 \\ 0 & 1.5275 & 0.2182 \\ 0 & 0 & 5.0780 \end{bmatrix}$$

31. We have $(\mathbf{u}, c\mathbf{v}) = c(\mathbf{u}, \mathbf{v}) = c \cdot 0 = 0$.

33. Let W be the subset of vectors in R^n that are orthogonal to $\mathbf{u}$. If $\mathbf{v}$ and $\mathbf{w}$ are in W, then $(\mathbf{u}, \mathbf{v}) = (\mathbf{u}, \mathbf{w}) = 0$. It follows that $(\mathbf{u}, \mathbf{v} + \mathbf{w}) = (\mathbf{u}, \mathbf{v}) + (\mathbf{u}, \mathbf{w}) = 0$ and for any scalar c, $(\mathbf{u}, c\mathbf{v}) = c(\mathbf{u}, \mathbf{v}) = 0$, so $\mathbf{u} + \mathbf{w}$ and $c\mathbf{v}$ are in W. Hence, W is a subspace of R^n.

35. Since S is an orthonormal basis for V, $\dim V = k$ and

$$(\mathbf{v}_i, \mathbf{v}_j) = \begin{cases} 0, & \text{if } i \neq j \\ 1, & \text{if } i = j \end{cases}$$

To show that T is a basis we need only show that it spans V and then use Theorem 4.12(b). Let $\mathbf{v}$ belong to V. Then there exist scalars c_i, $i = 1, 2, \ldots, k$ such that

$$\mathbf{v} = c_1\mathbf{v}_1 + c_2\mathbf{v}_2 + \cdots + c_k\mathbf{v}_k.$$

Since $a_j \neq 0$, we have

$$\mathbf{v} = \frac{c_1}{a_1}(a_1\mathbf{v}_1) + \frac{c_2}{a_2}(a_2\mathbf{v}_2) + \cdots + \frac{c_k}{a_k}(a_k\mathbf{v}_k)$$

so span $T = V$. Next we show that the members of T are orthogonal. Since S is orthonormal we have

$$(a_i\mathbf{v}_i, a_j\mathbf{v}_j) = a_i a_j (\mathbf{v}_i, \mathbf{v}_j) = \begin{cases} 0, & \text{if } i \neq j \\ a_i a_j, & \text{if } i = j \end{cases}$$

Hence T is an orthogonal set. In order for T to be an orthonormal set, we must have $a_i a_j = 1$ for all i and j. This is only possible if all $a_i = 1$.

37. If A is an $n \times n$ nonsingular matrix, then the columns of A are linearly independent, so by Theorem 4.8, A has a QR-factorization.

Section 5.5, p. 348

1. (a) Let

$$\mathbf{u} = \begin{bmatrix} a \\ b \\ c \end{bmatrix}$$

be a vector in $W^{\perp}$. Then $(\mathbf{u}, \mathbf{w}) = 2a - 3b + c = 0$. Therefore $a = \frac{3}{2}b - \frac{1}{2}c$ and hence the general vector $\mathbf{u}$ in $W^{\perp}$ has the form

$$\mathbf{u} = \begin{bmatrix} \frac{3}{2}b - \frac{1}{2}c \\ b \\ c \end{bmatrix} = b \begin{bmatrix} \frac{3}{2} \\ 1 \\ 0 \end{bmatrix} + c \begin{bmatrix} -\frac{1}{2} \\ 0 \\ 1 \end{bmatrix}.$$

Therefore the set

$$S = \left\{ \begin{bmatrix} \frac{3}{2} \\ 1 \\ 0 \end{bmatrix}, \begin{bmatrix} -\frac{1}{2} \\ 0 \\ 1 \end{bmatrix} \right\}$$

spans $W^{\perp}$. Since these two vectors are not multiples of each other, they are linearly independent and form a basis for $W^{\perp}$.

(b) $W^{\perp}$ is the plane through the origin determined by the vectors $\begin{bmatrix} \frac{3}{2} & 1 & 0 \end{bmatrix}^T$ and $\begin{bmatrix} -\frac{1}{2} & 0 & 1 \end{bmatrix}^T$. It consists of all points $P(x, y, z)$ such that

$$\mathbf{w} \cdot \mathbf{x} = \begin{bmatrix} 2 \\ -3 \\ 1 \end{bmatrix} \cdot \begin{bmatrix} x \\ y \\ z \end{bmatrix} = 2x - 3y + z = 0.$$

3. Let $\mathbf{u} = \begin{bmatrix} a_1 & a_2 & a_3 & a_4 & a_5 \end{bmatrix}$ be a vector in $W^{\perp}$. Then $\mathbf{u}$ is orthogonal to each of the vectors $\mathbf{w}_1$, ..., $\mathbf{w}_5$. Setting $(\mathbf{u}, \mathbf{w}_i) = 0$ for $i = 1, \ldots, 5$ leads to a system of homogeneous equations. We row reduce the coefficient matrix of the system:

$$\begin{bmatrix} 2 & -1 & 1 & 3 & 0 \\ 1 & 2 & 0 & 1 & -2 \\ 4 & 3 & 1 & 5 & -4 \\ 3 & 1 & 2 & -1 & 1 \\ 2 & -1 & 2 & -2 & 3 \end{bmatrix} \xrightarrow{\quad} \underset{\text{steps omitted}}{\cdots} \xrightarrow{\quad} \begin{bmatrix} 1 & 0 & 0 & \frac{17}{5} & -\frac{8}{5} \\ 0 & 1 & 0 & -\frac{6}{5} & -\frac{1}{5} \\ 0 & 0 & 1 & -5 & 3 \\ 0 & 0 & 0 & 0 & 0 \\ 0 & 0 & 0 & 0 & 0 \end{bmatrix}$$

The solution to the system is therefore $a_1 = -\frac{17}{5}r + \frac{8}{5}s$, $a_2 = \frac{6}{5}r + \frac{1}{5}s$, $a_3 = 5r - 3s$, $a_4 = r$, $a_5 = s$, where r and s are any real numbers. Hence

$$\mathbf{u} = \begin{bmatrix} -\frac{17}{5}r + \frac{8}{5}s & \frac{6}{5}r + \frac{1}{5}s & 5r - 3s & r & s \end{bmatrix} = r \begin{bmatrix} -\frac{17}{5} & \frac{6}{5} & 5 & 1 & 0 \end{bmatrix} + s \begin{bmatrix} \frac{8}{5} & \frac{1}{5} & -3 & 0 & 1 \end{bmatrix}.$$

Therefore the set

$$S = \left\{ \begin{bmatrix} -\frac{17}{5} & \frac{6}{5} & 5 & 1 & 0 \end{bmatrix}, \begin{bmatrix} \frac{8}{5} & \frac{1}{5} & -3 & 0 & 1 \end{bmatrix} \right\}$$

is a basis for $W^{\perp}$.

5. Let $p(t) = at^3 + bt^2 + ct + d$ be in $W^\perp$. Then $p(t)$ is orthogonal to $t - 1$ and t^2 and hence

$$(p(t), t - 1) = \int_0^1 (at^3 + bt^2 + ct + d)(t - 1)\, dt = -\frac{1}{20}a - \frac{1}{12}b - \frac{1}{6}c - \frac{1}{2}d = 0$$

$$(p(t), t^2) = \int_0^1 (at^3 + bt^2 + ct + d)(t^2)\, dt = \frac{1}{6}a + \frac{1}{5}b + \frac{1}{4}c + \frac{1}{3}d = 0$$

As in Exercise 3 we form the coefficient matrix of this system and row reduce it:

$$A = \begin{bmatrix} -\frac{1}{20} & -\frac{1}{12} & -\frac{1}{6} & -\frac{1}{2} \\ \frac{1}{6} & \frac{1}{5} & \frac{1}{4} & \frac{1}{3} \end{bmatrix} \underset{\text{steps omitted}}{\overset{\cdots}{\longrightarrow \longrightarrow}} \begin{bmatrix} 1 & 0 & -\frac{45}{14} & -\frac{130}{7} \\ 0 & 1 & \frac{55}{14} & \frac{120}{7} \end{bmatrix} = B$$

Thus the solution is $a = \frac{45}{14}r + \frac{130}{7}s$, $b = -\frac{55}{14}r - \frac{120}{7}s$, $c = r$, $d = s$, where r and s are any real numbers. Therefore

$$at^3 + bt^2 + ct + d = \left(\frac{45}{14}r + \frac{130}{7}s\right)t^3 + \left(-\frac{55}{14}r - \frac{120}{7}s\right)t^2 + rt + s$$

$$= r\left(\frac{45}{14}t^3 - \frac{55}{14}t^2 + t\right) + s\left(\frac{130}{7}t^3 - \frac{120}{7}t^2 + 1\right)$$

Hence the set $S = \left\{\frac{45}{14}t^3 - \frac{55}{14}t^2 + t, \frac{130}{7}t^3 - \frac{120}{7}t^2 + 1\right\}$ is a basis for $W^\perp$.

7. A point $(P(x, y, z)$ is on the plane W when $x = -\frac{2}{3}y + \frac{1}{3}z$. Thus,

$$\begin{bmatrix} x \\ y \\ z \end{bmatrix} = \begin{bmatrix} -\frac{2}{3}y + \frac{1}{3}z \\ y \\ z \end{bmatrix} = y\begin{bmatrix} -\frac{2}{3} \\ 1 \\ 0 \end{bmatrix} + z\begin{bmatrix} \frac{1}{3} \\ 0 \\ 1 \end{bmatrix}.$$

Hence a basis for the plane is

$$\{\mathbf{u}_1, \mathbf{u}_2\} = \left\{\begin{bmatrix} -\frac{2}{3} \\ 1 \\ 0 \end{bmatrix}, \begin{bmatrix} \frac{1}{3} \\ 0 \\ 1 \end{bmatrix}\right\}.$$

Now, a vector $\mathbf{v}$ is in $W^\perp$ when $(\mathbf{v}, \mathbf{u}_1) = (\mathbf{v}, \mathbf{u}_2) = 0$. Let $\mathbf{v} = \begin{bmatrix} a & b & c \end{bmatrix}^T$. Then

$$(\mathbf{v}, \mathbf{u}_1) = \begin{bmatrix} a \\ b \\ c \end{bmatrix} \cdot \begin{bmatrix} -\frac{2}{3} \\ 1 \\ 0 \end{bmatrix} = -\frac{2}{3}a + b = 0$$

$$(\mathbf{v}, \mathbf{u}_2) = \begin{bmatrix} a \\ b \\ c \end{bmatrix} \cdot \begin{bmatrix} \frac{1}{3} \\ 0 \\ 1 \end{bmatrix} = \frac{1}{3}a + c = 0$$

The solution to this system of equations is $a = r$, $b = \frac{2}{3}r$, $c = -\frac{1}{3}r$, where r is any real number. Thus

$$\begin{bmatrix} a \\ b \\ c \end{bmatrix} = \begin{bmatrix} r \\ \frac{2}{3}r \\ -\frac{1}{3}r \end{bmatrix} = r\begin{bmatrix} 1 \\ \frac{2}{3} \\ -\frac{1}{3} \end{bmatrix}.$$

Setting $r = -3$, we obtain $\left\{\begin{bmatrix} -3 \\ -2 \\ 1 \end{bmatrix}\right\}$ as a basis for $W^\perp$.

9. We follow the method used in Example 3. First transform the matrix A to its reduced row echelon form:

$$A = \begin{bmatrix} 1 & 5 & 3 & 7 \\ 2 & 0 & -4 & -6 \\ 4 & 7 & -1 & 2 \end{bmatrix} \underset{\text{steps omitted}}{\overset{\cdots}{\longrightarrow \longrightarrow}} \begin{bmatrix} 1 & 0 & -2 & -3 \\ 0 & 1 & 1 & 2 \\ 0 & 0 & 0 & 0 \end{bmatrix} = B$$

To find the null space of A we solve the system $B\mathbf{x} = \mathbf{0}$ and obtain

$$S = \left\{ \begin{bmatrix} 2 \\ -1 \\ 1 \\ 0 \end{bmatrix}, \begin{bmatrix} 3 \\ -2 \\ 0 \\ 1 \end{bmatrix} \right\}$$

as a basis for the null space of A. The rows of B form a basis for the row space of A:

$$T = \left\{ \begin{bmatrix} 1 & 0 & -2 & -3 \end{bmatrix}, \begin{bmatrix} 0 & 1 & 1 & 2 \end{bmatrix} \right\}.$$

Next, we have

$$A^T = \begin{bmatrix} 1 & 2 & 4 \\ 5 & 0 & 7 \\ 3 & -4 & -1 \\ 7 & -6 & 2 \end{bmatrix}.$$

We transform the matrix A^T to its reduced row echelon form:

$$A^T = \begin{bmatrix} 1 & 2 & 4 \\ 5 & 0 & 7 \\ 3 & -4 & -1 \\ 7 & -6 & 2 \end{bmatrix} \xrightarrow{\quad} \cdots \xrightarrow[\text{steps omitted}]{\quad} \begin{bmatrix} 1 & 0 & \frac{7}{5} \\ 0 & 1 & \frac{13}{10} \\ 0 & 0 & 0 \\ 0 & 0 & 0 \end{bmatrix} = C$$

To find the null space of A^T we solve the system $A^T\mathbf{x} = \mathbf{0}$ to obtain

$$S' = \left\{ \begin{bmatrix} -\frac{7}{5} \\ -\frac{13}{10} \\ 1 \end{bmatrix} \right\}$$

as a basis for the null space of A^T. The nonzero rows of C read vertically give a basis for the column space of A:

$$T' = \left\{ \begin{bmatrix} 1 \\ 0 \\ \frac{7}{5} \end{bmatrix}, \begin{bmatrix} 0 \\ 1 \\ \frac{13}{10} \end{bmatrix} \right\}.$$

11. Let

$$\mathbf{w}_1 = \begin{bmatrix} \frac{1}{\sqrt{5}} \\ 0 \\ \frac{2}{\sqrt{5}} \end{bmatrix}, \quad \mathbf{w}_2 = \begin{bmatrix} -\frac{2}{\sqrt{5}} \\ 0 \\ \frac{1}{\sqrt{5}} \end{bmatrix}.$$

Then $\{\mathbf{w}_1, \mathbf{w}_2\}$ is an orthonormal basis for W and hence, for any vector $\mathbf{v}$,

$$\text{proj}_W \mathbf{v} = \frac{(\mathbf{v}, \mathbf{w}_1)}{(\mathbf{w}_1, \mathbf{w}_1)} \mathbf{w}_1 + \frac{(\mathbf{v}, \mathbf{w}_2)}{(\mathbf{w}_2, \mathbf{w}_2)} \mathbf{w}_2 = (\mathbf{v}, \mathbf{w}_1)\mathbf{w}_1 + (\mathbf{v}, \mathbf{w}_2)\mathbf{w}_2.$$

(a) We have

$$(\mathbf{v}, \mathbf{w}_1) = \frac{1}{\sqrt{5}} \quad \text{and} \quad (\mathbf{v}, \mathbf{w}_2) = -\frac{7}{\sqrt{5}}.$$

Using the formula above, we obtain

$$\text{proj}_W \mathbf{v} = \left(\frac{1}{\sqrt{5}} \right) \begin{bmatrix} \frac{1}{\sqrt{5}} \\ 0 \\ \frac{2}{\sqrt{5}} \end{bmatrix} + \left(-\frac{7}{\sqrt{5}} \right) \begin{bmatrix} -\frac{2}{\sqrt{5}} \\ 0 \\ \frac{1}{\sqrt{5}} \end{bmatrix} = \begin{bmatrix} 3 \\ 0 \\ -1 \end{bmatrix}$$

(b) We have

$$(\mathbf{v}, \mathbf{w}_1) = \frac{8}{\sqrt{5}} \quad \text{and} \quad (\mathbf{v}, \mathbf{w}_2) = -\frac{1}{\sqrt{5}}.$$

Using the formula above, we obtain

$$\text{proj}_W \mathbf{v} = \left(\frac{8}{\sqrt{5}}\right) \begin{bmatrix} \frac{1}{\sqrt{5}} \\ 0 \\ \frac{2}{\sqrt{5}} \end{bmatrix} + \left(-\frac{1}{\sqrt{5}}\right) \begin{bmatrix} -\frac{2}{\sqrt{5}} \\ 0 \\ \frac{1}{\sqrt{5}} \end{bmatrix} = \begin{bmatrix} 2 \\ 0 \\ 3 \end{bmatrix}.$$

(c) We have

$$(\mathbf{v}, \mathbf{w}_1) = -\frac{3}{\sqrt{5}} \quad \text{and} \quad (\mathbf{v}, \mathbf{w}_2) = \frac{11}{\sqrt{5}}.$$

Using the formula above, we obtain

$$\text{proj}_W \mathbf{v} = \left(-\frac{3}{\sqrt{5}}\right) \begin{bmatrix} \frac{1}{\sqrt{5}} \\ 0 \\ \frac{2}{\sqrt{5}} \end{bmatrix} + \left(\frac{11}{\sqrt{5}}\right) \begin{bmatrix} -\frac{2}{\sqrt{5}} \\ 0 \\ \frac{1}{\sqrt{5}} \end{bmatrix} = \begin{bmatrix} -5 \\ 0 \\ 1 \end{bmatrix}.$$

13. Let

$$\mathbf{w}_1 = \frac{1}{\sqrt{2\pi}}, \quad \mathbf{w}_2 = \frac{1}{\sqrt{\pi}} \cos t, \quad \mathbf{w}_3 = \frac{1}{\sqrt{\pi}} \sin t.$$

Then $\{\mathbf{w}_1, \mathbf{w}_2, \mathbf{w}_3\}$ is an orthonormal basis for W. Hence, for any vector $\mathbf{v}$,

$$\text{proj}_W \mathbf{v} = (\mathbf{v}, \mathbf{w}_1)\mathbf{w}_1 + (\mathbf{v}, \mathbf{w}_2)\mathbf{w}_2 + (\mathbf{v}, \mathbf{w}_3)\mathbf{w}_3.$$

(a) For $\mathbf{v} = t$, we find that

$$(t, \mathbf{w}_1) = \int_{-\pi}^{\pi} (t) \frac{1}{\sqrt{2\pi}} \, dt = 0$$

$$(t, \mathbf{w}_2) = \int_{-\pi}^{\pi} (t) \frac{1}{\sqrt{\pi}} \cos t \, dt = 0$$

$$(t, \mathbf{w}_3) = \int_{-\pi}^{\pi} (t) \frac{1}{\sqrt{\pi}} \sin t \, dt = 2\sqrt{\pi}$$

Therefore

$$\text{proj}_W t = (0) \left(\frac{1}{\sqrt{2\pi}}\right) + (0) \left(\frac{1}{\sqrt{\pi}} \cos t\right) + (2\sqrt{\pi}) \frac{1}{\sqrt{\pi}} \sin t = 2 \sin t.$$

(b) For $\mathbf{v} = t^2$, we find that

$$(t^2, \mathbf{w}_1) = \int_{-\pi}^{\pi} (t^2) \frac{1}{\sqrt{2\pi}} \, dt = \frac{2\pi^3}{3\sqrt{2\pi}}$$

$$(t^2, \mathbf{w}_2) = \int_{-\pi}^{\pi} (t^2) \frac{1}{\sqrt{\pi}} \cos t \, dt = -4\sqrt{\pi}$$

$$(t^2, \mathbf{w}_3) = \int_{-\pi}^{\pi} (t^2) \frac{1}{\sqrt{\pi}} \sin t \, dt = 0$$

Therefore

$$\text{proj}_W t^2 = \left(\frac{2\pi^3}{3\sqrt{2\pi}}\right)\left(\frac{1}{\sqrt{2\pi}}\right) + (-4\sqrt{\pi})\left(\frac{1}{\sqrt{\pi}} \cos t\right) + (0)\left(\frac{1}{\sqrt{\pi}} \sin t\right) = \frac{\pi^2}{3} - 4 \cos t.$$

(c) For $\mathbf{v} = e^t$, we find that

$$(e^t, \mathbf{w}_1) = \int_{-\pi}^{\pi} (e^t) \frac{1}{\sqrt{2\pi}} \, dt = \frac{1}{\sqrt{2\pi}} (e^\pi - e^{-\pi})$$

$$(e^t, \mathbf{w}_2) = \int_{-\pi}^{\pi} (e^t) \frac{1}{\sqrt{\pi}} \cos t \, dt = \frac{1}{\sqrt{\pi}} \left(\frac{1}{2} e^{-\pi} - \frac{1}{2} e^\pi \right)$$

$$(e^t, \mathbf{w}_3) = \int_{-\pi}^{\pi} (e^t) \frac{1}{\sqrt{\pi}} \sin t \, dt = \frac{1}{\sqrt{\pi}} \left(-\frac{1}{2} e^{-\pi} + \frac{1}{2} e^\pi \right)$$

Therefore

$$\text{proj}_W e^t = \frac{1}{\sqrt{2\pi}} (e^\pi - e^{-\pi}) \left(\frac{1}{\sqrt{2\pi}} \right) + \frac{1}{\sqrt{\pi}} \left(\frac{1}{2} e^{-\pi} - \frac{1}{2} e^\pi \right) \frac{1}{\sqrt{\pi}} \cos t$$

$$+ \frac{1}{\sqrt{\pi}} \left(-\frac{1}{2} e^{-\pi} + \frac{1}{2} e^\pi \right) \frac{1}{\sqrt{\pi}} \sin t$$

$$= \frac{e^\pi - e^{-\pi}}{2\pi} + \frac{e^{-\pi} - e^\pi}{2\pi} \cos t + \frac{e^\pi - e^{-\pi}}{2\pi} \sin t.$$

15. The vector $\mathbf{w}$ is the projection of $\mathbf{v}$ onto the subspace W:

$$\mathbf{w} = (\mathbf{v}, \mathbf{w}_1)\mathbf{w}_1 + (\mathbf{v}, \mathbf{w}_2)\mathbf{w}_2 = 2\mathbf{w}_1 - \frac{1}{\sqrt{5}}\mathbf{w}_2 = \begin{bmatrix} -\frac{1}{5} \\ 2 \\ -\frac{2}{5} \end{bmatrix}$$

Let

$$\mathbf{u} = \mathbf{v} - \mathbf{w} = \begin{bmatrix} 1 \\ 2 \\ -1 \end{bmatrix} - \begin{bmatrix} -\frac{1}{5} \\ 2 \\ -\frac{2}{5} \end{bmatrix} = \begin{bmatrix} \frac{6}{5} \\ 0 \\ -\frac{3}{5} \end{bmatrix}.$$

Then $\mathbf{v} = \mathbf{w} + \mathbf{u}$, where $\mathbf{w}$ is in W and $\mathbf{u}$ is in $W^\perp$.

17. The vector $\mathbf{w}$ is the projection of $\mathbf{v}$ onto the subspace W. Thus:

$$\mathbf{w} = (\mathbf{v}, \mathbf{w}_1)\mathbf{w}_1 + (\mathbf{v}, \mathbf{w}_2)\mathbf{w}_2 + (\mathbf{v}, \mathbf{w}_3)\mathbf{w}_3.$$

For the inner products, we find:

$$(\mathbf{v}, \mathbf{w}_1) = \int_{-\pi}^{\pi} (t - 1) \frac{1}{\sqrt{2\pi}} \, dt = -\sqrt{2\pi}$$

$$(\mathbf{v}, \mathbf{w}_2) = \int_{-\pi}^{\pi} (t - 1) \frac{1}{\sqrt{\pi}} \cos t \, dt = 0$$

$$(\mathbf{v}, \mathbf{w}_3) = \int_{-\pi}^{\pi} (t - 1) \frac{1}{\sqrt{\pi}} \sin t \, dt = 2\sqrt{\pi}$$

Therefore

$$\mathbf{w} = -\sqrt{2\pi}\,\mathbf{w}_1 + (0)\mathbf{w}_2 + 2\sqrt{\pi}\,\mathbf{w}_3$$

$$= -\sqrt{2\pi} \left(\frac{1}{\sqrt{2\pi}} \right) + 2\sqrt{\pi} \left(\frac{1}{\sqrt{\pi}} \cos t \right) + 2\sqrt{\pi} \left(\frac{1}{\sqrt{\pi}} \sin t \right)$$

$$= -1 + 2\sin t$$

Let

$$\mathbf{u} = \mathbf{v} - \mathbf{w} = (t - 1) - (-1 + 2\sin t) = t - 2\sin t.$$

Then $\mathbf{v} = \mathbf{w} + \mathbf{u}$, where $\mathbf{w}$ is in W and $\mathbf{u}$ is in $W^\perp$.

19. Using the vectors $\mathbf{w}_1$, $\mathbf{w}_2$ from Exercise 15, the projection of $\mathbf{v}$ onto W is

$$\mathbf{w} = (\mathbf{v}, \mathbf{w}_1)\mathbf{w}_1 + (\mathbf{v}, \mathbf{w}_2)\mathbf{w}_2.$$

We find

$$(\mathbf{v}, \mathbf{w}_1) = \begin{bmatrix} -1 \\ 0 \\ 1 \end{bmatrix} \cdot \begin{bmatrix} 0 \\ 1 \\ 0 \end{bmatrix} = 0 \quad \text{and} \quad (\mathbf{v}, \mathbf{w}_2) = \begin{bmatrix} -1 \\ 0 \\ 1 \end{bmatrix} \cdot \begin{bmatrix} \frac{1}{\sqrt{5}} \\ 0 \\ \frac{2}{\sqrt{5}} \end{bmatrix} = \frac{1}{\sqrt{5}}.$$

Therefore

$$\mathbf{w} = (0)\mathbf{w}_1 + \frac{1}{\sqrt{5}}\mathbf{w}_2 = \begin{bmatrix} \frac{1}{5} \\ 0 \\ \frac{2}{5} \end{bmatrix}.$$

It follows that

$$\mathbf{v} - \text{proj}_W \mathbf{v} = \mathbf{v} - \mathbf{w} = \begin{bmatrix} -1 \\ 0 \\ 1 \end{bmatrix} - \begin{bmatrix} \frac{1}{5} \\ 0 \\ \frac{2}{5} \end{bmatrix} = \begin{bmatrix} -\frac{6}{5} \\ 0 \\ \frac{3}{5} \end{bmatrix}$$

and hence the distance from $\mathbf{v}$ to W is

$$\|\mathbf{v} - \text{proj}_W \mathbf{v}\| = \left\| \begin{bmatrix} -\frac{6}{5} \\ 0 \\ \frac{3}{5} \end{bmatrix} \right\| = \frac{\sqrt{45}}{5} = \frac{3\sqrt{5}}{5}.$$

21. From Exercise 13(a), the projection of $\mathbf{v} = t$ onto W is $\mathbf{w} = 2\sin t$. Hence the distance from $\mathbf{v}$ to W is

$$\|\mathbf{v} - \text{proj}_W \mathbf{v}\| = \|t - 2\sin t\| = \sqrt{\int_{-\pi}^{\pi} (t - 2\sin t)^2 \, dt} = \sqrt{\frac{2\pi^3}{3} - 4\pi}.$$

23. Let $\mathbf{v} = f(t) = e^t$. Then the Fourier polynomial of degree two for e^t is

$$\left(e^t, \frac{1}{\sqrt{2\pi}}\right) \frac{1}{\sqrt{2\pi}} + \left(e^t, \frac{1}{\sqrt{\pi}}\cos t\right) \frac{1}{\sqrt{\pi}}\cos t + \left(e^t, \frac{1}{\sqrt{\pi}}\sin t\right) \frac{1}{\sqrt{\pi}}\sin t$$
$$+ \left(e^t, \frac{1}{\sqrt{\pi}}\cos 2t\right) \frac{1}{\sqrt{\pi}}\cos 2t + \left(e^t, \frac{1}{\sqrt{\pi}}\sin 2t\right) \frac{1}{\sqrt{\pi}}\sin 2t.$$

Now,

$$\left(e^t, \frac{1}{\sqrt{2\pi}}\right) = \int_{\pi}^{\pi} e^t \frac{1}{\sqrt{2\pi}} \, dt = \frac{1}{\sqrt{2\pi}}(e^\pi - e^{-\pi})$$

$$\left(e^t, \frac{1}{\sqrt{\pi}}\cos t\right) = \int_{-\pi}^{\pi} e^t \frac{1}{\sqrt{\pi}}\cos t \, dt = -\frac{1}{2\sqrt{\pi}}(e^\pi - e^{-\pi})$$

$$\left(e^t, \frac{1}{\sqrt{\pi}}\sin t\right) = \int_{-\pi}^{\pi} e^t \frac{1}{\sqrt{\pi}}\sin t \, dt = \frac{1}{2\sqrt{\pi}}(e^\pi - e^{-\pi})$$

$$\left(e^t, \frac{1}{\sqrt{\pi}}\cos 2t\right) = \int_{-\pi}^{\pi} e^t \frac{1}{\sqrt{\pi}}\cos 2t \, dt = \frac{1}{5\sqrt{\pi}}(e^\pi - e^{-\pi})$$

$$\left(e^t, \frac{1}{\sqrt{\pi}}\sin 2t\right) = \int_{-\pi}^{\pi} e^t \frac{1}{\sqrt{\pi}}\sin 2t \, dt = -\frac{2}{5\sqrt{\pi}}(e^\pi - e^{-\pi})$$

Thus

$$\text{proj}_W e^t = \frac{1}{\sqrt{2\pi}}(e^\pi - e^{-\pi})\frac{1}{\sqrt{2\pi}} + \left(-\frac{1}{2\sqrt{\pi}}\right)(e^\pi - e^{-\pi})\frac{1}{\sqrt{\pi}}\cos t + \left(\frac{1}{2\sqrt{\pi}}\right)(e^\pi - e^{-\pi})\frac{1}{\sqrt{\pi}}\sin t$$

$$+ \left(\frac{1}{5\sqrt{\pi}}\right)(e^\pi - e^{-\pi})\frac{1}{\sqrt{\pi}}\cos 2t + \left(-\frac{2}{5\sqrt{\pi}}\right)(e^\pi - e^{-\pi})\frac{1}{\sqrt{\pi}}\sin 2t$$

$$= \frac{1}{2\pi}(e^\pi - e^{-\pi}) + \frac{1}{\pi}\left(-\frac{1}{2}e^\pi + \frac{1}{2}e^{-\pi}\right)\cos t + \frac{1}{\pi}\left(\frac{1}{2}e^\pi - \frac{1}{2}e^{-\pi}\right)\sin t$$

$$+ \frac{1}{\pi}\left(\frac{1}{5}e^\pi - \frac{1}{5}e^{-\pi}\right)\cos 2t + \frac{1}{\pi}\left(-\frac{2}{5}e^\pi + \frac{2}{5}e^{-\pi}\right)\sin 2t$$

25. If $\mathbf{v}$ is in $V^\perp$, then $(\mathbf{v}, \mathbf{v}) = 0$. By Definition 5.2, $\mathbf{v}$ must be the zero vector. If $W = \{\mathbf{0}\}$, then every vector $\mathbf{v}$ in V is in $W^\perp$ because $(\mathbf{v}, \mathbf{0}) = 0$. Thus $W^\perp = V$.

27. Let $\mathbf{v}$ be a vector in R^n. By Theorem 5.12(a), the column space of A^T is the orthogonal complement of the null space of A. This means that R^n is the direct sum of the null space of A and the column space of A^T:

$$R^n = \text{null space of } A \oplus \text{column space of } A^T.$$

Hence, there exist unique vectors $\mathbf{w}$ in the null space of A and $\mathbf{u}$ in the column space of A^T so that $\mathbf{v} = \mathbf{w} + \mathbf{u}$.

29. If $\{\mathbf{w}_1, \mathbf{w}_2, \ldots, \mathbf{w}_m\}$ is an orthogonal basis for W, then

$$\left\{ \frac{\mathbf{w}_1}{\|\mathbf{w}_1\|}, \frac{\mathbf{w}_2}{\|\mathbf{w}_2\|}, \ldots, \frac{\mathbf{w}_m}{\|\mathbf{w}_m\|} \right\}$$

is an orthonormal basis for W, so

$$\text{proj}_W \mathbf{v} = \left(\mathbf{v}, \frac{\mathbf{w}_1}{\|\mathbf{w}_1\|}\right)\frac{\mathbf{w}_1}{\|\mathbf{w}_1\|} + \left(\mathbf{v}, \frac{\mathbf{w}_2}{\|\mathbf{w}_2\|}\right)\frac{\mathbf{w}_2}{\|\mathbf{w}_2\|} + \cdots + \left(\mathbf{v}, \frac{\mathbf{w}_m}{\|\mathbf{w}_m\|}\right)\frac{\mathbf{w}_m}{\|\mathbf{w}_m\|}$$

$$= \frac{(\mathbf{v}, \mathbf{w}_1)}{(\mathbf{w}_1, \mathbf{w}_1)}\mathbf{w}_1 + \frac{(\mathbf{v}, \mathbf{w}_2)}{(\mathbf{w}_2, \mathbf{w}_2)}\mathbf{w}_2 + \cdots + \frac{(\mathbf{v}, \mathbf{w}_m)}{(\mathbf{w}_m, \mathbf{w}_m)}\mathbf{w}_m.$$

Section 5.6, p. 356

1. From Equation (1), the normal system of equations is $A^T A\hat{\mathbf{x}} = A^T \mathbf{b}$. Since A is nonsingular so is A^T and hence so is $A^T A$. It follows from matrix algebra that $(A^T A)^{-1} = A^{-1}(A^T)^{-1}$ and multiplying both sides of the preceding equation by $(A^T A)^{-1}$ gives

$$\hat{\mathbf{x}} = (A^T A)^{-1} A^T \mathbf{b} = A^{-1}(A^T)^{-1} A^T \mathbf{b} = A^{-1}\mathbf{b}.$$

3. In order to use Theorem 5.14 we must show that rank $A = 3$. We first obtain the reduced echelon form of A:

$$A = \begin{bmatrix} 1 & 2 & 1 \\ 1 & 3 & 2 \\ 2 & 5 & 3 \\ 2 & 0 & 1 \\ 3 & 1 & 1 \end{bmatrix} \xrightarrow{\quad \cdots \quad} \begin{bmatrix} 1 & 0 & 0 \\ 0 & 1 & 0 \\ 0 & 0 & 1 \\ 0 & 0 & 0 \\ 0 & 0 & 0 \end{bmatrix}$$
steps omitted

Hence rank $A = 3$. Then using Theorem 5.14 we proceed as follows. Form the normal system:

$$A^T A\hat{\mathbf{x}} = A^T \mathbf{b} \implies \begin{bmatrix} 19 & 18 & 14 \\ 18 & 39 & 24 \\ 14 & 24 & 16 \end{bmatrix} \hat{\mathbf{x}} = \begin{bmatrix} -3 \\ 2 \\ 2 \end{bmatrix}.$$

To solve the normal system we apply Gaussian elimination by reducing the augmented matrix to upper triangular form and then apply back substitution. (This is less work than determining $(A^T A)^{-1}$ and using the formula in Theorem 5.14.) For the upper triangular form we obtain:

$$
[\, A^T A \mid \mathbf{b} \,] = \begin{bmatrix} 19 & 18 & 14 & -3 \\ 18 & 39 & 24 & 2 \\ 14 & 24 & 16 & 2 \end{bmatrix} \xrightarrow{\underset{\text{steps omitted}}{\cdots}} \begin{bmatrix} 1 & -21 & -10 & -5 \\ 0 & 1 & \frac{204}{417} & \frac{92}{417} \\ 0 & 0 & 1 & \frac{768}{180} \end{bmatrix}
$$

Back substitution gives

$$
x_1 = -\frac{9591}{6255} \approx -1.5333, \quad x_2 = -\frac{35028}{18765} \approx -1.8667, \quad x_3 = \frac{768}{180} \approx 4.2667.
$$

Therefore $\hat{\mathbf{x}} \approx \begin{bmatrix} -1.5333 \\ -1.8667 \\ 4.2667 \end{bmatrix}$.

5. To find the QR-decomposition of A, let $\mathbf{u}_1$, $\mathbf{u}_2$, $\mathbf{u}_3$ be the column vectors of A:

$$
\mathbf{u}_1 = \begin{bmatrix} 1 \\ 1 \\ 2 \\ 2 \\ 3 \end{bmatrix}, \quad \mathbf{u}_2 = \begin{bmatrix} 2 \\ 3 \\ 5 \\ 0 \\ 1 \end{bmatrix}, \quad \mathbf{u}_3 = \begin{bmatrix} 1 \\ 2 \\ 3 \\ 1 \\ 1 \end{bmatrix}.
$$

Using the Gram-Schmidt process we find an orthogonal basis for the column space of A. Using the mathematics software package *Mathematica*, we find:

$$
\mathbf{v}_1 = \mathbf{u}_1, \quad \mathbf{v}_2 = \mathbf{u}_2 - \frac{(\mathbf{u}_2, \mathbf{v}_1)}{(\mathbf{v}_1, \mathbf{v}_1)} \mathbf{v}_1 = \begin{bmatrix} 1.0526 \\ 2.0526 \\ 3.1053 \\ -1.8947 \\ -1.8421 \end{bmatrix}, \quad \mathbf{v}_3 = \mathbf{u}_3 - \frac{(\mathbf{u}_3, \mathbf{v}_1)}{(\mathbf{v}_1, \mathbf{v}_1)} \mathbf{v}_1 - \frac{(\mathbf{u}_3, \mathbf{v}_2)}{(\mathbf{v}_2, \mathbf{v}_2)} \mathbf{v}_2 = \begin{bmatrix} -0.9629 \\ -2.1889 \\ -3.1518 \\ 1.5108 \\ 1.7740 \end{bmatrix}
$$

Next, find unit vectors in the same direction as the $\mathbf{v}_i$:

$$
\mathbf{w}_1 = \frac{\mathbf{v}_1}{\|\mathbf{v}_1\|} = \begin{bmatrix} 0.2294 \\ 0.2294 \\ 0.4588 \\ 0.4588 \\ 0.6882 \end{bmatrix}, \quad \mathbf{w}_2 = \frac{\mathbf{v}_2}{\|\mathbf{v}_2\|} = \begin{bmatrix} 0.2247 \\ 0.4381 \\ 0.6628 \\ -0.4044 \\ -0.3932 \end{bmatrix}, \quad \mathbf{w}_3 = \frac{\mathbf{v}_3}{\|\mathbf{v}_3\|} = \begin{bmatrix} -0.3833 \\ 0.3942 \\ 0.0110 \\ 0.6899 \\ -0.4709 \end{bmatrix}
$$

It now follows that

$$
Q = \begin{bmatrix} \mathbf{w}_1 & \mathbf{w}_2 & \mathbf{w}_3 \end{bmatrix} = \begin{bmatrix} 0.2294 & 0.2247 & -0.3833 \\ 0.2294 & 0.4381 & 0.3942 \\ 0.4588 & 0.6628 & 0.0110 \\ 0.4588 & -0.4044 & 0.6899 \\ 0.6882 & -0.3932 & -0.4709 \end{bmatrix}
$$

Next, we find the matrix $R = (r_{ij})$ using $r_{ji} = (\mathbf{u}_i, \mathbf{w}_j)$:

$$
R = \begin{bmatrix} 4.3589 & 4.1295 & 3.2118 \\ 0 & 4.6848 & 2.2918 \\ 0 & 0 & 0.6570 \end{bmatrix}
$$

Finally, to find the least squares solution to $A\mathbf{x} = \mathbf{b}$ we must solve the system $R\hat{\mathbf{x}} = Q^T\mathbf{b}$ by back substitution:

$$\begin{bmatrix} 4.3589 & 4.1295 & 3.2118 \\ 0 & 4.6848 & 2.2918 \\ 0 & 0 & 0.6570 \end{bmatrix} \begin{bmatrix} x_1 \\ x_2 \\ x_3 \end{bmatrix} = \begin{bmatrix} 0.2294 & 0.2247 & -0.3833 \\ 0.2294 & 0.4381 & 0.3942 \\ 0.4588 & 0.6628 & 0.0110 \\ 0.4588 & -0.4044 & 0.6899 \\ 0.6882 & -0.3932 & -0.4709 \end{bmatrix}^T \begin{bmatrix} -1 \\ 2 \\ 0 \\ 1 \\ -2 \end{bmatrix} = \begin{bmatrix} -0.6882 \\ 1.0335 \\ 2.8034 \end{bmatrix}$$

Back substitution then gives $x_3 = 4.2668$, $x_2 = -1.8668$, $x_1 = -1.5331$, which agrees with the solution found in Exercise 3 to within three decimal places of accuracy.

7. Minimizing E_2 amounts to searching over the vector space P_2 of all quadratics in order to determine the one whose coefficients give the smallest value in the expression E_2. Since P_1 is a subspace of P_2, the minimization of E_2 has already searched over P_1 and thus the minimum of E_1 cannot be smaller than the minimum of E_2.

9. (a) We row reduce the matrix, finding:

$$\begin{bmatrix} 1 & 3 & -3 \\ 2 & 4 & -2 \\ 0 & -1 & 2 \\ 1 & 2 & -1 \end{bmatrix} \xrightarrow{\;} \cdots \xrightarrow[\text{steps omitted}]{} \begin{bmatrix} 1 & 0 & 3 \\ 0 & 1 & -2 \\ 0 & 0 & 0 \\ 0 & 0 & 0 \end{bmatrix}$$

Therefore the rank of A is 2.

(b) To find a basis for the column space of A we determine the nonzero rows in the reduced row echelon form of A^T and convert them to columns. We find that

$$\begin{bmatrix} 1 & 2 & 0 & 1 \\ 3 & 4 & -1 & 2 \\ -3 & -2 & 2 & -1 \end{bmatrix} \xrightarrow{\;} \cdots \xrightarrow[\text{steps omitted}]{} \begin{bmatrix} 1 & 0 & -1 & 0 \\ 0 & 1 & \frac{1}{2} & \frac{1}{2} \\ 0 & 0 & 0 & 0 \end{bmatrix}$$

Then

$$\mathbf{u}_1 = \begin{bmatrix} 1 \\ 0 \\ -1 \\ 0 \end{bmatrix} \quad \text{and} \quad \mathbf{u}_2 = \begin{bmatrix} 0 \\ 1 \\ \frac{1}{2} \\ \frac{1}{2} \end{bmatrix}$$

form a basis for the column space of A. Applying the Gram-Schmidt process we obtain the orthonormal basis

$$\mathbf{w}_1 = \frac{1}{\sqrt{2}} \begin{bmatrix} 1 \\ 0 \\ -1 \\ 0 \end{bmatrix} \quad \text{and} \quad \mathbf{w}_2 = \frac{1}{\sqrt{22}} \begin{bmatrix} 1 \\ 4 \\ 1 \\ 2 \end{bmatrix}.$$

Hence we have

$$\text{proj}_W \mathbf{b} = (\mathbf{b}, \mathbf{w}_1)\mathbf{w}_1 + (\mathbf{b}, \mathbf{w}_2)\mathbf{w}_2 = \frac{1}{\sqrt{2}}\mathbf{w}_1 + \frac{3}{\sqrt{22}}\mathbf{w}_2 = \frac{1}{11} \begin{bmatrix} 7 \\ 6 \\ -4 \\ 3 \end{bmatrix}.$$

We now solve the system $A\hat{\mathbf{x}} = \text{proj}_W \mathbf{b}$ by row reducing the matrix $\begin{bmatrix} A & | & \text{proj}_W \mathbf{b} \end{bmatrix}$ to obtain

$$\begin{bmatrix} 1 & 0 & 3 & | & -\frac{5}{11} \\ 0 & 1 & -2 & | & \frac{4}{11} \\ 0 & 0 & 0 & | & 0 \\ 0 & 0 & 0 & | & 0 \end{bmatrix}$$

Thus we see that there are infinitely many solutions to this least squares problem. The general solution is

$$\hat{x}_1 = -3r - \frac{5}{11}, \quad \hat{x}_2 = 2r + \frac{4}{11}, \quad \hat{x}_3 = r,$$

where r is any real number. One possible answer is $\hat{\mathbf{x}} = \begin{bmatrix} -\frac{5}{11} \\ \frac{4}{11} \\ 0 \end{bmatrix}$.

11. (a) Let x represent the time in years from 1996 ($x = 0$ is 1996) and $y(x)$ the debt per capita in year $1996 + x$. We must find coefficients x_1, x_2 such that

$$y(x) = x_1 + x_2 x.$$

Let

$$A = \begin{bmatrix} 1 & 0 \\ 1 & 1 \\ 1 & 2 \\ 1 & 3 \\ 1 & 4 \\ 1 & 5 \\ 1 & 6 \\ 1 & 7 \\ 1 & 8 \\ 1 & 9 \end{bmatrix}, \quad \hat{\mathbf{x}} = \begin{bmatrix} \hat{x}_1 \\ \hat{x}_2 \end{bmatrix}, \quad \mathbf{b} = \begin{bmatrix} 20070 \\ 20548 \\ 20774 \\ 21182 \\ 20065 \\ 20874 \\ 22274 \\ 24077 \\ 25868 \\ 27430 \end{bmatrix}.$$

The rank of A is 2 and the normal system is

$$A^T A \hat{\mathbf{x}} = A^T \mathbf{b}$$

$$\begin{bmatrix} 1 & 1 & 1 & 1 & 1 & 1 & 1 & 1 & 1 & 1 \\ 0 & 1 & 2 & 3 & 4 & 5 & 6 & 7 & 8 & 9 \end{bmatrix} \begin{bmatrix} 1 & 0 \\ 1 & 1 \\ 1 & 2 \\ 1 & 3 \\ 1 & 4 \\ 1 & 5 \\ 1 & 6 \\ 1 & 7 \end{bmatrix} \hat{\mathbf{x}} = \begin{bmatrix} 1 & 1 & 1 & 1 & 1 & 1 & 1 & 1 & 1 & 1 \\ 0 & 1 & 2 & 3 & 4 & 5 & 6 & 7 & 8 & 9 \end{bmatrix} \begin{bmatrix} 20070 \\ 20548 \\ 20774 \\ 21182 \\ 20065 \\ 20874 \\ 22274 \\ 24077 \\ 25868 \\ 27430 \end{bmatrix}$$

$$\begin{bmatrix} 10 & 45 \\ 45 & 285 \end{bmatrix} \hat{\mathbf{x}} = \begin{bmatrix} 223162 \\ 1066269 \end{bmatrix}$$

Using Gaussian elimination to solve this system, we obtain

$$\hat{\mathbf{x}} = \begin{bmatrix} 18932.2 \\ 752 \end{bmatrix}.$$

Therefore the line of best fit to the given data is $y = 752x + 18932.2$.

(b) In the year 2008, $x = 12$. Then $y = 752(12) + 18932.2 = 27956.2$.

In the year 2010, $x = 14$. Then $y = 752(14) + 18932.2 = 29460.2$.

In the year 2015, $x = 19$. Then $y = 752(19) + 18932.2 = 33220.2$.

13. (a) Substituting the data from the given table into the equation $\ln y = c_1 + c_2 x$, we obtain the linear

system

$$c_1 + 6c_2 = \ln 23.34$$
$$c_1 + 10c_2 = \ln 19.67$$
$$c_1 + 12c_2 = \ln 18.52$$
$$c_1 + 14c_2 = \ln 17.60$$
$$c_1 + 16c_2 = \ln 16.81$$
$$c_1 + 20c_2 = \ln 15.90$$

The coefficiernt and augmented matrices

$$
\begin{bmatrix} 1 & 6 \\ 1 & 10 \\ 1 & 12 \\ 1 & 14 \\ 1 & 16 \\ 1 & 20 \end{bmatrix}, \qquad
\begin{bmatrix} 1 & 6 & \ln 23.34 \\ 1 & 10 & \ln 19.67 \\ 1 & 12 & \ln 18.52 \\ 1 & 14 & \ln 17.60 \\ 1 & 16 & \ln 16.81 \\ 1 & 20 & \ln 15.90 \end{bmatrix}
$$

have rank 2 and 3, respectively. Thus, by Theorem 4.21 the linear system is inconsistent.

(b) The normal system is

$$A^T A \widehat{\mathbf{x}} = A^T \mathbf{b}$$

$$
\begin{bmatrix} 1 & 1 & 1 & 1 & 1 & 1 \\ 6 & 10 & 12 & 14 & 16 & 20 \end{bmatrix}
\begin{bmatrix} 1 & 6 \\ 1 & 10 \\ 1 & 12 \\ 1 & 14 \\ 1 & 16 \\ 1 & 20 \end{bmatrix}
=
\begin{bmatrix} 1 & 1 & 1 & 1 & 1 & 1 \\ 6 & 10 & 12 & 14 & 16 & 20 \end{bmatrix}
\begin{bmatrix} \ln 23.34 \\ \ln 19.67 \\ \ln 18.52 \\ \ln 17.60 \\ \ln 17.60 \\ \ln 16.81 \\ \ln 15.90 \end{bmatrix}
$$

$$
\begin{bmatrix} 6 & 78 \\ 78 & 1132 \end{bmatrix} \widehat{\mathbf{x}} = \begin{bmatrix} 17.5049 \\ 224.356 \end{bmatrix}
$$

$$
\widehat{\mathbf{x}} = \begin{bmatrix} 3.2709 \\ -0.0272 \end{bmatrix}
$$

So $\ln y = -0.0272x + 3.2709$.

(c) $c_1 = \ln r \implies 3.2709 = \ln r \implies r = e^{3.2709} = 26.3350$
   $c_2 = s \implies s = -0.0272$

(d) Here $x = 18$, so $\ln y = -0.0272(18) + 3.2709 = 2.7813$ and hence $y = e^{2.7813} = 16.14$ mm.

Supplementary Exercises for Chapter 5, p. 358

1. Let

$$
\mathbf{u} = \begin{bmatrix} 1 \\ -2 \\ 1 \end{bmatrix}.
$$

To find the subspace W of vectors in R^3 that are orthogonal to $\mathbf{u}$, we determine all vectors $\mathbf{x} = \begin{bmatrix} x_1 & x_2 & x_3 \end{bmatrix}^T$ such that $\mathbf{u} \cdot \mathbf{x} = 0$. Using the standard inner product, this gives the equation $x_1 - 2x_2 + x_3 = 0$. Thus we have one equation in 3 unknowns so we can choose two of the unknowns arbitrarily. Let $x_2 = r$ and $x_3 = s$, then $x_1 = 2r - s$. Hence any vector $\mathbf{x}$ that is orthogonal to $\mathbf{u}$ has the form

$$
\mathbf{x} = \begin{bmatrix} x_1 \\ x_2 \\ x_3 \end{bmatrix} = \begin{bmatrix} 2r - s \\ r \\ s \end{bmatrix} = r \begin{bmatrix} 2 \\ 1 \\ 0 \end{bmatrix} + s \begin{bmatrix} -1 \\ 0 \\ 1 \end{bmatrix}.
$$

Let

$$\mathbf{u}_1 = \begin{bmatrix} 2 \\ 1 \\ 0 \end{bmatrix} \quad \text{and} \quad \mathbf{u}_2 = \begin{bmatrix} -1 \\ 0 \\ 1 \end{bmatrix}.$$

Then by the preceding expression $\{\mathbf{u}_1, \mathbf{u}_2\}$ spans W. We note that $\mathbf{u}_1$ and $\mathbf{u}_2$ are not scalar multiples of one another, hence they are linearly independent. Therefore $\{\mathbf{u}_1, \mathbf{u}_2\}$ is a basis for W. However, $\mathbf{u}_1 \cdot \mathbf{u}_2 \neq 0$ so we must use the Gram-Schmidt process to find an orthogonal basis for W. Let $\mathbf{v}_1 = \mathbf{u}_1$ and

$$\mathbf{v}_2 = \mathbf{u}_2 - \frac{(\mathbf{u}_2, \mathbf{v}_1)}{(\mathbf{v}_1, \mathbf{v}_1)}\mathbf{v}_1 = \mathbf{u}_2 - \frac{-2}{5}\mathbf{v}_1 = \begin{bmatrix} -\frac{1}{5} \\ \frac{2}{5} \\ 1 \end{bmatrix}.$$

Then $\{\mathbf{v}_1, \mathbf{v}_2\}$ is an orthogonal basis for W.

3. Let $S = \{\mathbf{u}_1, \mathbf{u}_2, \mathbf{u}_3\}$. Then using Theorem 5.5 we have

$$\mathbf{v} = (\mathbf{v}, \mathbf{u}_1)\mathbf{u}_1 + (\mathbf{v}, \mathbf{u}_2)\mathbf{u}_2 + (\mathbf{v}, \mathbf{u}_3)\mathbf{u}_3 = -\sqrt{2}\,\mathbf{u}_1 + 2\mathbf{u}_2 + 2\sqrt{2}\,\mathbf{u}_3.$$

5. We view the plane P as a two-dimensional subspace of R^3 which has basis $\{\mathbf{w}_1, \mathbf{w}_2\}$. The vector in P closest to $\mathbf{v}$ is just the projection of $\mathbf{v}$ onto plane P. To determine the projection we first find an orthonormal basis for P and then use Theorem 5.10. Let $\mathbf{v}_1 = \mathbf{w}_1$. Then by the Gram-Schmidt process we have

$$\mathbf{v}_2 = \mathbf{w}_2 - \frac{(\mathbf{w}_2, \mathbf{v}_1)}{(\mathbf{v}_1, \mathbf{v}_1)}\mathbf{v}_1 = \mathbf{w}_2 - \frac{12}{14}\mathbf{v}_1 = \frac{15}{7}\mathbf{i} + \frac{11}{7}\mathbf{j} - \frac{9}{7}\mathbf{k}.$$

Hence

$$\mathbf{u}_1 = \frac{\mathbf{v}_1}{\|\mathbf{v}_1\|} = \frac{1}{\sqrt{14}}\mathbf{v}_1 \quad \text{and} \quad \mathbf{u}_2 = \frac{\mathbf{v}_2}{\|\mathbf{v}_2\|} = \sqrt{\frac{7}{61}}\,\mathbf{v}_2$$

is an orthonormal basis for plane P. From Theorem 5.10

$$\text{proj}_P \mathbf{v} = (\mathbf{v}, \mathbf{u}_1)\mathbf{u}_1 + (\mathbf{v}, \mathbf{u}_1)\mathbf{u}_2 = -\frac{7}{\sqrt{14}}\mathbf{u}_1 + 4\sqrt{\frac{7}{61}}\mathbf{u}_2 = \frac{59}{122}\mathbf{i} + \frac{271}{122}\mathbf{j} + \frac{50}{61}\mathbf{k} = \frac{1}{122}\begin{bmatrix} 59 \\ 271 \\ 50 \end{bmatrix}.$$

The distance from $\mathbf{v}$ to P is therefore

$$\|\mathbf{v} - \text{proj}_P \mathbf{v}\| = \sqrt{\left(1 - \frac{59}{122}\right)^2 + \left(2 - \frac{271}{122}\right)^2 + \left(1 - \frac{25}{61}\right)^2} = \frac{9}{\sqrt{122}}.$$

7. To show that the functions $\sin nt$, for $n = 1, 2, \ldots$, is an orthogonal set, let k and m be integers with $k \geq 1$, $m \geq 1$, and $k \neq m$. Then from calculus or a table of integrals,

$$(\sin kt, \sin mt) = \int_0^\pi \sin kt \sin mt \, dt = \left. \frac{\sin(k-m)t}{2(k-m)} - \frac{\sin(k+m)t}{2(k+m)} \right|_0^\pi = 0.$$

9. (a) If we form the augmented matrix of the homogeneous system $A\mathbf{x} = \mathbf{0}$ it is easily seen that no row operations are needed to obtain the reduced row echelon form. The corresponding system can be expressed as

$$\begin{aligned} x_1 &= -5x_3 + 2x_4 \\ x_2 &= 2x_3 + 5x_4 \end{aligned}$$

Thus for $x_3 = s$ and $x_4 = t$ we have that the general solution of the system has the form

$$\mathbf{x} = \begin{bmatrix} -5s + 2t \\ 2s + 5t \\ s \\ t \end{bmatrix} = s\begin{bmatrix} -5 \\ 2 \\ 1 \\ 0 \end{bmatrix} + t\begin{bmatrix} 2 \\ 5 \\ 0 \\ 1 \end{bmatrix}.$$

It follows that $\mathbf{u}_1 = \begin{bmatrix} -5 & 2 & 1 & 0 \end{bmatrix}^T$ and $\mathbf{u}_2 = \begin{bmatrix} 2 & 5 & 0 & 1 \end{bmatrix}^T$ are a basis for the solution space. It also happens that $\mathbf{u}_1 \cdot \mathbf{u}_2 = 0$, so that $\mathbf{u}_1$ and $\mathbf{u}_2$ form an orthogonal basis. We determine unit vectors in the direction of $\mathbf{u}_1$ and $\mathbf{u}_2$ to obtain an orthonormal basis:

$$\mathbf{v}_1 = \frac{\mathbf{u}_1}{\|\mathbf{u}_1\|} = \frac{1}{\sqrt{30}} \begin{bmatrix} -5 \\ 2 \\ 1 \\ 0 \end{bmatrix}, \quad \mathbf{v}_2 = \frac{\mathbf{u}_2}{\|\mathbf{u}_2\|} = \frac{1}{\sqrt{30}} \begin{bmatrix} 2 \\ 5 \\ 0 \\ 1 \end{bmatrix}.$$

(b) As in part (a), the corresponding homogeneous system can be expressed as

$$\begin{aligned} x_1 &= -5x_3 + 2x_4 \\ x_2 &= 2x_3 - 4x_4 \end{aligned}$$

Thus for $x_3 = s$ and $x_4 = t$ we have that the general solution of the system has the form

$$\mathbf{x} = \begin{bmatrix} -5s + 2t \\ 2s - 4t \\ s \\ t \end{bmatrix} = s \begin{bmatrix} -5 \\ 2 \\ 1 \\ 0 \end{bmatrix} + t \begin{bmatrix} 2 \\ -4 \\ 0 \\ 1 \end{bmatrix}.$$

It follows that $\mathbf{u}_1 = \begin{bmatrix} -5 & 2 & 1 & 0 \end{bmatrix}^T$ and $\mathbf{u}_2 = \begin{bmatrix} 2 & -4 & 0 & 1 \end{bmatrix}^T$ form a basis for the solution space. However, in this case $\mathbf{u}_1 \cdot \mathbf{u}_2 \neq 0$, so that $\mathbf{u}_1$ and $\mathbf{u}_2$ are not orthogonal. To determine an orthogonal basis we use the Gram-Schmidt process. Let

$$\mathbf{v}_1 = \mathbf{u}_1 \quad \text{and} \quad \mathbf{v}_2 = \mathbf{u}_2 - \frac{(\mathbf{u}_2, \mathbf{v}_1)}{(\mathbf{v}_1, \mathbf{v}_1)} \mathbf{v}_1 = \mathbf{u}_2 - \frac{-18}{30} \mathbf{v}_1 = \begin{bmatrix} -1 \\ -\frac{14}{5} \\ \frac{3}{5} \\ 1 \end{bmatrix}.$$

Finally, we normalize $\mathbf{v}_1$ and $\mathbf{v}_2$ to obtain an orthonormal basis for the null space of A:

$$\mathbf{w}_1 = \frac{1}{\sqrt{30}} \begin{bmatrix} -5 \\ 2 \\ 1 \\ 0 \end{bmatrix} \quad \mathbf{w}_2 = \frac{1}{\sqrt{255}} \begin{bmatrix} -5 \\ -14 \\ 3 \\ 5 \end{bmatrix}.$$

11. Let $\mathbf{w}_1 = \begin{bmatrix} 1 & 0 & 1 \end{bmatrix}^T$ and $\mathbf{w}_2 = \begin{bmatrix} 0 & 1 & 0 \end{bmatrix}^T$. First note that $\mathbf{w}_1$ and $\mathbf{w}_2$ are orthogonal and since neither is the zero vector, they are linearly independent. Thus, $\{\mathbf{w}_1, \mathbf{w}_2\}$ is a basis for W.

(a) Any vector in the orthogonal complement of W must be orthogonal to both $\mathbf{w}_1$ and $\mathbf{w}_2$. If $\mathbf{x} = \begin{bmatrix} x_1 & x_2 & x_3 & x_4 \end{bmatrix}^T$ is in the orthogonal complement of W then we have using the standard inner product

$$(\mathbf{w}_1, \mathbf{x}) = x_1 + x_3 = 0 \quad \text{and} \quad (\mathbf{w}_2, \mathbf{x}) = x_2 = 0.$$

Hence we can choose one of x_1 and x_3 arbitrarily, so let $x_3 = t$ then $x_1 = -t$ and we also have $x_2 = 0$. Thus

$$\mathbf{x} = \begin{bmatrix} -t \\ 0 \\ t \end{bmatrix} = t \begin{bmatrix} -1 \\ 0 \\ 1 \end{bmatrix}$$

and it follows that the orthogonal complement of W consists of all the multiples of vector $\begin{bmatrix} -1 & 0 & 1 \end{bmatrix}^T$. Hence vector $\begin{bmatrix} -1 & 0 & 1 \end{bmatrix}^T$ is a basis for the orthogonal complement of W.

(b) Let

$$S = \left\{ \begin{bmatrix} 1 \\ 0 \\ 1 \end{bmatrix}, \begin{bmatrix} 0 \\ 1 \\ 0 \end{bmatrix}, \begin{bmatrix} -1 \\ 0 \\ 1 \end{bmatrix} \right\}.$$

Note that these three vectors are mutually orthogonal and since they are nonzero Theorem 3.4 guarantees that they are linearly independent. Hence S contains three linearly independent vectors in 3-dimensional space R^3 and so it is a basis for R^3.

(c) We express $\mathbf{v}$ in terms of the basis S in part (b).

i. For $\mathbf{v} = \begin{bmatrix} 1 & 0 & 0 \end{bmatrix}^T$ we have

$$\mathbf{v} = c_1 \begin{bmatrix} 1 \\ 0 \\ 1 \end{bmatrix} + c_2 \begin{bmatrix} 0 \\ 1 \\ 0 \end{bmatrix} + c_3 \begin{bmatrix} -1 \\ 0 \\ 1 \end{bmatrix} = \begin{bmatrix} 1 \\ 0 \\ 0 \end{bmatrix}.$$

Combining the matrices and equating corresponding elements gives $c_1 - c_3 = 1$, $c_2 = 0$, $c_1 + c_3 = 0$. Solving we have $c_1 = \frac{1}{2}$, $c_2 = 0$, $c_3 = -\frac{1}{2}$ and hence $\mathbf{v} = \mathbf{w} + \mathbf{u}$, where

$$\mathbf{w} = \frac{1}{2} \begin{bmatrix} 1 \\ 0 \\ 1 \end{bmatrix} \quad \text{and} \quad \mathbf{u} = -\frac{1}{2} \begin{bmatrix} -1 \\ 0 \\ 1 \end{bmatrix}.$$

ii. For $\mathbf{v} = \begin{bmatrix} 1 & 2 & 3 \end{bmatrix}^T$, we have

$$\mathbf{v} = c_1 \begin{bmatrix} 1 \\ 0 \\ 1 \end{bmatrix} + c_2 \begin{bmatrix} 0 \\ 1 \\ 0 \end{bmatrix} + c_3 \begin{bmatrix} -1 \\ 0 \\ 1 \end{bmatrix} = \begin{bmatrix} 1 \\ 2 \\ 3 \end{bmatrix}.$$

Combining the matrices and equating corresponding elements gives $c_1 - c_3 = 1$, $c_2 = 2$, $c_1 + c_3 = 3$. Solving we have $c_1 = 2$, $c_2 = 2$, $c_3 = 1$ and hence $\mathbf{v} = \mathbf{w} + \mathbf{u}$, where

$$\mathbf{w} = 2 \begin{bmatrix} 1 \\ 0 \\ 1 \end{bmatrix} + 2 \begin{bmatrix} 0 \\ 1 \\ 0 \end{bmatrix} = \begin{bmatrix} 2 \\ 2 \\ 2 \end{bmatrix} \quad \text{and} \quad \mathbf{u} = \begin{bmatrix} -1 \\ 0 \\ 1 \end{bmatrix}.$$

13. Let

$$\mathbf{v}_1 = 1$$

$$\mathbf{v}_2 = t - \frac{(t,t)}{(1,1)} 1 = t - \frac{\int_{-1}^{1} t \, dt}{\int_{-1}^{1} 1 \, dt} 1 = t - \frac{\frac{1}{2} t^2 \big|_{-1}^{1}}{t \big|_{-1}^{1}} 1 = t$$

$$\mathbf{v}_3 = t^2 - \frac{(t^2,1)}{(1,1)} 1 - \frac{(t^2,t)}{(t,t)} t = t^2 - \frac{\int_{-1}^{1} t^2 \, dt}{\int_{-1}^{1} 1 \, dt} 1 - \frac{\int_{-1}^{1} t^3 \, dt}{\int_{-1}^{1} t^2 \, dt} t = t^2 - \frac{\frac{1}{3} t^3 \big|_{-1}^{1}}{2} - \frac{\frac{1}{4} t^4 \big|_{-1}^{1}}{\frac{1}{3} t^3 \big|_{-1}^{1}} t = t^2 - \frac{1}{3}$$

Next normalize:

$$\mathbf{w}_1 = \frac{\mathbf{v}_1}{\|\mathbf{v}_1\|} = \frac{1}{\sqrt{(1,1)}} = \frac{1}{\sqrt{2}}$$

$$\mathbf{w}_2 = \frac{\mathbf{v}_2}{\|\mathbf{v}_2\|} = \frac{t}{\sqrt{(t,t)}} = \frac{t}{\sqrt{\int_{-1}^{1} t^2 \, dt}} = \frac{t}{\sqrt{\frac{2}{3}}} = \sqrt{\frac{3}{2}} t$$

$$\mathbf{w}_3 = \frac{\mathbf{v}_3}{\|\mathbf{v}_3\|} = \frac{t^2 - \frac{1}{3}}{\sqrt{\int_{-1}^{1} \left(t^2 - \frac{1}{3}\right)^2 \, dt}} = \frac{t^2 - \frac{1}{3}}{\sqrt{\frac{8}{45}}} = \sqrt{\frac{5}{8}} (3t^2 - 1)$$

From the Gram-Schmidt process $\{\mathbf{w}_1, \mathbf{w}_2, \mathbf{w}_3\}$ is an orthonormal basis for P_2.

15. Let $p_1(t) = \sqrt{\frac{1}{2}}$, $p_2(t) = \sqrt{\frac{3}{2}}\,t$, and $p_3(t) = \sqrt{\frac{5}{8}}\,(3t^2 - 1)$ be the polynomials obtained in Exercise 13 and let W be the subspace having p_1, p_2, and p_3 as basis. To find the distance of $\mathbf{v} = t^3 + 1$ from W we first compute the projection of $\mathbf{v}$ onto W:

$$\text{proj}_W \mathbf{v} = (\mathbf{v}, p_1)p_1 + (\mathbf{v}, p_2)p_2 + (\mathbf{v}, p_3)p_3$$

where we find

$$(\mathbf{v}, p_1) = (t^3 + 1, p_1) = \int_{-1}^{1} (t^3 + 1)\sqrt{\tfrac{1}{2}}\, dt = \sqrt{2}$$

$$(\mathbf{v}, p_2) = (t^3 + 1, p_2) = \int_{-1}^{1} (t^3 + 1)\sqrt{\tfrac{3}{2}}\,t\, dt = \frac{\sqrt{6}}{5}$$

$$(\mathbf{v}, p_3) = (t^3 + 1, p_3) = \int_{-1}^{1} (t^3 + 1)\sqrt{\tfrac{5}{8}}\,(3t^2 - 1)\, dt = 0$$

Therefore

$$\text{proj}_W \mathbf{v} = (\sqrt{2})\left(\sqrt{\tfrac{1}{2}}\right) + \left(\frac{\sqrt{6}}{5}\right)\left(\sqrt{\tfrac{3}{2}}\,t\right) + (0)\left(\sqrt{\tfrac{5}{8}}\,(3t^2 - 1)\right) = 1 + \frac{3}{5}t$$

and hence the distance from $\mathbf{v}$ to W is

$$\|\mathbf{v} - \text{proj}_W \mathbf{v}\| = \left\|(t^3 + 1) - \left(1 + \frac{3}{5}t\right)\right\| = \left\|t^3 - \frac{3}{5}t\right\| = \sqrt{\int_{-1}^{1}\left(t^3 - \frac{3}{5}t\right)^2 dt} = \sqrt{\frac{8}{175}}.$$

17. We show that $a_{ii} = 1$ and $a_{ij} = 0$ for $i \neq j$. We choose $\mathbf{u}$ in particular ways to accomplish this. First, let $\mathbf{u} = \mathbf{e}_i$ to get $(\mathbf{e}_i, \mathbf{e}_i) = 1$. Then $(\mathbf{e}_i, A\mathbf{e}_i) = a_{ii} = 1$. Next, let $\mathbf{u} = \mathbf{e}_i + \mathbf{e}_j$, with $i \neq j$. Then

$$(\mathbf{u}, \mathbf{u}) = a_{ii} + a_{jj} = 2 \quad \text{and} \quad (\mathbf{u}, A\mathbf{u}) = (\mathbf{u}, \text{col}_i(A) + \text{col}_j(A)) = a_{ii} + a_{ij} + a_{ji} + a_{jj} = 2 + 2a_{ij}.$$

Therefore $2a_{ij} = 0$, so that $a_{ij} = 0$ when $i \neq j$. Thus $A = I_n$.

19. First note that since the columns of P form an orthonormal set,

$$\mathbf{p}_i^T \mathbf{p}_j = \begin{cases} 0, & \text{for } i \neq j \\ 1, & \text{for } i = j \end{cases}$$

It follows that $P^T P = I_n$.

(a) $\|P\mathbf{x}\| = \sqrt{(P\mathbf{x}, P\mathbf{x})} = \sqrt{(P\mathbf{x})^T P\mathbf{x}} = \sqrt{\mathbf{x}^T P^T P\mathbf{x}} = \sqrt{\mathbf{x}^T I_n \mathbf{x}} = \sqrt{\mathbf{x}^T \mathbf{x}} = \|\mathbf{x}\|.$

(b) Let θ be the angle between $P\mathbf{x}$ and $P\mathbf{y}$. Then, using part (a), we have

$$\cos \theta = \frac{(P\mathbf{x}, P\mathbf{y})}{\|P\mathbf{x}\|\,\|P\mathbf{y}\|} = \frac{(P\mathbf{x})^T P\mathbf{y}}{\|\mathbf{x}\|\,\|\mathbf{y}\|} = \frac{\mathbf{x}^T P^T P\mathbf{y}}{\|\mathbf{x}\|\,\|\mathbf{y}\|} = \frac{\mathbf{x}^T \mathbf{y}}{\|\mathbf{x}\|\,\|\mathbf{y}\|}$$

But this last expression is the cosine of the angle between $\mathbf{x}$ and $\mathbf{y}$. Since the angle is restricted to be between 0 and π we have that the two angles are equal.

21. (a) The columns $\mathbf{b}_j$ are in R^m. Since the columns are orthonormal they are linearly independent vectors in R^m, thus $m \geq n$.

(b) We have

$$\mathbf{b}_i^T \mathbf{b}_j = \begin{cases} 0, & \text{for } i \neq j \\ 1, & \text{for } i = j. \end{cases}$$

It follows that $B^T B = I_n$, since the (i, j) element of $B^T B$ is computed by taking row i of B^T times column j of B. But row i of B^T is just $\mathbf{b}_i^T$ and column j of B is $\mathbf{b}_j$.

23. Let $\dim V = n$ and $\dim W = r$. Since $V = W \oplus W^\perp$ by Exercise 28, Section 5.5, $\dim W^\perp = n - r$. First observe that if $\mathbf{w}$ is in W, then $\mathbf{w}$ is orthogonal to every vector in $W^\perp$, so $\mathbf{w}$ is in $(W^\perp)^\perp$. Thus W is a subspace of $(W^\perp)^\perp$. Now, again by Exercise 28, $\dim(W^\perp)^\perp = n - (n - r) = r = \dim W$. Hence $(W^\perp)^\perp = W$.

25. We must show that the rows $\mathbf{v}_1, \mathbf{v}_2, \ldots, \mathbf{v}_m$ of AA^T are linearly independent. Consider

$$a_1\mathbf{v}_1 + a_2\mathbf{v}_2 + \cdots + c_m\mathbf{v}_m = \mathbf{0}$$

which can be written in matrix form as $\mathbf{x}A = \mathbf{0}$, where $\mathbf{x} = \begin{bmatrix} a_1 & a_2 & \cdots & a_m \end{bmatrix}$. Multiplying this equation by A^T we have $\mathbf{x}AA^T = \mathbf{0}$. Since AA^T is nonsingular, Theorem 2.9 implies that $\mathbf{x} = \mathbf{0}$, so $a_1 = a_2 = \cdots = a_m = 0$. Hence, rank $A = m$.

27. Let $\mathbf{v} = a_1\mathbf{v}_1 + a_2\mathbf{v}_2 + \cdots + a_n\mathbf{v}_n$ and $\mathbf{w} = b_1\mathbf{v}_1 + b_2\mathbf{v}_2 + \cdots + b_n\mathbf{v}_n$. Then $d(\mathbf{v}, \mathbf{w}) = \|\mathbf{v} - \mathbf{w}\|$ and hence

$$d(\mathbf{v}, \mathbf{w}) = \|\mathbf{v} - \mathbf{w}\| = \sqrt{(\mathbf{v} - \mathbf{w}, \mathbf{v} - \mathbf{w})}$$
$$= \sqrt{((a_1 - b_1)\mathbf{v}_1 + (a_2 - b_2)\mathbf{v}_2 + \cdots + (a_n - b_n)\mathbf{v}_n, (a_1 - b_1)\mathbf{v}_1 + (a_2 - b_2)\mathbf{v}_2 + \cdots + (a_n - b_n)\mathbf{v}_n)}$$
$$= \sqrt{(a_1 - b_1)^2 + (a_2 - b_2)^2 + \cdots + (a_n - b_n)^2}$$

since $(\mathbf{v}_i, \mathbf{v}_j) = 0$ if $i \neq j$ and 1 if $i = j$.

29. (a) $\left\| \begin{bmatrix} 2 & -2 & 3 \end{bmatrix} \right\|_1 = |2| + |-2| + |3| = 7$.
    $\left\| \begin{bmatrix} 2 & -2 & 3 \end{bmatrix} \right\|_2 = \sqrt{2^2 + (-2)^2 + 3^2} = \sqrt{17}$.
    $\left\| \begin{bmatrix} 2 & -2 & 3 \end{bmatrix} \right\|_\infty = \max\{|2|, |-2|, |3|\} = 3$.

    (b) $\left\| \begin{bmatrix} 0 & 3 & -2 \end{bmatrix} \right\|_1 = |0| + |3| + |-2| = 5$.
    $\left\| \begin{bmatrix} 0 & 3 & -2 \end{bmatrix} \right\|_2 = \sqrt{0^2 + 3^2 + (-2)^2} = \sqrt{13}$.
    $\left\| \begin{bmatrix} 0 & 3 & -2 \end{bmatrix} \right\|_\infty = \max\{|0|, |3|, |-2|\} = 3$.

    (c) $\left\| \begin{bmatrix} 2 & 0 & 0 \end{bmatrix} \right\|_1 = |2| + |0| + |0| = 2$.
    $\left\| \begin{bmatrix} 2 & 0 & 0 \end{bmatrix} \right\|_2 = \sqrt{2^2 + 0^2 + 0^2} = 2$.
    $\left\| \begin{bmatrix} 2 & 0 & 0 \end{bmatrix} \right\|_\infty = \max\{|2|, |0|, |0|\} = 2$.

31. (a) $\|\mathbf{x}\|_\infty = \max\{|x_1|, \ldots, |x_n|\} \geq 0$ since each of $|x_1|, \ldots, |x_n|$ is ≥ 0. Clearly, $\|\mathbf{x}\| = 0$ if and only if $\mathbf{x} = \mathbf{0}$.

    (b) If c is any real scalar

$$\|c\mathbf{x}\|_\infty = \max\{|cx_1|, \ldots, |cx_n|\} = \max\{|c| \, |x_1|, \ldots, |c| \, |x_n|\} = |c| \max\{|x_1|, \ldots, |x_n|\} = |c| \, \|\mathbf{x}\|_\infty.$$

    (c) Let $\mathbf{y} = \begin{bmatrix} y_1 & y_2 & \cdots & y_n \end{bmatrix}^T$ and let

$$\|\mathbf{x}\|_\infty = \max\{|x_1|, \ldots, |x_n|\} = |x_s|$$
$$\|\mathbf{y}\|_\infty = \max\{|y_1|, \ldots, |y_n|\} = |y_t|$$

for some s, t, where $1 \leq s \leq n$ and $1 \leq t \leq n$. Then for $i = 1, \ldots, n$, we have using the triangle inequality:

$$|x_i + y_i| \leq |x_i| + |y_i| \leq |x_s| + |y_t|.$$

Thus

$$\|\mathbf{x} + \mathbf{y}\| = \max\{|x_1 + y_1|, \ldots, |x_n + y_n|\} \leq |x_s| + |y_t| = \|\mathbf{x}\|_\infty + \|\mathbf{y}\|_\infty.$$

33. Let $\mathbf{x} = \begin{bmatrix} x_1 \\ x_2 \end{bmatrix}$ be a vector in R^2.

(a) $\|\mathbf{x}\|_1 = 1 \implies |x_1| + |x_2| = 1$. The graph of this equation is a diamond with vertices at $(1,0)$, $(0,1)$, $(-1,0)$, and $(0,-1)$; see figure (a) below.

(b) $\|\mathbf{x}\|_2 = 1 \implies \sqrt{x_1^2 + x_2^2} = 1$. The graph of this equation is a circle centered at the origin of radius 1 unit; see figure (b) below.

(c) $\|\mathbf{x}\|_\infty = 1 \implies \max\{|x_1|, |x_2|\} = 1$. If $|x_1| = 1$, then the graph consists of the points of the form $(1, x_2)$, $(-1, x_2)$, where $|x_2| \le 1$. If $|x_2| = 1$, the graph consists of the points of the form $(x_1, 1)$, $(x_1, -1)$, where $|x_1| \le 1$. Thus the graph is the square with vertices at $(1,1)$, $(-1,1)$, $(-1,-1)$, and $(1,-1)$; see figure (c) below.

(a) $\|\mathbf{x}\|_1 = 1$. \qquad (b) $\|\mathbf{x}\|_2 = 1$. \qquad (c) $\|\mathbf{x}\|_\infty = 1$.

Chapter Review for Chapter 4, p. 286

True or False

1. True. \qquad 2. False. \qquad 3. False. \qquad 4. False. \qquad 5. True. \qquad 6. True.

7. False. \qquad 8. False. \qquad 9. False. \qquad 10. False. \qquad 11. True. \qquad 12. True.

Quiz

1. Since $\cos 60° = \frac{1}{2}$, we have

$$\cos 60° = \frac{\mathbf{u} \cdot \mathbf{v}}{\|\mathbf{u}\|\|\mathbf{v}\|} = \frac{b}{\sqrt{2}\sqrt{b^2 + c^2}} = \frac{1}{2} \implies c^2 = b^2.$$

Since $\|\mathbf{v}\| = \sqrt{b^2 + c^2} = 1$, $c^2 = 1 - b^2$. Thus

$$1 - b^2 = b^2 \implies b = \pm\sqrt{\frac{1}{2}} \implies c = \pm\sqrt{\frac{1}{2}}.$$

However, $b > 0$ since

$$\cos 60° = \frac{b}{\sqrt{2}\sqrt{b^2 + c^2}} = \frac{1}{2} > 0.$$

Thus

$$b = \frac{1}{\sqrt{2}} = \frac{\sqrt{2}}{2}, \quad c = \pm\frac{1}{\sqrt{2}} = \pm\frac{\sqrt{2}}{2}.$$

2. Let $\mathbf{x} = \begin{bmatrix} a \\ b \\ c \\ d \end{bmatrix}$. Then

$$\mathbf{x} \cdot \mathbf{v}_1 = a + b + 2c - 2d = 0$$
$$\mathbf{x} \cdot \mathbf{v}_2 = 2a + b + c + 2d = 0.$$

So $x_1 = x_3 - 4x_4$, $x_2 = -3x_3 + 6x_4$, $x_3 = r$, $x_4 = s$, where r and s are any numbers. Thus

$$\mathbf{x} = \begin{bmatrix} r - 4s \\ -3r + 6s \\ r \\ s \end{bmatrix} = r \begin{bmatrix} 1 \\ -3 \\ 1 \\ 0 \end{bmatrix} + s \begin{bmatrix} -4 \\ 6 \\ 0 \\ 1 \end{bmatrix}.$$

3. Let $p(t) = a + bt$. Then

$$(p, f) = \int_0^1 p(t) f(t)\, dt = \int_0^1 (a + bt)(t + 1)\, dt$$

$$= \int_0^1 \left[bt^2 + (a + b)t + a \right]\, dt = \left(\tfrac{1}{3}bt^3 + \tfrac{1}{2}(a + b)t^2 + at \right)\big|_0^1 = 0$$

$$\implies \quad \tfrac{3}{2}a + \tfrac{5}{6}b = 0 \quad \implies \quad a = -\tfrac{5}{9}b.$$

So $p(t) = a + bt$, where $a = -\tfrac{5}{9}b$ and b is any number.

4. (a) The inner product of $\mathbf{u}$ and $\mathbf{v}$ is bounded by the product of the lengths of $\mathbf{u}$ and $\mathbf{v}$.
   (b) The cosine of the angle between $\mathbf{u}$ and $\mathbf{v}$ lies between -1 and 1.

5. (a) $\mathbf{v}_1 \cdot \mathbf{v}_2 = 0$, $\mathbf{v}_1 \cdot \mathbf{v}_3 = 0$, $\mathbf{v}_2 \cdot \mathbf{v}_3 = 0$.
   (b) Normalize the vectors in S: $\tfrac{1}{2}\mathbf{v}_1$, $\tfrac{1}{\sqrt{6}}\mathbf{v}_2$, $\tfrac{1}{\sqrt{12}}\mathbf{v}_3$.
   (c) Possible answer: $\mathbf{v}_4 = \begin{bmatrix} 0 \\ 0 \\ 1 \\ -1 \end{bmatrix}$.

6. (a) $\mathbf{u}_1 \cdot \mathbf{u}_1 = 1$, $\mathbf{u}_1 \cdot \mathbf{u}_2 = 0$, $\mathbf{u}_1 \cdot \mathbf{u}_3 = 0$, $\mathbf{u}_2 \cdot \mathbf{u}_2 = 1$, $\mathbf{u}_2 \cdot \mathbf{u}_3 = 0$, $\mathbf{u}_3 \cdot \mathbf{u}_3 = 1$. Thus S is an orthonormal set. Since the vectors are orthogonal, they are linearly independent. Thus S is an orthnormal basis for R^3.
   (b) Let $\mathbf{w} = a\mathbf{u}_1 + b\mathbf{u}_2 + c\mathbf{u}_3$. Then

$$a = (\mathbf{w}, \mathbf{u}_1) = \tfrac{5}{3}$$
$$b = (\mathbf{w}, \mathbf{u}_2) = \tfrac{1}{3}$$
$$c = (\mathbf{w}, \mathbf{u}_3) = \tfrac{1}{3}.$$

   So $\mathbf{w} = \tfrac{5}{3}\mathbf{u}_1 + \tfrac{1}{3}\mathbf{u}_2 + \tfrac{1}{3}\mathbf{u}_3$.

   (c) $\text{proj}_W \mathbf{w} = (\mathbf{w}, \mathbf{u}_1)\mathbf{u}_1 + (\mathbf{w}, \mathbf{u}_2)\mathbf{u}_2 = (1)\mathbf{u}_1 + (-1)\mathbf{u}_2 = \begin{bmatrix} -\tfrac{1}{3} \\ \tfrac{4}{3} \\ \tfrac{1}{3} \end{bmatrix}.$

   Distance from V to $\mathbf{w} = \|\mathbf{w} - \text{proj}_V \mathbf{w}\| = \left\| \begin{bmatrix} 1 \\ 2 \\ -1 \end{bmatrix} - \begin{bmatrix} -\tfrac{1}{3} \\ \tfrac{4}{3} \\ \tfrac{1}{3} \end{bmatrix} \right\| = \dfrac{2\sqrt{6}}{3}.$

7. Let $\mathbf{u}_1 = \begin{bmatrix} 1 \\ 0 \\ 1 \\ 0 \end{bmatrix}$, $\mathbf{u}_2 = \begin{bmatrix} 2 \\ 1 \\ -2 \\ 0 \end{bmatrix}$, and $\mathbf{u}_3 = \begin{bmatrix} 0 \\ 0 \\ 0 \\ 1 \end{bmatrix}$. Let $\mathbf{v}_1 = \mathbf{u}_1$. Then we found

$$\mathbf{v}_2 = \mathbf{u}_2 - \dfrac{(\mathbf{u}_2, \mathbf{v}_1)}{(\mathbf{v}_1, \mathbf{v}_1)}\mathbf{v}_1 = \begin{bmatrix} 2 \\ 1 \\ -2 \\ 0 \end{bmatrix} - \dfrac{0}{2}\mathbf{v}_1 = \begin{bmatrix} 2 \\ 1 \\ -2 \\ 0 \end{bmatrix}.$$

Finally,

$$\mathbf{v}_3 = \mathbf{u}_3 - \frac{(\mathbf{u}_3, \mathbf{v}_1)}{(\mathbf{v}_1, \mathbf{v}_1)}\mathbf{v}_1 - \frac{(\mathbf{u}_3, \mathbf{v}_2)}{(\mathbf{v}_2, \mathbf{v}_2)}\mathbf{v}_2 = \begin{bmatrix} 0 \\ 0 \\ 0 \\ 1 \end{bmatrix} - \frac{0}{2}\mathbf{v}_1 - \frac{0}{9}\mathbf{v}_2 = \begin{bmatrix} 0 \\ 0 \\ 0 \\ 1 \end{bmatrix}.$$

Multiplying each vector by the reciprocal of its length, we find that

$$\mathbf{w}_1 = \frac{\mathbf{v}_1}{\|\mathbf{v}_1\|} = \frac{1}{\sqrt{2}}\begin{bmatrix} 1 \\ 0 \\ 1 \\ 0 \end{bmatrix}, \quad \mathbf{w}_2 = \frac{\mathbf{v}_2}{\|\mathbf{v}_2\|} = \frac{1}{\sqrt{3}}\begin{bmatrix} 2 \\ 1 \\ -2 \\ 0 \end{bmatrix} \quad \mathbf{w}_3 = \frac{\mathbf{v}_3}{\|\mathbf{v}_3\|} = \begin{bmatrix} 0 \\ 0 \\ 0 \\ 1 \end{bmatrix}.$$

Thus, $\{\mathbf{w}_1, \mathbf{w}_2, \mathbf{w}_3\}$ is an orthonormal basis for W.

8. Let

$$A = \begin{bmatrix} 0 & 1 & 1 \\ 1 & 0 & 1 \\ 1 & 0 & 0 \\ 0 & 1 & 1 \end{bmatrix}^T = \begin{bmatrix} 0 & 1 & 1 & 0 \\ 1 & 0 & 0 & 1 \\ 1 & 1 & 0 & 1 \end{bmatrix}.$$

Then the reduced row echelon form of A^T is

$$\begin{bmatrix} 1 & 0 & 0 & 1 \\ 0 & 1 & 0 & 0 \\ 0 & 0 & 1 & 0 \end{bmatrix}.$$

The corresponding solution is $x_1 = -x_4 = -r$, $x_2 = 0$, $x_3 = 0$, $x_4 = r$, where r is any number. Hence

$$\mathbf{x} = \begin{bmatrix} -r \\ 0 \\ 0 \\ r \end{bmatrix} = r\begin{bmatrix} -1 \\ 0 \\ 0 \\ 1 \end{bmatrix}$$

and so a basis for $W^\perp$ is $\left\{ \begin{bmatrix} -1 \\ 0 \\ 0 \\ 1 \end{bmatrix} \right\}$.

9. Form the matrix A whose columns are the vectors in S. Find the row reduced echelon form of A. The columns of this matrix can be used to obtain a basis for W. The rows of this matrix give the solution to the homogeneous system $A\mathbf{x} = \mathbf{0}$ and from this we can find a basis for $W^\perp$.

10. We have

$$\begin{aligned} \mathrm{proj}_W(\mathbf{u} + \mathbf{v}) &= (\mathbf{u} + \mathbf{v}, \mathbf{w}_1) + (\mathbf{u} + \mathbf{v}, \mathbf{w}_2) + (\mathbf{u} + \mathbf{v}, \mathbf{w}_3) \\ &= (\mathbf{u}, \mathbf{w}_1) + (\mathbf{v}, \mathbf{w}_1) + (\mathbf{u}, \mathbf{w}_2) + (\mathbf{v}, \mathbf{w}_2) + (\mathbf{u}, \mathbf{w}_3) + (\mathbf{v}, \mathbf{w}_3) \\ &= (\mathbf{u}, \mathbf{w}_1) + (\mathbf{u}, \mathbf{w}_2) + (\mathbf{u}, \mathbf{w}_3) + (\mathbf{v}, \mathbf{w}_1) + (\mathbf{v}, \mathbf{w}_2) + (\mathbf{v}, \mathbf{w}_3) \\ &= \mathrm{proj}_W\mathbf{u} + \mathrm{proj}_W\mathbf{v}. \end{aligned}$$

Chapter 6

Linear Transformations and Matrices

Section 6.1, p. 372

1. (a) L is not a linear transformation; condition (a) of Definition 6.1 is not satisfied. To show this, let $\mathbf{u} = \begin{bmatrix} u_1 & u_2 \end{bmatrix}$ and $\mathbf{v} = \begin{bmatrix} b_1 & b_2 \end{bmatrix}$. Then

$$L(\mathbf{u} + \mathbf{v}) = L\left(\begin{bmatrix} u_1 + b_1 & u_2 + b_2 \end{bmatrix}\right) = \begin{bmatrix} u_1 + b_1 + 1 & u_2 + b_2 & u_1 + b_1 + u_2 + b_2 \end{bmatrix}$$

but

$$L(\mathbf{u}) + L(\mathbf{v}) = \begin{bmatrix} u_1 + 1 & u_2 & u_1 + u_2 \end{bmatrix} + \begin{bmatrix} b_1 + 1 & b_2 & b_1 + b_2 \end{bmatrix}$$
$$= \begin{bmatrix} u_1 + b_1 + 2 & u_2 + b_2 & u_1 + u_2 + b_1 + b_2 \end{bmatrix} \neq L(\mathbf{u} + \mathbf{v}).$$

(b) L is a linear transformation. To show this, let $\mathbf{u} = \begin{bmatrix} u_1 & u_2 \end{bmatrix}$, $\mathbf{v} = \begin{bmatrix} b_1 & b_2 \end{bmatrix}$, and let c be any scalar. Then

$$L(\mathbf{u} + \mathbf{v}) = L\left(\begin{bmatrix} u_1 + b_1 & u_2 + b_2 \end{bmatrix}\right)$$
$$= \begin{bmatrix} (u_1 + b_1) + (u_2 + b_2) & u_2 + b_2 & (u_1 + b_1) - (u_2 + b_2) \end{bmatrix}$$
$$= \begin{bmatrix} (u_1 + u_2) + (b_1 + b_2) & u_2 + b_2 & (u_1 - u_2) + (b_1 - b_2) \end{bmatrix}$$
$$= \begin{bmatrix} u_1 + u_2 & u_2 & u_1 - u_2 \end{bmatrix} + \begin{bmatrix} b_1 + b_2 & b_2 & b_1 - b_2 \end{bmatrix} = L(\mathbf{u}) + L(\mathbf{v})$$

and

$$L(c\mathbf{u}) = L\left(\begin{bmatrix} cu_1 & cu_2 \end{bmatrix}\right) = \begin{bmatrix} cu_1 + cu_2 & cu_2 & cu_1 - cu_2 \end{bmatrix} = c\begin{bmatrix} u_1 + u_2 & u_2 & u_1 - u_2 \end{bmatrix} = cL(\mathbf{u}).$$

3. (a) L is a linear transformation. To show this, let $p(t)$, $q(t)$ be polynomials in P_2 and let c be any scalar. Then

$$L(p(t) + q(t)) = t^3[p'(0) + q'(0)] + t^2[p(0) + q(0)]$$
$$= [t^3 p'(0) + t^2 p(0)] + [t^3 q'(0) + t^2 q(0)] = L(p(t)) + L(q(t))$$

and

$$L(cp(t)) = t^3[cp'(0)] + t^2[cp(0)] = c[t^3 p'(0) + t^2 p(0)] = cL(p(t)).$$

(b) L is a linear transformation. To show this, let $p(t)$, $q(t)$ be polynomials in P_1 and let c be any scalar. Then

$$L(p(t) + q(t)) = t[p(t) + q(t)] + [p(0) + q(0)]$$
$$= [tp(t) + p(0)] + [tq(t) + q(0)] = L(p(t)) + L(q(t))$$

and

$$L(cp(t)) = t[cp(t)] + cp(0) = c[tp(t) + p(0)] = cL(p(t)).$$

(c) L is not a linear transformation; condition (a) of Definition 6.1 is not satisfied. To show this, let $p(t)$, $q(t)$ be polynomials in P_1. Then

$$L(p(t) + q(t)) = t[p(t) + q(t)] + 1 = tp(t) + tq(t) + 1$$

but

$$L(p(t)) + L(q(t)) = [tp(t) + 1] + [tq(t) + 1] = tp(t) + tq(t) + 2 \neq L(p(t) + q(t)).$$

5. (a) L is not a linear transformation; condition (a) of Definition 6.1 is not satisfied. For example,

$$L\left(\begin{bmatrix} 1 & 0 \\ 0 & 2 \end{bmatrix} + \begin{bmatrix} 1 & 0 \\ 0 & 3 \end{bmatrix}\right) = L\left(\begin{bmatrix} 2 & 0 \\ 0 & 5 \end{bmatrix}\right) = \det\left(\begin{bmatrix} 2 & 0 \\ 0 & 5 \end{bmatrix}\right) = 10$$

but

$$L\left(\begin{bmatrix} 1 & 0 \\ 0 & 2 \end{bmatrix}\right) + L\left(\begin{bmatrix} 1 & 0 \\ 0 & 3 \end{bmatrix}\right) = \det\left(\begin{bmatrix} 1 & 0 \\ 0 & 2 \end{bmatrix}\right) + \det\left(\begin{bmatrix} 1 & 0 \\ 0 & 3 \end{bmatrix}\right) = 2 + 3 = 5 \neq 10.$$

(b) L is a linear transformation. To show this, we use properties of the trace from Exercise 43, Section 1.3. Let A and B be matrices in M_{nn} and let c be any scalar. Then

$$L(A + B) = \text{Tr}(A + B) = \text{Tr}(A) + \text{Tr}(B) = L(A) + L(B)$$
$$L(cA) = \text{Tr}(cA) = c\,\text{Tr}(A) = cL(A).$$

Therefore L is a linear transformation.

7. (a) We have

$$L(\mathbf{e}_1) = L\left(\begin{bmatrix} 1 \\ 0 \end{bmatrix}\right) = \begin{bmatrix} -1 \\ 0 \end{bmatrix}$$
$$L(\mathbf{e}_2) = L\left(\begin{bmatrix} 0 \\ 1 \end{bmatrix}\right) = \begin{bmatrix} 0 \\ 1 \end{bmatrix}.$$

Hence the standard matrix representing L is $\begin{bmatrix} -1 & 0 \\ 0 & 1 \end{bmatrix}$.

(b) We have

$$L(\mathbf{e}_1) = L\left(\begin{bmatrix} 1 \\ 0 \end{bmatrix}\right) = \begin{bmatrix} 0 \\ 1 \end{bmatrix}$$
$$L(\mathbf{e}_2) = L\left(\begin{bmatrix} 0 \\ 1 \end{bmatrix}\right) = \begin{bmatrix} -1 \\ 0 \end{bmatrix}.$$

Hence the standard matrix representing L is $\begin{bmatrix} 0 & -1 \\ 1 & 0 \end{bmatrix}$.

(c) We have

$$L(\mathbf{e}_1) = L\left(\begin{bmatrix} 1 \\ 0 \\ 0 \end{bmatrix}\right) = \begin{bmatrix} 1 \\ 0 \\ 0 \end{bmatrix}$$
$$L(\mathbf{e}_2) = L\left(\begin{bmatrix} 0 \\ 1 \\ 0 \end{bmatrix}\right) = \begin{bmatrix} 0 \\ 0 \\ 0 \end{bmatrix}$$
$$L(\mathbf{e}_3) = L\left(\begin{bmatrix} 0 \\ 0 \\ 1 \end{bmatrix}\right) = \begin{bmatrix} 0 \\ 0 \\ 0 \end{bmatrix}.$$

Hence the standard matrix representing L is $\begin{bmatrix} 1 & 0 & 0 \\ 0 & 0 & 0 \\ 0 & 0 & 0 \end{bmatrix}$.

9. (a) $L\left(\begin{bmatrix} 1 & 2 & 0 & -1 \\ 3 & 0 & 2 & 3 \\ 4 & 1 & -2 & 1 \end{bmatrix}\right) = \begin{bmatrix} 2 & 3 & 1 \\ 1 & 2 & -3 \end{bmatrix} \begin{bmatrix} 1 & 2 & 0 & -1 \\ 3 & 0 & 2 & 3 \\ 4 & 1 & -2 & 1 \end{bmatrix} = \begin{bmatrix} 15 & 5 & 4 & 8 \\ -5 & -1 & 10 & 2 \end{bmatrix}$

   (b) Let $M = \begin{bmatrix} 2 & 3 & 1 \\ 1 & 2 & -3 \end{bmatrix}$. Then $L(A) = MA$. Therefore

   $$L(A+B) = M(A+B) = MA + MB = L(A) + L(B) \quad \text{and} \quad L(cA) = M(cA) = c(MA) = cL(A).$$

   Therefore L is a transformation.

11. (a) We have
   $$L\left(\begin{bmatrix} 1 \\ 0 \end{bmatrix}\right) = \begin{bmatrix} 0 \\ 1 \end{bmatrix}, \quad L\left(\begin{bmatrix} 0 \\ 1 \end{bmatrix}\right) = \begin{bmatrix} 1 \\ 0 \end{bmatrix}.$$

   Thus the standard matrix representing L is $\begin{bmatrix} 0 & 1 \\ 1 & 0 \end{bmatrix}$.

   (b) We have
   $$L\left(\begin{bmatrix} 1 \\ 0 \end{bmatrix}\right) = \begin{bmatrix} 1 \\ 2 \\ 0 \end{bmatrix}, \quad L\left(\begin{bmatrix} 0 \\ 1 \end{bmatrix}\right) = \begin{bmatrix} -3 \\ -1 \\ 2 \end{bmatrix}.$$

   Thus the standard matrix representing L is $\begin{bmatrix} 1 & -3 \\ 2 & -1 \\ 0 & 2 \end{bmatrix}$.

   (c) We have
   $$L\left(\begin{bmatrix} 1 \\ 0 \\ 0 \end{bmatrix}\right) = \begin{bmatrix} 1 \\ 0 \\ 0 \end{bmatrix}, \quad L\left(\begin{bmatrix} 0 \\ 1 \\ 0 \end{bmatrix}\right) = \begin{bmatrix} 4 \\ 0 \\ 1 \end{bmatrix}, \quad L\left(\begin{bmatrix} 0 \\ 0 \\ 1 \end{bmatrix}\right) = \begin{bmatrix} 0 \\ -1 \\ 1 \end{bmatrix}.$$

   Thus the standard matrix representing L is $\begin{bmatrix} 1 & 4 & 0 \\ 0 & 0 & -1 \\ 0 & 1 & 1 \end{bmatrix}$.

13. (a) Since $\begin{bmatrix} 1 \\ -2 \\ 3 \end{bmatrix} = (1)\begin{bmatrix} 1 \\ 0 \\ 0 \end{bmatrix} + (-2)\begin{bmatrix} 0 \\ 1 \\ 0 \end{bmatrix} + (3)\begin{bmatrix} 0 \\ 0 \\ 1 \end{bmatrix}$ we obtain

   $$L\left(\begin{bmatrix} 1 \\ -2 \\ 3 \end{bmatrix}\right) = (1)L\left(\begin{bmatrix} 1 \\ 0 \\ 0 \end{bmatrix}\right) + (-2)L\left(\begin{bmatrix} 0 \\ 1 \\ 0 \end{bmatrix}\right) + (3)L\left(\begin{bmatrix} 0 \\ 0 \\ 1 \end{bmatrix}\right)$$

   $$= (1)\begin{bmatrix} 2 \\ -4 \end{bmatrix} + (-2)\begin{bmatrix} 3 \\ -5 \end{bmatrix} + (3)\begin{bmatrix} 2 \\ 3 \end{bmatrix} = \begin{bmatrix} 2 \\ 15 \end{bmatrix}$$

   (b) Using the same idea as in part (a), we obtain

   $$L\left(\begin{bmatrix} u_1 \\ u_2 \\ u_3 \end{bmatrix}\right) = u_1 L\left(\begin{bmatrix} 1 \\ 0 \\ 0 \end{bmatrix}\right) + u_2 L\left(\begin{bmatrix} 0 \\ 1 \\ 0 \end{bmatrix}\right) + u_3 L\left(\begin{bmatrix} 0 \\ 0 \\ 1 \end{bmatrix}\right)$$

   $$= u_1\begin{bmatrix} 2 \\ -4 \end{bmatrix} + u_2\begin{bmatrix} 3 \\ -5 \end{bmatrix} + u_3\begin{bmatrix} 2 \\ 3 \end{bmatrix} = \begin{bmatrix} 2u_1 + 3u_2 + 2u_3 \\ -4u_1 - 5u_2 + 3u_3 \end{bmatrix}$$

15. (a) $L(2t^2 - 5t + 3) = 2L(t^2) - 5L(t) + 3L(1) = 2(t^3 + t) - 5(t^2) + 3(1) = 2t^3 - 5t^2 + 2t + 3$

   (b) $L(at^2 + bt + c) = aL(t^2) + bL(t) + cL(1) = a(t^3 + t) + b(t^2) + c(1) = at^3 + bt^2 + at + c$

17. a can be any real number; $b = 0$. To show this, let $\mathbf{u}$, $\mathbf{v}$ be vectors in R. Then $L(\mathbf{u}+\mathbf{v}) = a(\mathbf{u}+\mathbf{v})+b = a\mathbf{u}+a\mathbf{v}+b$. For L to be a linear transformation, this expression must equal $L(\mathbf{u})+L(\mathbf{v}) = a\mathbf{u}+b+a\mathbf{v}+b$. This is possible only when $b = 0$.

19. (a) Reflection about the y axis. (b) Reflection about the origin. (c) Counterclockwise rotation through $\pi/2$ radians.

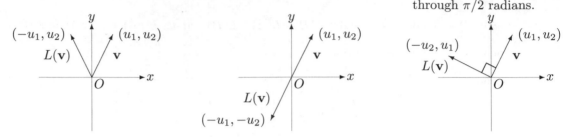
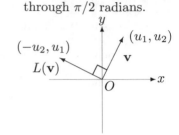

21. Let $\mathbf{u}$, $\mathbf{v}$, be vectors and c a scalar. Then we have $L(\mathbf{u} + \mathbf{v}) = \mathbf{0}_W = \mathbf{0}_W + \mathbf{0}_W = L(\mathbf{u}) + L(\mathbf{v})$ and $L(c\mathbf{u}) = \mathbf{0}_W = c\mathbf{0}_W = cL(\mathbf{u})$. Therefore O is a linear transformation.

23. Yes:

$$L\left(\begin{bmatrix} a_1 & b_1 \\ c_1 & d_1 \end{bmatrix} + \begin{bmatrix} a_2 & b_2 \\ c_2 & d_2 \end{bmatrix}\right) = L\left(\begin{bmatrix} a_1 + a_2 & b_1 + b_2 \\ c_1 + c_2 & d_1 + d_2 \end{bmatrix}\right)$$
$$= (a_1 + a_1) + (d_1 + d_2)$$
$$= (a_1 + d_1) + (a_2 + d_2)$$
$$= L\left(\begin{bmatrix} a_1 & b_1 \\ c_1 & d_1 \end{bmatrix}\right) + L\left(\begin{bmatrix} a_2 & b_2 \\ c_2 & d_2 \end{bmatrix}\right).$$

Also, if k is any real number

$$L\left(k\begin{bmatrix} a & b \\ c & d \end{bmatrix}\right) = L\left(\begin{bmatrix} ka & kb \\ kc & kd \end{bmatrix}\right) = ka + kd = k(a + d) = kL\left(\begin{bmatrix} a & b \\ c & d \end{bmatrix}\right).$$

25. L is a linear transformation. To prove this, let f, g be functions in V and let c be a real number. Then $L(f + g) = \int_a^b (f(x) + g(x))\, dx = \int_a^b f(x)\, dx + \int_a^b g(x)\, dx = L(f) + L(g)$ and $L(cf) = \int_a^b cf(x)\, dx = c\int_a^b f(x)\, dx = cL(f)$.

27. No. For example, consider $L: M_{22} \to R^1$, where

$$L\left(\begin{bmatrix} a_{11} & a_{12} \\ a_{21} & a_{22} \end{bmatrix}\right) = a_{11}a_{22}.$$

Then

$$L\left(\begin{bmatrix} 1 & 0 \\ 0 & 1 \end{bmatrix}\right) = 1 \cdot 1 = 1,$$

while

$$L\left(\begin{bmatrix} 1 & 0 \\ 0 & 1 \end{bmatrix} + \begin{bmatrix} 1 & 0 \\ 0 & 1 \end{bmatrix}\right) = L\left(\begin{bmatrix} 2 & 0 \\ 0 & 2 \end{bmatrix}\right) = 2 \cdot 2 = 4$$

but

$$L\left(\begin{bmatrix} 1 & 0 \\ 0 & 1 \end{bmatrix}\right) + L\left(\begin{bmatrix} 1 & 0 \\ 0 & 1 \end{bmatrix}\right) = 1 + 1 = 2.$$

29. We have by the properties of coordinate vectors discussed in Section 4.8, $L(\mathbf{u} + \mathbf{v}) = \begin{bmatrix} \mathbf{u} + \mathbf{v} \end{bmatrix}_S = \begin{bmatrix} \mathbf{u} \end{bmatrix}_S + \begin{bmatrix} \mathbf{v} \end{bmatrix}_S = L(\mathbf{u}) + L(\mathbf{v})$ and $L(c\mathbf{u}) = \begin{bmatrix} c\mathbf{u} \end{bmatrix}_S = c\begin{bmatrix} \mathbf{u} \end{bmatrix}_S = cL(\mathbf{u})$.

31. Let $L(\mathbf{v}_i) = \mathbf{w}_i$ for $i = 1, 2, \ldots, n$. Then for any $\mathbf{v}$ in V, express $\mathbf{v}$ in terms of the basis vectors of S, $\mathbf{v} = a_1\mathbf{v}_1 + a_2\mathbf{v}_2 + \cdots + a_n\mathbf{v}_n$, and define $L(\mathbf{v}) = \sum_{i=1}^{n} a_i\mathbf{w}_i$. If $\mathbf{v} = \sum_{i=1}^{n} a_i\mathbf{v}_i$ and $\mathbf{w} = \sum_{i=1}^{n} b_i\mathbf{v}_i$ are any vectors in V and c is any scalar, then

$$L(\mathbf{v} + \mathbf{w}) = L\left(\sum_{i=1}^{n}(a_i + b_i)\mathbf{v}_i\right) = \sum_{i=1}^{n}(a_i + b_i)\mathbf{w}_i = \sum_{i=1}^{n} a_i\mathbf{w}_i + \sum_{i=1}^{n} b_i\mathbf{w}_i = L(\mathbf{v}) + L(\mathbf{w})$$

and in a similar fashion

$$L(c\mathbf{v}) = \sum_{i=1}^{n} ca_i\mathbf{w}_i = c\sum_{i=1}^{n} a_i\mathbf{w}_i = cL(\mathbf{v})$$

for any scalar c, so L is a linear transformation.

33. Let $\mathbf{v}$ be any vector in V. Then $\mathbf{v} = c_1\mathbf{v}_1 + c_2\mathbf{v}_2 + \cdots + c_n\mathbf{v}_n$ for some real numbers $c_1, c_2, \ldots, c_n$. We now have

$$\begin{aligned} L_1(\mathbf{v}) &= L_1(c_1\mathbf{v}_1 + c_2\mathbf{v}_2 + \cdots + c_n\mathbf{v}_n) \\ &= c_1L_1(\mathbf{v}_1) + c_2L_1(\mathbf{v}_2) + \cdots + c_nL_1(\mathbf{v}_n) \\ &= c_1L_2(\mathbf{v}_1) + c_2L_2(\mathbf{v}_2) + \cdots + c_nL_2(\mathbf{v}_n) \\ &= L_2(c_1\mathbf{v}_1 + c_2\mathbf{v}_2 + \cdots + c_n\mathbf{v}_n) = L_2(\mathbf{v}). \end{aligned}$$

35. Let $\{\mathbf{e}_1, \ldots, \mathbf{e}_n\}$ be the natural basis for R^n. Then $O(\mathbf{e}_i) = \mathbf{0}$ for $i = 1, \ldots, n$. Hence the standard matrix representing O is the $n \times n$ zero matrix O.

37. Suppose there is another matrix B such that $L(\mathbf{x}) = B\mathbf{x}$ for all $\mathbf{x}$ in R^n. Then $L(\mathbf{e}_j) = B\mathbf{e}_j = \mathrm{Col}_j(B)$ for $j = 1, \ldots, n$. But by definition, $L(\mathbf{e}_j)$ is the jth column of A. Hence $\mathrm{Col}_j(B) = \mathrm{Col}_j(A)$ for $j = 1, \ldots, n$ and therefore $B = A$. Thus the matrix A is unique.

39. (a) Following the technique of Example 12, we determine the numbers associated with the letters in the message.

W	O	R	K		H	A	R	D
↕	↕	↕	↕		↕	↕	↕	↕
23	15	18	11		8	1	18	4

We break this string of numbers into four vectors in R^2 and multiply each vector by the matrix

$$A = \begin{bmatrix} 5 & 3 \\ 2 & 1 \end{bmatrix}.$$

We obtain

$$A\begin{bmatrix} 23 \\ 15 \end{bmatrix} = \begin{bmatrix} 13 \\ 11 \end{bmatrix}, \quad A\begin{bmatrix} 18 \\ 11 \end{bmatrix} = \begin{bmatrix} 123 \\ 47 \end{bmatrix}, \quad A\begin{bmatrix} 8 \\ 1 \end{bmatrix} = \begin{bmatrix} 43 \\ 17 \end{bmatrix}, \quad A\begin{bmatrix} 18 \\ 4 \end{bmatrix} = \begin{bmatrix} 102 \\ 40 \end{bmatrix}.$$

The final version of the code message is the string of numbers

$$13 \quad 11 \quad 123 \quad 47 \quad 43 \quad 17 \quad 102 \quad 40.$$

(b) We are given the coded message

$$93 \quad 36 \quad 60 \quad 21 \quad 159 \quad 60 \quad 110 \quad 43.$$

Breaking this into vectors in R^2, the decoded vector is obtained by multiplying each of the vectors by

$$A^{-1} = \begin{bmatrix} -1 & 3 \\ 2 & -5 \end{bmatrix}.$$

We obtain

$$A^{-1}\begin{bmatrix} 93 \\ 36 \end{bmatrix} = \begin{bmatrix} 15 \\ 6 \end{bmatrix}, \quad A^{-1}\begin{bmatrix} 60 \\ 21 \end{bmatrix} = \begin{bmatrix} 3 \\ 15 \end{bmatrix}, \quad A^{-1}\begin{bmatrix} 159 \\ 60 \end{bmatrix} = \begin{bmatrix} 21 \\ 18 \end{bmatrix}, \quad A^{-1}\begin{bmatrix} 110 \\ 43 \end{bmatrix} = \begin{bmatrix} 19 \\ 5 \end{bmatrix}.$$

Then the message is

$$
\begin{array}{cccccccc}
15 & 6 & 3 & 15 & 21 & 18 & 19 & 5 \\
\updownarrow & \updownarrow & \updownarrow & \updownarrow & \updownarrow & \updownarrow & \updownarrow & \updownarrow \\
O & F & C & O & U & R & S & E.
\end{array}
$$

Section 6.2, p. 387

1. (a) Yes; $L\left(\begin{bmatrix} 0 \\ 2 \end{bmatrix}\right) = \begin{bmatrix} 0 \\ 0 \end{bmatrix}$.

   (b) No; $L\left(\begin{bmatrix} 2 \\ 2 \end{bmatrix}\right) = \begin{bmatrix} 2 \\ 0 \end{bmatrix}$.

   (c) Yes; $\begin{bmatrix} 3 \\ 0 \end{bmatrix} = L\left(\begin{bmatrix} 3 \\ 0 \end{bmatrix}\right)$.

   (d) No; $\begin{bmatrix} 3 \\ 2 \end{bmatrix}$ does not have the form $\begin{bmatrix} u_1 \\ 0 \end{bmatrix}$

   (e) The kernel of L consists of all vectors $\begin{bmatrix} u_1 \\ u_2 \end{bmatrix}$ such that $L\left(\begin{bmatrix} u_1 \\ u_2 \end{bmatrix}\right) = \begin{bmatrix} u_1 \\ 0 \end{bmatrix} = \begin{bmatrix} 0 \\ 0 \end{bmatrix}$. This means that $u_1 = 0$, with u_2 equal to any real number. Therefore ker L consists of all vectors $\begin{bmatrix} 0 \\ a \end{bmatrix}$, where a is any real number. Geometrically, it is the y-axis.

   (f) The range of L consists of all vectors of the form $L\left(\begin{bmatrix} u_1 \\ u_2 \end{bmatrix}\right) = \begin{bmatrix} u_1 \\ 0 \end{bmatrix}$, where u_1 is any real number. Geometrically, it is the x-axis.

3. (a) No; $L\left(\begin{bmatrix} 2 & 3 & -2 & 3 \end{bmatrix}\right) = \begin{bmatrix} 0 & 6 \end{bmatrix}$

   (b) Yes; $L\left(\begin{bmatrix} 4 & -2 & -4 & 2 \end{bmatrix}\right) = \begin{bmatrix} 0 & 0 \end{bmatrix}$

   (c) Yes; $\begin{bmatrix} 1 & 2 \end{bmatrix} = L\left(\begin{bmatrix} 1 & 2 & 0 & 0 \end{bmatrix}\right)$

   (d) Yes; $\begin{bmatrix} 0 & 0 \end{bmatrix} = L\left(\begin{bmatrix} 0 & 0 & 0 & 0 \end{bmatrix}\right)$

   (e) The kernel of L consists of all vectors $\begin{bmatrix} u_1 & u_2 & u_3 & u_4 \end{bmatrix}$ such that

   $$L\left(\begin{bmatrix} u_1 & u_2 & u_3 & u_4 \end{bmatrix}\right) = \begin{bmatrix} u_1 + u_3 & u_2 + u_4 \end{bmatrix} = \begin{bmatrix} 0 & 0 \end{bmatrix}.$$

   This leads to the system of equations $u_1 + u_3 = u_2 + u_4 = 0$. The solution is $u_1 = -r$, $u_2 = -s$, $u_3 = r$, $u_4 = s$, where r, s are any real numbers. Therefore the kernel consists of all vectors of the form $\begin{bmatrix} -r & -s & r & s \end{bmatrix}$, where r and s are real numbers.

   (f) The range consists of all vectors of the form $\begin{bmatrix} u_1 + u_3 & u_2 + u_4 \end{bmatrix}$. Since $u_1 + u_3$ and $u_2 + u_4$ can assume any real value, the kernel is all vectors $\begin{bmatrix} a & b \end{bmatrix} = a\begin{bmatrix} 1 & 0 \end{bmatrix} + b\begin{bmatrix} 0 & 1 \end{bmatrix}$. A set of vectors spanning the range is $\left\{\begin{bmatrix} 1 & 0 \end{bmatrix}, \begin{bmatrix} 0 & 1 \end{bmatrix}\right\}$.

5. (a) $\begin{bmatrix} u_1 & u_2 & u_3 & u_4 \end{bmatrix}$ is in the kernel if and only if $u_1 + u_2 = u_3 + u_4 = u_1 + u_3 = 0$. The solution to this system of equations is $u_2 = u_3 = -u_1$, $u_4 = u_1$. Therefore the kernel consists of all vectors of the form $\begin{bmatrix} a & -a & -a & a \end{bmatrix}$, where a is any real number. A basis is $\left\{\begin{bmatrix} 1 & -1 & -1 & 1 \end{bmatrix}\right\}$.

   (b) $\dim \ker L = 1$.

   (c) The range consists of all vectors of the form $\begin{bmatrix} u_1 + u_2 & u_3 + u_4 & u_1 + u_3 \end{bmatrix} = u_1\begin{bmatrix} 1 & 0 & 1 \end{bmatrix} + u_2\begin{bmatrix} 1 & 0 & 0 \end{bmatrix} + u_3\begin{bmatrix} 0 & 1 & 1 \end{bmatrix} + u_4\begin{bmatrix} 0 & 1 & 0 \end{bmatrix}$. These four vectors are linearly dependent:

   $$\begin{bmatrix} 0 & 1 & 0 \end{bmatrix} = \begin{bmatrix} 0 & 1 & 1 \end{bmatrix} + \begin{bmatrix} 1 & 0 & 0 \end{bmatrix} - \begin{bmatrix} 1 & 0 & 1 \end{bmatrix},$$

   and these three vectors are independent. A basis is $\left\{\begin{bmatrix} 1 & 0 & 1 \end{bmatrix}, \begin{bmatrix} 0 & 1 & 1 \end{bmatrix}, \begin{bmatrix} 1 & 0 & 0 \end{bmatrix}\right\}$.

(d) $\dim \operatorname{range} L = 3$

7. (a) The kernel consists of all matrices $A = \begin{bmatrix} a_{11} & a_{12} & a_{13} \\ a_{21} & a_{22} & a_{23} \end{bmatrix}$ such that

$$L(A) = \begin{bmatrix} 2 & -1 \\ 1 & 2 \\ 3 & 1 \end{bmatrix} \begin{bmatrix} a_{11} & a_{12} & a_{13} \\ a_{21} & a_{22} & a_{23} \end{bmatrix} = \begin{bmatrix} 2a_{11} - a_{21} & 2a_{12} - a_{22} & 2a_{13} - a_{23} \\ a_{11} + 2a_{21} & a_{12} + 2a_{22} & a_{13} + 2a_{23} \\ 3a_{11} + a_{21} & 3a_{12} + a_{22} & 3a_{13} + a_{23} \end{bmatrix} = \begin{bmatrix} 0 & 0 & 0 \\ 0 & 0 & 0 \\ 0 & 0 & 0 \end{bmatrix}$$

To find these numbers, we row reduce the matrix of L to get $\begin{bmatrix} 1 & 0 \\ 0 & 1 \\ 0 & 0 \end{bmatrix}$. Therefore the kernel consists of only the zero vector. Hence $\dim \ker L = 0$.

(b) $\dim \operatorname{range} L = \dim M_{23} - \dim \ker L = 6 - 0 = 6$.

9. (a) The kernel consists of all polynomials $at^2 + bt + c$ such that $L(at^2 + bt + c) = \begin{bmatrix} a & b \end{bmatrix} = \begin{bmatrix} 0 & 0 \end{bmatrix}$. Therefore $a = b = 0$. Hence $\ker L$ consists of the polynomials c, c any number. A basis is $\{1\}$.

(b) The range consists of matrices $\begin{bmatrix} a & b \end{bmatrix} = a \begin{bmatrix} 1 & 0 \end{bmatrix} + b \begin{bmatrix} 0 & 1 \end{bmatrix}$. A basis is $\{ \begin{bmatrix} 1 & 0 \end{bmatrix}, \begin{bmatrix} 0 & 1 \end{bmatrix} \}$.

11. (a) The kernel consists of all matrices $\begin{bmatrix} a & b \\ c & d \end{bmatrix}$ such that

$$L\left(\begin{bmatrix} a & b \\ c & d \end{bmatrix} \right) = \begin{bmatrix} a+b & b+c \\ a+d & b+d \end{bmatrix} = \begin{bmatrix} 0 & 0 \\ 0 & 0 \end{bmatrix}.$$

This means that $a + b = b + c = a + d = b + d = 0$. The solution to these equations is $a = b = c = d = 0$. Therefore the kernel is $\left\{ \begin{bmatrix} 0 & 0 \\ 0 & 0 \end{bmatrix} \right\}$, which has no basis.

(b) The range consists of all matrices

$$\begin{bmatrix} a+b & b+c \\ a+d & b+d \end{bmatrix} = a \begin{bmatrix} 1 & 0 \\ 1 & 0 \end{bmatrix} + b \begin{bmatrix} 1 & 1 \\ 0 & 1 \end{bmatrix} + c \begin{bmatrix} 0 & 1 \\ 0 & 0 \end{bmatrix} + d \begin{bmatrix} 0 & 0 \\ 1 & 1 \end{bmatrix}.$$

A basis is $\left\{ \begin{bmatrix} 1 & 0 \\ 1 & 0 \end{bmatrix}, \begin{bmatrix} 1 & 1 \\ 0 & 1 \end{bmatrix}, \begin{bmatrix} 0 & 1 \\ 0 & 0 \end{bmatrix}, \begin{bmatrix} 0 & 0 \\ 1 & 1 \end{bmatrix} \right\}$.

13. (a) The kernel of L consists of the constant polynomials. Therefore $\dim \ker L = 1$. The range consists of the polynomials $tp'(t)$, where $p(t)$ has degree ≤ 2. So $\dim \operatorname{range} L = 2$. Hence $\dim \ker L + \dim \operatorname{range} L = 1 + 2 = 3 = \dim P_2$.

(b) The kernel of L consists of $\begin{bmatrix} u_1 & u_2 & u_3 \end{bmatrix}$ for which $u_1 + u_2 = u_1 + u_3 = 0$. The solution of these equations is $u_2 = -u_1$, $u_3 = -u_1$. So the kernel is the set of vectors $\begin{bmatrix} a & -a & -a \end{bmatrix}$, a any number, and hence $\dim \ker L = 1$. The range of L is all of R_2, so that $\dim \operatorname{range} L = 2$. Therefore $\dim \ker L + \dim \operatorname{range} L = 1 + 2 = 3 = \dim R_3$.

(c) The kernel of L is just the zero matrix (we omit the details), so that $\dim \ker L = 0$. The range of L is the set of matrices

$$\begin{bmatrix} a+2b & 2a+b & 3a+3b \\ c+2d & 2c+d & 3c+3d \end{bmatrix} = a \begin{bmatrix} 1 & 2 & 3 \\ 0 & 0 & 0 \end{bmatrix} + b \begin{bmatrix} 2 & 1 & 3 \\ 0 & 0 & 0 \end{bmatrix} + c \begin{bmatrix} 0 & 0 & 0 \\ 1 & 2 & 3 \end{bmatrix} + d \begin{bmatrix} 0 & 0 & 0 \\ 2 & 1 & 3 \end{bmatrix}$$

and hence $\dim \operatorname{range} L = 4$. Therefore $\dim \ker L + \dim \operatorname{range} L = 0 + 4 = 4 = \dim M_{22}$.

15. If $\mathbf{y}$ is in range L, then $\mathbf{y} = L(\mathbf{x}) = A\mathbf{x}$ for some $\mathbf{x}$ in R^n. This means that $\mathbf{y}$ is a linear combination of the columns of A, so $\mathbf{y}$ is in the column space of A. Conversely, if $\mathbf{y}$ is in the column space of A, then $\mathbf{y} = A\mathbf{x}$, so $\mathbf{y} = L(\mathbf{x})$ and $\mathbf{y}$ is in range L. Therefore range $L = $ column space of A.

17. No; the vectors are linearly dependent: $\begin{bmatrix} 2 & 1 & 1 \end{bmatrix} = \frac{1}{3}\begin{bmatrix} 3 & 0 & 0 \end{bmatrix} + \begin{bmatrix} 1 & 1 & 1 \end{bmatrix}$.

19. (a) The range of L is spanned by the vectors

$$\begin{bmatrix} 1 \\ 2 \\ 3 \end{bmatrix}, \begin{bmatrix} 0 \\ 1 \\ 1 \end{bmatrix}, \begin{bmatrix} 1 \\ 1 \\ 0 \end{bmatrix}.$$

Since this set of vectors is linearly independent, it is a basis for the range. Therefore L is one-to-one and onto and hence invertible.

(b) We obtain

$$L^{-1}\left(\begin{bmatrix} 2 \\ 3 \\ 4 \end{bmatrix}\right) = \begin{bmatrix} 1 & 0 & 1 \\ 2 & 1 & 1 \\ 3 & 1 & 0 \end{bmatrix}^{-1} \begin{bmatrix} 2 \\ 3 \\ 4 \end{bmatrix} = \frac{1}{2}\begin{bmatrix} 1 & -1 & 1 \\ -3 & 3 & -1 \\ 1 & 1 & -1 \end{bmatrix} \begin{bmatrix} 2 \\ 3 \\ 4 \end{bmatrix} = \begin{bmatrix} \frac{3}{2} \\ -\frac{1}{2} \\ \frac{1}{2} \end{bmatrix}.$$

21. We row reduce the matrix to obtain

$$\begin{bmatrix} 1 & 0 & 0 & -1 \\ 0 & 1 & 0 & \frac{8}{3} \\ 0 & 0 & 1 & -\frac{4}{3} \\ 0 & 0 & 0 & 0 \end{bmatrix}.$$

Therefore $x_1 = x_4$, $x_2 = -\frac{8}{3}x_4$, $x_3 = \frac{4}{3}x_4$, which means the solution space consists of the matrices

$$\begin{bmatrix} x_1 \\ x_2 \\ x_3 \\ x_4 \end{bmatrix} = \begin{bmatrix} x_4 \\ -\frac{8}{3}x_4 \\ \frac{4}{3}x_4 \\ x_4 \end{bmatrix} = x_4 \begin{bmatrix} 1 \\ -\frac{8}{3} \\ \frac{4}{3} \\ 1 \end{bmatrix}$$

and hence has dimension 1.

23. (a) The range of L is spanned by the vectors

$$\begin{bmatrix} 1 \\ 0 \\ 1 \end{bmatrix}, \begin{bmatrix} 1 \\ 1 \\ 2 \end{bmatrix}, \begin{bmatrix} 1 \\ 2 \\ 2 \end{bmatrix}.$$

Since this set of vectors is linearly independent, it is a basis for the range. Therefore L is one-to-one and onto and hence invertible.

(b) We obtain

$$L^{-1}\left(\begin{bmatrix} u_1 \\ u_2 \\ u_3 \end{bmatrix}\right) = \begin{bmatrix} 1 & 1 & 1 \\ 0 & 1 & 2 \\ 1 & 2 & 2 \end{bmatrix}^{-1} \begin{bmatrix} u_1 \\ u_2 \\ u_3 \end{bmatrix} = \begin{bmatrix} 2 & 0 & -1 \\ -2 & -1 & 2 \\ 1 & 1 & -1 \end{bmatrix} \begin{bmatrix} u_1 \\ u_2 \\ u_3 \end{bmatrix} = \begin{bmatrix} 2u_1 - u_3 \\ -2u_1 - u_2 + 2u_3 \\ u_1 + u_2 - u_3 \end{bmatrix}.$$

25. (a) $\dim \ker L + \dim \operatorname{range} L = 2 + \dim \operatorname{range} L = 4 \Longrightarrow \dim \operatorname{range} L = 2$.

(b) $\dim \ker L + \dim \operatorname{range} L = \dim \ker L + 3 = 4 \Longrightarrow \dim \ker L = 1$.

27. (a) L is not one-to-one. To prove this, let $f : V \to W$ be the function $f(x) = 1$. Then $L(f) = f'(x) = 0 = L(0)$ but $f \neq 0$, so L is not one-to-one.

(b) L is not onto. To prove this, let $g : V \to W$ be the function

$$g(x) = \begin{cases} 0 & \text{if } x < 0 \\ 1 & \text{if } x \geq 0 \end{cases}.$$

There is no differentiable function whose derivative is g. Therefore L is not onto.

29. Suppose that $\mathbf{x}_1$ and $\mathbf{x}_2$ are solutions to $L(\mathbf{x}) = \mathbf{b}$. We show that $\mathbf{x}_1 - \mathbf{x}_2$ is in $\ker L$:

$$L(\mathbf{x}_1 - \mathbf{x}_2) = L(\mathbf{x}_1) - L(\mathbf{x}_2) = \mathbf{b} - \mathbf{b} = \mathbf{0}.$$

31. From Theorem 6.6, we have $\dim \ker L + \dim \operatorname{range} L = \dim V$.

    (a) If L is one-to-one, then $\ker L = \{\mathbf{0}\}$, so $\dim \ker L = 0$. Hence $\dim \operatorname{range} L = \dim V = \dim W$ so L is onto.

    (b) If L is onto, then $\operatorname{range} L = W$, so $\dim \operatorname{range} L = \dim W = \dim V$. Hence $\dim \ker L = 0$ and L is one-to-one.

Section 6.3, p. 397

1. $L : R^2 \to R^2$ is given by

$$L\left(\begin{bmatrix} u_1 \\ u_2 \end{bmatrix}\right) = \begin{bmatrix} u_1 + 2u_2 \\ 2u_1 - u_2 \end{bmatrix}.$$

Let $S = \{\mathbf{e}_1, \mathbf{e}_2\}$ be the natural basis for R^2 and let $T = \left\{ \begin{bmatrix} -1 \\ 2 \end{bmatrix}, \begin{bmatrix} 2 \\ 0 \end{bmatrix} \right\}$ be another basis for R^2.

    (a) To find the representation of L with respect to S we assume that S is the basis used in both $V = R^2$ and $W = R^2$. We first compute the images of the members of S under L:

$$L(\mathbf{e}_1) = \begin{bmatrix} 1 \\ 2 \end{bmatrix}, \quad L(\mathbf{e}_2) = \begin{bmatrix} 2 \\ -1 \end{bmatrix}.$$

The columns of the matrix A that represents L in this case are the coordinates of the images of the S-basis vectors relative to the S-basis. That is

$$\operatorname{col}_1 A = \left[\ L(\mathbf{e}_1)\ \right]_S = \left[\ 1\mathbf{e}_1 + 2\mathbf{e}_2\ \right]_S = \begin{bmatrix} 1 \\ 2 \end{bmatrix}$$

$$\operatorname{col}_2 A = \left[\ L(\mathbf{e}_2)\ \right]_S = \left[\ 2\mathbf{e}_1 - 1\mathbf{e}_2\ \right]_S = \begin{bmatrix} 2 \\ -1 \end{bmatrix}$$

Hence $A = \begin{bmatrix} 1 & 2 \\ 2 & -1 \end{bmatrix}$. (Only because the natural basis is used in W do the columns of the matrix A become the images of the basis vectors.)

    (b) To find the representation of L with respect to S and T, we first compute the images of the members of S and T under L:

$$L(\mathbf{e}_1) = \begin{bmatrix} 1 \\ 2 \end{bmatrix}, \quad L(\mathbf{e}_2) = \begin{bmatrix} 2 \\ -1 \end{bmatrix}.$$

The columns of the matrix A that represent L in this case are the coordinates of the images of the S-basis vectors relative to the T-basis. That is,

$$\operatorname{col}_1 A = \left[\ L(\mathbf{e}_1)\ \right]_T \quad \text{and} \quad \operatorname{col}_2 A = \left[\ L(\mathbf{e}_2)\ \right]_T$$

as in part (a). But since we are not using the natural basis in W we must solve for these coordinates. That is, find coefficients c_1 and c_2 such that

$$c_1 \begin{bmatrix} -1 \\ 2 \end{bmatrix} + c_2 \begin{bmatrix} 2 \\ 0 \end{bmatrix} = L(\mathbf{e}_1) = \begin{bmatrix} 1 \\ 2 \end{bmatrix}$$

and coefficients k_1 and k_2 such that

$$k_1 \begin{bmatrix} -1 \\ 2 \end{bmatrix} + k_2 \begin{bmatrix} 2 \\ 0 \end{bmatrix} = L(\mathbf{e}_2) = \begin{bmatrix} 2 \\ -1 \end{bmatrix}.$$

In the first case we are led to a linear system with augmented matrix

$$\left[\begin{array}{cc|c} -1 & 2 & 1 \\ 2 & 0 & 2 \end{array}\right]$$

and in the second case to a linear system with augmented matrix

$$\left[\begin{array}{cc|c} -1 & 2 & 2 \\ 2 & 0 & -1 \end{array}\right].$$

We note that both systems have the same coefficient matrix so for the sake of efficiency we form the following partitioned matrix

$$\left[\begin{array}{cc|c|c} -1 & 2 & 1 & 2 \\ 2 & 0 & 2 & -1 \end{array}\right]$$

and proceed with row reduction to solve the systems involved.

$$\left[\begin{array}{cc|c|c} -1 & 2 & 1 & 2 \\ 2 & 0 & 2 & -1 \end{array}\right] \longrightarrow \left[\begin{array}{cc|c|c} 1 & -2 & -1 & -2 \\ 0 & 4 & 4 & 3 \end{array}\right] \longrightarrow \left[\begin{array}{cc|c|c} 1 & 0 & 1 & -\frac{1}{2} \\ 0 & 1 & 1 & \frac{3}{4} \end{array}\right]$$

add 2 times row 1 to row 2   multiply row 2 by $\frac{1}{2}$ reduced row echelon form
multiply row 1 by -1

Hence A is the last pair of columns in the preceding partitioned matrix. That is, $A = \left[\begin{array}{cc} 1 & -\frac{1}{2} \\ 1 & \frac{3}{4} \end{array}\right]$.

(c) To find the representation of L with respect to T and S we first compute the images of the members of T under L:

$$L\left(\left[\begin{array}{c} -1 \\ 2 \end{array}\right]\right) = \left[\begin{array}{c} 3 \\ -4 \end{array}\right], \quad L\left(\left[\begin{array}{c} 2 \\ 0 \end{array}\right]\right) = \left[\begin{array}{c} 2 \\ 4 \end{array}\right].$$

The columns of the matrix A that represents L in this case are the coordinates of the images of the T-basis vectors relative to the S-basis. That is,

$$\text{col}_1 A = \left[\begin{array}{c} 3 \\ -4 \end{array}\right]_S \quad \text{and} \quad \text{col}_2 A = \left[\begin{array}{c} 2 \\ 4 \end{array}\right]_S.$$

Since S is the natural basis we have that

$$\left[\begin{array}{c} 3 \\ -4 \end{array}\right]_S = \left[\begin{array}{c} 3 \\ -4 \end{array}\right] \quad \text{and} \quad \left[\begin{array}{c} 2 \\ 4 \end{array}\right]_S = \left[\begin{array}{c} 2 \\ 4 \end{array}\right]$$

and hence the matrix A of L with respect to T and S is $A = \left[\begin{array}{cc} 3 & 2 \\ -4 & 4 \end{array}\right]$.

(d) To find the representation of L with respect to T we assume that T is the basis used in both $V = R^2$ and $W = R^2$. We first compute the images of the members of T under L:

$$L\left(\left[\begin{array}{c} -1 \\ 2 \end{array}\right]\right) = \left[\begin{array}{c} 3 \\ -4 \end{array}\right], \quad L\left(\left[\begin{array}{c} 2 \\ 0 \end{array}\right]\right) = \left[\begin{array}{c} 2 \\ 4 \end{array}\right].$$

The columns of the matrix A that represents L in this case are the coordinates of the images of the T-basis vectors relative to the T-basis. That is,

$$\text{col}_1 A = \left[\begin{array}{c} 3 \\ -4 \end{array}\right]_T \quad \text{and} \quad \text{col}_2 A = \left[\begin{array}{c} 2 \\ 4 \end{array}\right]_T.$$

But since we are not using the natural basis in W we must solve for these coordinates. That is, find coefficients c_1, c_2 and k_1, k_2 such that

$$c_1\left[\begin{array}{c} -1 \\ 2 \end{array}\right] + c_2\left[\begin{array}{c} 2 \\ 0 \end{array}\right] = \left[\begin{array}{c} 3 \\ -4 \end{array}\right] \quad \text{and} \quad k_1\left[\begin{array}{c} -1 \\ 2 \end{array}\right] + k_2\left[\begin{array}{c} 2 \\ 0 \end{array}\right] = \left[\begin{array}{c} 2 \\ 4 \end{array}\right].$$

As in part (c) we create a partitioned matrix and proceed to row reduce it:

$$\begin{bmatrix} -1 & 2 & | & 3 & | & 2 \\ 2 & 0 & | & -4 & | & 4 \end{bmatrix} \longrightarrow \begin{bmatrix} 1 & -2 & | & -3 & | & -2 \\ 0 & 4 & | & 2 & | & 8 \end{bmatrix} \longrightarrow \begin{bmatrix} 1 & 0 & | & -2 & | & 2 \\ 0 & 1 & | & \frac{1}{2} & | & 2 \end{bmatrix}$$

add 2 times row 1 to row 2   multiply row 2 by $\frac{1}{4}$ reduced row echelon form

multiply row 1 by -1 add 2 times row 2 to row 1

Hence the matrix of L is $A = \begin{bmatrix} -2 & 2 \\ \frac{1}{2} & 2 \end{bmatrix}$.

(e) Let $\mathbf{x} = \begin{bmatrix} 1 \\ 2 \end{bmatrix}$ and $\mathbf{y} = L(\mathbf{x})$. From the definition of L, we have

$$\mathbf{y} = L(\mathbf{x}) = \begin{bmatrix} 5 \\ 0 \end{bmatrix}.$$

In order to use the matrices for L developed in the preceding parts of this exercise, recall that a matrix representing L connects coordinates in the bases involved. Thus we will first compute the following:

$$[\,\mathbf{x}\,]_S = \begin{bmatrix} 1 \\ 2 \end{bmatrix} \quad \text{and} \quad [\,\mathbf{x}\,]_T = \begin{bmatrix} 1 \\ 1 \end{bmatrix}.$$

From part (a): $[\,\mathbf{y}\,]_S = \begin{bmatrix} 1 & 2 \\ 2 & -1 \end{bmatrix}[\,\mathbf{x}\,]_S = \begin{bmatrix} 1 & 2 \\ 2 & -1 \end{bmatrix}\begin{bmatrix} 1 \\ 2 \end{bmatrix} = \begin{bmatrix} 5 \\ 0 \end{bmatrix}$

Thus $\mathbf{y} = 5\mathbf{e}_1 + 0\mathbf{e}_2 = \begin{bmatrix} 5 \\ 0 \end{bmatrix}$.

From part (b): $[\,\mathbf{y}\,]_T = \begin{bmatrix} 1 & -\frac{1}{2} \\ 1 & \frac{3}{4} \end{bmatrix}[\,\mathbf{x}\,]_S = \begin{bmatrix} 1 & -\frac{1}{2} \\ 1 & \frac{3}{4} \end{bmatrix}\begin{bmatrix} 1 \\ 2 \end{bmatrix} = \begin{bmatrix} 0 \\ \frac{5}{2} \end{bmatrix}$

Thus $\mathbf{y} = 0\begin{bmatrix} -1 \\ 2 \end{bmatrix} + \frac{5}{2}\begin{bmatrix} 2 \\ 0 \end{bmatrix} = \begin{bmatrix} 5 \\ 0 \end{bmatrix}$.

From part (c): $[\,\mathbf{y}\,]_S = \begin{bmatrix} 3 & 2 \\ -4 & 4 \end{bmatrix}[\,\mathbf{x}\,]_T = \begin{bmatrix} 3 & 2 \\ -4 & 4 \end{bmatrix}\begin{bmatrix} 1 \\ 1 \end{bmatrix} = \begin{bmatrix} 5 \\ 0 \end{bmatrix}$

Thus $\mathbf{y} = 5\mathbf{e}_1 + 0\mathbf{e}_2 = \begin{bmatrix} 5 \\ 0 \end{bmatrix}$.

From part (d): $[\,\mathbf{y}\,]_T = \begin{bmatrix} -2 & 2 \\ \frac{1}{2} & 2 \end{bmatrix}[\,\mathbf{x}\,]_T = \begin{bmatrix} -2 & 2 \\ \frac{1}{2} & 2 \end{bmatrix}\begin{bmatrix} 1 \\ 1 \end{bmatrix} = \begin{bmatrix} 0 \\ \frac{5}{2} \end{bmatrix}$

Thus $\mathbf{y} = 0\begin{bmatrix} -1 \\ 2 \end{bmatrix} + \frac{5}{2}\begin{bmatrix} 2 \\ 0 \end{bmatrix} = \begin{bmatrix} 5 \\ 0 \end{bmatrix}$.

3. Let

$$P = \begin{bmatrix} 1 & 0 & 1 & 1 \\ 0 & 1 & 2 & 1 \\ -1 & -2 & 1 & 0 \end{bmatrix} \quad \text{and} \quad \mathbf{x} = \begin{bmatrix} u_1 \\ u_2 \\ u_3 \\ u_4 \end{bmatrix}.$$

Then $L : R^4 \to R^3$ is defined as $L(\mathbf{x}) = P\mathbf{x}$. Next let $S = \{\mathbf{e}_1, \mathbf{e}_2, \mathbf{e}_3, \mathbf{e}_4\}$ be the natural basis for R^4 and $T = \{\mathbf{e}_1, \mathbf{e}_2, \mathbf{e}_3\}$ be the natural basis for R^3. Also let

$$S' = \{\mathbf{v}_1, \mathbf{v}_2, \mathbf{v}_3, \mathbf{v}_4\} = \left\{ \begin{bmatrix} 1 \\ 1 \\ 0 \\ 0 \end{bmatrix}, \begin{bmatrix} 0 \\ 1 \\ 0 \\ 0 \end{bmatrix}, \begin{bmatrix} 0 \\ 0 \\ 1 \\ 1 \end{bmatrix}, \begin{bmatrix} 0 \\ 1 \\ 1 \\ 0 \end{bmatrix} \right\} \quad \text{be a basis for } R^4$$

and

$$T' = \{\mathbf{w}_1, \mathbf{w}_2, \mathbf{w}_3\} = \left\{ \begin{bmatrix} 1 \\ 0 \\ 1 \end{bmatrix}, \begin{bmatrix} 0 \\ 1 \\ 1 \end{bmatrix}, \begin{bmatrix} 0 \\ 0 \\ 1 \end{bmatrix} \right\} \quad \text{be a basis for } R^3.$$

(a) To find the matrix A representing L relative to the S and T-bases we proceed as follows. Determine the images of the S-basis vectors and find their coordinates relative to the T-basis. Then A is given by

$$A = \left[\ \left[\ L(\mathbf{e}_1)\ \right]_T\quad \left[\ L(\mathbf{e}_2)\ \right]_T\quad \left[\ L(\mathbf{e}_3)\ \right]_T\quad \left[\ L(\mathbf{e}_4)\ \right]_T\ \right].$$

Since $L(\mathbf{x}) = P\mathbf{x}$ we compute the images of the S basis vectors by matrix multiplication:

$$L(\mathbf{e}_1) = \begin{bmatrix} 1 \\ 0 \\ -1 \end{bmatrix}, \quad L(\mathbf{e}_2) = \begin{bmatrix} 0 \\ 1 \\ -2 \end{bmatrix}, \quad L(\mathbf{e}_3) = \begin{bmatrix} 1 \\ 2 \\ 1 \end{bmatrix}, \quad L(\mathbf{e}_4) = \begin{bmatrix} 1 \\ 1 \\ 0 \end{bmatrix}.$$

Since T is the natural basis for R^3 we have that $\left[\ L(\mathbf{e}_j)\ \right]_T = L(\mathbf{e}_j)$ hence

$$A = \begin{bmatrix} 1 & 0 & 1 & 1 \\ 0 & 1 & 2 & 1 \\ -1 & -2 & 1 & 0 \end{bmatrix}.$$

(b) To find the matrix A representing L relative to the S' and T'-bases we proceed as follows. Determine the images of the S'-basis vectors and find their coordinates relative to the T'-basis. Then A is given by

$$A = \left[\ \left[\ L(\mathbf{e}_1)\ \right]_{T'}\quad \left[\ L(\mathbf{e}_2)\ \right]_{T'}\quad \left[\ L(\mathbf{e}_3)\ \right]_{T'}\quad \left[\ L(\mathbf{e}_4)\ \right]_{T'}\ \right].$$

Since $L(\mathbf{x}) = P\mathbf{x}$ we compute the images of the S'-basis vectors by matrix multiplication:

$$L(\mathbf{v}_1) = \begin{bmatrix} 1 \\ 1 \\ -3 \end{bmatrix}, \quad L(\mathbf{v}_2) = \begin{bmatrix} 0 \\ 1 \\ -2 \end{bmatrix}, \quad L(\mathbf{v}_3) = \begin{bmatrix} 2 \\ 3 \\ 1 \end{bmatrix}, \quad L(\mathbf{v}_4) = \begin{bmatrix} 1 \\ 3 \\ -1 \end{bmatrix}.$$

To determine the coordinates of $L(\mathbf{v}_j)$ relative to the T'-basis we have to determine constants so that $L(\mathbf{v}_j)$ is a linear combination of the members of T'. This leads to a linear system whose columns are the members of T' and the right hand side is the vector $L(\mathbf{v}_j)$. To determine the coordinates all at once we use the partitioned matrix device that appears in a number of preceding problems. In this case we reduce

$$\left[\begin{array}{ccc|c|c|c|c} 1 & 0 & 0 & 1 & 0 & 2 & 1 \\ 0 & 1 & 0 & 1 & 1 & 3 & 3 \\ 1 & 1 & 1 & -3 & -2 & 1 & -1 \end{array}\right] \longrightarrow \left[\begin{array}{ccc|c|c|c|c} 1 & 0 & 0 & 1 & 0 & 2 & 1 \\ 0 & 1 & 0 & 1 & 1 & 3 & 3 \\ 0 & 0 & 1 & -5 & -3 & -4 & -5 \end{array}\right]$$

add -1 times row 1 to row 3   reduced row echelon form
add -1 times row 2 to row 3

Thus $A = \begin{bmatrix} 1 & 0 & 2 & 1 \\ 1 & 1 & 3 & 3 \\ -5 & -3 & -4 & -5 \end{bmatrix}.$

5. We have $L : R^3 \rightarrow R^3$ defined on the natural basis for R^3 as

$$L(\mathbf{e}_1) = \begin{bmatrix} 1 \\ 1 \\ 0 \end{bmatrix}, \quad L(\mathbf{e}_2) = \begin{bmatrix} 2 \\ 0 \\ 1 \end{bmatrix}, \quad L(\mathbf{e}_3) = \begin{bmatrix} 1 \\ 0 \\ 1 \end{bmatrix}.$$

(a) The representation of L with respect to the natural basis S of R^3 is given by

$$A = \left[\ \left[\ L(\mathbf{e}_1)\ \right]_S\quad \left[\ L(\mathbf{e}_2)\ \right]_S\quad \left[\ L(\mathbf{e}_3)\ \right]_S\ \right].$$

The coordinates required are given by $\left[\, L(\mathbf{e}_j)\, \right]_S = L(\mathbf{e}_j)$ since S is the natural basis. Thus

$$A = \begin{bmatrix} 1 & 2 & 1 \\ 1 & 0 & 0 \\ 0 & 1 & 1 \end{bmatrix}.$$

(b) From part (a),

$$L\left(\begin{bmatrix} 1 \\ 2 \\ 3 \end{bmatrix}\right) = L(1\mathbf{e}_1 + 2\mathbf{e}_2 + 3\mathbf{e}_3) = 1L(\mathbf{e}_1) + 2L(\mathbf{e}_2) + 3L(\mathbf{e}_3)$$

$$= \begin{bmatrix} 1 \\ 1 \\ 0 \end{bmatrix} + 2\begin{bmatrix} 2 \\ 0 \\ 1 \end{bmatrix} + 3\begin{bmatrix} 1 \\ 0 \\ 1 \end{bmatrix} = \begin{bmatrix} 8 \\ 1 \\ 5 \end{bmatrix}.$$

Therefore

$$\left[L\left(\begin{bmatrix} 1 \\ 2 \\ 3 \end{bmatrix}\right) \right]_S = A\left[\begin{bmatrix} 1 \\ 2 \\ 3 \end{bmatrix}\right]_S = \begin{bmatrix} 1 & 2 & 1 \\ 1 & 0 & 0 \\ 0 & 1 & 1 \end{bmatrix}\begin{bmatrix} 1 \\ 2 \\ 3 \end{bmatrix} = \begin{bmatrix} 8 \\ 1 \\ 5 \end{bmatrix}.$$

7. For $L : R^3 \to R^3$ the matrix representing L with respect to the natural basis is

$$A = \begin{bmatrix} 1 & 3 & 1 \\ 1 & 2 & 0 \\ 0 & 1 & 1 \end{bmatrix}.$$

Since we are using the natural basis the coordinates of a vector are just the components of the vectors. Hence computing the image of a vector under L is obtained by multiplying the vector by A.

(a) $L\left(\begin{bmatrix} 1 \\ 2 \\ 3 \end{bmatrix}\right) = A\begin{bmatrix} 1 \\ 2 \\ 3 \end{bmatrix} = \begin{bmatrix} 10 \\ 5 \\ 5 \end{bmatrix}$ (b) $L\left(\begin{bmatrix} 0 \\ 1 \\ 1 \end{bmatrix}\right) = A\begin{bmatrix} 0 \\ 1 \\ 1 \end{bmatrix} = \begin{bmatrix} 4 \\ 2 \\ 2 \end{bmatrix}$

9. Vector space V has basis $S = \{1, t, e^t, te^t\}$ and $L : V \to V$ is given by $L(f) = f'$. To find the matrix representing L with respect to the S-basis, we find the images of the basis vectors and their coordinates with respect to the S-basis. Then the matrix A representing L is given by

$$A = \left[\; \left[\, L(1)\, \right]_S \;\; \left[\, L(t)\, \right]_S \;\; \left[\, L(e^t)\, \right]_S \;\; \left[\, L(te^t)\, \right]_S \;\right]$$

$$L(1) = 0, \quad \left[\, L(1)\, \right]_S = \begin{bmatrix} 0 \\ 0 \\ 0 \\ 0 \end{bmatrix}, \quad L(t) = 1, \quad \left[\, L(t)\, \right]_S = \begin{bmatrix} 1 \\ 0 \\ 0 \\ 0 \end{bmatrix}$$

$$L(e^t) = e^t, \quad \left[\, L(e^t)\, \right]_S = \begin{bmatrix} 0 \\ 0 \\ 1 \\ 0 \end{bmatrix}, \quad L(te^t) = e^t + te^t, \quad \left[\, L(te^t)\, \right]_S = \begin{bmatrix} 0 \\ 0 \\ 1 \\ 1 \end{bmatrix}$$

Thus $A = \begin{bmatrix} 0 & 1 & 0 & 0 \\ 0 & 0 & 0 & 0 \\ 0 & 0 & 1 & 1 \\ 0 & 0 & 0 & 1 \end{bmatrix}.$

11. Let

$$S = \{\mathbf{v}_1, \mathbf{v}_2, \mathbf{v}_3, \mathbf{v}_4\} = \left\{ \begin{bmatrix} 1 & 0 \\ 0 & 0 \end{bmatrix}, \begin{bmatrix} 0 & 1 \\ 0 & 0 \end{bmatrix}, \begin{bmatrix} 0 & 0 \\ 1 & 0 \end{bmatrix}, \begin{bmatrix} 0 & 0 \\ 0 & 1 \end{bmatrix} \right\}$$

and

$$T = \{\mathbf{w}_1, \mathbf{w}_2, \mathbf{w}_3, \mathbf{w}_4\} = \left\{ \begin{bmatrix} 1 & 0 \\ 0 & 1 \end{bmatrix}, \begin{bmatrix} 1 & 1 \\ 0 & 0 \end{bmatrix}, \begin{bmatrix} 1 & 0 \\ 1 & 0 \end{bmatrix}, \begin{bmatrix} 0 & 1 \\ 0 & 0 \end{bmatrix} \right\}$$

be bases for M_{22}. Let

$$A = \begin{bmatrix} 1 & 2 \\ 3 & 4 \end{bmatrix}$$

and let $L : M_{22} \to M_{22}$ be given by $L(X) = AX - XA$. To facilitate constructing the various representations of L that are requested we first record the images of the basis vectors in both S and T. (We omit the details.)

$$L(\mathbf{v}_1) = \begin{bmatrix} 0 & -2 \\ 3 & 0 \end{bmatrix}, \quad L(\mathbf{v}_2) = \begin{bmatrix} -3 & -3 \\ 0 & 3 \end{bmatrix}, \quad L(\mathbf{v}_3) = \begin{bmatrix} 2 & 0 \\ 3 & -2 \end{bmatrix}, \quad L(\mathbf{v}_4) = \begin{bmatrix} 0 & 2 \\ -3 & 0 \end{bmatrix}$$

$$L(\mathbf{w}_1) = \begin{bmatrix} 0 & 0 \\ 0 & 0 \end{bmatrix}, \quad L(\mathbf{w}_2) = \begin{bmatrix} -3 & -5 \\ 3 & 3 \end{bmatrix}, \quad L(\mathbf{w}_3) = \begin{bmatrix} 2 & -2 \\ 6 & -2 \end{bmatrix}, \quad L(\mathbf{w}_4) = \begin{bmatrix} -3 & -3 \\ 0 & 3 \end{bmatrix}$$

(a) The representation of L with respect to S is found by determining the coordinates of $L(\mathbf{v}_j)$, $j = 1$, 2, 3, 4 relative to the S-basis. Since the S-basis is the natural basis for M_{22} the coordinates we seek are easily obtained from the images:

$$[\, L(\mathbf{v}_1)\,]_S = \begin{bmatrix} 0 \\ -2 \\ 3 \\ 0 \end{bmatrix}, [\, L(\mathbf{v}_2)\,]_S = \begin{bmatrix} -3 \\ -3 \\ 0 \\ 3 \end{bmatrix}, [\, L(\mathbf{v}_3)\,]_S = \begin{bmatrix} 2 \\ 0 \\ 3 \\ -2 \end{bmatrix}, [\, L(\mathbf{v}_4)\,]_S = \begin{bmatrix} 0 \\ 2 \\ -3 \\ 0 \end{bmatrix}$$

The matrix of L is given by

$$A = \begin{bmatrix} 0 & -3 & 2 & 0 \\ -2 & -3 & 0 & 2 \\ 3 & 0 & 3 & -3 \\ 0 & 3 & -2 & 0 \end{bmatrix}.$$

(b) The representation of L with respect to T is found by determining the coordinates of $L(\mathbf{w}_j)$, $j = 1, 2, 3, 4$ relative to the T-basis. We must express each $L(\mathbf{w}_j)$ as a linear combination of the $\mathbf{w}_j$. As seen a number of times previously this leads to a linear system. In this case, we have

$$c_1 \mathbf{w}_1 + c_2 \mathbf{w}_2 + c_3 \mathbf{w}_3 + c_4 \mathbf{w}_4 = L(\mathbf{w}_j), \quad j = 1, 2, 3, 4.$$

Upon substitution of the matrices involved, combining terms on the left side, and equating corresponding elements from each side we obtain the partitioned linear system shown below on the left, and its reduced row echelon form on the right:

$$\left[\begin{array}{cccc|c|c|c|c} 1 & 1 & 1 & 0 & 0 & -3 & 2 & -3 \\ 0 & 1 & 0 & 1 & 0 & -5 & -2 & -3 \\ 0 & 0 & 1 & 0 & 0 & 3 & 6 & 0 \\ 1 & 0 & 0 & 0 & 0 & 3 & -2 & 3 \end{array} \right] \longrightarrow \left[\begin{array}{cccc|c|c|c|c} 1 & 0 & 0 & 0 & 0 & 3 & -2 & 3 \\ 0 & 1 & 0 & 0 & 0 & -9 & -2 & -6 \\ 0 & 0 & 1 & 0 & 0 & 3 & 6 & 0 \\ 0 & 0 & 0 & 1 & 0 & 4 & 0 & 3 \end{array} \right]$$

linear system   reduced row echelon form

Thus

$$[\, L(\mathbf{w}_1)\,]_T = \begin{bmatrix} 0 \\ 0 \\ 0 \\ 0 \end{bmatrix}, [\, L(\mathbf{w}_2)\,]_T = \begin{bmatrix} 3 \\ -9 \\ 3 \\ 4 \end{bmatrix}, [\, L(\mathbf{w}_3)\,]_T = \begin{bmatrix} -2 \\ -2 \\ 6 \\ 0 \end{bmatrix}, [\, L(\mathbf{w}_4)\,]_T = \begin{bmatrix} 3 \\ -6 \\ 0 \\ 3 \end{bmatrix}$$

and it follows that

$$A = \begin{bmatrix} 0 & 3 & -2 & 3 \\ 0 & -9 & -2 & -6 \\ 0 & 3 & 6 & 0 \\ 0 & 4 & 0 & 3 \end{bmatrix}.$$

(c) We repeat part (a) up until we are to determine the coordinates of $L(\mathbf{v}_j)$. Here we must write each of the $L(\mathbf{v}_j)$ in terms of the members of the T-basis. We follow the method of part (b). The array below on the left shows the linear system, the array on the right its reduced row echelon form:

$$\begin{bmatrix} 1 & 1 & 1 & 0 & 0 & -3 & 2 & 0 \\ 0 & 1 & 0 & 1 & -2 & -3 & 0 & 2 \\ 0 & 0 & 1 & 0 & 3 & 0 & 3 & -3 \\ 1 & 0 & 0 & 0 & 0 & 3 & -2 & 0 \end{bmatrix} \longrightarrow \begin{bmatrix} 1 & 0 & 0 & 0 & 0 & 3 & -2 & 0 \\ 0 & 1 & 0 & 0 & -3 & -6 & 1 & 3 \\ 0 & 0 & 1 & 0 & 3 & 0 & 3 & -3 \\ 0 & 0 & 0 & 1 & 1 & 3 & -1 & -1 \end{bmatrix}$$

linear system   reduced row echelon form

Therefore

$$[\,L(\mathbf{v}_1)\,]_T = \begin{bmatrix} 0 \\ -3 \\ 3 \\ 1 \end{bmatrix}, [\,L(\mathbf{v}_2)\,]_T = \begin{bmatrix} 3 \\ -6 \\ 0 \\ 3 \end{bmatrix}, [\,L(\mathbf{v}_3)\,]_T = \begin{bmatrix} -2 \\ 1 \\ 3 \\ -1 \end{bmatrix}, [\,L(\mathbf{v}_4)\,]_T = \begin{bmatrix} 0 \\ 3 \\ -3 \\ -1 \end{bmatrix}$$

and it follows that $A = \begin{bmatrix} 0 & 3 & -2 & 0 \\ -3 & -6 & 1 & 3 \\ 3 & 0 & 3 & -3 \\ 1 & 3 & -1 & -1 \end{bmatrix}.$

(d) We repeat part (b) up until we are to determine the coordinates of $L(\mathbf{w}_j)$. Here we must write each of the $L(\mathbf{w}_j)$ in terms of the members of the S-basis. But this is easy since S is the natural basis for M_{22}. We have

$$[\,L(\mathbf{w}_1)\,]_T = \begin{bmatrix} 0 \\ 0 \\ 0 \\ 0 \end{bmatrix}, [\,L(\mathbf{w}_2)\,]_T = \begin{bmatrix} -3 \\ -5 \\ 3 \\ 3 \end{bmatrix}, [\,L(\mathbf{w}_3)\,]_T = \begin{bmatrix} 2 \\ -2 \\ 6 \\ -2 \end{bmatrix}, [\,L(\mathbf{w}_4)\,]_T = \begin{bmatrix} -3 \\ -3 \\ 0 \\ 3 \end{bmatrix}$$

and it follows that

$$A = \begin{bmatrix} 0 & -3 & 2 & -3 \\ 0 & -5 & -2 & -3 \\ 0 & 3 & 6 & 0 \\ 0 & 3 & -2 & 3 \end{bmatrix}.$$

13. Let $L : R^2 \to R^2$ be given by

$$L\left(\begin{bmatrix} x \\ y \end{bmatrix}\right) = \begin{bmatrix} x \\ -y \end{bmatrix}$$

and let

$$S = \{\mathbf{e}_1, \mathbf{e}_2\} = \left\{\begin{bmatrix} 1 \\ 0 \end{bmatrix}, \begin{bmatrix} 0 \\ 1 \end{bmatrix}\right\}$$

be the natural basis for R^2 and

$$T = \{\mathbf{w}_1, \mathbf{w}_2\} = \left\{\begin{bmatrix} 1 \\ 1 \end{bmatrix}, \begin{bmatrix} -1 \\ 1 \end{bmatrix}\right\}$$

be another basis for R^2. To facilitate determining the matrix representations of L requested we first compute the images of the basis vectors in both S and T. We have

$$L(\mathbf{e}_1) = \begin{bmatrix} 1 \\ 0 \end{bmatrix}, \quad L(\mathbf{e}_2) = \begin{bmatrix} 0 \\ -1 \end{bmatrix}, \quad L(\mathbf{w}_1) = \begin{bmatrix} 1 \\ -1 \end{bmatrix}, \quad L(\mathbf{w}_2) = \begin{bmatrix} -1 \\ -1 \end{bmatrix}.$$

(a) The matrix representing L with respect to S is

$$\left[\ [\ L(\mathbf{e}_1)\ ]_S\quad [\ L(\mathbf{e}_2)\ ]_S\ \right] = \begin{bmatrix} 1 & 0 \\ 0 & -1 \end{bmatrix}.$$

This follows because S is the natural basis so the coordinates of a vector are just the components of the vector.

(b) The matrix representing L with respect to T is

$$\left[\ [\ L(\mathbf{w}_1)\ ]_T\quad [\ L(\mathbf{w}_2)\ ]_T\ \right]$$

Thus we must express both $L(\mathbf{w}_1)$ and $L(\mathbf{w}_2)$ in terms of $\mathbf{w}_1$ and $\mathbf{w}_2$. Constructing the appropriate equations leads to a pair of linear systems with the same coefficient matrix but different right-hand sides. As we did previously, we use a partitioned matrix for the system and find its row reduced echelon form:

$$\left[\begin{array}{rr|r|r} 1 & -1 & 1 & -1 \\ 1 & 1 & -1 & -1 \end{array}\right] \qquad\longrightarrow\qquad \left[\begin{array}{rr|r|r} 1 & 0 & 0 & -1 \\ 0 & 1 & -1 & 0 \end{array}\right]$$

$$\text{linear system} \qquad\qquad\qquad \text{reduced row echelon form}$$

Thus the matrix representing L is $\begin{bmatrix} 0 & -1 \\ -1 & 0 \end{bmatrix}.$

(c) The matrix representing L with respect to S and T is

$$\left[\ [\ L(\mathbf{e}_1)\ ]_T\quad [\ L(\mathbf{e}_2)\ ]_T\ \right].$$

Thus we must express both $L(\mathbf{e}_1)$ and $L(\mathbf{e}_2)$ in terms of $\mathbf{w}_1$ and $\mathbf{w}_2$. Constructing the appropriate equations leads to a pair of linear systems with the same coefficient matrix but different right-hand sides. As we did previously, we use a partitioned matrix:

$$\left[\begin{array}{rr|r|r} 1 & -1 & 1 & 0 \\ 1 & 1 & 0 & -1 \end{array}\right] \qquad\longrightarrow\qquad \left[\begin{array}{rr|r|r} 1 & 0 & \frac{1}{2} & -\frac{1}{2} \\ 0 & 1 & -\frac{1}{2} & -\frac{1}{2} \end{array}\right]$$

$$\text{linear system} \qquad\qquad\qquad \text{reduced row echelon form}$$

Thus the matrix representing L is $\begin{bmatrix} \frac{1}{2} & -\frac{1}{2} \\ -\frac{1}{2} & -\frac{1}{2} \end{bmatrix}.$

(d) The matrix representing L with respect to T and S is

$$\left[\ [\ L(\mathbf{w}_1)\ ]_S\quad [\ L(\mathbf{w}_2)\ ]_S\ \right].$$

Thus we must express both $L(\mathbf{w}_1)$ and $L(\mathbf{w}_2)$ in terms of $\mathbf{e}_1$ and $\mathbf{e}_2$. Since S is the natural basis we have that the matrix representing L is $\begin{bmatrix} 1 & -1 \\ -1 & -1 \end{bmatrix}.$

15. Let $O : V \to W$ be given by $O(\mathbf{v}) = \mathbf{0}_W$. Also let $S = \{\mathbf{v}_1, \mathbf{v}_2, \ldots, \mathbf{v}_n\}$ be a basis for V and $T = \{\mathbf{w}_1, \mathbf{w}_2, \ldots, \mathbf{w}_m\}$ be a basis for W. Then $O(\mathbf{v}_j) = \mathbf{0}_W$. Since the coordinates of the zero vector relative to any basis consist of all zero we have that

$$[\ O(\mathbf{v}_j)\ ]_T = \underbrace{\begin{bmatrix} 0 & 0 & \cdots & 0 \end{bmatrix}}_{m \text{ zeros}}{}^T$$

It follows that the matrix of O is $\left[\ [\ O(\mathbf{v}_1)\ ]_T\quad [\ O(\mathbf{v}_2)\ ]_T\quad \cdots\quad [\ O(\mathbf{v}_n)\ ]_T\ \right] = O.$

17. $I : R_2 \rightarrow R_2$ is the identity operator. That is, $I(\mathbf{x}) = \mathbf{x}$ for $\mathbf{x}$ in R^2. Let

$$S = \{\mathbf{e}_1, \mathbf{e}_2\} = \{[\,1\ \ 0\,], [\,0\ \ 1\,]\} \qquad \text{be the natural basis for } R^2$$

and let

$$T = \{\mathbf{w}_1, \mathbf{w}_2\} = \{[\,1\ \ -1\,], [\,2\ \ 3\,]\} \qquad \text{be another basis for } R^2.$$

To facilitate finding the matrix representations of I requested we first determine the images of the vectors in both S and T under I:

$$I(\mathbf{e}_1) = \mathbf{e}_1, \quad I(\mathbf{e}_2) = \mathbf{e}_2, \quad I(\mathbf{w}_1) = \mathbf{w}_1, \quad I(\mathbf{w}_2) = \mathbf{w}_2.$$

(a) The matrix representing I with respect to S is

$$\left[\ [\,I(\mathbf{e}_1)\,]_S \ \ [\,I(\mathbf{e}_2)\,]_S \ \right] = \begin{bmatrix} 1 & 0 \\ 0 & 1 \end{bmatrix}.$$

This follows from the fact that S is the natural basis.

(b) The matrix representing I with respect to T is

$$\left[\ [\,I(\mathbf{w}_1)\,]_T \ \ [\,I(\mathbf{w}_2)\,]_T \ \right].$$

We must express both $I(\mathbf{w}_1) = \mathbf{w}_1$ and $I(\mathbf{w}_2) = \mathbf{w}_2$ in terms of the vectors in T. It follows that

$$\left[\ [\,I(\mathbf{w}_1)\,]_T \ \ [\,I(\mathbf{w}_2)\,]_T \ \right] = \begin{bmatrix} 1 & 0 \\ 0 & 1 \end{bmatrix}.$$

(c) The matrix representing I with respect to S and T is

$$\left[\ [\,I(\mathbf{e}_1)\,]_T \ \ [\,I(\mathbf{e}_2)\,]_T \ \right].$$

We must express both $I(\mathbf{e}_1) = \mathbf{e}_1$ and $I(\mathbf{e}_2) = \mathbf{e}_2$ in terms of the vectors in T. Constructing the appropriate linear equations leads to two linear systems with the same coefficient matrix but different right-hand sides. We obtain the partitioned matrix and find its reduced row echelon form:

$$\begin{bmatrix} 1 & 2 & | & 1 & 0 \\ -1 & 3 & | & 0 & 1 \end{bmatrix} \qquad \longrightarrow \qquad \begin{bmatrix} -1 & 0 & | & \frac{3}{5} & | & -\frac{2}{5} \\ 0 & 1 & | & \frac{1}{5} & | & \frac{1}{5} \end{bmatrix}$$

$$\text{linear system} \qquad\qquad \text{reduced row echelon form}$$

Thus the matrix representing I is $\begin{bmatrix} \frac{3}{5} & -\frac{2}{5} \\ \frac{1}{5} & \frac{1}{5} \end{bmatrix}$.

(d) The matrix representing I with respect to T and S is

$$\left[\ [\,I(\mathbf{w}_1)\,]_S \ \ [\,I(\mathbf{w}_2)\,]_S \ \right].$$

We must express both $I(\mathbf{w}_1) = \mathbf{w}_1$ and $I(\mathbf{w}_2) = \mathbf{w}_2$ in terms of the vectors in S. It follows that

$$\left[\ [\,I(\mathbf{w}_1)\,]_S \ \ [\,I(\mathbf{w}_2)\,]_S \ \right] = \begin{bmatrix} 1 & 2 \\ -1 & 3 \end{bmatrix}$$

since S is the natural basis for R^2.

19. We compute the images of the vectors in S under L: $L(\sin t) = \cos t$, $L(\cos t) = -\sin t$. The matrix of L with respect to S is therefore

$$\left[\ [\,L(\sin t)\,]_S \ \ [\,L(\cos t)\,]_S \ \right] = \begin{bmatrix} 0 & -1 \\ 1 & 0 \end{bmatrix}.$$

21. Let $S = \{\mathbf{v}_1, \mathbf{v}_2, \ldots, \mathbf{v}_n\}$ be a basis for V. Then $L(\mathbf{v}_j) = c\mathbf{v}_j$ for $j = 1, 2, \ldots, n$. Hence $\left[L(\mathbf{v}_j) \right]_S = c\mathbf{e}_j$, where $\mathbf{e}_j$ is the jth natural basis vector for R^n. Hence the matrix representing L is

$$\left[\begin{array}{cccc} c\mathbf{e}_1 & c\mathbf{e}_2 & \cdots & c\mathbf{e}_n \end{array} \right] = cI_n.$$

Thus the matrix representing L is a scalar matrix (a constant multiple of the identity matrix).

23. Let $I : V \to V$ be the identity operator defined by $I(\mathbf{v}) = \mathbf{v}$ for $\mathbf{v}$ in V. The matrix of I with respect to S and T is obtained as follows. The jth column of A is $\left[I(\mathbf{v}_j) \right]_T = \left[\mathbf{v}_j \right]_T$. Therefore, as defined in Section 4.8, A is the transition matrix $P_{T \leftarrow S}$ from the S-basis to the T-basis.

Section 6.4, p. 405

1. (a) Let $\mathbf{u}$ and $\mathbf{v}$ be vectors in V and c_1 and c_2 scalars. Then

$$(L_1 \boxplus L_2)(c_1\mathbf{u} + c_2\mathbf{v}) = L_1(c_1\mathbf{u} + c_2\mathbf{v}) + L_2(c_1\mathbf{u} + c_2\mathbf{v})$$
$$\text{(from Definition 6.5)}$$
$$= c_1 L_1(\mathbf{u}) + c_2 L_1(\mathbf{v}) + c_1 L_2(\mathbf{u}) + c_2 L_2(\mathbf{v})$$
$$\text{(since } L_1 \text{ and } L_2 \text{ are linear transformations)}$$
$$= c_1(L_1(\mathbf{u}) + L_2(\mathbf{u})) + c_2(L_1(\mathbf{v}) + L_2(\mathbf{v}))$$
$$\text{(using properties of vector operations since the images are in } W)$$
$$= c_1(L_1 \boxplus L_2)(\mathbf{u}) + c_2(L_1 \boxplus L_2)(\mathbf{v})$$
$$\text{(from Definition 6.5)}$$

Thus by Exercise 6 in Section 6.1, $L_1 \boxplus L_2$ is a linear transformation.

(b) Let $\mathbf{u}$ and $\mathbf{v}$ be vectors in V and k_1 and k_2 be scalars. Then

$$(c \boxdot L)(k_1\mathbf{u} + k_2\mathbf{v}) = cL(k_1\mathbf{u} + k_2\mathbf{v})$$
$$\text{(from Definition 6.5)}$$
$$= c(k_1 L(\mathbf{u}) + k_2 L(\mathbf{v}))$$
$$\text{(since } L \text{ is a linear transformation)}$$
$$= ck_1 L(\mathbf{u}) + ck_2 L(\mathbf{v})$$
$$\text{(using properties of vector operations since the images are in } W)$$
$$= k_1 cL(\mathbf{u}) + k_2 cL(\mathbf{v})$$
$$\text{(using properties of vector operations)}$$
$$= k_1(c \boxdot L)(\mathbf{u}) + k_2(c \boxdot L)(\mathbf{v})$$
$$\text{(by Definition 6.5)}$$

(c) Let $S = \{\mathbf{v}_1, \mathbf{v}_2, \ldots, \mathbf{v}_n\}$. Then

$$A = \left[\begin{array}{cccc} \left[L(\mathbf{v}_1) \right]_T & \left[L(\mathbf{v}_2) \right]_T & \cdots & \left[L(\mathbf{v}_n) \right]_T \end{array} \right].$$

The matrix representing $c \boxdot L$ is given by

$$\left[\begin{array}{cccc} \left[L(\mathbf{v}_1) \right]_T & \left[L(\mathbf{v}_2) \right]_T & \cdots & \left[L(\mathbf{v}_n) \right]_T \end{array} \right]$$
$$= \left[\begin{array}{cccc} \left[c \boxdot L(\mathbf{v}_1) \right]_T & \left[c \boxdot L(\mathbf{v}_2) \right]_T & \cdots & \left[c \boxdot L(\mathbf{v}_n) \right]_T \end{array} \right]$$
$$= \left[\begin{array}{cccc} \left[cL(\mathbf{v}_1) \right]_T & \left[cL(\mathbf{v}_2) \right]_T & \cdots & \left[cL(\mathbf{v}_n) \right]_T \end{array} \right]$$

(by Definition 6.5)

$$= \left[\begin{array}{cccc} c \left[L(\mathbf{v}_1) \right]_T & c \left[L(\mathbf{v}_2) \right]_T & \cdots & c \left[L(\mathbf{v}_n) \right]_T \end{array} \right]$$

(by properties of coordinates)

$$= c \left[\begin{array}{cccc} \left[L(\mathbf{v}_1) \right]_T & \left[L(\mathbf{v}_2) \right]_T & \cdots & \left[L(\mathbf{v}_n) \right]_T \end{array} \right] = cA$$

(by matrix algebra)

3. (a) $(L_1 \boxplus L_2) \left(\left[\begin{array}{ccc} u_1 & u_2 & u_3 \end{array} \right] \right) = L_1 \left(\left[\begin{array}{ccc} u_1 & u_2 & u_3 \end{array} \right] \right) + L_2 \left(\left[\begin{array}{ccc} u_1 & u_2 & u_3 \end{array} \right] \right)$
   $= L_1(u_1\mathbf{e}_1^T + u_2\mathbf{e}_2^T + u_3\mathbf{e}_3^T) + L_2(u_1\mathbf{e}_1^T + u_2\mathbf{e}_2^T + u_3\mathbf{e}_3^T)$
   $= u_1 L_1(\mathbf{e}_1^T) + u_2 L_1(\mathbf{e}_2^T) + u_3 L_1(\mathbf{e}_3^T) + u_1 L_2(\mathbf{e}_1^T) + u_2 L_2(\mathbf{e}_2^T) + u_3 L_2(\mathbf{e}_3^T)$
   $= u_1 \left[\begin{array}{ccc} -1 & 2 & 1 \end{array} \right] + u_2 \left[\begin{array}{ccc} 0 & 1 & 2 \end{array} \right] + u_3 \left[\begin{array}{ccc} -1 & 1 & 3 \end{array} \right]$
   $\quad + u_1 \left[\begin{array}{ccc} 0 & 1 & 3 \end{array} \right] + u_2 \left[\begin{array}{ccc} 4 & -2 & 1 \end{array} \right] + u_3 \left[\begin{array}{ccc} 0 & 2 & 2 \end{array} \right]$
   $= u_1 \left[\begin{array}{ccc} -1 & 3 & 4 \end{array} \right] + u_2 \left[\begin{array}{ccc} 4 & -1 & 3 \end{array} \right] + u_3 \left[\begin{array}{ccc} -1 & 3 & 5 \end{array} \right]$
   $= \left[\begin{array}{ccc} -u_1 + 4u_2 - u_3 & 3u_1 - u_2 + 3u_3 & 4u_1 + 3u_2 + 5u_3 \end{array} \right]$

   (b) In part (a) set $u_1 = 2$, $u_2 = 1$, and $u_3 = -3$. Then $(L_1 \boxplus L_2) \left(\left[\begin{array}{ccc} 2 & 1 & -3 \end{array} \right] \right) = \left[\begin{array}{ccc} 5 & -4 & -4 \end{array} \right]$.

   (c) The representation of $L_1 \boxplus L_2$ with respect to S is

   $$\left[\begin{array}{ccc} \left[(L_1 \boxplus L_2)(\mathbf{e}_1^T) \right]_S & \left[(L_1 \boxplus L_2)(\mathbf{e}_2^T) \right]_S & \left[(L_1 \boxplus L_2)(\mathbf{e}_3^T) \right]_S \end{array} \right]$$

   $$= \left[\begin{array}{ccc} \left[\begin{array}{ccc} -1 & 3 & 4 \end{array} \right]_S & \left[\begin{array}{ccc} 4 & -1 & 3 \end{array} \right]_S & \left[\begin{array}{ccc} -1 & 3 & 5 \end{array} \right]_S \end{array} \right] = \left[\begin{array}{ccc} -1 & 4 & -1 \\ 3 & -1 & 3 \\ 4 & 3 & 5 \end{array} \right].$$

   (d) $(-2 \boxdot L_1) \left(\left[\begin{array}{ccc} u_1 & u_2 & u_3 \end{array} \right] \right) = -2L_1 \left(\left[\begin{array}{ccc} u_1 & u_2 & u_3 \end{array} \right] \right)$
   $= -2L_1(u_1\mathbf{e}_1^T + u_2\mathbf{e}_2^T + u_3\mathbf{e}_3^T)$
   $= -2u_1 L_1(\mathbf{e}_1^T) - 2u_2 L_1(\mathbf{e}_2^T) - 2u_3 L_1(\mathbf{e}_3^T)$
   $= -2u_1 \left[\begin{array}{ccc} -1 & 2 & 1 \end{array} \right] - 2u_2 \left[\begin{array}{ccc} 0 & 1 & 2 \end{array} \right] - 2u_3 \left[\begin{array}{ccc} -1 & 1 & 3 \end{array} \right]$
   $= \left[\begin{array}{ccc} 2u_1 + 2u_3 & -4u_1 - 2u_2 - 2u_3 & -2u_1 - 4u_2 - 6u_3 \end{array} \right]$

   (e) $((-2 \boxdot L_1) \boxplus (4 \boxdot L_2)) \left(\left[\begin{array}{ccc} 2 & 1 & -3 \end{array} \right] \right) = -2L_1 \left(\left[\begin{array}{ccc} 2 & 1 & -3 \end{array} \right] \right) + 4L_2 \left(\left[\begin{array}{ccc} 2 & 1 & -3 \end{array} \right] \right)$
   $= -2L_1(2\mathbf{e}_1^T + 1\mathbf{e}_2^T - 3\mathbf{e}_3^T) + 4L_2(2\mathbf{e}_1^T + 1\mathbf{e}_2^T - 3\mathbf{e}_3^T)$
   $= -4L_1(\mathbf{e}_1^T) - 2L_1(\mathbf{e}_2^T) + 6L_1(\mathbf{e}_3^T) + 8L_2(\mathbf{e}_1^T) + 4L_2(\mathbf{e}_2^T) - 12L_2(\mathbf{e}_3^T)$
   $= -4 \left[\begin{array}{ccc} -1 & 2 & 1 \end{array} \right] - 2 \left[\begin{array}{ccc} 0 & 1 & 2 \end{array} \right] + 6 \left[\begin{array}{ccc} -1 & 1 & 3 \end{array} \right]$
   $\quad + 8 \left[\begin{array}{ccc} 0 & 1 & 3 \end{array} \right] + 4 \left[\begin{array}{ccc} 4 & -2 & 1 \end{array} \right] - 12 \left[\begin{array}{ccc} 0 & 2 & 2 \end{array} \right]$
   $= \left[\begin{array}{ccc} 14 & -28 & 14 \end{array} \right]$

   (f) Since S is the natural basis, the representation of L_1 with respect to S is

   $$A = \left[\begin{array}{ccc} -1 & 0 & -1 \\ 2 & 1 & 1 \\ 1 & 2 & 3 \end{array} \right]$$

and the representation of L_2 with respect to S is

$$B = \begin{bmatrix} 0 & 4 & 0 \\ 1 & -2 & 2 \\ 3 & 1 & 2 \end{bmatrix}.$$

It follows that the representation of $(-2 \boxdot L_1) \boxplus (4 \boxdot L_2)$ with respect to S is

$$-2A + 4B = \begin{bmatrix} 2 & 16 & 2 \\ 0 & -10 & 6 \\ 10 & 0 & 2 \end{bmatrix}.$$

5. (a) For $\mathbf{x} = \begin{bmatrix} 3 & -2 & 1 \end{bmatrix}^T$, $(L_1 \boxplus L_2)(\mathbf{x}) = (A_1 + A_2)\mathbf{x} = \begin{bmatrix} 2 & 2 & 3 \\ 4 & 4 & 6 \\ 1 & 1 & -1 \end{bmatrix} \begin{bmatrix} 3 \\ -2 \\ 1 \end{bmatrix} = \begin{bmatrix} 5 \\ 10 \\ 0 \end{bmatrix}.$

(b) Ker L_1 consists of all solutions $\mathbf{x}$ of the homogeneous linear system $A_1\mathbf{x} = \mathbf{0}$ with coefficient matrix A_1. We form the reduced row echelon form of A_1 which is

$$\begin{bmatrix} 1 & 1 & 1 \\ 0 & 0 & 0 \\ 0 & 0 & 0 \end{bmatrix}.$$

Hence $\mathbf{x}$ is in ker L_1 only if $x_1 + x_2 + x_3 = 0$. That is, $x_1 = -r - s$, $x_2 = r$, $x_3 = s$, where r and s are any real numbers. Therefore

$$\mathbf{x} = \begin{bmatrix} -r - s \\ r \\ s \end{bmatrix} = r \begin{bmatrix} -1 \\ 1 \\ 0 \end{bmatrix} + s \begin{bmatrix} -1 \\ 0 \\ 1 \end{bmatrix}.$$

It follows that ker $L_1 = \text{span} \left\{ \begin{bmatrix} -1 & 1 & 0 \end{bmatrix}^T, \begin{bmatrix} -1 & 0 & 1 \end{bmatrix}^T \right\}$. Similarly, ker L_2 consists of all solutions $\mathbf{x}$ of the homogeneous linear system $A_2\mathbf{x} = \mathbf{0}$ with coefficient matrix A_2. We form the reduced row echelon form of A_2 which is

$$\begin{bmatrix} 1 & 1 & 2 \\ 0 & 0 & 0 \\ 0 & 0 & 0 \end{bmatrix}.$$

Hence $\mathbf{x}$ is in ker L_2 only if $x_1 + x_2 + 2x_3 = 0$. That is, $x_1 = -r - 2s$, $x_2 = r$, $x_3 = s$, where r and s are any real numbers. Therefore

$$\mathbf{x} = \begin{bmatrix} -r - 2s \\ r \\ s \end{bmatrix} = r \begin{bmatrix} -1 \\ 1 \\ 0 \end{bmatrix} + s \begin{bmatrix} -2 \\ 0 \\ 1 \end{bmatrix}.$$

It follows that ker $L_2 = \text{span} \left\{ \begin{bmatrix} -1 & 1 & 0 \end{bmatrix}^T, \begin{bmatrix} -2 & 0 & 1 \end{bmatrix}^T \right\}$. Thus

$$\text{ker } L_1 \cap \text{ker } L_2 = \text{span} \left\{ \begin{bmatrix} -1 & 1 & 0 \end{bmatrix}^T \right\}.$$

(c) To determined ker $(L_1 \boxplus L_2)$ we determine the general solution of the homogeneous linear system $(A_1 + A_2)\mathbf{x} = \mathbf{0}$. The reduced row echelon form of $A_1 + A_2$ is

$$\begin{bmatrix} 1 & 1 & 0 \\ 0 & 0 & 1 \\ 0 & 0 & 0 \end{bmatrix}.$$

Thus $x_1 = -r$, $x_2 = r$, $x_3 = 0$, where r is any real number. Then the general solution consists of all vectors of the form

$$r \begin{bmatrix} -1 \\ 1 \\ 0 \end{bmatrix}.$$

Hence $\ker (L_1 \boxplus L_2) = \text{span} \left\{ \begin{bmatrix} -1 & 1 & 0 \end{bmatrix}^T \right\}$.

(d) They are the same: $\ker (L_1 \boxplus L_2) = \ker L_1 \cap \ker L_2$

7. (a) We compute the images of the vectors in P using linear transformation $L_2 \circ L_1$:

$$(L_2 \circ L_1)(t+1) = L_2(L_1(t+1)) = L_2(t^2 + t) = t^2(2t + 1) = 2t^3 + t^2$$

$$(L_2 \circ L_1)(t-1) = L_2(L_1(t-1)) = L_2(t^2 - t) = t^2(2t - 1) = 2t^3 - t^2$$

Next we find the coordinates of the images relative to the T-basis. We have

$$a_1 t^3 + a_2(t^2 - 1) + a_3 t + a_4(t + 1) = 2t^3 + t^2$$

and upon collecting terms and equating coefficients of the like powers of t, the result is the linear system

$$a_1 = 2, \quad a_2 = 1, \quad a_3 + a_4 = 0, \quad -a_2 + a_4 = 0.$$

The solution is $a_1 = 2$, $a_2 = 1$, $a_3 = -1$, $a_4 = 1$ so

$$\begin{bmatrix} (L_2 \circ L_1)(t+1) \end{bmatrix}_T = \begin{bmatrix} 2t^3 + t^2 \end{bmatrix}_T = \begin{bmatrix} 2 & 1 & -1 & 1 \end{bmatrix}^T.$$

In a similar calculation we find that

$$\begin{bmatrix} (L_2 \circ L_1)(t-1) \end{bmatrix}_T = \begin{bmatrix} 2t^3 - t^2 \end{bmatrix}_T = \begin{bmatrix} 2 & -1 & 1 & -1 \end{bmatrix}^T.$$

Thus the matrix representing $L_2 \circ L_1$ is

$$C = \begin{bmatrix} 2 & 2 \\ 1 & -1 \\ -1 & 1 \\ 1 & -1 \end{bmatrix}.$$

(b) Following the same outline of steps as in part (a) we have

$$L_1(t+1) = t^2 + t \implies \begin{bmatrix} L_1(t+1) \end{bmatrix}_S = \begin{bmatrix} 1 & \frac{2}{3} & \frac{1}{3} \end{bmatrix}^T$$

$$L_1(t-1) = t^2 - t \implies \begin{bmatrix} L_1(t-1) \end{bmatrix}_S = \begin{bmatrix} 1 & -\frac{2}{3} & -\frac{1}{3} \end{bmatrix}^T.$$

Hence

$$A = \begin{bmatrix} 1 & 1 \\ \frac{2}{3} & -\frac{2}{3} \\ \frac{1}{3} & -\frac{1}{3} \end{bmatrix}.$$

In a similar fashion,

$$L_1(t+1) = t^2 + t \implies \begin{bmatrix} L_1(t+1) \end{bmatrix}_S = \begin{bmatrix} 1 & \frac{2}{3} & \frac{1}{3} \end{bmatrix}^T$$

$$L_1(t-1) = t^2 - t \implies \begin{bmatrix} L_1(t-1) \end{bmatrix}_S = \begin{bmatrix} 1 & -\frac{2}{3} & -\frac{1}{3} \end{bmatrix}^T$$

Hence

$$B = \begin{bmatrix} 2 & 0 & 0 \\ 0 & 1 & 1 \\ 0 & -1 & -1 \\ 0 & 1 & 1 \end{bmatrix} \quad \text{and therefore} \quad BA = \begin{bmatrix} 2 & 2 \\ 1 & -1 \\ -1 & 1 \\ 1 & -1 \end{bmatrix} = C.$$

9. Let $A = \begin{bmatrix} 1 & 4 & -1 \\ 2 & 1 & 3 \\ 1 & -1 & 2 \end{bmatrix}$.

    (a) The representation of $2 \square L$ is $2A = \begin{bmatrix} 2 & 8 & -2 \\ 4 & 2 & 6 \\ 2 & -2 & 4 \end{bmatrix}$.

    (b) The representation of $2 \square L \boxplus L \circ L$ is $2A + A^2 = \begin{bmatrix} 10 & 17 & 7 \\ 11 & 8 & 13 \\ 3 & -1 & 4 \end{bmatrix}$.

11. Use Theorem 6.9.

    (a) $\dim U = \dim M_{32} = 2 \times 3 = 6$

    (b) $\dim U = \dim M_{23} = 2 \times 3 = 6$

    (c) $\dim U = \dim M_{62} = 6 \times 2 = 12$

    (d) $\dim U = \dim M_{43} = 4 \times 3 = 12$

13. (a) We verify that L is a linear transformation by showing that Exercise 6 in Section 6.1 holds. Let $\mathbf{y} = k_1\mathbf{v}_1 + k_2\mathbf{v}_2 + \cdots + k_n\mathbf{v}_n$ and a and b be scalars. Then

$$L(a\mathbf{x} + b\mathbf{y}) = L\left[\sum_{i=1}^{n}(ac_i + bk_i)\mathbf{v}_i\right] \qquad \text{(by vector operations in } W\text{)}$$

$$= \left[\sum_{i=1}^{n}(ac_i + bk_i)L(\mathbf{v}_i)\right] \qquad \text{(definition of } L\text{)}$$

$$= \sum_{i=1}^{n}ac_iL(\mathbf{v}_i) + \sum_{i=1}^{n}bk_iL(\mathbf{v}_i) \qquad \text{(by vector operations)}$$

$$= a\sum_{i=1}^{n}c_iL(\mathbf{v}_i) + b\sum_{i=1}^{n}k_iL(\mathbf{v}_i) \qquad \text{(by vector operations)}$$

$$= aL(\mathbf{x}) + bL(\mathbf{y}) \qquad \text{(definition of } L\text{)}$$

    Thus L is a linear transformation.

    (b) The matrix representation of L with respect to S and T is

$$\begin{bmatrix} [L(\mathbf{v}_1)]_T & \cdots & [L(\mathbf{v}_n)]_T \end{bmatrix} = A$$

since $L(\mathbf{v}_i) = \sum_{k=1}^{m}a_{ki}\mathbf{w}_k$.

15. We are given that

$$A = \begin{bmatrix} 1 & 2 & -2 \\ 3 & 4 & -1 \end{bmatrix}$$

is the matrix representing L with respect to the bases $S = \{t^2, t, 1\}$ of P_2 and $T = \{t, 1\}$ of P_1. Hence we have

$$[L(t^2)]_T = \begin{bmatrix} 1 \\ 3 \end{bmatrix} \quad \Longrightarrow \quad L(t^2) = t + 3$$

$$[L(t)]_T = \begin{bmatrix} 2 \\ 4 \end{bmatrix} \quad \Longrightarrow \quad L(t) = 2t + 4$$

$$[L(1)]_T = \begin{bmatrix} -2 \\ -1 \end{bmatrix} \quad \Longrightarrow \quad L(1) = -2t - 1$$

(a) A linear transformation is specified by its action on a basis. That action is given above.

(b) $L(at^2+bt+c) = aL(t^2)+bL(t)+cL(1) = a(t+3)+b(2t+4)+c(-2t-1) = (a+2b-2c)t+(3a+4b-c)$

(c) In part (b) set $a = 2$, $b = -5$, $c = 4$ to obtain $L(2t^2 - 5t + 4) = -16t - 18$.

17. One possible answer is

$$L\left(\begin{bmatrix} u_1 \\ u_2 \end{bmatrix}\right) = \begin{bmatrix} u_2 \\ u_1 \end{bmatrix}.$$

That is, L interchanges the elements in the 2×1 matrix. Certainly $L \circ L = I$. Another answer is

$$L\left(\begin{bmatrix} u_1 \\ u_2 \end{bmatrix}\right) = (-1)\begin{bmatrix} u_1 \\ u_2 \end{bmatrix} = \begin{bmatrix} -u_1 \\ -u_2 \end{bmatrix}.$$

19. One possible answer is

$$L\left(\begin{bmatrix} u_1 \\ u_2 \end{bmatrix}\right) = \begin{bmatrix} u_1 \\ 0 \end{bmatrix}.$$

That is, L is the projection onto the x-axis. Certainly $L \circ L = I$. Another answer is

$$L\left(\begin{bmatrix} u_1 \\ u_2 \end{bmatrix}\right) = \begin{bmatrix} \frac{1}{2}(u_1 + u_2) \\ \frac{1}{2}(u_1 + u_2) \end{bmatrix}.$$

21. Just determine the inverse of $A = \begin{bmatrix} 1 & 1 & 1 \\ 0 & 1 & 2 \\ 1 & 2 & 2 \end{bmatrix}$. The matrix representing L^{-1} is $\begin{bmatrix} 2 & 0 & -1 \\ -2 & -1 & 2 \\ 1 & 1 & -1 \end{bmatrix}$.

23. The fact that A^2 represents L^2 follows directly from Theorem 6.11. Since $L^3 = L \circ L^2$, it follows from Theorem 6.11 that A^3 represents L^3. We continue this argument as long as necessary. A more formal proof can be given using induction.

Section 6.5, p. 413

1. (a) Since $A = I_n^{-1}AI_n$, A is similar to A.

    (b) If B is similar to A, then $B = P^{-1}AP$ for some matrix P. Then $A = PBP^{-1}$. Let $P^{-1} = Q$ so that $A = Q^{-1}BQ$. Therefore A is similar to B.

    (c) If C is similar to B and B is similar to A, then $C = P^{-1}BP$ and $B = Q^{-1}AQ$ for some matrices P and Q. Letting $M = QP$, it follows that $C = P^{-1}Q^{-1}AQP = M^{-1}AM$. Therefore C is similar to A.

3. From Exercise 1 in Section 6.3,

$$L\left(\begin{bmatrix} u_1 \\ u_2 \end{bmatrix}\right) = \begin{bmatrix} u_1 + 2u_2 \\ 2u_1 - u_2 \end{bmatrix}$$

with S the natural basis for R^2 and

$$T = \left\{\begin{bmatrix} -1 \\ 2 \end{bmatrix}, \begin{bmatrix} 2 \\ 0 \end{bmatrix}\right\}.$$

From part (a) the representation of L with respect to S is $A = \begin{bmatrix} 1 & 2 \\ 2 & -1 \end{bmatrix}$. Since S is the natural basis, the transition matrix from T to S is $P = \begin{bmatrix} -1 & 2 \\ 2 & 0 \end{bmatrix}$. Then we can compute $P^{-1} = \begin{bmatrix} 0 & \frac{1}{2} \\ \frac{1}{2} & \frac{1}{4} \end{bmatrix}$. Thus the representation of L with respect to T is $B = P^{-1}AP = \begin{bmatrix} -2 & 2 \\ \frac{1}{2} & 2 \end{bmatrix}$.

5. Let $L : R^2 \to R^2$ be given by $L\left(\begin{bmatrix} u_1 \\ u_2 \end{bmatrix}\right) = \begin{bmatrix} u_1 \\ -u_2 \end{bmatrix}$.

(a) For $S = \{e_1, e_2\}$, the natural basis for R^2,

$$L(e_1) = \begin{bmatrix} 1 \\ 0 \end{bmatrix} \quad \text{and} \quad L(e_2) = \begin{bmatrix} 0 \\ -1 \end{bmatrix}.$$

The coordinates of the images of the natural basis vectors relative to the natural basis are just the components of the image. Hence the representation of L with respect to the natural basis is

$$A = \begin{bmatrix} 1 & 0 \\ 0 & -1 \end{bmatrix}.$$

(b) Let $T = \{v_1, v_2\} = \left\{ \begin{bmatrix} 1 \\ -1 \end{bmatrix}, \begin{bmatrix} 1 \\ 2 \end{bmatrix} \right\}$ be a basis for R^2. Then

$$L(v_1) = \begin{bmatrix} 1 \\ 1 \end{bmatrix} \quad \text{and} \quad L(v_2) = \begin{bmatrix} 1 \\ -2 \end{bmatrix}.$$

Next we determine $\begin{bmatrix} L(v_1) \end{bmatrix}_T$ and $\begin{bmatrix} L(v_2) \end{bmatrix}_T$ by row reducing the partitioned matrix

$$\begin{bmatrix} 1 & 1 & 1 & 1 \\ -1 & 2 & 1 & -2 \end{bmatrix}.$$

We find that

$$\begin{bmatrix} L(v_1) \end{bmatrix}_T = \begin{bmatrix} \frac{1}{3} \\ \frac{2}{3} \end{bmatrix} \quad \text{and} \quad \begin{bmatrix} L(v_2) \end{bmatrix}_T = \begin{bmatrix} \frac{4}{3} \\ -\frac{1}{3} \end{bmatrix}.$$

Hence the matrix representing L with respect to T is

$$B = \begin{bmatrix} \frac{1}{3} & \frac{4}{3} \\ \frac{2}{3} & -\frac{1}{3} \end{bmatrix}.$$

(c) Since A and B are the representations of the same linear transformation with respect to different bases they are similar by Corollary 6.3. In fact Corollary 6.3 tells us that $B = P^{-1}AP$ where P is the transition matrix from T to S. Since S is the natural basis, it follows that

$$P = \begin{bmatrix} 1 & 1 \\ -1 & 2 \end{bmatrix}$$

and we can show that

$$P^{-1} = \begin{bmatrix} \frac{2}{3} & -\frac{1}{3} \\ \frac{1}{3} & \frac{1}{3} \end{bmatrix}.$$

We can verify that $P^{-1}AP = B$ so that A and B are shown directly to be similar.

(d) We compute the reduced row echelon form of both A and B. In each case the result is I_2, so they both have rank 2.

7. Since A and B are similar, there is a nonsingular matrix P such that $B = P^{-1}AP$. Taking the transpose of both sides, we obtain

$$B^T = (P^{-1}AP)^T = P^T A^T (P^{-1})^T = P^T A^T (P^T)^{-1}.$$

Since P is nonsingular, so is P^T. Letting $Q^{-1} = P^T$, then we have $B^T = Q^{-1}A^T Q$ and it follows that A^T and B^T are similar.

9. Use Theorem 6.11. Let S' be the natural basis in R_3 and T' be the natural basis in R_2. Since S' is the natural basis, it is easy to find P the transition matrix from S to S':

$$P = \begin{bmatrix} 1 & 0 & 1 \\ 0 & 2 & 2 \\ -1 & 0 & 3 \end{bmatrix}.$$

Similarly since T' is the natural basis it is easy to find Q the transition matrix from T to T':

$$Q = \begin{bmatrix} 1 & 2 \\ -1 & 0 \end{bmatrix}.$$

The following diagram depicts our strategy for this problem.

$$
\begin{array}{ccc}
 & \text{(find matrix } B) & \\
(S'\text{-base}) \quad R^2 & \longrightarrow & R^2 \quad (T'\text{-base}) \\
\Big\downarrow \begin{array}{l}\text{(move from } S' \\ \text{to } S \text{ using } P^{-1})\end{array} & & \Big\uparrow \begin{array}{l}\text{(move from } T \\ \text{to } T' \text{ using } Q)\end{array} \\
(S\text{-base}) \quad R^2 & \longrightarrow & R^2 \quad (T\text{-base}) \\
 & \text{(given matrix } A) &
\end{array}
$$

It follows that $B = QAP^{-1}$. To compute B we need only find P^{-1} and perform the multiplication. We obtain

$$P^{-1} = \begin{bmatrix} \frac{3}{4} & 0 & -\frac{1}{4} \\ -\frac{1}{4} & \frac{1}{2} & -\frac{1}{4} \\ \frac{1}{4} & 0 & \frac{1}{4} \end{bmatrix} \quad \text{and then} \quad B = \begin{bmatrix} \frac{13}{2} & \frac{1}{2} & -\frac{3}{2} \\ -\frac{5}{2} & \frac{1}{2} & -\frac{1}{2} \end{bmatrix}.$$

11. (a) Since A and B are similar, $B = P^{-1}AP$ for some nonsingular matrix P. Therefore B is a product of nonsingular matrices and hence is nonsingular.

(b) By part (a), $B = P^{-1}AP$. It follows that $B^{-1} = (P^{-1}AP)^{-1} = P^{-1}A^{-1}P$. Therefore A^{-1} and B^{-1} are similar.

13. Let $I : R_2 \to R_2$ be the identity linear operator,

$$S = \{ \begin{bmatrix} 1 & 0 \end{bmatrix}, \begin{bmatrix} 0 & 1 \end{bmatrix} \} \quad \text{and} \quad T = \{ \begin{bmatrix} 1 & -1 \end{bmatrix}, \begin{bmatrix} 2 & 3 \end{bmatrix} \}.$$

Since S is the natural basis for R_2, the representation of I with respect to S is just I_2, the 2×2 identity matrix. To find the matrix B representing I with respect to T using the techniques of this section we proceed as indicated in the following diagram.

$$
\begin{array}{ccc}
 & \text{(find matrix } B) & \\
(T\text{-basis}) \quad R^2 & \longrightarrow & R^2 \quad (T\text{-basis}) \\
\Big\downarrow \begin{array}{l}\text{(move from } T \\ \text{to } S \text{ using } P)\end{array} & & \Big\uparrow \begin{array}{l}\text{(move from } S \\ \text{to } T \text{ using } P^{-1})\end{array} \\
(S\text{-basis}) \quad R^2 & \longrightarrow & R^2 \quad (S\text{-basis}) \\
 & \text{(given matrix } I_2) &
\end{array}
$$

$$B = \begin{bmatrix} 1 & 0 \\ 0 & 1 \end{bmatrix}.$$

15. Let $L : V \to V$ be given by $L(f) = f'$, $S = \{\sin t, \cos t\}$, and $T = \{\sin t - \cos t, \sin t + \cos t\}$. The matrix representing L with respect to S is given by

$$A = \begin{bmatrix} 0 & -1 \\ 1 & 0 \end{bmatrix}.$$

(See Exercise 20(a) in Section 6.3) To find the matrix B representing L with respect to T we use the diagram given in Exercise 13 and construct the transition matrix P from T to S. It follows that

$$P = \left[\ \left[\ L(\sin t - \cos t)\ \right]_S\quad \left[\ L(\sin t + \cos t)\ \right]_S\ \right]$$

$$= \left[\ \left[\ \cos t + \sin t\ \right]_S\quad \left[\ \cos t - \sin t\ \right]_S\ \right] = \begin{bmatrix} 1 & -1 \\ 1 & 1 \end{bmatrix}.$$

It follows that $P^{-1} = \begin{bmatrix} \frac{1}{2} & \frac{1}{2} \\ -\frac{1}{2} & \frac{1}{2} \end{bmatrix}$ and then $B = P^{-1}AP = \begin{bmatrix} 0 & -1 \\ 1 & 0 \end{bmatrix}.$

17. Let $B = P^{-1}AP$. Then $\det(B) = \det(P^{-1}AP) = \det(P)^{-1}\det(A)\det(P) = \det(A)$.

Section 6.6, p. 425

1. (a) $M = RT = \begin{bmatrix} \cos(\pi/4) & -\sin(\pi/4) & 0 \\ \sin(\pi/4) & \cos(\pi/4) & 0 \\ 0 & 0 & 1 \end{bmatrix} \begin{bmatrix} 1 & 0 & 1 \\ 0 & 1 & 1 \\ 0 & 0 & 1 \end{bmatrix}$

$$= \begin{bmatrix} \sqrt{2}/2 & -\sqrt{2}/2 & 0 \\ \sqrt{2}/2 & \sqrt{2}/2 & 0 \\ 0 & 0 & 1 \end{bmatrix} \begin{bmatrix} 1 & 0 & 1 \\ 0 & 1 & 1 \\ 0 & 0 & 1 \end{bmatrix}$$

$$= \begin{bmatrix} \sqrt{2}/2 & -\sqrt{2}/2 & 0 \\ \sqrt{2}/2 & \sqrt{2}/2 & \sqrt{2} \\ 0 & 0 & 1 \end{bmatrix}.$$

(b)

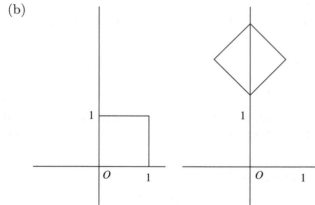

(c) No.

3. (a) The vertices are $(0,0)$, $(2,3)$, and $(4,4)$.

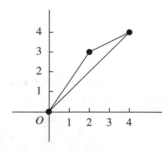

(b) $M = \begin{bmatrix} \cos(\pi/2) & -sin(\pi/2) & 0 \\ \sin(\pi/2) & \cos(\pi/2) & 0 \\ 0 & 0 & 1 \end{bmatrix} \begin{bmatrix} \cos(\pi/6) & -sin(\pi/6) & 0 \\ \sin(\pi/6) & \cos(\pi/6) & 0 \\ 0 & 0 & 1 \end{bmatrix}$

$= \begin{bmatrix} 0 & -1 & 0 \\ 1 & 0 & 0 \\ 0 & 0 & 1 \end{bmatrix} \begin{bmatrix} \sqrt{3}/2 & -1/2 & 0 \\ 1/2 & \sqrt{3}/2 & 0 \\ 0 & 0 & 1 \end{bmatrix} = \begin{bmatrix} -1/2 & -\sqrt{3}/2 & 0 \\ \sqrt{3}/2 & -1/2 & 0 \\ 0 & 0 & 1 \end{bmatrix}.$

(c)

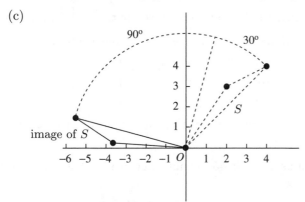

(d) $Q = RT = \begin{bmatrix} \cos(\pi/6) & -sin(\pi/6) & 0 \\ \sin(\pi/6) & \cos(\pi/6) & 0 \\ 0 & 0 & 1 \end{bmatrix} \begin{bmatrix} \cos(\pi/2) & -sin(\pi/2) & 0 \\ \sin(\pi/2) & \cos(\pi/2) & 0 \\ 0 & 0 & 1 \end{bmatrix}$

$= \begin{bmatrix} \sqrt{3}/2 & -1/2 & 0 \\ 1/2 & \sqrt{3}/2 & 0 \\ 0 & 0 & 1 \end{bmatrix} \begin{bmatrix} 0 & -1 & 0 \\ 1 & 0 & 0 \\ 0 & 0 & 1 \end{bmatrix} = \begin{bmatrix} -1/2 & -\sqrt{3}/2 & 0 \\ \sqrt{3}/2 & -1/2 & 0 \\ 0 & 0 & 1 \end{bmatrix}.$

(e)

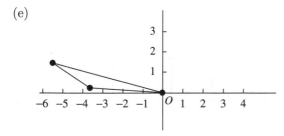

(f) Yes. They are the same since the product of the matrices involving the two rotations are the same for both orders of multiplication.

5. We have

$$A = \begin{bmatrix} 1 & 0 & 0 \\ 0 & -1 & 0 \\ 0 & 0 & 1 \end{bmatrix} \quad \text{and} \quad B = \begin{bmatrix} 1 & 0 & 4 \\ 0 & 1 & -2 \\ 0 & 0 & 1 \end{bmatrix}.$$

then

$$AB = \begin{bmatrix} 1 & 0 & 0 \\ 0 & -1 & 0 \\ 0 & 0 & 1 \end{bmatrix} \begin{bmatrix} 1 & 0 & 4 \\ 0 & 1 & -2 \\ 0 & 0 & 1 \end{bmatrix} = \begin{bmatrix} 1 & 0 & 4 \\ 0 & -1 & 2 \\ 0 & 0 & 1 \end{bmatrix}$$

$$BA = \begin{bmatrix} 1 & 0 & 4 \\ 0 & 1 & -2 \\ 0 & 0 & 1 \end{bmatrix} \begin{bmatrix} 1 & 0 & 0 \\ 0 & -1 & 0 \\ 0 & 0 & 1 \end{bmatrix} = \begin{bmatrix} 1 & 0 & 4 \\ 0 & -1 & -2 \\ 0 & 0 & 1 \end{bmatrix}.$$

Since $BA \neq AB$, the images will not be the same.

7. Observe that the image rectangle has half the width and half the height of the original rectangle and is reflected about the y-axis. Thus, the matrix M that performs these operations is given by

$$M = \begin{bmatrix} \frac{1}{2} & 0 & 0 \\ 0 & \frac{1}{2} & 0 \\ 0 & 0 & 1 \end{bmatrix} \begin{bmatrix} -1 & 0 & 0 \\ 0 & 1 & 0 \\ 0 & 0 & 1 \end{bmatrix} = \begin{bmatrix} -\frac{1}{2} & 0 & 0 \\ 0 & \frac{1}{2} & 0 \\ 0 & 0 & 1 \end{bmatrix}.$$

9. Observe that the image semicircle is translated to the right 1 unit so that $\mathbf{t} = \begin{bmatrix} 1 \\ 0 \end{bmatrix}$, and is then reflected about the x-axis. Thus, the matrix M that performs these operations is given by

$$M = \begin{bmatrix} 1 & 0 & 0 \\ 0 & -1 & 0 \\ 0 & 0 & 1 \end{bmatrix} \begin{bmatrix} 1 & 0 & 1 \\ 0 & -1 & 1 \\ 0 & 0 & 1 \end{bmatrix} = \begin{bmatrix} 1 & 0 & 1 \\ 0 & -1 & -1 \\ 0 & 0 & 1 \end{bmatrix}.$$

Note that another solution is to first reflect the semicircle about the x-axis and then translate it to the right 1 unit.

11. The image semicircle is the original semicircle rotated $45°$ counterclockwise. The matrix M that performs this operation is

$$M = \begin{bmatrix} \cos(\pi/4) & -\sin(\pi/4) & 0 \\ \sin(\pi/45) & \cos(\pi/4) & 0 \\ 0 & 0 & 1 \end{bmatrix} = \begin{bmatrix} \frac{1}{2}\sqrt{2} & -\frac{1}{2}\sqrt{2} & 0 \\ \frac{1}{2}\sqrt{2} & \frac{1}{2}\sqrt{2} & 0 \\ 0 & 0 & 1 \end{bmatrix}.$$

13. (a)

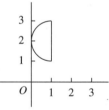

(b) $$M = \begin{bmatrix} 1 & 0 & 1 \\ 0 & 1 & 1 \\ 0 & 0 & 1 \end{bmatrix} \begin{bmatrix} \cos(\pi/2) & -\sin(\pi/2) & 0 \\ \sin(\pi/2) & \cos(\pi/2) & 0 \\ 0 & 0 & 1 \end{bmatrix} \begin{bmatrix} 1 & 0 & -1 \\ 0 & 1 & -1 \\ 0 & 0 & 1 \end{bmatrix}.$$

$$= \begin{bmatrix} 1 & 0 & 1 \\ 0 & 1 & 1 \\ 0 & 0 & 1 \end{bmatrix} \begin{bmatrix} 0 & -1 & 0 \\ 1 & 0 & 0 \\ 0 & 0 & 1 \end{bmatrix} \begin{bmatrix} 1 & 0 & -1 \\ 0 & 1 & -1 \\ 0 & 0 & 1 \end{bmatrix}$$

$$= \begin{bmatrix} 0 & -1 & 2 \\ 1 & 0 & 0 \\ 0 & 0 & 1 \end{bmatrix}$$

(c) $M \begin{bmatrix} 1 \\ 1 \\ 1 \end{bmatrix} = \begin{bmatrix} 0 & -1 & 2 \\ 1 & 0 & 0 \\ 0 & 0 & 1 \end{bmatrix} \begin{bmatrix} 1 \\ 1 \\ 1 \end{bmatrix} = \begin{bmatrix} 1 \\ 1 \\ 1 \end{bmatrix}.$

$M \begin{bmatrix} 3 \\ 1 \\ 1 \end{bmatrix} = \begin{bmatrix} 0 & -1 & 2 \\ 1 & 0 & 0 \\ 0 & 0 & 1 \end{bmatrix} \begin{bmatrix} 3 \\ 1 \\ 1 \end{bmatrix} = \begin{bmatrix} 1 \\ 3 \\ 1 \end{bmatrix}.$

15. (a) We form a composition of a rotation about the z-axis by $\theta_j = s_j\theta$ and translation by the vector in homogeneous form given by

$$\begin{bmatrix} 0 \\ 0 \\ k\theta_j \\ 1 \end{bmatrix},$$

where $k = -1$. We obtain

$$
\begin{bmatrix} \cos\theta_j & -\sin\theta_j & 0 & 0 \\ \sin\theta_j & \cos\theta_j & 0 & 0 \\ 0 & 0 & 1 & 0 \\ 0 & 0 & 0 & 1 \end{bmatrix}
\begin{bmatrix} 1 & 0 & 0 & 0 \\ 0 & 1 & 0 & 0 \\ 0 & 0 & 1 & k\theta_j \\ 0 & 0 & 0 & 1 \end{bmatrix}
=
\begin{bmatrix} \cos\theta_j & -\sin\theta_j & 0 & 0 \\ \sin\theta_j & \cos\theta_j & 0 & 0 \\ 0 & 0 & 1 & k\theta_j \\ 0 & 0 & 0 & 1 \end{bmatrix}.
$$

(b) It appears that the length of the vector being "screwed" is decreasing as we move down the z-axis. Thus a scaling in the x- and y-direction by a factor smaller than 1 is included.

Supplementary Exercises for Chapter 6, p. 430

1. From Exercise 43 in Section 1.3 we have that $\text{Tr}(A + B) = \text{Tr}(A) + \text{Tr}(B)$ and $\text{Tr}(cA) = c\,\text{Tr}(A)$. Thus Definition 6.1 is satisfied and it follows that the trace of A is a linear transformation.

3. L is not a linear transformation; condition (a) of Definition 6.1 is not satisfied. To show this, let

$$
A = \begin{bmatrix} 1 & 0 \\ 0 & 0 \end{bmatrix} \quad \text{and} \quad B = \begin{bmatrix} 0 & 0 \\ 0 & 1 \end{bmatrix}.
$$

Then $A + B = I_2$ so that $L(A + B) = I_2^{-1} = I_2$. But A and B are singular so that $L(A) = L(B) = O$. Therefore $L(A) + L(B) = O + O = O \neq L(A + B)$. Thus L is not a linear transformation.

5. (a) Since $5t + 1 = 2(t - 1) + 3(t + 1)$ and since L is a linear transformation, we have

$$
\begin{aligned}
L(5l + 1) &= L(2(t - 1) + 3(t + 1)) = 2L(t - 1) + 3L(t - 1) \\
&= 2(t + 2) + 3(2t + 1) = 8t + 7
\end{aligned}
$$

(b) Proceeding as in part (a) we write $at + b$ as a linear combination of $t - 1$ and $t + 1$:

$$
at + b = c_1(t - 1) + c_2(t + 1).
$$

Expanding the right side and collecting terms gives $at + b = (c_1 + c_2)t + (c_2 - c_1)$. Equating coefficients of like terms leads to the linear system $a = c_1 + c_2$, $b = c_2 - c_1$ whose solution is

$$
c_1 = \frac{a - b}{2} \quad \text{and} \quad c_2 = \frac{a + b}{2}.
$$

Thus

$$
L(at + b) = \frac{a - b}{2} L(t - 1) + \frac{a + b}{2} L(t + 1) = \frac{3a + b}{2} t + \frac{3a - b}{2}.
$$

7. Let $L : P_3 \to P_3$ be the linear transformation given by $L(at^3 + bt^2 + ct + d) = (a - b)t^3 + (c - d)t$.

(a) $L(t^3 + t^2 + t - 1) = (1 - 1)t^3 + (1 - (-1))t = 2t \neq 0$ so $t^3 + t^2 + t - 1$ is not in ker L.

(b) $L(t^3 - t^2 + t - 1) = (1 - (-1))t^3 + (1 - (-1))t = 2t^3 + 2t \neq 0$ so $t^3 - t^2 + t - 1$ is not in ker L.

(c) Vector $3t^3 + t$ is in range L only if there is some polynomial $at^3 + bt^2 + ct + d$ such that

$$
L(at^3 + bt^2 + ct + d) = 3t^3 + t.
$$

But

$$
L(at^3 + bt^2 + ct + d) = (a - b)t^3 + (c - d)t = 3t^3 + t
$$

only if $a - b = 3$ and $c - d = 1$. For $a = 5$, $b = 2$, $c = 1$, $d = 0$ these relations are satisfied, hence $3t^3 + t$ is in range L. (The choices for a, b, c, and d are not unique. There are many vectors whose image is $3t^3 + t$.)

(d) Vector $3t^3 - t^2$ is in range L only if there is some polynomial $at^3 + bt^2 + ct + d$ such that

$$L(at^3 + bt^2 + ct + d) = 3t^3 - t^2.$$

But

$$L(at^3 + bt^2 + ct + d) = (a - b)t^3 + (c - d)t \neq 3t^3 - t^2$$

since no choice of a, b, c, d gives an image containing a t^2 term. Thus $3t^3 - t^2$ is not in range L.

(e) $L(at^3 + bt^2 + ct + d) = 0 \iff (a - b)t^3 + (c - d)t = 0 \iff a - b = 0,\ c - d = 0$.
The general solution of this linear system is $a = b = r$, $c = d = s$ which in vector form is

$$r\begin{bmatrix} 1 \\ 1 \\ 0 \\ 0 \end{bmatrix} + s\begin{bmatrix} 0 \\ 0 \\ 1 \\ 1 \end{bmatrix}.$$

The preceding column vectors represent coordinates of vectors $t^3 + t^2$ and $t + 1$, respectively, which form a basis for ker L.

(f) Any vector in range L is a linear combination of t^2 and t, hence these vectors span range L. Since they are different powers of t it follows that they are also linearly independent. Thus $\{t^3, t\}$ is a basis for range L.

9. $L : M_{22} \to M_{22}$ is given by $L(A) = A^T$.

(a) To find the matrix of L with respect to S we compute the images of the base vectors in S under L and write them as linear combinations of the vectors in S:

$$L\left(\begin{bmatrix} 1 & 0 \\ 0 & 0 \end{bmatrix}\right) = \begin{bmatrix} 1 & 0 \\ 0 & 0 \end{bmatrix}^T = \begin{bmatrix} 1 & 0 \\ 0 & 0 \end{bmatrix} = 1\begin{bmatrix} 1 & 0 \\ 0 & 0 \end{bmatrix} + 0\begin{bmatrix} 0 & 1 \\ 0 & 0 \end{bmatrix} + 0\begin{bmatrix} 0 & 0 \\ 1 & 0 \end{bmatrix} + 0\begin{bmatrix} 0 & 0 \\ 0 & 1 \end{bmatrix}$$

$$L\left(\begin{bmatrix} 0 & 1 \\ 0 & 0 \end{bmatrix}\right) = \begin{bmatrix} 0 & 1 \\ 0 & 0 \end{bmatrix}^T = \begin{bmatrix} 0 & 0 \\ 1 & 0 \end{bmatrix} = 0\begin{bmatrix} 1 & 0 \\ 0 & 0 \end{bmatrix} + 0\begin{bmatrix} 0 & 1 \\ 0 & 0 \end{bmatrix} + 1\begin{bmatrix} 0 & 0 \\ 1 & 0 \end{bmatrix} + 0\begin{bmatrix} 0 & 0 \\ 0 & 1 \end{bmatrix}$$

$$L\left(\begin{bmatrix} 0 & 0 \\ 1 & 0 \end{bmatrix}\right) = \begin{bmatrix} 0 & 0 \\ 1 & 0 \end{bmatrix}^T = \begin{bmatrix} 0 & 1 \\ 0 & 0 \end{bmatrix} = 0\begin{bmatrix} 1 & 0 \\ 0 & 0 \end{bmatrix} + 1\begin{bmatrix} 0 & 1 \\ 0 & 0 \end{bmatrix} + 0\begin{bmatrix} 0 & 0 \\ 1 & 0 \end{bmatrix} + 0\begin{bmatrix} 0 & 0 \\ 0 & 1 \end{bmatrix}$$

$$L\left(\begin{bmatrix} 0 & 0 \\ 0 & 1 \end{bmatrix}\right) = \begin{bmatrix} 0 & 0 \\ 0 & 1 \end{bmatrix}^T = \begin{bmatrix} 0 & 0 \\ 0 & 1 \end{bmatrix} = 0\begin{bmatrix} 1 & 0 \\ 0 & 0 \end{bmatrix} + 0\begin{bmatrix} 0 & 1 \\ 0 & 0 \end{bmatrix} + 0\begin{bmatrix} 0 & 0 \\ 1 & 0 \end{bmatrix} + 1\begin{bmatrix} 0 & 0 \\ 0 & 1 \end{bmatrix}$$

Therefore the matrix of L with respect to S is $\begin{bmatrix} 1 & 0 & 0 & 0 \\ 0 & 0 & 1 & 0 \\ 0 & 1 & 0 & 0 \\ 0 & 0 & 0 & 1 \end{bmatrix}$.

(b) Find the image of each vector in S and write it as a linear combination of vectors in T:

$$L\left(\begin{bmatrix} 1 & 0 \\ 0 & 0 \end{bmatrix}\right) = \begin{bmatrix} 1 & 0 \\ 0 & 0 \end{bmatrix} = 1\begin{bmatrix} 1 & 1 \\ 0 & 0 \end{bmatrix} + (-1)\begin{bmatrix} 0 & 1 \\ 0 & 0 \end{bmatrix} + 0\begin{bmatrix} 0 & 0 \\ 1 & 1 \end{bmatrix} + 0\begin{bmatrix} 1 & 0 \\ 0 & 1 \end{bmatrix}$$

$$L\left(\begin{bmatrix} 0 & 1 \\ 0 & 0 \end{bmatrix}\right) = \begin{bmatrix} 0 & 0 \\ 1 & 0 \end{bmatrix} = 1\begin{bmatrix} 1 & 1 \\ 0 & 0 \end{bmatrix} + (-1)\begin{bmatrix} 0 & 1 \\ 0 & 0 \end{bmatrix} + 1\begin{bmatrix} 0 & 0 \\ 1 & 1 \end{bmatrix} + (-1)\begin{bmatrix} 1 & 0 \\ 0 & 1 \end{bmatrix}$$

$$L\left(\begin{bmatrix} 0 & 0 \\ 1 & 0 \end{bmatrix}\right) = \begin{bmatrix} 0 & 1 \\ 0 & 0 \end{bmatrix} = 0\begin{bmatrix} 1 & 1 \\ 0 & 0 \end{bmatrix} + 1\begin{bmatrix} 0 & 1 \\ 0 & 0 \end{bmatrix} + 0\begin{bmatrix} 0 & 0 \\ 1 & 1 \end{bmatrix} + 0\begin{bmatrix} 1 & 0 \\ 0 & 1 \end{bmatrix}$$

$$L\left(\begin{bmatrix} 0 & 0 \\ 0 & 1 \end{bmatrix}\right) = \begin{bmatrix} 0 & 0 \\ 0 & 1 \end{bmatrix} = (-1)\begin{bmatrix} 1 & 1 \\ 0 & 0 \end{bmatrix} + 1\begin{bmatrix} 0 & 1 \\ 0 & 0 \end{bmatrix} + 0\begin{bmatrix} 0 & 0 \\ 1 & 1 \end{bmatrix} + 1\begin{bmatrix} 1 & 0 \\ 0 & 1 \end{bmatrix}$$

Therefore the matrix of L with respect to S and T is $\begin{bmatrix} 1 & 1 & 0 & -1 \\ -1 & -1 & 1 & 1 \\ 0 & 1 & 0 & 0 \\ 0 & -1 & 0 & 1 \end{bmatrix}$.

(c) Find the image of each vector in T and write it as a linear combination of vectors in S:

$$L\left(\begin{bmatrix} 1 & 1 \\ 0 & 0 \end{bmatrix}\right) = \begin{bmatrix} 1 & 0 \\ 1 & 0 \end{bmatrix} = 1\begin{bmatrix} 1 & 0 \\ 0 & 0 \end{bmatrix} + 0\begin{bmatrix} 0 & 1 \\ 0 & 0 \end{bmatrix} + 1\begin{bmatrix} 0 & 0 \\ 1 & 0 \end{bmatrix} + 0\begin{bmatrix} 0 & 0 \\ 0 & 1 \end{bmatrix}$$

$$L\left(\begin{bmatrix} 0 & 1 \\ 0 & 0 \end{bmatrix}\right) = \begin{bmatrix} 0 & 0 \\ 1 & 0 \end{bmatrix} = 0\begin{bmatrix} 1 & 0 \\ 0 & 0 \end{bmatrix} + 0\begin{bmatrix} 0 & 1 \\ 0 & 0 \end{bmatrix} + 1\begin{bmatrix} 0 & 0 \\ 1 & 0 \end{bmatrix} + 0\begin{bmatrix} 0 & 0 \\ 0 & 1 \end{bmatrix}$$

$$L\left(\begin{bmatrix} 0 & 0 \\ 1 & 1 \end{bmatrix}\right) = \begin{bmatrix} 0 & 1 \\ 0 & 1 \end{bmatrix} = 0\begin{bmatrix} 1 & 0 \\ 0 & 0 \end{bmatrix} + 1\begin{bmatrix} 0 & 1 \\ 0 & 0 \end{bmatrix} + 0\begin{bmatrix} 0 & 0 \\ 1 & 0 \end{bmatrix} + 1\begin{bmatrix} 0 & 0 \\ 0 & 1 \end{bmatrix}$$

$$L\left(\begin{bmatrix} 1 & 0 \\ 0 & 1 \end{bmatrix}\right) = \begin{bmatrix} 1 & 0 \\ 0 & 1 \end{bmatrix} = 1\begin{bmatrix} 1 & 0 \\ 0 & 0 \end{bmatrix} + 0\begin{bmatrix} 0 & 1 \\ 0 & 0 \end{bmatrix} + 0\begin{bmatrix} 0 & 0 \\ 1 & 0 \end{bmatrix} + 1\begin{bmatrix} 0 & 0 \\ 0 & 1 \end{bmatrix}$$

Therefore the matrix of L with respect to T and S is $\begin{bmatrix} 1 & 0 & 0 & 1 \\ 0 & 0 & 1 & 0 \\ 1 & 1 & 0 & 0 \\ 0 & 0 & 1 & 1 \end{bmatrix}$.

(d) Find the image of each vector in T and write it as a linear combination of vectors in T:

$$L\left(\begin{bmatrix} 1 & 1 \\ 0 & 0 \end{bmatrix}\right) = \begin{bmatrix} 1 & 0 \\ 1 & 0 \end{bmatrix} = 2\begin{bmatrix} 1 & 1 \\ 0 & 0 \end{bmatrix} + (-2)\begin{bmatrix} 0 & 1 \\ 0 & 0 \end{bmatrix} + 1\begin{bmatrix} 0 & 0 \\ 1 & 1 \end{bmatrix} + (-1)\begin{bmatrix} 1 & 0 \\ 0 & 1 \end{bmatrix}$$

$$L\left(\begin{bmatrix} 0 & 1 \\ 0 & 0 \end{bmatrix}\right) = \begin{bmatrix} 0 & 0 \\ 1 & 0 \end{bmatrix} = 1\begin{bmatrix} 1 & 1 \\ 0 & 0 \end{bmatrix} + (-1)\begin{bmatrix} 0 & 1 \\ 0 & 0 \end{bmatrix} + 1\begin{bmatrix} 0 & 0 \\ 1 & 1 \end{bmatrix} + (-1)\begin{bmatrix} 1 & 0 \\ 0 & 1 \end{bmatrix}$$

$$L\left(\begin{bmatrix} 0 & 0 \\ 1 & 1 \end{bmatrix}\right) = \begin{bmatrix} 0 & 1 \\ 0 & 1 \end{bmatrix} = (-1)\begin{bmatrix} 1 & 1 \\ 0 & 0 \end{bmatrix} + 2\begin{bmatrix} 0 & 1 \\ 0 & 0 \end{bmatrix} + 0\begin{bmatrix} 0 & 0 \\ 1 & 1 \end{bmatrix} + 1\begin{bmatrix} 1 & 0 \\ 0 & 1 \end{bmatrix}$$

$$L\left(\begin{bmatrix} 1 & 0 \\ 0 & 1 \end{bmatrix}\right) = \begin{bmatrix} 1 & 0 \\ 0 & 1 \end{bmatrix} = 0\begin{bmatrix} 1 & 1 \\ 0 & 0 \end{bmatrix} + 0\begin{bmatrix} 0 & 1 \\ 0 & 0 \end{bmatrix} + 0\begin{bmatrix} 0 & 0 \\ 1 & 1 \end{bmatrix} + 1\begin{bmatrix} 1 & 0 \\ 0 & 1 \end{bmatrix}$$

Therefore the matrix of L with respect to T is $\begin{bmatrix} 2 & 1 & -1 & 0 \\ -2 & -1 & 2 & 0 \\ 1 & 1 & 0 & 0 \\ -1 & -1 & 1 & 1 \end{bmatrix}$.

11. (a) Let f and g be functions in V and let c be a scalar. Then

$$L(f + g) = (f + g)(0) = f(0) + g(0) = L(f) + L(g)$$

and

$$L(cf) = (cf)(0) = cf(0) = cL(f).$$

Therefore f is a linear transformation.

(b) The kernel of L consists of any continuous function $f(x)$ such that $L(f) = f(0) = 0$. That is, f is in ker L provided the value of f at $x = 0$ is zero. The following functions are in ker L:

$$x, \quad x^2, \quad x\cos x, \quad \sin x, \quad \frac{x}{x^2 + 1}, \quad xe^x.$$

(c) Yes. In this case

$$L(f + g) = (f + g)(\tfrac{1}{2}) = f(\tfrac{1}{2}) + g(\tfrac{1}{2}) = L(f) + L(g)$$

and

$$L(cf) = (cf)(\tfrac{1}{2}) = cf(\tfrac{1}{2}) = cL(f).$$

13. $L : P_2 \rightarrow P_2$ is defined by $L(at^2 + bt + c) = (a + 2c)t^2 + (b - c)t + (a - c)$. We have ordered bases $S = \{1, t, t^2\}$ and $T = \{t^2 - 1, t, t - 1\}$.

(a) We have

$$L(1) = 2t^2 - t - 1 = 2(t^2 - 1) - (t - 1)$$
$$L(t) = t$$
$$L(t^2) = t^2 + 1 = (t^2 - 1) + 2t - 2(t - 1)$$

(In the last case we really constructed a linear system to obtain the expression given, but we have omitted the details here.) Thus the matrix of L with respect to S and T is

$$A = \left[\ [\ L(1)\ ]_T\quad [\ L(t)\ ]_T\quad [\ L(t^2)\ ]_T\ \right] = \begin{bmatrix} 2 & 0 & 1 \\ 0 & 1 & 2 \\ -1 & 0 & -2 \end{bmatrix}.$$

(b) We have that

$$[\ L(p(t))\ ]_T = A\ [\ p(t)\ ]_S = A\begin{bmatrix} 1 \\ -3 \\ 2 \end{bmatrix} = \begin{bmatrix} 4 \\ 1 \\ -5 \end{bmatrix}.$$

Thus $L(p(t)) = 4(t^2 - 1) + t - 5(t - 1) = 4t^2 - 4t + 1$.

15. $L : P_3 \to P_3$ is given by $L(at^3 + bt^2 + ct + d) = 3at^2 + 2bt + c$ and S is the basis $\{t^3, t^2, t, 1\}$. We first find the images of the basis vectors:

$$L(t^3) = 3t^2, \quad L(t^2) = 2t, \quad L(t) = 1, \quad L(1) = 0.$$

Then

$$[\ L(t^3)\ ]_S = \begin{bmatrix} 0 \\ 3 \\ 0 \\ 0 \end{bmatrix}, \quad [\ L(t^2)\ ]_S = \begin{bmatrix} 0 \\ 0 \\ 2 \\ 0 \end{bmatrix}, \quad [\ L(t)\ ]_S = \begin{bmatrix} 0 \\ 0 \\ 0 \\ 1 \end{bmatrix}, \quad [\ L(1)\ ]_S = \begin{bmatrix} 0 \\ 0 \\ 0 \\ 0 \end{bmatrix}$$

hence the matrix representing L with respect to S is

$$\begin{bmatrix} 0 & 0 & 0 & 0 \\ 3 & 0 & 0 & 0 \\ 0 & 2 & 0 & 0 \\ 0 & 0 & 1 & 0 \end{bmatrix}.$$

17. Assume that $(L_1 + L_2)^2 = L_1^2 + 2L_2 \circ L_2 + L_2^2$. Then

$$L_1^2 + L_1 \circ L_2 + L_2 \circ L_1 + L_2^2 = L_1^2 + 2L_1 \circ L_2 + L_2^2,$$

and simplifying gives $L_1 \circ L_2 = L_2 \circ L_1$. The steps are reversible.

19. (a) Suppose that $L(\mathbf{v}) = \mathbf{0}$. Then $0 = (0,0) = (L(\mathbf{v}), L(\mathbf{v})) = (\mathbf{v}, \mathbf{v})$. But then from the definition of an inner product, $\mathbf{v} = \mathbf{0}$. Hence ker $L = \{\mathbf{0}\}$.

(b) $(\mathbf{v}, \mathbf{v}) = (L(\mathbf{v}), L(\mathbf{v}))$ (since L preserves inner products)

$$= (A\mathbf{v}, A\mathbf{v}) \qquad \text{(since } L(\mathbf{v}) = A\mathbf{v})$$
$$= (\mathbf{v}, A^T A\mathbf{v}) \qquad \text{(by Equation 3 in Section 5.3)}$$

21. (a) We use Exercise 6 in Section 6.1 to show that L is a linear transformation. Let

$$\mathbf{u} = \begin{bmatrix} u_1 \\ u_2 \\ \vdots \\ u_n \end{bmatrix} \quad \text{and} \quad \mathbf{v} = \begin{bmatrix} v_1 \\ v_2 \\ \vdots \\ v_n \end{bmatrix}$$

be vectors in R^n and let a and b be scalars. Then

$$L(a\mathbf{u} + b\mathbf{v}) = L\left(a\begin{bmatrix} u_1 \\ u_2 \\ \vdots \\ u_n \end{bmatrix} + b\begin{bmatrix} v_1 \\ v_2 \\ \vdots \\ v_n \end{bmatrix}\right) = L\left(\begin{bmatrix} au_1 + bv_1 \\ au_2 + bv_2 \\ \vdots \\ au_n + bv_n \end{bmatrix}\right)$$

$$= (au_1 + bv_1)\mathbf{v}_1 + (au_2 + bv_2)\mathbf{v}_2 + \cdots + (au_n + bv_n)\mathbf{v}_n$$
$$= a(u_1\mathbf{v}_1 + u_2\mathbf{v}_2 + \cdots + u_n\mathbf{v}_n) + b(v_1\mathbf{v}_1 + v_2\mathbf{v}_2 + \cdots + v_n\mathbf{v}_n)$$
$$= aL(\mathbf{u}) + bL(\mathbf{v})$$

Therefore L is a linear transformation.

(b) We show that $\ker L = \{\mathbf{0}_V\}$. Let $\mathbf{v}$ be in the kernel of L. Then $L(\mathbf{v}) = a_1\mathbf{v}_1 + a_2\mathbf{v}_2 + \cdots a_n\mathbf{v}_n = \mathbf{0}$. Since the vectors $\mathbf{v}_1$, $\mathbf{v}_2$, $\ldots$, $\mathbf{v}_n$ form a basis for V, they are linearly independent. Therefore $a_1 = 0$, $a_2 = 0$, $\ldots$, $a_n = 0$. Hence $\mathbf{v} = \mathbf{0}$. Therefore $\ker L = \{\mathbf{0}\}$ and hence L is one-to-one by Theorem 6.4.

(c) Since both R^n and V have dimension n, it follows from Corollary 6.2 that L is onto.

23. Since A is nonsingular, A^{-1} exists. Therefore, $BA = A^{-1}(BA)A$ and hence AB and BA are similar.

Chapter Review for Chapter 6, p. 432

True or False

1. True.	2. False.	3. True.	4. False.	5. False.	6. True.
7. True.	8. True.	9. True.	10. False.	11. True.	12. False.

Quiz

1. Yes. If $\begin{bmatrix} a & b \\ c & d \end{bmatrix}$ and $\begin{bmatrix} a' & b' \\ c' & d' \end{bmatrix}$ are any matrices in M_{22}, then

$$L\left(\begin{bmatrix} a & b \\ c & d \end{bmatrix} + \begin{bmatrix} a' & b' \\ c' & d' \end{bmatrix}\right) = L\left(\begin{bmatrix} a + a' & b + b' \\ c + c' & d + d' \end{bmatrix}\right)$$
$$= (a + a') - (d + d') + (b + b') - (c + c')$$
$$= (a - d + b - c) + (a' - d' + b' - c')$$
$$= L\left(\begin{bmatrix} a & b \\ c & d \end{bmatrix}\right) + L\left(\begin{bmatrix} a' & b' \\ c' & d' \end{bmatrix}\right)$$

and, for any scalar k,

$$L\left(k\begin{bmatrix} a & b \\ c & d \end{bmatrix}\right) = L\left(\begin{bmatrix} ka & kb \\ kc & kd \end{bmatrix}\right)$$
$$= ka - kd + kb - kc$$
$$= k(a - d + b - c)$$
$$= kL\left(\begin{bmatrix} a & b \\ c & d \end{bmatrix}\right).$$

2. (a) Let $\begin{bmatrix} a_1 \\ b_1 \end{bmatrix}$ and $\begin{bmatrix} a_2 \\ b_2 \end{bmatrix}$ be any vectors and c_1, c_2 any scalars. Then

$$L_k\left(c_1\begin{bmatrix} a_1 \\ b_1 \end{bmatrix} + c_2\begin{bmatrix} a_2 \\ b_2 \end{bmatrix}\right) = L_k\left(\begin{bmatrix} c_1a_1 + c_2a_2 \\ c_1b_1 + c_2b_2 \end{bmatrix}\right)$$

$$= \begin{bmatrix} c_1a_1 + c_2a_2 \\ k(c_1a_1 + c_2a_2) + (c_1b_1 + c_2b_2) \end{bmatrix}$$

$$= \begin{bmatrix} c_1a_1 \\ k(c_1a_1) + c_1b_1 \end{bmatrix} + \begin{bmatrix} c_2a_2 \\ k(c_2a_2) + c_2b_2 \end{bmatrix}$$

$$= c_1L_k\left(\begin{bmatrix} a_1 \\ b_1 \end{bmatrix}\right) + c_2L_k\left(\begin{bmatrix} a_2 \\ b_2 \end{bmatrix}\right).$$

Therefore, L is a linear transformation.

(b) We have

$$L\left(\begin{bmatrix} 1 \\ 0 \end{bmatrix}\right) = \begin{bmatrix} 1 \\ k \end{bmatrix} \quad \text{and} \quad L\left(\begin{bmatrix} 0 \\ 1 \end{bmatrix}\right) = \begin{bmatrix} 0 \\ 1 \end{bmatrix}.$$

So the standard matrix representing L_k is $\begin{bmatrix} 1 & 0 \\ k & 1 \end{bmatrix}$.

3. (a) Possible answer: $\begin{bmatrix} 1 & 1 & -1 \\ 1 & 2 & -1 \\ 1 & 3 & -1 \end{bmatrix}$.

(b) No. If the nonzero vector $\begin{bmatrix} 1 \\ 0 \\ 1 \end{bmatrix}$ is in the kernel, then L cannot be nonsingular.

4. Since

$$\begin{bmatrix} 3 \\ 1 \\ -5 \end{bmatrix} = 2\begin{bmatrix} 1 \\ 1 \\ 0 \end{bmatrix} + \begin{bmatrix} 1 \\ 2 \\ 1 \end{bmatrix} - 3\begin{bmatrix} 0 \\ 1 \\ 2 \end{bmatrix}$$

we have

$$L\left(\begin{bmatrix} 3 \\ 1 \\ -5 \end{bmatrix}\right) = 2L\left(\begin{bmatrix} 1 \\ 1 \\ 0 \end{bmatrix}\right) + L\left(\begin{bmatrix} 1 \\ 2 \\ 1 \end{bmatrix}\right) - 3L\left(\begin{bmatrix} 0 \\ 1 \\ 2 \end{bmatrix}\right)$$

$$= 2\begin{bmatrix} 0 \\ 1 \\ 1 \end{bmatrix} + \begin{bmatrix} 2 \\ 1 \\ 2 \end{bmatrix} - 3\begin{bmatrix} 2 \\ 0 \\ 0 \end{bmatrix}$$

$$= \begin{bmatrix} -4 \\ 3 \\ 4 \end{bmatrix}.$$

5. We have

$$L\left(\begin{bmatrix} 1 \\ 1 \end{bmatrix}\right) = \begin{bmatrix} 0 \\ 3 \end{bmatrix} = (0)\mathbf{e}_1 + 3\mathbf{e}_2$$

$$L\left(\begin{bmatrix} 2 \\ 1 \end{bmatrix}\right) = \begin{bmatrix} -1 \\ 5 \end{bmatrix} = (-1)\mathbf{e}_1 + 5\mathbf{e}_2$$

so the representation of L with respect to T and S is $\begin{bmatrix} 0 & -1 \\ 3 & 5 \end{bmatrix}$.

6. (a) Since

$$L\left(\begin{bmatrix} 1 \\ 0 \end{bmatrix}\right) = \begin{bmatrix} 1 \\ 1 \\ 2 \end{bmatrix} = 1\mathbf{e}_1 + 1\mathbf{e}_2 + 2\mathbf{e}_3$$

$$L\left(\begin{bmatrix} 0 \\ 1 \end{bmatrix}\right) = \begin{bmatrix} -1 \\ 1 \\ 0 \end{bmatrix} = (-1)\mathbf{e}_1 + 1\mathbf{e}_2 + 0\mathbf{e}_3,$$

the matrix of L with respect to S and T is $\begin{bmatrix} 1 & -1 \\ 1 & 1 \\ 2 & 0 \end{bmatrix}$.

(b) We need to write each vector in S' as a linear combination of vectors in S. This is easy to do since S is is the natural basis for R^2:

$$\begin{bmatrix} 1 \\ 2 \end{bmatrix} = 1\mathbf{e}_1 + 2\mathbf{e}_2 \quad \text{and} \quad \begin{bmatrix} 0 \\ 1 \end{bmatrix} = 0\mathbf{e}_1 + 1\mathbf{e}_2.$$

Therefore, $P = \begin{bmatrix} 1 & 0 \\ 2 & 1 \end{bmatrix}$.

(c) As in part (b), we find that

$$\begin{bmatrix} 1 \\ 1 \\ 0 \end{bmatrix} = 1\mathbf{e}_1 + 1\mathbf{e}_2 + 0\mathbf{e}_3, \quad \begin{bmatrix} 1 \\ 2 \\ 1 \end{bmatrix} = 1\mathbf{e}_1 + 2\mathbf{e}_2 + 1\mathbf{e}_3, \quad \text{and} \quad \begin{bmatrix} 2 \\ 0 \\ 0 \end{bmatrix} = 2\mathbf{e}_1 + 0\mathbf{e}_2 + 0\mathbf{e}_3.$$

Therefore, $Q = \begin{bmatrix} 1 & 1 & 2 \\ 1 & 2 & 0 \\ 0 & 1 & 0 \end{bmatrix}$.

(d) $B = Q^{-1}AP = \begin{bmatrix} 0 & 1 & -2 \\ 0 & 0 & 1 \\ \frac{1}{2} & -\frac{1}{2} & \frac{1}{2} \end{bmatrix} \begin{bmatrix} 1 & -1 \\ 1 & 1 \\ 2 & 0 \end{bmatrix} \begin{bmatrix} 1 & 0 \\ 2 & 1 \end{bmatrix} = \begin{bmatrix} -1 & 1 \\ 2 & 0 \\ -1 & -1 \end{bmatrix}$.

Chapter 7

Eigenvalues and Eigenvectors

Section 7.1, p. 450

1. Let $L : R^2 \to R^2$ be the counterclockwise rotation through an angle π. That is, using Example 8 in Section 1.6

$$L\left(\begin{bmatrix} x_1 \\ x_2 \end{bmatrix}\right) = A\mathbf{x} = \begin{bmatrix} \cos\pi & -\sin\pi \\ \sin\pi & \cos\pi \end{bmatrix} \begin{bmatrix} x_1 \\ x_2 \end{bmatrix} = \begin{bmatrix} -1 & 0 \\ 0 & -1 \end{bmatrix} \begin{bmatrix} x_1 \\ x_2 \end{bmatrix}.$$

It follows that $L(\mathbf{x}) = A\mathbf{x} = \lambda\mathbf{x}$ if and only if $(\lambda I_2 - A)\mathbf{x} = \mathbf{0}$. An eigenvector $\mathbf{x}$ must be a nonzero vector, hence the coefficient matrix of the homogeneous system must be singular. This is the case provided that

$$\det(\lambda I_2 - A) = \begin{vmatrix} \lambda+1 & 0 \\ 0 & \lambda+1 \end{vmatrix} = (\lambda+1)^2 = 0.$$

Hence we obtain an eigenvector if and only if $\lambda = -1$. That is, we have that L has eigenvalue -1. To find the associated eigenvector(s) we solve the homogeneous linear system $(\lambda I_2 - A)\mathbf{x} = \mathbf{0}$ with $\lambda = -1$. We have

$$(-1I_2 - A)\mathbf{x} = \mathbf{0} \quad \Longleftrightarrow \quad \begin{bmatrix} 0 & 0 \\ 0 & 0 \end{bmatrix} \begin{bmatrix} x_1 \\ x_2 \end{bmatrix} = \begin{bmatrix} 0 \\ 0 \end{bmatrix}.$$

The general solution is

$$\mathbf{x} = \begin{bmatrix} r \\ s \end{bmatrix} = r\begin{bmatrix} 1 \\ 0 \end{bmatrix} + s\begin{bmatrix} 0 \\ 1 \end{bmatrix}$$

where r and s are any real numbers, but not both zero. Hence every nonzero vector in R^2 is an eigenvector. In particular,

$$\begin{bmatrix} 1 \\ 0 \end{bmatrix} \quad \text{and} \quad \begin{bmatrix} 0 \\ 1 \end{bmatrix}$$

are eigenvectors associated with eigenvalue $\lambda = -1$.

3. Since

$$L(t^2 + 1) = 1 - t^2 = (-1)(t^2 + 1) + (0)t + (2)1$$
$$L(t) = 0 = (0)(t^2 + 1) + (0)t + (0)1$$
$$L(1) = 1 = (0)(t^2 + 1) + (0)t + (1)1$$

the matrix of L with respect to the basis $\{t^2 + 1, t, 1\}$ is

$$A = \begin{bmatrix} -1 & 0 & 0 \\ 0 & 0 & 0 \\ 2 & 0 & 1 \end{bmatrix}.$$

The characteristic polynomial of A is

$$p(\lambda) = \det(\lambda I_3 - A) = \begin{vmatrix} \lambda + 1 & 0 & 0 \\ 0 & \lambda & 0 \\ -2 & 0 & \lambda - 1 \end{vmatrix} = \lambda(\lambda + 1)(\lambda - 1).$$

Hence the eigenvalues of L are $\lambda_1 = 1$, $\lambda_2 = -1$, and $\lambda_3 = 0$.

To find the associated eigenvectors for $\lambda_1 = 1$, we solve the system

$$((1)I_3 - A)\mathbf{x} = \begin{bmatrix} 2 & 0 & 0 \\ 0 & 1 & 0 \\ 1 & 0 & 0 \end{bmatrix} \begin{bmatrix} x_1 \\ x_2 \\ x_3 \end{bmatrix} = \begin{bmatrix} 0 \\ 0 \\ 0 \end{bmatrix}$$

to obtain the general solution

$$\mathbf{x} = \begin{bmatrix} 0 \\ 0 \\ r \end{bmatrix} = r \begin{bmatrix} 0 \\ 0 \\ 1 \end{bmatrix}$$

where $r \neq 0$ is any real number. Choosing $r = 1$, we find a particular eigenvector associated with $\lambda_3 = 1$ is

$$\begin{bmatrix} 0 \\ 0 \\ 1 \end{bmatrix} \quad \text{or, using the basis } \{t^2 + 1, t, 1\}, \text{ the vector} \quad (0)(t^2 + 1) + (0)t + (1)1 = 1.$$

For $\lambda_2 = -1$, we have

$$((-1)I_3 - A)\mathbf{x} = \begin{bmatrix} 0 & 0 & 0 \\ 0 & -1 & 0 \\ -2 & 0 & -2 \end{bmatrix} \begin{bmatrix} x_1 \\ x_2 \\ x_3 \end{bmatrix} = \begin{bmatrix} 0 \\ 0 \\ 0 \end{bmatrix}$$

whose general solution is

$$\mathbf{x} = \begin{bmatrix} r \\ 0 \\ -r \end{bmatrix} = r \begin{bmatrix} 1 \\ 0 \\ -1 \end{bmatrix}$$

where $r \neq 0$ is any real number. Choosing $r = 1$, we find a particular eigenvector associated with $\lambda_2 = -1$ is

$$\begin{bmatrix} 1 \\ 0 \\ -1 \end{bmatrix} \quad \text{or, using the basis } \{t^2 + 1, t, 1\}, \text{ the vector} \quad (1)(t^2 + 1) + (0)t + (-1)1 = t^2$$

For $\lambda_3 = 0$, we have

$$(0I_3 - A)\mathbf{x} = \begin{bmatrix} 1 & 0 & 0 \\ 0 & 0 & 0 \\ 2 & 0 & -1 \end{bmatrix} \begin{bmatrix} x_1 \\ x_2 \\ x_3 \end{bmatrix} = \begin{bmatrix} 0 \\ 0 \\ 0 \end{bmatrix}$$

to obtain the general solution

$$\mathbf{x} = \begin{bmatrix} 0 \\ r \\ 0 \end{bmatrix} = r \begin{bmatrix} 0 \\ 1 \\ 0 \end{bmatrix}$$

where $r \neq 0$ is any real number. Choosing $r = 1$, we find a particular eigenvector associated with $\lambda_1 = 0$ is

$$\begin{bmatrix} 0 \\ 1 \\ 0 \end{bmatrix} \quad \text{or, using the basis } \{t^2 + 1, t, 1\}, \text{ the vector} \quad (0)(t^2 + 1) + (1)t + (0)1 = t.$$

5. Let A be the given matrix. In each case the characteristic polynomial is $p(\lambda) = \det(\lambda I_n - A)$.

(a) $p(\lambda) = \begin{vmatrix} \lambda - 2 & -1 \\ 1 & \lambda - 3 \end{vmatrix} = \lambda^2 - 5\lambda + 7.$

(b) $p(\lambda) = \begin{vmatrix} \lambda - 1 & -2 & -1 \\ 0 & \lambda - 1 & -2 \\ 1 & -3 & \lambda - 2 \end{vmatrix} = \lambda^3 - 4\lambda^2 + 7.$

(c) $p(\lambda) = \begin{vmatrix} \lambda - 4 & 1 & -3 \\ 0 & \lambda - 2 & -1 \\ 0 & 0 & \lambda - 3 \end{vmatrix} = \lambda^3 - 9\lambda^2 + 26\lambda - 24.$

(d) $p(\lambda) = \begin{vmatrix} \lambda - 4 & -2 \\ -3 & \lambda - 3 \end{vmatrix} = \lambda^2 - 7\lambda + 6.$

7. (a) Let $A = \begin{bmatrix} 1 & -1 \\ 2 & 4 \end{bmatrix}$. Then the characteristic polynomial is

$$p(\lambda) = \det(\lambda I_2 - A) = \begin{vmatrix} \lambda - 1 & 1 \\ -2 & \lambda - 4 \end{vmatrix} = (\lambda - 1)(\lambda - 4) + 2 = \lambda^2 - 5\lambda + 6 = (\lambda - 2)(\lambda - 3).$$

The eigenvalues are $\lambda_1 = 2$, $\lambda_2 = 3$. To determine the eigenvectors corresponding to $\lambda_1 = 2$ we solve the homogeneous linear system

$$(2I_2 - A)\mathbf{x} = \begin{bmatrix} 1 & 1 \\ -2 & -2 \end{bmatrix} \begin{bmatrix} x_1 \\ x_2 \end{bmatrix} = \begin{bmatrix} 0 \\ 0 \end{bmatrix}.$$

Row reducing the coefficient matrix we have the equivalent linear system

$$\begin{bmatrix} 1 & 1 \\ 0 & 0 \end{bmatrix} \begin{bmatrix} x_1 \\ x_2 \end{bmatrix} = \begin{bmatrix} 0 \\ 0 \end{bmatrix}$$

whose solution is $x_1 = -r$, $x_2 = r$, or in matrix form

$$\mathbf{x} = r \begin{bmatrix} -1 \\ 1 \end{bmatrix}$$

where r is any nonzero real number. Thus a particular eigenvector for eigenvalue $\lambda_1 = 2$ is

$$\mathbf{x}_1 = \begin{bmatrix} 1 \\ -1 \end{bmatrix}$$

where we chose $r = -1$. To determine the eigenvectors corresponding to $\lambda_2 = 3$ we solve the homogeneous linear system

$$(3I_2 - A)\mathbf{x} = \begin{bmatrix} 2 & 1 \\ -2 & -1 \end{bmatrix} \begin{bmatrix} x_1 \\ x_2 \end{bmatrix} = \begin{bmatrix} 0 \\ 0 \end{bmatrix}.$$

Row reducing the coefficient matrix we have the equivalent linear system

$$\begin{bmatrix} 1 & \frac{1}{2} \\ 0 & 0 \end{bmatrix} \begin{bmatrix} x_1 \\ x_2 \end{bmatrix} = \begin{bmatrix} 0 \\ 0 \end{bmatrix}$$

whose solution is $x_1 = -\frac{1}{2}r$, $x_2 = r$, or in matrix form

$$\mathbf{x} = r \begin{bmatrix} -\frac{1}{2} \\ 1 \end{bmatrix}$$

where r is any nonzero real number. Thus a particular eigenvector for eigenvalue $\lambda_2 = 3$ is

$$\mathbf{x}_2 = \begin{bmatrix} 1 \\ -2 \end{bmatrix}$$

where we chose $r = -2$.

(b) Let $A = \begin{bmatrix} 2 & -2 & 3 \\ 0 & 3 & -2 \\ 0 & -1 & 2 \end{bmatrix}$. Then the characteristic polynomial is

$$p(\lambda) = \det(\lambda I_3 - A) = \begin{vmatrix} \lambda - 2 & 2 & -3 \\ 0 & \lambda - 3 & 2 \\ 0 & 1 & \lambda - 2 \end{vmatrix} = (\lambda - 2)[(\lambda - 3)(\lambda - 2) - 2] = (\lambda - 1)(\lambda - 2)(\lambda - 4).$$

The eigenvalues are $\lambda_1 = 1$, $\lambda_2 = 2$, $\lambda_3 = 4$. To determine the eigenvectors corresponding to $\lambda_1 = 1$ we solve the homogeneous linear system

$$(1I_3 - A)\mathbf{x} = \begin{bmatrix} -1 & 2 & -3 \\ 0 & -2 & 2 \\ 0 & 1 & -1 \end{bmatrix} \begin{bmatrix} x_1 \\ x_2 \\ x_3 \end{bmatrix} = \begin{bmatrix} 0 \\ 0 \\ 0 \end{bmatrix}.$$

Row reducing the coefficient matrix we have the equivalent linear system

$$\begin{bmatrix} 1 & 0 & 1 \\ 0 & 1 & -1 \\ 0 & 0 & 0 \end{bmatrix} \begin{bmatrix} x_1 \\ x_2 \\ x_3 \end{bmatrix} = \begin{bmatrix} 0 \\ 0 \\ 0 \end{bmatrix}$$

whose solution is $x_1 = -r$, $x_2 = r$, $x_3 = r$, or in matrix form

$$\mathbf{x} = \begin{bmatrix} -r \\ r \\ r \end{bmatrix} = r \begin{bmatrix} -1 \\ 1 \\ 1 \end{bmatrix}$$

where r is any nonzero real number. Thus a particular eigenvector for eigenvalue $\lambda_1 = 1$ is

$$\mathbf{x}_1 = \begin{bmatrix} -1 \\ 1 \\ 1 \end{bmatrix}$$

where we chose $r = 1$. To determine the eigenvectors corresponding the $\lambda_2 = 2$ we solve the homogeneous linear system

$$(2I_3 - A)\mathbf{x} = \begin{bmatrix} 0 & 2 & -3 \\ 0 & -1 & 2 \\ 0 & 1 & 0 \end{bmatrix} \begin{bmatrix} x_1 \\ x_2 \\ x_3 \end{bmatrix} = \begin{bmatrix} 0 \\ 0 \\ 0 \end{bmatrix}.$$

Row reducing the coefficient matrix we have the equivalent linear system

$$\begin{bmatrix} 0 & 1 & 0 \\ 0 & 0 & 1 \\ 0 & 0 & 0 \end{bmatrix} \begin{bmatrix} x_1 \\ x_2 \\ x_3 \end{bmatrix} = \begin{bmatrix} 0 \\ 0 \\ 0 \end{bmatrix}$$

whose solution is $x_1 = r$, $x_2 = 0$, $x_3 = 0$, or in matrix form

$$\mathbf{x} = \begin{bmatrix} r \\ 0 \\ 0 \end{bmatrix} = r \begin{bmatrix} 1 \\ 0 \\ 0 \end{bmatrix}$$

where r is any nonzero real number. Thus a particular eigenvector for eigenvalue $\lambda_2 = 2$ is

$$\mathbf{x}_2 = \begin{bmatrix} 1 \\ 0 \\ 0 \end{bmatrix}$$

where we chose $r = 2$. To determine eigenvectors corresponding to $\lambda_3 = 4$ we solve the homogeneous linear system

$$(4I_3 - A)\mathbf{x} = \begin{bmatrix} 2 & 2 & -3 \\ 0 & 1 & 2 \\ 0 & 1 & 2 \end{bmatrix} \begin{bmatrix} x_1 \\ x_2 \\ x_3 \end{bmatrix} = \begin{bmatrix} 0 \\ 0 \\ 0 \end{bmatrix}.$$

Row reducing the coefficient matrix we have the equivalent linear system

$$\begin{bmatrix} 1 & 0 & -\frac{7}{2} \\ 0 & 1 & 2 \\ 0 & 0 & 0 \end{bmatrix} \begin{bmatrix} x_1 \\ x_2 \\ x_3 \end{bmatrix} = \begin{bmatrix} 0 \\ 0 \\ 0 \end{bmatrix}$$

whose solution is $x_1 = \frac{7}{2}r$, $x_2 = -2r$, $x_3 = r$, or in matrix form

$$\mathbf{x} = r \begin{bmatrix} \frac{7}{2} \\ -2 \\ 1 \end{bmatrix}$$

where r is any nonzero real number. Thus a particular eigenvector for eigenvalue $\lambda_3 = 4$ is

$$\mathbf{x}_3 = \begin{bmatrix} 7 \\ -4 \\ 2 \end{bmatrix}$$

where we chose $r = 2$.

(c) Let $A = \begin{bmatrix} 2 & 2 & 3 \\ 1 & 2 & 1 \\ 2 & -2 & 1 \end{bmatrix}$. Then the characteristic polynomial is

$$\begin{aligned} p(\lambda) = \det(\lambda I_3 - A) &= \begin{vmatrix} \lambda - 2 & -2 & -3 \\ -1 & \lambda - 2 & -1 \\ -2 & 2 & \lambda - 1 \end{vmatrix} \\ &= (\lambda - 2)(\lambda - 2)(\lambda - 1) - 4 + 6 - 6(\lambda - 2) + 2(\lambda - 2) - 2(\lambda - 1) \\ &= \lambda^3 - 5\lambda^2 + 2\lambda + 8 \\ &= (\lambda + 1)(\lambda - 2)(\lambda - 4) \end{aligned}$$

The eigenvalues are $\lambda_1 = -1$, $\lambda_2 = 2$, $\lambda_3 = 4$. To determine the eigenvectors corresponding to $\lambda_1 = -1$ we solve the homogeneous linear system

$$(-1I_3 - A)\mathbf{x} = \begin{bmatrix} -3 & -2 & -3 \\ -1 & -3 & -1 \\ -2 & 2 & -2 \end{bmatrix} \begin{bmatrix} x_1 \\ x_2 \\ x_3 \end{bmatrix} = \begin{bmatrix} 0 \\ 0 \\ 0 \end{bmatrix}.$$

Row reducing the coefficient matrix we have the following equivalent linear system

$$\begin{bmatrix} 1 & 0 & 1 \\ 0 & 1 & 0 \\ 0 & 0 & 0 \end{bmatrix} \begin{bmatrix} x_1 \\ x_2 \\ x_3 \end{bmatrix} = \begin{bmatrix} 0 \\ 0 \\ 0 \end{bmatrix}$$

whose solution is $x_1 = -r$, $x_2 = 0$, $x_3 = r$, or in matrix form

$$\mathbf{x} = \begin{bmatrix} -r \\ 0 \\ r \end{bmatrix} = r \begin{bmatrix} -1 \\ 0 \\ 1 \end{bmatrix}$$

where r is any nonzero real number. Thus a particular eigenvector for eigenvalue $\lambda_1 = -1$ is

$$\mathbf{x}_1 = \begin{bmatrix} 1 \\ 0 \\ -1 \end{bmatrix}$$

where we chose $r = -1$. To determine eigenvectors corresponding to $\lambda_2 = 2$ we solve the homogeneous linear system

$$(2I_3 - A)\mathbf{x} = \begin{bmatrix} 0 & -2 & -3 \\ -1 & 0 & -1 \\ -2 & 2 & 1 \end{bmatrix} \begin{bmatrix} x_1 \\ x_2 \\ x_3 \end{bmatrix} = \begin{bmatrix} 0 \\ 0 \\ 0 \end{bmatrix}.$$

Row reducing the coefficient matrix we have the equivalent linear system

$$\begin{bmatrix} 1 & 0 & 1 \\ 0 & 1 & \frac{3}{2} \\ 0 & 0 & 0 \end{bmatrix} \begin{bmatrix} x_1 \\ x_2 \\ x_3 \end{bmatrix} = \begin{bmatrix} 0 \\ 0 \\ 0 \end{bmatrix}$$

whose solution is $x_1 = -r$, $x_2 = -\frac{3}{2}r$, $x_3 = r$, or in matrix form

$$\mathbf{x} = r \begin{bmatrix} -1 \\ -\frac{3}{2} \\ 1 \end{bmatrix}$$

where r is any nonzero real number. Thus a particular eigenvector for eigenvalue $\lambda_2 = 2$ is

$$\mathbf{x}_2 = \begin{bmatrix} -2 \\ -3 \\ 2 \end{bmatrix}$$

where we chose $r = 2$. To determine the eigenvectors corresponding to $\lambda_3 = 4$ we solve the homogeneous linear system

$$(4I_3 - A)\mathbf{x} = \begin{bmatrix} 2 & -2 & -3 \\ -1 & 2 & -1 \\ -2 & 2 & 3 \end{bmatrix} \begin{bmatrix} x_1 \\ x_2 \\ x_3 \end{bmatrix} = \begin{bmatrix} 0 \\ 0 \\ 0 \end{bmatrix}.$$

Row reducing the coefficient matrix we have the equivalent linear system

$$\begin{bmatrix} 1 & 0 & -4 \\ 0 & 1 & -\frac{5}{2} \\ 0 & 0 & 0 \end{bmatrix} \begin{bmatrix} x_1 \\ x_2 \\ x_3 \end{bmatrix} = \begin{bmatrix} 0 \\ 0 \\ 0 \end{bmatrix}$$

whose solution is $x_1 = 4r$, $x_2 = \frac{5}{2}r$, $x_3 = r$, or in matrix form

$$\mathbf{x} = \begin{bmatrix} 4r \\ \frac{5}{2}r \\ r \end{bmatrix} = r \begin{bmatrix} 4 \\ \frac{5}{2} \\ 1 \end{bmatrix}$$

where r is any nonzero real number. Thus a particular eigenvector for the eigenvalue $\lambda_3 = 4$ is

$$\mathbf{x}_3 = \begin{bmatrix} 8 \\ 5 \\ 2 \end{bmatrix}$$

where we chose $r = 2$.

(d) Let $A = \begin{bmatrix} -2 & -2 & 3 \\ 0 & 3 & -2 \\ 0 & -1 & 2 \end{bmatrix}$. Then the characteristic polynomial is

$$p(\lambda) = \det(\lambda I_2 - A) = \begin{vmatrix} \lambda + 2 & -2 & 3 \\ 0 & \lambda - 3 & 2 \\ 0 & 1 & \lambda - 2 \end{vmatrix} = \lambda^3 - 3\lambda^2 - 6\lambda + 8 = (\lambda + 2)(\lambda - 4)(\lambda - 1).$$

The eigenvalues are $\lambda_1 = -2$, $\lambda_2 = 4$, $\lambda_3 = 1$. To determine the eigenvectors associated with $\lambda_1 = -2$, we solve the homogeneous linear system

$$(-2I_2 - A)\mathbf{x} = \begin{bmatrix} 0 & 2 & -3 \\ 0 & -5 & 2 \\ 0 & 1 & -4 \end{bmatrix} \begin{bmatrix} x_1 \\ x_2 \\ x_3 \end{bmatrix} = \begin{bmatrix} 0 \\ 0 \\ 0 \end{bmatrix}.$$

Row reducing the coefficient matrix we obtain the solution $x_1 = r$, $x_2 = 0$, $x_3 = 0$, or

$$\mathbf{x} = \begin{bmatrix} r \\ 0 \\ 0 \end{bmatrix} = r \begin{bmatrix} 1 \\ 0 \\ 0 \end{bmatrix},$$

where r is any nonzero real number. Then a particular eigenvector associated with $\lambda_1 = -2$ is

$$\mathbf{x}_1 = \begin{bmatrix} 1 \\ 0 \\ 0 \end{bmatrix}.$$

To determine the eigenvectors associated with $\lambda_2 = 4$ we solve the homogeneous system

$$(4I_2 - A)\mathbf{x} = \begin{bmatrix} 6 & 2 & -3 \\ 0 & 1 & 2 \\ 0 & 1 & 2 \end{bmatrix} \begin{bmatrix} x_1 \\ x_2 \\ x_3 \end{bmatrix} = \begin{bmatrix} 0 \\ 0 \\ 0 \end{bmatrix}.$$

Row reducing the coefficient matrix, we obtain the solution $x_1 = \frac{7}{6}r$, $x_2 = -2r$, $x_3 = r$, or

$$\mathbf{x} = \begin{bmatrix} \frac{7}{6}r \\ -2r \\ r \end{bmatrix} = r \begin{bmatrix} \frac{7}{6} \\ -2 \\ 1 \end{bmatrix},$$

where r is any nonzero real number. Choosing $r = 6$, we obtain the particular eigenvector

$$\mathbf{x}_2 = \begin{bmatrix} 7 \\ -12 \\ 6 \end{bmatrix}.$$

To determine the eigenvectors associated with $\lambda_3 = 1$, we solve the homogeneous system

$$(1I_3 - A)\mathbf{x} = \begin{bmatrix} 3 & 2 & -3 \\ 0 & -2 & 2 \\ 0 & 1 & -1 \end{bmatrix} \begin{bmatrix} x_1 \\ x_2 \\ x_3 \end{bmatrix} = \begin{bmatrix} 0 \\ 0 \\ 0 \end{bmatrix}.$$

Row reducing the coefficient matrix, we obtain the solution $x_1 = \frac{1}{3}r$, $x_2 = r$, $x_3 = r$, or

$$\mathbf{x} = \begin{bmatrix} \frac{1}{3}r \\ r \\ r \end{bmatrix} = r \begin{bmatrix} \frac{1}{3} \\ 1 \\ 1 \end{bmatrix},$$

where r is any nonzero real number. Choosing $r = 3$, we obtain the particular eigenvector

$$\mathbf{x}_3 = \begin{bmatrix} 1 \\ 3 \\ 3 \end{bmatrix}.$$

9. (a) $p(\lambda) = \det(\lambda I_2 - A) = \begin{vmatrix} \lambda & -1 \\ 1 & \lambda \end{vmatrix} = \lambda^2 + 1$. The eigenvalues are $\lambda_1 = i$ and $\lambda_2 = -i$.

To find an associated eigenvector for $\lambda_1 = i$, we solve the homogeneous linear system

$$(iI_2 - A)\mathbf{x} = \begin{bmatrix} i & -1 \\ 1 & i \end{bmatrix} \begin{bmatrix} x_1 \\ x_2 \end{bmatrix} = \begin{bmatrix} 0 \\ 0 \end{bmatrix}.$$

Row reducing the coefficient matrix, we obtain the equivalent linear system

$$\begin{bmatrix} 1 & i \\ 0 & 0 \end{bmatrix} \begin{bmatrix} x_1 \\ x_2 \end{bmatrix} = \begin{bmatrix} 0 \\ 0 \end{bmatrix}$$

whose solution is $x_1 = -ix_2$, $x_2 = r$, or in matrix form

$$\mathbf{x} = r \begin{bmatrix} -i \\ 1 \end{bmatrix}.$$

Choosing $r = i$, we obtain the eigenvector

$$\mathbf{x}_1 = \begin{bmatrix} 1 \\ i \end{bmatrix}$$

associated with $\lambda_1 = i$. To find an associated eigenvector for $\lambda_2 = -i$, we solve the homogeneous linear system

$$(-iI_2 - A)\mathbf{x} = \begin{bmatrix} -i & -1 \\ 1 & -i \end{bmatrix} \begin{bmatrix} x_1 \\ x_2 \end{bmatrix} = \begin{bmatrix} 0 \\ 0 \end{bmatrix}.$$

Row reducing the coefficient matrix, we obtain the equivalent linear system

$$\begin{bmatrix} 1 & -i \\ 0 & 0 \end{bmatrix} \begin{bmatrix} x_1 \\ x_2 \end{bmatrix} = \begin{bmatrix} 0 \\ 0 \end{bmatrix}$$

whose solution is $x_1 = ix_2$, $x_2 = r$, or in matrix form

$$\mathbf{x} = r \begin{bmatrix} i \\ 1 \end{bmatrix}.$$

Choosing $r = -i$, we obtain the eigenvector

$$\mathbf{x}_1 = \begin{bmatrix} 1 \\ -i \end{bmatrix}$$

associated with $\lambda_2 = -i$.

(b) $p(\lambda) = \det(\lambda I_3 - A) = \begin{vmatrix} \lambda+2 & 4 & 8 \\ -1 & \lambda & 0 \\ 0 & -1 & \lambda \end{vmatrix} = (\lambda+2)\lambda^2 + 8 + 4\lambda = (\lambda+2)(\lambda-2i)(\lambda+2i)$. The eigenvalues are $\lambda_1 = -2$, $\lambda_2 = 2i$, and $\lambda_3 = -2i$.

To find an associated eigenvector for $\lambda_1 = -2$, we solve the homogeneous linear system

$$(iI_3 - A)\mathbf{x} = \begin{bmatrix} 0 & 4 & 8 \\ -1 & -2 & 0 \\ 0 & -1 & -2 \end{bmatrix} \begin{bmatrix} x_1 \\ x_2 \\ x_3 \end{bmatrix} = \begin{bmatrix} 0 \\ 0 \\ 0 \end{bmatrix}.$$

Row reducing the coefficient matrix, we obtain the equivalent linear system

$$\begin{bmatrix} 1 & 0 & -4 \\ 0 & 1 & 2 \\ 0 & 0 & 0 \end{bmatrix} \begin{bmatrix} x_1 \\ x_2 \\ x_3 \end{bmatrix} = \begin{bmatrix} 0 \\ 0 \\ 0 \end{bmatrix}$$

whose solution is $x_1 = -4x_3$, $x_2 = -2x_3$, $x_3 = r$ or in matrix form

$$\mathbf{x} = r \begin{bmatrix} 4 \\ -2 \\ 1 \end{bmatrix}.$$

Choosing $r = 1$, we obtain the eigenvector

$$\mathbf{x}_1 = \begin{bmatrix} 4 \\ -2 \\ 1 \end{bmatrix}$$

associated with $\lambda_1 = -2$.

To find an associated eigenvector for $\lambda_2 = 2i$, we solve the homogeneous linear system

$$(2iI_3 - A)\mathbf{x} = \begin{bmatrix} 2i+2 & 4 & 8 \\ -1 & 2i & 0 \\ 0 & -1 & 2i \end{bmatrix} \begin{bmatrix} x_1 \\ x_2 \\ x_3 \end{bmatrix} = \begin{bmatrix} 0 \\ 0 \\ 0 \end{bmatrix}.$$

Row reducing the coefficient matrix, we obtain the equivalent linear system

$$\begin{bmatrix} 1 & 0 & 4 \\ 0 & 1 & -2i \\ 0 & 0 & 0 \end{bmatrix} \begin{bmatrix} x_1 \\ x_2 \\ x_3 \end{bmatrix} = \begin{bmatrix} 0 \\ 0 \\ 0 \end{bmatrix}$$

whose solution is $x_1 = 4x_3$, $x_2 = 2ix_3$, $x_3 = r$ or in matrix form

$$\mathbf{x} = r \begin{bmatrix} -4 \\ 2i \\ 1 \end{bmatrix}.$$

Choosing $r = 1$, we obtain the eigenvector

$$\mathbf{x}_1 = \begin{bmatrix} -4 \\ 2i \\ 1 \end{bmatrix}$$

associated with $\lambda_1 = 2i$.

To find an associated eigenvector for $\lambda_3 = -2i$, we solve the homogeneous linear system

$$(-2iI_3 - A)\mathbf{x} = \begin{bmatrix} -2i+2 & 4 & 8 \\ -1 & -2i & 0 \\ 0 & -1 & -2i \end{bmatrix} \begin{bmatrix} x_1 \\ x_2 \\ x_3 \end{bmatrix} = \begin{bmatrix} 0 \\ 0 \\ 0 \end{bmatrix}.$$

Row reducing the coefficient matrix, we obtain the equivalent linear system

$$\begin{bmatrix} 1 & 0 & 4 \\ 0 & 1 & 2i \\ 0 & 0 & 0 \end{bmatrix} \begin{bmatrix} x_1 \\ x_2 \\ x_3 \end{bmatrix} = \begin{bmatrix} 0 \\ 0 \\ 0 \end{bmatrix}$$

whose solution is $x_1 = -4x_3$, $x_2 = -2ix_3$, $x_3 = r$ or in matrix form

$$\mathbf{x} = r \begin{bmatrix} -4 \\ -2i \\ 1 \end{bmatrix}.$$

Choosing $r = 1$, we obtain the eigenvector

$$\mathbf{x}_1 = \begin{bmatrix} -4 \\ -2i \\ 1 \end{bmatrix}$$

associated with $\lambda_1 = -2i$.

(c) $p(\lambda) = \det(\lambda I_3 - A) = \begin{vmatrix} \lambda - 2 + i & -2i & 0 \\ -1 & \lambda & 0 \\ 0 & -1 & \lambda \end{vmatrix} = (\lambda - 2 + i)\lambda^2 - 2i\lambda = \lambda(\lambda + i)(\lambda - 2).$ The eigenvalues are $\lambda_1 = 0$, $\lambda_2 = -i$, and $\lambda_3 = 2$.

To find an associated eigenvector for $\lambda_1 = 0$, we solve the homogeneous linear system

$$(0I_3 - A)\mathbf{x} = \begin{bmatrix} -2+i & -2i & 0 \\ -1 & 0 & 0 \\ 0 & -1 & 0 \end{bmatrix} \begin{bmatrix} x_1 \\ x_2 \\ x_3 \end{bmatrix} = \begin{bmatrix} 0 \\ 0 \\ 0 \end{bmatrix}.$$

Row reducing the coefficient matrix, we obtain the equivalent linear system

$$\begin{bmatrix} 1 & 0 & 0 \\ 0 & 1 & 0 \\ 0 & 0 & 0 \end{bmatrix} \begin{bmatrix} x_1 \\ x_2 \\ x_3 \end{bmatrix} = \begin{bmatrix} 0 \\ 0 \\ 0 \end{bmatrix}$$

whose solution is $x_1 = 0$, $x_2 = 0$, $x_3 = r$ or in matrix form

$$\mathbf{x} = r \begin{bmatrix} 0 \\ 0 \\ 1 \end{bmatrix}.$$

Choosing $r = -1$, we obtain the eigenvector

$$\mathbf{x}_1 = \begin{bmatrix} 0 \\ 0 \\ 1 \end{bmatrix}$$

associated with $\lambda_1 = 0$.

To find an associated eigenvector for $\lambda_2 = -i$, we solve the homogeneous linear system

$$(-iI_3 - A)\mathbf{x} = \begin{bmatrix} -2 & -2i & 0 \\ -1 & -i & 0 \\ 0 & -1 & -i \end{bmatrix} \begin{bmatrix} x_1 \\ x_2 \\ x_3 \end{bmatrix} = \begin{bmatrix} 0 \\ 0 \\ 0 \end{bmatrix}.$$

Row reducing the coefficient matrix, we obtain the equivalent linear system

$$\begin{bmatrix} 1 & 0 & 1 \\ 0 & 1 & i \\ 0 & 0 & 0 \end{bmatrix} \begin{bmatrix} x_1 \\ x_2 \\ x_3 \end{bmatrix} = \begin{bmatrix} 0 \\ 0 \\ 0 \end{bmatrix}$$

whose solution is $x_1 = -x_3$, $x_2 = -ix_3$, $x_3 = r$ or in matrix form

$$\mathbf{x} = r \begin{bmatrix} -1 \\ -i \\ 1 \end{bmatrix}.$$

Choosing $r = -1$, we obtain the eigenvector

$$\mathbf{x}_2 = \begin{bmatrix} -1 \\ -i \\ 1 \end{bmatrix}$$

associated with $\lambda_1 = 0$.

To find an associated eigenvector for $\lambda_3 = 2$, we solve the homogeneous linear system

$$(2I_3 - A)\mathbf{x} = \begin{bmatrix} i & -2i & 0 \\ -1 & 2 & 0 \\ 0 & -1 & 2 \end{bmatrix} \begin{bmatrix} x_1 \\ x_2 \\ x_3 \end{bmatrix} = \begin{bmatrix} 0 \\ 0 \\ 0 \end{bmatrix}.$$

Row reducing the coefficient matrix, we obtain the equivalent linear system

$$\begin{bmatrix} 1 & 0 & -4 \\ 0 & 1 & -2 \\ 0 & 0 & 0 \end{bmatrix} \begin{bmatrix} x_1 \\ x_2 \\ x_3 \end{bmatrix} = \begin{bmatrix} 0 \\ 0 \\ 0 \end{bmatrix}$$

whose solution is $x_1 = 4x_3$, $x_2 = 2x_3$, $x_3 = r$ or in matrix form

$$\mathbf{x} = r \begin{bmatrix} 4 \\ 2 \\ 1 \end{bmatrix}.$$

Choosing $r = -1$, we obtain the eigenvector

$$\mathbf{x}_3 = \begin{bmatrix} 4 \\ 2 \\ 1 \end{bmatrix}$$

associated with $\lambda_1 = 0$.

(d) $p(\lambda) = \det(\lambda I_3 - A) = \begin{vmatrix} \lambda - 5 & -2 \\ 1 & \lambda - 3 \end{vmatrix} = (\lambda - 5)(\lambda - 3) + 2 = (\lambda - 4 - i)(\lambda - 4 + i)$. The eigenvalues are $\lambda_1 = 4 + i$ and $\lambda_2 = 4 - i$.

To find an associated eigenvector for $\lambda_1 = 4 + i$, we solve the homogeneous linear system

$$((4+i)I_3 - A)\mathbf{x} = \begin{bmatrix} -1+i & -2 \\ 1 & 1+i \end{bmatrix} \begin{bmatrix} x_1 \\ x_2 \\ x_3 \end{bmatrix} = \begin{bmatrix} 0 \\ 0 \\ 0 \end{bmatrix}.$$

Row reducing the coefficient matrix, we obtain the equivalent linear system

$$\begin{bmatrix} 1 & 1+i \\ 0 & 0 \end{bmatrix} \begin{bmatrix} x_1 \\ x_2 \end{bmatrix} = \begin{bmatrix} 0 \\ 0 \end{bmatrix}$$

whose solution is $x_1 = -(1 + i)x_2$, $x_2 = r$, or in matrix form

$$\mathbf{x} = r \begin{bmatrix} -1 - i \\ 1 \end{bmatrix}.$$

Choosing $r = -1 + i$, we obtain the eigenvector

$$\mathbf{x}_1 = \begin{bmatrix} 2 \\ -1 + i \end{bmatrix}$$

associated with $\lambda_1 = 4 + i$.

To find an associated eigenvector for $\lambda_2 = 4 - i$, we solve the homogeneous linear system

$$((4-i)I_3 - A)\mathbf{x} = \begin{bmatrix} -1-i & -2 \\ 1 & 1-i \end{bmatrix} \begin{bmatrix} x_1 \\ x_2 \\ x_3 \end{bmatrix} = \begin{bmatrix} 0 \\ 0 \\ 0 \end{bmatrix}.$$

Row reducing the coefficient matrix, we obtain the equivalent linear system

$$\begin{bmatrix} 1 & 1-i \end{bmatrix} \begin{bmatrix} x_1 \\ x_2 \end{bmatrix} = \begin{bmatrix} 0 \\ 0 \end{bmatrix}$$

whose solution is $x_1 = -(1 - i)x_2$, $x_2 = r$, or in matrix form

$$\mathbf{x} = r \begin{bmatrix} -1 + i \\ 1 \end{bmatrix}.$$

Choosing $r = -1 - i$, we obtain the eigenvector

$$\mathbf{x}_2 = \begin{bmatrix} 2 \\ -1 - i \end{bmatrix}$$

associated with $\lambda_2 = 4 - i$.

11. Let $A = \begin{bmatrix} a_{ij} \end{bmatrix}$ be an $n \times n$ upper triangular matrix, that is, $a_{ij} = 0$ for $i > j$. Then the characteristic polynomial of A is

$$p(\lambda) = \det(\lambda I_n - A) = \begin{vmatrix} \lambda - a_{11} & -a_{12} & \cdots & -a_{1n} \\ 0 & \lambda - a_{22} & \cdots & -a_{2n} \\ \vdots & \vdots & \ddots & \vdots \\ 0 & 0 & 0 & \lambda - a_{nn} \end{vmatrix} = (\lambda - a_{11})(\lambda - a_{22}) \cdots (\lambda - a_{nn}),$$

which we obtain by expanding along the cofactors of the first column repeatedly. Thus the eigenvalues of A are $a_{11}, \ldots, a_{nn}$, which are the elements on the main diagonal of A. A similar proof shows the same result if A is lower triangular.

13. The characteristic polynomial of L is

$$p(\lambda) = \det(\lambda I_4 - A) = \begin{vmatrix} \lambda - 1 & -2 & -3 & -4 \\ 0 & \lambda + 1 & -3 & -2 \\ 0 & 0 & \lambda - 3 & -3 \\ 0 & 0 & 0 & \lambda - 2 \end{vmatrix} = (\lambda - 1)(\lambda + 1)(\lambda - 3)(\lambda - 2).$$

Thus the eigenvalues of L are $\lambda_1 = 1$, $\lambda_2 = -1$, $\lambda_3 = 3$, and $\lambda_4 = 2$.

To find the eigenvectors associated with $\lambda_1 = 1$, we solve the system

$$(1I_4 - A)\mathbf{x} = \begin{bmatrix} 0 & -2 & -3 & -4 \\ 0 & 2 & -3 & -2 \\ 0 & 0 & -2 & -3 \\ 0 & 0 & 0 & -1 \end{bmatrix} \begin{bmatrix} x_1 \\ x_2 \\ x_3 \\ x_4 \end{bmatrix} = \begin{bmatrix} 0 \\ 0 \\ 0 \\ 0 \end{bmatrix}$$

to obtain the general solution

$$\mathbf{x} = \begin{bmatrix} r \\ 0 \\ 0 \\ 0 \end{bmatrix} = r \begin{bmatrix} 1 \\ 0 \\ 0 \\ 0 \end{bmatrix}$$

where $r \neq 0$ is any real number. Choosing $r = 1$, we obtain the particular eigenvector

$$\begin{bmatrix} 1 \\ 0 \\ 0 \\ 0 \end{bmatrix}$$

associated with $\lambda_1 = 1$. Using the basis S, this is the matrix

$$1 \begin{bmatrix} 1 & 0 \\ 0 & 0 \end{bmatrix} + 0 \begin{bmatrix} 0 & 1 \\ 0 & 0 \end{bmatrix} + 0 \begin{bmatrix} 0 & 0 \\ 1 & 0 \end{bmatrix} + 0 \begin{bmatrix} 0 & 0 \\ 0 & 1 \end{bmatrix} = \begin{bmatrix} 1 & 0 \\ 0 & 0 \end{bmatrix}.$$

To find the eigenvectors associated with $\lambda_2 = -1$, we solve the system

$$((-1)I_4 - A)\mathbf{x} = \begin{bmatrix} -2 & -2 & -3 & -4 \\ 0 & 0 & -3 & -2 \\ 0 & 0 & -4 & -3 \\ 0 & 0 & 0 & -3 \end{bmatrix} \begin{bmatrix} x_1 \\ x_2 \\ x_3 \\ x_4 \end{bmatrix} = \begin{bmatrix} 0 \\ 0 \\ 0 \\ 0 \end{bmatrix}$$

to obtain the general solution

$$\mathbf{x} = \begin{bmatrix} r \\ -r \\ 0 \\ 0 \end{bmatrix} = r \begin{bmatrix} 1 \\ -1 \\ 0 \\ 0 \end{bmatrix}$$

where $r \neq 0$ is any real number. Choosing $r = 1$, we obtain the particular eigenvector

$$\begin{bmatrix} 1 \\ -1 \\ 0 \\ 0 \end{bmatrix}$$

associated with $\lambda_2 = -1$. Using the basis S, this is the matrix

$$1 \begin{bmatrix} 1 & 0 \\ 0 & 0 \end{bmatrix} + (-1) \begin{bmatrix} 0 & 1 \\ 0 & 0 \end{bmatrix} + 0 \begin{bmatrix} 0 & 0 \\ 1 & 0 \end{bmatrix} + 0 \begin{bmatrix} 0 & 0 \\ 0 & 1 \end{bmatrix} = \begin{bmatrix} 1 & -1 \\ 0 & 0 \end{bmatrix}.$$

To find the eigenvectors associated with $\lambda_3 = 3$, we solve the system

$$(3I_4 - A)\mathbf{x} = \begin{bmatrix} 2 & -2 & -3 & -4 \\ 0 & 4 & -3 & -2 \\ 0 & 0 & 0 & -3 \\ 0 & 0 & 0 & 1 \end{bmatrix} \begin{bmatrix} x_1 \\ x_2 \\ x_3 \\ x_4 \end{bmatrix} = \begin{bmatrix} 0 \\ 0 \\ 0 \\ 0 \end{bmatrix}$$

to obtain the general solution

$$\mathbf{x} = \begin{bmatrix} \frac{9}{4}r \\ \frac{3}{4}r \\ r \\ 0 \end{bmatrix} = r \begin{bmatrix} \frac{9}{4} \\ \frac{3}{4} \\ 1 \\ 0 \end{bmatrix}$$

where $r \neq 0$ is any real number. Choosing $r = 4$, we obtain the particular eigenvector

$$\begin{bmatrix} 9 \\ 3 \\ 4 \\ 0 \end{bmatrix}$$

associated with $\lambda_3 = 3$. Using the basis S, this is the matrix

$$9 \begin{bmatrix} 1 & 0 \\ 0 & 0 \end{bmatrix} + 3 \begin{bmatrix} 0 & 1 \\ 0 & 0 \end{bmatrix} + 4 \begin{bmatrix} 0 & 0 \\ 1 & 0 \end{bmatrix} + 0 \begin{bmatrix} 0 & 0 \\ 0 & 1 \end{bmatrix} = \begin{bmatrix} 9 & 3 \\ 4 & 0 \end{bmatrix}.$$

To find the eigenvectors associated with $\lambda_4 = 2$, we solve the system

$$(2I_4 - A)\mathbf{x} = \begin{bmatrix} 1 & -2 & -3 & -4 \\ 0 & 3 & -3 & -2 \\ 0 & 0 & -1 & -3 \\ 0 & 0 & 0 & 0 \end{bmatrix} \begin{bmatrix} x_1 \\ x_2 \\ x_3 \\ x_4 \end{bmatrix} = \begin{bmatrix} 0 \\ 0 \\ 0 \\ 0 \end{bmatrix}$$

to obtain the general solution

$$\mathbf{x} = \begin{bmatrix} -\frac{29}{3}r \\ -\frac{7}{3}r \\ -3r \\ r \end{bmatrix} = r \begin{bmatrix} -\frac{29}{3} \\ -\frac{7}{3} \\ -3 \\ 1 \end{bmatrix}$$

where $r \neq 0$ is any real number. Choosing $r = 3$, we obtain the particular eigenvector

$$\begin{bmatrix} -29 \\ -7 \\ -9 \\ 3 \end{bmatrix}$$

associated with $\lambda_4 = 2$. Using the basis S, this is the matrix

$$(-29) \begin{bmatrix} 1 & 0 \\ 0 & 0 \end{bmatrix} + (-7) \begin{bmatrix} 0 & 1 \\ 0 & 0 \end{bmatrix} + (-9) \begin{bmatrix} 0 & 0 \\ 1 & 0 \end{bmatrix} + (3) \begin{bmatrix} 0 & 0 \\ 0 & 1 \end{bmatrix} = \begin{bmatrix} -29 & -7 \\ -9 & 3 \end{bmatrix}.$$

15. We use Exercise 14 as follows. Let $L : R^n \to R^n$ be defined by $L(\mathbf{x}) = A\mathbf{x}$. Then we saw in Chapter 4 that L is a linear transformation and matrix A represents this transformation. Hence Exercise 14 implies that all the eigenvectors of A with associated eigenvalue λ, together with the zero vector, form a subspace of V.

17. (a) To find the eigenvectors associated with $\lambda = 1$, we solve

$$(1I_3 - A)\mathbf{x} = \begin{bmatrix} 1 & 0 & -1 \\ 0 & 0 & 0 \\ -1 & 0 & 1 \end{bmatrix} \begin{bmatrix} x_1 \\ x_2 \\ x_3 \end{bmatrix} = \begin{bmatrix} 0 \\ 0 \\ 0 \end{bmatrix}.$$

The general solution to this system is $x_1 = s$, $x_2 = r$, $x_3 = s$, or

$$\begin{bmatrix} r \\ s \\ r \end{bmatrix} = r \begin{bmatrix} 1 \\ 0 \\ 1 \end{bmatrix} + s \begin{bmatrix} 0 \\ 1 \\ 0 \end{bmatrix}$$

where r and s are any nonzero real numbers. Since the vectors $\begin{bmatrix} 1 \\ 0 \\ 1 \end{bmatrix}$ and $\begin{bmatrix} 0 \\ 1 \\ 0 \end{bmatrix}$ are linearly independent, a basis for the eigenspace of A associated with $\lambda = 1$ is

$$\left\{ \begin{bmatrix} 1 \\ 0 \\ 1 \end{bmatrix}, \begin{bmatrix} 0 \\ 1 \\ 0 \end{bmatrix} \right\}.$$

(b) To find the eigenvectors associated with $\lambda = 2$, we solve

$$(2I_3 - A)\mathbf{x} = \begin{bmatrix} 0 & -1 & 0 \\ -1 & 0 & -1 \\ 0 & -1 & 0 \end{bmatrix} \begin{bmatrix} x_1 \\ x_2 \\ x_3 \end{bmatrix} = \begin{bmatrix} 0 \\ 0 \\ 0 \end{bmatrix}.$$

The general solution to this system is $x_1 = -r$, $x_2 = 0$, $x_3 = r$, or

$$\begin{bmatrix} -r \\ 0 \\ r \end{bmatrix} = r \begin{bmatrix} -1 \\ 0 \\ 1 \end{bmatrix}$$

where $r \neq 0$ is any real number. Thus a basis for the eigenspace of A associated with $\lambda = 2$ is

$$\left\{ \begin{bmatrix} -1 \\ 0 \\ 1 \end{bmatrix} \right\}.$$

19. (a) To find the eigenvectors associated with $\lambda = 2i$, we solve

$$(2iI_3 - A)\mathbf{x} = \begin{bmatrix} 2i & 4 & 0 \\ -1 & 2i & 0 \\ 0 & -1 & 2i \end{bmatrix} \begin{bmatrix} x_1 \\ x_2 \\ x_3 \end{bmatrix} = \begin{bmatrix} 0 \\ 0 \\ 0 \end{bmatrix}.$$

The general solution to this system is $x_1 = -4r$, $x_2 = 2ir$, $x_3 = r$, or

$$\begin{bmatrix} -4r \\ 2ir \\ r \end{bmatrix} = r \begin{bmatrix} -4 \\ 2i \\ 1 \end{bmatrix}$$

where $r \neq 0$ is any number. Thus a basis for the eigenspace of A associated with $\lambda = 2$ is

$$\left\{ \begin{bmatrix} -4 \\ 2i \\ 1 \end{bmatrix} \right\}.$$

(b) To find the eigenvectors associated with $\lambda = -2i$, we solve

$$(-2iI_3 - A)\mathbf{x} = \begin{bmatrix} -2i & 4 & 0 \\ -1 & -2i & 0 \\ 0 & -1 & -2i \end{bmatrix} \begin{bmatrix} x_1 \\ x_2 \\ x_3 \end{bmatrix} = \begin{bmatrix} 0 \\ 0 \\ 0 \end{bmatrix}.$$

The general solution to this system is $x_1 = -4r$, $x_2 = -2ir$, $x_3 = r$, or

$$\begin{bmatrix} -4r \\ -2ir \\ r \end{bmatrix} = r \begin{bmatrix} -4 \\ -2i \\ 1 \end{bmatrix}$$

where $r \neq 0$ is any number. Thus a basis for the eigenspace of A associated with $\lambda = 2$ is

$$\left\{ \begin{bmatrix} -4 \\ -2i \\ 1 \end{bmatrix} \right\}.$$

21. If λ is an eigenvalue of A with associated eigenvector $\mathbf{x}$, then $A\mathbf{x} = \lambda\mathbf{x}$. Therefore

$$A^2\mathbf{x} = A(A\mathbf{x}) = A(\lambda\mathbf{x}) = \lambda A\mathbf{x} = \lambda(\lambda\mathbf{x}) = \lambda^2\mathbf{x}.$$

Continuing in this manner, it follows by induction that $A^k\mathbf{x} = \lambda^k\mathbf{x}$ for all positive integers k. Thus, λ^k is an eigenvalue of the matrix A^k with associated eigenvector $\mathbf{x}$ for all positive integers k.

23. If A is nilpotent, then $A^k = O$ for some positive integer k. If λ is an eigenvalue of A with associated eigenvector $\mathbf{x}$, then by Exercise 21 we have $O = A^k\mathbf{x} = \lambda^k\mathbf{x}$. Since $\mathbf{x} \neq \mathbf{0}$, $\lambda^k = 0$ so $\lambda = 0$. Thus, 0 is the only eigenvalue of A.

25. (a) Since $L(\mathbf{x}) = \lambda\mathbf{x}$ and since L is invertible, we have $\mathbf{x} = L^{-1}(\lambda\mathbf{x}) = \lambda L^{-1}(\mathbf{x})$. Therefore $L^{-1}(\mathbf{x}) = (1/\lambda)\mathbf{x}$. Hence $1/\lambda$ is an eigenvalue of L^{-1} with associated eigenvector $\mathbf{x}$.

(b) Let A be a nonsingular matrix with eigenvalue λ and associated eigenvector $\mathbf{x}$. Then $1/\lambda$ is an eigenvalue of A^{-1} with associated eigenvector $\mathbf{x}$. For if $A\mathbf{x} = \lambda\mathbf{x}$, then $A^{-1}\mathbf{x} = (1/\lambda)\mathbf{x}$.

27. If $A\mathbf{x} = \lambda\mathbf{x}$, then, for any scalar r,

$$(A + rI_n)\mathbf{x} = A\mathbf{x} + r\mathbf{x} = \lambda\mathbf{x} + r\mathbf{x} = (\lambda + r)\mathbf{x}.$$

Thus $\lambda + r$ is an eigenvalue of $A + rI_n$ with associated eigenvector $\mathbf{x}$.

29. (a) $(A + B)\mathbf{x} = A\mathbf{x} + B\mathbf{x} = \lambda\mathbf{x} + \mu\mathbf{x} = (\lambda + \mu)\mathbf{x}$

   (b) $(AB)\mathbf{x} = A(B\mathbf{x}) = A(\mu\mathbf{x}) = \mu(A\mathbf{x}) = \mu\lambda\mathbf{x} = (\lambda\mu)\mathbf{x}$

31. Let A be an $n \times n$ nonsingular matrix with characteristic polynomial

$$p(\lambda) = \lambda^n + a_1\lambda^{n-1} + \cdots + a_{n-1}\lambda + a_n.$$

By the Cayley-Hamilton Theorem (see Exercise 30)

$$p(A) = A^n + a_1 A^{n-1} + \cdots + a_{n-1}A + a_n I_n = O.$$

Multiply the preceding expression by A^{-1} to obtain

$$A^{n-1} + a_1 A^{n-2} + \cdots + a_{n-1}I_n + a_n A^{-1} = O.$$

Rearranging terms we have

$$a_n A^{-1} = -A^{n-1} - a_1 A^{n-2} - \cdots - a_{n-1}I_n.$$

Since A is nonsingular $\det(A) \neq 0$. From the discussion prior to Example 11, $a_n = (-1)^n \det(A)$, so $a_n \neq 0$. Hence we have

$$A^{-1} = -\frac{1}{a_n}\left(A^{n-1} + a_1 A^{n-2} + \cdots + a_{n-1}I_n\right).$$

33. Let A be an $n \times n$ matrix all of whose columns add up to 1 and let $\mathbf{x}$ be the $m \times 1$ matrix

$$\mathbf{x} = \begin{bmatrix} 1 \\ \vdots \\ 1 \end{bmatrix}.$$

Then

$$A^T\mathbf{x} = \begin{bmatrix} 1 \\ \vdots \\ 1 \end{bmatrix} = \mathbf{x} = 1\mathbf{x}.$$

Therefore $\lambda = 1$ is an eigenvalue of A^T. By Exercise 12, $\lambda = 1$ is an eigenvalue of A.

35. (a) Since $A\mathbf{u} = \mathbf{0} = 0\mathbf{u}$, it follows that 0 is an eigenvalue of A with associated eigenvector $\mathbf{u}$.

   (b) Since $A\mathbf{v} = 0\mathbf{v} = \mathbf{0}$, it follows that $A\mathbf{x} = \mathbf{0}$ has a nontrivial solution, namely $\mathbf{x} = \mathbf{v}$.

Section 7.2, p. 461

1. Let $L : P_2 \to P_2$ be the linear transformation defined by $L(p(t)) = p'(t)$. That is, for $p(t) = c_2 t^2 + c_1 t + c_0$, $L(p(t)) = 2c_2 t + c_1$. To determine if L is diagonalizable we first find a matrix A representing L. We use the basis $T = \{t^2, t, 1\}$ for P_2. We have $L(t^2) = 2t$, $L(t) = 1$, $L(1) = 0$. Hence

$$A = \begin{bmatrix} [\ L(t^2)\ ]_T & [\ L(t)\ ]_T & [\ L(1)\ ]_T \end{bmatrix} = \begin{bmatrix} 0 & 0 & 0 \\ 2 & 0 & 0 \\ 0 & 1 & 0 \end{bmatrix}.$$

By Theorem 7.4 A is diagonalizable if and only if A has three linearly independent eigenvectors. Thus we find the eigenvalues and eigenvectors of A.

$$\det(\lambda I_3 - A) = \begin{vmatrix} \lambda & 0 & 0 \\ -2 & \lambda & 0 \\ 0 & -1 & \lambda \end{vmatrix} = \lambda^3 = 0.$$

Then A has eigenvalue $\lambda = 0$ of multiplicity 3. To determine the corresponding eigenvectors for $\lambda = 0$ we must find nonzero solutions of

$$(0I_3 - A)\mathbf{x} = \begin{bmatrix} 0 & 0 & 0 \\ -2 & 0 & 0 \\ 0 & -1 & 0 \end{bmatrix} \begin{bmatrix} x_1 \\ x_2 \\ x_3 \end{bmatrix} = \begin{bmatrix} 0 \\ 0 \\ 0 \end{bmatrix}.$$

It follows that $x_1 = x_2 = 0$, $x_3 = r$. Hence there is only one linearly independent eigenvector of A. Therefore L is not diagonalizable.

3. We observe that $L(t^2) = t^2$, $L(t) = 0 = (0)t$, and $L(1) = -1 = (-1)1$. The vectors in $S = \{t^2, t, 1\}$ form a basis for P_2 and the matrix of L with respect to S is the diagonal matrix

$$\begin{bmatrix} 1 & 0 & 0 \\ 0 & 0 & 0 \\ 0 & 0 & -1 \end{bmatrix}.$$

5. Let $L : P_2 \to P_2$ be defined by

$$L(at^2 + bt + c) = (2a + b + c)t^2 + (2c - 3b)t + 4c.$$

We first find a matrix A representing L. It is convenient to use the natural basis $S = \{t^2, t, 1\}$. We have

$$L(t^2) = 2t^2, \quad L(t) = t^2 - 3t, \quad L(1) = t^2 + 2t + 4.$$

Then

$$A = \begin{bmatrix} [\ L(t^2)\ ]_S & [\ L(t)\ ]_S & [\ L(1)\ ]_S \end{bmatrix} = \begin{bmatrix} 2 & 1 & 1 \\ 0 & -3 & 2 \\ 0 & 0 & 4 \end{bmatrix}.$$

Since A is upper triangular its eigenvalues are its diagonal entries. Thus $\lambda_1 = 2$, $\lambda_2 = -3$, $\lambda_3 = 4$. Since the eigenvalues are different, Theorem 7.5 implies that A is diagonalizable, hence L is diagonalizable. To complete the problem we find the eigenvectors of A and then the eigenvectors of L. Solving the appropriate homogeneous linear systems we obtain the following pair of eigenvalues and eigenvectors:

$$\lambda_1 = 2, \mathbf{x}_1 = \begin{bmatrix} 1 \\ 0 \\ 0 \end{bmatrix}; \quad \lambda_2 = -3, \mathbf{x}_2 = \begin{bmatrix} 1 \\ -5 \\ 0 \end{bmatrix}; \quad \lambda_3 = 4, \mathbf{x}_3 = \begin{bmatrix} 9 \\ 4 \\ 14 \end{bmatrix}.$$

Note, your eigenvectors may be different. The corresponding eigenvectors of L are obtained by considering $\mathbf{x}_j$ as coordinates of vectors in P_2 relative to basis S. Thus we have eigenvalues and eigenvectors of L as given below:

$$\lambda_1 = 2, \quad p_1(t) = t^2$$
$$\lambda_2 = -3, \quad p_2(t) = t^2 - 5t$$
$$\lambda_3 = 4, \quad p_3(t) = 9t^2 + 4t + 14$$

7. Let A be the given matrix.

(a) Not diagonalizable. The characteristic polynomial is

$$p(\lambda) = \det(\lambda I_3 - A) = \begin{vmatrix} \lambda - 3 & -1 & 0 \\ 0 & \lambda - 3 & -1 \\ 0 & 0 & \lambda - 3 \end{vmatrix} = (\lambda - 3)^3.$$

Thus there is one eigenvalue of multiplicity 3: $\lambda_1 = \lambda_2 = \lambda_3 = 3$. To find the associated eigenvectors, we solve the system

$$(3I_3 - A)\mathbf{x} - \begin{bmatrix} 0 & -1 & 0 \\ 0 & 0 & -1 \\ 0 & 0 & 0 \end{bmatrix} \begin{bmatrix} x_1 \\ x_2 \\ x_3 \end{bmatrix} = \begin{bmatrix} 0 \\ 0 \\ 0 \end{bmatrix}$$

to obtain the eigenvectors

$$\begin{bmatrix} r \\ 0 \\ 0 \end{bmatrix} = r \begin{bmatrix} 1 \\ 0 \\ 0 \end{bmatrix}$$

where $r \neq 0$ is any real number. Since the eigenvectors do not form a basis for R^3, A is not diagonalizable.

(b) Diagonalizable. The characteristic polynomial is

$$p(\lambda) = \det(\lambda I_2 - A) = \begin{vmatrix} \lambda + 2 & -2 \\ -5 & \lambda - 1 \end{vmatrix} = \lambda^2 + \lambda - 12 = (\lambda + 4)(\lambda - 3).$$

Hence the eigenvalues of A are $\lambda_1 = -4$, $\lambda_2 = 3$. By Theorem 7.5, A is diagonalizable.

(c) Not diagonalizable. The characteristic polynomial is

$$p(\lambda I_3 - A) = \begin{vmatrix} \lambda - 2 & 0 & -3 \\ 0 & \lambda - 1 & 0 \\ 0 & -1 & \lambda - 2 \end{vmatrix} = (\lambda - 2)(\lambda - 1)(\lambda - 2)$$

so the eigenvalues are $\lambda_1 = 2$, $\lambda_2 = 2$, $\lambda_3 = 1$. To find the eigenvectorsd associated with $\lambda_1 = \lambda_2 = 2$, we solve the system

$$(2I_3 - A)\mathbf{x} = \begin{bmatrix} 0 & 0 & -3 \\ 0 & 1 & 0 \\ 0 & -1 & 0 \end{bmatrix} \begin{bmatrix} x_1 \\ x_2 \\ x_3 \end{bmatrix} = \begin{bmatrix} 0 \\ 0 \\ 0 \end{bmatrix}$$

to obtain the eigenvectors

$$\begin{bmatrix} r \\ 0 \\ 0 \end{bmatrix} = r \begin{bmatrix} 1 \\ 0 \\ 0 \end{bmatrix}$$

where $r \neq 0$ is any scalar. Choose

$$\mathbf{x}_1 = \begin{bmatrix} 1 \\ 0 \\ 0 \end{bmatrix}$$

as an eigenvectors associated with $\lambda_1 = \lambda_2 = 2$.
To find eigenvectors associated with $\lambda_3 = 1$, we solve the system

$$(1I_3 - A)\mathbf{x} = \begin{bmatrix} -1 & 0 & -3 \\ 0 & 0 & 0 \\ 0 & -1 & -1 \end{bmatrix} \begin{bmatrix} x_1 \\ x_2 \\ x_3 \end{bmatrix} = \begin{bmatrix} 0 \\ 0 \\ 0 \end{bmatrix}$$

to obtain the eigenvectors

$$\begin{bmatrix} -3r \\ r \\ r \end{bmatrix} = r \begin{bmatrix} -3 \\ 1 \\ 1 \end{bmatrix}$$

where $r \neq 0$ is any real number. Choose

$$\mathbf{x}_2 = \begin{bmatrix} -3 \\ 1 \\ 1 \end{bmatrix}.$$

Since $\{\mathbf{x}_1, \mathbf{x}_2\}$ is not a basis for R^3, A is not diagonalizable.

(d) Not diagonalizable. The characteristic polynomial is

$$p(\lambda) = \det(\lambda I_4 - A) = \begin{vmatrix} \lambda - 2 & -3 & -3 & -5 \\ -3 & \lambda - 2 & -2 & -3 \\ 0 & 0 & \lambda - 2 & -2 \\ 0 & 0 & 0 & \lambda - 2 \end{vmatrix} = \lambda^4 - 8\lambda^3 + 15\lambda^2 + 4\lambda - 20 = (\lambda+1)(\lambda-2)^2(\lambda-5)$$

so the eigenvalues are $\lambda_1 = -1$, $\lambda_2 = 2$, $\lambda_3 = 2$, and $\lambda_4 = 5$.

To find eigenvectors associated with $\lambda_1 = -1$, we solve the system

$$((-1)I_4 - A)\mathbf{x} = \begin{bmatrix} -3 & -3 & -3 & -5 \\ -3 & -3 & -2 & -3 \\ 0 & 0 & -3 & -2 \\ 0 & 0 & 0 & -3 \end{bmatrix} \begin{bmatrix} x_1 \\ x_2 \\ x_3 \\ x_4 \end{bmatrix} = \begin{bmatrix} 0 \\ 0 \\ 0 \\ 0 \end{bmatrix}$$

to obtain the eigenvectors

$$\begin{bmatrix} r \\ -r \\ 0 \\ 0 \end{bmatrix} = r \begin{bmatrix} 1 \\ -1 \\ 0 \\ 0 \end{bmatrix}$$

where $r \neq 0$ is any real number. Choose

$$\mathbf{x}_1 = \begin{bmatrix} 1 \\ -1 \\ 0 \\ 0 \end{bmatrix}$$

as an eigenvector associated with eigenvalue $\lambda_1 = -1$.

To find eigenvectors associated with $\lambda_2 = \lambda_3 = 2$, we solve the system

$$(2I_4 - A)\mathbf{x} = \begin{bmatrix} 0 & -3 & -3 & -5 \\ -3 & 0 & -2 & -3 \\ 0 & 0 & 0 & -2 \\ 0 & 0 & 0 & 0 \end{bmatrix} \begin{bmatrix} x_1 \\ x_2 \\ x_3 \\ x_4 \end{bmatrix} = \begin{bmatrix} 0 \\ 0 \\ 0 \\ 0 \end{bmatrix}$$

to obtain the eigenvectors

$$\begin{bmatrix} -\frac{2}{3}r \\ -r \\ r \\ 0 \end{bmatrix} = r \begin{bmatrix} -\frac{2}{3} \\ -1 \\ 1 \\ 0 \end{bmatrix}.$$

Choose

$$x_2 = \begin{bmatrix} -2 \\ -3 \\ 3 \\ 0 \end{bmatrix}$$

as an eigenvector associated with eigenvalue $\lambda_2 = \lambda_3 = 2$.

To find eigenvectors associated with $\lambda_4 = 5$, we solve the system

$$(5I_4 - A)\mathbf{x} = \begin{bmatrix} 3 & -3 & -3 & -5 \\ -3 & 3 & -2 & -3 \\ 0 & 0 & 3 & -2 \\ 0 & 0 & 0 & 3 \end{bmatrix} \begin{bmatrix} x_1 \\ x_2 \\ x_3 \\ x_4 \end{bmatrix} = \begin{bmatrix} 0 \\ 0 \\ 0 \\ 0 \end{bmatrix}$$

to obtain the eigenvectors

$$\begin{bmatrix} r \\ r \\ 0 \\ 0 \end{bmatrix} = r \begin{bmatrix} 1 \\ 1 \\ 0 \\ 0 \end{bmatrix}.$$

At this point we may stop and conclude that A is not diagonalizable since there are only three eigenvectors for A and they cannot form a basis for R^4.

9. Let

$$D = \begin{bmatrix} -2 & 0 & 0 \\ 0 & -2 & 0 \\ 0 & 0 & 3 \end{bmatrix} \quad \text{and} \quad P = \begin{bmatrix} 1 & 0 & 1 \\ 0 & 1 & 1 \\ 1 & 1 & 1 \end{bmatrix}.$$

Then $P^{-1}AP = D$, so

$$A = PDP^{-1} = \begin{bmatrix} 3 & 5 & -5 \\ 5 & 3 & -5 \\ 5 & 5 & -7 \end{bmatrix}$$

is a matrix whose eigenvalues and associated eigenvectors are as given.

11. If such a matrix P exists, then the given matrix is diagonalizable and P is the transition matrix to the basis of eigenvectors.

(a) The matrix is upper triangular so its eigenvalues are $\lambda_1 = 3$, $\lambda_2 = 2$, and $\lambda_3 = 0$. To find the eigenvectors associated with $\lambda_1 = 3$, we solve the system

$$(3I_3 - A)\mathbf{x} = \begin{bmatrix} 0 & 2 & -1 \\ 0 & 1 & 0 \\ 0 & 0 & 3 \end{bmatrix} \begin{bmatrix} x_1 \\ x_2 \\ x_3 \end{bmatrix} = \begin{bmatrix} 0 \\ 0 \\ 0 \end{bmatrix}$$

obtaining the solution

$$\begin{bmatrix} r \\ 0 \\ 0 \end{bmatrix} = r \begin{bmatrix} 1 \\ 0 \\ 0 \end{bmatrix}$$

where $r \neq 0$ is any real number. Choose the eigenvector

$$\mathbf{x}_1 = \begin{bmatrix} 1 \\ 0 \\ 0 \end{bmatrix}.$$

For the eigenvectors associated with $\lambda_2 = 2$, we solve the system

$$(2I_3 - A)\mathbf{x} = \begin{bmatrix} -1 & 2 & -1 \\ 0 & 0 & 0 \\ 0 & 0 & 2 \end{bmatrix} \begin{bmatrix} xz_1 \\ x_2 \\ x_3 \end{bmatrix} = \begin{bmatrix} 0 \\ 0 \\ 0 \end{bmatrix}$$

obtaining the solution

$$\begin{bmatrix} 2r \\ r \\ 0 \end{bmatrix} = r \begin{bmatrix} 2 \\ 1 \\ 0 \end{bmatrix}$$

where $r \neq 0$ is any real number. Choose the eigenvector

$$\mathbf{x}_2 = \begin{bmatrix} 2 \\ 1 \\ 0 \end{bmatrix}.$$

For the eigenvectors associated with $\lambda_3 = 0$, we solve the system

$$(0I_3 - A)\mathbf{x} = \begin{bmatrix} 3 & -2 & 1 \\ 0 & 2 & 0 \\ 0 & 0 & 0 \end{bmatrix} \begin{bmatrix} x_1 \\ x_2 \\ x_3 \end{bmatrix} = \begin{bmatrix} 0 \\ 0 \\ 0 \end{bmatrix}$$

obtaining the solution

$$\begin{bmatrix} -\frac{1}{3}r \\ 0 \\ r \end{bmatrix} = r \begin{bmatrix} -\frac{1}{3} \\ 0 \\ 1 \end{bmatrix}$$

where $r \neq 0$ is any real number. Choose the eigenvector

$$x_3 = \begin{bmatrix} 1 \\ 0 \\ -3 \end{bmatrix}.$$

The eigenvectors

$$\left\{ \begin{bmatrix} 1 \\ 0 \\ 0 \end{bmatrix}, \begin{bmatrix} 2 \\ 1 \\ 0 \end{bmatrix}, \begin{bmatrix} 1 \\ 0 \\ -3 \end{bmatrix} \right\}$$

form a basis for R^3, so A is diagonalizable. The transition matrix P is

$$P = \begin{bmatrix} 1 & 2 & 1 \\ 0 & 1 & 0 \\ 0 & 0 & -3 \end{bmatrix}$$

and we have

$$P^{-1}AP = \begin{bmatrix} 3 & 0 & 0 \\ 0 & 2 & 0 \\ 0 & 0 & 0 \end{bmatrix}.$$

(b) The characteristic polynomial is

$$p(\lambda) = \det(\lambda I_3 - A) = \begin{vmatrix} \lambda - 2 & -2 & -2 \\ -2 & \lambda - 2 & -2 \\ -2 & -2 & \lambda - 2 \end{vmatrix} = \lambda^3 - 6\lambda^2 = \lambda^2(\lambda - 6)$$

so the eigenvalues are $\lambda_1 = \lambda_2 = 0$, $\lambda_3 = 6$. To find the eigenvectors associated with $\lambda_1 = \lambda_2 = 0$, we solve the system

$$(0I_3 - A)\mathbf{x} = \begin{bmatrix} -2 & -2 & -2 \\ -2 & -2 & -2 \\ -2 & -2 & -2 \end{bmatrix} \begin{bmatrix} x_1 \\ x_2 \\ x_3 \end{bmatrix} = \begin{bmatrix} 0 \\ 0 \\ 0 \end{bmatrix}$$

obtaining

$$\begin{bmatrix} -r - s \\ r \\ s \end{bmatrix} = r \begin{bmatrix} -1 \\ 1 \\ 0 \end{bmatrix} + s \begin{bmatrix} -1 \\ 0 \\ 1 \end{bmatrix}$$

where r, s are nonzero real numbers. Choose

$$\mathbf{x}_1 = \begin{bmatrix} -1 \\ 1 \\ 0 \end{bmatrix}, \quad \mathbf{x}_2 = \begin{bmatrix} -1 \\ 0 \\ 1 \end{bmatrix}.$$

To find the eigenvectors associated with $\lambda_3 = 6$, we solve the system

$$(6I_3 - A)\mathbf{x} = \begin{bmatrix} 4 & -2 & -2 \\ -2 & 4 & -2 \\ -2 & -2 & 4 \end{bmatrix} \begin{bmatrix} x_1 \\ x_2 \\ x_3 \end{bmatrix} = \begin{bmatrix} 0 \\ 0 \\ 0 \end{bmatrix}$$

obtaining

$$\begin{bmatrix} r \\ r \\ r \end{bmatrix} = r \begin{bmatrix} 1 \\ 1 \\ 1 \end{bmatrix}$$

where $r \neq 0$ is any real number. Choose

$$\mathbf{x}_3 = \begin{bmatrix} 1 \\ 1 \\ 1 \end{bmatrix}.$$

Then the eigenvectors

$$\left\{ \begin{bmatrix} -1 \\ 1 \\ 0 \end{bmatrix}, \begin{bmatrix} -1 \\ 0 \\ 1 \end{bmatrix}, \begin{bmatrix} 1 \\ 1 \\ 1 \end{bmatrix} \right\}$$

form a basis for R^3, so A is diagonalizable. The transition matrix P is

$$P = \begin{bmatrix} -1 & -1 & 1 \\ 1 & 0 & 1 \\ 0 & 1 & 1 \end{bmatrix}$$

and

$$P^{-1}AP = \begin{bmatrix} 0 & 0 & 0 \\ 0 & 0 & 0 \\ 0 & 0 & 6 \end{bmatrix}.$$

(c) The characteristic polynomial is

$$p(\lambda) = \det(\lambda I_3 - A) = \begin{vmatrix} \lambda - 3 & 0 & 0 \\ -2 & \lambda - 3 & 0 \\ 0 & 0 & \lambda - 3 \end{vmatrix} = (\lambda - 3)^3$$

so the eigenvalues are $\lambda_1 = \lambda_2 = \lambda_3 = 3$. To find the eigenvectors associated with this eigenvalue, we solve

$$(3I_3 - A)\mathbf{x} = \begin{bmatrix} 0 & 0 & 0 \\ -2 & 0 & 0 \\ 0 & 0 & 0 \end{bmatrix} \begin{bmatrix} x_1 \\ x_2 \\ x_3 \end{bmatrix} = \begin{bmatrix} 0 \\ 0 \\ 0 \end{bmatrix}$$

to obtain the eigenvectors

$$\begin{bmatrix} 0 \\ r \\ s \end{bmatrix} = r \begin{bmatrix} 0 \\ 1 \\ 0 \end{bmatrix} + s \begin{bmatrix} 0 \\ 0 \\ 1 \end{bmatrix}$$

where r, s are any nonzero real numbers. Since the eigenvectors

$$\left\{ \begin{bmatrix} 0 \\ 1 \\ 0 \end{bmatrix}, \begin{bmatrix} 0 \\ 0 \\ 1 \end{bmatrix} \right\}$$

do not form a basis for R^3, A is not diagonalizable so no such matrix P exists.

(d) The characteristic polynomial is

$$p(\lambda) = \det(\lambda I_3 - A) = \begin{vmatrix} \lambda - 1 & 0 & -1 \\ 0 & \lambda - 1 & 0 \\ 0 & -1 & 2 - \lambda \end{vmatrix} = (\lambda - 1)^2 (2 - \lambda)$$

so the eigenvalues are $\lambda_1 = \lambda_2 = 1$, $\lambda_3 = 2$. For $\lambda_1 = \lambda_2 = 1$, we solve the system

$$(1I_3 - A)\mathbf{x} = \begin{bmatrix} 0 & 0 & -1 \\ 0 & 0 & 0 \\ 0 & -1 & 1 \end{bmatrix} \begin{bmatrix} x_1 \\ x_2 \\ x_3 \end{bmatrix} = \begin{bmatrix} 0 \\ 0 \\ 0 \end{bmatrix}$$

to obtain the eigenvectors

$$\begin{bmatrix} r \\ 0 \\ 0 \end{bmatrix} = r \begin{bmatrix} 1 \\ 0 \\ 0 \end{bmatrix}.$$

Choose

$$\mathbf{x}_1 = \begin{bmatrix} 1 \\ 0 \\ 0 \end{bmatrix}.$$

For $\lambda_3 = 2$, we solve the system

$$(2I_3 - A)\mathbf{x} = \begin{bmatrix} 1 & 0 & -1 \\ 0 & 1 & 0 \\ 0 & -1 & 0 \end{bmatrix} \begin{bmatrix} x_1 \\ x_2 \\ x_3 \end{bmatrix} = \begin{bmatrix} 0 \\ 0 \\ 0 \end{bmatrix}$$

to obtain the eigenvectors

$$\begin{bmatrix} r \\ 0 \\ r \end{bmatrix} = r \begin{bmatrix} 1 \\ 0 \\ 1 \end{bmatrix}$$

where $r \neq 0$ is any real number. Since the eigenvectors

$$\left\{ \begin{bmatrix} 1 \\ 0 \\ 0 \end{bmatrix}, \begin{bmatrix} 1 \\ 0 \\ 1 \end{bmatrix} \right\}$$

do not form a basis for R^3, A is not diagonalizable so no such matrix P exists.

13. P is the transition matrix to the basis of eigenvectors:

$$P = \begin{bmatrix} -1 & 0 & 0 \\ 0 & 0 & 1 \\ 1 & 1 & 1 \end{bmatrix}, \quad D = P^{-1}AP = \begin{bmatrix} -3 & 0 & 0 \\ 0 & 4 & 0 \\ 0 & 0 & 4 \end{bmatrix}.$$

15. Let A be the given matrix.

(a) The characteristic polynomial is

$$p(\lambda) = \det(\lambda I_2 - A) = \begin{vmatrix} \lambda - 4 & -2 \\ -3 & \lambda - 3 \end{vmatrix} = \lambda^2 - 7\lambda + 6 = (\lambda - 6)(\lambda - 1)$$

so the eigenvalues are $\lambda_1 = 6$, $\lambda_2 = 1$. To find the eigenvectors associated with $\lambda_1 = 6$, we solve the system

$$(6I_2 - A)\mathbf{x} = \begin{bmatrix} 2 & -2 \\ -3 & 2 \end{bmatrix} \begin{bmatrix} x_1 \\ x_2 \end{bmatrix} = \begin{bmatrix} 0 \\ 0 \end{bmatrix}$$

to obtain the eigenvectors

$$\begin{bmatrix} r \\ r \end{bmatrix} = r \begin{bmatrix} 1 \\ 1 \end{bmatrix}.$$

Choose

$$\mathbf{x}_1 = \begin{bmatrix} 1 \\ 1 \end{bmatrix}.$$

To find the eigenvectors associated with $\lambda_2 = 1$, we solve the system

$$(1I_2 - A)\mathbf{x} = \begin{bmatrix} -3 & -2 \\ -3 & -2 \end{bmatrix} \begin{bmatrix} x_1 \\ x_2 \end{bmatrix} = \begin{bmatrix} 0 \\ 0 \end{bmatrix}$$

to obtain

$$\begin{bmatrix} -\frac{2}{3}r \\ r \end{bmatrix} = r \begin{bmatrix} -\frac{2}{3} \\ 1 \end{bmatrix}$$

where $r \neq 0$ is any real number. Choose

$$\mathbf{x}_2 = \begin{bmatrix} -2 \\ 3 \end{bmatrix}.$$

Since

$$\left\{ \begin{bmatrix} 1 \\ 1 \end{bmatrix}, \begin{bmatrix} -2 \\ 3 \end{bmatrix} \right\}$$

is a basis for R^2, A is diagonalizable. We have

$$P = \begin{bmatrix} 1 & -2 \\ 1 & 3 \end{bmatrix} \quad \text{and} \quad D = P^{-1}AP = \begin{bmatrix} 6 & 0 \\ 0 & 1 \end{bmatrix}.$$

(b) The characteristic polynomial is

$$p(\lambda) = \det(\lambda I_2 - A) = \begin{vmatrix} \lambda - 3 & -2 \\ -6 & \lambda - 4 \end{vmatrix} = \lambda^2 - 7\lambda + 12 - 12 = \lambda(\lambda - 7)$$

so the eigenvalues are $\lambda_1 = 0$, $\lambda_2 = 7$. To find the eigenvectors associated with $\lambda_1 = 0$, we solve

$$(0I_2 - A)\mathbf{x} = \begin{bmatrix} -3 & -2 \\ -6 & -4 \end{bmatrix} \begin{bmatrix} x_1 \\ x_2 \end{bmatrix} = \begin{bmatrix} 0 \\ 0 \end{bmatrix}$$

to obtain the eigenvectors

$$\begin{bmatrix} -\frac{2}{3}r \\ r \end{bmatrix} = r \begin{bmatrix} -\frac{2}{3} \\ 1 \end{bmatrix}.$$

Choose $\mathbf{x}_1 = \begin{bmatrix} -2 \\ 3 \end{bmatrix}.$

To find the eigenvectors associated with $\lambda_2 = 7$, we solve the system

$$(7I_2 - A)\mathbf{x} = \begin{bmatrix} 4 & -2 \\ -6 & 3 \end{bmatrix} \begin{bmatrix} x_1 \\ x_2 \end{bmatrix} = \begin{bmatrix} 0 \\ 0 \end{bmatrix}$$

to obtain

$$\begin{bmatrix} \frac{1}{2}r \\ r \end{bmatrix} = r \begin{bmatrix} \frac{1}{2} \\ 1 \end{bmatrix}$$

where $r \neq 0$ is any real number. Choose $\mathbf{x}_2 = \begin{bmatrix} 1 \\ 2 \end{bmatrix}.$ Since

$$\left\{ \begin{bmatrix} -2 \\ 3 \end{bmatrix}, \begin{bmatrix} 1 \\ 2 \end{bmatrix} \right\}$$

is a basis for R^2, A is diagonalizable. We have

$$P = \begin{bmatrix} -2 & 1 \\ 3 & 2 \end{bmatrix} \quad \text{and} \quad D = P^{-1}AP = \begin{bmatrix} 0 & 0 \\ 0 & 7 \end{bmatrix}.$$

(c) The characteristic polynomial is

$$p(\lambda) = \det(\lambda I_3 - A) = \begin{vmatrix} \lambda - 2 & 2 & -3 \\ 0 & \lambda - 3 & 2 \\ 0 & 1 & \lambda - 2 \end{vmatrix} = \lambda^3 - 7\lambda^2 + 14\lambda - 8 = (\lambda - 2)(\lambda - 4)(\lambda - 1)$$

so the eigenvalues are $\lambda_1 = 2$, $\lambda_2 = 4$, $\lambda_3 = 1$. To find the eigenvectors associated with $\lambda_1 = 2$, we solve the system

$$(2I_3 - A)\mathbf{x} = \begin{bmatrix} 0 & 2 & -3 \\ 0 & -1 & 2 \\ 0 & 1 & 0 \end{bmatrix} \begin{bmatrix} x_1 \\ x_2 \\ x_3 \end{bmatrix} = \begin{bmatrix} 0 \\ 0 \\ 0 \end{bmatrix}$$

to obtain

$$\begin{bmatrix} r \\ 0 \\ 0 \end{bmatrix} = r \begin{bmatrix} 1 \\ 0 \\ 0 \end{bmatrix}$$

where $r \neq 0$ is any real number. Choose $\mathbf{x}_1 = \begin{bmatrix} 1 \\ 0 \\ 0 \end{bmatrix}$.

To find the eigenvectors associated with $\lambda_2 = 4$, we solve the system

$$(4I_3 - A)\mathbf{x} = \begin{bmatrix} 2 & 2 & -3 \\ 0 & 1 & 2 \\ 0 & 1 & 2 \end{bmatrix} \begin{bmatrix} x_1 \\ x_2 \\ x_3 \end{bmatrix} = \begin{bmatrix} 0 \\ 0 \\ 0 \end{bmatrix}$$

to obtain

$$\begin{bmatrix} \frac{7}{2}r \\ -2r \\ r \end{bmatrix} = r \begin{bmatrix} \frac{7}{2} \\ -2 \\ 1 \end{bmatrix}.$$

Choose $\mathbf{x}_3 = \begin{bmatrix} 7 \\ -4 \\ 2 \end{bmatrix}$.

To find the eigenvectors associated with $\lambda_3 = 1$, we solve the system

$$(1I_3 - A)\mathbf{x} = \begin{bmatrix} -1 & 2 & -3 \\ 0 & -2 & 2 \\ 0 & 1 & -1 \end{bmatrix} \begin{bmatrix} x_1 \\ x_2 \\ x_3 \end{bmatrix} = \begin{bmatrix} 0 \\ 0 \\ 0 \end{bmatrix}$$

to obtain the eigenvectors

$$\begin{bmatrix} -r \\ r \\ r \end{bmatrix} = r \begin{bmatrix} -1 \\ 1 \\ 1 \end{bmatrix}$$

where $r \neq 0$ is any real number. Choose $\mathbf{x}_3 = \begin{bmatrix} -1 \\ 1 \\ 1 \end{bmatrix}$. The eigenvectors $\{\mathbf{x}_1, \mathbf{x}_2, \mathbf{x}_3\}$ form a basis

for R^3 so A is diagonalizable. We have

$$P = \begin{bmatrix} 1 & 7 & -1 \\ 0 & -4 & 1 \\ 0 & 2 & 1 \end{bmatrix} \quad \text{and} \quad D = P^{-1}AP = \begin{bmatrix} 2 & 0 & 0 \\ 0 & 4 & 0 \\ 0 & 0 & 1 \end{bmatrix}.$$

(d) The characteristic polynomial is

$$p(\lambda) = \det(\lambda I_3 - A) = \begin{vmatrix} \lambda & 2 & -1 \\ -1 & \lambda - 3 & 1 \\ 0 & 0 & \lambda - 1 \end{vmatrix} = \lambda^3 - 4\lambda^2 + 5\lambda - 2 = (\lambda - 1)^2(\lambda - 2)$$

so the eigenvalues are $\lambda_1 = \lambda_2 = 1$, $\lambda_3 = 2$. To find the eigenvectors associated with $\lambda_1 = \lambda_2 = 1$, we solve the system

$$(1I_3 - A)\mathbf{x} = \begin{bmatrix} 1 & 2 & -1 \\ -1 & -2 & 1 \\ 0 & 0 & 0 \end{bmatrix} \begin{bmatrix} x_1 \\ x_2 \\ x_3 \end{bmatrix} = \begin{bmatrix} 0 \\ 0 \\ 0 \end{bmatrix}$$

to obtain the eigenvectors

$$\begin{bmatrix} -2r + s \\ r \\ s \end{bmatrix} = r \begin{bmatrix} -2 \\ 1 \\ 0 \end{bmatrix} + s \begin{bmatrix} 1 \\ 0 \\ 1 \end{bmatrix}$$

where r, s are any nonzero real numbers. Choose

$$\mathbf{x}_1 = \begin{bmatrix} -2 \\ 1 \\ 0 \end{bmatrix}, \quad \mathbf{x}_2 = \begin{bmatrix} 1 \\ 0 \\ 1 \end{bmatrix}.$$

To find the eigenvectors associated with $\lambda_3 = 2$, we solve the system

$$(2I_3 - A)\mathbf{x} = \begin{bmatrix} 2 & 2 & -1 \\ -1 & -1 & 1 \\ 0 & 0 & 1 \end{bmatrix} \begin{bmatrix} x_1 \\ x_2 \\ x_3 \end{bmatrix} = \begin{bmatrix} 0 \\ 0 \\ 0 \end{bmatrix}$$

to obtain

$$\begin{bmatrix} -r \\ r \\ 0 \end{bmatrix} = r \begin{bmatrix} -1 \\ 1 \\ 0 \end{bmatrix}$$

where $r \neq 0$ is any real number. Choose

$$\mathbf{x}_3 = \begin{bmatrix} -1 \\ 1 \\ 0 \end{bmatrix}.$$

Since the eigenvectors $\{\mathbf{x}_1, \mathbf{x}_2, \mathbf{x}_3\}$ form a basis for R^3, A is diagonalizable. We have

$$P = \begin{bmatrix} -2 & 1 & -1 \\ 1 & 0 & 1 \\ 0 & 1 & 0 \end{bmatrix} \quad \text{and} \quad D = P^{-1}AP = \begin{bmatrix} 1 & 0 & 0 \\ 0 & 1 & 0 \\ 0 & 0 & 2 \end{bmatrix}.$$

We note that if the basis of eigenvectors is reordered as $\{\mathbf{x}_1, \mathbf{x}_3, \mathbf{x}_2\}$, then the transition matrix is

$$P = \begin{bmatrix} -2 & -1 & 1 \\ 1 & 1 & 0 \\ 0 & 0 & 1 \end{bmatrix} \quad \text{and} \quad D = P^{-1}AP = \begin{bmatrix} 1 & 0 & 0 \\ 0 & 2 & 0 \\ 0 & 0 & 1 \end{bmatrix}.$$

17. Let A be the given matrix. We must determine the dimension of the solution space to the system $(\lambda I_n - A)\mathbf{x} = \mathbf{0}$.

(a) $8I_2 - A = \begin{bmatrix} 0 & -7 \\ 0 & 0 \end{bmatrix} \implies$ solution space has dimension 1 so A is defective.

(b) $3I_3 - A = \begin{bmatrix} 0 & 0 & 0 \\ 2 & 0 & 2 \\ -2 & 0 & -2 \end{bmatrix} \implies$ solution space has dimension 2 so A is not defective since $\lambda = 3$ has multiplicity 2.

(c) $0I_3 - A = \begin{bmatrix} -3 & -3 & -3 \\ -3 & -3 & -3 \\ 3 & 3 & 3 \end{bmatrix} \implies$ solution space has dimension 2 so A is not defective since $\lambda = 0$ has multiplicity 2.

(d) $\lambda = 1:$ $1I_4 - A = \begin{bmatrix} 1 & 0 & -1 & 0 \\ 0 & 1 & 0 & 1 \\ -1 & 0 & 1 & 0 \\ 0 & 1 & 0 & 1 \end{bmatrix}$ $\implies$ solution space has dimension 2

$\lambda = -1:$ $(-1)I_4 - A = \begin{bmatrix} -1 & 0 & -1 & 0 \\ 0 & -1 & 0 & 1 \\ -1 & 0 & -1 & 0 \\ 0 & 1 & 0 & -1 \end{bmatrix}$ $\implies$ solution space has dimension 2.

Therefore A is not defective.

19. Following the hint we find that A has eigenvalues and eigenvectors given by

$$\lambda_1 = 2, \mathbf{x}_1 = \begin{bmatrix} 5 \\ 1 \end{bmatrix}; \quad \lambda_2 = -2, \mathbf{x}_2 = \begin{bmatrix} 1 \\ 1 \end{bmatrix}.$$

Then for $P = \begin{bmatrix} \mathbf{x}_1 & \mathbf{x}_2 \end{bmatrix} = \begin{bmatrix} 5 & 1 \\ 1 & 1 \end{bmatrix}$ we have $A = PDP^{-1}$, where $D = \begin{bmatrix} 2 & 0 \\ 0 & -2 \end{bmatrix}$. Hence

$$A^9 = (PDP^{-1})^9 = PD^9P^{-1} = P \begin{bmatrix} 2^9 & 0 \\ 0 & (-2)^9 \end{bmatrix} P^{-1} = P \begin{bmatrix} 512 & 0 \\ 0 & -512 \end{bmatrix} P^{-1}.$$

Computing P^{-1} we have $P^{-1} = \begin{bmatrix} \frac{1}{4} & -\frac{1}{4} \\ -\frac{1}{4} & \frac{5}{4} \end{bmatrix}$. Then forming the product above we obtain

$$A^9 = \begin{bmatrix} 768 & -1280 \\ 256 & -768 \end{bmatrix}.$$

21. Since A and B are nonsingular, A^{-1} and B^{-1} exist. Then $BA = A^{-1}(AB)A$. Therefore AB and BA are similar and hence by Theorem 7.5 they have the same eigenvalues.

23. Let A be diagonalizable with $A = PDP^{-1}$, where D is diagonal.

   (a) $A^T = (PDP^{-1})^T = (P^{-1})^T D^T P^T = QDQ^{-1}$, where $Q = (P^{-1})^T$. Thus A^T is similar to a diagonal matrix and hence is diagonalizable.

   (b) $A^k = (PDP^{-1})^k = PD^kP^{-1}$. Since D^k is diagonal we have A^k is similar to a diagonal matrix and hence diagonalizable.

25. First observe the difference between this result and Theorem 7.5. Theorem 7.5 shows that if *all* the eigenvalues of A are distinct, then the associated eigenvectors are linearly independent. In the present exercise, we are asked to show that if any subset of k eigenvalues are distinct, then the associated eigenvectors are linearly independent. To prove this result, we basically imitate the proof of Theorem 7.5

   Suppose that $S = \{\mathbf{x}_1, \ldots, \mathbf{x}_k\}$ is linearly dependent. Then Theorem 4.7 implies that some vector $\mathbf{x}_j$ is a linear combination of the preceding vectors in S. We can assume that $S_1 = \{\mathbf{x}_1, \mathbf{x}_2, \ldots, \mathbf{x}_{j-1}\}$ is linearly independent, for otherwise one of the vectors in S_1 is a linear combination of the preceding ones, and we can choose a new set S_2, and so on. We thus have that S_1 is linearly independent and that

$$\mathbf{x}_j = a_1\mathbf{x}_1 + a_2\mathbf{x}_2 + \cdots + a_{j-1}\mathbf{x}_{j-1}, \tag{7.1}$$

   where $a_1, a_2, \ldots, a_{j-1}$ are real numbers. This means that

$$A\mathbf{x}_j = A(a_1\mathbf{x}_1 + a_2\mathbf{x}_2 + \cdots + a_{j-1}\mathbf{x}_{j-1}) = a_1A\mathbf{x}_1 + a_2A\mathbf{x}_2 + \cdots + a_{j-1}A\mathbf{x}_{j-1}. \tag{7.2}$$

Since $\lambda_1, \lambda_2, \ldots, \lambda_j$ are eigenvalues and $\mathbf{x}_1, \mathbf{x}_2, \ldots, \mathbf{x}_j$ are associated eigenvectors, we know that $A\mathbf{x}_i = \lambda_i \mathbf{x}_i$ for $i = 1, 2, \ldots, n$. Substituting in (7.2), we have

$$\lambda_j \mathbf{x}_j = a_1 \lambda_1 \mathbf{x}_1 + a_2 \lambda_2 \mathbf{x}_2 + \cdots + a_{j-1} \lambda_{j-1} \mathbf{x}_{j-1}. \tag{7.3}$$

Multiplying (7.1) by λ_j, we get

$$\lambda_j \mathbf{x}_j = \lambda_j a_1 \mathbf{x}_1 + \lambda_j a_2 \mathbf{x}_2 + \cdots + \lambda_j a_{j-1} \mathbf{x}_{j-1}. \tag{7.4}$$

Subtracting (7.4) from (7.3), we have

$$\mathbf{0} = \lambda_j \mathbf{x}_j - \lambda_j \mathbf{x}_j = a_1 (\lambda_1 - \lambda_j) \mathbf{x}_1 + a_2 (\lambda_2 - \lambda_j) \mathbf{x}_2 + \cdots + a_{j-1} (\lambda_{j-1} - \lambda_j) \mathbf{x}_{j-1}.$$

Since S_1 is linearly independent, we must have

$$a_1 (\lambda_1 - \lambda_j) = 0, \quad a_2 (\lambda_2 - \lambda_j) = 0, \quad \ldots, \quad a_{j-1} (\lambda_{j-1} - \lambda_j) = 0.$$

Now $(\lambda_1 - \lambda_j) \neq 0$, $(\lambda_2 - \lambda_j) \neq 0$, $\ldots$, $(\lambda_{j-1} - \lambda_j) \neq 0$, since the λ's are distinct, which implies that

$$a_1 = a_2 = \cdots = a_{j-1} = 0.$$

This means that $\mathbf{x}_j = \mathbf{0}$, which is impossible if $\mathbf{x}_j$ is an eigenvector. Hence S is linearly independent, so A is diagonalizable.

27. Let P be a nonsingular matrix such that $P^{-1}AP = D$. Then

$$\mathrm{Tr}(D) = \mathrm{Tr}(P^{-1}AP) = \mathrm{Tr}(P^{-1}(AP)) = \mathrm{Tr}((AP)P^{-1}) = \mathrm{Tr}(APP^{-1}) = \mathrm{Tr}(AI_n) = \mathrm{Tr}(A).$$

Section 7.3, p. 475

1. We use Definition 7.4 and show that $P^T P = I_3$.

$$P^T P = \begin{bmatrix} \frac{2}{3} & \frac{2}{3} & \frac{1}{3} \\ -\frac{2}{3} & \frac{1}{3} & \frac{2}{3} \\ \frac{1}{3} & -\frac{2}{3} & \frac{2}{3} \end{bmatrix} \begin{bmatrix} \frac{2}{3} & -\frac{2}{3} & \frac{1}{3} \\ \frac{2}{3} & \frac{1}{3} & -\frac{2}{3} \\ \frac{1}{3} & \frac{2}{3} & \frac{2}{3} \end{bmatrix} = \begin{bmatrix} 1 & 0 & 0 \\ 0 & 1 & 0 \\ 0 & 0 & 1 \end{bmatrix} = I_3.$$

3. Since A and B are orthogonal matrices, we have $A^T A = I_n$ and $B^T B = I_n$. Hence

$$(AB)^T AB = B^T A^T AB = B^T (A^T A)B = B^T I_n B = B^T B = I_n.$$

Thus by Definition 7.4 AB is an orthonormal matrix.

5. Let the columns of A be denoted by $\mathbf{u}_1, \mathbf{u}_2, \ldots, \mathbf{u}_n$. Suppose that A is an orthogonal matrix. Then $A^T A = I_n$. Therefore

$$I_n = A^T A = \begin{bmatrix} \mathbf{u}_1 & \mathbf{u}_2 & \cdots & \mathbf{u}_n \end{bmatrix}^T \begin{bmatrix} \mathbf{u}_1 & \mathbf{u}_2 & \cdots & \mathbf{u}_n \end{bmatrix}$$

$$= \begin{bmatrix} \mathbf{u}_1^T \\ \mathbf{u}_2^T \\ \vdots \\ \mathbf{u}_n^T \end{bmatrix} \begin{bmatrix} \mathbf{u}_1 & \mathbf{u}_2 & \cdots & \mathbf{u}_n \end{bmatrix} = \begin{bmatrix} \mathbf{u}_1^T \mathbf{u}_1 & \mathbf{u}_1^T \mathbf{u}_2 & \cdots & \mathbf{u}_1^T \mathbf{u}_n \\ \mathbf{u}_2^T \mathbf{u}_1 & \mathbf{u}_2^T \mathbf{u}_2 & \cdots & \mathbf{u}_2^T \mathbf{u}_n \\ \vdots & \vdots & \cdots & \vdots \\ \mathbf{u}_n^T \mathbf{u}_1 & \mathbf{u}_n^T \mathbf{u}_2 & \cdots & \mathbf{u}_n^T \mathbf{u}_n \end{bmatrix}$$

Equating corresponding entries gives

$$\mathbf{u}_i^T \mathbf{u}_j = \begin{cases} 1, & \text{for } i = j \\ 0, & \text{for } i \neq j. \end{cases}$$

Hence the columns of A form an orthonormal set. The preceding steps are reversible and show that if the columns of A form an orthonormal set then A is an orthogonal matrix. Thus, A is orthogonal if and only if the columns of A form an orthonormal set. A similar proof shows that A is orthogonal if and only if the rows of A form an orthonormal set.

7. Let $P = \begin{bmatrix} 0 & -\frac{1}{\sqrt{5}} & \frac{2}{\sqrt{5}} \\ 1 & 0 & 0 \\ 0 & \frac{2}{\sqrt{5}} & \frac{1}{\sqrt{5}} \end{bmatrix} = \frac{1}{\sqrt{5}} \begin{bmatrix} 0 & -1 & 2 \\ \sqrt{5} & 0 & 0 \\ 0 & 2 & 1 \end{bmatrix}$. Then

$$P^T P = \frac{1}{\sqrt{5}} \begin{bmatrix} 0 & \sqrt{5} & 0 \\ -1 & 0 & 2 \\ 2 & 0 & 1 \end{bmatrix} \frac{1}{\sqrt{5}} \begin{bmatrix} 0 & -1 & 2 \\ \sqrt{5} & 0 & 0 \\ 0 & 2 & 1 \end{bmatrix} = \frac{1}{5} \begin{bmatrix} 5 & 0 & 0 \\ 0 & 5 & 0 \\ 0 & 0 & 5 \end{bmatrix} = I_3.$$

For $A = \begin{bmatrix} 0 & 0 & -2 \\ 0 & -2 & 0 \\ -2 & 0 & 3 \end{bmatrix}$ we have

$$P^{-1}AP = P^T AP = \frac{1}{\sqrt{5}} \begin{bmatrix} 0 & \sqrt{5} & 0 \\ -1 & 0 & 2 \\ 2 & 0 & 1 \end{bmatrix} \begin{bmatrix} 0 & 0 & -2 \\ 0 & -2 & 0 \\ -2 & 0 & 3 \end{bmatrix} \frac{1}{\sqrt{5}} \begin{bmatrix} 0 & -1 & 2 \\ \sqrt{5} & 0 & 0 \\ 0 & 2 & 1 \end{bmatrix}$$

$$= \frac{1}{5} \begin{bmatrix} 0 & -2\sqrt{5} & 0 \\ -4 & 0 & 8 \\ -2 & 0 & -1 \end{bmatrix} \begin{bmatrix} 0 & -1 & 2 \\ \sqrt{5} & 0 & 0 \\ 0 & 2 & 1 \end{bmatrix}$$

$$= \frac{1}{5} \begin{bmatrix} -10 & 0 & 0 \\ 0 & 20 & 0 \\ 0 & 0 & -5 \end{bmatrix} = \begin{bmatrix} -2 & 0 & 0 \\ 0 & 4 & 0 \\ 0 & 0 & -1 \end{bmatrix}.$$

9. Let $B = \begin{bmatrix} \cos\phi & -\sin\phi \\ \sin\phi & \cos\phi \end{bmatrix}$.

(a) Since

$$B^T B = \begin{bmatrix} \cos\phi & \sin\phi \\ -\sin\phi & \cos\phi \end{bmatrix} \begin{bmatrix} \cos\phi & -\sin\phi \\ \sin\phi & \cos\phi \end{bmatrix} = \begin{bmatrix} \cos^2\phi + \sin^2\phi & 0 \\ 0 & \cos^2\phi + \sin^2\phi \end{bmatrix} = I_2$$

it follows that B is an orthogonal matrix.

(b) Let $A = \begin{bmatrix} a & c \\ b & d \end{bmatrix}$. Then, since A is orthogonal, we have

$$A^T A = \begin{bmatrix} a^2 + b^2 & ac + bd \\ ac + bd & c^2 + d^2 \end{bmatrix} = I_2.$$

Equating corresponding entries and using the fact that $\det(A) = \pm 1$ since A is orthogonal (see Exercise 8), we have the following four equations:

$$a^2 + b^2 = 1 \tag{7.5}$$
$$c^2 + d^2 = 1 \tag{7.6}$$
$$ac + bd = 0 \tag{7.7}$$
$$ad - bc = \pm 1 \tag{7.8}$$

From (7.5) there exists a ϕ_1 such that $a = \cos\phi_1$ and $b = \sin\phi_1$.
From (7.6) there exists a ϕ_2 such that $c = \cos\phi_2$ and $d = \sin\phi_2$.
Using (7.7) and a trigonometric identity we have

$$\cos\phi_1 \cos\phi_2 + \sin\phi_1 \sin\phi_2 = \cos(\phi_2 - \phi_1) = 0.$$

Using (7.8) and a trigonometric identity we have

$$\cos\phi_1 \sin\phi_2 - \sin\phi_1 \cos\phi_2 = \sin(\phi_2 - \phi_1) = \pm 1.$$

These last two expressions imply that $\phi_2 - \phi_1 = \pm\frac{\pi}{2}$, or equivalently that $\phi_2 = \phi_1 \pm \frac{\pi}{2}$. Hence we have $\cos\phi_2 = \mp\sin\phi_1$ and $\sin\phi_2 = \pm\cos\phi_1$. Thus either

$$A = \begin{bmatrix} \cos\phi & -\sin\phi \\ \sin\phi & \cos\phi \end{bmatrix} \quad \text{or} \quad A = \begin{bmatrix} \cos\phi & \sin\phi \\ \sin\phi & -\cos\phi \end{bmatrix}.$$

11. If θ is the angle between $\mathbf{x}$ and $\mathbf{y}$, then

$$\begin{aligned}
\cos\theta &= \frac{(\mathbf{x}, \mathbf{y})}{\|\mathbf{x}\|\,\|\mathbf{y}\|} = \frac{(\mathbf{x}, \mathbf{y})}{\sqrt{(\mathbf{x}, \mathbf{x})}\,\sqrt{(\mathbf{y}, \mathbf{y})}} = \frac{(\mathbf{x}, I_n\mathbf{y})}{\sqrt{(\mathbf{x}, I_n\mathbf{x})}\,\sqrt{(\mathbf{y}, I_n\mathbf{y})}} \\
&= \frac{(\mathbf{x}, A^T A\mathbf{y})}{\sqrt{(\mathbf{x}, A^T A\mathbf{x})}\,\sqrt{(\mathbf{y}, A^T A\mathbf{y})}} \qquad \text{(since } A \text{ is orthogonal } A^T A = I_n) \\
&= \frac{(A\mathbf{x}, A\mathbf{y})}{\sqrt{(A\mathbf{x}, A\mathbf{x})}\,\sqrt{(A\mathbf{y}, A\mathbf{y})}} \qquad \text{(by Equation (3) in Section 5.3)} \\
&= \frac{(L(\mathbf{x}), L(\mathbf{y}))}{\|L(\mathbf{x})\|\,\|L(\mathbf{y})\|}
\end{aligned}$$

Hence it follows that the angle between $L(\mathbf{x})$ and $L(\mathbf{y})$ is also θ.

13. From Example 8, Section 1.6, the matrix representing a linear transformation that performs a counterclockwise rotation through $\pi/4$ is

$$A = \begin{bmatrix} \cos\pi/4 & -\sin\pi/4 \\ \sin\pi/4 & \cos\pi/4 \end{bmatrix} = \begin{bmatrix} \frac{1}{2}\sqrt{2} & -\frac{1}{2}\sqrt{2} \\ \frac{1}{2}\sqrt{2} & \frac{1}{2}\sqrt{2} \end{bmatrix}.$$

Then

$$A^T A = \begin{bmatrix} \frac{1}{2}\sqrt{2} & \frac{1}{2}\sqrt{2} \\ -\frac{1}{2}\sqrt{2} & \frac{1}{2}\sqrt{2} \end{bmatrix} \begin{bmatrix} \frac{1}{2}\sqrt{2} & -\frac{1}{2}\sqrt{2} \\ \frac{1}{2}\sqrt{2} & \frac{1}{2}\sqrt{2} \end{bmatrix} = \begin{bmatrix} 1 & 0 \\ 0 & 1 \end{bmatrix}.$$

Therefore A is an orthogonal matrix.

15. Let $A = \begin{bmatrix} 2 & 2 \\ 2 & 2 \end{bmatrix}$. The characteristic polynomial of A is

$$\det(\lambda I_3 - A) = \begin{vmatrix} \lambda - 2 & -2 \\ -2 & \lambda - 2 \end{vmatrix} = (\lambda - 2)^2 - 4 = \lambda^2 - 4\lambda = \lambda(\lambda - 4).$$

Hence A has eigenvalues $\lambda_1 = 0$, and $\lambda_2 = 4$. Since the eigenvalues are distinct the corresponding eigenvectors will be orthogonal. For $\lambda_1 = 0$ we have the system

$$\begin{bmatrix} -2 & -2 \\ -2 & -2 \end{bmatrix} \begin{bmatrix} x_1 \\ x_2 \end{bmatrix} = \begin{bmatrix} 0 \\ 0 \end{bmatrix}.$$

Row reducing the coefficient matrix we have the equivalent system

$$\begin{bmatrix} 1 & 1 \\ 0 & 0 \end{bmatrix} \begin{bmatrix} x_1 \\ x_2 \end{bmatrix} = \begin{bmatrix} 0 \\ 0 \end{bmatrix}.$$

The solution is $x_1 = -r$, $x_2 = r$ and hence

$$\mathbf{x}_1 = \begin{bmatrix} 1 \\ -1 \end{bmatrix}$$

is an eigenvector. For $\lambda_2 = 4$ we have the system

$$\begin{bmatrix} 2 & -2 \\ -2 & 2 \end{bmatrix} \begin{bmatrix} x_1 \\ x_2 \end{bmatrix} = \begin{bmatrix} 0 \\ 0 \end{bmatrix}.$$

Row reducing the coefficient matrix we have the equivalent system

$$\begin{bmatrix} 1 & -1 \\ 0 & 0 \end{bmatrix} \begin{bmatrix} x_1 \\ x_2 \end{bmatrix} = \begin{bmatrix} 0 \\ 0 \end{bmatrix}.$$

The solution is $x_1 = x_2 = r$ and hence

$$\mathbf{x}_2 = \begin{bmatrix} 1 \\ 1 \end{bmatrix}$$

is an eigenvector. To form an orthogonal matrix P we normalize the vectors $\mathbf{x}_1$ and $\mathbf{x}_2$:

$$\mathbf{y}_1 = \frac{\mathbf{x}_1}{\|\mathbf{x}_1\|} = \begin{bmatrix} \frac{1}{2}\sqrt{2} \\ -\frac{1}{2}\sqrt{2} \end{bmatrix} \quad \text{and} \quad \mathbf{y}_2 = \frac{\mathbf{x}_2}{\|\mathbf{x}_2\|} = \begin{bmatrix} \frac{1}{2}\sqrt{2} \\ \frac{1}{2}\sqrt{2} \end{bmatrix}.$$

Then

$$P = \begin{bmatrix} \mathbf{y}_1 & \mathbf{y}_2 \end{bmatrix} = \begin{bmatrix} \frac{1}{2}\sqrt{2} & \frac{1}{2}\sqrt{2} \\ -\frac{1}{2}\sqrt{2} & \frac{1}{2}\sqrt{2} \end{bmatrix} \quad \text{and} \quad P^T A P = \begin{bmatrix} 0 & 0 \\ 0 & 4 \end{bmatrix}.$$

17. Let $A = \begin{bmatrix} 0 & 0 & 0 \\ 0 & 2 & 2 \\ 0 & 2 & 2 \end{bmatrix}$. The characteristic polynomial is

$$\det(\lambda I_3 - A) = \begin{vmatrix} \lambda & 0 & 0 \\ 0 & \lambda - 2 & -2 \\ 0 & -2 & \lambda - 2 \end{vmatrix} = \lambda^2(\lambda - 4).$$

Hence A has eigenvalues $\lambda_1 = 0$, $\lambda_2 = 0$, and $\lambda_3 = 4$. For $\lambda_1 = \lambda_2 = 0$ we have the linear system

$$\begin{bmatrix} 0 & 0 & 0 \\ 0 & -2 & -2 \\ 0 & -2 & -2 \end{bmatrix} \begin{bmatrix} x_1 \\ x_2 \\ x_3 \end{bmatrix} = \begin{bmatrix} 0 \\ 0 \\ 0 \end{bmatrix}.$$

Row reducing the coefficient matrix we have the equivalent linear system

$$\begin{bmatrix} 0 & 1 & 1 \\ 0 & 0 & 0 \\ 0 & 0 & 0 \end{bmatrix} \begin{bmatrix} x_1 \\ x_2 \\ x_3 \end{bmatrix} = \begin{bmatrix} 0 \\ 0 \\ 0 \end{bmatrix}.$$

The solution is $x_1 = r$, $x_2 = -s$, $x_3 = s$ and hence

$$\mathbf{x}_1 = \begin{bmatrix} 1 \\ 0 \\ 0 \end{bmatrix} \quad \text{and} \quad \mathbf{x}_2 = \begin{bmatrix} 0 \\ -1 \\ 1 \end{bmatrix}$$

are eigenvectors. Note that $\mathbf{x}_1$ and $\mathbf{x}_2$ are orthogonal in this case. For $\lambda_3 = 4$ we have the linear system

$$\begin{bmatrix} 4 & 0 & 0 \\ 0 & 2 & -2 \\ 0 & -2 & 2 \end{bmatrix} \begin{bmatrix} x_1 \\ x_2 \\ x_3 \end{bmatrix} = \begin{bmatrix} 0 \\ 0 \\ 0 \end{bmatrix}.$$

Row reducing the coefficient matrix we have the equivalent linear system

$$\begin{bmatrix} 1 & 0 & 0 \\ 0 & 1 & -1 \\ 0 & 0 & 0 \end{bmatrix} \begin{bmatrix} x_1 \\ x_2 \\ x_3 \end{bmatrix} = \begin{bmatrix} 0 \\ 0 \\ 0 \end{bmatrix}.$$

The solution is $x_1 = 0$, $x_2 = r$, $x_3 = r$ and hence

$$\mathbf{x}_3 = \begin{bmatrix} 0 \\ 1 \\ 1 \end{bmatrix}$$

is an eigenvector. Vector $\mathbf{x}_3$ is guaranteed to be orthogonal to both $\mathbf{x}_1$ and $\mathbf{x}_2$ since the corresponding eigenvalues are distinct. To form P we need only normalize the eigenvectors. Let

$$\mathbf{y}_1 = \mathbf{x}_1, \quad \mathbf{y}_2 = \frac{\mathbf{x}_2}{\sqrt{2}}, \quad \mathbf{y}_3 = \frac{\mathbf{x}_3}{\sqrt{2}}.$$

Then

$$P = \begin{bmatrix} \mathbf{y}_1 & \mathbf{y}_2 & \mathbf{y}_3 \end{bmatrix} = \begin{bmatrix} 1 & 0 & 0 \\ 0 & -\frac{1}{2}\sqrt{2} & \frac{1}{2}\sqrt{2} \\ 0 & \frac{1}{2}\sqrt{2} & \frac{1}{2}\sqrt{2} \end{bmatrix} \quad \text{and} \quad P^T A P = \begin{bmatrix} 0 & 0 & 0 \\ 0 & 0 & 0 \\ 0 & 0 & 4 \end{bmatrix}.$$

19. Let $A = \begin{bmatrix} 0 & -1 & -1 \\ -1 & 0 & -1 \\ -1 & -1 & 0 \end{bmatrix}$. The characteristic polynomial is

$$\det(\lambda I_3 - A) = \begin{vmatrix} \lambda & 1 & 1 \\ 1 & \lambda & 1 \\ 1 & 1 & \lambda \end{vmatrix} = (\lambda - 1)^2 (\lambda + 2).$$

Hence A has eigenvalues $\lambda_1 = -2$, $\lambda_2 = 1$, and $\lambda_3 = 1$. For $\lambda_1 = -2$, we have the linear system

$$\begin{bmatrix} -2 & 1 & 1 \\ 1 & -2 & 1 \\ 1 & 1 & -2 \end{bmatrix} \begin{bmatrix} x_1 \\ x_2 \\ x_3 \end{bmatrix} = \begin{bmatrix} 0 \\ 0 \\ 0 \end{bmatrix}.$$

Row reducing the coefficient matrix we have the equivalent system

$$\begin{bmatrix} 1 & 0 & -1 \\ 0 & 1 & -1 \\ 0 & 0 & 0 \end{bmatrix} \begin{bmatrix} x_1 \\ x_2 \\ x_3 \end{bmatrix} = \begin{bmatrix} 0 \\ 0 \\ 0 \end{bmatrix}.$$

The solution is $x_1 = x_2 = x_3 = r$ and hence

$$\mathbf{x}_1 = \begin{bmatrix} 1 \\ 1 \\ 1 \end{bmatrix}$$

is an eigenvector. For $\lambda_2 = \lambda_3 = 1$ we have the linear system

$$\begin{bmatrix} 1 & 1 & 1 \\ 1 & 1 & 1 \\ 1 & 1 & 1 \end{bmatrix} \begin{bmatrix} x_1 \\ x_2 \\ x_3 \end{bmatrix} = \begin{bmatrix} 0 \\ 0 \\ 0 \end{bmatrix}.$$

Row reducing the coefficient matrix we have the equivalent linear system

$$\begin{bmatrix} 1 & 1 & 1 \\ 0 & 0 & 0 \\ 0 & 0 & 0 \end{bmatrix} \begin{bmatrix} x_1 \\ x_2 \\ x_3 \end{bmatrix} = \begin{bmatrix} 0 \\ 0 \\ 0 \end{bmatrix}.$$

The solution is $x_1 = -r - s$, $x_2 = r$, $x_3 = s$ and hence

$$\mathbf{x}_2 = \begin{bmatrix} -1 \\ 1 \\ 0 \end{bmatrix} \quad \text{and} \quad \mathbf{x}_3 = \begin{bmatrix} -1 \\ 0 \\ 1 \end{bmatrix}$$

are eigenvectors. The vectors $\mathbf{x}_2$ and $\mathbf{x}_3$ are orthogonal to $\mathbf{x}_1$ since the corresponding eigenvalues are distinct. But $\mathbf{x}_2$ and $\mathbf{x}_3$ are not orthogonal. Thus we use the Gram-Schmidt process to determine a pair of orthogonal eigenvectors corresponding to eigenvalue 1. Let $\mathbf{y}_2 = \mathbf{x}_2$ and then

$$\mathbf{y}_3 = \mathbf{x}_3 - \frac{(\mathbf{x}_3, \mathbf{y}_2)}{(\mathbf{y}_2, \mathbf{y}_2)} \mathbf{x}_2 = \begin{bmatrix} -\frac{1}{2} \\ -\frac{1}{2} \\ 1 \end{bmatrix}.$$

To form P we need only normalize the eigenvectors. Let

$$\mathbf{z}_1 = \frac{\mathbf{x}_1}{\sqrt{3}}, \quad \mathbf{z}_2 = \frac{\mathbf{y}_2}{\sqrt{2}}, \quad \mathbf{z}_3 = \frac{\mathbf{y}_3}{\sqrt{\frac{3}{2}}} = \frac{2}{\sqrt{6}} \mathbf{y}_3.$$

Then

$$P = \begin{bmatrix} \mathbf{z}_1 & \mathbf{z}_2 & \mathbf{z}_3 \end{bmatrix} = \begin{bmatrix} \frac{1}{\sqrt{3}} & -\frac{1}{\sqrt{2}} & -\frac{1}{\sqrt{6}} \\ \frac{1}{\sqrt{3}} & \frac{1}{\sqrt{2}} & -\frac{1}{\sqrt{6}} \\ \frac{1}{\sqrt{3}} & 0 & \frac{2}{\sqrt{6}} \end{bmatrix} \quad \text{and} \quad P^T A P = \begin{bmatrix} -2 & 0 & 0 \\ 0 & 1 & 0 \\ 0 & 0 & 1 \end{bmatrix}.$$

21. Let $A = \begin{bmatrix} 2 & 1 \\ 1 & 2 \end{bmatrix}$. The characteristic polynomial is

$$\det(\lambda I_2 - A) = \begin{vmatrix} \lambda - 2 & -1 \\ -1 & \lambda - 2 \end{vmatrix} = (\lambda - 2)^2 - 1 = \lambda^2 - 4\lambda + 3 = (\lambda - 1)(\lambda - 3).$$

Hence A has eigenvalues $\lambda_1 = 3$ and $\lambda_2 = 1$. Since the eigenvalues are distinct the corresponding eigenvectors will be orthogonal. For $\lambda_1 = 3$ we have the system

$$\begin{bmatrix} 1 & -1 \\ -1 & 1 \end{bmatrix} \begin{bmatrix} x_1 \\ x_2 \end{bmatrix} = \begin{bmatrix} 0 \\ 0 \end{bmatrix}$$

Row reducing the coefficient matrix we have the equivalent system

$$\begin{bmatrix} 1 & -1 \\ 0 & 0 \end{bmatrix} \begin{bmatrix} x_1 \\ x_2 \end{bmatrix} = \begin{bmatrix} 0 \\ 0 \end{bmatrix}.$$

The solution is $x_1 = x_2 = r$ and hence

$$\mathbf{x}_1 = \begin{bmatrix} 1 \\ 1 \end{bmatrix}$$

is an eigenvector. For $\lambda_2 = 1$ we have the system

$$\begin{bmatrix} -1 & -1 \\ -1 & -1 \end{bmatrix} \begin{bmatrix} x_1 \\ x_2 \end{bmatrix} = \begin{bmatrix} 0 \\ 0 \end{bmatrix}.$$

Row reducing the coefficient matrix we have the equivalent system

$$\begin{bmatrix} 1 & 1 \\ 0 & 0 \end{bmatrix} \begin{bmatrix} x_1 \\ x_2 \end{bmatrix} = \begin{bmatrix} 0 \\ 0 \end{bmatrix}.$$

The solution is $x_1 = -r$, $x_2 = r$ and hence

$$\mathbf{x}_2 = \begin{bmatrix} -1 \\ 1 \end{bmatrix}$$

is an eigenvector. To form the orthogonal matrix P we normalize the vectors $\mathbf{x}_1$ and $\mathbf{x}_2$. Let

$$\mathbf{y}_1 = \frac{\mathbf{x}_1}{\|\mathbf{x}_1\|} = \begin{bmatrix} \frac{1}{\sqrt{2}} \\ \frac{1}{\sqrt{2}} \end{bmatrix} \quad \text{and} \quad \mathbf{y}_2 = \frac{\mathbf{x}_2}{\|\mathbf{x}_2\|} = \begin{bmatrix} -\frac{1}{\sqrt{2}} \\ \frac{1}{\sqrt{2}} \end{bmatrix}.$$

Then

$$P = \begin{bmatrix} \mathbf{y}_1 & \mathbf{y}_2 \end{bmatrix} = \begin{bmatrix} \frac{1}{\sqrt{2}} & -\frac{1}{\sqrt{2}} \\ \frac{1}{\sqrt{2}} & \frac{1}{\sqrt{2}} \end{bmatrix} \quad \text{and} \quad P^T A P = \begin{bmatrix} 3 & 0 \\ 0 & 1 \end{bmatrix}.$$

23. Let $A = \begin{bmatrix} 1 & 1 & 0 \\ 1 & 1 & 0 \\ 0 & 0 & 1 \end{bmatrix}$. The characteristic polynomial is

$$\det(\lambda I_3 - A) = \begin{vmatrix} \lambda - 1 & -1 & 0 \\ -1 & \lambda - 1 & 0 \\ 0 & 0 & \lambda - 1 \end{vmatrix} = \lambda(\lambda - 1)(\lambda - 2).$$

Hence A has eigenvalues $\lambda_1 = 1$, $\lambda_2 = 2$, and $\lambda_3 = 0$. For $\lambda_1 = 1$ we have the linear system

$$\begin{bmatrix} 0 & -1 & 0 \\ -1 & 0 & 0 \\ 0 & 0 & 0 \end{bmatrix} \begin{bmatrix} x_1 \\ x_2 \\ x_3 \end{bmatrix} = \begin{bmatrix} 0 \\ 0 \\ 0 \end{bmatrix}.$$

Row reducing the coefficient matrix we have the equivalent linear system

$$\begin{bmatrix} 1 & 0 & 0 \\ 0 & 1 & 0 \\ 0 & 0 & 0 \end{bmatrix} \begin{bmatrix} x_1 \\ x_2 \\ x_3 \end{bmatrix} = \begin{bmatrix} 0 \\ 0 \\ 0 \end{bmatrix}.$$

The solution is $x_1 = x_2 = 0$, $x_3 = r$ and hence

$$\mathbf{x}_1 = \begin{bmatrix} 0 \\ 0 \\ 1 \end{bmatrix}$$

is an eigenvector. For $\lambda_2 = 2$ we have the linear system

$$\begin{bmatrix} 1 & -1 & 0 \\ -1 & 1 & 0 \\ 0 & 0 & 1 \end{bmatrix} \begin{bmatrix} x_1 \\ x_2 \\ x_3 \end{bmatrix} = \begin{bmatrix} 0 \\ 0 \\ 0 \end{bmatrix}.$$

Row reducing the coefficient matrix we have the equivalent linear system

$$\begin{bmatrix} 1 & -1 & 0 \\ 0 & 0 & 1 \\ 0 & 0 & 0 \end{bmatrix} \begin{bmatrix} x_1 \\ x_2 \\ x_3 \end{bmatrix} = \begin{bmatrix} 0 \\ 0 \\ 0 \end{bmatrix}.$$

The solution is $x_1 = x_2 = r$, $x_3 = 0$ and hence

$$\mathbf{x}_2 = \begin{bmatrix} 1 \\ 1 \\ 0 \end{bmatrix}$$

is an eigenvector. For $\lambda_3 = 0$ we have the linear system

$$\begin{bmatrix} -1 & -1 & 0 \\ -1 & -1 & 0 \\ 0 & 0 & -1 \end{bmatrix} \begin{bmatrix} x_1 \\ x_2 \\ x_3 \end{bmatrix} = \begin{bmatrix} 0 \\ 0 \\ 0 \end{bmatrix}.$$

Row reducing the coefficient matrix we have the equivalent linear system

$$\begin{bmatrix} 1 & 1 & 0 \\ 0 & 0 & 1 \\ 0 & 0 & 0 \end{bmatrix} \begin{bmatrix} x_1 \\ x_2 \\ x_3 \end{bmatrix} = \begin{bmatrix} 0 \\ 0 \\ 0 \end{bmatrix}.$$

The solution is $x_1 = -r$, $x_2 = r$, $x_3 = 0$ and hence

$$\mathbf{x}_3 = \begin{bmatrix} -1 \\ 1 \\ 0 \end{bmatrix}$$

is an eigenvector. Vectors $\mathbf{x}_1$, $\mathbf{x}_2$, and $\mathbf{x}_3$ are guaranteed to be orthogonal to each other since the corresponding eigenvalues are distinct. To form P we need only normalize the eigenvectors. Let

$$\mathbf{y}_1 = \mathbf{x}_1, \quad \mathbf{y}_2 = \frac{\mathbf{x}_2}{\sqrt{2}}, \quad \mathbf{y}_3 = \frac{\mathbf{x}_3}{\sqrt{2}}.$$

Then

$$P = \begin{bmatrix} \mathbf{y}_1 & \mathbf{y}_2 & \mathbf{y}_3 \end{bmatrix} = \begin{bmatrix} 0 & \frac{1}{\sqrt{2}} & -\frac{1}{\sqrt{2}} \\ 0 & \frac{1}{\sqrt{2}} & \frac{1}{\sqrt{2}} \\ 1 & 0 & 0 \end{bmatrix} \quad \text{and} \quad P^T A P = \begin{bmatrix} 1 & 0 & 0 \\ 0 & 2 & 0 \\ 0 & 0 & 0 \end{bmatrix}.$$

25. Let $A = \begin{bmatrix} 1 & 0 & 0 \\ 0 & 1 & 1 \\ 0 & 1 & 1 \end{bmatrix}$. The characteristic polynomial is

$$\det(\lambda I_3 - A) = \begin{vmatrix} \lambda - 1 & 0 & 0 \\ 0 & \lambda - 1 & -1 \\ 0 & -1 & \lambda - 1 \end{vmatrix} = \lambda(\lambda - 1)(\lambda - 2).$$

Hence A has eigenvalues $\lambda_1 = 1$, $\lambda_2 = 0$, $\lambda_3 = 2$. For $\lambda_1 = 1$ we have the linear system

$$\begin{bmatrix} 0 & 0 & 0 \\ 0 & 0 & -1 \\ 0 & -1 & 0 \end{bmatrix} \begin{bmatrix} x_1 \\ x_2 \\ x_3 \end{bmatrix} = \begin{bmatrix} 0 \\ 0 \\ 0 \end{bmatrix}.$$

Row reducing the coefficient matrix we have the equivalent linear system

$$\begin{bmatrix} 0 & 1 & 0 \\ 0 & 0 & 1 \\ 0 & 0 & 0 \end{bmatrix} \begin{bmatrix} x_1 \\ x_2 \\ x_3 \end{bmatrix} = \begin{bmatrix} 0 \\ 0 \\ 0 \end{bmatrix}.$$

The solution is $x_1 = r$, $x_2 = x_3 = 0$ and hence

$$\mathbf{x}_1 = \begin{bmatrix} 1 \\ 0 \\ 0 \end{bmatrix}$$

is an eigenvector. For $\lambda_2 = 0$ we have the linear system

$$\begin{bmatrix} -1 & 0 & 0 \\ 0 & -1 & -1 \\ 0 & -1 & -1 \end{bmatrix} \begin{bmatrix} x_1 \\ x_2 \\ x_3 \end{bmatrix} = \begin{bmatrix} 0 \\ 0 \\ 0 \end{bmatrix}.$$

Row reducing the coefficient matrix we have the equivalent linear system

$$\begin{bmatrix} 1 & 0 & 0 \\ 0 & 1 & 1 \\ 0 & 0 & 0 \end{bmatrix} \begin{bmatrix} x_1 \\ x_2 \\ x_3 \end{bmatrix} = \begin{bmatrix} 0 \\ 0 \\ 0 \end{bmatrix}.$$

The solution is $x_1 = 0$, $x_2 = -r$, $x_3 = r$ and hence

$$\mathbf{x}_2 = \begin{bmatrix} 0 \\ -1 \\ 1 \end{bmatrix}$$

is an eigenvector. For $\lambda_3 = 2$ we have the linear system

$$\begin{bmatrix} 1 & 0 & 0 \\ 0 & 1 & -1 \\ 0 & -1 & 1 \end{bmatrix} \begin{bmatrix} x_1 \\ x_2 \\ x_3 \end{bmatrix} = \begin{bmatrix} 0 \\ 0 \\ 0 \end{bmatrix}.$$

Row reducing the coefficient matrix we have the equivalent linear system

$$\begin{bmatrix} 1 & 0 & 0 \\ 0 & 1 & -1 \\ 0 & 0 & 0 \end{bmatrix} \begin{bmatrix} x_1 \\ x_2 \\ x_3 \end{bmatrix} = \begin{bmatrix} 0 \\ 0 \\ 0 \end{bmatrix}.$$

The solution is $x_1 = 0$, $x_2 = x_3 = r$ and hence

$$\mathbf{x}_3 = \begin{bmatrix} 0 \\ 1 \\ 1 \end{bmatrix}$$

is an eigenvector. Vectors $\mathbf{x}_1$, $\mathbf{x}_2$, and $\mathbf{x}_3$ are guaranteed to be orthogonal to each other since the corresponding eigenvalues are distinct. To form P we need only normalize the eigenvectors. Let

$$\mathbf{y}_1 = \mathbf{x}_1, \quad \mathbf{y}_2 = \frac{\mathbf{x}_2}{\sqrt{2}}, \quad \mathbf{y}_3 = \frac{\mathbf{x}_3}{\sqrt{2}}.$$

Then

$$P = \begin{bmatrix} \mathbf{y}_1 & \mathbf{y}_2 & \mathbf{y}_3 \end{bmatrix} = \begin{bmatrix} 1 & 0 & 0 \\ 0 & -\frac{1}{\sqrt{2}} & \frac{1}{\sqrt{2}} \\ 0 & \frac{1}{\sqrt{2}} & \frac{1}{\sqrt{2}} \end{bmatrix} \quad \text{and} \quad P^T A P = \begin{bmatrix} 1 & 0 & 0 \\ 0 & 0 & 0 \\ 0 & 0 & 2 \end{bmatrix}.$$

27. Let $A = \begin{bmatrix} 1 & -1 & 2 \\ -1 & 1 & 2 \\ 2 & 2 & 2 \end{bmatrix}$. The characteristic polynomial is

$$\det(\lambda I_3 - A) = \begin{vmatrix} \lambda - 1 & 1 & -2 \\ 1 & \lambda - 1 & -2 \\ -2 & -2 & \lambda - 2 \end{vmatrix} = (\lambda - 2)(\lambda + 2)(\lambda - 4).$$

Hence A has eigenvalues $\lambda_1 = 2$, $\lambda_2 = -2$, and $\lambda_3 = 4$. For $\lambda_1 = 2$ we have the linear system

$$\begin{bmatrix} 1 & 1 & -2 \\ 1 & 1 & -2 \\ -2 & -2 & 0 \end{bmatrix} \begin{bmatrix} x_1 \\ x_2 \\ x_3 \end{bmatrix} = \begin{bmatrix} 0 \\ 0 \\ 0 \end{bmatrix}.$$

Row reducing the coefficient matrix we have the equivalent linear system

$$\begin{bmatrix} 1 & 0 & 0 \\ 0 & 0 & 1 \\ 0 & 0 & 0 \end{bmatrix} \begin{bmatrix} x_1 \\ x_2 \\ x_3 \end{bmatrix} = \begin{bmatrix} 0 \\ 0 \\ 0 \end{bmatrix}.$$

The solution is $x_1 = -r$, $x_2 = r$, $x_3 = 0$ and hence

$$\mathbf{x}_1 = \begin{bmatrix} -1 \\ 1 \\ 0 \end{bmatrix}$$

is an eigenvector. For $\lambda_2 = -2$ we have the linear system

$$\begin{bmatrix} -3 & 1 & -2 \\ 1 & -3 & -2 \\ -2 & -2 & -4 \end{bmatrix} \begin{bmatrix} x_1 \\ x_2 \\ x_3 \end{bmatrix} = \begin{bmatrix} 0 \\ 0 \\ 0 \end{bmatrix}.$$

Row reducing the coefficient matrix we have the equivalent linear system

$$\begin{bmatrix} 1 & 0 & 1 \\ 0 & 1 & 1 \\ 0 & 0 & 0 \end{bmatrix} \begin{bmatrix} x_1 \\ x_2 \\ x_3 \end{bmatrix} = \begin{bmatrix} 0 \\ 0 \\ 0 \end{bmatrix}.$$

The solution is $x_1 = -r$, $x_2 = -r$, $x_3 = r$ and hence

$$\mathbf{x}_2 = \begin{bmatrix} -1 \\ -1 \\ 1 \end{bmatrix}$$

is an eigenvector. For $\lambda_3 = 4$ we have the linear system

$$\begin{bmatrix} 3 & 1 & -2 \\ 1 & 3 & -2 \\ -2 & -2 & 2 \end{bmatrix} \begin{bmatrix} x_1 \\ x_2 \\ x_3 \end{bmatrix} = \begin{bmatrix} 0 \\ 0 \\ 0 \end{bmatrix}.$$

Row reducing the coefficient matrix we have the equivalent linear system

$$\begin{bmatrix} 1 & 0 & -\frac{1}{2} \\ 0 & 1 & -\frac{1}{2} \\ 0 & 0 & 0 \end{bmatrix} \begin{bmatrix} x_1 \\ x_2 \\ x_3 \end{bmatrix} = \begin{bmatrix} 0 \\ 0 \\ 0 \end{bmatrix}.$$

The solution is $x_1 = \frac{1}{2}r$, $x_2 = \frac{1}{2}r$, $x_3 = r$ and hence

$$\mathbf{x}_3 = \begin{bmatrix} 1 \\ 1 \\ 2 \end{bmatrix}$$

is an eigenvector. Vectors $\mathbf{x}_1$, $\mathbf{x}_2$, and $\mathbf{x}_3$ are guaranteed to be orthogonal to each other since the corresponding eigenvalues are distinct. To form P we need only normalize the eigenvectors. Let

$$\mathbf{y}_1 = \frac{\mathbf{x}_1}{\sqrt{2}}, \quad \mathbf{y}_2 = \frac{\mathbf{x}_2}{\sqrt{3}}, \quad \mathbf{y}_3 = \frac{\mathbf{x}_3}{\sqrt{6}}.$$

Then

$$P = \begin{bmatrix} \mathbf{y}_1 & \mathbf{y}_2 & \mathbf{y}_3 \end{bmatrix} = \begin{bmatrix} -\frac{1}{\sqrt{2}} & -\frac{1}{\sqrt{3}} & \frac{1}{\sqrt{6}} \\ \frac{1}{\sqrt{2}} & -\frac{1}{\sqrt{3}} & \frac{1}{\sqrt{6}} \\ 0 & \frac{1}{\sqrt{3}} & \frac{2}{\sqrt{6}} \end{bmatrix} \quad \text{and} \quad P^T A P = \begin{bmatrix} 2 & 0 & 0 \\ 0 & -2 & 0 \\ 0 & 0 & 4 \end{bmatrix}.$$

29. To prove Theorem 7.9 for the 2×2 case we proceed as follows. Let

$$A = \begin{bmatrix} a & b \\ b & c \end{bmatrix}.$$

Then the characteristic polynomial is

$$\det(\lambda I_2 - A) = (\lambda - a)(\lambda - c) - b^2 = \lambda^2 - (a + c)\lambda + (ac - b^2).$$

The roots of the characteristic polynomial are

$$\lambda_1 = \frac{(a + c) + \sqrt{(a + c)^2 - 4(ac - b^2)}}{2} \quad \text{and} \quad \lambda_2 = \frac{(a + c) - \sqrt{(a + c)^2 - 4(ac - b^2)}}{2}.$$

Suppose $\lambda_1 \neq \lambda_2$. Since eigenvectors corresponding to distinct eigenvalues of a symmetric matrix are orthogonal, we can normalize the eigenvectors and thus form an orthogonal matrix P to diagonalize A. Now suppose $\lambda_1 = \lambda_2$. In this case $(a+c)^2 - 4(ac - b^2) = 0$. Rearranging terms we have $(a-c)^2 + 4b^2 = 0$ which implies that $a = c$ and $b = 0$. That is, matrix A is already diagonal.

31. To show that L is an isometry we verify Equation (7). First note that matrix A satisfies $A^T A = I_2$. Then

$$(L(\mathbf{u}), L(\mathbf{v})) = (A\mathbf{u}, A\mathbf{v}) = (\mathbf{u}, A^T A\mathbf{v}) = (\mathbf{u}, \mathbf{v})$$

so L is an isometry.

33. (a) Let L be an isometry. Then $(L(\mathbf{x}), L(\mathbf{x})) = (\mathbf{x}, \mathbf{x})$ so $\|L(\mathbf{x})\| = \|\mathbf{x}\|$.

    (b) Let L be an isometry. Then the angle θ between $L(\mathbf{x})$ and $L(\mathbf{y})$ is determined by

    $$\cos\theta = \frac{(L(\mathbf{x}), L(\mathbf{y}))}{\|L(\mathbf{x})\|\,\|L(\mathbf{y})\|} = \frac{(\mathbf{x}, \mathbf{y})}{\|\mathbf{x}\|\,\|\mathbf{y}\|},$$

    which is the cosine of the angle between $\mathbf{x}$ and $\mathbf{y}$.

35. Suppose that L is an isometry. Then $(L(\mathbf{x}), L(\mathbf{y})) = (\mathbf{x}, \mathbf{y})$ for any vectors $\mathbf{x}$, $\mathbf{y}$. Since S is an orthonormal basis,

    $$(\mathbf{v}_i, \mathbf{v}_j) = \begin{cases} 1, & \text{for } i = j \\ 0, & \text{for } i \neq j. \end{cases}$$

    We have $(L(\mathbf{v}_i), L(\mathbf{v}_j)) = (\mathbf{v}_i, \mathbf{v}_j)$ so T is a set of n orthonormal vectors in the n-dimensional space R^n. Hence T is an orthonormal basis for R^n. Conversely, suppose that T is an orthonormal basis for R^n. Let $\mathbf{x}$ and $\mathbf{y}$ be any vectors in R^n. Then

    $$\mathbf{x} = \sum_{i=1}^{n} a_i \mathbf{v}_i \quad \text{and} \quad \mathbf{y} = \sum_{i=1}^{n} b_i \mathbf{v}_i$$

    for some scalars a_i and b_i. Since S is an orthonormal basis,

    $$(\mathbf{x}, \mathbf{y}) = \sum_{i=1}^{n} a_i b_i.$$

    Since L is a linear transformation,

    $$L(\mathbf{x}) = \sum_{i=1}^{n} a_i L(\mathbf{v}_i) \quad \text{and} \quad L(\mathbf{y}) = \sum_{i=1}^{n} b_i L(\mathbf{v}_i).$$

    Hence, since T is an orthonormal basis,

    $$(L(\mathbf{x}), L(\mathbf{y})) = \sum_{i=1}^{n} a_i b_i.$$

    Thus $(\mathbf{x}, \mathbf{y}) = (L(\mathbf{x}), L(\mathbf{y}))$ so L is an isometry.

37. If A is orthogonal, then $A^T = A^{-1}$. Since

    $$(A^T)^T = (A^{-1})^T = (A^T)^{-1},$$

    we have that A^T is orthogonal.

Supplementary Exercises for Chapter 7, p. 477

1. (a) Let

    $$A = \begin{bmatrix} 1 & 4 & 0 \\ 0 & 4 & 0 \\ 0 & 0 & 1 \end{bmatrix}.$$

    Since A is upper triangular its eigenvalues are its diagonal elements. Hence $\lambda_1 = 1$, $\lambda_2 = 1$, $\lambda_3 = 4$. To find the corresponding eigenvectors we proceed as follows. The linear system

    $$(\lambda_1 I_3 - A)\mathbf{x} = \begin{bmatrix} 0 & -4 & 0 \\ 0 & -3 & 0 \\ 0 & 0 & 0 \end{bmatrix} \begin{bmatrix} x_1 \\ x_2 \\ x_3 \end{bmatrix} = \begin{bmatrix} 0 \\ 0 \\ 0 \end{bmatrix}$$

is equivalent to the system

$$\begin{bmatrix} 0 & 1 & 0 \\ 0 & 0 & 0 \\ 0 & 0 & 0 \end{bmatrix} \begin{bmatrix} x_1 \\ x_2 \\ x_3 \end{bmatrix} = \begin{bmatrix} 0 \\ 0 \\ 0 \end{bmatrix}$$

which has the general solution

$$\mathbf{x} = \begin{bmatrix} r \\ 0 \\ s \end{bmatrix}.$$

Hence we have two linearly independent eigenvectors:

$$\mathbf{x}_1 = \begin{bmatrix} 1 \\ 0 \\ 0 \end{bmatrix}, \quad \mathbf{x}_2 = \begin{bmatrix} 0 \\ 0 \\ 1 \end{bmatrix}.$$

For λ_3, the linear system

$$(\lambda_3 I_3 - A)\mathbf{x} = \begin{bmatrix} 3 & -4 & 0 \\ 0 & 0 & 0 \\ 0 & 0 & 3 \end{bmatrix} \begin{bmatrix} x_1 \\ x_2 \\ x_3 \end{bmatrix} = \begin{bmatrix} 0 \\ 0 \\ 0 \end{bmatrix}$$

is equivalent to the system

$$\begin{bmatrix} 1 & -\frac{4}{3} & 0 \\ 0 & 0 & 1 \\ 0 & 0 & 0 \end{bmatrix} \begin{bmatrix} x_1 \\ x_2 \\ x_3 \end{bmatrix} = \begin{bmatrix} 0 \\ 0 \\ 0 \end{bmatrix}$$

which has the solution

$$\mathbf{x}_3 = \begin{bmatrix} 4 \\ 3 \\ 0 \end{bmatrix}.$$

Since $\lambda_1 \ne \lambda_3$, the eigenvectors $\mathbf{x}_1$, $\mathbf{x}_2$, and $\mathbf{x}_3$ are linearly independent, so A is similar to a diagonal matrix.

(b) Let $A = \begin{bmatrix} 2 & -2 & 0 \\ -3 & 1 & 0 \\ 0 & 0 & 3 \end{bmatrix}$. The characteristic polynomial of A is

$$\det(\lambda I_3 - A) = \begin{vmatrix} \lambda - 2 & 2 & 0 \\ 3 & \lambda - 1 & 0 \\ 0 & 0 & \lambda - 3 \end{vmatrix} = (\lambda - 3)[(\lambda - 2)(\lambda - 1) - 6] = (\lambda - 3)(\lambda - 4)(\lambda + 1).$$

Hence the eigenvalues are $\lambda_1 = 3$, $\lambda_2 = 4$, $\lambda_3 = -1$. These are distinct so we can immediately say that A is similar to a diagonal matrix. Solving the appropriate homogeneous linear systems we can show that the corresponding eigenvectors are

$$\mathbf{x}_1 = \begin{bmatrix} 0 \\ 0 \\ 1 \end{bmatrix}, \quad \mathbf{x}_2 = \begin{bmatrix} 1 \\ -1 \\ 0 \end{bmatrix}, \quad \mathbf{x}_3 = \begin{bmatrix} 2 \\ 3 \\ 0 \end{bmatrix}.$$

(c) Let $A = \begin{bmatrix} 0 & 1 & 0 \\ 0 & 0 & 1 \\ 6 & -11 & 6 \end{bmatrix}$. The characteristic polynomial of A is

$$\det(\lambda I_3 - A) = \begin{vmatrix} \lambda & -1 & 0 \\ 0 & \lambda & -1 \\ -6 & 11 & \lambda - 6 \end{vmatrix} = \lambda^2(\lambda - 6) - 6 + 11\lambda = (\lambda - 1)(\lambda - 2)(\lambda - 3).$$

Hence the eigenvalues are $\lambda_1 = 1$, $\lambda_2 = 2$, $\lambda_3 = 3$. These are distinct so we can immediately say that A is similar to a diagonal matrix. Solving the appropriate homogeneous linear systems the corresponding eigenvectors are

$$\mathbf{x}_1 = \begin{bmatrix} 1 \\ 1 \\ 1 \end{bmatrix}, \quad \mathbf{x}_2 = \begin{bmatrix} 1 \\ 2 \\ 4 \end{bmatrix}, \quad \mathbf{x}_3 = \begin{bmatrix} 1 \\ 3 \\ 9 \end{bmatrix}.$$

(d) Let $A = \begin{bmatrix} 0 & 1 & 0 \\ 0 & 0 & 1 \\ 3 & 1 & -3 \end{bmatrix}$. The characteristic polynomial of A is

$$\det(\lambda I_3 - A) = \begin{vmatrix} \lambda & -1 & 0 \\ 0 & \lambda & -1 \\ -3 & -1 & \lambda + 3 \end{vmatrix} = \lambda^2(\lambda + 3) - 3 - \lambda = (\lambda + 3)(\lambda - 1)(\lambda + 1).$$

Hence the eigenvalues are $\lambda_1 = -3$, $\lambda_2 = 1$, $\lambda_3 = -1$. These are distinct so we can immediately say that A is similar to a diagonal matrix. Solving the appropriate homogeneous linear systems the corresponding eigenvectors are

$$\mathbf{x}_1 = \begin{bmatrix} 1 \\ -3 \\ 9 \end{bmatrix}, \quad \mathbf{x}_2 = \begin{bmatrix} 1 \\ 1 \\ 1 \end{bmatrix}, \quad \mathbf{x}_3 = \begin{bmatrix} -1 \\ 1 \\ -1 \end{bmatrix}.$$

3. Let A be any $n \times n$ matrix.

(a) To find the coefficient of λ^{n-1} in the characteristic polynomial of A, first note that the characteristic polynomial is

$$\det(\lambda I_n - A) = \begin{vmatrix} \lambda - a_{11} & -a_{12} & \cdots & \cdots & -a_{1n} \\ -a_{21} & \lambda - a_{22} & \cdots & \cdots & -a_{2n} \\ \vdots & \vdots & \vdots & \vdots & \vdots \\ -a_{n1} & -a_{n2} & \cdots & -a_{n\,n-1} & \lambda - a_{nn} \end{vmatrix}.$$

Any product in $\det(\lambda I_n - A)$, other than the product of the diagonal entries, can contain at most $n - 2$ of the diagonal entries of $\lambda I_n - A$. This follows because at least two of the column indices must be out of natural order in every other product appearing in $\det(\lambda I_n - A)$. This implies that the coefficient of λ^{n-1} is formed by the expansion of the product of the diagonal entries. The coefficient of λ^{n-1} is the sum of the coefficients of λ^{n-1} from each of the products

$$-a_{ii}(\lambda - a_{11}) \cdots (\lambda - a_{i-1\,i-1})(\lambda - a_{i+1\,i+1}) \cdots (\lambda - a_{nn})$$

for $i = 1, 2, \ldots, n$. The coefficient of λ^{n-1} in each such term is $-a_{ii}$ and so the coefficient of λ^{n-1} in the characteristic polynomial is

$$(-a_{11}) + (-a_{22}) + \cdots + (-a_{nn}) = -\operatorname{Tr}(A).$$

(b) If $\lambda_1, \lambda_2, \ldots, \lambda_n$ are the eigenvalues of A then $\lambda - \lambda_i$, for $i = 1, 2, \ldots, n$, are factors of the characteristic polynomial $\det(\lambda I_n - A)$. It follows that

$$\det(\lambda I_n - A) = (\lambda - \lambda_1)(\lambda - \lambda_2) \cdots (\lambda - \lambda_n).$$

Proceeding as in part (a), the coefficient of λ^{n-1} is the sum of the coefficients of λ^{n-1} from each of the products

$$-\lambda_i(\lambda - \lambda_1)(\lambda - \lambda_2) \cdots (\lambda - \lambda_{i-1})(\lambda - \lambda_{i+1}) \cdots (\lambda - \lambda_n)$$

for $i = 1, 2, \ldots, n$. The coefficient of λ^{n-1} in each such term is $-\lambda_i$, so the coefficient of λ^{n-1} in the characteristic polynomial is

$$(-\lambda_1) + (-\lambda_2) + \cdots + (-\lambda_n) = -\operatorname{Tr}(A)$$

by (a). Thus, $\operatorname{Tr}(A)$ is the sum of the eigenvalues of A.

5. Let $p(x) = a_0 + a_1 x + a_2 x^2 + \cdots + a_k x^k$ be a polynomial in x. In Exercise 21 of Section 7.1 we showed that if λ is an eigenvalue of A with associated eigenvector $\mathbf{x}$, then λ^k is an eigenvalue of A^k, k a positive integer. For any positive integers j and k and any scalars a and b, the eigenvalues of $aA^j + bA^k$ are $a\lambda^j + b\lambda^k$ since

$$(aA^j + bA^k)\mathbf{x} = aA^j\mathbf{x} + bA^k\mathbf{x} = a\lambda^j\mathbf{x} + b\lambda^k\mathbf{x} = (a\lambda^j + b\lambda^k)\mathbf{x}.$$

This result generalizes to finite linear combinations of powers of A and to scalar multiples of the identity matrix. Thus,

$$\begin{aligned} p(A)\mathbf{x} &= (a_0 I_n + a_1 A + a_2 A^2 + \cdots + a_k A^k)\mathbf{x} \\ &= a_0 I_n \mathbf{x} + a_1 A\mathbf{x} + a_2 A^2\mathbf{x} + \cdots + a_k A^k\mathbf{x} \\ &= a_0\mathbf{x} + a_1\lambda\mathbf{x} + a_2\lambda^2\mathbf{x} + \cdots + a_k\lambda^k\mathbf{x} \\ &= (a_0 + a_1\lambda + a_2\lambda^2 + \cdots + a_k\lambda^k)\mathbf{x} \\ &= p(\lambda)\mathbf{x} \end{aligned}$$

Thus, if $\lambda_1, \ldots, \lambda_n$ are eigenvalues of A, then $p(\lambda_1), \ldots, p(\lambda_n)$ are eigenvalues of $p(A)$.

7. Let $L : P_1 \to P_1$ be the linear transformation defined by

$$L(at + b) = \frac{a + b}{2} t$$

and let $S = \{2 - t, 3 + t\}$ be a basis for P_1.

(a) $L(2 - t) = \frac{1}{2} t$. To find the coordinates of $L(2 - t)$ with respect to the S-basis we proceed as follows. Determine constants a_1 and a_2 such that

$$a_1(2 - t) + a_2(3 + t) = \frac{1}{2} t.$$

Expanding and collecting like terms, we have

$$(-a_1 + a_2)t + (2a_1 + 3a_2) = \frac{1}{2} t.$$

Equating coefficients of like terms gives the linear system

$$\begin{aligned} -a_1 &+ a_2 &= \tfrac{1}{2} \\ 2a_1 &+ 3a_2 &= 0. \end{aligned}$$

Solving the system we have $a_1 = -\frac{3}{10}$ and $a_2 = \frac{1}{5}$. Thus

$$\left[\, L(2 - t) \,\right]_S = \begin{bmatrix} -\frac{3}{10} \\ \frac{1}{5} \end{bmatrix}.$$

For $3 + t$ we have $L(3 + t) = 2t$. To find the coordinates of $L(3 + t)$ with respect to the S-basis we proceed as follows. Determine constants a_1 and a_2 such that

$$a_1(2 - t) + a_2(3 + t) = 2t.$$

Expanding and collecting like terms we have

$$(-a_1 + a_2)t + (2a_1 + 3a_2) = 2t.$$

Equating coefficients of like terms gives the linear system

$$
\begin{array}{rcrcl}
-a_1 & + & a_2 & = & 2 \\
2a_1 & + & 3a_2 & = & 0
\end{array}
$$

Solving the system we have $a_1 = -\frac{6}{5}$ and $a_2 = \frac{4}{54}$. Thus

$$\left[\, L(2-t) \,\right]_S = \begin{bmatrix} -\frac{6}{5} \\ \frac{4}{5} \end{bmatrix}.$$

(b) A matrix representing L with respect to S is

$$A = \left[\ \left[\, L(2-t)\, \right]_S \quad \left[\, L(3+t)\, \right]_S \ \right] = \begin{bmatrix} -\frac{3}{10} & -\frac{6}{5} \\ \frac{1}{5} & \frac{4}{5} \end{bmatrix} = \frac{1}{10}\begin{bmatrix} -3 & -12 \\ 2 & 8 \end{bmatrix}.$$

(c) The characteristic polynomial of A is

$$\det(\lambda I_2 - A) = \begin{vmatrix} \lambda + \frac{3}{10} & \frac{6}{5} \\ -\frac{1}{5} & \lambda - \frac{4}{5} \end{vmatrix} = \left(\lambda + \frac{3}{10}\right)\left(\lambda - \frac{4}{5}\right) + \frac{6}{25} = \lambda\left(\lambda - \frac{1}{2}\right).$$

Thus the eigenvalues of A are $\lambda_1 = 0$ and $\lambda_2 = \frac{1}{2}$. To find the corresponding eigenvectors we solve the following homogeneous systems:

$$(\lambda_1 I_2 - A)\mathbf{x} = \begin{bmatrix} \frac{3}{10} & \frac{6}{5} \\ -\frac{1}{5} & -\frac{4}{5} \end{bmatrix}\begin{bmatrix} x_1 \\ x_2 \end{bmatrix} = \begin{bmatrix} 0 \\ 0 \end{bmatrix} \implies \mathbf{x} = \begin{bmatrix} -4r \\ r \end{bmatrix}.$$

Choosing $r = 1$, we obtain the eigenvector

$$\mathbf{x}_1 = \begin{bmatrix} -4 \\ 1 \end{bmatrix}.$$

For λ_2,

$$(\lambda_2 I_2 - A)\mathbf{x} = \begin{bmatrix} \frac{4}{5} & \frac{6}{5} \\ -\frac{1}{5} & -\frac{3}{10} \end{bmatrix}\begin{bmatrix} x_1 \\ x_2 \end{bmatrix} = \begin{bmatrix} 0 \\ 0 \end{bmatrix} \implies \mathbf{x} = \begin{bmatrix} -\frac{3}{2}r \\ r \end{bmatrix}.$$

Choosing $r = 2$, we obtain the eigenvector

$$\mathbf{x}_2 = \begin{bmatrix} -3 \\ 2 \end{bmatrix}.$$

(d) The eigenvalues of A are also the eigenvalues of L. The eigenvectors of A represent the coordinates of the eigenvectors of L with respect to S. Thus the eigenvectors of L are

$$-4(2-t) + (3+t) = 5t - 5 \quad \text{and} \quad -3(2-t) + 2(3+t) = 5t.$$

(e) The eigenspace for $\lambda_1 = 0$ is the subspace of P_1 with basis $\{5t - 5\}$. The eigenspace for $\lambda_2 = \frac{1}{2}$ is the subspace of P_1 with basis $\{5t\}$.

9. Let V be the real vector space of (trigonometric) polynomials of the form $a + b\sin x + c\cos x$. Let $L : V \to V$ be the linear transformation defined by

$$L(\mathbf{v}) = \frac{d}{dx}[\mathbf{v}].$$

To find the eigenvalues and eigenvectors of L we first determine a matrix representation of L. The set $S = \{1, \sin x, \cos x\}$ is a basis for V. We find the matrix representing L with respect to S. We first find the images of the members of S under L:

$$L(1) = 0, \quad L(\sin x) = \cos x, \quad L(\cos x) = -\sin x.$$

Then a matrix A representing L with respect to S is

$$A = \begin{bmatrix} [\, L(1)\,]_S & [\, L(\sin x)\,]_S & [\, L(\cos x)\,]_S \end{bmatrix} = \begin{bmatrix} 0 & 0 & 0 \\ 0 & 0 & -1 \\ 0 & 1 & 0 \end{bmatrix}.$$

The characteristic polynomial of A is

$$\det(\lambda I_3 - A) = \begin{vmatrix} \lambda & 0 & 0 \\ 0 & \lambda & 1 \\ 0 & -1 & \lambda \end{vmatrix} = \lambda(\lambda^2 + 1).$$

Since all the eigenvalues must be real, the only eigenvalue is $\lambda = 0$. To find the corresponding eigenvector we solve the system

$$(0I_3 - A)\mathbf{x} = \begin{bmatrix} 0 & 0 & 0 \\ 0 & 0 & 1 \\ 0 & -1 & 0 \end{bmatrix} \begin{bmatrix} x_1 \\ x_2 \\ x_3 \end{bmatrix} = \begin{bmatrix} 0 \\ 0 \\ 0 \end{bmatrix} \implies \mathbf{x} = \begin{bmatrix} r \\ 0 \\ 0 \end{bmatrix}.$$

Choosing $r = 1$ gives the eigenvector

$$\begin{bmatrix} 1 \\ 0 \\ 0 \end{bmatrix}.$$

This vector gives the coordinates of the eigenvector of L for eigenvalue $\lambda = 0$. Thus, the eigenvector for L is $p_1(x) = 1$.

11. Since A is similar to a diagonal matrix D, there exists a nonsingular matrix P such that $P^{-1}AP = D$. Thus it follows that

$$D = D^T = (P^{-1}AP)^T = P^T A^T (P^{-1})^T = ((P^T)^{-1})^{-1} A^T (P^T)^{-1}.$$

Let $Q = (P^T)^{-1}$. Then we have $Q^{-1}A^T Q = D$. That is, both A and A^T are similar to D. Hence A and A^T are similar to one another.

Chapter Review for Chapter 7, p. 492

True or False

1. True.	2. False.	3. True.	4. True.	5. False.
6. True.	7. True.	8. True.	9. True.	10. True.
11. False.	12. True.	13. True.	14. True.	15. True.
16. True.	17. True.	18. True.	19. True.	20. True.

Quiz

1. We solve the characteristic polynomial:

$$p(\lambda) = \det\left(\begin{bmatrix} \lambda+1 & 2 \\ -4 & \lambda-5 \end{bmatrix} \right) = (\lambda+1)(\lambda-5) + 8 = \lambda^2 - 4\lambda + 3 = (\lambda-3)(\lambda-1).$$

So the eigenvalues are $\lambda_1 = 1$, $\lambda_2 = 3$. Now,

$$\lambda_1 I_2 - A = \begin{bmatrix} 2 & 2 \\ -4 & -4 \end{bmatrix} \quad \rightarrow \quad \begin{bmatrix} 1 & 1 \\ 0 & 0 \end{bmatrix} \quad \text{choose } \mathbf{x}_1 = \begin{bmatrix} 1 \\ -1 \end{bmatrix}$$

$$\lambda_2 I_2 - A = \begin{bmatrix} 4 & 2 \\ -4 & -2 \end{bmatrix} \quad \rightarrow \quad \begin{bmatrix} 1 & \frac{1}{2} \\ 0 & 0 \end{bmatrix} \quad \text{chosse } \mathbf{x}_2 = \begin{bmatrix} -1 \\ 2 \end{bmatrix}.$$

2. (a) Computing images, we find that

$$L(t^2) = 0 = (0)t^2 + (0)(2+t) + (0)(2-t)$$
$$L(2+t) = 2t+1 = (0)t^2 + \tfrac{5}{4}(2+t) - \tfrac{3}{4}(2-t)$$
$$L(2-t) = 2t-1 = (0)t^2 + \tfrac{3}{4}(2+t) - \tfrac{5}{4}(2-t).$$

Therefore the matrix of L with respect to S is $\begin{bmatrix} 0 & 0 & 0 \\ 0 & \frac{5}{4} & \frac{3}{4} \\ 0 & -\frac{3}{4} & -\frac{5}{4} \end{bmatrix}$.

(b) $p(\lambda) = \det\left(\begin{bmatrix} \lambda & 0 & 0 \\ 0 & \lambda - \frac{5}{4} & -\frac{3}{4} \\ 0 & \frac{3}{4} & \lambda + \frac{5}{4} \end{bmatrix} \right) = \lambda\left(\lambda - \frac{5}{4}\right)\left(\lambda + \frac{5}{4}\right) + \frac{9}{16}\lambda = \lambda(\lambda^2 - 1)$. So $\lambda_1 = 0$, $\lambda_2 = 1$, $\lambda_3 = -1$. Now,

$$\lambda_1 = 0: \begin{bmatrix} 0 & 0 & 0 \\ 0 & -\frac{5}{4} & -\frac{3}{4} \\ 0 & \frac{3}{4} & \frac{5}{4} \end{bmatrix} \quad \rightarrow \quad \begin{bmatrix} 0 & 0 & 0 \\ 0 & 1 & 0 \\ 0 & 0 & 1 \end{bmatrix} \quad \text{choose } \mathbf{x}_1 = \begin{bmatrix} 1 \\ 0 \\ 0 \end{bmatrix}$$

$$\lambda_2 = 1: \begin{bmatrix} 1 & 0 & 0 \\ 0 & -\frac{1}{4} & -\frac{3}{4} \\ 0 & \frac{3}{4} & \frac{9}{4} \end{bmatrix} \quad \rightarrow \quad \begin{bmatrix} 1 & 0 & 0 \\ 0 & 1 & 3 \\ 0 & 0 & 0 \end{bmatrix} \quad \text{choose } \mathbf{x}_2 = \begin{bmatrix} 0 \\ -3 \\ 1 \end{bmatrix}.$$

$$\lambda_3 = -1: \begin{bmatrix} -1 & 0 & 0 \\ 0 & -\frac{9}{4} & -\frac{3}{4} \\ 0 & \frac{3}{4} & \frac{1}{4} \end{bmatrix} \quad \rightarrow \quad \begin{bmatrix} 1 & 0 & 0 \\ 0 & 1 & \frac{1}{3} \\ 0 & 0 & 0 \end{bmatrix} \quad \text{choose } \mathbf{x}_3 = \begin{bmatrix} 0 \\ -1 \\ 3 \end{bmatrix}$$

3. We find that $P(-1) = 0$. Therefore, $p(\lambda) = (\lambda+1)(\lambda^2 - 4\lambda + 4) = (\lambda+1)(\lambda-2)^2$. Hence the eigenvalues are $\lambda_1 = -1$, $\lambda_2 = 2$, and $\lambda_3 = 2$.

4. An eigenvalue of B is $\lambda = 9$ and an associated eigenvector is $\mathbf{x}$.

5. We have

$$2I_3 - A = \begin{bmatrix} 0 & -1 & -1 \\ 0 & 0 & 0 \\ 0 & 1 & 1 \end{bmatrix} \quad \rightarrow \quad \begin{bmatrix} 0 & 1 & 1 \\ 0 & 0 & 0 \\ 0 & 0 & 0 \end{bmatrix}$$

so the eigenspace associated with $\lambda = 2$ has basis $\left\{ \begin{bmatrix} 1 \\ 0 \\ 0 \end{bmatrix}, \begin{bmatrix} 0 \\ 1 \\ -1 \end{bmatrix} \right\}$.

6. $p(\lambda) = \det\left(\begin{bmatrix} \lambda-2 & -1 & 3 \\ -3 & \lambda & -3 \\ 1 & -1 & \lambda \end{bmatrix}\right) = (\lambda-2)\lambda^2 + 3 + 9 - 3\lambda - 3(\lambda-2) - 3\lambda = (\lambda)(\lambda^2 - 9)$ so $\lambda_1 = -3$,

$\lambda_2 = 2$, and $\lambda_3 = 3$. Now,

$$\lambda_1 = -3: \begin{bmatrix} -5 & -1 & 3 \\ -3 & -3 & -3 \\ 1 & -1 & -3 \end{bmatrix} \rightarrow \begin{bmatrix} 1 & 0 & -1 \\ 0 & 1 & 2 \\ 0 & 0 & 0 \end{bmatrix} \quad \text{chhose } \mathbf{x}_1 = \begin{bmatrix} 1 \\ -2 \\ 1 \end{bmatrix}$$

$$\lambda_2 = 2: \begin{bmatrix} 0 & -1 & 3 \\ -3 & 2 & -3 \\ 1 & -1 & 2 \end{bmatrix} \rightarrow \begin{bmatrix} 1 & 0 & -1 \\ 0 & 1 & -3 \\ 0 & 0 & 0 \end{bmatrix} \quad \text{choose } \mathbf{x}_2 = \begin{bmatrix} 1 \\ 3 \\ 1 \end{bmatrix}$$

$$\lambda_3 = 3: \begin{bmatrix} 1 & -1 & 3 \\ -3 & 3 & -3 \\ 1 & -1 & 3 \end{bmatrix} \rightarrow \begin{bmatrix} 1 & -1 & 0 \\ 0 & 0 & 1 \\ 0 & 0 & 0 \end{bmatrix} \quad \text{choose } \mathbf{x}_3 = \begin{bmatrix} 1 \\ 1 \\ 0 \end{bmatrix}.$$

Then $\{\mathbf{x}_1, \mathbf{x}_2, \mathbf{x}_3\}$ is a basis for R^3, so a diagonal matrix similar to A is

$$\begin{bmatrix} -3 & 0 & 0 \\ 0 & 2 & 0 \\ 0 & 0 & 3 \end{bmatrix}.$$

7. The matrix is upper triangular. So the eigenvalues are $\lambda_1 = \lambda_2 = 1$ and $\lambda_3 = 0$. Now,

$$\lambda_1 = \lambda_2 = 1: \begin{bmatrix} 0 & -1 & -1 \\ 0 & 0 & -1 \\ 0 & 0 & 0 \end{bmatrix} \rightarrow \begin{bmatrix} 0 & 1 & 0 \\ 0 & 0 & 1 \\ 0 & 0 & 0 \end{bmatrix} \quad \text{choose } \mathbf{x}_1 = \begin{bmatrix} 1 \\ 0 \\ 0 \end{bmatrix}$$

$$\lambda_3 = 0: \begin{bmatrix} 1 & -1 & -1 \\ 0 & 1 & -1 \\ 0 & 0 & 0 \end{bmatrix} \rightarrow \begin{bmatrix} 1 & 0 & -2 \\ 0 & 1 & -1 \\ 0 & 0 & 0 \end{bmatrix} \quad \text{choose } \mathbf{x}_3 = \begin{bmatrix} 2 \\ 1 \\ 1 \end{bmatrix}.$$

Since the two eigenvectors do not form a basis for R^3, A is not diagonalizable.

8. No. For example, let

$$A = \begin{bmatrix} 1 & 1 \\ -1 & 1 \end{bmatrix}.$$

Then

$$A^{-1} = \begin{bmatrix} \frac{1}{2} & -\frac{1}{2} \\ \frac{1}{2} & \frac{1}{2} \end{bmatrix} \neq A^T$$

but

$$\begin{bmatrix} 1 \\ -1 \end{bmatrix}^T \begin{bmatrix} 1 \\ 1 \end{bmatrix} = 0.$$

9. (a) Let $\mathbf{z} = \begin{bmatrix} a \\ b \\ c \end{bmatrix}$. Then

$$\mathbf{z} \cdot \mathbf{x} = a - b + 2c = 0$$
$$\mathbf{z} \cdot \mathbf{y} = 3a + 3b = 0.$$

We find $a = b$, $c = 0$, so a possible answer is

$$\mathbf{z} = \begin{bmatrix} -1 \\ 1 \\ 1 \end{bmatrix}.$$

(b) $A = \begin{bmatrix} 1 & 3 & -1 \\ -1 & 3 & 1 \\ 2 & 0 & 1 \end{bmatrix}$. Thus

$$A^T A = \begin{bmatrix} 1 & -1 & 2 \\ 3 & 3 & 0 \\ -1 & 1 & 1 \end{bmatrix} \begin{bmatrix} 1 & 3 & -1 \\ -1 & 3 & 1 \\ 2 & 0 & 1 \end{bmatrix} = \begin{bmatrix} 6 & 0 & 0 \\ 0 & 18 & 0 \\ 0 & 0 & 3 \end{bmatrix}.$$

In general,

$$A^T A = \begin{bmatrix} \mathbf{x}^T \\ \mathbf{y}^T \\ \mathbf{z}^T \end{bmatrix} \begin{bmatrix} \mathbf{x} & \mathbf{y} & \mathbf{z} \end{bmatrix} = \begin{bmatrix} \mathbf{x}^T\mathbf{x} & \mathbf{x}^T\mathbf{y} & \mathbf{x}^T\mathbf{z} \\ \mathbf{y}^T\mathbf{x} & \mathbf{y}^T\mathbf{y} & \mathbf{y}^T\mathbf{z} \\ \mathbf{z}^T\mathbf{x} & \mathbf{z}^T\mathbf{y} & \mathbf{z}^T\mathbf{z} \end{bmatrix}.$$

Since $\mathbf{z}$ is orthogonal to $\mathbf{x}$ and $\mathbf{y}$, and $\mathbf{x}$ and $\mathbf{y}$ are orthogonal, all entries not on the diagonal of this matrix are zero. The diagonal entries are the squares of the magnitudes of the vectors: $\|\mathbf{x}\|^2 = 6$, $\|\mathbf{y}\|^2 = 18$, and $\|\mathbf{z}\|^2 = 3$.

(c) Normalize each vector from part (b).

(d) diagonal

(e) Since

$$A^T A = \begin{bmatrix} \mathbf{x}^T \\ \mathbf{y}^T \\ \mathbf{z}^T \end{bmatrix} \begin{bmatrix} \mathbf{x} & \mathbf{y} & \mathbf{z} \end{bmatrix} = \begin{bmatrix} \mathbf{x}^T\mathbf{x} & \mathbf{x}^T\mathbf{y} & \mathbf{x}^T\mathbf{z} \\ \mathbf{y}^T\mathbf{x} & \mathbf{y}^T\mathbf{y} & \mathbf{y}^T\mathbf{z} \\ \mathbf{z}^T\mathbf{x} & \mathbf{z}^T\mathbf{y} & \mathbf{z}^T\mathbf{z} \end{bmatrix},$$

it follows that if the columns of A are mutually orthogonal, then all entries of $A^T A$ not on the diagonal are zero. Thus, $A^T A$ is a diagonal matrix.

10. False. Let $A = \begin{bmatrix} 1 & 0 \\ 0 & 1 \end{bmatrix}$. The eigenvalues of A are $\lambda_1 = \lambda_2 = 1$. Let $B = \begin{bmatrix} 0 & 1 \\ 1 & 0 \end{bmatrix}$. The eigenvalues of B are $\lambda_1 = -1$ and $\lambda_2 = 1$.

11. Let

$$A = \begin{bmatrix} k & 0 & 0 \\ a_{21} & a_{22} & a_{23} \\ a_{31} & a_{32} & a_{33} \end{bmatrix}.$$

Then $kI_3 - A$ has its first row all zero and hence $\det(kI_3 - A) = 0$. Therefore, $\lambda = k$ is an eigenvalue of A.

12. (a) $\det(4I_3 - A) = \det\left(\begin{bmatrix} -5 & 1 & 2 \\ 1 & -5 & 2 \\ 2 & 2 & -2 \end{bmatrix} \right) = 0$; $\det(10I_3 - A) = \det\left(\begin{bmatrix} 1 & 1 & 2 \\ 1 & 1 & 2 \\ 2 & 2 & 4 \end{bmatrix} \right) = 0$.

Now:

$$\lambda = 4: \quad \begin{bmatrix} -5 & 1 & 2 \\ 1 & -5 & 2 \\ 2 & 2 & -2 \end{bmatrix} \rightarrow \begin{bmatrix} 1 & 0 & -\frac{1}{2} \\ 0 & 1 & -\frac{1}{2} \\ 0 & 0 & 0 \end{bmatrix} \quad \text{choose } \mathbf{x}_1 = \begin{bmatrix} 1 \\ 1 \\ 2 \end{bmatrix}$$

$$\lambda = 10: \quad \begin{bmatrix} 1 & 1 & 2 \\ 1 & 1 & 2 \\ 2 & 2 & 4 \end{bmatrix} \rightarrow \begin{bmatrix} 1 & 1 & 2 \\ 0 & 0 & 0 \\ 0 & 0 & 0 \end{bmatrix} \quad \text{choose } \mathbf{x}_2 = \begin{bmatrix} -1 \\ 1 \\ 0 \end{bmatrix}, \mathbf{x}_3 = \begin{bmatrix} -2 \\ 0 \\ 1 \end{bmatrix}.$$

Observe that $\mathbf{x}_1$ is orthogonal to $\mathbf{x}_2$ and $\mathbf{x}_3$. But $\mathbf{x}_2$ and $\mathbf{x}_3$ are not orthogonal. so we must first find an orthogonal basis for the eigenspace of $\lambda = 10$. Let

$$\mathbf{v}_1 = \mathbf{x}_2 = \begin{bmatrix} -1 \\ 1 \\ 0 \end{bmatrix}$$

and

$$\mathbf{v}_2 = \mathbf{x}_3 - \frac{(\mathbf{x}_3, \mathbf{v}_1)}{(\mathbf{v}_1, \mathbf{v}_1)}\mathbf{v}_1 = \begin{bmatrix} -2 \\ 0 \\ 1 \end{bmatrix} - \frac{2}{2}\begin{bmatrix} -1 \\ 1 \\ 0 \end{bmatrix} = \begin{bmatrix} -1 \\ -1 \\ 1 \end{bmatrix}.$$

Then $\{\mathbf{v}_1, \mathbf{v}_2\}$ is an orthogonal basis for the eigenspace of $\lambda = 10$.

(b) Let

$$P = \begin{bmatrix} \frac{1}{\sqrt{6}} & -\frac{1}{\sqrt{2}} & -\frac{1}{\sqrt{3}} \\ \frac{1}{\sqrt{6}} & \frac{1}{\sqrt{2}} & -\frac{1}{\sqrt{3}} \\ \frac{2}{\sqrt{6}} & 0 & \frac{1}{\sqrt{3}} \end{bmatrix}.$$

Then

$$P^T A P = \begin{bmatrix} 4 & 0 & 0 \\ 0 & 10 & 0 \\ 0 & 0 & 10 \end{bmatrix}.$$

Chapter 8

Applications of Eigenvalues and Eigenvectors (Optional)

Section 8.1, p. 486

1. Suppose that $\mathbf{u} = \begin{bmatrix} a_1 \\ a_2 \\ a_3 \end{bmatrix}$ is a stable age distribution. Then

$$A\mathbf{u} = \mathbf{u} \implies \begin{bmatrix} 0 & 0 & 8 \\ \frac{1}{4} & 0 & 0 \\ 0 & \frac{1}{2} & 0 \end{bmatrix} \begin{bmatrix} a_1 \\ a_2 \\ a_3 \end{bmatrix} = \begin{bmatrix} a_1 \\ a_2 \\ a_3 \end{bmatrix} \implies \begin{array}{l} 8a_3 = a_1 \\ \frac{1}{4}a_1 = a_2 \\ \frac{1}{2}a_2 = a_3 \end{array}$$

We find that $a_1 = 8r$, $a_2 = 2r$, $a_3 = r$, where $r > 0$ is a real number. Choosing $r = 1$, we obtain the stable age distribution

$$\begin{bmatrix} 8 \\ 2 \\ 1 \end{bmatrix}.$$

3. T is a transition matrix if and only if $0 \le t_{ij} \le 1$ for $1 \le i, j \le n$ and the sum of the entries in each column is 1.

   (a) Let $T = \begin{bmatrix} 0.3 & 0.7 \\ 0.4 & 0.6 \end{bmatrix}$. T is not a transition matrix because the sum of the entries in each column is not 1.

   (b) Let $T = \begin{bmatrix} 0.2 & 0.3 & 0.1 \\ 0.8 & 0.5 & 0.7 \\ 0 & 0.2 & 0.2 \end{bmatrix}$. T is a transition matrix.

   (c) Let $T = \begin{bmatrix} 0.55 & 0.33 \\ 0.45 & 0.67 \end{bmatrix}$. T is a transition matrix.

   (d) Let $T = \begin{bmatrix} 0.3 & 0.4 & 0.2 \\ 0.2 & 0 & 0.8 \\ 0.1 & 0.3 & 0.6 \end{bmatrix}$. T is not a transition matrix because the sum of the entries in each column is not 1.

5. Let $T = \begin{bmatrix} 0.7 & 0.4 \\ 0.3 & 0.6 \end{bmatrix}$.

(a) For $\mathbf{x}^{(0)} = \begin{bmatrix} 1 \\ 0 \end{bmatrix}$,

$$\mathbf{x}^{(1)} = T\mathbf{x}^{(0)} = \begin{bmatrix} 0.700 \\ 0.300 \end{bmatrix}, \qquad \mathbf{x}^{(2)} = T\mathbf{x}^{(1)} = \begin{bmatrix} 0.610 \\ 0.390 \end{bmatrix}, \qquad \mathbf{x}^{(3)} = T\mathbf{x}^{(2)} = \begin{bmatrix} 0.583 \\ 0.417 \end{bmatrix}.$$

(b) All the entries of T are positive thus T is regular. Following the method in Example 2, we solve for a vector $\mathbf{u}$ such that $T\mathbf{u} = \mathbf{u}$. Hence solve the homogeneous system

$$(I_3 - T)\mathbf{u} = \begin{bmatrix} 0.3 & -0.4 \\ -0.3 & 0.4 \end{bmatrix} \begin{bmatrix} u_1 \\ u_2 \end{bmatrix} = \mathbf{0}.$$

Row reduce the coefficient matrix to obtain

$$\begin{bmatrix} 1 & -\frac{4}{3} \\ 0 & 0 \end{bmatrix}.$$

The solution is $u_1 = \frac{4}{3}r$, $u_2 = r$, where r is any real number. Since $\mathbf{u}$ is to be a probability vector, we require that

$$u_1 + u_2 = \tfrac{4}{3}r + r = \tfrac{7}{3}r = 1.$$

Set $r = \frac{3}{7}$. Then the steady-state vector is

$$\mathbf{u} = \begin{bmatrix} \frac{4}{7} \\ \frac{3}{7} \end{bmatrix} \approx \begin{bmatrix} 0.571 \\ 0.429 \end{bmatrix}.$$

7. (a) $T = \begin{bmatrix} 0 & \frac{1}{2} \\ 1 & \frac{1}{2} \end{bmatrix}$ is regular since $T^2 = \begin{bmatrix} \frac{1}{2} & \frac{1}{4} \\ \frac{1}{2} & \frac{3}{4} \end{bmatrix} > \mathbf{0}.$

(b) $T = \begin{bmatrix} \frac{1}{2} & 0 & 0 \\ 0 & 1 & \frac{1}{2} \\ \frac{1}{2} & 0 & \frac{1}{2} \end{bmatrix}$ is not regular since T^k will always have $\begin{bmatrix} 0 \\ 1 \\ 0 \end{bmatrix}$ as its second column.

(c) $T = \begin{bmatrix} 1 & \frac{1}{3} & 0 \\ 0 & \frac{1}{3} & 1 \\ 0 & \frac{1}{3} & 0 \end{bmatrix}$ is not regular since T^k will always have $\begin{bmatrix} 1 \\ 0 \\ 0 \end{bmatrix}$ as its first column.

(d) $T = \begin{bmatrix} \frac{1}{4} & \frac{3}{5} & \frac{1}{2} \\ \frac{1}{2} & 0 & 0 \\ \frac{1}{4} & \frac{2}{5} & \frac{1}{2} \end{bmatrix}$ is regular since $T^2 = \begin{bmatrix} 0.4875 & 0.35 & 0.375 \\ 0.1250 & 0.30 & 0.250 \\ 0.3875 & 0.35 & 0.375 \end{bmatrix} > \mathbf{0}.$

9. Let T be the given matrix. In each case we determine a solution $\mathbf{u}$ to the system $(I - A)\mathbf{u} = \mathbf{0}$ whose components add to 1.

(a) $I - A = \begin{bmatrix} \frac{2}{3} & -\frac{1}{2} \\ -\frac{2}{3} & \frac{1}{2} \end{bmatrix} \rightarrow \begin{bmatrix} 1 & -\frac{3}{4} \\ 0 & 0 \end{bmatrix}$

Thus $u_1 = \frac{3}{4}u_2$. Since $u_1 + u_2 = 1$, we obtain the solution $u_1 = \frac{3}{7}$, $u_2 = \frac{4}{7}$. Therefore the steady state vector is

$$\begin{bmatrix} \frac{3}{7} \\ \frac{4}{7} \end{bmatrix}.$$

(b) $I - A = \begin{bmatrix} 0.7 & -0.1 \\ -0.7 & 0.1 \end{bmatrix} \rightarrow \begin{bmatrix} 1 & -\frac{1}{7} \\ 0 & 0 \end{bmatrix}$

Thus $u_1 = \frac{1}{7}u_2$. Since $u_1 + u_2 = 1$, we obtain the solution $u_1 = \frac{1}{8}$, $u_2 = \frac{7}{8}$. Therefore the steady state vector is

$$\begin{bmatrix} \frac{1}{8} \\ \frac{7}{8} \end{bmatrix}.$$

(c) $I - A = \begin{bmatrix} \frac{3}{4} & -\frac{1}{2} & -\frac{1}{3} \\ 0 & \frac{1}{2} & -\frac{2}{3} \\ -\frac{3}{4} & 0 & 1 \end{bmatrix} \rightarrow \begin{bmatrix} 1 & 0 & -\frac{4}{3} \\ 0 & 1 & -\frac{4}{3} \\ 0 & 0 & 0 \end{bmatrix}$

Thus $u_1 = \frac{4}{3}u_3$, $u_2 = \frac{4}{3}u_3$. Since $u_1 + u_2 + u_3 = 1$, we obtain the solution $u_1 = \frac{4}{11}$, $u_2 = \frac{4}{11}$, $u_3 = \frac{3}{11}$. Therefore the steady state vector is

$$\begin{bmatrix} \frac{4}{11} \\ \frac{4}{11} \\ \frac{3}{11} \end{bmatrix}.$$

(d) $I - A = \begin{bmatrix} 0.6 & 0.0 & -0.1 \\ -0.2 & 0.5 & -0.3 \\ -0.4 & -0.5 & 0.4 \end{bmatrix} \rightarrow \begin{bmatrix} 1 & 0 & -\frac{1}{6} \\ 0 & 1 & -\frac{2}{3} \\ 0 & 0 & 0 \end{bmatrix}$

Thus $u_1 = \frac{1}{6}u_3$, $u_2 = \frac{2}{3}u_3$. Since $u_1 + u_2 + u_3 = 1$, we obtain the solution $u_1 = \frac{1}{11}$, $u_2 = \frac{4}{11}$, $u_3 = \frac{6}{11}$. THerefore the steady state vector is

$$\begin{bmatrix} \frac{1}{11} \\ \frac{4}{11} \\ \frac{6}{11} \end{bmatrix}.$$

11. Let $T = \begin{bmatrix} 0.8 & 0.3 & 0.2 \\ 0.1 & 0.5 & 0.2 \\ 0.1 & 0.2 & 0.6 \end{bmatrix}$.

(a) We have that an individual is a professional. Thus the current state is

$$\mathbf{x}^{(0)} = \begin{bmatrix} 1 \\ 0 \\ 0 \end{bmatrix}.$$

We are asked to determine the probability that a grandson will be a professional. This represents a transition through two stages:

$$\mathbf{x}^{(2)} = T\mathbf{x}^{(1)} = T(T\mathbf{x}^{(0)}) = T \begin{bmatrix} 0.8 \\ 0.1 \\ 0.1 \end{bmatrix} = \begin{bmatrix} 0.69 \\ 0.15 \\ 0.16 \end{bmatrix}.$$

Thus the probability that a grandson of a professional will be a professional is 0.69.

(b) The "long-run" distribution is the steady-state vector. Thus solve

$$(I_3 - T)\mathbf{u} = \begin{bmatrix} 0.2 & -0.3 & -0.2 \\ -0.1 & 0.5 & -0.2 \\ -0.1 & -0.2 & 0.4 \end{bmatrix} \begin{bmatrix} u_1 \\ u_2 \\ u_3 \end{bmatrix} = \mathbf{0}.$$

Row reduce the coefficient matrix to obtain

$$\begin{bmatrix} 1 & 0 & -\frac{16}{7} \\ 0 & 1 & -\frac{6}{7} \\ 0 & 0 & 0 \end{bmatrix}.$$

The solution is $u_1 = \frac{16}{7}r$, $u_2 = \frac{6}{7}r$, $u_3 = r$, where r is any real number. Since the steady-state vector is a probability vector we have

$$u_1 + u_2 + u_3 = \frac{16}{7}r + \frac{6}{7}r + r = 1.$$

Thus $r = \frac{7}{29}$ and

$$\mathbf{u} = \begin{bmatrix} \frac{16}{29} \\ \frac{6}{29} \\ \frac{7}{29} \end{bmatrix} \approx \begin{bmatrix} 0.552 \\ 0.207 \\ 0.241 \end{bmatrix}.$$

The proportion of the population that will be farmers is .207 or 20.7%.

13. Let

$$T = \begin{bmatrix} 0.7 & 0.2 \\ 0.3 & 0.8 \end{bmatrix}.$$

The initial state is that 30% of commuters use mass transit and 70% use their automobiles. Hence the initial state is

$$\mathbf{x}^{(0)} = \begin{bmatrix} 0.3 \\ 0.7 \end{bmatrix}.$$

(a) The state 1 year from now is calculated as

$$\mathbf{x}^{(1)} = T\mathbf{x}^{(0)} = \begin{bmatrix} 0.35 \\ 0.65 \end{bmatrix}.$$

The state 2 years from now is

$$\mathbf{x}^{(2)} = T\mathbf{x}^{(1)} = \begin{bmatrix} 0.375 \\ 0.625 \end{bmatrix}.$$

Hence after one year 35% of the commuters will be using mass transit and at the end of two years 37.5% will be using mass transit.

(b) Find the steady-state vector. Solve the homogeneous linear system

$$(I_2 - T)\mathbf{u} = \begin{bmatrix} 0.3 & -0.2 \\ -0.3 & 0.2 \end{bmatrix} \begin{bmatrix} u_1 \\ u_2 \end{bmatrix} = \mathbf{0}.$$

Row reduce the coefficient matrix to obtain

$$\begin{bmatrix} 1 & -\frac{2}{3} \\ 0 & 0 \end{bmatrix}.$$

The solution is $u_1 = \frac{2}{3}r$, $u_2 = r$, where r is any real number. Since the steady-state vector is a probability vector, $u_1 + u_2 = \frac{5}{3}r = 1$. Thus $r = \frac{3}{5}$ and

$$\mathbf{u} = \begin{bmatrix} \frac{2}{5} \\ \frac{3}{5} \end{bmatrix}.$$

In the long run 40% of commuters will be using mass transit.

Section 8.2, p. 500

1. (a) $A^T A = \begin{bmatrix} 25 & 0 \\ 0 & 1 \end{bmatrix}$ with eigenvalues 25 and 1. It follows that the singular values of A are 5 and 1, the square root of the eigenvalues.

(b) $A^T A = \begin{bmatrix} 2 & 2 \\ 2 & 2 \end{bmatrix}$ with eigenvalues 4 and 0. It follows that the singular values of A are 2 and 0, the square root of the eigenvalues.

(c) $A^T A = \begin{bmatrix} 5 & 0 \\ 0 & 6 \end{bmatrix}$ with eigenvalues 5 and 6. It follows that the singular values of A are $\sqrt{5}$ and $\sqrt{6}$, the square root of the eigenvalues.

(d) $A^T A = \begin{bmatrix} 1 & 0 & 1 & -1 \\ 0 & 1 & -1 & 1 \\ 1 & -1 & 2 & -2 \\ -1 & 1 & -2 & 2 \end{bmatrix}$ with eigenvalues 0, 0, 1, and 5. It follows that the singular values of A are 0, 0, 1, and $\sqrt{5}$, the square root of the eigenvalues.

3. We follow the method of Example 5. First compute

$$AA^T = \begin{bmatrix} 1 & 1 \\ -1 & 1 \end{bmatrix} \begin{bmatrix} 1 & -1 \\ 1 & 1 \end{bmatrix} = \begin{bmatrix} 2 & 0 \\ 0 & 2 \end{bmatrix}.$$

Thus the eigenvalues of AA^T are 2 and 2. Associated eigenvectors are $\begin{bmatrix} 1 \\ 0 \end{bmatrix}$ and $\begin{bmatrix} 0 \\ 1 \end{bmatrix}$. Thus

$$V = \begin{bmatrix} 1 & 0 \\ 0 & 1 \end{bmatrix}, \quad s_{11} = s_{22} = \sqrt{2}, \quad \text{and} \quad S = \begin{bmatrix} \sqrt{2} & 0 \\ 0 & \sqrt{2} \end{bmatrix}.$$

Next we find the matrix $U = \begin{bmatrix} \mathbf{u}_1 & \mathbf{u}_2 \end{bmatrix}$:

$$\mathbf{u}_1 = \frac{1}{\sqrt{2}} A \begin{bmatrix} 1 \\ 0 \end{bmatrix} = \frac{1}{\sqrt{2}} \begin{bmatrix} 1 & 1 \\ -1 & 1 \end{bmatrix} \begin{bmatrix} 1 \\ 0 \end{bmatrix} = \frac{1}{\sqrt{2}} \begin{bmatrix} 1 \\ -1 \end{bmatrix}$$

$$\mathbf{u}_2 = \frac{1}{\sqrt{2}} A \begin{bmatrix} 0 \\ 1 \end{bmatrix} = \frac{1}{\sqrt{2}} \begin{bmatrix} 1 & 1 \\ -1 & 1 \end{bmatrix} \begin{bmatrix} 0 \\ 1 \end{bmatrix} = \frac{1}{\sqrt{2}} \begin{bmatrix} 1 \\ 1 \end{bmatrix}$$

so

$$U = \frac{1}{\sqrt{2}} \begin{bmatrix} 1 & 1 \\ -1 & 1 \end{bmatrix}.$$

Observe that the vectors $\{\mathbf{u}_1, \mathbf{u}_2\}$ form an orthonormal basis for R^2. Therefore the singular value decomposition of A is

$$A = USV^T = \frac{1}{\sqrt{2}} \begin{bmatrix} 1 & 1 \\ -1 & 1 \end{bmatrix} \begin{bmatrix} \sqrt{2} & 0 \\ 0 & \sqrt{2} \end{bmatrix} \begin{bmatrix} 1 & 0 \\ 0 & 1 \end{bmatrix}^T.$$

Section 8.3, p. 514

1. (a) The eigenvalues of this matrix were obtained in Example 10, Section 7.1: $\lambda_1 = 2$, $\lambda_2 = 3$. Hence the dominant eigenvalue is 3.

   (b) The eigenvalues of this matrix were obtained in Examples 11 and 12, Section 7.1: $\lambda_1 = 1$, $\lambda_2 = 2$, and $\lambda_3 = 3$. Hence the dominant eigenvalue is 3.

3. (a) $\left\| \begin{bmatrix} 3 & -5 \\ 2 & 2 \end{bmatrix} \right\|_1 = \max \left\{ \left\| \begin{bmatrix} 3 \\ 2 \end{bmatrix} \right\|_1, \left\| \begin{bmatrix} -5 \\ 2 \end{bmatrix} \right\|_1 \right\} = \max\{5, 7\} = 7.$

   (b) $\left\| \begin{bmatrix} 4 & 1 & 0 \\ 2 & 5 & 0 \\ -4 & -4 & 7 \end{bmatrix} \right\|_1 = \max \left\{ \left\| \begin{bmatrix} 4 \\ 2 \\ -4 \end{bmatrix} \right\|_1, \left\| \begin{bmatrix} 1 \\ 5 \\ -4 \end{bmatrix} \right\|_1, \left\| \begin{bmatrix} 0 \\ 0 \\ 7 \end{bmatrix} \right\|_1 \right\} = \max\{10, 10, 7\} = 10.$

   (c) $\left\| \begin{bmatrix} 2 & -1 & 1 & 0 \\ 4 & -2 & 2 & 3 \\ 1 & 0 & 2 & 6 \\ -3 & 4 & 8 & 1 \end{bmatrix} \right\|_1 = \max \left\{ \left\| \begin{bmatrix} 2 \\ 4 \\ 1 \\ -3 \end{bmatrix} \right\|_1, \left\| \begin{bmatrix} -1 \\ -2 \\ 0 \\ 4 \end{bmatrix} \right\|_1, \left\| \begin{bmatrix} 1 \\ 2 \\ 2 \\ 8 \end{bmatrix} \right\|_1, \left\| \begin{bmatrix} 0 \\ 3 \\ 6 \\ 1 \end{bmatrix} \right\|_1 \right\}$

   $= \max\{10, 7, 13, 10\} = 13.$

5. (a) $\left\|\begin{bmatrix} 1 & 1 \\ -2 & 4 \end{bmatrix}\right\| = \max\{3,5\} = 5.$

(b) $\left\|\begin{bmatrix} 1 & 2 & -1 \\ 1 & 0 & 1 \\ 4 & -4 & 5 \end{bmatrix}\right\| = \max\{6,6,7\} = 7.$

7. If A is symmetric, $A^T = A$. Therefore, $\|A\|_1 = \|A^T\|_1$.

9. First recall tht $A^k \to \mathbf{0}$ for any vector $\mathbf{x}$ if $|\lambda_1| < 1$, where λ_1 is a unique dominant eigenvalue of A. To show that $\lambda_1| < 1$, recall that $|\lambda_1| \le \|A\|_1$. Hence if $\|A\|_1 < 1$, we have $|\lambda_1| \le \|A\|_1 < 1$, so $|\lambda_1| < 1$ and therefore $A^k\mathbf{x} \to \mathbf{0}$.

11. Sample mean $= \frac{1}{5}(56 + 62 + 59 + 73 + 75) = 65$.
Sample variance $= \frac{1}{5}[(56-65)^2 + (62-65)^2 + (59-65)^2 + (73-65)^2 + (75-65)^2] = 58$.
Standard deviation $= \sqrt{58} \approx 7.6158$.

13. The sample means are

$$x_1 = \tfrac{1}{5}(337 + 449 + 631 + 550 + 582) = 509.80$$

$$x_2 = \tfrac{1}{5}(425 + 847 + 846 + 617 + 647) = 676.4$$

Thus, the vector of sample means is

$$\overline{\mathbf{x}} = \begin{bmatrix} 509.80 \\ 676.40 \end{bmatrix}.$$

For the sample covariances, we find:

$$s_{11} = 11014.96, \quad s_{21} = 9822.88, \quad s_{22} = 25092.64.$$

Hence the covariance matrix is

$$\begin{bmatrix} 11014.96 & 9822.88 \\ 9822.88 & 25092.64 \end{bmatrix}.$$

15. Let

$$S = \begin{bmatrix} 11014.96 & 9822.88 \\ 9822.88 & 25092.64 \end{bmatrix}$$

be the covariance matrix obtained in Exercise 13. For the eigenvalues and eigenvectors of S, we obtain

$$\lambda_1 = 30138.3, \quad \mathbf{u}_1 = \begin{bmatrix} 0.456908 \\ 0.889514 \end{bmatrix}$$

$$\lambda_2 = 5969.34, \quad \mathbf{u}_2 = \begin{bmatrix} 0.889514 \\ -0.456908 \end{bmatrix}$$

Using Theorem 8.7, the first principal component is

$$Y_1 = X\mathbf{u}_1 = 0.456908\mathrm{col}_1(X) + 0.889514\mathrm{col}_2(X)$$

$$= 0.456908 \begin{bmatrix} 337 \\ 449 \\ 631 \\ 550 \\ 582 \end{bmatrix} + 0.889514 \begin{bmatrix} 425 \\ 847 \\ 846 \\ 617 \\ 647 \end{bmatrix}$$

$$= \begin{bmatrix} 550.298 \\ 958.57 \\ 1040.84 \\ 800.13 \\ 841.436 \end{bmatrix}.$$

17. Let λ be an eigenvalue of a square symmetric matrix C with associated eigenvector $\mathbf{x}$. Then $C\mathbf{x} = \lambda\mathbf{x}$ and hence

$$\mathbf{x}^T C\mathbf{x} > 0 \implies \mathbf{x}^T(\lambda\mathbf{x}) > 0 \implies \lambda(\mathbf{x}^T\mathbf{x}) > 0 \implies \lambda\|\mathbf{x}\|^2 > 0.$$

since $\mathbf{x} \neq \mathbf{0}$, $\|\mathbf{x}\|^2 > 0$. Therefore, $\lambda > 0$.

Section 8.4 , p. 524

1. The linear system of differential equations

$$\begin{bmatrix} x_1' \\ x_2' \\ x_3' \end{bmatrix} = \begin{bmatrix} -3 & 0 & 0 \\ 0 & 4 & 0 \\ 0 & 0 & 2 \end{bmatrix} \begin{bmatrix} x_1 \\ x_2 \\ x_3 \end{bmatrix}$$

is a diagonal system, thus we follow the method in Example 2.

(a) The system can be written as

$$x_1' = -3x_1, \quad x_2' = 4x_2, \quad x_3' = 2x_3.$$

Integrating these equations we obtain

$$x_1 = b_1 e^{-3t}, \quad x_2 = b_2 e^{4t}, \quad x_3 = b_3 e^{2t},$$

where b_1, b_2, and b_3 are arbitrary constants. Thus

$$\mathbf{x}(t) = \begin{bmatrix} b_1 e^{-3t} \\ b_2 e^{4t} \\ b_3 e^{2t} \end{bmatrix} = b_1 \begin{bmatrix} 1 \\ 0 \\ 0 \end{bmatrix} e^{-3t} + b_2 \begin{bmatrix} 0 \\ 1 \\ 0 \end{bmatrix} e^{4t} + b_3 \begin{bmatrix} 0 \\ 0 \\ 1 \end{bmatrix} e^{2t}$$

is the general solution.

(b) To find the solution of the initial value problem with initial conditions $x_1(0) = 3$, $x_2(0) = 4$, $x_3(0) = 5$, we set $t = 0$ in the expressions

$$x_1 = b_1 e^{-3t}, \quad x_2 = b_2 e^{4t}, \quad x_3 = b_3 e^{2t},$$

and use the given values for the left sides. We obtain $b_1 = 3$, $b_2 = 4$, and $b_3 = 5$. Thus the solution of the initial value problem is

$$\mathbf{x}(t) = \begin{bmatrix} 3e^{-3t} \\ 4e^{4t} \\ 5e^{2t} \end{bmatrix} = 3 \begin{bmatrix} 1 \\ 0 \\ 0 \end{bmatrix} e^{-3t} + 4 \begin{bmatrix} 0 \\ 1 \\ 0 \end{bmatrix} e^{4t} + 5 \begin{bmatrix} 0 \\ 0 \\ 1 \end{bmatrix} e^{2t}.$$

3. Since the linear system of differential equations

$$\begin{bmatrix} x_1' \\ x_2' \\ x_3' \end{bmatrix} = \begin{bmatrix} 4 & 0 & 0 \\ 3 & -5 & 0 \\ 2 & 1 & 2 \end{bmatrix} \begin{bmatrix} x_1 \\ x_2 \\ x_3 \end{bmatrix}$$

is not diagonal, we follow the procedure outlined prior to Theorem 8.8 and used in Examples 3 and 4. The eigenvalues are the diagonal entries since the matrix is lower triangular. Let $\lambda_1 = 4$, $\lambda_2 = -5$, and $\lambda_3 = 2$. Using the methods from Chapter 7 we find that the corresponding eigenvectors are respectively

$$\mathbf{x}_1 = \begin{bmatrix} 6 \\ 2 \\ 7 \end{bmatrix}, \quad \mathbf{x}_2 = \begin{bmatrix} 0 \\ -7 \\ 1 \end{bmatrix}, \quad \mathbf{x}_3 = \begin{bmatrix} 0 \\ 0 \\ 1 \end{bmatrix}.$$

Since the eigenvalues are distinct the eigenvectors are linearly independent. Hence by Theorem 8.8 the general solution is

$$\mathbf{x}(t) = b_1 \begin{bmatrix} 6 \\ 2 \\ 7 \end{bmatrix} e^{4t} + b_2 \begin{bmatrix} 0 \\ -7 \\ 1 \end{bmatrix} e^{-5t} + b_3 \begin{bmatrix} 0 \\ 0 \\ 1 \end{bmatrix} e^{2t}$$

where the b's are arbitrary constants.

5. Let

$$\begin{bmatrix} x_1' \\ x_2' \\ x_3' \end{bmatrix} = \begin{bmatrix} 5 & 0 & 0 \\ 0 & -4 & 3 \\ 0 & 3 & 4 \end{bmatrix} \begin{bmatrix} x_1 \\ x_2 \\ x_3 \end{bmatrix}.$$

We proceed as in Exercise 3. In this case we must determine the eigenvalues. We have

$$\det(\lambda I_3 - A) = \begin{vmatrix} \lambda - 5 & 0 & 0 \\ 0 & \lambda + 4 & -3 \\ 0 & -3 & \lambda - 4 \end{vmatrix} = (\lambda - 5)(\lambda^2 - 16 - 9) = (\lambda - 5)^2(\lambda + 5).$$

It follows that the eigenvalues are $\lambda_1 = \lambda_2 = 5$, $\lambda_3 = -5$. Since the matrix of the system is symmetric we are assured that there will be three linearly independent eigenvectors. Using the techniques developed in Chapter 7 we find that the corresponding linearly independent eigenvectors can be taken as

$$\mathbf{x}_1 = \begin{bmatrix} 1 \\ 0 \\ 0 \end{bmatrix}, \quad \mathbf{x}_2 = \begin{bmatrix} 0 \\ 1 \\ 3 \end{bmatrix}, \quad \mathbf{x}_3 = \begin{bmatrix} 0 \\ -3 \\ 1 \end{bmatrix}.$$

Hence by Theorem 8.8 the general solution is

$$\mathbf{x}(t) = b_1 \begin{bmatrix} 1 \\ 0 \\ 0 \end{bmatrix} e^{5t} + b_2 \begin{bmatrix} 0 \\ 1 \\ 3 \end{bmatrix} e^{5t} + b_3 \begin{bmatrix} 0 \\ -3 \\ 1 \end{bmatrix} e^{-5t}$$

where the b's are arbitrary constants.

7. Let

$$\begin{bmatrix} x_1' \\ x_2' \\ x_3' \end{bmatrix} = \begin{bmatrix} -2 & -2 & 3 \\ 0 & -2 & 2 \\ 0 & 2 & 1 \end{bmatrix} \begin{bmatrix} x_1 \\ x_2 \\ x_3 \end{bmatrix}.$$

We proceed as in Exercise 5. We must determine the eigenvalues. We have

$$\det(\lambda I_3 - A) = \begin{vmatrix} \lambda + 2 & 2 & -3 \\ 0 & \lambda + 2 & -2 \\ 0 & -2 & \lambda - 1 \end{vmatrix} = (\lambda + 2)(\lambda^2 + \lambda - 6) = (\lambda + 2)(\lambda - 2)(\lambda + 3).$$

It follows that the eigenvalues are $\lambda_1 = -2$, $\lambda_2 = 2$, $\lambda_3 = -3$. Since the eigenvalues are distinct we are assured that there will be three linearly independent eigenvectors. Using the techniques developed in Chapter 7 we find that the corresponding linearly independent eigenvectors can be taken as

$$\mathbf{x}_1 = \begin{bmatrix} 1 \\ 0 \\ 0 \end{bmatrix}, \quad \mathbf{x}_2 = \begin{bmatrix} 1 \\ 1 \\ 2 \end{bmatrix}, \quad \mathbf{x}_3 = \begin{bmatrix} -7 \\ -2 \\ 1 \end{bmatrix}.$$

Hence by Theorem 8.8 the general solution is

$$\mathbf{x}(t) = b_1 \begin{bmatrix} 1 \\ 0 \\ 0 \end{bmatrix} e^{-2t} + b_2 \begin{bmatrix} 1 \\ 1 \\ 2 \end{bmatrix} e^{2t} + b_3 \begin{bmatrix} -7 \\ -2 \\ 1 \end{bmatrix} e^{-3t}$$

where the b's are arbitrary constants.

9. The initial value problem that models this situation is given by

$$\mathbf{x}'(t) = \begin{bmatrix} x_1'(t) \\ x_2'(t) \end{bmatrix} = \begin{bmatrix} -3 & 6 \\ 1 & -2 \end{bmatrix} \begin{bmatrix} x_1(t) \\ x_2(t) \end{bmatrix}, \quad \begin{matrix} x_1(0) = 500 \\ x_2(0) = 200 \end{matrix}$$

We find the eigenvalues and corresponding eigenvectors of the coefficient matrix. They are

$$\lambda_1 = 0, \ \mathbf{x}_1 = \begin{bmatrix} 2 \\ 1 \end{bmatrix}; \quad \lambda_2 = -5, \ \mathbf{x}_2 = \begin{bmatrix} -3 \\ 1 \end{bmatrix}.$$

It follows that the general solution is given by

$$\mathbf{x}(t) = b_1 \begin{bmatrix} 2 \\ 1 \end{bmatrix} + b_2 \begin{bmatrix} -3 \\ 1 \end{bmatrix} e^{-5t}.$$

We apply the initial conditions to determine the constants b_1 and b_2.

$$\begin{matrix} x_1(0) = 500 \\ x_2(0) = 200 \end{matrix} \implies \mathbf{x}(0) = \begin{bmatrix} 500 \\ 200 \end{bmatrix} \implies \begin{bmatrix} 2 & -3 \\ 1 & 1 \end{bmatrix} \begin{bmatrix} b_1 \\ b_2 \end{bmatrix} = \begin{bmatrix} 500 \\ 200 \end{bmatrix}.$$

Solving for b_1 and b_2 we obtain $b_1 = 220$ and $b_2 = -20$. Hence the solution of the initial value problem is

$$\mathbf{x}(t) = \begin{bmatrix} 440 + 60e^{-5t} \\ 220 - 20e^{-5t} \end{bmatrix}.$$

Section 8.5, p. 534

1. The eigenvalues of the coefficient matrix are $\lambda_1 = -1$ and $\lambda_2 = -3$ with associated eigenvectors

$$\mathbf{p}_1 = \begin{bmatrix} 1 \\ 0 \end{bmatrix} \quad \text{and} \quad \mathbf{p}_2 = \begin{bmatrix} 0 \\ 1 \end{bmatrix}.$$

Thus the origin is a stable equilibrium. The phase portrait shows all trajectories tending toward the origin.

3. The eigenvalues of the coefficient matrix are $\lambda_1 = -1$ and $\lambda_2 = -1$ with one linearly independent eigenvector $\mathbf{p}_1 = \begin{bmatrix} 1 \\ 0 \end{bmatrix}$. Thus the origin is a stable equilibrium. The phase portrait shows all trajectories tending toward the origin with those passing through points not on the eigenvector aligning themselves to be tangent to the eigenvector at the origin.

5. The eigenvalues of the coefficient matrix are $\lambda_1 = 2$ and $\lambda_2 = -2$ with associated eigenvectors

$$\mathbf{p}_1 = \begin{bmatrix} 1 \\ 1 \end{bmatrix} \quad \text{and} \quad \mathbf{p}_2 = \begin{bmatrix} -1 \\ 3 \end{bmatrix}.$$

Thus the origin is a saddle point. The phase portrait shows trajectories not in the direction of an eigenvector heading towards the origin, but bending away as $t \to \infty$.

7. The eigenvalues of the coefficient matrix are $\lambda_1 = -1$ and $\lambda_2 = -4$ with one linearly independent eigenvector

$$\mathbf{p}_1 = \begin{bmatrix} 1 \\ 1 \end{bmatrix} \quad \text{and} \quad \mathbf{p}_2 = \begin{bmatrix} -1 \\ 2 \end{bmatrix}.$$

Thus the origin is a stable equilibrium. The phase portrait shows all trajectories tending towards the origin.

9. The eigenvalues of the coefficient matrix are $\lambda_1 = 2i$ and $\lambda_2 = -2i$ with associated eigenvectors

$$\mathbf{p}_1 = \begin{bmatrix} -0.5345 + 0.8018i \\ 0.2673i \end{bmatrix} \quad \text{and} \quad \mathbf{p}_2 = \begin{bmatrix} -0.5345 - 0.8018i \\ -0.2673i \end{bmatrix}.$$

Since the eigenvalues have real part zero, the trajectories are ellipses whose major and minor axes are determined by the eigenvectors. The origin is called marginally stable in this case.

Section 8.6, p. 542

1. We use Equations (1) and (2) along with Examples 1 and 2.

(a) $-3x^2 + 5xy - 2y^2 = \begin{bmatrix} x & y \end{bmatrix} \begin{bmatrix} -3 & \frac{5}{2} \\ \frac{5}{2} & -2 \end{bmatrix} \begin{bmatrix} x \\ y \end{bmatrix}$

(b) $2x_1^2 + 3x_1x_2 - 5x_1x_3 + 7x_2x_3 = \begin{bmatrix} x_1 & x_2 & x_3 \end{bmatrix} \begin{bmatrix} 2 & \frac{3}{2} & -\frac{5}{2} \\ \frac{3}{2} & 0 & \frac{7}{2} \\ -\frac{5}{2} & \frac{7}{2} & 0 \end{bmatrix} \begin{bmatrix} x_1 \\ x_2 \\ x_3 \end{bmatrix}$

(c) $3x_1^2 + x_2^2 - 2x_3^2 + x_1x_2 - x_1x_3 - 4x_2x_3 = \begin{bmatrix} x_1 & x_2 & x_3 \end{bmatrix} \begin{bmatrix} 3 & \frac{1}{2} & -\frac{1}{2} \\ \frac{1}{2} & 1 & -2 \\ -\frac{1}{2} & -2 & -2 \end{bmatrix} \begin{bmatrix} x_1 \\ x_2 \\ x_3 \end{bmatrix}$

3. Since each of the matrices A is symmetric we use Theorem 7.9. The diagonal matrix D we seek has the eigenvalues of A as diagonal entries. Thus we compute the eigenvalues of A and form matrix D.

(a) Let $A = \begin{bmatrix} -1 & 0 & 0 \\ 0 & 1 & 1 \\ 0 & 1 & 1 \end{bmatrix}$. The characteristic polynomial of A is

$$\det(\lambda I_3 - A) = \begin{vmatrix} \lambda + 1 & 0 & 0 \\ 0 & \lambda - 1 & -1 \\ 0 & -1 & \lambda - 1 \end{vmatrix} = (\lambda + 1)((\lambda - 1)^2 - 1) = (\lambda + 1)(\lambda - 2)\lambda.$$

Hence the eigenvalues of A are $\lambda_1 = -1$, $\lambda_2 = 2$, and $\lambda_3 = 0$. Thus a diagonal matrix D congruent to A is

$$D = \begin{bmatrix} -1 & 0 & 0 \\ 0 & 2 & 0 \\ 0 & 0 & 0 \end{bmatrix}.$$

There is more than one diagonal matrix D congruent to A. Others are found by reordering the eigenvalues of A on the diagonal.

(b) Let $A = \begin{bmatrix} 1 & 1 & 1 \\ 1 & 1 & 1 \\ 1 & 1 & 1 \end{bmatrix}$. The characteristic polynomial of A is

$$\det(\lambda I_3 - A) = \begin{vmatrix} \lambda - 1 & -1 & -1 \\ -1 & \lambda - 1 & -1 \\ -1 & -1 & \lambda - 1 \end{vmatrix} = (\lambda - 1)^3 - 1 - 1 - 3(\lambda - 1) = (\lambda - 3)\lambda^2.$$

Hence the eigenvalues of A are $\lambda_1 = 3$, $\lambda_2 = \lambda_3 = 0$. Thus a diagonal matrix D congruent to A is

$$D = \begin{bmatrix} 3 & 0 & 0 \\ 0 & 0 & 0 \\ 0 & 0 & 0 \end{bmatrix}.$$

There is more than one diagonal matrix D congruent to A. Others are found by reordering the eigenvalues of A on the diagonal.

(c) Let $A = \begin{bmatrix} 0 & 2 & 2 \\ 2 & 0 & 2 \\ 2 & 2 & 0 \end{bmatrix}$. The characteristic polynomial of A is

$$\det(\lambda I_3 - A) = \begin{vmatrix} \lambda & -2 & -2 \\ -2 & \lambda & -2 \\ -2 & -2 & \lambda \end{vmatrix} = \lambda^3 - 8 - 8 - 12\lambda = (\lambda - 4)(\lambda + 2)^2.$$

Hence the eigenvalues of A are $\lambda_1 = 4$, $\lambda_2 = -2$, and $\lambda_3 = -2$. Thus a diagonal matrix D congruent to A is

$$D = \begin{bmatrix} 4 & 0 & 0 \\ 0 & -2 & 0 \\ 0 & 0 & -2 \end{bmatrix}.$$

There is more than one diagonal matrix D congruent to A. Others are found by reordering the eigenvalues of A on the diagonal.

5. Let $g(\mathbf{x}) = 2x^2 - 4xy - y^2$. Then the matrix of the quadratic form is

$$A = \begin{bmatrix} 2 & -2 \\ -2 & -1 \end{bmatrix}.$$

The characteristic polynomial of A is

$$\det(\lambda I_2 - A) = \begin{vmatrix} \lambda - 2 & 2 \\ 2 & \lambda + 1 \end{vmatrix} = (\lambda - 2)(\lambda + 1) - 4 = \lambda^2 - \lambda - 6 = (\lambda - 3)(\lambda + 2).$$

Thus the eigenvalues of A are $\lambda_1 = 3$, $\lambda_2 = -2$. Hence $g(\mathbf{x})$ is equivalent to $h(\mathbf{y}) = 3x'^2 - 2y'^2$, where $\mathbf{y} = \begin{bmatrix} x' & y' \end{bmatrix}^T$. Note that if the eigenvalues had been labeled $\lambda_1 = -2$, $\lambda_2 = 3$, then $h(\mathbf{y}) = -2x'^2 + 3y'^2$. Hence $h(\mathbf{y})$ is not unique.

7. Let $g(\mathbf{x}) = 2x_1 x_3$. Then the matrix of the quadratic form is

$$A = \begin{bmatrix} 0 & 0 & 1 \\ 0 & 0 & 0 \\ 1 & 0 & 0 \end{bmatrix}.$$

The characteristic polynomial of A is

$$\det(\lambda I_3 - A) = \begin{vmatrix} \lambda & 0 & -1 \\ 0 & \lambda & 0 \\ -1 & 0 & \lambda \end{vmatrix} = \lambda^3 - \lambda = \lambda(\lambda^2 - 1) = (\lambda - 1)(\lambda)(\lambda + 1).$$

Thus the eigenvalues of A are $\lambda_1 = 1$, $\lambda_2 = 0$, $\lambda_3 = -1$. Hence $g(\mathbf{x})$ is equivalent to $h(\mathbf{y}) = y_1^2 - y_3^2$, where $\mathbf{y} = \begin{bmatrix} y_1 & y_2 & y_3 \end{bmatrix}^T$. Note that using the eigenvalues in a different order produces another equivalent quadratic form of the desired type.

9. Let $g(\mathbf{x}) = -2x_1^2 - 4x_2^2 + 4x_3^2 - 6x_2 x_3$. Then the matrix of the quadratic form is

$$A = \begin{bmatrix} -2 & 0 & 0 \\ 0 & -4 & -3 \\ 0 & -3 & 4 \end{bmatrix}.$$

The characteristic polynomial of A is

$$\det(\lambda I_3 - A) = \begin{vmatrix} \lambda + 2 & 0 & 0 \\ 0 & \lambda + 4 & 3 \\ 0 & 3 & \lambda - 4 \end{vmatrix} = (\lambda + 2)(\lambda + 4)(\lambda - 4) - 9(\lambda + 2) = (\lambda + 2)(\lambda - 5)(\lambda + 5).$$

Thus the eigenvalues of A are $\lambda_1 = -2$, $\lambda_2 = 5$, $\lambda_3 = -5$. Hence $g(\mathbf{x})$ is equivalent to $h(\mathbf{y}) = -2y_1^2 + 5y_2^2 - 5y_3^2$, where $\mathbf{y} = \begin{bmatrix} y_1 & y_2 & y_3 \end{bmatrix}^T$. Note that using the eigenvalues in a different order produces another equivalent quadratic form of the desired type.

11. Let $g(\mathbf{x}) = 2x^2 + 4xy + 2y^2$. Then the matrix of the quadratic form is

$$A = \begin{bmatrix} 2 & 2 \\ 2 & 2 \end{bmatrix}.$$

The characteristic polynomial of A is

$$\det(\lambda I_2 - A) = \begin{vmatrix} \lambda - 2 & -2 \\ -2 & \lambda - 2 \end{vmatrix} = (\lambda - 2)^2 - 4 = \lambda^2 - 4\lambda = \lambda(\lambda - 4).$$

Thus the eigenvalues of A are $\lambda_1 = 0$, $\lambda_2 = 4$. In this case $n = 2$, $r = 1$, $p = 1$, and $r - p = 0$. Hence $g(\mathbf{x})$ is equivalent to $h(\mathbf{y}) = y'^2$, where $\mathbf{y} = \begin{bmatrix} x' & y' \end{bmatrix}^T$.

13. Let $g(\mathbf{x}) = 2x_1^2 + 4x_2^2 + 4x_3^2 + 10x_2x_3$. Then the matrix of the quadratic form is

$$A = \begin{bmatrix} 2 & 0 & 0 \\ 0 & 4 & 5 \\ 0 & 5 & 4 \end{bmatrix}.$$

The characteristic polynomial of A is

$$\det(\lambda I_3 - A) = \begin{vmatrix} \lambda - 2 & 0 & 0 \\ 0 & \lambda - 4 & -5 \\ 0 & -5 & \lambda - 4 \end{vmatrix} = (\lambda - 2)(\lambda - 4)^2 - 25(\lambda - 2) = (\lambda - 2)(\lambda - 9)(\lambda + 1).$$

Thus the eigenvalues are $\lambda_1 = 2$, $\lambda_2 = 9$, $\lambda_3 = -1$. In this case $n = 3$, $r = 3$, $p = 2$, and $r - p = 1$. Hence $g(\mathbf{x})$ is equivalent to $h(\mathbf{y}) = y_1^2 + y_2^2 - y_3^2$, where $\mathbf{y} = \begin{bmatrix} y_1 & y_2 & y_3 \end{bmatrix}^T$.

15. Let $g(\mathbf{x}) = -3x_1^2 + 2x_2^2 + 2x_3^2 + 4x_2x_3$. Then the matrix of the quadratic form is

$$A = \begin{bmatrix} -3 & 0 & 0 \\ 0 & 2 & 2 \\ 0 & 2 & 2 \end{bmatrix}.$$

The characteristic polynomial of A is

$$\det(\lambda I_3 - A) = \begin{vmatrix} \lambda + 3 & 0 & 0 \\ 0 & \lambda - 2 & 2 \\ 0 & 2 & \lambda - 2 \end{vmatrix} = (\lambda + 3)(\lambda - 2)^2 - 4(\lambda + 3) = (\lambda + 3)\lambda(\lambda - 4).$$

Thus the eigenvalues of A are $\lambda_1 = -3$, $\lambda_2 = 0$, $\lambda_3 = 4$. In this case $n = 3$, $r = 2$, $p = 1$, and $r - p = 1$. Hence $g(\mathbf{x})$ is equivalent to $h(\mathbf{y}) = y_1^2 - y_2^2$, where $\mathbf{y} = \begin{bmatrix} y_1 & y_2 & y_3 \end{bmatrix}^T$.

17. Let $g(\mathbf{x}) = 4x_2^2 + 4x_3^2 - 10x_2x_3$. Then the matrix of the quadratic form is

$$A = \begin{bmatrix} 0 & 0 & 0 \\ 0 & 4 & -5 \\ 0 & -5 & 4 \end{bmatrix}.$$

The characteristic polynomial of A is

$$\det(\lambda I_3 - A) = \begin{vmatrix} \lambda & 0 & 0 \\ 0 & \lambda - 4 & 5 \\ 0 & 5 & \lambda - 4 \end{vmatrix} = \lambda(\lambda - 4)2 - 25\lambda = \lambda(\lambda - 9)(\lambda + 1).$$

Thus the eigenvalues of A are $\lambda_1 = 0$, $\lambda_2 = 9$, $\lambda_3 = -1$. In this case $n = 3$, $r = 2$, $p = 1$, and $r - p = 1$. Hence $g(\mathbf{x})$ is equivalent to $h(\mathbf{y}) = y_1^2 - y_2^2$, where $\mathbf{y} = \begin{bmatrix} y_1 & y_2 & y_3 \end{bmatrix}^T$. The rank of g is 2 and the signature of g is 0.

19. For a 2×2 matrix A of a quadratic form $g(\mathbf{x}) = \mathbf{x}^T A \mathbf{x}$ we have the following possibilities for its eigenvalues with the corresponding quadratic form of the type described in Theorem 8.10:

eigenvalues	quadratic form
both positive	$y_1^2 + y_2^2$
both negative	$-y_1^2 - y_2^2$
one positive, one negative	$y_1^2 - y_2^2$
one positive, one zero	y_1^2
one negative, one zero	$-y_1^2$

The conics for the equations $\mathbf{x}^T A \mathbf{x} = 1$ are given next.

$$y_1^2 + y_2^2 = 1 \qquad \text{is a circle}$$
$$-y_1^2 - y_2^2 = 1 \qquad \text{is empty; it represents no conic}$$
$$y_1^2 - y_2^2 = 1 \qquad \text{is a hyperbola}$$
$$y_1^2 = 1 \qquad \text{is a pair of lines; } y_1 = 1,\ y_1 = -1$$
$$-y_1^2 = 1 \qquad \text{is empty; it represents no concic}$$

21. In the discussion preceding Example 7 it is stated that quadratic forms g and h are equivalent if and only if they have equal ranks and the same signature. We determine the rank and signature for each of the quadratic forms listed.

- Let $g_1(\mathbf{x}) = x_1^2 + x_2^2 + x_3^2 + 2x_1 x_2$. Then the matrix of the quadratic form is

$$A = \begin{bmatrix} 1 & 1 & 0 \\ 1 & 1 & 0 \\ 0 & 0 & 1 \end{bmatrix}.$$

The characteristic polynomial of A is

$$\det(\lambda I_3 - A) = \begin{vmatrix} \lambda - 1 & -1 & 0 \\ -1 & \lambda - 1 & 0 \\ 0 & 0 & \lambda - 1 \end{vmatrix} = (\lambda - 1)^3 - (\lambda - 1) = \lambda(\lambda - 1)(\lambda - 2).$$

Thus the eigenvalues of A are $\lambda_1 = 0$, $\lambda_2 = 1$, $\lambda_3 = 2$. In this case $n = 3$, $r = 2$, $p = 2$, and $r - p = 0$. The rank of g_1 is 2 and the signature of g_1 is 2.

- Let $g_2(\mathbf{x}) = 2x_2^2 + 2x_3^2 + 2x_2 x_3$. Then the matrix of the quadratic form is

$$A = \begin{bmatrix} 0 & 0 & 0 \\ 0 & 2 & 1 \\ 0 & 1 & 2 \end{bmatrix}.$$

The characteristic polynomial of A is

$$\det(\lambda I_3 - A) = \begin{vmatrix} \lambda & 0 & 0 \\ 0 & \lambda - 2 & 1 \\ 0 & 1 & \lambda - 2 \end{vmatrix} = \lambda(\lambda - 2)^2 - \lambda = \lambda(\lambda - 1)(\lambda - 3).$$

Thus the eigenvalues of A are $\lambda_1 - 0$, $\lambda_2 = 1$, $\lambda_3 = 3$. In this case $n = 3$, $r = 2$, $p = 2$, and $r - p = 0$. The rank of g_2 is 2 and the signature of g_2 is 2. Thus $g_1(\mathbf{x})$ and $g_2(\mathbf{x}_2)$ are equivalent.

• Let $g_3(\mathbf{x}) = 3x_2^2 - 3x_3^2 + 8x_2x_3$. Then the matrix of the quadratic form is

$$A = \begin{bmatrix} 0 & 0 & 0 \\ 0 & 3 & 4 \\ 0 & 4 & -3 \end{bmatrix}.$$

The characteristic polynomial of A is

$$\det(\lambda I_3 - A) = \begin{vmatrix} \lambda & 0 & 0 \\ 0 & \lambda - 3 & -4 \\ 0 & -4 & \lambda + 3 \end{vmatrix} = \lambda(\lambda - 3)(\lambda + 3) - 16\lambda = \lambda(\lambda - 5)(\lambda + 5).$$

Thus the eigenvalues of A are $\lambda_1 = 0$, $\lambda_2 = 5$, $\lambda_3 = -5$. In this case $n = 3$, $r = 2$, $p = 1$, and $r - p = 1$. The rank of g_3 is 2 and the signature of g_3 is 0. Thus $g_3(\mathbf{x})$ is not equivalent to either $g_1(\mathbf{x})$ or $g_2(\mathbf{x})$.

• Let $g_4(\mathbf{x}) = 3x_2^2 + 3x_3^2 - 4x_2x_3$. Then the matrix of the quadratic form is

$$A = \begin{bmatrix} 0 & 0 & 0 \\ 0 & 3 & -2 \\ 0 & -2 & 3 \end{bmatrix}.$$

The characteristic polynomial of A is

$$\det(\lambda I_3 - A) = \begin{vmatrix} \lambda & 0 & 0 \\ 0 & \lambda - 3 & 2 \\ 0 & 2 & \lambda - 3 \end{vmatrix} = \lambda(\lambda - 3)^2 - 4\lambda = \lambda(\lambda - 5)(\lambda - 1).$$

Thus the eigenvalues of A are $\lambda_1 = 0$, $\lambda_2 = 5$, $\lambda_3 = 1$. In this case $n = 3$, $r = 2$, $p = 2$, and $r - p = 0$. The rank of g_4 is 2 and the signature of g_4 is 2. Thus $g_4(\mathbf{x})$ is equivalent to both $g_1(\mathbf{x})$ and $g_2(\mathbf{x})$.

23. We compute the eigenvalues of each matrix and use Theorem 8.11.

(a) Let $A = \begin{bmatrix} 2 & -1 \\ -1 & 2 \end{bmatrix}$. The characteristic polynomial of A is

$$\det(\lambda I_2 - A) = \begin{vmatrix} \lambda - 2 & 1 \\ 1 & \lambda - 2 \end{vmatrix} = (\lambda - 2)^2 - 1 = \lambda^2 - 4\lambda + 3 = (\lambda - 1)(\lambda - 3).$$

Thus the eigenvalues of A are $\lambda_1 = 1$ and $\lambda_2 = 3$. Since all the eigenvalues are positive, A is positive definite.

(b) Let $A = \begin{bmatrix} 2 & 1 \\ 1 & 2 \end{bmatrix}$. The characteristic polynomial of A is

$$\det(\lambda I_2 - A) = \begin{vmatrix} \lambda - 2 & -1 \\ -1 & \lambda - 2 \end{vmatrix} = (\lambda - 2)^2 - 1 = \lambda^2 - 4\lambda + 3 = (\lambda - 1)(\lambda - 3).$$

Thus the eigenvalues of A are $\lambda_1 = 1$ and $\lambda_2 = 3$. Since all the eigenvalues are positive, A is positive definite.

(c) Let $A = \begin{bmatrix} 3 & 1 & 0 \\ 1 & 3 & 0 \\ 0 & 0 & 3 \end{bmatrix}$. The characteristic polynomial of A is

$$\det(\lambda I_3 - A) = \begin{vmatrix} \lambda - 3 & -1 & 0 \\ -1 & \lambda - 3 & 0 \\ 0 & 0 & \lambda - 3 \end{vmatrix} = (\lambda - 3)^3 - (\lambda - 3) = (\lambda - 3)(\lambda - 4)(\lambda - 2).$$

Thus the eigenvalues of A are $\lambda_1 = 3$, $\lambda_2 = 4$, and $\lambda_3 = 2$. Since all the eigenvalues are positive, A is positive definite.

(d) Let $A = \begin{bmatrix} 1 & 0 & 0 \\ 0 & 2 & 0 \\ 0 & 0 & -3 \end{bmatrix}$. Since A is diagonal, its eigenvalues are its diagonal entries. It follows that A is not positive definite.

(e) Let $A = \begin{bmatrix} 2 & 2 \\ 2 & 2 \end{bmatrix}$. Matrix A is singular, so one of its eigenvalues is zero. Thus A is not positive definite.

25. Let A be symmetric. Then $A^T = A$ and hence

$$(P^T A P)^T = P^T A^T (P^T)^T = P^T A^T P = P^T A P.$$

Therefore $P^T A P$ is symmetric.

27. If A is symmetric then by Theorem 7.9 in Section 7.4 there exists an orthogonal matrix P such that $P^{-1} A P = D$, a diagonal matrix. Since $P^{-1} = P^T$ we have $P^T A P = D$. Thus A is congruent to a diagonal matrix.

29. Let A be the matrix of the quadratic form $g(\mathbf{x}) = \mathbf{x}^T A \mathbf{x}$. Then by Theorem 7.13 $g(\mathbf{x})$ is equivalent to

$$h(\mathbf{y}) = y_1^2 + y_2^2 + \cdots + y_p^2 - y_{p+1}^2 - \cdots - y_r^2.$$

Assume A is positive definite. Since g and h are equivalent $h(\mathbf{y}) > 0$ for each $\mathbf{y} \neq \mathbf{0}$. However, this can happen if and only if all the summands in $h(\mathbf{y})$ are positive, that is, A is congruent to I_n. Thus it follows that there exists a nonsingular matrix P such that

$$A = P^T I_n P = P^T P.$$

The preceding steps are reversible to show that if $A = P^T P$ for some nonsingular matrix P, then A is positive definite.

Section 8.7, p. 551

1. Rewrite the equation $x^2 + 9y^2 - 9 = 0$ as

$$\frac{x^2}{9} + \frac{9y^2}{9} = \frac{9}{9} \quad \Longrightarrow \quad \frac{x^2}{9} + \frac{y^2}{1} = 1.$$

This is the equation of an ellipse in standard position with $a = 3$ and $b = 1$. The x-intercepts are $(-3, 0)$ and $3, 0)$ and the y-intercepts are $(0, -1)$ and $(0, 1)$.

3. Rewrite the equation $25y^2 - 4x^2 = 100$ as

$$\frac{25y^2}{100} - \frac{4x^2}{100} = \frac{100}{100} \quad \Longrightarrow \quad \frac{y^2}{4} - \frac{x^2}{25} = 1.$$

This is the equation of a hyperbola in standard position with $a = 2$ and $b = 5$. The y-intercepts are $(0, -2$ and $(0, 2)$.

5. Rewrite the equation $3x^2 - y^2 = 0$ as

$$y^2 = 3x^2 \quad \Longrightarrow \quad y = \sqrt{3}\, x \quad \text{and} \quad y = -\sqrt{3}\, x$$

This represents the graph of a pair of intersecting lines which is a degenerate conic section.

7. Rewrite the equation $4x^2 + 4y^2 - 9 = 0$ as

$$\frac{4x^2}{9} + \frac{4y^2}{9} = \frac{9}{9} \quad \Longrightarrow \quad \frac{x^2}{\left(\frac{3}{2}\right)^2} + \frac{y^2}{\left(\frac{3}{2}\right)^2} = 1.$$

This is the equation of a circle in standard position with $a = \frac{3}{2}$.

9. Upon inspection, the equation $4x^2 + y^2 = 0$ is satisfied if and only if $x = y = 0$. Hence this is a degenerate conic section which represents the single point $(0,0)$.

11. In the equation $x^2 + 2y^2 - 4x - 4y + 4 = 0$ we note that the x^2 and x and y^2 and y terms appear. Hence we complete the square in both x and y. Rewrite the equation as

$$\underbrace{x^2 - 4x + 4}_{} + \underbrace{2y^2 - 4y}_{} = 0$$

$$(x-2)^2 + 2(y^2 - 2y \quad) = 0$$

$$(x-2)^2 + 2(y^2 - 2y + 1) = 2$$

$$(x-2)^2 + 2(y-1)^2 = 2$$

Let $x' = x - 2$ and $y' = y - 1$. Then the preceding equation is written as $x'^2 + 2y'^2 = 2$. Next we transform this equation to

$$\frac{x'^2}{2} + y'^2 = 1.$$

If we translate the xy-coordinate system to the $x'y'$-coordinate system, whose origin is at $(2,1)$, then the graph is an ellipse in standard position with respect to the $x'y'$-coordinate system.

13. In the equation $x^2 + y^2 - 8x - 6y = 0$ we note that the x^2 and x and y^2 and y terms appear. Hence we complete the square in both x and y. Rewrite the equation as

$$\underbrace{x^2 - 8x}_{} + \underbrace{y^2 - 6y}_{} = 0$$

$$x^2 - 8x + 16 + y^2 - 6y + 9 = 16 + 9$$

$$(x-4)^2 + (y-3)^2 = 25$$

Let $x' = x - 4$ and $y' = y - 3$. Then the preceding equation can be written as $x'^2 + y'^2 = 25$. Next we transform this equation to

$$\frac{x'^2}{25} + \frac{y'^2}{25} = 1 \implies \frac{x'^2}{5^2} + \frac{y'^2}{5^2} = 1.$$

If we translate the xy-coordinate system to the $x'y'$-coordinate system, whose origin is at $(4,3)$, then the graph is a circle in standard position with respect to the $x'y'$-coordinate system.

15. In the equation $y^2 - 4y = 0$ we note that only y^2 and y terms appear. Hence we complete the square only in y. Rewrite the equation as

$$y^2 - 4y + 4 = 4 \implies (y-2)^2 = 4.$$

Let $y' = y - 2$. Then we have

$$y'^2 = 4 \implies y' = 2 \quad \text{and} \quad y' = -2.$$

If we translate the xy-coordinate system to the $x'y'$-coordinate system, whose origin is at $(0,2)$, then the graph is a pair of parallel lines.

17. In the equation $x^2 + y^2 - 2x - 6y + 10 = 0$ we note that x^2 and x and y^2 and y terms appear. Hence we complete the square in both x and y. Rewrite the equation as

$$\underbrace{x^2 - 2x}_{} + \underbrace{y^2 - 6y}_{} = -10$$

$$x^2 - 2x + 1 + y^2 - 6y + 9 = -10 + 1 + 9$$

$$(x-1)^2 + (y-3)^2 = 0$$

Let $x' = x - 1$ and $y' = y - 3$. Then the preceding equation can be written as $x'^2 + y'^2 = 0$. If we translate the xy-coordinate system to the $x'y'$-coordinate system, whose origin is at $(1, 3)$, then the graph is a single point which is the origin of the $x'y'$-coordinate system.

19. For the equation $x^2 + xy + y^2 = 6$, we have $A = \begin{bmatrix} 1 & \frac{1}{2} \\ \frac{1}{2} & 1 \end{bmatrix}$. The characteristic polynomial is

$$\det(\lambda I_2 - A) = \begin{vmatrix} \lambda - 1 & -\frac{1}{2} \\ -\frac{1}{2} & \lambda - 1 \end{vmatrix} = \lambda^2 - 2\lambda + \frac{3}{4} = \left(\lambda - \frac{1}{2}\right)\left(\lambda - \frac{3}{2}\right).$$

Thus the eigenvalues of A are $\lambda_1 = \frac{1}{2}$ and $\lambda_2 = \frac{3}{2}$. Next we determine an eigenvector associated with eigenvalue λ_1 by solving the equation

$$(\lambda_1 I_2 - A)\mathbf{x} = \begin{bmatrix} -\frac{1}{2} & -\frac{1}{2} \\ -\frac{1}{2} & -\frac{1}{2} \end{bmatrix}\begin{bmatrix} x_1 \\ x_2 \end{bmatrix} = \begin{bmatrix} 0 \\ 0 \end{bmatrix} \implies \mathbf{x} = \begin{bmatrix} r \\ -r \end{bmatrix}.$$

For $r = 1$, we have the eigenvector

$$\mathbf{x}_1 = \begin{bmatrix} 1 \\ -1 \end{bmatrix}.$$

To find an eigenvector associated with $\lambda_2 = \frac{3}{2}$, we have

$$(\lambda_2 I_2 - A)\mathbf{x}) = \begin{bmatrix} \frac{1}{2} & -\frac{1}{2} \\ -\frac{1}{2} & \frac{1}{2} \end{bmatrix}\begin{bmatrix} x_1 \\ x_2 \end{bmatrix} = \begin{bmatrix} 0 \\ 0 \end{bmatrix} \implies \mathbf{x} = \begin{bmatrix} r \\ r \end{bmatrix}.$$

For $r = 1$, we have the eigenvector

$$\mathbf{x}_2 = \begin{bmatrix} 1 \\ 1 \end{bmatrix}.$$

Normalizing the eigenvectors and forming the matrix P we have

$$P = \begin{bmatrix} \frac{1}{\sqrt{2}} & \frac{1}{\sqrt{2}} \\ -\frac{1}{\sqrt{2}} & \frac{1}{\sqrt{2}} \end{bmatrix}.$$

Let $\mathbf{x} = P\mathbf{y}$, where $\mathbf{y} = \begin{bmatrix} x' & y' \end{bmatrix}^T$. Then we can rewrite the original equation to obtain

$$\frac{1}{2}x'^2 + \frac{3}{2}y'^2 = 6 \implies \frac{x'^2}{12} + \frac{y'^2}{4} = 1.$$

Thus this conic section is an ellipse. The preceding is only a possible answer. The roles of x' and y' would be reversed if the eigenvalues of A were used in a different order.

21. For the equation $9x^2 + y^2 + 6xy = 4$, we have $A = \begin{bmatrix} 9 & 3 \\ 3 & 1 \end{bmatrix}$. The characteristic polynomial is

$$\det(\lambda I_2 - A) = \begin{vmatrix} \lambda - 9 & -3 \\ -3 & \lambda - 1 \end{vmatrix} = \lambda^2 - 10\lambda = \lambda(\lambda - 10).$$

Thus the eigenvalues are $\lambda_1 = 0$ and $\lambda_2 = 10$. Next we determine an eigenvector $\mathbf{x}_1$ associated with eigenvalue $\lambda_1 = 0$ by solving the equation

$$(\lambda_1 I_2 - A)\mathbf{x} = \begin{bmatrix} -9 & -3 \\ -3 & -1 \end{bmatrix}\begin{bmatrix} x_1 \\ x_2 \end{bmatrix} = \begin{bmatrix} 0 \\ 0 \end{bmatrix} \implies \mathbf{x} = \begin{bmatrix} -\frac{1}{3}r \\ r \end{bmatrix}.$$

For $r = 3$, we have the eigenvector

$$\mathbf{x}_1 = \begin{bmatrix} -1 \\ 3 \end{bmatrix}.$$

For the eigenvalue $\lambda_2 = 10$, we find:

$$(\lambda_2 I_2 - A)\mathbf{x} = \begin{bmatrix} 1 & -3 \\ -3 & 9 \end{bmatrix} \begin{bmatrix} x_1 \\ x_2 \end{bmatrix} = \begin{bmatrix} 0 \\ 0 \end{bmatrix} \implies \mathbf{x} = \begin{bmatrix} 3r \\ r \end{bmatrix}.$$

For $r = 1$, we obtain the eigenvector

$$\mathbf{x}_2 = \begin{bmatrix} 3 \\ 1 \end{bmatrix}.$$

Normalizing the eigenvectors and forming the matrix P we have

$$P = \begin{bmatrix} -\frac{1}{\sqrt{10}} & \frac{3}{\sqrt{10}} \\ \frac{3}{\sqrt{10}} & \frac{1}{\sqrt{10}} \end{bmatrix}.$$

Note that $\det(P) = -1$. For a counterclockwise rotation we require $\det(P) = 1$. Since any nonzero multiple of an eigenvector is still an eigenvector for the same eigenvalue we replace $\mathbf{x}_1$ by $-\mathbf{x}_1$ (an alternate procedure is to interchange columns of P) which results in redefining P as

$$P = \begin{bmatrix} \frac{1}{\sqrt{10}} & \frac{3}{\sqrt{10}} \\ -\frac{3}{\sqrt{10}} & \frac{1}{\sqrt{10}} \end{bmatrix}.$$

Let $\mathbf{x} = P\mathbf{y}$, where $\mathbf{y} = \begin{bmatrix} x' & y' \end{bmatrix}^T$. Then we can rewrite the original equation as in Equation (6) to obtain

$$0x'^2 + 10y'^2 = 4 \implies y'^2 = \frac{4}{10} \implies y' = \pm\frac{2}{\sqrt{10}}.$$

Thus this conic section consists of two parallel lines. The preceding is only a possible answer. The roles of x' and y' would be reversed if the eigenvalues of A were used in a different order.

23. For the equation $4x^2 + 4y^2 - 10xy = 0$, we have $A = \begin{bmatrix} 4 & -5 \\ -5 & 4 \end{bmatrix}$. The characteristic polynomial is

$$\det(\lambda I_2 - A) = \begin{vmatrix} \lambda - 4 & 5 \\ 5 & \lambda - 4 \end{vmatrix} = \lambda^2 - 8\lambda - 9 = (\lambda - 9)(\lambda + 1).$$

Thus the eigenvalues are $\lambda_1 = 9$ and $\lambda_2 = -1$. Next we determine an eigenvector $\mathbf{x}_1$ associated with eigenvalue $\lambda_1 = 0$ by solving the equation

$$(\lambda_1 I_2 - A)\mathbf{x} = \begin{bmatrix} 5 & 5 \\ 5 & 5 \end{bmatrix} \begin{bmatrix} x_1 \\ x_2 \end{bmatrix} = \begin{bmatrix} 0 \\ 0 \end{bmatrix}.$$

We find

$$\mathbf{x} = \begin{bmatrix} -r \\ r \end{bmatrix}.$$

For $r = 1$, we have the eigenvector

$$\mathbf{x}_1 = \begin{bmatrix} -1 \\ 1 \end{bmatrix}.$$

We determine an eigenvector $\mathbf{x}_2$ associated with eigenvalue λ_2 by solving

$$(\lambda_2 I_2 - A)\mathbf{x} = \begin{bmatrix} -5 & 5 \\ 5 & -5 \end{bmatrix} \begin{bmatrix} x_1 \\ x_2 \end{bmatrix} = \begin{bmatrix} 0 \\ 0 \end{bmatrix}.$$

We find

$$\mathbf{x} = \begin{bmatrix} r \\ r \end{bmatrix}.$$

For $r = 1$, we have the eigenvector

$$\mathbf{x}_2 = \begin{bmatrix} 1 \\ 1 \end{bmatrix}.$$

Normalizing the eigenvectors and forming the matrix P we obtain

$$P = \begin{bmatrix} -\frac{1}{\sqrt{2}} & \frac{1}{\sqrt{2}} \\ \frac{1}{\sqrt{2}} & \frac{1}{\sqrt{2}} \end{bmatrix}.$$

Note that $\det(P) = -1$. For a counterclockwise rotation we require $\det(P) = 1$. Since any nonzero multiple of an eigenvector is still an eigenvector for the same eigenvalue we replace $\mathbf{x}_1$ by $-\mathbf{x}_1$ which results in redefining P as

$$P = \begin{bmatrix} \frac{1}{\sqrt{2}} & \frac{1}{\sqrt{2}} \\ -\frac{1}{\sqrt{2}} & \frac{1}{\sqrt{2}} \end{bmatrix}.$$

Let $\mathbf{x} = P\mathbf{y}$, where $\mathbf{y} = \begin{bmatrix} x' & y' \end{bmatrix}^T$. Then we can rewrite the original equation as in Equation (6) to obtain

$$9x'^2 - y'^2 = 0 \quad \Longrightarrow \quad y'^2 = 9x'^2 \quad \Longrightarrow \quad y' = \pm 3x'.$$

Thus this conic section is a pair of intersecting lines. The preceding is only a possible answer. The roles of x' and y' would be reversed if the eigenvalues of A were used in a different order.

25. For equation $9x^2 + y^2 + 6xy - 10\sqrt{10}\,x + 10\sqrt{10}\,y + 90 = 0$, we have from Equation (5) that

$$A = \begin{bmatrix} 9 & 3 \\ 3 & 1 \end{bmatrix}, \quad B = \begin{bmatrix} -10\sqrt{10} & 10\sqrt{10} \end{bmatrix}, \quad f = 90.$$

The characteristic polynomial for matrix A is

$$\det(\lambda I_2 - A) = \begin{vmatrix} \lambda - 9 & -3 \\ -3 & \lambda - 1 \end{vmatrix} = \lambda^2 - 10\lambda = \lambda(\lambda - 10).$$

Thus the eigenvalues are $\lambda_1 = 0$ and $\lambda_2 = 10$. Next we determine an eigenvector $\mathbf{x}_1$ associated with eigenvalue λ_1 by solving

$$(\lambda_1 I_2 - A)\mathbf{x} = \begin{bmatrix} -9 & -3 \\ -3 & -1 \end{bmatrix} \begin{bmatrix} x_1 \\ x_2 \end{bmatrix} = \begin{bmatrix} 0 \\ 0 \end{bmatrix} \quad \Longrightarrow \quad \mathbf{x} = \begin{bmatrix} -\frac{1}{3}r \\ r \end{bmatrix}.$$

For $r = 3$ we obtain the eigenvector

$$\mathbf{x}_1 = \begin{bmatrix} -1 \\ 3 \end{bmatrix}.$$

We determine an eigenvector $\mathbf{x}_2$ associated with eigenvalue λ_2 by solving

$$(\lambda_2 I_2 - A)\mathbf{x} = \begin{bmatrix} 1 & -3 \\ -3 & 9 \end{bmatrix} \begin{bmatrix} x_1 \\ x_2 \end{bmatrix} = \begin{bmatrix} 0 \\ 0 \end{bmatrix} \quad \Longrightarrow \quad \mathbf{x} = \begin{bmatrix} 3r \\ r \end{bmatrix}.$$

For $r = 1$, we have the eigenvector

$$\mathbf{x}_2 = \begin{bmatrix} 3 \\ 1 \end{bmatrix}.$$

Normalizing the eigenvectors and forming the matrix P we have

$$P = \begin{bmatrix} -\frac{1}{\sqrt{10}} & \frac{3}{\sqrt{10}} \\ \frac{3}{\sqrt{10}} & \frac{1}{\sqrt{10}} \end{bmatrix}.$$

Note that $\det(P) = -1$. For a counterclockwise rotation we require $\det(P) = 1$. Since any nonzero multiple of an eigenvector is still an eigenvector for the same eigenvalue we replace $\mathbf{x}_1$ by $-\mathbf{x}_1$ which results in redefining P as

$$P = \begin{bmatrix} \frac{1}{\sqrt{10}} & \frac{3}{\sqrt{10}} \\ -\frac{3}{\sqrt{10}} & \frac{1}{\sqrt{10}} \end{bmatrix}.$$

Let $\mathbf{x} = P\mathbf{y}$, where $\mathbf{y} = \begin{bmatrix} x' & y' \end{bmatrix}^T$. Then we can rewrite the original equation as in (6) to obtain

$$\begin{bmatrix} x' & y' \end{bmatrix} \begin{bmatrix} 0 & 0 \\ 0 & 10 \end{bmatrix} \begin{bmatrix} x' \\ y' \end{bmatrix} + \begin{bmatrix} -10\sqrt{10} & 10\sqrt{10} \end{bmatrix} \begin{bmatrix} \frac{1}{\sqrt{10}} & \frac{3}{\sqrt{10}} \\ -\frac{3}{\sqrt{10}} & \frac{1}{\sqrt{10}} \end{bmatrix} \begin{bmatrix} x' \\ y' \end{bmatrix} + 90 = 0.$$

Performing the matrix operations and simplifying gives

$$10y'^2 - 20y' + 40x' + 90 = 0.$$

Dividing by 10 and completing the square in y' we obtain

$$(y' - 1)^2 + 4(x' + 2) = 0.$$

Let $x'' = x' + 2$ and $y'' = y' - 1$. Then we can rewrite the equation as

$$y''^2 = -4x''.$$

Thus this conic section is a parabola. The preceding is only a possible answer. The roles of x'' and y'' would be reversed if the eigenvalues of A were used in a different order.

27. For equation $5x^2 + 12xy - 12\sqrt{13}\,x = 36$ we have from Equation (5)

$$A = \begin{bmatrix} 5 & 6 \\ 6 & 0 \end{bmatrix}, \quad B = \begin{bmatrix} -12\sqrt{13} & 0 \end{bmatrix}, \quad f = -36.$$

The characteristic polynomial is

$$\det(\lambda I_2 - A) = \begin{vmatrix} \lambda - 5 & -6 \\ -6 & \lambda \end{vmatrix} = \lambda^2 - 5\lambda - 36 = (\lambda - 9)(\lambda + 4).$$

Thus the eigenvalues are $\lambda_1 = 9$ and $\lambda_2 = -4$. Next we determine an eigenvector $\mathbf{x}_1$ associated with eigenvalue λ_1 by solving

$$(\lambda_1 I_2 - A)\mathbf{x} = \begin{bmatrix} 4 & -6 \\ -6 & 9 \end{bmatrix} \begin{bmatrix} x_1 \\ x_2 \end{bmatrix} = \begin{bmatrix} 0 \\ 0 \end{bmatrix} \implies \mathbf{x} = \begin{bmatrix} \frac{3}{2}r \\ r \end{bmatrix}.$$

For $r = 2$ we have the eigenvector

$$\mathbf{x}_1 = \begin{bmatrix} 3 \\ 2 \end{bmatrix}.$$

We determine an eigenvector $\mathbf{x}_2$ associated with eigenvalue λ_2 by solving

$$(\lambda_2 I_2 - A)\mathbf{x} = \begin{bmatrix} -9 & -6 \\ -6 & -4 \end{bmatrix} \begin{bmatrix} x_1 \\ x_2 \end{bmatrix} = \begin{bmatrix} 0 \\ 0 \end{bmatrix} \implies \mathbf{x} = \begin{bmatrix} -\frac{2}{3}r \\ r \end{bmatrix}.$$

For $r = 3$ we have the eigenvector

$$\mathbf{x}_2 = \begin{bmatrix} -2 \\ 3 \end{bmatrix}.$$

Normalizing the eigenvectors and forming the matrix P we obtain

$$P = \begin{bmatrix} \frac{3}{\sqrt{13}} & -\frac{2}{\sqrt{13}} \\ \frac{2}{\sqrt{13}} & \frac{3}{\sqrt{13}} \end{bmatrix}.$$

Let $\mathbf{x} = P\mathbf{y}$, where $\mathbf{y} = \begin{bmatrix} x' & y' \end{bmatrix}^T$. Then we can rewrite the original equation as in (6) to obtain

$$\begin{bmatrix} x' & y' \end{bmatrix} \begin{bmatrix} 9 & 0 \\ 0 & -4 \end{bmatrix} \begin{bmatrix} x' \\ y' \end{bmatrix} + \begin{bmatrix} -12\sqrt{13} & 0 \end{bmatrix} \begin{bmatrix} \frac{3}{\sqrt{13}} & -\frac{2}{\sqrt{13}} \\ \frac{2}{\sqrt{13}} & \frac{3}{\sqrt{13}} \end{bmatrix} \begin{bmatrix} x' \\ y' \end{bmatrix} - 36 = 0.$$

Performing the matrix operations and simplifying we obtain

$$9x'^2 - 4y'^2 - 36x' + 24y' - 36 = 0.$$

Completing the square in x' and y' gives

$$9(x' - 2)^2 - 4(y' - 3)^2 = 36.$$

Let $x'' = x' - 2$ and $y'' = y' - 3$. Then we can write the equation as

$$9x''^2 - 4y''^2 = 36 \quad \text{or equivalently} \quad \frac{x''^2}{4} - \frac{y''^2}{9} = 1.$$

Thus this conic section is a hyperbola. The preceding is only a possible answer. The roles of x'' and y'' would be reversed if the eigenvalues of A were used in a different order.

29. For equation $x^2 - y^2 + 2\sqrt{3}\,xy + 6x = 0$ we have from Equation (5)

$$A = \begin{bmatrix} 1 & \sqrt{3} \\ \sqrt{3} & -1 \end{bmatrix}, \quad B = \begin{bmatrix} 6 & 0 \end{bmatrix}, \quad f = 0.$$

The characteristic polynomial for the matrix A is

$$\det(\lambda I_2 - A) = \begin{vmatrix} \lambda - 1 & -\sqrt{3} \\ -\sqrt{3} & \lambda + 1 \end{vmatrix} = \lambda^2 - 4 = (\lambda - 2)(\lambda + 2).$$

Thus the eigenvalues are $\lambda_1 = 2$ and $\lambda_2 = -2$. Next we determine an eigenvector $\mathbf{x}_1$ associated with eigenvalue λ_1 by solving

$$(\lambda_1 I_2 - A)\mathbf{x} = \begin{bmatrix} 1 & -\sqrt{3} \\ -\sqrt{3} & 3 \end{bmatrix} \begin{bmatrix} x_1 \\ x_2 \end{bmatrix} = \begin{bmatrix} 0 \\ 0 \end{bmatrix} \implies \mathbf{x} = \begin{bmatrix} \sqrt{3}\,r \\ r \end{bmatrix}.$$

For $r = 1$, we have the eigenvector

$$\mathbf{x}_1 = \begin{bmatrix} \sqrt{3} \\ 1 \end{bmatrix}.$$

We determine an eigenvector $\mathbf{x}_2$ associated with eigenvalue λ_2 by solving

$$(\lambda_2 I_2 - A)\mathbf{x} = \begin{bmatrix} -3 & -\sqrt{3} \\ -\sqrt{3} & -1 \end{bmatrix} \begin{bmatrix} x_1 \\ x_2 \end{bmatrix} = \begin{bmatrix} 0 \\ 0 \end{bmatrix} \implies \mathbf{x} = \begin{bmatrix} -\frac{1}{3}\sqrt{3}\,r \\ r \end{bmatrix}.$$

For $r = 1$ we have the eigenvector

$$\mathbf{x}_2 = \begin{bmatrix} -\frac{1}{3}\sqrt{3} \\ 1 \end{bmatrix}.$$

Normalizing the eigenvectors and forming the matrix P we have

$$P = \begin{bmatrix} \frac{1}{2}\sqrt{3} & -\frac{1}{2} \\ \frac{1}{2} & \frac{1}{2}\sqrt{3} \end{bmatrix}.$$

Let $\mathbf{x} = P\mathbf{y}$, where $\mathbf{y} = \begin{bmatrix} x' & y' \end{bmatrix}^T$. Then we can rewrite the original equation as in (6) to obtain

$$\begin{bmatrix} x' & y' \end{bmatrix} \begin{bmatrix} 2 & 0 \\ 0 & -2 \end{bmatrix} \begin{bmatrix} x' \\ y' \end{bmatrix} + \begin{bmatrix} 6 & 0 \end{bmatrix} \begin{bmatrix} \frac{1}{2}\sqrt{3} & -\frac{1}{2} \\ \frac{1}{2} & \frac{1}{2}\sqrt{3} \end{bmatrix} \begin{bmatrix} x' \\ y' \end{bmatrix} = 0.$$

Performing the matrix operations and simplifying we obtain

$$2x'^2 - 2y'^2 + 3\sqrt{3}\,x' - 3y' = 0.$$

Completing the square in x' and y' gives

$$2\left(x'^2 + \frac{3}{2}\sqrt{3}\,x' + \frac{27}{16}\right) - 2\left(y'^2 + \frac{3}{2}y' + \frac{9}{16}\right) = \frac{9}{4}$$

or equivalently

$$2\left(x' + \frac{3\sqrt{3}}{4}\right)^2 - 2\left(y' + \frac{3}{4}\right)^2 = \frac{9}{4}.$$

Let $x'' = x' + \frac{3}{4}\sqrt{3}$ and $y'' = y' + \frac{3}{4}$. Then we can write the equation as

$$2x''^2 - 2y''^2 = \frac{9}{4} \quad \text{or equivalently} \quad \frac{x''^2}{\frac{9}{8}} - \frac{y''^2}{\frac{9}{8}} = 1.$$

Thus this conic section is a hyperbola. The preceding is only a possible answer. The roles of x'' and y'' would be reversed if the eigenvalues of A were used in a different order.

Section 8.8, p. 560

1. For the quadric surface $x^2 + y^2 + 2z^2 - 2xy - 4xz - 4yz + 4x = 8$, we have

$$A = \begin{bmatrix} 1 & -1 & -2 \\ -1 & 1 & -2 \\ -2 & -2 & 2 \end{bmatrix}.$$

The characteristic equation is

$$\det(\lambda I_3 - A) = \begin{vmatrix} \lambda - 1 & 1 & 2 \\ 1 & \lambda - 1 & 2 \\ 2 & 2 & \lambda - 2 \end{vmatrix} = \lambda^3 - 4\lambda^2 - 4\lambda + 16 = (\lambda - 4)(\lambda - 2)(\lambda + 2) = 0$$

and hence the eigenvalues are $\lambda = 4$, 2, and -2. Thus the inertia of A is $(2,1,0)$ and it follows from Table 8.2 that this quadric surface is a hyperboloid of one sheet.

3. For the quadric surface $z = 4xy$, or equivalently $-4xy + z = 0$, we have

$$A = \begin{bmatrix} 0 & -2 & 0 \\ -2 & 0 & 0 \\ 0 & 0 & 0 \end{bmatrix}.$$

The characteristic equation is

$$\det(\lambda I_3 - A) = \begin{vmatrix} \lambda & 2 & 0 \\ 2 & \lambda & 0 \\ 0 & 0 & \lambda \end{vmatrix} = \lambda^3 - 4\lambda = \lambda(\lambda^2 - 4) = 0$$

and hence the eigenvalues are $\lambda = 2$, -2, and 0. Thus the inertia of A is $(1,1,1)$ and it follows from Table 8.2 that this quadric surface is a hyperbolic paraboloid.

5. For the quadric surface $x^2 - y = 0$, $A = \begin{bmatrix} 1 & 0 & 0 \\ 0 & 0 & 0 \\ 0 & 0 & 0 \end{bmatrix}$. The characteristic equation is

$$\det(\lambda I_3 - A) = \begin{vmatrix} \lambda - 1 & 0 & 0 \\ 0 & \lambda & 0 \\ 0 & 0 & \lambda \end{vmatrix} = \lambda^2(\lambda - 1) = 0$$

and hence the eigenvalues are $\lambda = 1$, 0, 0. Thus the inertia of A is $(1,0,2)$ and it follows from Table 8.2 that this quadric surface is a parabolic cylinder.

7. For the quadric surface $5y^2 + 20y + z - 23 = 0$, we have

$$A = \begin{bmatrix} 0 & 0 & 0 \\ 0 & 5 & 0 \\ 0 & 0 & 0 \end{bmatrix}.$$

The characteristic equation is

$$\det(\lambda I_3 - A) = \begin{vmatrix} \lambda & 0 & 0 \\ 0 & \lambda - 5 & 0 \\ 0 & 0 & \lambda \end{vmatrix} = \lambda^2(\lambda - 5) = 0$$

and hence the eigenvalues are $\lambda = 5, 0, 0$. Thus the inertia of A is $(1, 0, 2)$ and it follows from Table 8.2 that this quadric surface is a parabolic cylinder.

9. For the quadric surface $4x^2 + 9y^2 + z^2 + 8x - 18y - 4z - 19 = 0$, we have

$$A = \begin{bmatrix} 4 & 0 & 0 \\ 0 & 9 & 0 \\ 0 & 0 & 1 \end{bmatrix}.$$

Since A is diagonal, its eigenvalues are $\lambda = 9, 4, 1$. Thus the inertia of A is $(3, 0, 0)$ and it follows from Table 8.2 that this quadric surface is an ellipsoid.

11. For the quadric surface $x^2 + 4y^2 + 4x + 16y - 16z - 4 = 0$, we have

$$A = \begin{bmatrix} 1 & 0 & 0 \\ 0 & 4 & 0 \\ 0 & 0 & 0 \end{bmatrix}.$$

Since A is diagonal, its eigenvalues are $\lambda = 4, 1, 0$. Thus the inertia of A is $(2, 0, 1)$ and it follows from Table 8.2 that this quadric surface is an elliptic paraboloid.

13. For the quadric surface $x^2 - 4z^2 - 4x + 8z = 0$, we have

$$A = \begin{bmatrix} 1 & 0 & 0 \\ 0 & 0 & 0 \\ 0 & 0 & -4 \end{bmatrix}.$$

Since A is diagonal, its eigenvalues are $\lambda = 1, -4, 0$. Thus the inertia of A is $(1, 1, 1)$ and it follows from Table 8.2 that this quadric surface is an hyperbolic paraboloid.

15. For the quadric surface $x^2 + 2y^2 + 2z^2 + 2yz = 1$, we have

$$A = \begin{bmatrix} 1 & 0 & 0 \\ 0 & 2 & 1 \\ 0 & 1 & 2 \end{bmatrix}.$$

The characteristic equation of A is

$$\det(\lambda I_3 - A) = \begin{vmatrix} \lambda - 1 & 0 & 0 \\ 0 & \lambda - 2 & -1 \\ 0 & -1 & \lambda - 2 \end{vmatrix} = (\lambda - 1)(\lambda^2 - 4\lambda + 3) = (\lambda - 1)(\lambda - 1)(\lambda - 3) = 0$$

and hence the eigenvalues are $\lambda = 3, 1, 1$. Thus the inertia of A is $(3, 0, 0)$ and it follows from Table 8.2 that this quadric surface is an ellipsoid. Since there is a cross product term we must perform a rotation (and possibly a translation) to obtain the standard form. Hence we require the eigenvectors. Solving the appropriate homogeneous systems we have eigenvalue and eigenvector pairs

$$\lambda_1 = 3, \ \mathbf{x}_1 = \begin{bmatrix} 0 \\ 1 \\ 1 \end{bmatrix}; \quad \lambda_2 = 1, \ \mathbf{x}_2 = \begin{bmatrix} 1 \\ 0 \\ 0 \end{bmatrix}; \quad \lambda_3 = 1, \ \mathbf{x}_3 = \begin{bmatrix} 0 \\ -1 \\ 1 \end{bmatrix}.$$

Normalizing the eigenvectors we have

$$\mathbf{u}_1 = \begin{bmatrix} 0 \\ \frac{1}{2} \\ \frac{1}{\sqrt{2}} \end{bmatrix}, \quad \mathbf{u}_2 = \mathbf{x}_2, \quad \mathbf{u}_3 = \begin{bmatrix} 0 \\ -\frac{1}{\sqrt{2}} \\ \frac{1}{\sqrt{2}} \end{bmatrix}.$$

Let $P = \begin{bmatrix} \mathbf{u}_1 & \mathbf{u}_2 & \mathbf{u}_3 \end{bmatrix}$. Then $\det(P) = -1$, so to have a counterclockwise rotation we will redefine the eigenvector $\mathbf{u}_2 = -\mathbf{x}_2$. (This is valid since any nonzero multiple of an eigenvector is another eigenvector.) Thus

$$P = \begin{bmatrix} 0 & -1 & 0 \\ \frac{1}{\sqrt{2}} & 0 & -\frac{1}{\sqrt{2}} \\ \frac{1}{\sqrt{2}} & 0 & \frac{1}{\sqrt{2}} \end{bmatrix}.$$

Let $\mathbf{x} = P\mathbf{y}$, where $\mathbf{y} = \begin{bmatrix} x' & y' & z' \end{bmatrix}^T$. Then the quadric surface can be written as

$$(P\mathbf{y})^T A (P\mathbf{y}) = x'^2 + y'^2 + 3z'^2 = 1.$$

Hence the standard form is

$$\frac{x'^2}{1} + \frac{y'^2}{1} + \frac{z'^2}{\frac{1}{3}} = 1.$$

Note for this problem no translations were required. The expression for the standard form is not unique. The roles of x', y', and z' may be interchanged depending upon the order of the eigenvectors used as columns in P.

17. For the quadric surface $2xz - 2x - 4y - 4z + 8 = 0$, we have

$$A = \begin{bmatrix} 0 & 0 & 1 \\ 0 & 0 & 0 \\ 1 & 0 & 0 \end{bmatrix}.$$

The characteristic equation of A is

$$\det(\lambda I_3 - A) = \begin{vmatrix} \lambda & 0 & -1 \\ 0 & \lambda & 0 \\ -1 & 0 & \lambda \end{vmatrix} = \lambda^3 - \lambda = \lambda(\lambda^2 - 1) = 0$$

and hence the eigenvalues are $\lambda = 1, -1, 0$. Thus the inertia of A is $(1, 1, 1)$ and it follows from Table 8.2 that this quadric surface is a hyperbolic paraboloid. Since there is a cross product term we must perform a rotation (and possibly a translation) to obtain the standard form. Hence we require the eigenvectors. Solving the appropriate homogeneous systems we have eigenvalue and eigenvector pairs

$$\lambda_1 = 1, \ \mathbf{x}_1 = \begin{bmatrix} 1 \\ 0 \\ 1 \end{bmatrix}; \quad \lambda_2 = -1, \ \mathbf{x}_2 = \begin{bmatrix} 1 \\ 0 \\ -1 \end{bmatrix}; \quad \lambda_3 = 0, \ \mathbf{x}_3 = \begin{bmatrix} 0 \\ 1 \\ 0 \end{bmatrix}.$$

Normalizing the eigenvectors we have

$$\mathbf{u}_1 = \begin{bmatrix} \frac{1}{\sqrt{2}} \\ 0 \\ \frac{1}{\sqrt{2}} \end{bmatrix}, \quad \mathbf{u}_2 = \begin{bmatrix} \frac{1}{\sqrt{2}} \\ 0 \\ -\frac{1}{\sqrt{2}} \end{bmatrix}, \quad \mathbf{u}_3 = \begin{bmatrix} 0 \\ 1 \\ 0 \end{bmatrix}.$$

Let $P = \begin{bmatrix} \mathbf{u}_1 & \mathbf{u}_2 & \mathbf{u}_3 \end{bmatrix}$. Then $\det(P) = 1$ so we have a counterclockwise rotation. Let $\mathbf{x} = P\mathbf{y}$, where $\mathbf{y} = \begin{bmatrix} x' & y' & z' \end{bmatrix}^T$. Then he quadric surface can be written as

$$(P\mathbf{y})^T A (P\mathbf{y}) + \begin{bmatrix} -2 & -4 & -4 \end{bmatrix} P\mathbf{y} + 8 = 0$$

which simplifies to

$$x'^2 - y'^2 - \frac{6}{\sqrt{2}} x' + \frac{2}{\sqrt{2}} y' - 4z' + 8 = 0.$$

Completing the square in x' and y' we have

$$\left(x'^2 - \frac{6}{\sqrt{2}} x' + \frac{9}{2} \right) - \left(y'^2 - \frac{2}{\sqrt{2}} y' + \frac{1}{2} \right) - 4z' + 8 = \frac{9}{2} - \frac{1}{2}$$

$$\implies \quad \left(x' - \frac{3}{\sqrt{2}} \right)^2 - \left(y' - \frac{1}{\sqrt{2}} \right)^2 - 4(z' - 1) = 0.$$

Let $x'' = x' - \frac{3}{\sqrt{2}}$, $y'' = y' - \frac{1}{\sqrt{2}}$, and $z'' = z' - 1$. Then we have $x''^2 - y''^2 - 4z'' = 0$ and the standard form is

$$\frac{x''^2}{4} - \frac{y''^2}{4} = z''.$$

The expression for the standard form is not unique. The roles of x'', y'', and z'' may be interchanged depending upon the order of the eigenvectors used as columns in P.

19. For the quadric surface $x^2 + y^2 + z^2 + 2xy = 8$, we have

$$A = \begin{bmatrix} 1 & 1 & 0 \\ 1 & 1 & 0 \\ 0 & 0 & 1 \end{bmatrix}.$$

The characteristic equation of A is

$$\det(\lambda I_2 - A) = \begin{vmatrix} \lambda - 1 & -1 & 0 \\ -1 & \lambda - 1 & 0 \\ 0 & 0 & \lambda - 1 \end{vmatrix} = (\lambda - 1)^3 - (\lambda - 1) = (\lambda - 1)(\lambda - 2)\lambda = 0$$

and hence the eigenvalues are $\lambda = 2, 1, 0$. Thus the inertia of A is $(2, 0, 1)$ and it follows from Table 8.2 that this quadric surface is an elliptic paraboloid. Since there is a cross product term we must perform a rotation (and possibly a translation) to obtain the standard form. Hence we require the eigenvectors. Solving the appropriate homogeneous systems we have eigenvalue and eigenvector pairs

$$\lambda_1 = 2, \mathbf{x}_1 = \begin{bmatrix} 1 \\ 1 \\ 0 \end{bmatrix}; \quad \lambda_2 = 1, \mathbf{x}_2 = \begin{bmatrix} 0 \\ 0 \\ 1 \end{bmatrix}; \quad \lambda_3 = 0, \mathbf{x}_3 = \begin{bmatrix} -1 \\ 1 \\ 0 \end{bmatrix}.$$

Normalizing the eigenvectors we have

$$\mathbf{u}_1 = \begin{bmatrix} \frac{1}{\sqrt{2}} \\ \frac{1}{\sqrt{2}} \\ 0 \end{bmatrix}, \quad \mathbf{u}_2 = \mathbf{x}_2, \quad \mathbf{u}_3 = \begin{bmatrix} -\frac{1}{\sqrt{2}} \\ \frac{1}{\sqrt{2}} \\ 0 \end{bmatrix}.$$

Let $P = \begin{bmatrix} \mathbf{u}_1 & \mathbf{u}_2 & \mathbf{u}_3 \end{bmatrix}$. Then $\det(P) = -1$, so to have a counterclockwise rotation we will redefine the eigenvector $\mathbf{u}_2 = -\mathbf{x}_2$. (This is valid since any nonzero multiple of an eigenvector is another eigenvector.) Thus

$$P = \begin{bmatrix} \frac{1}{\sqrt{2}} & 0 & -\frac{1}{\sqrt{2}} \\ \frac{1}{\sqrt{2}} & 0 & \frac{1}{\sqrt{2}} \\ 0 & 1 & 0 \end{bmatrix}.$$

Let $\mathbf{x} = P\mathbf{y}$, where $\mathbf{y} = \begin{bmatrix} x' & y' & z' \end{bmatrix}^T$. Then the quadric surface can be written as

$$(P\mathbf{y})^T A(P\mathbf{y}) = 2x'^2 + y'^2 = 8.$$

Hence the standard form is

$$\frac{x'^2}{4} + \frac{y'^2}{8} = 1.$$

Note that for this problem no translations were required. The expression for the standard form is not unique. The roles of x', y', and z' may be interchanged depending upon the order of the eigenvectors used as columns in P.

21. For the quadric surface $2x^2 + 2y^2 + 4z^2 - 4xy - 8xz - 8yz + 8x = 15$, we have

$$A = \begin{bmatrix} 2 & -2 & -4 \\ -2 & 2 & -4 \\ -4 & -4 & 4 \end{bmatrix}.$$

The characteristic equation of A is

$$\det(\lambda I_2 - A) = \begin{vmatrix} \lambda - 2 & 2 & 4 \\ 2 & \lambda - 2 & 4 \\ 4 & 4 & \lambda - 4 \end{vmatrix} = \lambda^3 - 8\lambda^2 - 16\lambda + 128$$

$$= \lambda^2(\lambda - 8) - 16(\lambda - 8)$$
$$= (\lambda - 8)(\lambda^2 - 16)$$
$$= (\lambda - 8)(\lambda - 4)(\lambda + 4) = 0$$

and hence the eigenvalues are $\lambda = 8,\ 4,\ -4$. Thus the inertia of A is $(2, 1, 0)$ and it follows from Table 8.2 that this quadric surface is an hyperboloid of one sheet. From the terms present we see that we must perform a rotation and a translation to obtain the standard form. Hence we require the eigenvectors. Solving the appropriate homogeneous systems we have eigenvalue and eigenvector pairs

$$\lambda_1 = 8,\ \mathbf{x}_1 = \begin{bmatrix} -\frac{1}{2} \\ -\frac{1}{2} \\ 1 \end{bmatrix}; \quad \lambda_2 = 4,\ \mathbf{x}_2 = \begin{bmatrix} -1 \\ 1 \\ 0 \end{bmatrix}; \quad \lambda_3 = -4,\ \mathbf{x}_3 = \begin{bmatrix} 1 \\ 1 \\ 1 \end{bmatrix}.$$

Normalizing the eigenvectors we have

$$\mathbf{u}_1 = \begin{bmatrix} -\frac{1}{\sqrt{6}} \\ -\frac{1}{\sqrt{6}} \\ \frac{2}{\sqrt{6}} \end{bmatrix}, \quad \mathbf{u}_2 = \begin{bmatrix} -\frac{1}{\sqrt{2}} \\ \frac{1}{\sqrt{2}} \\ 0 \end{bmatrix}, \quad \mathbf{u}_3 = \begin{bmatrix} \frac{1}{\sqrt{3}} \\ \frac{1}{\sqrt{3}} \\ \frac{1}{\sqrt{3}} \end{bmatrix}.$$

Let $P = \begin{bmatrix} \mathbf{u}_1 & \mathbf{u}_2 & \mathbf{u}_3 \end{bmatrix}$. Then $\det(P) = -1$, so to have a counterclockwise rotation. Here we replace $\mathbf{u}_1$ by $-\mathbf{u}_1$ and redefine P to be

$$P = \begin{bmatrix} \frac{1}{\sqrt{6}} & -\frac{1}{\sqrt{2}} & \frac{1}{\sqrt{3}} \\ \frac{1}{\sqrt{6}} & \frac{1}{\sqrt{2}} & \frac{1}{\sqrt{3}} \\ -\frac{2}{\sqrt{6}} & 0 & \frac{1}{\sqrt{3}} \end{bmatrix}.$$

Then $\det(P) = 1$ (verify). Let $\mathbf{x} = P\mathbf{y}$, where $\mathbf{y} = \begin{bmatrix} x' & y' & z' \end{bmatrix}^T$. Then the quadric surface can be written as

$$(P\mathbf{y})^T A(P\mathbf{y}) + \begin{bmatrix} 8 & 0 & 0 \end{bmatrix} P\mathbf{y} = 15$$

which simplifies to

$$8x'^2 + 4y'^2 - 4z'^2 + \frac{8}{\sqrt{6}} x' - \frac{8}{\sqrt{2}} y' + \frac{8}{\sqrt{3}} z' = 15.$$

Completing the square in x', y', and z' we have

$$8\left(x'^2 + \frac{1}{\sqrt{6}}x' + \frac{1}{24}\right) + 4\left(y'^2 - \frac{2}{\sqrt{2}}y' + \frac{1}{2}\right) - 4\left(z'^2 - \frac{2}{\sqrt{3}}z' + \frac{1}{3}\right) = \frac{8}{24} + 2 - \frac{4}{3} + 15$$

$$\implies \quad 8\left(x' + \frac{1}{\sqrt{24}}\right)^2 + 4\left(y' - \frac{1}{\sqrt{2}}\right)^2 - 4\left(z' - \frac{1}{\sqrt{3}}\right)^2 = 16.$$

Let $x'' = x' + \frac{1}{\sqrt{24}}$, $y'' = y' - \frac{1}{\sqrt{2}}$, and $z'' = z' - \frac{1}{\sqrt{3}}$. Then we have

$$8x''^2 + 4y''^2 - 4z''^2 = 16$$

and the standard form is

$$\frac{x''^2}{2} + \frac{y''^2}{4} - \frac{z''^2}{4} = 1.$$

The expression for the standard form is not unique. The roles of x'', y'', and z'' may be interchanged depending upon the order of the eigenvectors used as columns in P.

23. For the quadric surface $2y^2 + 2z^2 + 4yz + \frac{16}{\sqrt{2}}x + 4 = 0$, we have

$$A = \begin{bmatrix} 0 & 0 & 0 \\ 0 & 2 & 2 \\ 0 & 2 & 2 \end{bmatrix}.$$

The characteristic equation of A is

$$\det(\lambda I_3 - A) \begin{vmatrix} \lambda & 0 & 0 \\ 0 & \lambda - 2 & -2 \\ 0 & -2 & \lambda - 2 \end{vmatrix} = \lambda[(\lambda - 2)^2 - 4] = \lambda^2(\lambda - 4) = 0$$

and thus the eigenvalues are $\lambda = 4, 0, 0$. Thus the inertial of A is $(1, 0, 2)$ and it follows from Table 8.2 that this quadric surface is a parabolic cylinder. Since there is a cross product term we must perform a rotation to obtain the standard form. Hence we require the eigenvectors. Solving the appropriate homogeneous systems we have eigenvalue and eigenvector pairs

$$\lambda_1 = 4, \ \mathbf{x}_1 = \begin{bmatrix} 0 \\ 1 \\ 1 \end{bmatrix}; \quad \lambda_2 = 0, \ \mathbf{x}_2 = \begin{bmatrix} 1 \\ 0 \\ 0 \end{bmatrix}; \quad \lambda_3 = 0, \ \mathbf{x}_3 = \begin{bmatrix} 0 \\ -1 \\ 1 \end{bmatrix}.$$

Normalizing the eigenvectors we have

$$\mathbf{u}_1 = \begin{bmatrix} 0 \\ \frac{1}{\sqrt{2}} \\ \frac{1}{\sqrt{2}} \end{bmatrix}, \quad \mathbf{u}_2 = \mathbf{x}_2, \quad \mathbf{u}_3 = \begin{bmatrix} 0 \\ -\frac{1}{\sqrt{2}} \\ \frac{1}{\sqrt{2}} \end{bmatrix}.$$

Let $P = \begin{bmatrix} \mathbf{u}_1 & \mathbf{u}_2 & \mathbf{u}_3 \end{bmatrix}$. Then $\det(P) = -1$, so to have a counterclockwise rotation we will redefine the eigenvector $\mathbf{u}_2 = -\mathbf{x}_2$. (This is valid since any nonzero multiple of an eigenvector is another eigenvector.) Thus

$$P = \begin{bmatrix} 0 & -1 & 0 \\ \frac{1}{\sqrt{2}} & 0 & -\frac{1}{\sqrt{2}} \\ \frac{1}{\sqrt{2}} & 0 & \frac{1}{\sqrt{2}} \end{bmatrix}.$$

Let $\mathbf{x} = P\mathbf{y}$, where $\mathbf{y} = \begin{bmatrix} x' & y' & z' \end{bmatrix}^T$. Then the quadric surface can be written as

$$(P\mathbf{y})^T A(P\mathbf{y}) + \begin{bmatrix} \frac{16}{\sqrt{2}} & 0 & 0 \end{bmatrix} P\mathbf{y} + 4 = 0$$

or equivalently

$$4x'^2 - \frac{16}{\sqrt{2}} y' + 4 = 0.$$

Next we rearrange the equation to determine the translation required. We have

$$4x'^2 - \frac{16}{\sqrt{2}} \left(y' - \frac{\sqrt{2}}{4} \right) = 0.$$

Let $x'' = x'$ and $y'' = y' - \frac{\sqrt{2}}{4}$ and it follows that the equation is

$$4x''^2 - \frac{16}{\sqrt{2}} y'' = 0.$$

Hence the standard form is

$$x''^2 = \frac{4}{\sqrt{2}} y''.$$

The expression for the standard form is not unique. The roles of x'', y'', and z'' may be interchanged depending upon the order of the eigenvectors used as columns in P.

25. For the quadric surface $-x^2 - y^2 - z^2 + 4xy + 4xz + 4yz + \frac{3}{\sqrt{2}} x - \frac{3}{\sqrt{2}} y = 6$, we have

$$A = \begin{bmatrix} -1 & 2 & 2 \\ 2 & -1 & 2 \\ 2 & 2 & -1 \end{bmatrix}.$$

The characteristic equation of A is

$$\det(\lambda I_3 - A) = \begin{vmatrix} \lambda+1 & -2 & -2 \\ -2 & \lambda+1 & -2 \\ -2 & -2 & \lambda+1 \end{vmatrix} = \lambda^3 + 3\lambda^2 - 9\lambda - 27 = (\lambda - 3)(\lambda + 3)^2 = 0$$

and hence the eigenvalues are $\lambda = 3, -3, -3$. Thus the inertia of A is $(1, 2, 0)$ and it follows from Table 8.2 that this quadric surface is a hyperboloid of two sheets. From the terms present we see that we must perform a rotation and a translation to obtain the standard form. Hence we require the eigenvectors. Solving the appropriate homogeneous systems we have eigenvalue and eigenvector pairs

$$\lambda_1 = 3, \mathbf{x}_1 = \begin{bmatrix} 1 \\ 1 \\ 1 \end{bmatrix}; \quad \lambda_2 = -3, \mathbf{x}_2 = \begin{bmatrix} -1 \\ 1 \\ 0 \end{bmatrix}; \lambda_3 = -3, \mathbf{x}_3 = \begin{bmatrix} -1 \\ 0 \\ 1 \end{bmatrix}.$$

Unfortunately the eigenvectors corresponding to the eigenvalue $\lambda = -3$, which has multiplicity two, are not orthogonal. Thus we need to use the Gram-Schmidt process. For consistency of notation, let $\mathbf{y}_1 = \mathbf{x}_1$ and proceed as follows. Let $\mathbf{y}_2 = \mathbf{x}_2$. Then define

$$\mathbf{y}_3 = \mathbf{x}_3 - \frac{(\mathbf{x}_3, \mathbf{y}_2)}{(\mathbf{y}_2, \mathbf{y}_2)} \mathbf{y}_2 = \begin{bmatrix} -\frac{1}{2} \\ -\frac{1}{2} \\ 1 \end{bmatrix}.$$

Normalizing the eigenvectors $\mathbf{y}_j$ we have

$$\mathbf{u}_1 = \begin{bmatrix} \frac{1}{\sqrt{3}} \\ \frac{1}{\sqrt{3}} \\ \frac{1}{\sqrt{3}} \end{bmatrix}, \quad \mathbf{u}_2 = \begin{bmatrix} -\frac{1}{\sqrt{2}} \\ \frac{1}{\sqrt{2}} \\ 0 \end{bmatrix}, \quad \mathbf{u}_3 = \begin{bmatrix} -\frac{1}{\sqrt{6}} \\ -\frac{1}{\sqrt{6}} \\ \frac{2}{\sqrt{6}} \end{bmatrix}.$$

Let $P = \begin{bmatrix} \mathbf{u}_1 & \mathbf{u}_2 & \mathbf{u}_3 \end{bmatrix}$. Then $\det(P) = 1$ so we do have a counterclockwise rotation. Let $\mathbf{x} = P\mathbf{y}$, where $\mathbf{y} = \begin{bmatrix} x' & y' & z' \end{bmatrix}^T$. Then the quadric surface can be written as

$$(P\mathbf{y})^T A (P\mathbf{y}) + \begin{bmatrix} \frac{3}{\sqrt{2}} & -\frac{3}{\sqrt{2}} & 0 \end{bmatrix} P\mathbf{y} = 6$$

which simplifies to

$$3x'^2 - 3y'^2 - 3z'^2 - 3y' = 6.$$

Completing the square in y' we have

$$3x'^2 - 3\left(y'^2 + y' + \frac{1}{4}\right) - 3z'^2 = 6 - \frac{3}{4}$$

$$\implies \quad 3x'^2 - 3\left(y' + \frac{1}{2}\right)^2 - 3z'^2 = \frac{21}{4}.$$

Let $x'' = x'$, $y'' = y' + \frac{1}{2}$, and $z'' = z'$. Then we have

$$3x''^2 - 3y''^2 - 3z''^2 = \frac{21}{4}$$

and the standard form is

$$\frac{x''^2}{\frac{7}{4}} - \frac{y''^2}{\frac{7}{4}} - \frac{z''^2}{\frac{7}{4}} = 1.$$

The expression for the standard form is not unique. The roles of x'', y'', and z'' may be interchanged depending upon the order of the eigenvectors used in the columns in P.

27. For the quadric surface $x^2 + y^2 - z^2 - 2x - 4y - 4z + 1 = 0$, we have

$$A = \begin{bmatrix} 1 & 0 & 0 \\ 0 & 1 & 0 \\ 0 & 0 & -1 \end{bmatrix}.$$

The characteristic equation is

$$\det(\lambda I_3 - A) = \begin{vmatrix} \lambda - 1 & 0 & 0 \\ 0 & \lambda - 1 & 0 \\ 0 & 0 & \lambda + 1 \end{vmatrix} = (\lambda - 1)^2(\lambda + 1) = 0$$

and hence the eigenvalues are $\lambda = 1, 1, -1$. Thus the inertia of A is $(2, 1, 0)$ and it follows from Table 8.2 that this quadric surface is a hyperboloid of one sheet. Since there are no cross product terms we do not need the eigenvectors. We merely complete the square in each of the variables:

$$(x^2 - 2x + 1) + (y^2 - 4y + 4) - (z^2 + 4z + 4) = -1 + 1 + 4 - 4$$

$$\implies \quad (x - 1)^2 + (y - 2)^2 - (z + 2)^2 = 0.$$

Let $x' = x - 1$, $y' = y - 2$, and $z' = z + 2$. Then we have the standard form

$$x''^2 + y''^2 - z''^2 = 0.$$

This is really a cone which is a special case of a hyperboloid of one sheet.

Chapter 10

MATLAB Exercises

Section 10.1, p. 597

Basic Matrix Properties, p. 598

ML.1. (a) Commands: $\mathbf{A(2,3)}$, $\mathbf{B(3,2)}$, $\mathbf{B(1,2)}$.

(b) For $\text{row}_1(\mathbf{A})$, use command $\mathbf{A(1,:)}$.
For $\text{col}_3(\mathbf{A})$, use command $\mathbf{A(:,3)}$.
For $\text{row}_2(\mathbf{B})$, use command $\mathbf{B(2,:)}$.
(In this context the colon means "all.")

(c) Matrix B in **format long** is

$$\begin{bmatrix} 8.00000000000000 & 0.666666666666667 \\ 0.00497512437811 & -3.200000000000000 \\ 0.00001000000000 & 4.333333333333333 \end{bmatrix}.$$

Matrix Operations, p. 598

ML.1. (a) $\begin{bmatrix} 4.5000 & 2.2500 & 3.7500 \\ 1.5833 & 0.9167 & 1.5000 \\ 0.9667 & 0.5833 & 0.9500 \end{bmatrix}.$

(b) `??? Error using ==> *`
`Inner matrix dimensions must agree.`

(c) $\begin{bmatrix} 5.0000 & 1.5000 \\ 1.5833 & 2.2500 \\ 2.4500 & 3.1667 \end{bmatrix}.$

(d) `??? Error using ==> *`
`Inner matrix dimensions must agree.`

(e) `??? Error using ==> *`
`Inner matrix dimensions must agree.`

(f) `??? Error using ==> −`
`Inner matrix dimensions must agree.`

(g) $\begin{bmatrix} 18.2500 & 7.4583 & 12.2833 \\ 7.4583 & 5.7361 & 8.9208 \\ 12.2833 & 8.9208 & 14.1303 \end{bmatrix}.$

ML.3. $\begin{bmatrix} 4 & -3 & 2 & -1 & -5 \\ 2 & 1 & -3 & 0 & 7 \\ -1 & 4 & 1 & 2 & 8 \end{bmatrix}.$

ML.5. (a) $\begin{bmatrix} 1 & 0 & 0 & 0 \\ 0 & 2 & 0 & 0 \\ 0 & 0 & 3 & 0 \\ 0 & 0 & 0 & 4 \end{bmatrix}$.

(b) $\begin{bmatrix} 0 & 0 & 0 & 0 & 0 \\ 0 & 1.0000 & 0 & 0 & 0 \\ 0 & 0 & 0.5000 & 0 & 0 \\ 0 & 0 & 0 & 0.3333 & 0 \\ 0 & 0 & 0 & 0 & 0.2500 \end{bmatrix}$.

(c) $\begin{bmatrix} 5 & 0 & 0 & 0 & 0 & 0 \\ 0 & 5 & 0 & 0 & 0 & 0 \\ 0 & 0 & 5 & 0 & 0 & 0 \\ 0 & 0 & 0 & 5 & 0 & 0 \\ 0 & 0 & 0 & 0 & 5 & 0 \\ 0 & 0 & 0 & 0 & 0 & 5 \end{bmatrix}$.

Powers of a Matrix, p. 599

ML.1. (a) $k = 3$.   (b) $k = 5$.

ML.3. (a) $\begin{bmatrix} 0 & -2 & 4 \\ 4 & 0 & -2 \\ -2 & 4 & 0 \end{bmatrix}$.   (b) $\begin{bmatrix} 0 & 0 & 0 \\ 0 & 0 & 0 \\ 0 & 0 & 0 \end{bmatrix}$.

ML.5. The sequence seems to be converging to $\begin{bmatrix} 1.0000 & 0.7500 \\ 0 & 0 \end{bmatrix}$.

ML.7. (a) $A^T A = \begin{bmatrix} 2 & -3 & -1 \\ -3 & 9 & 2 \\ -1 & 2 & 6 \end{bmatrix}$, $AA^T = \begin{bmatrix} 6 & -1 & -3 \\ -1 & 6 & 4 \\ -3 & 4 & 5 \end{bmatrix}$.

(b) $B = \begin{bmatrix} 2 & -3 & 1 \\ -3 & 2 & 4 \\ 1 & 4 & 2 \end{bmatrix}$, $C = \begin{bmatrix} 0 & -1 & 1 \\ 1 & 0 & 0 \\ -1 & 0 & 0 \end{bmatrix}$.

(c) $B + C = \begin{bmatrix} 2 & -4 & 2 \\ -2 & 2 & 4 \\ 0 & 4 & 2 \end{bmatrix}$, $B + C = 2A$.

Row Operations and Echelon Forms, p. 600

ML.1. (a) $\begin{bmatrix} 1.0000 & 0.5000 & 0.5000 \\ -3.0000 & 1.0000 & 4.0000 \\ 1.0000 & 0 & 3.0000 \\ 5.0000 & -1.0000 & 5.0000 \end{bmatrix}$.

(b) $\begin{bmatrix} 1.0000 & 0.5000 & 0.5000 \\ 0 & 2.5000 & 5.5000 \\ 1.0000 & 0 & 3.0000 \\ 5.0000 & -1.0000 & 5.0000 \end{bmatrix}$.

(c) $\begin{bmatrix} 1.0000 & 0.5000 & 0.5000 \\ 0 & 2.5000 & 5.5000 \\ 0 & -0.5000 & 2.5000 \\ 5.0000 & -1.0000 & 5.0000 \end{bmatrix}$.

$$(d) \begin{bmatrix} 1.0000 & 0.5000 & 0.5000 \\ 0 & 2.5000 & 5.5000 \\ 0 & -0.5000 & 2.5000 \\ 0 & -3.5000 & 2.5000 \end{bmatrix}.$$

$$(e) \begin{bmatrix} 1.0000 & 0.5000 & 0.5000 \\ 0 & -3.5000 & 2.5000 \\ 0 & -0.5000 & 2.5000 \\ 0 & 2.5000 & 5.5000 \end{bmatrix}.$$

ML.3. $\begin{bmatrix} 1 & 0 & 0 \\ 0 & 1 & 0 \\ 0 & 0 & 1 \\ 0 & 0 & 0 \end{bmatrix}.$

ML.5. $x = -2 + r$, $y = -1$, $z = 8 - 2r$, $w = r$, $r =$ any real number.

ML.7. $x_1 = -r + 1$, $x_2 = r + 2$, $x_3 = r - 1$, $x_4 = r$,
$r =$ any real number.

ML.9. $\mathbf{x} = \begin{bmatrix} 0.5r \\ r \end{bmatrix}.$

ML.11. Exercise 15:

(a) Unique solution: $x = -1$, $y = 4$, $z = -3$.

(b) The only solution is the trivial one.

Exercise 16:

(a) $x = r$, $y = -2r$, $z - r$, where r is any real number.

(b) Unique solution: $x = 1$, $y = 2$, $z = 2$.

ML.13. The \ command yields a matrix showing that the system is inconsistent. The **rref** command leads to the display of a warning that the result may contain large roundoff errors.

LU-Factorization, p. 601

ML.1. $L = \begin{bmatrix} 1 & 0 & 0 \\ 1 & 1 & 0 \\ 0.5 & 0.3333 & 1 \end{bmatrix}$, $U = \begin{bmatrix} 2 & 8 & 0 \\ 0 & -6 & -3 \\ 0 & 0 & 8 \end{bmatrix}.$

ML.3. $L = \begin{bmatrix} 1.0000 & 0 & 0 & 0 \\ 0.5000 & 1.0000 & 0 & 0 \\ -2.0000 & -2.0000 & 1.0000 & 0 \\ -1.0000 & 1.0000 & -2.0000 & 1.0000 \end{bmatrix}$, $U = \begin{bmatrix} 6 & -2 & -4 & 4 \\ 0 & -2 & -4 & -1 \\ 0 & 0 & 5 & -2 \\ 0 & 0 & 0 & 8 \end{bmatrix}$,

$\mathbf{z} = \begin{bmatrix} 2 \\ -5 \\ 2 \\ -32 \end{bmatrix}$, $\mathbf{x} = \begin{bmatrix} 4.5000 \\ 6.9000 \\ -1.2000 \\ -4.0000 \end{bmatrix}.$

Matrix Inverses, p. 601

ML.1. (a) and (c).

ML.3. (a) $\begin{bmatrix} -2 & 3 \\ 1 & -1 \end{bmatrix}$. (b) $\begin{bmatrix} -\frac{1}{4} & \frac{3}{4} & -\frac{1}{4} \\ -\frac{1}{4} & -\frac{1}{4} & \frac{3}{4} \\ \frac{3}{4} & -\frac{1}{4} & -\frac{1}{4} \end{bmatrix}$.

ML.5. (a) $t = 4$. (b) $t = 3$.

Determinants by Row Reduction, p. 601

ML.1. (a) -18. (b) 5.

ML.3. (a) 4. (b) 0.

ML.5. $t = 3$, $t = 4$.

Determinants by Cofactor Expansion, p. 602

ML.1. $A_{11} = -11$, $A_{23} = -2$, $A_{31} = 2$.

ML.3. 0.

ML.5. (a) $\dfrac{1}{28} \begin{bmatrix} 30 & 5 & -9 & -46 \\ 32 & -4 & -4 & -36 \\ 12 & 2 & 2 & -24 \\ -16 & 2 & 2 & 32 \end{bmatrix}$. (b) $\dfrac{1}{14} \begin{bmatrix} 3 & -6 & 2 \\ 2 & 10 & -8 \\ -1 & 2 & 4 \end{bmatrix}$. (c) $\dfrac{1}{18} \begin{bmatrix} 4 & -2 \\ 3 & 3 \end{bmatrix}$.

Subspaces, p. 603

ML.3. (a) No. (b) Yes.

ML.5. (a) $0\mathbf{v}_1 + \mathbf{v}_2 - \mathbf{v}_3 - \mathbf{v}_4 = \mathbf{v}$.

(b) $p_1(t) + 2p_2(t) + 2p_3(t) = p(t)$.

ML.7. (a) Yes. (b) Yes. (c) Yes.

Linear Independence/Dependence, p. 604

ML.1. (a) Linearly dependent.

(b) Linearly independent.

(c) Linearly independent.

(d) Linearly dependent.

ML.3. (a) $\mathbf{v} = \frac{1}{2}\mathbf{v}_1 + \frac{2}{5}\mathbf{v}_2 - \frac{1}{10}\mathbf{v}_3$.

(b) $\mathbf{v} = \mathbf{v}_1 + \mathbf{v}_2$.

(c) $\mathbf{v} = 2\mathbf{v}_1$.

Bases and Dimension, p. 604

ML.1. Basis.

ML.3. Basis.

ML.5. Basis.

ML.7. dim span $S = 3$, span $S \neq R^4$.

ML.9. dim span $S = 3$, span $S = P_2$.

ML.11. $\{t^3 - t + 1, t^3 + 2, t, 1\}$.

Coordinates and Change of Basis, p. 605

ML.1. (a) $\begin{bmatrix} 1 \\ 2 \\ 3 \end{bmatrix}$.　(b) $\begin{bmatrix} -1 \\ 2 \\ -1 \end{bmatrix}$.　(c) $\begin{bmatrix} 1 \\ 1 \\ 1 \end{bmatrix}$.

ML.3. (a) $\begin{bmatrix} 0.5000 \\ -0.5000 \\ 0 \\ -0.5000 \end{bmatrix}$.　(b) $\begin{bmatrix} 1.0000 \\ 0.5000 \\ 0.3333 \\ 0 \end{bmatrix}$.　(c) $\begin{bmatrix} 0.5000 \\ 0.1667 \\ -0.3333 \\ -1.5000 \end{bmatrix}$.

ML.5. $\begin{bmatrix} -0.5000 & -1.0000 & -0.5000 & 0 \\ -0.5000 & 0 & 1.5000 & 0 \\ 1.0000 & 0 & -1.0000 & 1.0000 \\ 0 & 0 & 0 & 1.0000 \end{bmatrix}$.

ML.7. (a) $\begin{bmatrix} 1.0000 & -1.6667 & 2.3333 \\ 1.0000 & 0.6667 & -1.3333 \\ 0 & 1.3333 & -0.6667 \end{bmatrix}$.　(b) $\begin{bmatrix} 2 & 0 & 1 \\ -1 & 1 & -1 \\ 0 & -1 & 2 \end{bmatrix}$.　(c) $\begin{bmatrix} 2 & -2 & 4 \\ 0 & 1 & -3 \\ -1 & 2 & 0 \end{bmatrix}$.　(d) QP.

Homogeneous Linear Systems, p. 605

ML.1. $\left\{ \begin{bmatrix} -2 \\ 0 \\ 1 \\ 0 \\ 0 \end{bmatrix}, \begin{bmatrix} -1 \\ -1 \\ 0 \\ 1 \\ 0 \end{bmatrix}, \begin{bmatrix} -2 \\ 1 \\ 0 \\ 0 \\ 1 \end{bmatrix} \right\}$.

ML.3. $\left\{ \begin{bmatrix} 1 \\ -2 \\ 1 \\ 0 \end{bmatrix}, \begin{bmatrix} \frac{4}{3} \\ -\frac{1}{3} \\ 0 \\ 1 \end{bmatrix} \right\}$.

ML.5. $\mathbf{x} = \begin{bmatrix} t \\ t \\ t \end{bmatrix}$, where t is any nonzero real number.

Rank of a Matrix, p. 606

ML.3. (a) The original columns of A and

$$\left\{ \begin{bmatrix} 1 \\ 0 \\ 2 \\ 0 \end{bmatrix}, \begin{bmatrix} 0 \\ 1 \\ 1 \\ 0 \end{bmatrix}, \begin{bmatrix} 0 \\ 0 \\ 0 \\ 1 \end{bmatrix} \right\}.$$

(b) The first two columns of A and

$$\left\{ \begin{bmatrix} 1 \\ 0 \\ 0 \\ 1 \\ 1 \end{bmatrix}, \begin{bmatrix} 0 \\ 0 \\ 1 \\ 2 \\ 1 \end{bmatrix} \right\}.$$

ML.5. (a) Consistent.　(b) Inconsistent.　(c) Inconsistent.

Standard Inner Product, p. 607

ML.1. (a) (i) 15.   (i) 0.

(b) $k = -\frac{2}{3}$.

(c) (i) 29.   (ii) 127.   (iii) 39.

ML.3. (a) 2.2361. (b) 5.4772. (c) 3.1623.

ML.5. (a) 19. (b) −11. (c) −55.

ML.9. (a) $\begin{bmatrix} 0.6667 \\ 0.6667 \\ -0.3333 \end{bmatrix}$ or in rational form $\begin{bmatrix} \frac{2}{3} \\ \frac{2}{3} \\ -\frac{1}{3} \end{bmatrix}$.

(b) $\begin{bmatrix} 0 \\ 0.8000 \\ -0.6000 \\ 0 \end{bmatrix}$ or in rational form $\begin{bmatrix} 0 \\ \frac{4}{5} \\ -\frac{3}{5} \\ 0 \end{bmatrix}$. (c) $\begin{bmatrix} 0.3015 \\ 0 \\ 0.3015 \\ 0 \end{bmatrix}$.

Cross Product, p. 608

ML.1. (a) $\begin{bmatrix} -11 & 2 & 5 \end{bmatrix}$.   (b) $\begin{bmatrix} 3 & 1 & -1 \end{bmatrix}$.   (c) $\begin{bmatrix} 1 & -8 & -5 \end{bmatrix}$.

ML.5. $\cos\theta = -0.4655$, $\theta = 2.0550$ rad $= 117.7409°$.

The Gram–Schmidt Process, p. 608

ML.1. $\left\{ \begin{bmatrix} 0.7071 \\ 0.7071 \\ 0 \end{bmatrix}, \begin{bmatrix} 0.7071 \\ -0.7071 \\ 0 \end{bmatrix}, \begin{bmatrix} 0 \\ 0 \\ 1.0000 \end{bmatrix} \right\} = \left\{ \begin{bmatrix} \frac{\sqrt{2}}{2} \\ \frac{\sqrt{2}}{2} \\ 0 \end{bmatrix}, \begin{bmatrix} \frac{\sqrt{2}}{2} \\ -\frac{\sqrt{2}}{2} \\ 0 \end{bmatrix}, \begin{bmatrix} 0 \\ 0 \\ 1 \end{bmatrix} \right\}$.

ML.3. (a) $\begin{bmatrix} -1.4142 \\ 1.4142 \\ 1.0000 \end{bmatrix}$.   (b) $\begin{bmatrix} 0 \\ 1.4142 \\ 1.0000 \end{bmatrix}$.   (c) $\begin{bmatrix} 0.7071 \\ 0.7071 \\ -1.0000 \end{bmatrix}$.

ML.5. (a) $\mathbf{w} = \begin{bmatrix} 1 \\ -2 \end{bmatrix}$.

(b) $\mathbf{u}_1 = \frac{1}{\sqrt{5}} \begin{bmatrix} 2 \\ 1 \end{bmatrix}$, $\mathbf{u}_2 = \frac{1}{\sqrt{5}} \begin{bmatrix} 1 \\ -2 \end{bmatrix}$.

Projections, p. 609

ML.1. (a) $\begin{bmatrix} 0 \\ \frac{5}{6} \\ \frac{5}{3} \\ \frac{5}{6} \end{bmatrix}$. (b) $\begin{bmatrix} \frac{3}{5} \\ \frac{3}{5} \\ \frac{3}{5} \\ \frac{3}{5} \\ \frac{3}{5} \end{bmatrix}$.

ML.3. (a) $\begin{bmatrix} 2.4286 \\ 3.9341 \\ 7.9011 \end{bmatrix}$.

(b) $\sqrt{(2.4286 - 2)^2 + (3.9341 - 4)^2 + (7.9011 - 8)^2} \approx 0.4448$.

ML.5. $\mathbf{p} = \begin{bmatrix} 0.8571 \\ 0.5714 \\ 1.4286 \\ 0.8571 \\ 0.8571 \end{bmatrix}$.

Least Squares, p. 609

ML.1. $y = 1.87 + 1.345t$.

ML.3. (a) $T = -8.278t + 188.1$, where $t = $ time.

(b) $T(1) = 179.7778° \text{F}$.
$T(6) = 138.3889° \text{F}$.
$T(8) = 121.8333° \text{F}$.

(c) 3.3893 minutes.

ML.5. $y = 1.0204x^2 + 3.1238x + 1.0507$, when $x = 7$, $y = 72.9169$.

Kernel and Range of Linear Transformations, p. 611

ML.1. Basis for ker L: $\left\{ \begin{bmatrix} -1 \\ -2 \\ 1 \\ 0 \end{bmatrix}, \begin{bmatrix} 1 \\ -3 \\ 0 \\ 1 \end{bmatrix} \right\}$. Basis for range L: $\left\{ \begin{bmatrix} 1 \\ 0 \end{bmatrix}, \begin{bmatrix} 0 \\ 1 \end{bmatrix} \right\}$.

ML.3. Basis for ker L: $\left\{ \begin{bmatrix} -2 \\ 0 \\ 1 \\ -2 \\ 1 \end{bmatrix}, \begin{bmatrix} -1 \\ 1 \\ 0 \\ 0 \\ 0 \end{bmatrix} \right\}$. Basis for range L: $\left\{ \begin{bmatrix} 1 \\ 0 \\ 0 \end{bmatrix}, \begin{bmatrix} 0 \\ 1 \\ 0 \end{bmatrix}, \begin{bmatrix} 0 \\ 0 \\ 1 \end{bmatrix} \right\}$.

Matrix of a Linear Transformation, p. 611

ML.1. $A = \begin{bmatrix} -1 & 0 & 3 \\ 1 & 0 & -2 \end{bmatrix}$.

ML.3. (a) $A = \begin{bmatrix} 1.3333 & -0.3333 \\ -1.6667 & -3.3333 \end{bmatrix}$.

(b) $B = \begin{bmatrix} -3.6667 & 0.3333 \\ -3.3333 & 1.6667 \end{bmatrix}$.

(c) $P = \begin{bmatrix} -0.3333 & 0.6667 \\ 1.6667 & -0.3333 \end{bmatrix}$.

Eigenvalues and Eigenvectors, p. 612

ML.1. (a) $\lambda^2 - 5$. (b) $\lambda^3 - 6\lambda^2 + 4\lambda + 8$. (c) $\lambda^4 - 3\lambda^3 - 3\lambda^2 + 11\lambda - 6$.

ML.3. (a) $\begin{bmatrix} 1 \\ 1 \end{bmatrix}$. (b) $\begin{bmatrix} 0 \\ 0 \\ 1 \end{bmatrix}$. (c) $\begin{bmatrix} 1 \\ -2 \\ 1 \end{bmatrix}$.

ML.5. $\begin{bmatrix} 1 & -1 & 1 \\ 0 & 0 & 1 \\ 0 & 0 & 1 \end{bmatrix}.$

ML.7. The sequence $A, A^3, A^5, \ldots$ converges to $\begin{bmatrix} -1 & 1 & -1 \\ -2 & 2 & -1 \\ -2 & 2 & -1 \end{bmatrix}.$

The sequence $A^2, A^4, A^6, \ldots$ converges to $\begin{bmatrix} 1 & -1 & 1 \\ 0 & 0 & 1 \\ 0 & 0 & 1 \end{bmatrix}.$

Diagonalization, p. 613

ML.1. (a) $\lambda_1 = 0$, $\lambda_2 = 12$; $P = \begin{bmatrix} 0.7071 & 0.7071 \\ -0.7071 & 0.7071 \end{bmatrix}.$

(b) $\lambda_1 = -1$, $\lambda_2 = -1$, $\lambda_3 = 5$;

$$P = \begin{bmatrix} 0.7743 & -0.2590 & 0.5774 \\ -0.6115 & -0.5411 & 0.5774 \\ -0.1629 & 0.8001 & 0.5774 \end{bmatrix}.$$

(c) $\lambda_1 = 5.4142$, $\lambda_2 = 4.0000$, $\lambda_3 = 2.5858$.

$$P = \begin{bmatrix} 0.5000 & -0.7071 & -0.5000 \\ 0.7071 & -0.0000 & 0.7071 \\ 0.5000 & 0.7071 & -0.5000 \end{bmatrix}.$$

Dominant Eigenvalue, p. 568

ML.1. (a) $\lambda_1 \approx 7.6904$, $\mathbf{x}_1 \approx \begin{bmatrix} -0.7760 \\ -0.6308 \end{bmatrix}$

$\lambda_2 \approx -1.6904$, $\mathbf{x}_2 \approx \begin{bmatrix} -0.8846 \\ 0.4664 \end{bmatrix}.$

λ_1 is dominant.

(b) $\lambda_1 \approx 8.8655$, $\mathbf{x}_1 \approx \begin{bmatrix} -0.6065 \\ -0.7951 \end{bmatrix}$

$\lambda_2 \approx -5.8655$, $\mathbf{x}_2 \approx \begin{bmatrix} -0.6581 \\ 0.7530 \end{bmatrix}.$

λ_1 is dominant.

Complex Numbers

Appendix B.1, p. A-11

1. Let $c_1 = 3 + 4i$, $c_2 = 1 - 2i$, and $c_3 = -1 + i$.

   (a) $c_1 + c_2 = (3 + 4i) + (1 - 2i) = 4 + 2i$

   (b) $c_3 - c_1 = (-1 + i) - (3 + 4i) = -4 - 3i$

   (c) $c_1 c_2 = (3 + 4i)(1 - 2i) = 11 - 2i$

   (d) $c_2 \overline{c_3} = (1 - 2i)(-1 - i) = -3 + i$

   (e) $4c_3 + \overline{c_2} = (-4 + 4i) + (1 + 2i) = -3 + 6i$

   (f) $(-i)c_2 = (-i)(1 - 2i) = -2 - i$

   (g) $\overline{3c_1 - ic_2} = \overline{(9 + 12i) - (i - 2i^2)} = \overline{7 + 11i} = 7 - 11i$

   (h) $c_1 c_2 c_3 = (c_1 c_2)c_3 = (11 - 2i)(-1 + i) = -9 + 13i$

3.

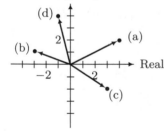

5. (a) $\mathrm{Re}(c_1 + c_2) = \mathrm{Re}((a_1 + a_2) + (b_1 + b_2)i) = a_1 + a_2 = \mathrm{Re}(c_1) + \mathrm{Re}(c_2)$
   $\mathrm{Im}(c_1 + c_2) = \mathrm{Im}((a_1 + a_2) + (b_1 + b_2)i) = b_1 + b_2 = \mathrm{Im}(c_1) + \mathrm{Im}(c_2)$

   (b) $\mathrm{Re}(kc) = \mathrm{Re}(ka + kbi) = ka = k\mathrm{Re}(c)$
   $\mathrm{Im}(kc) = \mathrm{Im}(ka + kbi) = kb = k\mathrm{Im}(c)$

   (c) No.

   (d) $\mathrm{Re}(c_1 c_2) = \mathrm{Re}((a_1 + b_1 i)(a_2 + b_2 i)) = \mathrm{Re}((a_1 a_2 - b_1 b_2) + (a_1 b_2 + a_2 b_1)i) = a_1 a_2 - b_1 b_2 \neq \mathrm{Re}(c_1)\mathrm{Re}(c_2)$

7. (a) $A + B = \begin{bmatrix} 2 + 4i & 5i \\ -2 & 4 - 2i \end{bmatrix}$

   (b) $(1 - 2i)C = \begin{bmatrix} (1 - 2i)(2 + i) \\ (1 - 2i)(-i) \end{bmatrix} = \begin{bmatrix} 4 - 3i \\ -2 - i \end{bmatrix}$

   (c) $AB = \begin{bmatrix} 2 + 2i & -1 + 3i \\ -2 & 1 - i \end{bmatrix}\begin{bmatrix} 2i & 1 + 2i \\ 0 & 3 - i \end{bmatrix} = \begin{bmatrix} -4 + 4i & -2 + 16i \\ -4i & -8i \end{bmatrix}$

   (d) $BC = \begin{bmatrix} 2i(2 + i) + (1 + 2i)(-i) \\ 0(2 + i) + (3 - i)(-i) \end{bmatrix} = \begin{bmatrix} 3i \\ -1 - 3i \end{bmatrix}$

(e) $A - 2I_2 = \begin{bmatrix} (2+2i) - 2 & -1 + 3i \\ -2 & (1-i) - 2 \end{bmatrix} = \begin{bmatrix} 2i & -1 + 3i \\ -2 & -1 - i \end{bmatrix}$

(f) $\overline{B} = \begin{bmatrix} \overline{2i} & \overline{1+2i} \\ \overline{0} & \overline{3-i} \end{bmatrix} = \begin{bmatrix} -2i & 1 - 2i \\ 0 & 3 + i \end{bmatrix}$

(g) $A\overline{C} = \begin{bmatrix} 2+2i & -1+3i \\ -2 & 1-i \end{bmatrix} \begin{bmatrix} \overline{2+i} \\ \overline{-i} \end{bmatrix} = \begin{bmatrix} 2+2i & -1+3i \\ -2 & 1-i \end{bmatrix} \begin{bmatrix} 2-i \\ i \end{bmatrix} = \begin{bmatrix} 3+i \\ -3+3i \end{bmatrix}$

(h) $(\overline{A+B})C = \begin{bmatrix} \overline{2+4i} & \overline{5i} \\ \overline{-2} & \overline{4-2i} \end{bmatrix} \begin{bmatrix} 2+i \\ -i \end{bmatrix} = \begin{bmatrix} 2-4i & -5i \\ -2 & 4+2i \end{bmatrix} \begin{bmatrix} 2+i \\ -i \end{bmatrix} = \begin{bmatrix} 3-6i \\ -2-6i \end{bmatrix}$

9. We have

$$A^2 = \begin{bmatrix} -1 & 0 \\ 0 & -1 \end{bmatrix} = -I_2, \quad A^3 = A^2 A = -I_2 A = -A = \begin{bmatrix} 0 & -i \\ -i & 0 \end{bmatrix}, \quad A^4 = A^2 A^2 = I_2 = \begin{bmatrix} 1 & 0 \\ 0 & 1 \end{bmatrix}$$

and, in general, $A^{4n} = I_2$, $A^{4n+1} = A$, $A^{4n+2} = A^2 = -I_2$, and $A^{4n+3} = A^3 = -A$.

11. Let $A = \begin{bmatrix} a_{ij} \end{bmatrix}$ be a Hermitian matrix.

   (a) Since $\overline{A^T} = A$, $\overline{a_{ii}} = a_{ii}$. Hence a_{ii} is real. Thus, the diagonal entries of a Hermitian matrix are real.

   (b) Since $\overline{A^T} = A$, $A^T = \overline{A}$. Let $B = \dfrac{1}{2}(A + \overline{A})$. Then

$$\overline{B} = \overline{\frac{1}{2}\left(A + \overline{A}\right)} = \frac{1}{2}\left(\overline{A} + \overline{\overline{A}}\right) = \frac{1}{2}\left(\overline{A} + A\right) = \frac{1}{2}\left(A + \overline{A}\right) = B.$$

   Therefore B is a real matrix. Also,

$$B^T = \left[\frac{1}{2}\left(A + \overline{A}\right)\right]^T = \frac{1}{2}\left(A^T + \overline{A}^T\right) = \frac{1}{2}\left(\overline{A} + A\right) = \frac{1}{2}\left(A + \overline{A}\right) = B.$$

   Therefore B is symmetric. Next, let $C = \dfrac{1}{2i}\left(A - \overline{A}\right)$. Then

$$\overline{C} = \overline{\frac{1}{2i}\left(A - \overline{A}\right)} = \frac{1}{-2i}\left(\overline{A} - \overline{\overline{A}}\right) = \frac{1}{2i}(A - \overline{A}) = C$$

   so C is a real matrix. Also,

$$C^T = \left(\frac{1}{2i}(A - \overline{A})\right)^T = \frac{1}{2i}\left(A^T - \overline{A}^T\right) = \frac{1}{2i}\left(A^T - \overline{A^T}\right) = \frac{1}{2i}\left(\overline{A} - A\right) = -\frac{1}{2i}\left(A - \overline{A}\right) = -C$$

   so C is also skew symmetric. Moreover, $A = B + iC$.

   (c) If A is real symmetric, then $A = A^T$ and $A = \overline{A}$. Therefore $\overline{A^T} = \overline{A} = A$. Hence A is Hermitian.

13. (a) Let $B = \dfrac{1}{2}\left(A + \overline{A^T}\right)$ and $C = \dfrac{1}{2i}\left(A - \overline{A^T}\right)$. Then

$$\overline{B^T} = \overline{\frac{1}{2}\left(A + \overline{A^T}\right)^T} = \frac{1}{2}\left(\overline{A^T} + \overline{\overline{A^T}}^T\right) = \frac{1}{2}\left(\overline{A^T} + A\right) = \frac{1}{2}\left(A + \overline{A^T}\right) = B$$

   so B is Hermitian. Also,

$$\overline{C^T} = \overline{\frac{1}{2i}\left(A - \overline{A^T}\right)^T} = \frac{1}{-2i}\left(\overline{A^T} - \overline{\overline{A^T}}^T\right) = \frac{1}{2i}\left(A - \overline{A^T}\right) = C$$

   so C is Hermitian. Moreover, $A = B + iC$.

(b) We have

$$\overline{A^T} A = \left(\overline{B^T + iC^T} \right)(B + iC) = \left(\overline{B^T} + \overline{iC^T} \right)(B + iC)$$
$$= (B - iC)(B + iC)$$
$$= B^2 - iCB + iBC - i^2 C^2$$
$$= (B^2 + C^2) + i(BC - CB)$$

Similarly,

$$A \overline{A^T} = (B + iC)\overline{(B^T + iC)^T} = (B + iC)\left(\overline{B^T} + \overline{iC^T} \right)$$
$$= (B + iC)(B - iC)$$
$$= B^2 - iBC + iCB - i^2 C^2$$
$$= (B^2 + C^2) + i(CB - BC)$$

Since $\overline{A^T} A = A\overline{A^T}$, we equate imaginary parts obtaining $BC - CB = CB - BC$ which implies that $BC = CB$. The steps are reversible, establishing the converse.

15. Let $A = B + iC$ be skew Hermitian. Then $\overline{A^T} = -A$ so that $B^T - iC^T = -B - iC$. It follows that $B^T = -B$ and $C^T = C$. Thus, B is skew symmetric and C is symmetric. Conversely, if B is skew symmetric and C is symmetric, then $B^T = -B$ and $C^T = C$ so that $B^T - iC^T = -B - iC$, or equivalently $\overline{A^T} = -A$. Hence A is skew Hermitian.

17. (a) $A^2 = \begin{bmatrix} 9 & 0 \\ 0 & 9 \end{bmatrix}$. Then $p(A) = 2 \begin{bmatrix} 9 & 0 \\ 0 & 9 \end{bmatrix} + 5 \begin{bmatrix} -3 & 0 \\ 0 & -3 \end{bmatrix} - 3 \begin{bmatrix} 1 & 0 \\ 0 & 1 \end{bmatrix} = \begin{bmatrix} 0 & 0 \\ 0 & 0 \end{bmatrix}$.

   (b) $A^2 = \begin{bmatrix} 1 & 4 \\ 0 & 1 \end{bmatrix}$. Then $p(A) = 2 \begin{bmatrix} 1 & 4 \\ 0 & 1 \end{bmatrix} + 5 \begin{bmatrix} 1 & 2 \\ 0 & 1 \end{bmatrix} - \begin{bmatrix} 1 & 0 \\ 0 & 1 \end{bmatrix} = \begin{bmatrix} 4 & 18 \\ 0 & 4 \end{bmatrix}$.

   (c) $A^2 = \begin{bmatrix} -1 & 0 \\ 0 & -1 \end{bmatrix}$. Then $p(A) = 2 \begin{bmatrix} -1 & 0 \\ 0 & -1 \end{bmatrix} + 5 \begin{bmatrix} 0 & i \\ i & 0 \end{bmatrix} - 3 \begin{bmatrix} 1 & 0 \\ 0 & 1 \end{bmatrix} = \begin{bmatrix} -5 & 5i \\ 5i & -5 \end{bmatrix}$.

   (d) $A^2 = \begin{bmatrix} 1 & i \\ 0 & 0 \end{bmatrix}$. Then $p(A) = 2 \begin{bmatrix} 1 & i \\ 0 & 0 \end{bmatrix} + 5 \begin{bmatrix} 1 & i \\ 0 & 0 \end{bmatrix} - 3 \begin{bmatrix} 1 & 0 \\ 0 & 1 \end{bmatrix} = \begin{bmatrix} 4 & 7i \\ 0 & -3 \end{bmatrix}$.

19. $p(kI_2) = (kI_2)^2 - (kI_2) - 2I_2 = (k^2 - k - 2)I_2 = O_2$ if and only if $k^2 - k - 2 = (k - 2)(k + 1) = 0$. Thus $k = 2$ or $k = -1$. Hence

$$\begin{bmatrix} 2 & 0 \\ 0 & 2 \end{bmatrix} = 2I_2 \quad \text{and} \quad \begin{bmatrix} -1 & 0 \\ 0 & -1 \end{bmatrix} = -1I_2$$

are the only solutions.

Appendix B.2, p. A-20

1. Form the augmented matrix and use row operations

   (a) $\begin{bmatrix} 1 + 2i & -2 + i & | & 1 - 3i \\ 2 + i & -1 + 2i & | & -1 - i \end{bmatrix} \longrightarrow \begin{bmatrix} 1 & i & | & -1 - i \\ 2 + i & -1 + 2i & | & -1 - i \end{bmatrix} \longrightarrow \begin{bmatrix} 1 & i & | & -1 - i \\ 0 & 0 & | & 2i \end{bmatrix}$

   $\frac{1}{1+2i}\mathbf{r}_1 \qquad\qquad -(2 + i)\mathbf{r}_1 + \mathbf{r}_2$

   The system is inconsistent; there is no solution.

   (b) $\begin{bmatrix} 2i & -(1 - i) & | & 1 + i \\ 1 - i & 1 & | & 1 - i \end{bmatrix} \longrightarrow \begin{bmatrix} 1 & \frac{1}{2} + \frac{1}{2}i & | & \frac{1}{2} - \frac{1}{2}i \\ 1 - i & 1 & | & 1 - i \end{bmatrix} \longrightarrow \begin{bmatrix} 1 & \frac{1}{2} + \frac{1}{2}i & | & \frac{1}{2} - \frac{1}{2}i \\ 0 & 0 & | & 1 \end{bmatrix}$

   $\frac{1}{2i}\mathbf{r}_1 \qquad\qquad -(1 - i)\mathbf{r}_1 + \mathbf{r}_2$

   The system is inconsistent; there is no solution.

(c) $\begin{bmatrix} 1+i & -1 & -2+i \\ 2i & 1-i & i \end{bmatrix} \longrightarrow \begin{bmatrix} 1 & -\frac{1}{2}+\frac{1}{2}i & -\frac{1}{2}+\frac{3}{2}i \\ 2i & 1-i & i \end{bmatrix} \longrightarrow \begin{bmatrix} 1 & -\frac{1}{2}+\frac{1}{2}i & -\frac{1}{2}+\frac{3}{2}i \\ 0 & 2 & 3+2i \end{bmatrix}$

$\quad\quad\quad\quad \frac{1}{1+i}\mathbf{r}_1 \quad\quad\quad\quad\quad\quad\quad -2i\mathbf{r}_1+\mathbf{r}_2 \quad\quad\quad\quad\quad\quad \frac{1}{2}\mathbf{r}_2$

$\begin{bmatrix} 1 & -\frac{1}{2}+\frac{1}{2}i & -\frac{1}{2}+\frac{3}{2}i \\ 0 & 1 & \frac{3}{2}+i \end{bmatrix} \longrightarrow \begin{bmatrix} 1 & 0 & \frac{3}{4}+\frac{5}{4}i \\ 0 & 1 & \frac{3}{2}+i \end{bmatrix}$

$\quad \frac{1}{2}(1-i)\mathbf{r}_2+\mathbf{r}_1$

Thus the solution is $x_1 = \frac{3}{4}+\frac{5}{4}i$ and $x_2 = \frac{3}{2}+i$.

3. Form the augmented matrix and apply row operations to obtain row echelon form, then apply back substitution.

(a) $\begin{bmatrix} i & 1+i & 0 & i \\ 1-i & 1 & -i & 1 \\ 0 & i & 1 & 1 \end{bmatrix} \longrightarrow \begin{bmatrix} 1 & 1-i & 0 & 1 \\ 0 & 1+2i & -i & i \\ 0 & i & 1 & 1 \end{bmatrix} \longrightarrow \begin{bmatrix} 1 & 1-i & 0 & 1 \\ 0 & 1 & -\frac{2}{5}-\frac{1}{5}i & \frac{2}{5}+\frac{1}{5}i \\ 0 & 0 & \frac{4}{5}+\frac{2}{5}i & \frac{6}{5}-\frac{2}{5}i \end{bmatrix} \longrightarrow$

$\quad\quad \frac{1}{i}\mathbf{r}_1 \quad\quad\quad\quad\quad\quad\quad\quad \frac{1}{1+2i}\mathbf{r}_2 \quad\quad\quad\quad\quad\quad\quad\quad \frac{5}{4+2i}\mathbf{r}_3$

$\quad -(1+i)\mathbf{r}_1+\mathbf{r}_2 \quad\quad\quad\quad\quad -1\mathbf{r}_2+\mathbf{r}_3$

$\begin{bmatrix} 1 & 1-i & 0 & 1 \\ 0 & 1 & -\frac{2}{5}-\frac{1}{5}i & \frac{2}{5}+\frac{1}{5}i \\ 0 & 0 & 1 & 1-i \end{bmatrix}$

Use back substitution; the solution is

$$x_3 = 1-i$$
$$x_2 = \frac{2}{5}+\frac{1}{5}i - \left(-\frac{2}{5}-\frac{1}{5}i\right)x_3 = 1$$
$$x_1 = 1-(1-i)x_2 = i$$

(b) $\begin{bmatrix} 1 & i & 1-i & 2+i \\ i & 0 & 1+i & -1+i \\ 0 & 2i & -1 & 2-i \end{bmatrix} \longrightarrow \begin{bmatrix} 1 & i & 1-i & 2+i \\ 0 & 1 & 0 & -i \\ 0 & 2i & -1 & 2-i \end{bmatrix} \longrightarrow \begin{bmatrix} 1 & i & 1-i & 2+i \\ 0 & 1 & 0 & -i \\ 0 & 0 & -1 & -i \end{bmatrix} \longrightarrow \begin{bmatrix} 1 & 1 & 1-i & 2+i \\ 0 & 1 & 0 & -i \\ 0 & 0 & 1 & i \end{bmatrix}$

$\quad -i\mathbf{r}_1+\mathbf{r}_2 \quad\quad\quad\quad\quad -2i\mathbf{r}_2+\mathbf{r}_3 \quad\quad\quad\quad\quad -1\mathbf{r}_3$

Use back substitution to get $x_3 = i$, $x_2 = -i$, $x_1 = (2+i)-(1-i)x_3-x_2 = 0$.

5. (a) $\begin{bmatrix} i & 2 & 1 & 0 \\ 1+i & -i & 0 & 1 \end{bmatrix} \longrightarrow \begin{bmatrix} 1 & -2i & -i & 0 \\ 0 & -2+i & -1+i & 1 \end{bmatrix} \longrightarrow \begin{bmatrix} 1 & 0 & \frac{2}{5}+\frac{1}{5}i & \frac{2}{5}-\frac{4}{5}i \\ 0 & 1 & \frac{3}{5}-\frac{1}{5}i & -\frac{2}{5}-\frac{1}{5}i \end{bmatrix}$

$\quad\quad \frac{1}{i}\mathbf{r}_1 \quad\quad\quad\quad\quad\quad\quad \frac{1}{-2+i}\mathbf{r}_2$
$\quad -(1+i)\mathbf{r}_1+\mathbf{r}_2 \quad\quad\quad 2i\mathbf{r}_2+\mathbf{r}_1$

Thus the inverse is $\begin{bmatrix} \frac{2}{5}+\frac{1}{5}i & \frac{2}{5}-\frac{4}{5}i \\ \frac{3}{5}-\frac{1}{5}i & -\frac{2}{5}-\frac{1}{5}i \end{bmatrix} = \frac{1}{5}\begin{bmatrix} 2+i & 2-4i \\ 3-i & -2-i \end{bmatrix}$.

(b) $\begin{bmatrix} 2 & i & 3 & 1 & 0 & 0 \\ 1+i & 0 & 1-i & 0 & 1 & 0 \\ 2 & 1 & 2+i & 0 & 0 & 1 \end{bmatrix} \longrightarrow \begin{bmatrix} 1 & \frac{1}{2}i & \frac{3}{2} & \frac{1}{2} & 0 & 0 \\ 0 & \frac{1}{2}-\frac{1}{2}i & -\frac{1}{2}-\frac{5}{2}i & -\frac{1}{2}-\frac{1}{2}i & 1 & 0 \\ 0 & 1-i & -1+i & -1 & 0 & 1 \end{bmatrix} \longrightarrow$

$\quad\quad \frac{1}{2}\mathbf{r}_1 \quad\quad\quad\quad\quad\quad\quad\quad\quad\quad \frac{2}{1-i}\mathbf{r}_2$
$\quad -(1+i)\mathbf{r}_1+\mathbf{r}_2 \quad\quad\quad\quad\quad -\frac{1}{2}i\mathbf{r}_2+\mathbf{r}_1$
$\quad\quad -2\mathbf{r}_1+\mathbf{r}_3 \quad\quad\quad\quad\quad\quad -(1-i)\mathbf{r}_2+\mathbf{r}_3$

$\begin{bmatrix} 1 & 0 & -i & 0 & \frac{1}{2}-\frac{1}{2}i & 0 \\ 0 & 1 & 2-3i & -i & 1+i & 0 \\ 0 & 0 & 6i & i & -2 & 1 \end{bmatrix} \longrightarrow \begin{bmatrix} 1 & 0 & 0 & \frac{1}{6}i & \frac{1}{6}-\frac{1}{2}i & \frac{1}{6} \\ 0 & 1 & 0 & -\frac{1}{3}-\frac{1}{2}i & \frac{1}{3}i & \frac{1}{2}+\frac{1}{3}i \\ 0 & 0 & 1 & \frac{1}{6} & \frac{1}{3}i & -\frac{1}{6}i \end{bmatrix}$

$\quad\quad \frac{1}{6i}\mathbf{r}_3$
$\quad -(2-3i)\mathbf{r}_3+\mathbf{r}_2$
$\quad\quad i\mathbf{r}_3+\mathbf{r}_1$

$$\text{Thus the inverse is } \begin{bmatrix} \frac{1}{6}i & \frac{1}{6}-\frac{1}{2}i & \frac{1}{6} \\ -\frac{1}{3}-\frac{1}{2}i & \frac{1}{3}i & \frac{1}{2}+\frac{1}{3}i \\ \frac{1}{6} & \frac{1}{3}i & -\frac{1}{6}i \end{bmatrix} = \frac{1}{6}\begin{bmatrix} i & 1-3i & 1 \\ -2-3i & 2i & 3+2i \\ 1 & 2i & -i \end{bmatrix}$$

7. (a) Let A and B be Hermitian and let k be a complex scalar. Then

$$\overline{(A+B)^T} = \overline{A^T + B^T} = \overline{A^T} + \overline{B^T} = A+B$$

so the sum of Hermitian matrices is Hermitian. Next,

$$\overline{(kA)^T} = \overline{kA^T} = \overline{k}\,\overline{A^T} = \overline{k}A \neq kA,$$

so the set of Hermitian matrices is not closed under scalar multiplication and hence is not a complex subspace of the complex vector space of $n \times n$ complex matrices.

(b) From part (a), we have closure of addition and since the scalars are real in this case, $\overline{k} = k$ and hence $\overline{(kA)^T} = kA$. Thus W is a real subspace of the real vector space of all $n \times n$ complex matrices.

9. (a) We must determine if there are complex scalars c_1, c_2, c_3 such that

$$c_1\mathbf{v}_1 + c_2\mathbf{v}_2 + c_3\mathbf{v}_3 = \begin{bmatrix} c_1(-1+i) + c_2 + c_3(-5+2i) \\ c_1(2) + c_2(1+i) + c_3(-1-3i) \\ c_1 + c_2(i) + c_3(2-3i) \end{bmatrix} = \begin{bmatrix} i \\ 0 \\ 0 \end{bmatrix}.$$

Equating corresponding entries from both sides of the equation gives the linear system

$$\begin{array}{rcrcrcl} (-1+i)c_1 & + & c_2 & + & (-5+2i)c_3 & = & i \\ 2c_1 & + & (1+i)c_2 & + & (-1-3i)c_3 & = & 0 \\ c_1 & + & ic_2 & + & (2-3i)c_3 & = & 0 \end{array}$$

Form the augmented matrix

$$\left[\begin{array}{ccc|c} -1+i & 1 & -5+2i & i \\ 2 & 1+i & -1-3i & 0 \\ 1 & i & 2-3i & 0 \end{array}\right]$$

and row reduce the matrix to obtain the reduced row echelon form

$$\left[\begin{array}{ccc|c} 1 & 0 & 0 & -\frac{2}{5}-\frac{4}{5}i \\ 0 & 1 & 0 & \frac{11}{10}+\frac{7}{10}i \\ 0 & 0 & 1 & \frac{3}{10}+\frac{1}{10}i \end{array}\right]$$

Thus the system is consistent which implies that $\mathbf{v}$ belongs to W.

(b) Form the expression

$$c_1\mathbf{v}_1 + c_2\mathbf{v}_2 + c_3\mathbf{v}_3 = \mathbf{0}.$$

Expand, add the vectors, and equate corresponding entries from both sides of the equation. We obtain the homogeneous system

$$\begin{array}{rcrcrcl} (-1+i)c_1 & + & c_2 & + & (-5+2i)c_3 & = & 0 \\ 2c_1 & + & (1+i)c_2 & + & (-1-3i)c_3 & = & 0 \\ c_1 & + & ic_2 & + & (2-3i)c_3 & = & 0 \end{array}$$

Form the coefficient matrix and apply row operations to obtain the reduced row echelon form which is I_3. Hence the only solution to the linear system is the trivial solution. Thus $\{\mathbf{v}_1, \mathbf{v}_2, \mathbf{v}_3\}$ is linearly independent.

11. (a) $A = \begin{bmatrix} 1 & 1 \\ -1 & 1 \end{bmatrix}$. Then

$$\det(\lambda I_2 - A) = \begin{vmatrix} \lambda - 1 & -1 \\ 1 & \lambda - 1 \end{vmatrix} = \lambda^2 - 2\lambda + 2.$$

The eigenvalues are $\lambda_1 = 1 + i$ and $\lambda_2 = 1 - i$. For $\lambda_1 = 1 + i$ we obtain

$$(\lambda_1 I_2 - A)\mathbf{x} = \begin{bmatrix} i & -1 \\ 1 & i \end{bmatrix} \begin{bmatrix} x_1 \\ x_2 \end{bmatrix} = \begin{bmatrix} 0 \\ 0 \end{bmatrix}.$$

The reduced row echelon form of the coefficient matrix is

$$\begin{bmatrix} 1 & i \\ 0 & 0 \end{bmatrix}.$$

Thus $x_1 = -ic$, $x_2 = c$, where c is any complex number. Choosing $c = 1$, we obtain the eigenvector

$$\mathbf{x}_1 = \begin{bmatrix} -i \\ 1 \end{bmatrix}$$

corresponding to eigenvalue $\lambda_1 = 1 + i$. For $\lambda_2 = 1 - i$ we have the system

$$(\lambda_2 I_2 - A)\mathbf{x} = \begin{bmatrix} -i & -1 \\ 1 & -i \end{bmatrix} \begin{bmatrix} x_1 \\ x_2 \end{bmatrix} = \begin{bmatrix} 0 \\ 0 \end{bmatrix}.$$

The reduced row echelon form of the coefficient matrix is

$$\begin{bmatrix} 1 & -i \\ 0 & 0 \end{bmatrix}.$$

This $x_1 = ic$, $x_2 = c$, where c is any complex number. Choosing $c = 1$, we obtain the eigenvector

$$\mathbf{x}_2 = \begin{bmatrix} i \\ 1 \end{bmatrix}$$

corresponding to eigenvalue $\lambda_2 = 1 - i$.

(b) $A = \begin{bmatrix} 1 & i \\ -i & 1 \end{bmatrix}$. Then

$$\det(\lambda I_2 - A) = \begin{vmatrix} \lambda - 1 & -i \\ i & \lambda - 1 \end{vmatrix} = \lambda(\lambda - 2).$$

The eigenvalues are $\lambda_1 = 0$ and $\lambda_2 = 2$. For $\lambda_1 = 0$ we have the system

$$(\lambda_1 I_2 - A)\mathbf{x} = \begin{bmatrix} -1 & -i \\ i & -1 \end{bmatrix} \begin{bmatrix} x_1 \\ x_2 \end{bmatrix} = \begin{bmatrix} 0 \\ 0 \end{bmatrix}.$$

The reduced echelon form of the coefficient matrix is

$$\begin{bmatrix} 1 & i \\ 0 & 0 \end{bmatrix}.$$

Thus $x_1 = -ic$, $x_2 = c$, where c is any complex number. Choosing $c = 1$ we have the eigenvector

$$\mathbf{x}_1 = \begin{bmatrix} -i \\ 1 \end{bmatrix}$$

corresponding to eigenvalue $\lambda_1 = 0$. For $\lambda_2 = 2$ we have the system

$$(\lambda_2 I_2 - A)\mathbf{x} = \begin{bmatrix} 1 & -i \\ i & 1 \end{bmatrix} \begin{bmatrix} x_1 \\ x_2 \end{bmatrix} = \begin{bmatrix} 0 \\ 0 \end{bmatrix}.$$

The reduced row echelon form of the coefficient matrix is

$$\begin{bmatrix} 1 & -i \\ 0 & 0 \end{bmatrix}.$$

Thus $x_1 = ic$, $x_2 = c$, where c is any complex number. Choosing $c = 1$ we have the eigenvector

$$\mathbf{x}_2 = \begin{bmatrix} i \\ 1 \end{bmatrix}$$

corresponding to eigenvalue $\lambda_2 = 2$.

(c) $A - \begin{bmatrix} 2 & 0 & 0 \\ 0 & 2 & i \\ 0 & -i & 2 \end{bmatrix}$. Then

$$\det(\lambda I_3 - A) = \begin{vmatrix} \lambda - 2 & 0 & 0 \\ 0 & \lambda - 2 & -i \\ 0 & i & \lambda - 2 \end{vmatrix} = (\lambda - 1)(\lambda - 2)(\lambda - 3).$$

The eigenvalues are $\lambda_1 = 1$, $\lambda_2 = 2$, $\lambda_3 = 3$. For $\lambda_1 = 1$, we have the system

$$(\lambda_1 I_3 - A)\mathbf{x} = \begin{bmatrix} -1 & 0 & 0 \\ 0 & -1 & -i \\ 0 & i & -1 \end{bmatrix} \begin{bmatrix} x_1 \\ x_2 \\ x_3 \end{bmatrix} = \begin{bmatrix} 0 \\ 0 \\ 0 \end{bmatrix}.$$

The reduced row echelon form of the coefficient matrix is

$$\begin{bmatrix} 1 & 0 & 0 \\ 0 & 1 & i \\ 0 & 0 & 0 \end{bmatrix}.$$

Thus $x_1 = 0$, $x_2 = -ic$, $x_3 = c$, where c is any complex number. Choosing $c = 1$ we obtain the eigenvector

$$\mathbf{x}_1 = \begin{bmatrix} 0 \\ -i \\ 1 \end{bmatrix}$$

corresponding to eigenvalue $\lambda_1 = 1$. For $\lambda_2 = 2$, we have the system

$$(\lambda_2 I_3 - A)\mathbf{x} = \begin{bmatrix} 0 & 0 & 0 \\ 0 & 0 & -i \\ 0 & i & 0 \end{bmatrix} \begin{bmatrix} x_1 \\ x_2 \\ x_3 \end{bmatrix} = \begin{bmatrix} 0 \\ 0 \\ 0 \end{bmatrix}.$$

The reduced row echelon form of the coefficient matrix is

$$\begin{bmatrix} 0 & 1 & 0 \\ 0 & 0 & 1 \\ 0 & 0 & 0 \end{bmatrix}.$$

Thus $x_1 = c$, $x_2 = 0$, $x_3 = 0$, where c is any complex number. Choosing $c = 1$ we obtain the eigenvector

$$\mathbf{x}_2 = \begin{bmatrix} 1 \\ 0 \\ 0 \end{bmatrix}$$

corresponding to eigenvalue $\lambda_2 = 2$. For $\lambda_3 = 3$ we have the system

$$(\lambda_3 I_3 - A)\mathbf{x} = \begin{bmatrix} 1 & 0 & 0 \\ 0 & 1 & -i \\ 0 & i & 1 \end{bmatrix} \begin{bmatrix} x_1 \\ x_2 \\ x_3 \end{bmatrix} = \begin{bmatrix} 0 \\ 0 \\ 0 \end{bmatrix}.$$

The reduced row echelon form of the coefficient matrix is

$$\begin{bmatrix} 1 & 0 & 0 \\ 0 & 1 & -i \\ 0 & 0 & 0 \end{bmatrix}.$$

Thus $x_1 = 0$, $x_2 = ic$, $x_3 = c$, where c is any complex number. Choosing $c = 1$ we obtain the eigenvector

$$\mathbf{x}_3 = \begin{bmatrix} 0 \\ i \\ 1 \end{bmatrix}$$

corresponding to eigenvalue $\lambda_3 = 3$.

13. (a) Let A be Hermitian and suppose that $A\mathbf{x} = \lambda\mathbf{x}$, $\lambda \neq 0$. We show that $\lambda = \overline{\lambda}$. Since A is Hermitian,

$$\overline{(A\mathbf{x})^T} = \overline{\mathbf{x}^T A^T} = \overline{\mathbf{x}^T}\,\overline{A^T} = \overline{\mathbf{x}^T} A.$$

But $A\mathbf{x} = \lambda\mathbf{x}$, so that $\overline{(A\mathbf{x})^T} = \overline{(\lambda\mathbf{x})^T} = \overline{\lambda}\,\overline{\mathbf{x}^T}$. Thus $\overline{\mathbf{x}^T} A = \overline{\lambda}\,\overline{\mathbf{x}^T}$. Multiplying both sides of this equation on the right by $\mathbf{x}$ and using the fact that $A\mathbf{x} = \lambda\mathbf{x}$, we obtain

$$\overline{\mathbf{x}^T} A\mathbf{x} = \overline{\lambda}\,\overline{\mathbf{x}^T}\mathbf{x} \quad\Longrightarrow\quad \overline{\mathbf{x}^T}\lambda\mathbf{x} = \overline{\lambda}\,\overline{\mathbf{x}^T}\mathbf{x} \quad\Longrightarrow\quad (\lambda - \overline{\lambda})\,\overline{\mathbf{x}^T}\mathbf{x} = \mathbf{0}.$$

Since $\overline{\mathbf{x}^T}\mathbf{x} > 0$, it follows that $\lambda = \overline{\lambda}$. Thus the eigenvalues of a Hermitian matrix are real.

(b) $\overline{A^T} = \begin{bmatrix} 2 & 0 & 0 \\ 0 & 2 & \overline{-i} \\ 0 & \overline{i} & 2 \end{bmatrix} = \begin{bmatrix} 2 & 0 & 0 \\ 0 & 2 & i \\ 0 & -i & 2 \end{bmatrix} = A$

(c) No, see Exercise 11(c). An eigenvector $\mathbf{x}$ associated with a real eigenvalue λ of a complex matrix A is, in general, complex, because $A\mathbf{x}$ is, in general, complex. Thus $\lambda\mathbf{x}$ must also be complex.

15. Let A be a skew symmetric matrix, so that $\overline{A^T} = -A$, and let λ be an eigenvalue of A with corresponding eigenvector $\mathbf{x}$. We show that $\overline{\lambda} = -\lambda$. We have $A\mathbf{x} = \lambda\mathbf{x}$. Multiplying both sides of this equation by $\overline{\mathbf{x}^T}$ on the left we have

$$\overline{\mathbf{x}^T} A\mathbf{x} = \overline{\mathbf{x}^T}\lambda\mathbf{x}.$$

Taking the conjugate transpose of both sides gives

$$\overline{\mathbf{x}^T}\,\overline{A^T}\mathbf{x} = \overline{\lambda}\,\overline{\mathbf{x}^T}\mathbf{x} \quad\Longrightarrow\quad -\overline{\mathbf{x}^T} A\mathbf{x} = \overline{\lambda}\,\overline{\mathbf{x}^T}\mathbf{x} \quad\Longrightarrow\quad -\lambda\overline{\mathbf{x}^T}\mathbf{x} = \overline{\lambda}\,\overline{\mathbf{x}^T}\mathbf{x}.$$

Therefore $(\lambda + \overline{\lambda})\,\overline{\mathbf{x}^T}\mathbf{x} = 0$. Since $\mathbf{x} \neq \mathbf{0}$, $\overline{\mathbf{x}^T}\mathbf{x} \neq 0$, so $\overline{\lambda} = -\lambda$. Hence the real part of λ is zero.